Field Ecology

Michael Mühlenberg • Thomas Waßmer

Field Ecology

Concepts, Issues and Practice

Springer

Michael Mühlenberg
Department of Conservation Biology
University of Göttingen
Göttingen, Germany

Thomas Waßmer
Department of Biology
Siena Heights University
Adrian, MI, USA

ISBN 978-3-031-87546-5 ISBN 978-3-031-87547-2 (eBook)
https://doi.org/10.1007/978-3-031-87547-2

This Springer imprint is published by the registered company Springer Nature Switzerland AG
The registered company address is: Gewerbestrasse 11, 6330 Cham, Switzerland

Foreword

Ecological science requires reliable data to answer questions about conservation, community, and ecosystem dynamics, and these data can only be obtained with the best field methods. All our progress in answering current ecological questions and solving ecological problems must rely on the finest available methods. In this book, Michael Muehlenberg and Thomas Wassmer have brought field ecology up to date with all the details and cross-references for methods of study in aquatic and terrestrial animal ecology, whether for sampling plankton, predator–prey dynamics, estimating population density, or devising methods to measure habitat fragmentation. This book is a thorough introduction to field ecology with the concepts and methods needed by students, researchers, and practitioners of animal ecology. It is wholly readable and well-illustrated. Precise methods and explicit research questions for field studies are more and more essential in this era of climate change and population growth when more and more citizens request guidance for biodiversity conservation. This is an outstanding guide to the best field ecological methods every animal ecologist requires.

Charles J. Krebs

Department of Zoology
University of British Columbia
Vancouver, BC, Canada

Acknowledgments

A previous version of this book, *Freilandökologie*, appeared in three editions in Germany and was written in response to the wide-ranging interest Michael Muehlenberg (MM) received from students, doctoral candidates, and conservation professionals. He wants to thank his mentors, Prof. F. Schremmer and Prof. K.-E. Linsenmair. On the student side, MM would particularly like to thank Lect. Dr. Julian Schrader, Dr. Christoph Parsch, Lect. Dr. Choimaa Dulamsuren, and Benjamin Neusel, M.Sc., for their suggestions and encouragement. Thanks also to Prof. A. Bogenrieder, who contributed abiotic and botanical measurements in an earlier edition. MM's activities abroad have been consistently supported by the German Academic Exchange Service (DAAD), for which he is very grateful. Individual partners in other countries always supported him, and he wants to highlight Prof. H.H. Roth and Dr. E. Waitkuwait from the Ivory Coast, Prof. R. Samiya from Mongolia, and particularly Enkhmaa Ayush, MSc. for her management of the Khonin Nuga Ecological Station in Northern Mongolia for decades. Through tireless efforts, she made numerous field studies in the diverse natural landscape possible for many students and scientists. In Indonesia and West Papua, MM was supported by Samuel Renyaan, MSc., Dr. Konstantina Kameubun, John D. Kalor, M.Sc., and Paulus Savyai. Many thanks also go to Dr. Simone Pfeiffer, who coordinated the work in Papua from Germany. In New Zealand, MM's colleagues, Prof. I. Spellerberg and Dr. K-J. Wilson, supported him in friendship. With them, MM established a new Master's program with a double degree between Göttingen, Germany, and Christchurch, New Zealand.

Thomas Wassmer (TW) wants to thank his mentors Guenther Osche, Ottie Wilmanns, Klaus Peschke, Peter Sowig, Franziska Wollnik, Serge Daan, Hubert Markl, Ireneus Eibl-Eibesfeldt, and his collaborators Roberto Refinetti, Jorge Ari Noriega, Joaquín Hortal, Andreas Fahlmann, Dennis Murray, Stan Boutin, Frants Havmand Jensen, and Peter Dunn. His work was generously supported by the German Research Foundation (DFG), Siena Heights University, and the Oak Foundation. Several Siena Heights University undergraduate students contributed to the research showcased in this book, including Olivia Smith, Crystal Wilcoxen, Kristin Delclos, Ciara Ginn, Rosemary Hanna, Elise Armstrong, Noah Gascon, and Thomas Bakoway.

Many years ago, TW contacted MM, suggesting that he translate his book *Freilandökologie*, as such a book was missing on the international market. The translation did not go forward then, but the two researchers have stayed in touch over the years. With the current cooperation of the two seasoned authors from Germany and the USA, the idea of a book on field ecology has matured to an international level, merging their decades-long experiences into a substantially extended and contemporary manual on field ecological methods.

We would like to thank the Springer Verlag, especially Dr. Lars Koerner and his staff, who made it possible to compile our ambitious intentions into a book, even though they had to forego many of their wishes. Finally, we are incredibly grateful to our patient life partners, families, and friends, who had to give up many days of quality time with us that we both enthusiastically invested in this book that is so close to our hearts.

About the Book

This book is a theoretical and practical guide to ecological work in the field, focusing on animal ecology. By highlighting examples, it provides students, researchers, and professionals with the tools to develop ecological questions and corresponding working hypotheses. It offers guidelines to choose the appropriate methods for successful data collection and analysis. This book focuses on methods for assessing biological diversity and habitats in a changing world, relating specifically to conservation issues and concerns.

Introduction

Ecology, the science of the interactions between organisms and their environment, finds its practical application in field ecology. This term encapsulates the practice of conducting ecological work in the natural environment, where the true dynamics of these interactions can be observed and studied.

This book serves as a comprehensive guide, introducing the various areas a field ecologist may encounter. Through its rich illustrations, it aids in the understanding of fieldwork and interactions with organisms. Its primary aim is to encourage a range of topics while also serving as a practical guide for conducting one's own ecological investigations.

Field ecologists are confronted with many questions: Why is a particular organism in a specific place at a given time? Why is a species missing from its expected location? If populations or the species community is of interest, additional questions arise: What determines the number of species that live together in a habitat? How are the densities of the populations regulated? What role do abiotic factors, resource availability, competition, predation pressure, and other aspects play? Is there interspecific competition for resources? Do predators control herbivores? How can several similar species live together? How do the coexisting species differ in terms of their resource use? Are there temporal changes or disturbances in the observed system? What can be done to ensure that declining populations do not disappear locally or globally?

This catalog of questions could be expanded, but field ecologists must limit the number of questions they ask when planning an investigation. In most cases, you need to be content with clarifying only partial aspects of overarching questions. It would be best to use the following principle as guidance: Before beginning an investigation, familiarize yourself with the issue. First, a question needs to be defined; then, you must identify data to answer it. The next step is to plan how the necessary data should be collected, including assessing the corresponding effort, time, and costs. It is also helpful to consider the types of diagrams that will be used to present the data. Histograms, for example, are suitable for visualizing frequency distributions, whereas scatter plots help break down relationships between two variables. Tables are helpful if precise numerical values are essential, and maps can be used to explain the procedures in the field.

In most areas of field ecology, state-of-the-art equipment and techniques are available to collect data quickly and comprehensively. Nevertheless, in this book we emphasize on simple and inexpensive methods and the most recent literature while also providing essential older references ("oldies but goldies"). In appropriate cases, we refer to the use of more expensive modern techniques. Some ecological analyses must be carried out in the laboratory. However, as this is not the focus of our book, such analyses are only mentioned in a few individual cases.

As a science, ecology has advanced far enough that general answers to many questions already exist. However, it always remains to be seen whether these general findings are valid in every specific case. With this book, we want to encourage field ecologists to explore the interactions in nature through their own investigations and to engage with the fascinating biodiversity. We believe it is crucial to prioritize investigations in the context of nature conservation whenever possible. Many of the research questions raised in this book are highly relevant to nature conservation, and the results are eminently applicable for this purpose. The consequences of various management measures affect nature conservation, and scientific studies are the most promising preparation for fact-based decision-making in this area.

The activities of field ecologists are almost always subject to three limitations: (1) Lack of taxonomic knowledge, (2) exceedingly short project timelines, and (3) too small areas.

Lack of Taxonomic Knowledge

Unfortunately, taxonomic training is increasingly taking a back seat to theoretical learning, with the number of taxonomists, i.e., experts familiar with a group of organisms, dwindling worldwide. This causes complications in conducting surveys, data collection, and inquiries. We have included numerous illustrations in this book to encourage people to make their own observations and gain an understanding of the various groups of organisms. It is widely acknowledged that we can only truly appreciate what we know, and only care for what we appreciate. We, therefore, believe that ecologists must experience nature's biodiversity directly and first hand by themselves.

Short Timelines

Climate change and habitat loss present significant challenges for nature conservation in real time. However, corresponding studies require monitoring over many years. This book provides instructions for conducting studies with results that can be analyzed in shorter time spans. Ideally, long-term studies should incorporate multiple approaches to record the impact of climate change and habitat loss on various species groups.

Small Areas

Many research projects are limited by the need to study small areas. The available area is typically divided into different plots (e.g., treatment A, treatment B,..., and control), and the fauna is analyzed at great expense. Differences between plots observed in this way often cannot be clearly attributed to the different treatments, and it is usually impossible to repeat the data collections or experiments. The area required for successful experiments and data collections is almost always underestimated. A model for a sufficiently large scaled area is the Kluane project in North America (see Chap. 8), in which plots of one square kilometer were compared. Many animals are highly mobile, and their home range is so extensive that they can roam through various study plots regardless of the treatment (usually smaller than 1 ha, only rarely slightly more extensive). In study areas that are too small, the results can be additionally blurred by immigration and emigration processes.

We have divided this book into three parts: Part I mainly consists of descriptive studies, Part II covers analyses of ecological relationships, and Part III focuses on management and planning. Almost all animal ecology questions require knowledge of the populations and the abundances of individual species. Therefore, the first two chapters of Part I (Chapter 1, Recording Vertebrates, and Chapter 2, Sampling Invertebrates) deal with sampling methods for various animal groups and selected species. Chapter 3 , Recording Abiotic Parameters, presents methods and instrumentation to record abiotic parameters in terrestrial areas. Subsequently (in Chapter 4, Mapping Landscape and Locating Individuals), we point out ways to classify vegetation types and show how to quantify the structural parameters of vegetation. A description of GPS and UAV (drone) -based data acquisation then leads to the application of radiotelemetry methods to study the movement of individuals. Chapter 5, Estimating Populations, introduces various methods for estimating the abundances and densities of animal populations, regardless of whether they belong to vertebrates or invertebrates. Methods and questions in limnology (freshwater ecology), separated into flowing (rivers) and standing waters (lakes), are presented in Chapter 6, Freshwater Ecology.

Questions concerning what animals have ingested as food are central to ecological relationships (Part II). Chapter 7, Foraging Ecology, deals with the possibilities of deciphering this. With the food intake of carnivores comes the more general problem of predation pressure. How can we estimate the magnitude and impact of predation pressure on selected animal populations and vegetation (Chapter 8, Estimating Predation Pressure)? The fundamental question of ecology, where and why a species resides in a particular area, is addressed in the following chapter (Chapter 9, Habitat Selection). Once we establish that a particular species favors a habitat with specific characteristics, we arrive at Chapter, 10, Habitat Evaluation. The evaluation of habitat plays a crucial role in nature conservation when selecting areas worthy of protection and proposing management measures to preserve biotopes. Habitat loss and fragmentation are among the leading causes of species decline and biodiversity loss on every scale, locally to globally, explained by Landscape

Ecology (Chapter 11, Landscape Ecology). When resources become scarce in small habitats, it becomes clearer that species with ecologically similar requirements compete and can only coexist if they differ in some aspects. This problem is dealt with (in Chapter 12, Niche Differentiation Between Related Species).

Part III focuses on management and planning. The management of landscape components includes both the assessment of anthropogenic impact and the planning of measures in response to disturbances in the ecosystem (Chapter 13, Managing Landscape and Populations). Harvesting forest products and interventions in wildlife populations must be managed sustainably. Approaches for the extensive restoration of altered ecosystems are outlined in Chapter 14, Restoration Ecology.

Since ecological research and data collections generally require scientific quantification, Chapter 15, Designing Data Collection, offers guidelines for the design of experiments and data collections and facilitates an understanding of statistics and insight into the corresponding analysis methods. Since numerous software programs and books are available, we have mostly restricted us to explain how to process ecological data and how to use standard statistical methods independently. We also provide references to locate more specific methods of analysis and the use of relevant software programs. Field ecology research typically requires a significant investment of both time and money. Preparing scientific reports and publications can help increase your chances of securing financial project support or grants. Chapter 16 ,Writing Research and Grant Proposals, summarizes critical basic principles for scientific reports and proposals.

Contents

About the Authors

Michael Mühlenberg received his doctorate in 1972 and habilitation in 1977 from the University of Heidelberg. He was then appointed to the University of Würzburg, where he founded the Ecological Station Fabrikschleichach in Steigerwald, Northern Bavaria. In 1994, he joined the University of Göttingen and established the Conservation Biology program. He supervised more than 40 Ph.D. theses. After many years of wildlife surveys in West Africa, he worked, e.g., in New Zealand, Mongolia, and Papua Indonesia, focusing equally on the biodiversity and conservation of invertebrates and vertebrates. He has received honorary professorships in Mongolia and Indonesia. The study area around his research site in Northern Mongolia (Khonin Nuga) was declared a protected area with several thousand square kilometers and incorporated into the existing Strictly Protected Area of Khan Khentey. He was awarded the Order of Merit by the Mongolian Ministry of Environment and Nature Conservation.

Thomas Wassmer (in German: Waßmer) received his Diplom-Biologe degree from the University of Freiburg in 1991 and his Dr. rer. nat. degree from the University of Konstanz in 1998. He currently works at Siena Heights University in Adrian, Michigan, and holds the rank of Full Professor of Biology. He teaches courses in terrestrial ecology and freshwater ecology, focusing on field biology, and vertebrate physiology, evolution, human anatomy and physiology, and a liberal arts class on environmental and earth justice and activism. His research centers on the population, community, and landscape ecology of dung-inhabiting beetles, the ecophysiology and chronoecology of tree squirrels and hamsters, and on using drones and acoustic recording devices for biodiversity conservation.

Part I

Focus on Description

1 Estimating and Recording Abundance and Occurrence of Vertebrate Populations

Abstract

Knowledge of species' occurrences and population sizes is a prerequisite for practically all ecological questions. Direct and indirect methods of assessing larger vertebrates (mammals and birds) are presented. Some survey methods can only determine the presence or absence of species. In addition to traditional on-site methods, e.g., counts along line transects, other detection tools such as camera traps, detection dogs, and sound decoys are also used depending on the animal group. Bird territories can be mapped using a bird species' territorial behavior, and nest search can supplement visual and auditory bird detections. Smaller birds can be recorded in a standardized way using mist nets. Community-based long-term monitoring conducted by indigenous and local people is introduced.

1.1 Inventory Surveys of Larger Wildlife

To protect a species from extirpation, ecologists need to locate their populations and estimate their numbers, preferences for specific habitat characteristics, and their susceptibility to influencing factors, especially habitat loss and fragmentation due to changes in human land use such as deforestation, poaching, and road construction, as well as climate change. Depending on the animal species that comes into focus, one must select or use different methods side by side to answer the questions about frequency, and spatial and temporal distribution (see also Krebs 1999; Caughley 2004).

We start with inventory surveys of wildlife and birds. After the animals are discovered using direct observation, distance sampling is used to estimate population density (Buckland et al. 2015). This method is based on the fact that more distant animals are more easily overlooked. The method is presented in detail in Sect. 5.1. Here, we will deal with detecting larger animals using alternative tools.

- Camera trapping in connection with mark-recapture methods (Rowcliffe et al. 2008)
- Genetic methods with the amplification of DNA extracted from field samples (hairs or feces) (Bellemain and Swenson 2005)
- Use of detection dogs (Grimm-Seyfarth et al. 2021)
- Indirect assessments using tracks, like snow tracking (Stephens et al. 2006)
- Occurrence modeling for wildlife assessments (MacKenzie et al. 2017)
- Sound decoys and automatic sound recognition (e.g., Kiffner et al. 2008; Cornell University 2024)
- Mist-netting birds (Euring 2007)

M. Mühlenberg, T. Waßmer, *Field Ecology*, https://doi.org/10.1007/978-3-031-87547-2_1

1.2 Camera Traps

Camera traps are systems that automatically take pictures of animals in the wild without requiring an operator's presence (Fig. 1.1). They usually use a passive infrared sensor to detect motion and trigger a camera. At night and low light conditions, a infrared LED flash does not blind the animals as a white-light flash would do. Camera traps allow the study of animal populations and behavior with minimal disturbance (e.g., O'Connell et al. 2011; Soria-Díaz and Monroy-Vilchis 2015; Wearn and Glover-Kapfer 2017; Searle et al. 2025). First, camera traps can document the occurrence of species (Fig. 1.2).

Fig. 1.1 Camera trap fixed to a tree for automatic recordings

Fig. 1.2 A motion-triggered automatic camera has captured a wolverine (*Gulo gulo*) with prey in Northern Mongolia. Wolverines are shy and difficult to observe. (Picture credit: Khan Khentii Park Management. Reproduced by permission)

In most cases, a camera trapping system is not triggered explicitly by the target species and detects a range of species. Camera traps can, therefore, also be used to capture segments of the diversity of specific communities (*community diversity*), mainly medium to large, ground-dwelling mammals, birds, and reptiles (Wearn & Glover-Kapfer 2019). If the cameras are set up randomly, stratified, or mounted in a grid over a larger area, the rate at which wildlife is discovered can be assumed to be directly proportional to their frequency (*local abundance*). Under certain conditions, this rate can also be converted into absolute density values (Moeller et al. 2018), e.g., camera trapping can be combined with the method of *distance sampling* and allow for abundance estimation (Howe et al. 2017).

Camera trapping is encouraging in studies of the activity patterns and behaviors of elusive species in forests. The analytical approach of capture-recapture relies on individual recognition of camera-trapped animals but is challenging in most ungulates. An alternative approach is the Random Encounter Model (REM). It describes contact rates between camera traps and animals. Required information: animal group density, camera days/camera station; daily distance walked (km/day, e.g., 3.36 km/day for deer, Pépin et al. 2009; Carranza et al. 1991); detection distance (r, in km, e.g., 0.012 for deer, depending on camera type), detection angle (Θ, 0.175°, depending on camera type), group size (individuals/group, e.g., 2.78 in North Iran National Park). The most influential parameters are group size and animal movement (daily distance walked).

Camera traps have been used to estimate white-tailed deer densities in the USA at artificial salt licks (see Figs. 1.3 and 1.4 as an example), with

Fig. 1.3 Maral habitat in the Khentey region, northern Mongolia

Fig. 1.4 A natural salt lick with tracks of Maral

one camera being sufficient to estimate deer densities on 165 ha.

In tropical forests, camera traps and distance sampling (see Sect 5.1) are used together for various species different in size and behavior (Bessone et al. 2020). This combination of methods has enormous potential to study terrestrial wildlife in inaccessible areas and, in this way, also contributes to reassessing the conservation status of individual species in the IUCN Red List of Threatened Species (https://www.iucnredlist.org).

Case Study: Camera Trapping for Activity Tracking in Squirrels

Research conducted by Kristin Delclos, Rosemary Hanna, Ciara Ginn, Noah Gascon, and Thomas Wassmer—Siena Heights University.

North American red squirrels (*Tamiasciurus hudsonicus*) and fox squirrels (*Sciurus niger*) are prevalent throughout the northern United States and along southern Canada (Koprowski 1994; Steele 1998). Both species are believed to be diurnal with an unimodal to bimodal activity pattern and variable activity duration depending on season and latitude (Hicks 1949; Amspacher et al. 2019; Kramm 1975; Wassmer and Refinetti 2016; Williams et al. 2014). However, both tree squirrel species were also shown to exhibit substantial individual variation in both activity duration and pattern (Boon et al. 2008; Wassmer and Refinetti 2019). Our first objective was to verify if camera trapping can describe and contrast activity patterns between species and seasons. Our second goal was to compare the activity patterns of individual fox squirrels. We established 4 camera traps facing a tree-mounted feeding platform from 1 to 2 m away (Fig. 1.5). Four individual fox squirrels were tagged using differently colored neck collars. We recorded each picture's date/time, species, and collar color. For some analyses, we only included the first/last daily frames as indicators for the onsets and offsets of the activity phase. We determined the activity durations as the time difference between the first and last frames. The statistical significance of the timing and differences between species and seasons was tested using circular statistics. In contrast, differences in the activity duration were examined using a two-way ANOVA, both at p-levels below 0.05.

In the second year of this ongoing study, we recorded 1460 trapping days, totaling 350 first/last frames for fox squirrels but only 126 first/last frames for red squirrels, as the species was almost absent in the winter and spring of 2020/2021. In summer, the first and last frames were almost indifferent between the two species and occurred at about 7:20 and 16:10 ET, respec-

Fig. 1.5 Setup of camera traps and portraits of the two studied tree squirrel species. (**a**) A Moultrie Trace MCS-12639 camera is attached to a fence post using hardware that allows rotation around all three space dimensions to properly align with the Tomahawk life trap used as a bait station. (**b**) Camera trap picture of an American red squirrel. (**c**) Camera trap picture of a fox squirrel

tively (Fig. 1.6). In contrast, in the fall, the first frames of fox squirrels were taken about an hour later than red squirrels (8:35). The last frames were taken about an hour earlier than red squirrels (14:57). First frames for fox squirrels were taken significantly earlier during the spring and summer as compared to fall (+1 h) and winter (+2 h). In contrast, the last frames occurred significantly earlier in fall and winter (−2 h) than in spring and summer. Individual fox squirrels showed significantly different timing and activity durations compared to each other and the unmarked population. The duration of the activity phase was significantly influenced by species, season, and their interaction. Our results compare well with the findings of other studies using more laborious, expensive, or obstructive methods (Amlaner 1987; Anders et al. 2017; Koprowski and Corse 2005; Osgood 1972; Wassmer and Refinetti 2019; Wauters and Dhondt 1987), suggesting that camera trapping should be more commonly used in chronoecological research.

1.3 Genetic Methods

Researchers have recently developed DNA-based methods to identify species, individuals, and their sex and estimate their population size (Bellemain et al. 2005). The methods have the advantage that genetic samples (i.e., hairs or feces) are easily collected without the need for direct observations, and compared to conservative indirect methods, result in a surplus of information about population structure, sex ratio, and genetic variation. As a unique multilocus genotype can characterize every individual, it is pos-

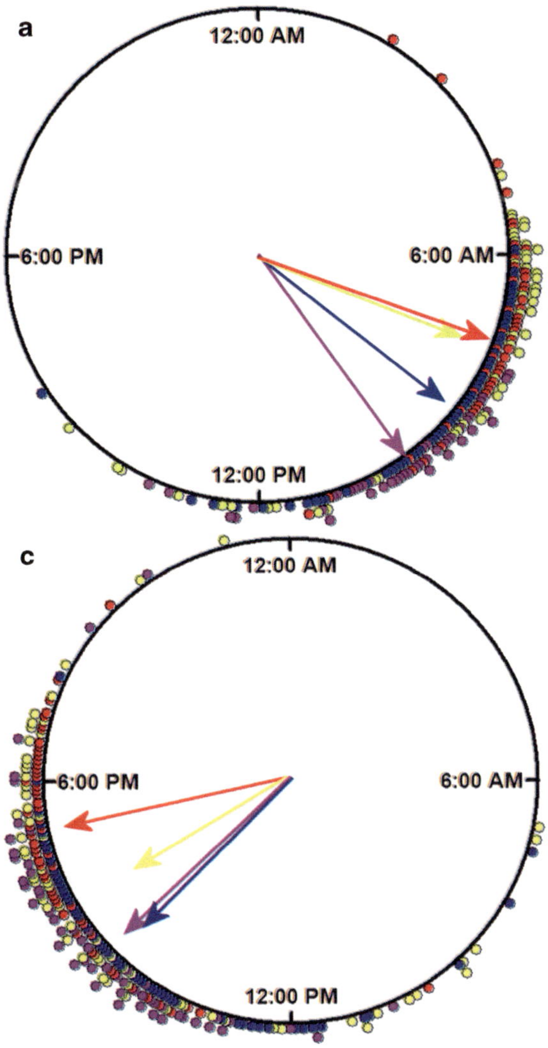

Fig. 1.6 Circular plots of the timing of both squirrel species' first and last frames during the year's four seasons. Circles and arrows are color-coded by season: blue (Fall), red (Winter), yellow (Spring), and pink (Summer). The lengths of the mean vectors are a measure of the concentration of events around the mean time. (**a**) First frame of fox squirrels—each symbol represents nine observations. (**b**) First frames of red squirrels—each symbol represents five observations. (**c**) Last frames of fox squirrels—each symbol represents eight observations. (**d**) Last frames of red squirrels—each symbol represents four observations. (Picture credit: Kristin Delclos)

sible to determine the number and the sex of animals sampled.

Another approach is to use the cumulative number of unique multilocus genotypes expressed as a function of the number of feces sampled. By determining the asymptote of this curve, you can estimate the local population size. The amplification of DNA extracted from field samples is usually problematic because the DNA is often degraded and in low quantity. Fecal DNA genotyping functions better when sampled in winter.

The higher costs of genetic sampling and the considerable sampling effort compared to conservative methods significantly restrict its application.

The great advantage of genetic methods is that they are noninvasive in estimating population size, especially for elusive animals. Good application examples in the Mongolian Taiga, would be the brown bear, wolves, or musk deer. The most significant method is based on fecal samples. According to experiences in Scandinavia, the number of samples should exceed the expected/estimated population size by about three to four times. Articles describing the methods and comparing different approaches are Solberg et al. (2006), Bellemain et al. (2005), Maudet et al. (2004), Smallwood and Schonewald (1998), Wilson et al. (2003), Creel et al. (2003), Flagstad et al. (2004), Mowat and Paetkau (2003), Schwartz et al. (1999), Foran et al. (1997), McDaniel et al. (2000), Wasser et al. (1997), Dalén et al. (2004), Piggott (2004), Frantzen et al. (1998), Kohn et al. (1999), Lucchini et al. (2002), Murphy et al. (2002), Waits et al. (2001), Lehman et al. (1992), Palomares et al. (2002), Stephens et al. (2006) and Goode et al. (2014).

The search for feces, pellets, or dung becomes particularly effective when trained search dogs are used.

1.4 The Use of Detection Dogs in Nature Conservation

The use of search dogs (also "scent detection dogs") has now been developed internationally (Grimm-Seyfarth et al. 2021). *Specially trained conservation detection dogs sniff out odors and search for things related to conservation biology.* There are also platforms of associations, e.g., https://k9conservationists.org/, Wildlife Detection Dogs e.V. (https://www.wildlifedetectiondogs.org/en/homepage/), and in Australia, the *USC Detection Dogs for Conservation team* (https://www.usc.edu.au/about/structure/schools/school-of-science-technology-and-engineering/detection-dogs-for-conservation). Search dogs can achieve much more than human visual counting methods. Through intensive and enduring training of search dogs (e.g., Simon et al. 2020), it is possible to expand the detection probabilities of target objects many times over with their sense of smell. However, the training usually requires several years of education (see Bottaro 2023). Various breeds can be successfully trained (Beebe et al. 2016; Hewings 2019). Their use can be crucial in conservation, for example, when proving the existence of rare species or animals with hidden or nocturnal lifestyles (Bennett et al. 2019; Karp 2020; Thomas et al. 2020). Even in terrain that is difficult for field biologists to traverse and rich in cover, search dogs are superior. Even in the limnic area, search dogs can be successfully used (Richards 2018). For many target species, evidence of tracks, feces, pellets, hair, feathers, sleeping places, and nests is possible (Grimm-Seyfarth et al. 2019). Even the individual distinction of feces is possible with *detection dogs* and can even replace expensive and complex DNA analyses. Although the current focus of using search dogs is on placental mammals, there are already successful case examples for marsupials (Hollerbach 2012; Cristescu et al. 2015), birds (Wasser et al. 2012; Beaumont et al. 2022), reptiles, amphibians (Glover et al. 2023), and insects (Lehnert and Weeks 2016; Aviles-Rosa et al. 2023).

1.5 Indirect Assessments Using Tracks

Several books on trace detection are available, such as Olson (2013), Baker (2013), and https://naturetracking.com/mammal-tracks/ for North America (also offering the smartphone app iTrack Wildlife for North America and S-Africa) and https://naturestudyaustralia.com.au/wp-content/uploads/2020/06/Australian-Animal-Track-and-Scat-Montessori-Cards.pdf for Australia.

Some sources also refer to snow tracking (Figs. 1.7 and 1.8) and recording the location of each track or trail that crosses a transect using GPS, allowing for studying the species' spatial distribution and habitat use over broad areas (D'Eon et al. 2006).

It "requires an estimate of the daily travel distance of the surveyed species to estimate the

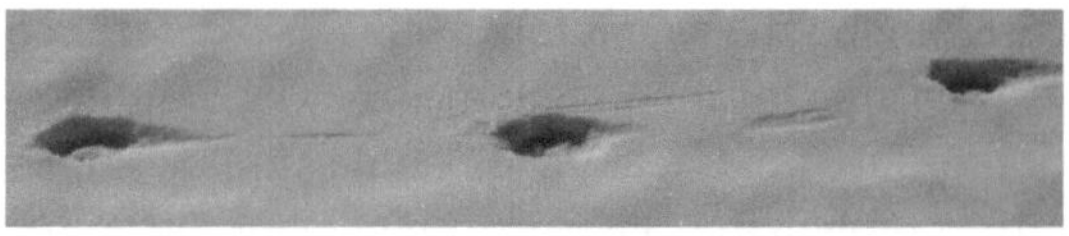

Fig. 1.7 A wolf track in the snow, Mongolia

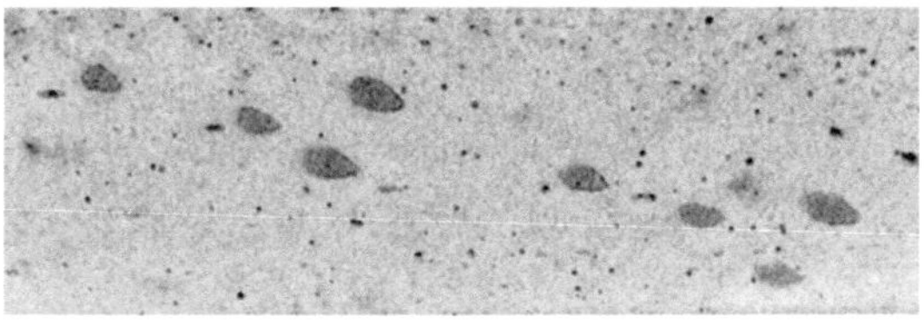

Fig. 1.8 Track of a running hare, Germany

probability of an individual crossing a transect within the 24 h before the survey" (Stephens et al. 2006: 340). The authors found surprisingly consistent travel distances for red deer during their research. Additionally, group sizes have not been recorded in the past, which can lead to serious bias, especially when red deer traveling in groups use each other's tracks when the snow level exceeds ca. 20 cm. "It is unlikely that very low-density ungulate populations (e.g., of less than 0.5/km^2) can be surveyed effectively by this method" (Stephens et al. 2006: 343–344). Mooty and Karns (1984) compared the count results of tracks and pellet groups. Mandujano (2005) calibrated track counts by comparing them with observations on the line transects.

Tracking in the snow is only possible to a limited extent due to the weather conditions. For this reason, attempts have been made to lay out surface-treated panels for recording tracks (Reynolds et al. 2010). Reynolds et al. (2010) used track-recording rafts to detect the mink (*Neovison vison*). This method is particularly suitable for recording marten-like species (Mustelidae), which are difficult to detect using camera traps. Only camera traps in combination with scent lures can be used successfully (see also the wildcat example, Fig. 13.3), whereby pine martens (*Martes martes*) can be attracted relatively well (Lyra-Jorge et al. 2008; Burki et al. 2010; Kriegs et al. 2012). Pine martens have individualized chest markings (yellow spots), allowing for the identification of individual animals in the photos. In species where this is possible, camera traps can be used to determine abundance (see photo-ID in Sect. 5.4). Harrington et al. (2008) provide instructions on measuring tracks of similar marten species. Carbon-sooted track plates can be laid out to obtain traces of the mustelids. This technique is inexpensive compared to photo traps with bait. Zielinski and Truex (1995) give instructions on how to distinguish between the closely related American marten (*Martes americana*) and Fisher's marten (*Martes pennanti*). Similarly, evidence of dormice (Gliridae) can also be obtained (see Chap. 2.18). Stier et al. (2015) provide a detailed overview of the various survey methods and their effectiveness for pine martens and polecats.

1.6 Recording Occurrence

One can also make population estimates by logging the presence or absence of a species in specific area units. While such data is relatively easy to collect, some prerequisites must be met for the evaluation. One distinguishes so-called "open" or "closed" populations. Closed populations are not subject to undetected changes like migration, mortality, or reproduction. It is easy to understand that most naturally occurring populations are open populations.

The following table (Table 1.1) compares the advantages and disadvantages of *Occurrence* models and population estimation with *distance sampling*:

Gopalaswamy et al. (2012) compare the results of density estimates between occupancy and distance sampling for five ungulate species in Southeast Asian forests and demonstrate similar results through modeling. Davis et al. (2022) compared mammalian occurrence indices for monitoring herbivores derived from camera trapping and pellet counts, and found that camera trapping provided higher statistical power to detect changes in abundance and habitat use than pellet counts.

1.7 Counting Birds by Line Transects

Ornithologists have used standardized inventory surveys for a long time. Several internationally recognized recommendations exist (Emlen 1971; Svensson 1992; Oelke 1980; Berthold 1976; Ralph and Scott 1981; Marchant 1983; Bibby et al. 2000). A complete breeding bird inventory

Table 1.1 Comparison of occurrence modeling and distance sampling with respective advantages and disadvantages

Occurrence modeling	Distance sampling
Occupancy state is "closed": species are present at occupied sites for the sampling season. Occupancy does not change at a site within the sampling season, but it can change between sampling seasons	100% detection on/near line transect
Sites are independent: detecting the target species at one site is independent of detecting the species at other sites. Animals should not move between sites to avoid duplicate detections	Transects are independent samples, but you can repeat them
No unexplained heterogeneity in occupancy: the probability of occupancy is the same across sites, or differences in occupancy can be explained with site characteristics (covariates) that have been quantified	"Given the fact that visibility varies substantially, distance sampling is potentially well suited to estimating ungulate population density in forest habitats" (Gill et al. 1997)
No unexplained heterogeneity in detectability: detectability at occupied sites is the same across all surveys and sites, or differences can be explained with site or survey characteristics that have been quantified for inclusion in the model	Before detection, there was no movement in response to observers. Many study objects may remain undetected since there is post-survey control of strip width
A large sample size is required if only frequency (proportion of sample units) is recorded. Relatively insensitive to seasonal or management changes	Should in total have at least 40 detections of the target species (pooling data possible)
Difficult to quantify heterogeneity in occupancy and detectability	Exact measurements for distances
Required data (detection/non-detection information) are relatively simple and inexpensive to collect	The assumptions are mild

consists of three parts: line transect, determination with sound lures, and nest search.

Transect methods for breeding bird mapping:

The first step is mapping (Oelke 1980): During control walks through the test area in the morning hours on randomly selected routes (the transect method is vital at the peak of the breeding season), the researcher records all optical and acoustic findings of bird individuals. In forests, the test area should be 10–30 ha (in a cultural landscape, at least 60 ha), and no part of the test area should be more than 50 m away from the control route (in open terrain <100 m). Overall, 8–10 systematic walks are required, and the time required is 10–15 min/ha. In Central Europe, the recording should take place in April and May. However, recording should start in March for woodpeckers, tree creepers, many birds of prey, and owl species (Andretzke et al. 2005). It is best to start before 9 am or after 5 pm for a 3-h census.

For each walk, such a breeding bird inventory maps out all territory-indicating bird individuals on a daily map. The varying detectability of individual bird species and the weather-dependent observation result in an error rate of up to 50% (Berthold 1976). Ellenberg et al. (1989) checked the reproducibility of registering "singing males" in forests. They concluded that the species inventories of independently registering observers in the same areas matched each other by about 90%. In the relative abundances of the different species, the observers matched each other by 80%. The errors become smaller with larger study areas (over 60 ha). Mattes (1977) recommends using test areas of at least 30–35 ha in coniferous forests to avoid exceeding the 10% error limit. Edge areas are the leading cause of errors. The results can be improved for certain species by performing an absolute inventory in a delimited test area (e.g., a 3–4 ha large section of the study area).

For the mapping of individual species, the researcher records precisely each observed activity of a bird species on the map of the control area. From the ten control maps of the visits, species-specific maps are then finally drawn, where the observed activities provide information, including territory boundaries and nest locations (Example, Fig. 1.9). For the bird species, abbreviations are suitable, preferably consisting of two to three let-

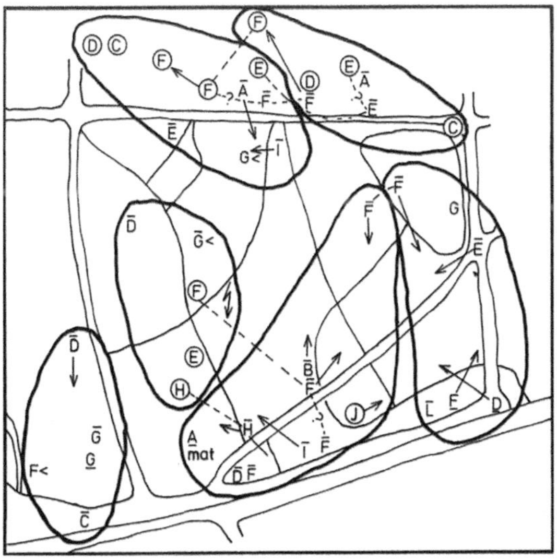

Fig. 1.9 Example of a summarized species map for a songbird (after Marchant 1983). All letters represent encounters with the same species, with registrations at consecutive checks indicated by alphabetically increasing letters. Bold lines delimit the identified territories

ters of the scientific genus and species name (e.g., Tomialojc 1976). Also, codes can be used to register different activities (suggestions can be found at the *British Trust for Ornithology* under https://www.bto.org/get-involved/volunteer/projects/bbs/taking-part/download-forms-instructions – or by Google "BTO forms". There is an app: The BirdTrack app: https://www.bto.org/get-involved/volunteer/projects/birdtrack and in Sutherland (2006). These resources also provide examples of recording forms for *bird surveys*, habitat recording forms, and observations of mammals (mammal *summary sheet).*

Population studies on individual bird species have shown that many species have polygamous mating systems while other species are monogamous. Moreover, depending on the species, in almost all areas, a certain proportion of unpaired males reside who currently do not occupy a territory. This is particularly true in Central Europe's densely populated anthropogenically shaped secondary forests, but are also found under primary forest conditions (Wesolowski et al. 1987). Therefore, one cannot simply infer from the males identified during the mapping of breeding pairs (one female for each male) the actual individual density of a bird species.

1.8 Point-Stop Counts

Instead of line transects, you can also carry out point-stop counts (point transects). They are initially simpler and can be applied to large areas. They are often used for monitoring purposes in comparisons between countries (e.g., Ralph et al. 1995a, b).

Compared to the mapping method through control walks, point registration methods by counting individuals perceived at a certain distance and within a specific time (point transects) are particularly suitable outside the breeding season for systematic bird population recording in forests (Blondel 1975; Reynolds et al. 1980).

A researcher records all bird contacts within a 50 m radius for 15 min at fixed marked points. Use individual trees to mark the survey radius. Distribute the point transects across different types of forest. Since some birds might leave the area upon the observer's arrival and are not rediscovered within the counting time, counting bird contacts should begin right upon arrival. For a large-scale inventory, increase the radius in which you record bird sightings and vocalizations. For example, Svensson (1977) recommends an observation radius of 125 m so that 8 counting points cover 1 km^2. He suggested a counting duration of only 5 min per point but only recorded 62% of the species that he recorded during a 30-min observation duration. However, longer observation duration also leads to recording duplicate bird contacts. You can record far more birds if you obtain all vocalizations and sightings at each point, regardless of the distance. Especially cuckoos, carrion crows, wood pigeons, redstarts, song thrushes, skylarks, chaffinches, gardens, and lesser whitethroats, and many water birds are often heard or seen outside of the 100 m limit (70%, Svensson 1977). However, despite the "tempting" registration of all birds one hears or sees, only recordings within a narrow area yield quantitatively comparable results.

It is best to combine the point registrations with the *distance sampling* method (Buckland

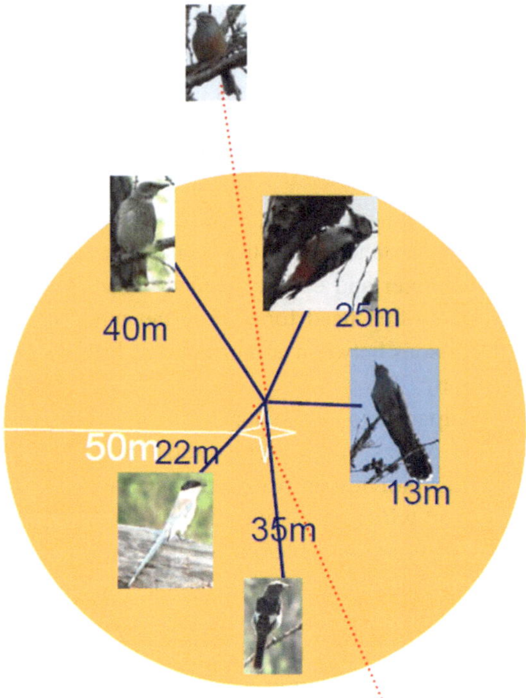

Fig. 1.10 Bird point counts. The calculation example shows the process of comparing four habitats. Sample effort: one point count at the center of a study site (habitat) for 20 min between 06:00 and 09:00 h. Four replicate sites are selected for each of the four studied habitats. Each site is visited alternatively, with eight visits per site. 1 point count per site × 4 replicate sites × 4 habitat types × 8 visits each = 128 bird point counts. A rangefinder is used for distance estimation within the 50 m circle, and birds beyond 50 m are discarded

et al. 2015, see Sect. 5.1), as this approach also allows you to estimate the density of individual bird species. The software Distance (https://distancesampling.org) offers the corresponding calculations. Point counts are also well suited for comparing different habitats with each other in terms of bird populations. Figure 1.10 provides an example of a point count.

The decision to use either a transect or point count method depends on the habitat type. The transect method is preferred in extensive habitats with good possibilities for marking the control routes; in dense and less clear habitats with patchy distribution of different vegetation types, the point count is to be preferred (Anderson and Ohmart 1981; Verner 1985). The mapping of territories provides more accurate information on specific study areas. However, due to the higher effort, it is limited to smaller areas and only possible at certain times of the year. For a recurring annual monitoring program, transect or point counts without fixed boundaries are best suited (Ralph et al. 1995a, b). In Central European forests, about 30–40 breeding bird species can be expected in a study area of less than 1 km^2. The density of breeding pairs total per 1 km^2 usually varies between 300 and 1100 (Scherner 1977; Bezzel 1982; Klafs and Stübs 1987; Tomiałojć et al. 1984; Erard 1989; Hogstad 2013).

1.9 Sound Decoys and Listening

Territory owners can be determined by playing species-specific songs from a portable tape recorder (e.g., Parker 1991). Most bird voices are available on cassettes or CDs (for Europe: Palmér and Boswall 1981; Roché 1990; Schulze 2003) or smartphone apps (Cornell University 2024; mydigitalearth.com 2024; National Audubon Society 2024). Source worthless without pointing to the exact publication! The Merlin app (Cornell University 2024) provides pretty good, automated bird identification using artificial intelligence for many areas worldwide. Wildlife Acoustics offers virtual courses on using bioacoustic applications, recorder features and functionality, and sound analysis (https://www.wildlifeacoustics.com/resources/training-courses/virtual-community-training).

Standard gray partridge (*Perdix perdix*) mapping is also carried out with sound dummies along 1–1.5 km transects in the cultivated landscape (Gottschalk 2024. In Germany, the mapping period is between the end of February and the end of March. Mapping takes place ½ to 1 h after sunset. Partridges respond from a distance of 100–300 m, so transects should be 500 m apart. The recordings are entered directly onto a field map with the transects marked. A small external speaker is needed in addition to the mobile phone to ensure that the call can be heard. You can test whether the playback call can still be heard by moving 200–300 m away from the speaker. The call is played back every 150 m for

Fig. 1.11 Partridges (*Perdix*) occur in great parties' during the non-breeding season and seek cover in tall grass or verges in the cultivated landscape

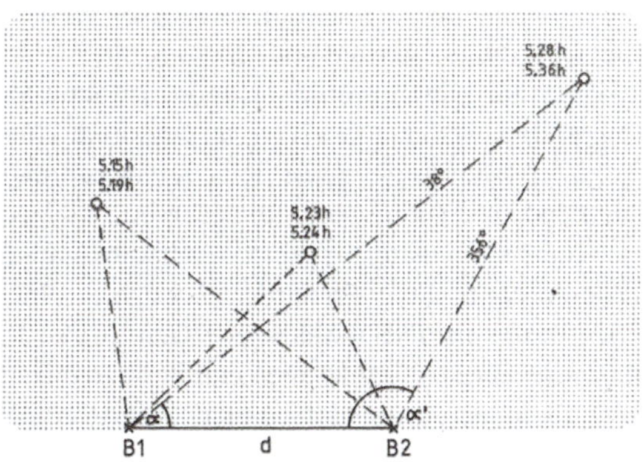

Fig. 1.12 Diagram to illustrate the bird call triangulation and the associated area estimation. Two observers (**B1** and **B2**) are at a known distance (*d*) from each other, and each records the bearing angle and the exact time for each interesting bird sound. In a terrain map, the location and distance of the calling bird can then be determined by triangulation. The recorded time facilitates the identification of the heard birds

30 s, if possible, broadcasting in all directions. You can find precise instructions (in German) at https://www.rebhuhnschutzprojekt.de/kartierung.html (Wiedenmann and Purkart 2022). Katzenberger (2022) provides a detailed presentation on gray partridge monitoring in an online video (in German) (Fig. 1.11).

With a "playback" equipment (excellent directional microphone, two recorders, and a separate speaker box), you can also identify unknown singers: You record the song and play it back immediately via the speaker box. Almost all singing or calling birds will come closer, and the song or call recording can be improved. Due to the improved, immediate call playback, the birds usually fly so closely that they can be visually identified. The playback method is beneficial in regions new for the investigations (e.g., tropics), where not all bird sounds are available, and many birds are challenging to see in the foliage.

While the distance of bird calls is difficult to estimate, you can determine the direction with a compass with only a few degrees of deviation. This limitation can be overcome by working in groups of two people listening to birds within the area. You will need to know the exact distance between the listeners who are precisely targeting the specific bird calls or songs and accurately note the times and direction when they hear the target bird sounds. From the records of at least two observers, the location of the individual bird singers can then be determined cartographically by triangulating (see Fig. 1.12). A rough habitat mapping completes the desired result, determining how many birds of a selected species have been called in a particular area and how extensive the corresponding habitat is. Analyzing many triangulations, you also get a value for the maximum hearing distance for the examined species. Suppose the "listened" area cannot be spatially defined. In that case, comparing the average distance between the located bird calls can measure the relative density of the singing males.

In individual cases, you can also use special sound-generating devices for the detection and estimation of the population density of selected species that are difficult to track, such as the decoy whistles for the hazel grouse (*Bonasa bonasia*), which can be found in hunting equipment stores. Bergmann et al. (1996) describe their application.

For conservation reasons, decoys must not be used to catch animals.

1.10 Luring Top Carnivores

Researchers can also use decoys to estimate the population of larger carnivores. For example, lions and spotted hyenas can be attracted by the calls of young antelopes at night when the predators are otherwise difficult to spot (Kiffner et al. 2008). Particularly effective have been various calls representing the distress bleating of a wildebeest (*Connochaetes taurinus*) calf, the distress call of a buffalo calf, or the sound of hyenas competing over a kill (Cozzi et al. 2013). Playing the calls at full volume with an MP3 player has proven effective. The player was attached to a mobile 120 W amplifier, and the sound was broadcast via two 60 W horn speakers. In order to estimate response probability and the maximum distance from which lions respond to the playback, researchers must conduct calibration experiments. Again, having two sets of observers/cars with radio communication is necessary. Lions are best detected when systematically searched for in the afternoon. One set of observers stays with the lions while the second car (with the speakers) drives a few km away.

After obtaining the three required results—(a) maximum response distance, (b) response probability, and (c) the results of the survey counts—it is possible to estimate the expected number of responding carnivores within the response range $(\hat{\mu})$ by the formula: $\hat{\mu} = y - / \hat{\pi}$ with $y-$ being the mean number of lions/hyenas responding per station and $\hat{\pi}$ the probability of the carnivores to approach the station (Mills et al. 2001).

1.11 Automated Birdsong Recognition

Bird songs and bird calls can also be automatically recorded by sound recorders, e.g., AudioMoth (Open Acoustic Devices 2023, https://www.openacousticdevices.info/audiomoth) and Song Meter (e.g., Song Meter Mini 2[1]; Wildlife Acoustics, Inc. 2024a) (Fig. 1.13). Such monitoring has many advantages. First, after the sound recorders have been distributed to selected points in the study area (e.g., by attaching them to trees), the study area only needs to be entered to check the equipment and retrieve the recordings, usually on SD or microSD cards. As each visit by one or more observers causes an unpredictable disturbance of the animal community in the study area, this method is much less invasive. Second, the recordings register sounds over the entire 24-h cycle, which would be more difficult or impossible if an observer is required. For extended battery life, some devices can be time-triggered. The sound recordings can be analyzed in the lab using special software, e.g., Kaleidoscope Pro (Wildlife Acoustics, Inc. 2024b) and BirdNET-Analyzer (Kahl 2024). In this way, certain bird species can be recognized and their activities determined (Katz et al. 2016; Leach et al. 2016; Ovaskainen et al. 2018; Priyadarshani et al. 2018; Shonfield and Bayne 2017; Darras et al. 2018a, b).

Fig. 1.13 Acoustic recording device attached to a beech tree in the forest with weatherproof housing (Singer 2020)

The Haiku box (https://haikubox.com) is available for citizen scientists and bird enthusiasts in the US. It uses the artificial intelligence Merlin at its backend (Cornell University 2024) to automatically identify bird sounds—the box records continuously and allows data sharing as long as it is connected to Wi-Fi and a power source.

1.12 Recording Ultrasonic Calls of Bats

Sound recorders can also be used to record ultrasonic calls of bats (Russo and Voigt 2016). Monitoring bat species is vital for conservation in Europe, North America, and worldwide. In principle, the Habitats Directive requires monitoring

[1] https://www.wildlifeacoustics.com/products/song-meter-mini-2-aa $499 (August 2024).

all bat species' conservation status in Central Europe (Dietz and Simon 2005). Direct field determination of ultrasonic calls can be done with simple bat detectors.[2] However, there is the difficulty of most species being highly mobile and nocturnal, and some bat species cannot be conclusively identified with a detector.

Passive acoustic monitoring (PAM) refers to using acoustic sensors to monitor vocalizing animals and environmental effects (Browning et al. 2017). PAM can be widely applied in ecology and nature conservation studies. PAM is also suitable for recording anurans (frogs and toads) (e.g., Stevenson et al. 2015). Especially in the tropics, anurans are hard to discover in forests, and many species are nocturnal.

With PAM, it is also possible to investigate disturbance effects, such as logging in forests, agricultural expansion, or energy production (Deichmann et al. 2017). To more effectively capture disturbances from human activities on wildlife, PAM can also be combined with camera traps (Buxton et al. 2018). Acoustic monitoring of specific species, taxonomic groups, or entire "soundscapes" (Grinfeder et al. 2022; Pijanowski et al. 2011; Ratcliffe 2021; Turlington et al. 2024) is the foundation of a new discipline coined "ecoacoustics" (Abrahams et al. 2023; Calanchi 2023; Farina et al. 2022; Farina and Gage 2017).

1.13 Nest Search

In addition to territory mapping with sound lures, a systematic nest search of open nesters should be carried out; only this combination of methods can determine the actual number of breeding pairs (Berthold 1976; Oelke 1977). Nests rarely need to be checked up close: building, breeding, or feeding adult birds can also be located and recognized from a considerable distance. The risk of disturbance is generally low, but one should proceed with particular caution at the beginning of the breeding season. Before a nest search, everyone should familiarize themselves thoroughly with the breeding biology of the target species using the relevant literature. For cavity nesters like great tits, blue tits, and pied flycatchers, nest box checks are easy to carry out. There are field guidebooks for identifying nests in Europe, e.g., Harrison (2002), and in North America, e.g., Harrison (1998), also for monitoring bird nests, the *British Trust for Ornithology* Recommendations (Ferguson-Lees et al. 2011).

When the leaves have fallen in winter, there is also the possibility of recording the distribution of bird nests in hedges or at forest edges. The nests can usually be assigned to bird species with the help of field guide identification aids (Fitter and Richardson 1974; Harrison 1998, 2002; McFarland et al. 2021). We suggest you record each nest's type, height, and plant species. With these data, the value of a hedge or forest edge for breeding birds can be estimated, and recommendations for the creation or maintenance of hedges and forest edges can be developed (see the corresponding evaluation of hedges by Heusinger in Zwölfer et al. 1981, 1984), see also Sect. 10.14 (Fig. 1.14).

A particular difficulty is the recording of birds of prey in extensive forests. While in Central Europe, most forests are not very extensive; mapping the nests in early spring, when the deciduous trees have no leaves yet, is more accessible (Mammen and Stubbe 2003). An orienting hint is the presence or absence of corvids, as these avoid goshawk territories (Ellenberg 1983; Ellenberg et al. 1984). An identification guide for raptor nests in Germany is available from Hoffmann and Lösekrug (2020). The tree crowns in tropical forests are often not visible. Thiollay (1989), therefore, combined transect checks (visual observations and listening) with mapping the birds of prey circling above the tree crowns and conducted these observations regularly from suitable hills with an expansive view. The key was monitoring an extensive area (approx. 100 km^2) and adapting the estimation method to the different behaviors of the species in terms of daily activity, circling, and calling.

---

[2] A professional bat detector is, for example, the hand-held "Echo Meter Touch 2 Pro" from Wildlife Acoustics, Inc. It costs approx. $350 (Wildlife Acoustics, Inc. 2024c). The company also offers stationary bat monitors like the Song Meter Mini Bat 2 that run for approx. $800.

Fig. 1.14 (**a**) The red kite (*Milvus milvus*) was observed building its nest in spring (March). Lower Saxony, Germany. (**b**) The same red kite nest as in left in June is challenging to discover. (**c**) A red kite (*Milvus milvus)* sits on its nest. It is the same nest as in (**b**)

Fig. 1.15 (**a**) A mist-net set up between riparian shrubs. (**b**) Several birds are trapped alive in a mist net in Mongolia. (**c**) Right: if a bird flies against the mist net, it is captured in a pocket without injuries

1.14 Mist-Netting for Recording Understory Birds

A standardized catch of songbirds with so-called Japan-nets (Lockley and Russell 1953; Mac Arthur and Mac Arthur 1974; Bub 1976; Bub and Oelke 1980; Dorsch 1971; Bird Ringing Scheme | British Trust for Ornithology BTO; Euring 2007) is only suitable for specially trained ornithologists. It requires both permission from the conservation authority and a ringing permit from the bird observatory institution responsible. In Germany, for the federal states of Bremen, Hamburg, Hessen, Lower Saxony, North Rhine-Westphalia, and Schleswig-Holstein, the bird observatory Helgoland, 2940 Wilhelmshaven-Rüstersiel, is responsible. For Baden-Württemberg, Bavaria, Saarland, and Rhineland-Palatinate, as well as for Austria, the Radolfzell Bird Observatory, 7760 Radolfzell-Möggingen, is responsible (Berthold et al. 1980). In the new federal states of Germany and Berlin, one should contact the Hiddensee Bird Observatory, D-17493 Greifswald.

Birds in the understory of forests are difficult to observe. This difficulty is especially true in tropical forests, where species differentiation is complicated. Live trapping with Japan-nets allows for a standardized and observer-independent recording, which also allows for monitoring relative frequencies in temporal and locational comparison. Large-scale population dynamic parameters of many songbird species were recorded in this way in a long-term project of the three German bird observatories (Köppen 2003; IMS 2013) (Figs. 1.15, 1.16, 1.17 and 1.18).

Fig. 1.16 (**a**) The equipment for bird banding in the field: (from left to right) a small bag to transport and keep captured birds, a paper funnel for weighing the bird, an electronic scale, rulers, and rings of a specific size; in front left, a red pair of pliers (arrow) to fasten the ring on the feet of the bird. (**b**) The bird is placed headfirst into a paper funnel, causing it to stay quiet while weighing

Fig. 1.17 Azur-winged magpie (*Cyanopica cyanus*) marked with colored rings for individual identification in the field observations. Northern Mongolia

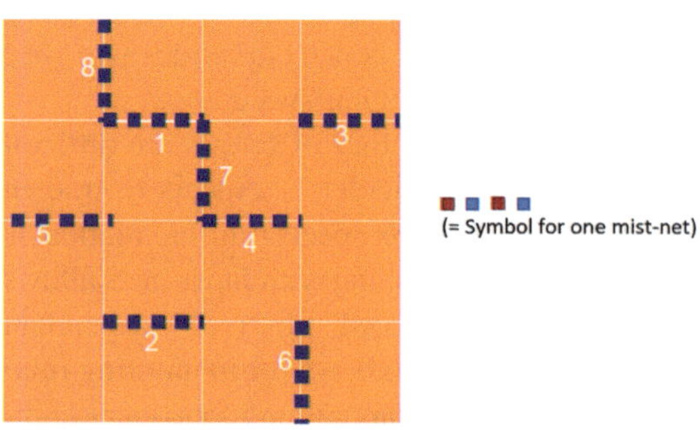

Fig. 1.18 Design of mist nets in a habitat. A mist net (indicated by a row of four blue squares) is most commonly 12 m long. However, there are also mist nets of 6 m in length

In the following example, we want to show how mist-netting can be installed to record the bird survey in a habitat. We also calculate the time required for monitoring if we want to compare five habitats in this way.

The sample effort can be calculated as follows:

- Eight mist-net sample lines per study site, five study sites (habitats surveyed).
- One mist-net line is open for 10 h, 15:00–18:00 and 06:00–13:00 h, = 2 days' work.
- 8 repetitions for mist-netting = 8 samples.
- 8 samples × 8 nets × 10 h = 640 netting-hours per site, 5 sites × 640 netting-hours = 3200 net-hours in total.
- 5 sites × 8 mist-net samples × 2 days = 80 days of fieldwork.

1.15 Community-Based Monitoring

Almost everywhere in the tropics, people use forests in some way (Vancutsem et al. 2021). If there has been logging, many larger vertebrates may have disappeared from the area. Hunting mammals and larger birds by the local residents to cover their protein needs might also lead to extirpation (see Sect. 13.8) (e.g., Pangau-Adam et al. 2012). In the case of wood exploitation and so-called "sustainable" bushmeat use, the question arises of whether the species that should be present in the tropical forest are still present in sufficient numbers, i.e., are ecologically still in viable populations. Even when a forest is to be restored or rehabilitated, we must know the corresponding animal populations. They can also serve as a target for forest management (Mühlenberg et al. 1995). "Sustainable" in an ecological sense can only mean that biodiversity is preserved.

1.15.1 Monitoring of Mammals and Large Birds in the Tropics

How can one establish monitoring of mammals and large birds in a tropical forest? The first step is evaluating the forest: what is missing compared to a primary forest? Is it the thick trees, the sparse undergrowth, or the overall basal area covered with wood? A one-time vegetation survey (see Chap. 4) is sufficient here. It would be advantageous if a local expert in recognizing tree species participated in the field to describe the vegetation. The local expert can also align the local tree species' names with their scientific names. The botanical description of the forest can be carried out either on a chosen plot or along a transect. A transect has the advantage that animal population surveys are faster collected along a transect line (see Sect. 5.1).

The main question arises: who should carry out the animal population surveys? One should carry out long-term monitoring using the same method. (e.g., McComb et al. 2010). It should identify trends in the development of populations, comparing specific management methods. The experts for individual animal groups usually come from developed countries ("outside experts"), who are only available temporarily and expensive to fly in and employ for a limited time on site.

In contrast, a "community-based approach" aims to involve indigenous local people (see also Spellerberg 2005). Local forces are financially cheaper and bring several advantages: they come from neighboring villages and thus guarantee the inclusion of the local community, which is particularly necessary for conservation efforts (Sheil 2001). Local people can also guarantee continuity for extended periods, such as several years. As a rule, they are also more "resilient" to adverse weather conditions, such as heavy rain. However, the local workforce might require more extended training.

For the selection, we recommend holding a village meeting and identifying those interested who can write almost flawlessly and recognize the species of interest in the field. Often, current or former hunters are particularly suitable for this, and a second person with writing skills can be added. Difficulties for a later protocol can arise because the villagers know the species by local names, and you will need to compile an adapted dictionary before the start of the monitoring program. The process can, therefore, take a long time, sometimes a year. Books like field guides of mammals or birds for the corresponding region are suitable for the training. Intensive training of the ecological staff incorporating the Indigenous knowledge of local hunters for identification of game species and their footprints, as well as bird calls and regular follow-up, are prerequisites for successful program implementation (Fig. 1.19).

What can be achieved with standardized biomonitoring?

The continuous monitoring of the species results in a temporal sequence of the frequency of encounters of the different species over the years. Several ways exist for interpreting these phenology curves: as long as the sampling effort is the same, you can deduce a population trend. Fluctuations within a year can have biological causes or can be caused by your methodology. Individual species can, for example, leave the forest seasonally or remain silent or particularly secretive during their life cycle at a particular time of year. Methodologically, discovering animals can be complicated in the rainy season. The comparison of the phenology curves between different forests or different recording groups of villagers ("ecological teams") usually allows the correct interpretation.

For observations to be carried out and recorded continuously, e.g., once a month throughout the year, the transect in the forest must be marked clearly, including the marking of the route, e.g., every 500 m. This has the advantage that future observations can be assigned to corresponding sectors. The habitat characteristics often change in a natural forest, even over smaller distances. A transect should not be longer than about 2 km in a tropical forest, as the observers' attention decreases over time when walking the transect. Experience shows that observations along a 2 km transect take at least an hour. After this time, you also have to return to the starting point. For an adequate return, it is advantageous to design the transect as a more or less equilateral triangle (see example at the pellet counts, Sect. 5.2). In addition to the transect marking, a GPS device helps. Section 4.4 explains its use and application.

Fig. 1.19 Monitoring by trained villagers, Southwest Cameroon

Which species to select for monitoring?

Usually, you cannot record all occurring species in a limited time. We distinguish between a "*specialist approach*" and a "*community-based approach*" (Table 1.2). Danielsen et al. (2009) provide a detailed summary of monitoring approaches with different levels of involvement of local people.

An expert will think of a "best indicator species" like true habitat species specialists or sensitive species. However, actual indicators are rare and often difficult to identify, and their data is usually insufficiently recorded for a statistical analysis (Sect. 9.8). Species reliably identified by the village hunters can also be significant for con-

Table 1.2 Advantages and disadvantages of two alternative strategies of biomonitoring data collection

	Specialist approach	Community-based approach
Data collection	Scientists are expensive; field workers come from outside (strangers), and one scientist per taxon	Trained villagers: Cheaper but longer training is necessary; the village supports the conservation; continuity is guaranteed/more likely; local people are used to working in adverse field conditions (e.g., rainy season)
Selection of species for monitoring	The number of species increased by forest specialists using "best indicators," but true habitat specialists or sensitive species may be rare. Therefore, a database for statistical analyses of trends is insufficient. They are only identified by experts	Hunters know local species and species of conservation concern: sufficient data for statistical analyses (at least in 1-year samples), an essential component of biodiversity, includes charismatic species (e.g., in Africa: forest elephant, chimpanzee, hornbills)

Survey of the fauna in West Khentei

Date______

Area ____

Observer ____

Field trip by horse ___ car ___ foot ___

Cloud cover clear cloudy rainy **Wind** (N/NE/SE/S/SW/W) weak strong __

Time I ________ Way across____________GPS_______________

Time II ________ Way across ____________ GPS_______________

Time III ________ Way across____________ GPS_______________

Vertical distance of the object to the way________________________________

Distance to Khonin Nuga _______________________________________

Total distance covered for this trip ________________________________

Duration of the tour (hrs) _____________________________________

Pellets/Traces/Signs/Number of findings

Bear __________________ Wolf __________________ Otter ________

Maral __________________ Wild pig __________________________

Roe-Deer ___________________ Musk deer ______________________

Biotope types

Dark Taiga ___ Bright Taiga___Floodplain vegetation ___

Meadow steppe __ Herb steppe (river valley terrace) ___ Mountain dry steppe __

C. Further information

Sights (time I or II or III ?)___________________________________

Adult= **A** , Juvenile= **J** Male= **M** , Female= **F**

Age/Sex/Number/Biotope type__ (Example: **BearA/F/1, J/?/2/river valley** = female bear with two cubs of unknown sex, seen in river valley)

Bear __________________ Wolf __________________ Lynx ________

Red fox ______________________ Badger __________________________

Roe deer __________________ Wild pig ____________ Moose __________

Deer(Maral) __________ Musk deer ____________ Wolverine ________ Otter ________

Sable ______________ Squirrel ________________ Weasel ____________

BIRDS (number, sex F or M for grouse (*Tetraonidae*)

Capercaillie ________ Black grouse ______________ Hazel grouse ______________

Ural owl ______________ Lappland Owl ______________ Black vulture __________

White-tailed eagle __________ Osprey ________ Golden/Spotted Eagle ___________

Goshawk ____________ Sparrow hawk ____________ Common buzzard ___________

Fig. 1.20 Example of a "seen sheet" for monitoring purposes servation, such as forest elephants, certain primate species, and giant hornbills. The decline of such charismatic species indicates a likewise declining state of the forest under consideration. It is possible to include up to 20 species (mammals and birds) (Mühlenberg et al. 1995, 1999). The monitoring teams are trained in species identification and distance estimation (with the help of range finders) for distance sampling.

The use of "distance sampling" is strongly recommended for the surveys. Section 5.1 provides a more detailed presentation and implementation in the field. Experts in their home country can evaluate the data after entering the observations on site into pre-made tables (see Sect. 5.1) (Mühlenberg and Waltert 2001; Lien et al. 2001). Distance sampling also offers the advantage that the data collection can be interpreted later as sufficient or precise (Buckland et al. 2015). For individual bird species, one could also consider "spot-mapping" (Bibby et al. 2000) to review the *distance* results.

A tricky part is assessing the effects of management on forest management itself, which can also be the basis for evaluating a designated protection zone. For this purpose, reference areas are necessary for selected forest parcels, which are analyzed using the same monitoring method. In addition, replicates should be carried out, see Sect. 1.14 case study. The problem is that, strictly speaking, different areas on a larger scale (over 1 ha, aimed between 10 and 100 ha) never ecologically match all parameters. It is essential to filter out decisive parameters during vegetation recording (see above) and select comparable forest pieces based on their commonality. Statistically, it would be desirable to reach about 5–6 replicates. However, this is usually not realistic in the tropics. However, a significant change in the inventory data after the onset of a particular management would be significant, e.g., after logging. However, we would have to ensure that the changes in the population data are not seasonally conditioned and do not fluctuate due to biotic factors independent of management. When comparing a treated area with a reference area, not only is the number of endangered species essential for the evaluation in nature conservation but also their affiliation to different categories of endemism and habitat choice (e.g., forest species or savannah species) (cf. Sects. 10.3 and 10.7).

Regarding nature conservation, we would have two options to react: (A) We could demand a resource that guarantees the preservation of selected species (or at least not jeopardizes their survival by reducing the population). (B) We could also demand the designation of a protected area or a refuge where no management measures may occur. (A) would correspond to an integrative (B) to a segregated approach. In (B), we could add up the weakened population in the management zone and the refuge's subpopulation for their viability. There are also examples of resource use in a management area that can increase due to the neighborhood being a fully protected area. Concrete evidence has been provided several times in marine conservation management (Roberts et al. 2001; Abesamis and Russ 2005; Hughes et al. 2006).

1.15.2 Seen Sheets

To get information on uncommon species in the vertebrate community, "seen sheets" shall be used. Such forms provide a way of monitoring populations by recording random encounters with animals seen while doing other jobs in the area. Indeed, the observers vary significantly in their sighting abilities, but because some team members represent "high-quality" observers, their results can be used as a baseline for calibration (Hochachka et al. 2000).

The following form may serve as an example of the patterns of "seen sheets" for observations in a wilderness area of the Mongolian Taiga (Fig. 1.20), compare also Clark et al. (2006):

References

Abesamis R.A. & Russ G.R. 2005: Density-Dependent Spillover from a Marine Reserve: Long-Term Evidence. – *Ecological Applications* 15(5), 1798-1812

Abrahams, C., Ashington, B., Baker, E., Bradfer-Lawrence, T., Browning, E., Carruthers-Jones, J., et al. (2023). *Good practice guidelines for long-term ecoacoustic monitoring in the UK.*

Amlaner, C. (1987). 24 Hour Activity Patterns of the Douglas Squirrel (Tamiasciurus douglasii). *Biotelemetry IX - Editors: P. Kimmich and M. R Neuman, Doring-Druck Ltd, Germany*. https://www.academia.edu/21235917/24Hour_Activity_Patterns_of_the_Douglas_Squirrel_Tamiasciurus_douglasii_. Accessed 17 April 2022

Amspacher, K., Bauer, B., Waldron, J., Wiggers, E., & Welch, S. (2019). Sciurus niger niger (Southern Fox Squirrel) Density and the Diurnal Patterns, Occupancy, and Detection of Sympatric Southern Fox Squirrels and S. carolinensis (Eastern Gray Squirrel) on Spring Island, South Carolina. Southeastern Naturalist, 18(2), 321–333. https://doi.org/10.1656/058.018.0214

Anderson B.W. & Ohmart R.D. 1981: Comparisons of avian census results using variable distance transect and variable circular plot techniques. – *Studies Avian Biol.* 6, 186–192

Anders, J. L., Uchida, K., Watanabe, M., Tanio, I., Shimamoto, T., Hamada, M., et al. (2017). Usefulness and limitation of a tiny light-temperature logger to monitor daily activity levels of arboreal squirrels in temperate areas. *Mammal Research*, *62*(4), 397–404. https://doi.org/10.1007/s13364-017-0326-0

Andretzke H., Schikore T. & Schröder K. (2005): Species profiles. In: Südbeck P. et al. (eds.): *Method standards for recording breeding birds in Germany*. p. 135-695. Radolfzell.

Aviles-Rosa E.O., Nita M., Feuerbacher E. & Hall N.J. 2023: An evaluation of Spotted Lanternfly (*Lycorma delicatula*) detection dog training and performance. – *Applied Animal Behaviour Science*, 10.1016/j.applanim.2022.105816, 258, (105816)

Baker N. 2013: The Nature Tracker's Handbook. - Bloomsbury Publishing, 288 p.

Beaumont S.N., Ehlers-Smith D.A., Ehlers Smith Y.C. & Downs C.T. 2022: The Development and Application of a Method for Estimating Red-Winged Francolin Population Density with Pointing Dogs in the Midlands, Kwazulu-Natal, South Africa. – *African Journal of Wildlife Research*, 10.3957/056.052.0122, 52, 1. https://doi.org/10.3957/056.052.0122

Beebe S.C., Howell T.J. & Bennett P.C. 2016: Using scent detection dogs in conservation settings: A review of scientific literature regarding their selection. – *Frontiers in Veterinary Science* 3, 96. https://doi.org/10.3389/fvets.2016.00096

Bellemain, E., Swenson, J.E. 2005: Estimating Population Size of Elusive Animals with DNA from hunter collected Feces: Four Methods for Brown Bears. – *Conservation Biology* (2005) 19

Bellemain, E., Swenson, J.E., Tallmon, D., Brunberg, S. and Taberlet, P. 2005: Estimating population size of elusive animals with DNA from hunter-collected feces: four methods for brown bears. – *Conservation Biology* 19, 150-161.

Bennett E.M., Hauser C.E. & Moore J.L. 2019: Evaluating conservation dogs in the search for rare species. – *Conservation Biology*, 34(2), 314–325. https://doi.org/10.1111/cobi.13431

Bergmann H-H, Klaus S., Müller F., Scherzinger W., Swenson J.E. & Wiesner J. 1996: *The Hazel Grouse*. – The New Brehm Library 77, Magdeburg, 278 p.

Berthold P. 1976: Methoden der Bestandserfassung in der Ornithologie: Übersicht und kritische Betrachtung. – *J. Orn*. 117, 1–69

Berthold P., Bezzel E. & Thielcke G. 1980: *Praktische Vogelkunde. Ein Leitfaden für Feldornithologen*. – Kilda-Verl., Greven/Westfalen, 159 p.

Bessone M., Kühl H.S., Hohmann G., Herbinger I., N'Goran K.P., Asanzi P., Da Costa P.B., Dérozier V., Fotsing E.D.B.Beka B.I., Iyomi M.D., Iyatshi I.B., Kafaando P., Kambere M.A., Moundzoho D.B., Wanzalire M.L.K. & Fruth B. 2020: Drawn out of the shadows: Surveying secretive forest species with camera trap distance sampling. – *Journal of Applied Ecology* 57: 963-974

Bezzel E. 1982: *Vögel in der Kulturlandschaft*. – Verlag Eugen Ulmer, Stuttgart, 350 p.

Bibby C, Burgess N, Hill D & Mustoe S. 2000: *Bird census technique* s, 2nd ed. – Academic Press, London, UK, 302 p.

Blondel J. 1975: L'analyse des peuplements d'oiseaux, elements d'un diagnostic ecologique. I. La methode des echantillonnages frequentiels progressifs (E. F. P.). – *Terre et la Vie* 29, 533–589

Boon, A. K., Réale, D., & Boutin, S. (2008). Personality, habitat use, and their consequences for survival in North American red squirrels Tamiasciurus hudsonicus. *Oikos*, *117*(9), 1321–1328. https://doi.org/10.1111/j.0030-1299.2008.16567.x

Bottaro R. 2023: *Detection dog: theoretical and practical handbook*. – Rose Canine Editions, 158 p.

Browning E., Gibb R., Glover-Kapfer P. & Jones K.E. 2017: Passive acoustic monitoring in ecology and conservation. - *Technical Report WWF 75p.*, https://doi.org/10.13140/RG.2.2.18158.46409

Bub H. & Oelke H. 1980: *Markierungsmethoden für Vögel*. – A. Ziemsen-Verlag Wittenberg Lutherstadt, 152 p.

Bub H. 1976: *Vogelfang und Vogelberingung zur Brutzeit*. – A. Ziemsen-Verl., Wittenberg Lutherstadt, 112 p.

Buckland S.T., Rexstad E.A., Marques T.A. & Oedekoven C.S. 2015: *Distance Sampling: Methods and Applications*. – Springer International Publishing Switzerland, 277 p.

Burki S., Roth T., Robin K. & Weber D. 2010: Lure sticks as a method to detect pine martens *Martes martes*. – *Acta Theriol* 55, 223–230. https://doi.org/10.4098/j.at.0001-7051.023.2009

Buxton, R.T., Lendrum, P.E., Crooks, K.R., & Wittemyer, G. (2018). Pairing camera traps and acoustic recorders to monitor the ecological impact of human disturbance. – *Global Ecology and Conservation*, 16, e00493.

Calanchi, A. (2023). Methods in Ecoacoustics. The acoustic complexity indices. *LINGUAE &-RIVISTA DI LINGUE E CULTURE MODERNE*, *23*(1), 163–165.

Caughley G. 2004: Analysis of Vertebrate Populations. - Caldwell, NJ: The Blackburn Press., 234 p.

Carranza J., de Trucios S.J.H., Medina R., Valencia J. & Delgado J. 1991: Space use by red deer in a Mediterranean ecosystem as determined by radio-tracking. - *Appl. Anim. Beh. Science* 30, 363-371

Clark E.L., Munkhbat J., Dulamtseren S., Baillie J.E.M., Batsaikhan N., Samiya R. & Stubbe M. 2006: Mongolian Red List of Mammals. – Regional Red List Series Vol. I. Zoological Society of London, London. (In English and Mongolian)

Cornell University. (2024, May 22). Merlin Bird ID by Cornell Lab. https://apps.apple.com/us/app/merlin-bird-id-by-cornell-lab/id773457673. Accessed 15 June 2024

Cozzi G., Broekhuis F., McNutt JW & Schmid B. 2013: Density and habitat use of lions and spotted hyenas in northern Botswana and the influence of survey and ecological variables on call-in survey estimation. – *Biodiversity and Conservation* 22 (12): 2937-2956

Creel, S., Spong, G., Sands, J., Rotella, J., Zeigle, J., Joe, L., Murphy, K.M. and Smith, D. 2003: Population size estimation in Yellowstone wolves with error-prone noninvasive genotypes. – *Molecular Ecology* 12, 2003-009.

Cristescu R.H., Foley E., Markula A., Jackson G., Jones D. & Frère C. 2015: Accuracy and efficiency of detection dogs: a powerful new tool for koala conservation and management. – *Scientific Reports* 5, Article number: 8349. https://doi.org/10.1038/srep08349

D'Eon R.G., Wilson S.F. & Hamilton D. 2006: *Ground-based Inventory Methods for Ungulate Snow-track Surveys*. Standards for Components of British Columbia's Biodiversity No. 33a. – Published by the Resources Information Standards Committee, The Province of British Columbia. 19 p.

Dalén, L., Götherström, A. and Angerbjörn, A. 2004: Identifying species from pieces of faeces. – *Conservation Genetics* 5, 109-111

Danielsen F., Burgess N.D., Balmford A., Donald P.F., Funder M., Jones J.P.G., Alviola P., Balete D.S., Blomley T., Brashares J., Child B., Enghoff M., Fjeldså J., Holt S., Hübertz H., Jensen A.E., Jensen P.M., Massao J., Mendoza M.M., Ngaga Y., Poulsen M.K., Rueda R., Sam M., Skielboe T., Stuart-Hill G., Topp-Jørgensen E. & Yonten D. 2009: Local Participation in Natural Resource Monitoring: a Characterization of Approaches. – *Conservation Biology* 23, 31-42. https://doi.org/10.1111/j.1523-1739.2008.01063.x

Darras K., Batáry P., Furnas B., Celis-Murillo A., Van Wilgenburg S.L., Mulyani Y.A. & Tscharntke T. 2018a: Comparing the sampling performance of sound recorders versus point counts in bird surveys: A meta-analysis. – *Journal of Applied Ecology* 55(6), 2575-2586

Darras K., Furnas B., Fitriawan I., Mulyani Y. & Tscharntke T. 2018b: Estimating bird detection distances in sound recordings for standardizing detection ranges and distance sampling. – *Methods Ecol. Evol.* 9, 1928–1938

Davis N.E., Di Stefano J., Whelan J. Wright J., Taylor L., Coulson G. & Sitters H. 2022: Power of faecal pellet count and camera trapping indices to monitor mammalian herbivore activity. – *Wildlife Research*, https://doi.org/10.1071/WR21135, 49, 8, 686-697

Deichmann J.L., Hernández-Serna A., Delgado C., Campos-Cerqueira M. & Aide T.M. 2017: Soundscape analysis and acoustic monitoring document impacts of natural gas exploration on biodiversity in a tropical forest. – *Ecological Indicators* 74: 39-48

Dietz M & Simon M. 2005: General notes on the recording of bats (Chiroptera). – In: Doerpinghaus A., Eichen C., Gunnemann H., Leopold P., Neukirchen M., Petermann J. & Schröder E. (Eds.): Methods for recording species of Annexes IV and V of the Fauna-Flora-Habitat Directive. – *Nature Conservation and Biological Diversity* 20. 318-330.

Dorsch H. 1971: Vogelberingung und Faunistik. – *Falke* 18, 340–350

Ecology Letters 2018: Animal Sound Identifier (ASI): software for automated identification of vocal animals. – *Ecology Letters* 21: 1244-1254

Ellenberg H. 1983: Wechselwirkungen im Ökosystem Kulturlandschaft: Habicht und Beute. – *Allgem. Forstzeitschrift* 44, 1195–1201

Ellenberg H., Gast F. & Dietrich J. 1984: Elster, Krähe und Habicht. Ein Beziehungsgefüge aus Territorialität, Konkurrenz und Prädation. – *Verh. Ges. Ökol.* 12, 319–329

Ellenberg H., Hudeczek H. & Lade H.J. 1989: Zur Reproduzierbarkeit der Erfassung von Vogelgemeinschaften in den Wäldern durch Registrierung „singender Männchen". – *Die Vogelwelt* 110 (5), 166–171

Emlen JT 1971: Population Densities of Birds Derived from Transect Counts. – *The Auk*, Boston, 88, 323–342

Erard C. 1989: Bird Community Structure in Two Rainforests: Africa (Gabon) and South America (French Guiana)-A Comparison. – Vertebrates in Complex Tropical Systems, 89-122. – In: *Ecological Studies* 69.

Euring 2007: *Bird Ringing for Science and Conservation*. – The European Union for Bird Ringing. 33 p. www.euring.org

Farina, A., Eldridge, A., Fuller, S., & Pavan, G. (2022). Editorial: Advances in ecoacoustics. *Frontiers in Ecology and Evolution, 10*. https://doi.org/10.3389/fevo.2022.978516

Farina, A., & Gage, S. H. (2017). Ecoacoustics: A New Science. In *Ecoacoustics* (pp. 1–11). John Wiley & Sons, Ltd. https://doi.org/10.1002/9781119230724.ch1

Ferguson-Lees J., Castell R. & Leech D. 2011: *A Field Guide to Monitoring Nests*. – British Trust for Ornithology (BTO).

Fitter RSR & Richardson R.A. 1974 (reprinted): *Collins Pocket Guide to Nests and Eggs*. – Collins, London, 175p.

Flagstad, O., Hedmark, E., Landa, A., Broseth, H., Persson, J., Andersen, R., Segerström, P. and Ellegren,

H. 2004: Colonization history and noninvasive monitoring of a re-established wolverine population. – *Conservation Biology* 18, 676-688.

Foran, D.R.; Minta, S.C., Heinemeyer, K.S. 1997: DNA-based analysis of hair to identify species and individuals for population research and monitoring. – *Wildlife Society Bulletin* 25, 840-847.

Frantzen, M.A.J., Silk, J.B., Ferguson, J.W.H., Wayne, R.K. and Kohn, M.H. 1998: Empirical evaluation of preservation methods for faecal DNA. – *Molecular Ecology* 7, 1423-1428

Gill R.M.A., Thomas M.L. and Stocker D. 1997: The Use of Portable Thermal Imaging for Estimating Deer Population Density in Forest Habitats. - *Journal of Applied Ecology* 34(5), 1273-1286

Glover N.J., Wilson L.E., Leedale A. & Jehle R. 2023: An experimental assessment of detection dog ability to locate great crested newts (*Triturus cristatus*) at distance and through soil. – *PLOS ONE*, 10.1371/journal.pone.0285084, 18, 6, (e0285084)

Goode M.J., Beaver J.T., Muller L.I., Clark J.D., van Manen F.T., Harper C.A. & Basinger P.S. 2014: Capture – recapture of white-tailed deer using DNA from fecal pellet groups. – *Wildlife Biology* 20, 270–278

Gopalaswamy A.M., Karanth K.U., Kumar N.S. & Macdonald D.W. 2012: Estimating tropical forest ungulate densities from sign surveys using abundance models of occupancy. – *Animal Conservation* 15 (6), 669-679

Gottschalk E. 2024: www.rebhuhnschutzprojekt.de, University of Göttingen. 60 p. (in German).

Grimm-Seyfarth A., Harms W. & Berger A. 2021: Detection dogs in nature conservation: A database on their worldwide deployment with a review on breeds used and their performance compared to other methods. – *Methods in Ecology and Evolution* 12(4), 568-579

Grimm-Seyfarth A., Zarzycka A., Nitz T., Heynig L., Weissheimer N., Lampa S. & Klenke R. 2019: Performance of detection dogs and visual searches for scat detection and discrimination amongst related species with identical diets. – *Nature Conservation* 37, 81–98.

Grinfeder, E., Lorenzi, C., Haupert, S., & Sueur, J. (2022). What Do We Mean by "Soundscape"? A Functional Description. *Frontiers in Ecology and Evolution, 10.* https://doi.org/10.3389/fevo.2022.894232

Harrington L.A., Harrington A.L. & Macdonald D. 2008: Distinguishing tracks of mink Mustela vison and polecat M. putorius. – *European Journal of Wildlife Research* 54(2), 367-371

Harrison C. 2002: *Bird Nests, Eggs and Nestlings (Collins Field Guide) of Britain and Europe*. 2nd Edition, Collins, 488p.

Harrison H.H. 1998: *Peterson Field Guide: Eastern Bird' Nests*. – Mariner Books. 288p.

Hewings R. 2019: *Scent Training for Every Dog*. – First Stone Publishing, 175 p.

Hicks, E. A. (1949). Ecological Factors Affecting the Activity of the Western Fox Squirrel, Sciurus niger rufiventer (Geoffroy). *Ecological Monographs*, *19*(4), 287–302. https://doi.org/10.2307/1943272

Hochachka W.M., Martin K., Doyle F. & Krebs C.J. 2000: Monitoring vertebrate populations using observational data. – *Canadian Journal of Zoology* 78: 521-529

Hoffmann M. & Lösekrug R-G. 2020: Bestimmungshilfe für Horste im Wald. – Hessisches Landesamt für Naturschutz, Wiesbaden, www.hlnug.de, 27 p.

Hogstad O. 2013: Species richness and structure of a breeding passerine bird community in a spruce-dominated boreal forest in central Norway: stability from 1960s to 2013. – *Ornis Norvegica* 36, 52–60

Hollerbach L. 2012: Canine detection of free-ranging koalas (*Phascolarctos cinereus*). – Master's Thesis Dept. Conservation Biology, University of Göttingen. 123 p.

Hondong H.: Conservation Detection Canines (CDK9), Georg-August-University Göttingen, Dept. of Conservation Biology

Howe, E. J., Buckland, S. T., Després-Einspenner, M., & Kühl, H. S. (2017). Distance sampling with camera traps. – *Methods in Ecology and Evolution*, 8, 1558–1565.

Hughes T.P., Bellwood D.R., Folke C.S., McCook L.J. & Pandolfi J.M. 2006: No-take areas, herbivory and coral reef resilience. – *TRENDS in Ecology and Evolution* 22 (1), 1-3

Integrated Monitoring of Songbird Populations. -

Kahl, S. (2024, June 17). BirdNET-Analyzer. Python. https://github.com/kahst/BirdNET-Analyzer. Accessed 17 June 2024

Karp D. 2020: Detecting small and cryptic animals by combining thermography and a wildlife detection dog. – *Scientific Reports* 10, 5220. https://doi.org/10.1038/s41598-020-61594-y

Katz J., Hafner S.D. & Donovan T. 2016: Tools for automated acoustic monitoring within the R package monitor. – *Bioacoustics* 25 (2): 197-210

Katzenberger J. 2022: Schulung zum Rebhuhn-Monitoring. - YouTube Video. Universität Göttingen, DDA.

Kiffner C., Waltert M., Meyer B. & Mühlenberg M. 2008: Response of lions (*Panthera leo* LINNAEUS 1758) and spotted hyenas (*Crocuta crocuta* ERXLEBEN 1777) to sound playbacks. – *African Journal of Ecology* 46: 223-226

Klafs G. & Stübs J. (eds.) 1987: Die Vogelwelt Mecklenburgs: Bezirke Rostock, Schwerin, Neubrandenburg. – Gustav Fischer Verlag, 426 p.

Kohn, M.H., York, E.C., Kamradt, D.A., Haught, G., Sauvajot, R.M. and Wayne, R.K. 1999: Estimating population size by genotyping faeces. – *Proceedings of the Royal society of London, Series B. Biological Sciences* 266, 657-663

Köppen U. 2003: The "*Integrated Monitoring Songbird Populations*" (IMS) – Potentials for a national bird monitoring concept and current status in Germany. –

Ber. State Office for Environmental Protection Sachsen-Anhalt, Special Issue 1/2003: 56-61

Koprowski, J. L. (1994). Sciurus niger. *Mammalian Species*, (479), 1–9. https://doi.org/10.2307/3504263

Koprowski, J. L., & Corse, M. C. (2005). Time budgets, activity periods, and behavior of Mexican fox squirrels. *Journal of Mammalogy*, *86*(5), 947–952. https://doi.org/10.1644/1545-1542(2005)86[947:Tbapab]2.0.Co;2

Kramm, K. R. (1975). Circadian activity of the red squirrel, Tamiasciurus hudsonicus, in continuous darkness and continuous illumination. *International Journal of Biometeorology*, *19*(4), 232–245. https://doi.org/10.1007/BF01451034

Krebs C.J. 1999: *Ecological Methodology.* 2nd ed. – Benjamin/Cummings, California, 620 p.

Kriegs J.O., Eversmann N., Hachmann H-U., Lindenschmidt M., Pickel T. & Rehage H-O. 2012: Eine Methode zur Kartierung des Baummarders *Martes martes* (Linnaeus, 1758) am Beispiel der Westfälischen Bucht. – *Natur und Heimat* 72(4), 107-116

Leach E.C., Burwell C.J., Ashton L.A., Jones D.N. &Kitching R.L. 2016: Comparison of point counts and automated acoustic monitoring: detecting birds in a rainforest biodiversity survey. – *Emu* 116 (3), 305-309

Lehman N., Clarckson, P., Mech L.D., Meyer, T.J., and Wayne R.K. 1992: A study of the genetic relationships within and among wolf packs using DNA fingerprinting and mitochondrial DNA. – *Behavioral Ecology and Sociobiology* 30, 83-94

Lehnert M.P. & Weeks ENI 2016: Trained dogs in insect detection. – In: T. Jezierski, J. Ensminger, & L. E. Papet (Eds.), Canine olfaction in science and law. *Advances in forensic science, medicine, conservation, and environmental remediation,* 321–333. CRC Press.

Lien, Waltert M., Faber K., von Loebenstein K. & Mühlenberg M. 2001: Community-based wildlife population assessment in the Korup Project Area, SW Cameroon. *ETFRN News* 32, 59-60.

Lockley R.M. & Russell R. 1953: *Bird-Ringing*. The Art of Bird Study by Individual Marking. – Crosby Lockwood & Son Ltd., London, 119 p.

Lucchini, V., Fabri, E., Marucco, F., Ricci, S., Biotani, L. And Randi, E. 2002: Noninvasive molecular tracking of colonising wolf (*Canis lupus*) packs in the western Italian Alps. – *Molecular Ecology* 11, 857-868

Lyra-Jorge M.C., Ciocheti G., Pivello V.R. & Meirelles ST 2008: Comparing methods for sampling large- and medium-sized mammals: camera traps and track plots. – *Eur. J. Wildl. Res.* 54, 739-744.

Mac Arthur R.H. & Mac Arthur A.T. 1974: On the use of mist nets for population studies of birds. – *Proc. Nat. Acad. Sei. USA* 71 (8), 3230–3233

MacKenzie, D., Nichols, J., Royle, J, Pollock, K., Bailey, L., & Hines, J. (2017). *Occupancy Estimation and Modeling,* 2nd edn. – Academic Press, San Diego, CA.

Mammen U. & Stubbe M. 2003: Monitoring Raptors and Owls of Europe. – *Ber. State Office for Environmental Protection Sachsen-Anhalt,* Special Issue 1/2003: 50-55

Mandujano S. 2005: Track count calibration to estimate density of White-tailed deer (*Odocoileus virginianus*) in Mexican dry tropical forest. – *The Southwestern Naturalist* 50(2), 223-229

Marchant J. 1983: *Common Birds Census of the BTO* (British Trust for Ornithology). – Maund & Irvine Ltd., Tring. Herts., England, 12 p.

Mattes, H. (1977): Erfahrungen mit der Kartierungsmethode zur Brutvogelbestandsaufnahme in Nadelwäldern. – *Die Vogelwelt* 98 (1), 1–15

Maudet, C., Luikart, G., Dubray, D., Von Hardenberg, A., Taberlet, P. 2004: Low genotyping error rates in wild ungulate feces sampled in winter. – *Mol. Ecol. Notes* 4, 772-775.

McFarland C., Monjello M. & Moskowitz D. 2021: *Peterson Field Guide to North American Bird Nests*. – Mariner Books, 500p.

McComb B., Zuckerberg B., Vesely D. & Jordan C. 2010: *Monitoring Animal Populations and their Habitats*. – CRC Press, Taylor & Francis Group, 277 p.

McDaniel, G.W., McKelvey, K.S., Squires, J.R. and Ruggiero, L.F. 2000: Efficacy of lures and hair snares to detect lynx. – *Wildlife Society Bulletin* 28, 119-123

Mills M.G.L., Juritz J.M. & Zucchini W. 2001: Estimating the size of spotted hyena (*Crocuta crocuta*) populations through playback recordings allowing for non response. – *Anim. Cons.* 4, 335–343

Moeller, A. K., Lukacs, P. M., & Horne, J. S. (2018). Three novel methods to estimate abundance of unmarked animals using remote cameras. – *Ecosphere*, 9, e02331.

Mooty J.J. & Karns P.D. 1984: The relationship between white-tailed deer track counts and pellet-group surveys. – *J. Wildl. Manag.* 48, 275-279

Mowat, G. and Paetkau, D. 2003: Estimating marten *Martes americana* population size using hair capture and genetic tagging. – *Wildlife Biology* 8, 201-209.

Mühlenberg M. & Waltert M. 2001: Influence of behavior on monitoring methods: Examples of vertebrates in West Africa. p. 123 – 134. – In: *Nature Conservation and Behavior.* International Symposium at the Center for Nature Conservation of the University of Göttingen (Gottschalk, E., Barkow, A., Mühlenberg, M. & Settele, J., Edts.) Environmental Research Center Leipzig.

Mühlenberg M., Slowik J., Waltert M., Fermon H., Bamba S. & Waitkuwait W.E. 1999: Biomonitoring in tropical forests. A contribution to the evaluation of Sustainable Use. – *Göttingen Contributions to Land and Forestry in the Tropics and Subtropics* 133, 220 - 225

Mühlenberg M., Slowik J., Wöll H. & Waitkuwait W.E. 1995: Strategies for the restoration of tropical forest, that incorporate Wildlife Protection: an example from the Ivory Coast, West Africa. p. 413-417 – In: Bissonette J.A. & Krausman P.R. (eds.): *Integrating people and wildlife for a sustainable future*. Proc. 1st

Internat. Wildl. Management Congr., The Wildlife Society, Bethesda, Maryland.

Murphy, M.A., Waits, L.P., Kendall, K.C., Wasser, S.K., Higbee, J.A. and Bogden, R. 2002: An evaluation of long-term preservation methods for brown bear (*Ursus arctos*) faecal samples. – *Conservation Genetics* 3, 435-440

mydigitalearth.com. (2024, May 8). Sibley Birds 2nd Edition. https://apps.apple.com/us/app/sibley-birds-2nd-edition/id1236011411. Accessed 15 June 2024

National Audubon Society. (2024, June 6). Audubon Bird Guide. https://apps.apple.com/us/app/audubon-bird-guide/id333227386. Accessed 15 June 2024

O'Connell A.F., Nichols J.D. & Karanth K.U. 2011: *Camera Traps in Animal Ecology. Methods and Analyses*. – Springer, 271p.

Oelke H. 1977: Vogelsiedlungsdichten und ornithoökologische Differenzierungen der Laubwälder und Laubholzanlagen des Westharzes. – *Ber. Int. Symp. Int. Vergg. Veg.kunde Rinteln*, 353–396

Oelke H. 1980: 4. Quantitative Untersuchungen. 4.1 Siedlungsdichte. – In: Berthold P., Bezzel E. & Thielke G. (eds.): *Praktische Vogelkunde*, 34–45. – Kilda-Verlag, Greven Westfalen, 158 p.

Olson L.-H. 2013: *Tracks and Signs of the Animals and Birds of Britain and Europe*. - Princeton University Press, 273 p.

Open Acoustic Devices. (2023). AudioMoth | Open Acoustic Devices. *openacousticdevices*. https://www.openacousticdevices.info/audiomoth. Accessed 17 June 2024

Osgood, D. W. (1972). Monitoring Activity of Small Mammals by Temperature-Telemetry. *Ecology*, *53*(4), 738-. https://doi.org/10.2307/1934794

Ovaskainen O., Moliterno de Camargo U. & Somervuo P. 2018: Animal Sound Identifier (ASI): software for automated identification of vocal animals. - *Ecology Letters* 21 (8), 1244-1254

Palmér S. & Boswall J. 1981: *A Field Guide to the Bird Songs of Britain and Europe*. – Swedish Radio Company, Stockholm, Kassetten 1–12 (dtsch. Bearbeitung 1982).

Palomares, F., Godoy, J.A., Piriz, A., O'Brien S.J., and Johnson, W.E. 2002: Faecal genetic analysis to determine the presence and distribution of elusive carnivores: design and feasibility for the Iberian lynx. – *Molecular Ecology* 11, 2171-2182

Pangau-Adam M., Noske R. & Muehlenberg, M. 2012: Wildmeat or Bushmeat? Subsistence Hunting and Commercial Harvesting in Papua (West New Guinea), Indonesia. – *Hum. Ecol.* https://doi.org/10.1007/s10745-012-9492-5; published online Springer.

Parker T.A. 1991: On the use of tape recorders in avifaunal surveys. – *The Auk* 108, 443-444

Pépin D., Morellet N. & Goulard M. 2009: Seasonal and daily walking activity patterns of free-ranging adult red deer (*Cervus elaphus*) at the individual level. - *Eur J Wildl Res*., https://doi.org/10.1007/s10344-009-0267-2

Piggott, M.P. 2004: Effect of sample age and season of collection on the reliability of microsatellite genotyping of faecal DNA. – *Wildlife Research* 31, 485-493

Pijanowski, B. C., Villanueva-Rivera, L. J., Dumyahn, S. L., Farina, A., Krause, B. L., Napoletano, B. M., et al. (2011). Soundscape Ecology: The Science of Sound in the Landscape. *BioScience*, *61*(3), 203–216. https://doi.org/10.1525/bio.2011.61.3.6

Priyadarshani N., Marsland S. & Castro I. 2018: Automated birdsong recognition in complex acoustic environments: a review. – *Journal of Avian Biology* 49 (5), https://doi.org/10.1111/jav.01447

Ralph C.J. & Scott J.M. (eds.) 1981: Estimating the numbers of terrestrial birds. – *Studies in Avian Biology* 6, Lawrence, Kansas, 1–630

Ralph C.J., Droege S. & Sauer J.R. 1995a: Managing and monitoring birds using point counts: Standards and applications. – In: Ralph C.J., Sauer J.R. & Droege S. (eds.). *Monitoring Bird Populations by Point Counts*. USDA Forest Service General Technical Reports PSW-GTR-149, p.161-168

Ralph C.J., Sauer J.R. & Droege S. 1995b: *Monitoring bird populations by point counts*. – Gen. Tech. Rep. PSW-GTR-149. Albany, CA: US Department of Agriculture, Forest Service, Pacific Southwest Research Station. 187 p.

Ratcliffe, E. (2021). Sound and Soundscape in Restorative Natural Environments: A Narrative Literature Review. *Frontiers in Psychology*, *12*, 570563. https://doi.org/10.3389/fpsyg.2021.570563

Reynolds J.C., Porteus T.A., Richardson S.M., Leigh R.J. & Short MJ 2010: Detectability of American Mink Using Rafts to Solicit Field Signs in a Population Control Context. – *Wildlife* 74 (7), 1601-1606

Reynolds R.T., Scott J.M. & Nussbaum R.A. 1980: A variable circular-plot method for estimating bird numbers. – *Condor* 82, 309–313

Richards N.I. (ed.) 2018: *Using Detection Dogs to Monitor Aquatic Ecosystem Health and Protect Aquatic Resources*. – Palgrave Macmillan, 396 p.

Roberts C.M., Bohnsack J.A., Gell F.R., Hawkins J.P. & Goodridge R. 2001: Effect of Marine Reserves on Adjacent Fisheries. – *Science* 294 (5548), 1920-1923

Roché J.C. 1990: All the birds of Europe. (All the bird songs of Britain and Europe) – 4 CD, Sittelle 01202.

Rowcliffe J.M., Field J., Turvey S.T. & Carbone C. 2008: Estimating animal density using camera traps without the need for individual recognition. - *Journal of Applied Ecology* 45, 1228-1236. https://doi.org/10.1111/j.1365-2664.2008.01473.x

Russo D. & Voigt C. 2016: The use of automated identification of bat echolocation calls in acoustic monitoring: A cautionary note for a sound analysis. – *Ecological Indicators* 66: 598-602

Scherner E.R. 1977: Struktur und Dynamik der Avifauna des Sollings. – *Verhandlungen der Gesellschaft für Ökologie*, Göttingen 1976, 145-160

Schulze A. 2003: The bird voices of Europe, North Africa and the Middle East. – 819 bird species on 17 audio CDs. – Music publishing Edition AMPLE

Schwartz, M.C., Tallmon, D. and Luikart, G. 1999: Using genetics to estimate the size of wild populations: many methods, much potential, uncertain utility. – *Animal Conservation* 2, 321-323.

Searle C.E., Strampelli P., Parsais S.N. and Dickman, A. 2025: Spatially explicit camera trap-based lion monitoring in Tanzania's Selous–Nyerere landscape. - *Journal of Zoology*, p. 12. https://doi.org/10.1111/jzo.70019

Sheil D. 2001: Conservation and biodiversity monitoring in the tropics: Realities, priorities, and distractions. – *Conservation Biology* 15(4), 1179-1182

Shonfield J. and Bayne E.M. 2017: Autonomous recording units in avian ecological research: current use and future applications. – *Avian Conservation & Ecology* 12 (1), Article 14

Simon A., Lazarowski L., Singletary M., Barrow J., Van Arsdale K., Angle T., Waggoner P. & Giles K. 2020: A review of the types of training aids used for canine detection training. – *Frontiers in Veterinary Science* 7, 313. https://doi.org/10.3389/fvets.2020.00313

Singer D. 2020: Ongoing test projects for a NWE basic monitoring. – Presentation at the Northwest German Forestry Research Institute, Göttingen

Smallwood, K.S. and Schonewald, C. 1998: Study design and interpretation of mammalian carnivore density estimates. – *Oecologia* 113, 474-491.

Solberg, K.H., Bellemain, E., Drageset, O-M., Taberlet, P. and Swenson, J.E. 2006: An evaluation of field and noninvasive genetic methods to estimate brown bear (*Ursus arctos*) population size. – *Biological Conservation* 128, 158-168.

Soria-Díaz L. & Monroy-Vilchis O. 2015: Monitoring population density and activity pattern of white-tailed deer (*Odocoileus virginianus*) in Central Mexico, using camera trapping. – *Mammalia*, 79 (1), 43-50 https://doi.org/10.1515/mammalia-2013-0107

Spellerberg I. 2005: Community-based ecological monitoring, p. 249-273 – In: Spellerberg I. 2005: *Monitoring Ecological Change*. 2e[nd] ed.. – Cambridge University Press, 391 p.

Steele, M. A. (1998). Tamiasciurus hudsonicus. *Mammalian Species*, (586), 1–9. https://doi.org/10.2307/3504443

Stephens P.A., Zaumyslova O.Y., Miquelle D.G., Myslenkov A.I. & Hayward G.D. 2006: Estimating population density from indirect sign: track counts and the Formozov-Malyshev-Pereleshin formula. – *Animal Conservation* 9, 339-348.

Stevenson B.C., Borchers D.L., Altwegg R., Swift R.J., Gillespie D.M. & Measey G.J. 2015: A general framework for animal density estimation from acoustic detections across a fixed microphone array. – *Methods in Ecology and Evolution* 6 (1), 38-48

Stier N., Borchert M., Meißner-Hylanova V., Pinnecke J., Schmüser H., Hoffmann D., Eckern S., Häger B. & Roth M. 2015: *Erfassungsmethoden von Baummarder und Iltis zur Beurteilung ihrer Populationszustände*. – Abschlussbericht BMLE und BLE, TU Dresden, 152 p.

Sutherland W.J. 2006: *Ecological Census Techniques*, a handbook. 2nd ed. – Cambridge University Press, 432 p.

Svensson L. 1992: *Identification guide to European passerines*. – British Trust for Ornithology, 368 p.

Svensson S. 1977: Land Use Planning and Bird Census work with Particular Reference to the Application of the Point Sampling Method. – *Pol. Ecol. Stud.* 3 (4), 99–117

Thiollay J.M. 1989: Area Requirements for the Conservation of Rain Forest Raptors and Game Birds in French Guiana. – *Conservation Biology* 3 (2), 128-137

Thomas M.L., Baker L., Beattie J.R. & Baker A.M. 2020: Determining the efficacy of camera traps, live capture traps, and detection dogs for locating cryptic small mammal species. – *Ecology and Evolution* 10(3), 1054-1068

Tomiałojć L., Wesołowski T. & Walankiewicz W. 1984: Breeding bird community of a primaeval temperate forest (Bialowieza National Park, Poland). – *Acta Ornithologica* 20, 241-310

Tomialojc, L. 1976: *Birds of Poland. A List of Species and Their Distribution* (Birds of Poland. List of species and distribution). - Fish and Wildlife Service, US Dpt. Washington DC, 254p.

Turlington, K., Suarez-Castro, A. F., Teixeira, D., Linke, S., & Sheldon, F. (2024). Exploring the relationship between the soundscape and the environment: A systematic review. *Ecological Indicators*, *166*, 112388. https://doi.org/10.1016/j.ecolind.2024.112388

University of the Sunshine Coast, Australia, https://www.usc.edu.au/about/structure/schools/school-of-science-technology-and-engineering/detection-dogs-for-conservation

Vancutsem C., Achard F., Pekel J., Vieilledent G., Carboni S., Simonetti D., Gallego Pinilla F., Aragao L. & Nasi R. 2021: Long-term (1990-2019) monitoring of forest cover changes in the humid tropics, SCIENCE ADVANCES, ISSN 2375-2548, 7 (10), 2021, p. eabe1603, JRC121591.

Verner J. 1985: Assessment of Counting Techniques. – *Current Ornithology* 2, 247-302

Waits, L.P., Luikart, G. and Taberlet, P. 2001: Estimating the probability of identity among genotypes in natural populations: cautions and guidelines. – *Molecular Ecology* 10, 249-256

Wasser S.K., Hayward L.S., Hartman J., Booth R.K., Broms K., Berg J., Seely E., Lewis L. & Smith H. 2012: Using Detection Dogs to Conduct Simultaneous Surveys of Northern Spotted (*Strix occidentalis caurina*) and Barred Owls (*Strix varia*). – PLOS One https://doi.org/10.1371/journal.pone.0042892

Wasser, S.K., Houston, C.S., Koehler, G.M., Caidd, G.G. and Fain, S.R. 1997: Techniques for application of fecal DNA methods to field studies of Ursids. – *Molecular Ecology* 6, 1091-1097

Wassmer, T., & Refinetti, R. (2016). Daily activity and nest occupation patterns of fox squirrels (Sciurus niger) throughout the year. *PLOS One*, *11*(3), e0151249. https://doi.org/10.1371/journal.pone.0151249

Wassmer, T., & Refinetti, R. (2019). Individual daily and seasonal activity patterns in fox squirrels (Sciurus niger) quantified by temperature-sensitive data loggers. *Frontiers in Ecology and Evolution*, *7*, 179. https://doi.org/10.3389/fevo.2019.00179

Wauters, L. A., & Dhondt, A. A. (1987). Activity budget and foraging behaviour of the Red squirrel (Sciurus vulgaris Linnaeus, 1758) in a coniferous habitat. *Zeitschrift Fur Saugetierkunde-International Journal of Mammalian Biology*, *52*(6), 341–353.

Wearn, O. R., & Glover-Kapfer, P. (2019). Snap happy: camera traps are an effective sampling tool when compared with alternative methods. – *Royal Society Open Science*, 6, 181748.

Wearn, O. R., & Glover-Kapfer, P. 2017: *Camera-trapping for conservation: a guide to best-practices.* WWF Conservation Technology Series. – Available at: https://www.wwf.org.uk/sites/default/files/2019-04/CameraTraps-WWF-guidelines.pdf

Wesolowski T., Tomialojc L. & Stawarczyk T. 1987: Why low numbers of *Parus major* in Bialowieza Forest – removal experiments. – *Acta ornithologica* 23 (2), 303–316

Wiedenmann A. & Purkart L. 2022: Rebhuhnkartierung. - YouTube Video. leben.natur.vielfalt. BfN

Wildlife Acoustics, Inc. (2024a). Song Meter Mini 2 AA. *Wildlife Acoustics.* https://www.wildlifeacoustics.com/products/song-meter-mini-2-aa. Accessed 17 June 2024

Wildlife Acoustics, Inc. (2024b). Kaleidoscope Pro Analysis Software. https://www.wildlifeacoustics.com/products/kaleidoscope-pro. Accessed 17 June 2024

Wildlife Acoustics, Inc. (2024c). Echo Meter Touch 2 PRO Active Bat Detector. *Wildlife Acoustics.* https://www.wildlifeacoustics.com/products/echo-meter-touch-2-pro. Accessed 17 June 2024

Williams, C. T., Wilsterman, K., Kelley, A. D., Breton, A. R., Stark, H., Humphries, M. M., et al. (2014). Light loggers reveal weather-driven changes in the daily activity patterns of arboreal and semifossorial rodents. *Journal of Mammalogy*, *95*(6), 1230–1239. https://doi.org/10.1644/14-MAMM-A-062

Wilson, G.J., Frantz, A.C., Pope, LC et al. 2003: Estimation of badger abundance using fecal DNA typing. – *J. Appl. Ecol.* 40, 658-666

Zielinski W.J. & Truex R.L. 1995: Distinguishing Tracks of Marten and Fisher at Track-Plate Stations. – *The Journal of Wildlife Management* 59(3), 571-579

Zwölfer H., Bauer G. & Heusinger G. 1981: *Ökologische Funktionsanalyse von Feldhecken* – Tierökologische Untersuchungen über Struktur und Funktion biozönotischer Komplexe. – Lehrstuhl Tierökologie, Universität Bayreuth, 422 p.

Zwölfer H., Bauer G. Heusinger G. & Stechmann D. 1984: *Die tierökologische Bedeutung und Bewertung von Hecken.* – Akademie für Naturschutz u. Landespflege, Beiheft **3** (2), 155 p.

2 Sampling Invertebrates and Small Vertebrates

Abstract

Invertebrates make up most species in organismic biodiversity. Their life forms are diverse, and they are found in almost every microhabitat. Their movement patterns are also highly diverse. Therefore, it is necessary to adapt the sampling methods to the respective taxa and locations. Various methods and techniques are presented. Some methods depend on the observer's efforts, but many are automated. This chapter also emphasizes whether the survey method records occurrences and frequencies or whether it is also able to determine abundance. In some cases, high species richness requires rapid recording, whereby species are differentiated as morphospecies by working names. This approach also opens the possibility of involving volunteers and nonscientists in large-scale surveys.

2.1 Species Inventories

The identification of arthropods in the field is challenging. Most amateurs overestimate their taxonomic knowledge. Even with conspicuous animal groups, such as butterflies, dragonflies, grasshoppers, and, for example, large ground beetles, there are often groups of species that can only be distinguished after careful identification. However, when the value of a habitat is to be assessed for conservation issues (see Chap. 10) based on the species inventory, it is often the rarer and less conspicuous species that are important (cf. Zwölfer 1980). Therefore, identifying preserved animals using a magnifying glass or stereomicroscope, in addition to specialist literature and a reference collection, are essential. If in doubt, the specimen should be sent to a specialist for confirmation.

The inventory process should not threaten the survival of individual species. Therefore, in all assessment programs, careful consideration must be exercised to determine whether animals should be killed when caught and whether the collection can be minimized despite the need for extensive sampling for statistical evaluation (Rümer and Mühlenberg 1988). Preference should be given to protocols based on the catch and release of animals.

The Federal Nature Conservation Act in Germany (2009, improved 2022) fundamentally prohibits any capture based on the killing of animals. § 39 (1) states: "It is forbidden to disturb wild animals deliberately or to catch, injure or kill them without reasonable cause." § 44 (1) adds: "It is forbidden to pursue, catch, injure or kill wild animals of particularly protected species or to remove, damage or destroy their developmental forms in nature."

The regulation for protecting wild animal and plant species (Federal Species Protection Regulation 2005) lists particularly protected species for Germany. These include, for example, all native dragonfly species unless they are already

M. Mühlenberg, T. Waßmer, *Field Ecology*, https://doi.org/10.1007/978-3-031-87547-2_2

listed separately as endangered species. The same applies, for example, to all large ground beetles (*Carabus* spp.), all native bee and bumblebee species, and many butterflies and moth species. Therefore, all scientific actions or research programs associated with animal capture require a permit or consultation with the respective responsible nature conservation authority (§ 45 Federal Nature Conservation Act deals with exceptions). Permits can be granted according to § 13 BArtSchV. Paragraph 3.3 states: "The authority responsible under state law can allow further exceptions from the prohibitions of paragraph 1 in individual cases, insofar as this is necessary for research or teaching purposes or breeding for one of these purposes, ...". The application for a permit must be justified and should be submitted early, as the decision often takes longer. Before collecting, researchers must inform themselves about the legal regulations in the countries they intend to research.

2.2 Estimation of Abundances of Small Animals

There are two different qualities of frequency data: relative and absolute. From the recording methods described below, only the suction trap, the "biocoenometer," quadrat, and extraction methods provide absolute catch data, i.e., data related to a known unit area. The effective catchment area of Barber (pitfall) traps and Malaise or window traps (flight interception traps) is unknown. Baars and van Dijk (1984) attempted to estimate the catchment area for a ground beetle species. Even a standardized net catch in open terrain can hardly be related to a precisely defined area. Suppose the catch result depends on the activity of the animals. In that case, we evaluate the encountered frequency as "activity density" as opposed to measures that indicate animal numbers per unit of area or volume ("individual density" or abundance).

For the processing of ecological problems, an abundance determination of the populations should always be attempted: fluctuations in the abundance of individual populations are studied for predator-prey relationships, interspecific competition or stability of ecosystems, and for estimates of the survival probability of a species (essential in nature conservation). Knowledge of the population sizes is also required to determine ecosystem energy flow. Many populations are regulated by density-dependent factors at relatively high and possibly relatively low densities. However, most observed ranges of possible densities are regulated independently of density (May 1986). Therefore, studying the density dependence of populations is particularly important at extremes. To better understand population fluctuations, data on the density variation in different localities within a generation are more important than comparing the overall average density in successive generations (May 1986). The behavior of individual animals can significantly influence population dynamics (Dubois and Binning 2022; Modlmeier et al. 2015; Wolf and Weissing 2012). The variability of population density can be expressed as the standard deviation of the logarithm of the annual density and is strongly dependent on how many years in a row a population is monitored (Pimm and Redfearn 1988). In long-term studies on insects (more than 10 years), about 80% showed a density-dependent regulation of abundance. Other studies highlighted other factors, such as weather as density-independent; for a discussion, see Begon et al. (2006).

Only in rare cases can the individuals of a population be counted directly. The distinction between absolute and relative abundances is crucial here. Examples of absolute estimates are quadrate samples, line transects, extraction methods, catching with a soil or ground photoeclector, catching with a suction device covering a defined sample area (biocoenometers, D-Vac), counting animal individuals on plant parts, and mark-recapture methods.

Most methods for recording small animals provide data on relative abundances. The mark-recapture method is also essential for this (see Sect. 5.3).

The fundamental question "How common is a species?" can be broken down into a series of specific questions:

- How is a population structured?
- What proportion is made up of young animals?
- What is the ratio between males and females?
- Is the population growing, or is it declining?

The study of population changes is critical in the applied ecology of phytophagous insects, which can occur as plant pests (e.g., Ohnesorge 1976). If field data on the class frequency of the individual developmental stages in clearly distinguishable generations are available, then so-called life tables can be created. One goal is to quantitatively specify the impacts of successive events, such as various mortality factors in the individual developmental stages. With these data, one can then check whether a population is in equilibrium or which density regulation, measured by the different survival rates in each developmental stage, is decisively changed during an outbreak of abundance (mass production) (Varley et al. 1974, 1980; Dempster and McLean 1999; Liebhold and Tobin 2008; Morris 1960; Murray and Sandercock 2020; Rockwood 2015). For further explanation of life tables, see also general ecology books like (Begon et al. 2006). In some insects, males occur in several morphs, e.g., major and minor forms, with different behavioral strategies that can influence their relative abundance in a population. An interesting case study is described in Wassmer and Armstrong (2023) and showcased in Sect. 5.3.1.

Another complex of possible research questions cover the relationship between resource supply and use by the animals.

- Could higher individual densities of certain species live in the same area?
- What is a meadow's carrying capacity for herbivorous animals?
- What is the relationship between a grassland's phytomass and biomass of phytophagous animals?
- Does fertilizing a meadow increase or decrease the abundance of certain animal species?
- How does the abundance of two populations of the same species differ in two different areas?

Productivity estimates are also important information for the management of agricultural land. While carrying capacity calculations are usually performed for large herbivores (cf. Gossow 1976), the recording of arthropods (density of individual developmental stages and their energy content) plays a significant role in determining energy flow. Reasonable estimates exist in Germany for the beech forest ecosystem (overview and summary in Ellenberg et al. 1986). For species protection in Central Europe, the problem of eutrophication is becoming increasingly severe, as even in strictly protected areas, nitrogen input through the air leads to a shift in species composition (Ellenberg 1987, 1988; Kratochwil 1989), which forces changes in population density to follow.

Even if your research focuses on the relationships between different species, you need solid population density data.

- What numerical relationship exists between predators and their prey?
- How does the supply of prey vary depending on location or over time?
- Does a predator effectively limit the prey population?
- To what extent do ecologically similar species co-occur in a community?

Due to the many parameters one is confronted within the field, it is not easy to work on the questions hinted above. Examples of analyses of the food supply and its use are comparisons of the prey supply (flying insects) for flycatchers or web-building spiders at different locations and at different times using Malaise traps (Davies 1977) and comparisons of the prey supply for roaming carnivorous arthropods through standardized sweep net and pitfall trap catches (e.g., Nentwig 1982). Some predators change their hunting strategy depending on the density and quality of the prey supply between more prolonged waiting times at the same place, waiting at different places, or roaming (for flycatchers, see Davies 1977; for herons, see Krebs 1974; for web-building spiders, see Gillespie and Caraco 1987). Newton (1980) provides an overview of the role of food supply as a density-regulating factor in birds.

Competition between different species can only be experimentally demonstrated by changing the resource supply and controlling the population densities of the studied species (see Schoener 1983). As an example of artificial density changes in spiders, the works of Schaefer (1975, 1978) and Wise (1981) may serve.

2.3 Brief Overview of Different Recording Methods for Different Taxa

The most reliable method to determine the **frequency of small animals** in a meadow is directly counting the animals in unit areas or possibly collecting them if this does not disturb the collection of further animals.

For ground beetles or spiders in the herb layer, you mark off a square area and search it systematically. Random coordinate points determine the **quadrats** to be checked within the study area. The quadrat's size should be chosen to maximize the number of unit areas to be fully sampled (collected or counted) in the available time. They should, therefore, be kept relatively small, e.g., for beetles or spiders between ¼ and 1 m^2. A lower limit follows the expectation that, ideally, several representatives of the studied species should be found in each quadrat. Early studies attempted to define the minimal area objectively by gradually increasing the sampling area until no more new species were found by further increasing the area (species-area curve, minimal area concept, e.g., Cain 1938). However, this idea is still under debate, with some evidence refuting it (Hopkins 1957), while others supporting it (Hao et al. 2020). If the results of the first counted quadrats vary greatly, the number of unit areas to be collected must be increased. The choice for the number of quadrat samples follows the general rules of determining the sample size (Sect. 15.6).

Shell snails can be counted relatively easily using quadrate sampling. However, one must be careful in dry weather as many snails hide in cracks in the ground or other hiding places. The number of snails is also easily underestimated. In leaf litter, densities of up to 2000 animals per m^2, and in some forest locations in Central Europe, up to 30 species can occur (Martin 2002).

The frequency of ants is determined by the number of nests per area (Lévieux 1969; du Merle 1978).

Instead of directly collecting soil organisms from a quadrat, dig up the top soil from a standard surface area (e.g., 25 cm × 25 cm), pack it into plastic bags, and take them to the lab for **heat extraction**. Heat, e.g., from an iridescent light bulb, drives out almost all of the animals from the soil surface and top layer (over 90%). Suitable devices are described further down in Sect. 2.20. For example, Workman (1978) found the highest densities of the wolf spider (*Trochosa terricola*) from meadows using heat extraction (approx. 30 individuals per m^2). In contrast, he calculated only 12 individuals per m^2 by directly collecting from a ¼ m^2 sample area.

A significant group of animals that determine the nutrient cycle and the productivity of meadows and forests as decomposers are the **earthworms** (Lumbricidae). In meadows and pastures, their densities were estimated to be between 8 and 300 individuals per m^2, weighing 20–80 g/m^2 (Kühnelt 1950; Dunger 2008). In English pastures, even higher densities of up to 524 animals/m^2 with a biomass of 52–120 g/m^2 have been reported (Edwards and Lofty 1977). Their species number can reach six to eight in Central European meadows. In limestone beech forests near Göttingen, Lower Saxony, the number of Lumbricidae species reached 11, with a density of up to 205 individuals per m^2 (Schaefer 1996). Section 2.22 shows how to determine earthworm density when sampling the forest floor fauna.

Web-building spiders are temporarily stationary and can, therefore, be counted reliably. To make recognizing their capture webs in the vegetation easier, you can spray the spider webs with water (Tolbert 1977).

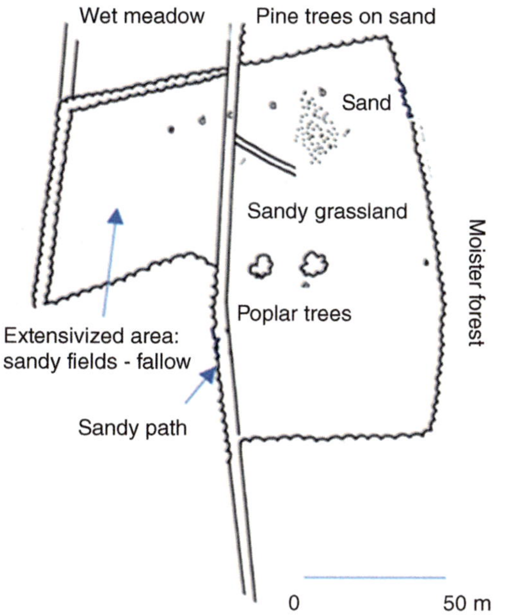

Fig. 2.1 Sketch of a study area for the recording of grasshoppers

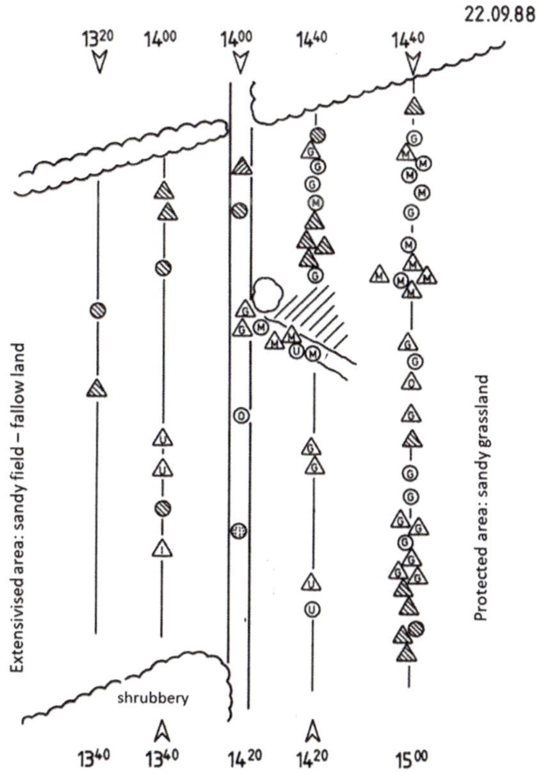

Fig. 2.2 Result of a line census on five transects recording grasshopper species. The encountered grasshoppers were marked by capital letters and round symbols (for females) and triangular symbols (for males) at the locations where they were encountered. The letters stand for the following species: (M) *Myrmeleotettix maculatus,* (O) *Omocestus haemorrhoidalis, Omocestus ventralis;* (G) *Glyptobothrus mollis, Glyptobothrus brunneus, Glyptobothrus biguttulus;* (U) *Tetrix undulata; Phaneroptera falcata;* (I) *Chorthippus dorsatus, Chorthippus parallelus.* After Hess and Ritschel-Kandel (1988)

Grasshoppers, especially field grasshoppers (Caelifera), can be detected by startling them: You mark a **transect strip** of 1.5–2 m width (depending on grass height) and about 30 m length and count all jumping grasshoppers with a hand counter while you are slowly and continuously walking through the strip, see the case study of line transects for grasshoppers below, and Figs. 2.1 and 2.2. The study also shows how individual grasshopper species are bound to certain small-scale site factors.

After some preliminary studies (trial catches), you can learn to distinguish the adult forms of the common species relatively easily through visual observation. Keys to the German species and their juvenile and adult stages are provided by Richards and Waloff (1954), Ingrisch (1977), and Oschmann (1969). For a more in-depth population study, breeding several species in the lab is almost unavoidable in developing a good knowledge of their different juvenile stages. With field data on the distribution of age classes of a species, a population can also be described demographically (Bradley 1985).

Case Study: Line Census for Grasshoppers

Repeat your line census using the same methodology after 2–3 days with similar weather conditions. If several observers are available, you can mark several strips parallel to each other with cords and check them simultaneously by walking through the field at the same pace, forming a straight counting front. For more extended studies, it might be worth enclosing a test area of, e.g., 20 m × 20 m with a 1 m high, vertically erected plastic sheet. Grasshoppers usually do not flee from the test area during the count.

Quadrat sampling methods are also suitable for determining the density of grasshoppers (Thompson 2019). In the field, at high density, rings with 0.1 m^2 (simple aluminum wire rings made from 114 cm long pieces with 2 cm overlap bent together resulting in 35.7 cm diameter) are laid out along a transect at approx. 8 m intervals and only checked the next day (Onsager and Henry 1977). One approaches carefully up to 2 m to a ring and scares all grasshoppers out of the ring area with a 2–3 m long pointer. You encounter four or fewer grasshoppers per ring at a density of about 20 animals per m^2. This method provides quick and accurate determinations of the density for a larger area and over many sub-samples with only a tiny counting error. At lower densities, you can increase the ring area to 0.25 or 0.5 m^2 (maximum diameter 79.8 cm).

Since many **butterflies** can easily be recognized by observation in the field, you can also use the transect method (line transect) for their density determination and for recording their distribution in space. The method was described in detail in Frazer (1973), Owen (1975), Moore (1975), Pollard (1977, 1982), Thomas (1983), and Pollard (1991). The British Butterfly Monitoring Scheme was described in detail by Pollard and Yates (1993). Transect counts should be performed once a week from April to September but do not provide absolute population figures. However, trends in population development become visible. Transect monitoring requires the weather to be not too windy or cold. According to Pollard (1977), standardization is achieved by carrying out counts between 10:45 and 15:45 local time and only at temperatures above 13 °C. If it is between 13 and 17 °C warm, there must be at least 60% sunshine; above 17 °C, it can be sunny or cloudy. Butterflies depend heavily on weather conditions for their entire population development (Pollard 1988). As there are not that many species in the UK (approx. 50), it is possible for non-experts to learn to distinguish between species. A prepared recording form, dictaphone, or smartphone can be used to record in the field. One slowly walks a path or a marked route and counts all butterflies within a strip width of up to 5 m. The annual sum of the observed numbers of butterfly species per weekly controlled transect is provided as a relative frequency measure. The respective weekly average counts if the transect was walked several times weekly. The European-wide application of the Butterfly Monitoring Scheme (in ten countries) was evaluated in terms of its informative value and methodology by Van Swaay et al. (2008). Compare also Schmucki et al. (2016) on the use of indices.

According to the transect method, a density indication is also possible (see Sect. 5.1). Isaac et al. (2011) compared the results for abundance estimates using the simple Pollard method (described above) with the distance sampling method. Although detectability differs between species and regions, the density estimates from the Pollard walk correlate well with distance sampling calculations.

Small cicadas and spiders are best collected with the **"D-Vac" vacuum sampler** (see Sect. 2.8, Duffey 1974; Harper and Guynn 1998). A box is placed over the habitat volume to be sampled beforehand so that the small arthropods cannot escape during vacuuming and to ensure you always collect the same volume size.

We call the demarcation of a meadow piece with a box and the subsequent collection of the arthropods within such a gaze cage the **"Biocoenometer"** method (described in detail in Sect. 2.8). It is the most elaborate but also the best method for the quantitative recording of small insects (Diptera, Hymenoptera, Auchenorrhyncha) and the determination of the percentage shares of different groups in the biocoenosis (Gromadzka and Trojan 1967). Olechowicz (1970) compared different sizes of biocoenometers in terms of the catch result. Converted to insect number per m^2, he obtained the highest values using small biocoenometers (20 × 20 cm^2 base area), but most species were caught with large biocoenometers (80 × 80 cm ^{2} base area). To concentrate many insects, Olechowicz first used the biocoenometer like a ground photoeclector (Sect. 2.20.2), darkened the sample volume for about 2 h, and then vacuumed the remaining insects. Calculating the necessary sample size for insect collections using the biocoenometer or the D-Vac vacuum follows the usual statistical rules (see Sect. 15.6).

During the vegetation period, **ground photoeclectors** (ground eclectors) must be moved in the meadow every 4 weeks, and these methods are not as effective as other methods (see also Haas 1980). A detailed description can be found in Sect. 2.20.2.

Catching insects and spiders from the herb layer with the **net** is simple, cheap, and yields high catch numbers. Various authors have written about the possibility of comparing the catches quantitatively, e.g., Witsack (1975) and Regensburg (1977). Sweep net results are influenced by weather and time of day (Lewis and Taylor 1964; Lowrie 1971). Compared to the biocoenometer, larger insects are more easily captured with the sweep net. However, the transect strip method (see above) offers the best solution for their quantitative determination.

Much work has been done on how the convenient and productive **pitfall trap catches** can be evaluated quantitatively, i.e., in terms of individual densities (Skuhravy 1957; Weidemann 1971; Gist and Crossley 1973; Bonkowska and Ryszkowski 1975; Fechter 1977; Baars 1979; Thomas and Sleeper 1977; Uetz and Unzicker 1976). The results are arbitrary: while Weidemann (1971) found no significant correlations between the activity density (activity-abundance assessed by pitfall traps) and the abundance (assessed by quadrat methods) over the year in spiders, Chilopoda, and the ground beetle, *Pterostichus oblongopunctatus* (Carabidae), Uetz and Unzicker (1976) and Baars (1979) report sound correspondences, so that changes in the pitfall trap catches indeed reflect similar changes in population density.

To use pitfall trap in a spatially related manner, one can enclose smaller areas (1–4 m^2) with a vertical plastic wall about 10 cm into the ground and at least 30 cm protruding and use the pitfall trap for **empty trapping** (removal trapping, see Sect. 5.5) of the enclosed sample area (Tanier 1977; Bonkowska and Ryszkowski 1975). Gist and Crossley (1973) bordered a study area of 5 m × 5 m with a resin-coated aluminum band so that epigaeic arthropods could no longer enter or run out of the area. They obtained the same result through extrapolation of the decreasing catch rate (daily collection!) as through quadrat methods. The ground trap should be placed for empty catch experiments at the partition, as most arthropods run along it after hitting the wall.

Another way to determine individual densities with known trapping methods (e.g., pitfall trap and net catches, see below) is offered by the **recapture methods** (mark-release-recapture, MRR, or capture–mark-recapture, CMR, methods), see Sect. 5.3 for a detailed description.

The following presents the devices mentioned above in detail. Applications of various collection techniques, especially for the tropics, are also presented in Grootaert et al. (2010). Zou et al. (2012) also briefly overview different sampling methods for terrestrial arthropods.

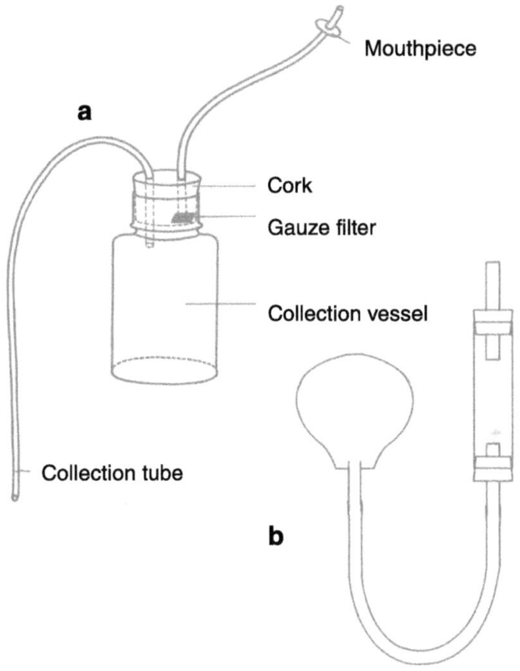

Fig. 2.3 Exhaustor, (**a**) for self-suction, (**b**) for suction with a balloon

2.4 Hand Collection with an Exhaustor and Tweezers: Catching Through Own Effort

An exhaustor is used to identify individually collected arthropods. The principle is that a suddenly created vacuum sucks in small arthropods into a transparent collection vessel.

Polyethylene wide-mouth bottles (250 mL) are suitable as a collection vessel for a self-suction exhaustor (Fig. 2.3). They are quickly interchangeable, closable with a lid, and unbreakable. A deflection plate at the inner end of the intake hose or a piece of pulp at the bottom of the vessel is recommended to prevent the squashing of sucked-in animals. Especially with ants (Formicinae spray formic acid) and when sucking up small insects on animal carcasses and dung heaps, an exhaustor with a balloon is recommended (Fig. 2.3b). Smaller arthropods are selectively picked up from the ground, leaves, bark, and other objects.

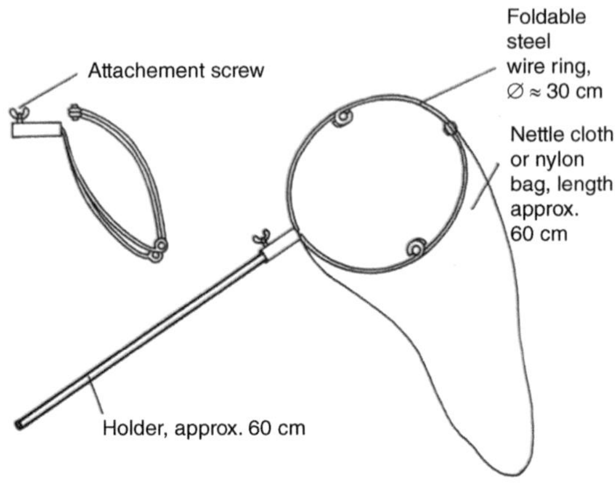

Fig. 2.4 Insect sweep net

Advantages and Disadvantages of Exhaustors They mainly collect agile arthropods, which are difficult to grasp with tweezers. With a longer intake hose, you can reach into crevices and under the bark of trees. Practically only selective catching or collecting is possible.

2.5 Insect Sweep Net

With an insect net, you sweep off herb and shrub layers and catch flying insects and sitting arthropods in this way. You can determine species spectra, relative frequencies, and the horizontal and vertical distributions of certain species in this way (Figs. 2.4 and 2.5).

Advantages and Disadvantages of Sweep Netting Handy and easy to use and assemble in a DIY manner. The catch result depends on the skill and individual handling of the catcher. Ground-level layers are underrepresented in the catch result. The catch result is strongly dependent on the weather.

Sweep-net captures are used for standardized sampling of butterflies in a specific study site within a fixed time period (e.g., 1 h). That corresponds to "catch per unit effort" if the same observer always acts in standardized time and weather conditions.

The group of animals of interest is determined from the net bag and, for a secure species deter-

Fig. 2.5 Insect net in use: in this case, specific species are targeted and then caught with a short sweep of the net

mination, transferred to 70% alcohol or, for a dry preparation, killed in acetic acid ethyl ester and later prepared (instructions see, e.g., Piechocki 1985). Sweep-net catches depend on both the expertise of the investigator and the activity of the arthropods and thus on the weather or the current food supply (e.g., Lewis and Taylor 1964; Maier and Waldbauer 1979). For a species inventory, therefore, samples must be taken at different times of the day and, depending on the question, at different seasons. Since many animals drop immediately when the plant stems are shaken or fly up, the net strokes must be quick (Luczak 1958) and spatially well executed in front of the walking observer.

The best species inventories of a meadow through net catches are achieved on grasshoppers and cicadas. In the smooth oat meadow in Germany, up to 70 breeding species of cicadas live, with a per-plant association of 30–50 species (Müller 1978).

Many native jumping insects (Saltatoria) can be recognized by their species-specific songs. "Listen" to your study area in favorable weather conditions (little wind, warm temperatures). You will be able to compile a species inventory with the help of the calls of leaf and short-horned grasshoppers and crickets. Many animals start their songs in May. Crickets and leaf grasshoppers usually call at dusk and night. The voices of 61 German grasshopper species are available on CD (Bellmann 2004). "Songs of Insects" provides the voices of 77 species of crickets, locusts and cicadas native to eastern and central North America (Elliot and Hershberger (2007) and online at https://songsofinsects.com).

Since the determination of the juvenile stages in grasshoppers is not easy (for identification literature, refer to Richards and Waloff 1954; Oschmann 1969; Ingrisch 1977), systematic sweep-net catches for field grasshoppers (Caelifera) are usually only worthwhile in the summer to late summer.

Sweep-net catches are never independent of the personal commitment of the observer and, therefore, are not entirely free from systematic errors. A better way to objectively capture arthropods in the herb layer is the suction devices mentioned below (Johnson et al. 1957), presented in Sect. 2.8.

2.6 Knock-Down Methods

The knock-down method captures arthropods from the trunk and canopy layer of trees. Insects and spiders of the crown region are otherwise difficult to access. Assessing what proportion of the fauna drops down and what proportion is not captured remains difficult.

(a) Knock-Down from a Tree (Fig. 2.6a)

With a strong branch or wooden hammer, younger trees can be suddenly shaken so that many arthropods fall down. A laid-out bright linen cloth serves to catch them, and they must be quickly collected.

(b) Beating Tray (Fig. 2.6b)

The beating tray method is used to collect arthropods, mainly from the shrub layer, at forest edges and hedges. It can also be applied when one wants to estimate individual numbers in relation to leaf volume.

With a standardized beating stick thickened as much as possible at the end and wrapped with a rubber hose, branches and twigs of shrubs are hit. The beating tray is held under the beaten part of

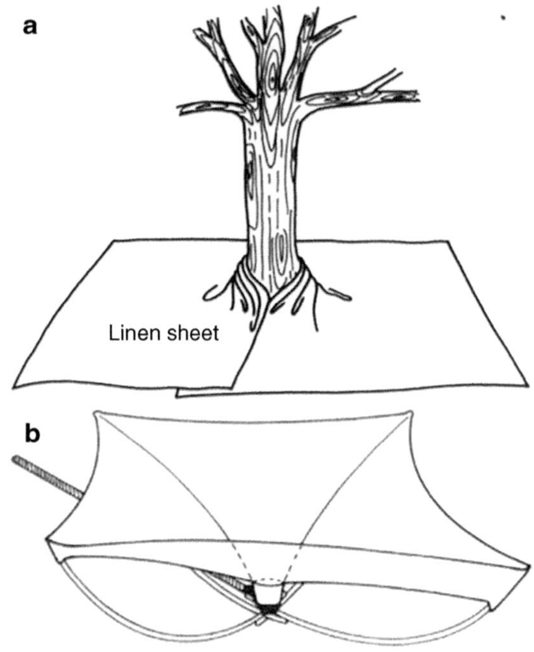

Fig. 2.6 Beating methods, (**a**) beating a tree, (**b**) beating tray

the shrub with the other hand during the hits. The arthropods let themselves fall due to the vibration and are caught in the tray. Every sample is collected using the same number of beats (e.g., 10×). The depicted beating tray can be disassembled into the rigid holding stick and four tensioning sticks. A catch glass in the center safely collects the falling small animals.

The beating tray method can also be used for quantitative comparisons. It replaces nets in stiff and thorny branches. Not all parts of a hedge can be beaten equally well, and there are also differences between the types of woody plants.

2.7 Canopy Fogging

Arthropods living freely in the canopy layer of trees are best collected quantitatively by fogging the canopy with an insecticide (Simberloff and Wilson 1970; Floren 2010; Floren et al. 2022; Allison et al. 1993, 1997; Paarmann and Stork 1987; Basset et al. 1997; Marques et al. 2006) (Fig. 2.7).

Insecticides based on pyrethrum are considered environmentally compatible as they have relatively low mammalian toxicity. As a wide array of pyrethroids are available, it is possible to choose one with high toxicity for pest insects but relatively low toxicity to beneficial insects such as pollinators (Johnson et al. 2006). However, pyrethrum-based pesticides were found to be highly toxic to fish and other aquatic organisms (Farag et al. 2021). Stunned or killed arthropods fall from the sprayed canopy layer and are collected on the ground on a spread tarp. Marques et al. (2006) installed funnel-shaped trays (1 m^2) made of nylon under the canopy. In this way, new insights into the high biodiversity of canopy faunas have been gained in the tropics and temperate latitudes. Allison et al. (1993) obtained 4840 beetles representing 633 species in 54 families by canopy fogging of 8 trees in New Guinea. Marques et al. (2006) sprayed three trees in the Pantanal, Brazil, and gained 15,744 arthropods representing 20 taxonomic orders (87% Insecta, including 2197 adult Coleoptera, and 13% Arachnida). Floren et al. (2022) collected approximately 3.2 million arthropods in 32 taxa (order, suborder, or family) through 1152 foggings of 24 tree species in Central Europe.

Allison et al. (1993) describe the method in detail. They used Pyranone® mixed with kerosene to form a 5% solution that was delivered by a Curtis Dynafog model 2610 ("Golden Eagle") fogger with the fog density dial set to "5." The fogger was operated manually at canopy level from an adjacent tree. The trees were fogged for 15 min in the early morning. Marques et al. (2006) sprayed the insecticide from the ground with an agricultural sprayer (Bomba Hatsuda Costal, type Pulverisador UBL, FP-4, JD22-2.T) or a thermonebulizer Swing Fog SN50. Floren (2010) used the fogging machine SN-50 by Swingtec GmbH, Germany. Floren et al. (2022) used a natural pyrethrum with a concentration below 1%, where the arthropods are mostly only temporarily stunned, and the insecticide is rapidly degraded in the environment.

Fig. 2.7 Canopy fogging in rubber monoculture. On the right in the photo are collecting funnels. (Photo credit: Yohanes Bayu Rohaditomo, from Pollierer et al. 2023. Reproduced by permission)

2.8 Vacuum-Insect Net, D-Vac

The D-Vac (Fig. 2.8) combines covering the vegetation and suction and works like a vacuum cleaner. A two-stroke engine generates a constant suction airflow, and the arthropods are caught in an exchangeable net bag. The unit area to be vacuumed is quickly delimited with a metal frame or covered with an eclector-like "biocoenometer" (see Fig. 2.9). The enclosed vegetation and soil surface are thoroughly vacuumed with the "trunk" of the vacuum device in a standardized time duration (e.g., always 2 min long). If the vegetation is too high, it is cut in the delimited area and read out separately. There are exchangeable, differently shaped parts for the suction nozzle. The showcased D-Vac is manufactured in the USA (https://www.rinconvitova.com/d-vac.htm) and sold in Europe by ecoTech, Environmental Measurement Systems (https://www.ecotech.de/en/) as "ecoVac—Insect Vacuum."

A D-Vac determines absolute animal densities (number of animals per unit area). D-Vac collection methods are advantageous for pest and predator control in agricultural cultivation areas (Elliott et al. 2006).

The method allows frequent and extensive sampling without significant habitat impairment. It is particularly suitable for narrow-edge strips (meadow and field edges, see Sect. 10.17). Since the animals are sucked in alive, a selection of the net bags can be made in the field, and unneeded animal groups can be rereleased. Due to the standardization, the results are more reliable than net catches (see above). The catch efficiency is reduced in dense vegetation. Animals in crevices and soil depressions may not be recorded; therefore, controlling them by cutting out the soil is

Fig. 2.8 Vacuum-insect net ("D-Vac"). Left panel: schematic setup: SM—two-stroke engine for generating the suction air stream, TG—carrying frame, SR—suction pipe, HG—handle, AR—suction trunk with a suction nozzle. Right panel: use of the *vacuum-insect net* in the field. Ecological Research Station Fabrikschleichach, University Würzburg, Northern Bavaria

necessary. If the soil is too loose or in the leaf litter, too much foreign material may be sucked in. In such cases, heat extraction might be better suited.

The volume of meadow intended to be sampled for insects and spiders is covered by the biocoenometer (Fig. 2.9). A gauze hose is used to insert the vacuum cleaner pipe.

The collecting tube (catch can, Fig. 2.9, lower panel) is closed at the front and back by a lid. The inlet and outlet pipes protrude about 3 cm into the cylinder. The suction pipe is flexibly connected to the suction opening of the vacuum cleaner, and a 95 cm long pipe with a 2.9 cm inner diameter is firmly connected to the front end (lid). This pipe is inserted into the biocoenometer for collection. The gauze cone at the entrance of the rear pipe prevents the animals from being sucked onto the closing gauze. The plates in the cylinder greatly slow down the airflow and thus prevent too much damage to the sucked-in animals. The vacuum cleaner used by Haas (1980) operated at 550 W and had a suction speed of 14.8 m/s (Fig. 2.9).

In the following, we want to introduce capture techniques and inventories of small animals with automatic catching devices.

2.9 Light Traps for Moths

Moths develop from phytophagous caterpillars, which usually only feed on specific plants. A forest stand with more plant species should also host a more diverse fauna of moths. We can check this with automated light traps.

While it is known that insectivorous birds during the breeding season depend more on the vegetation structure than on individual plant species (e.g., Erdelen 1978; Oelke 1977), we can investigate whether moths, as phytophagous insects, have a closer relationship with the woody flora of the forest. We can test such a question by catching moths at night with light traps in the same test areas where we conducted the bird population surveys (see Sect. 1.8). The light traps (Fig. 2.10) are preferably set up during the day in the centers

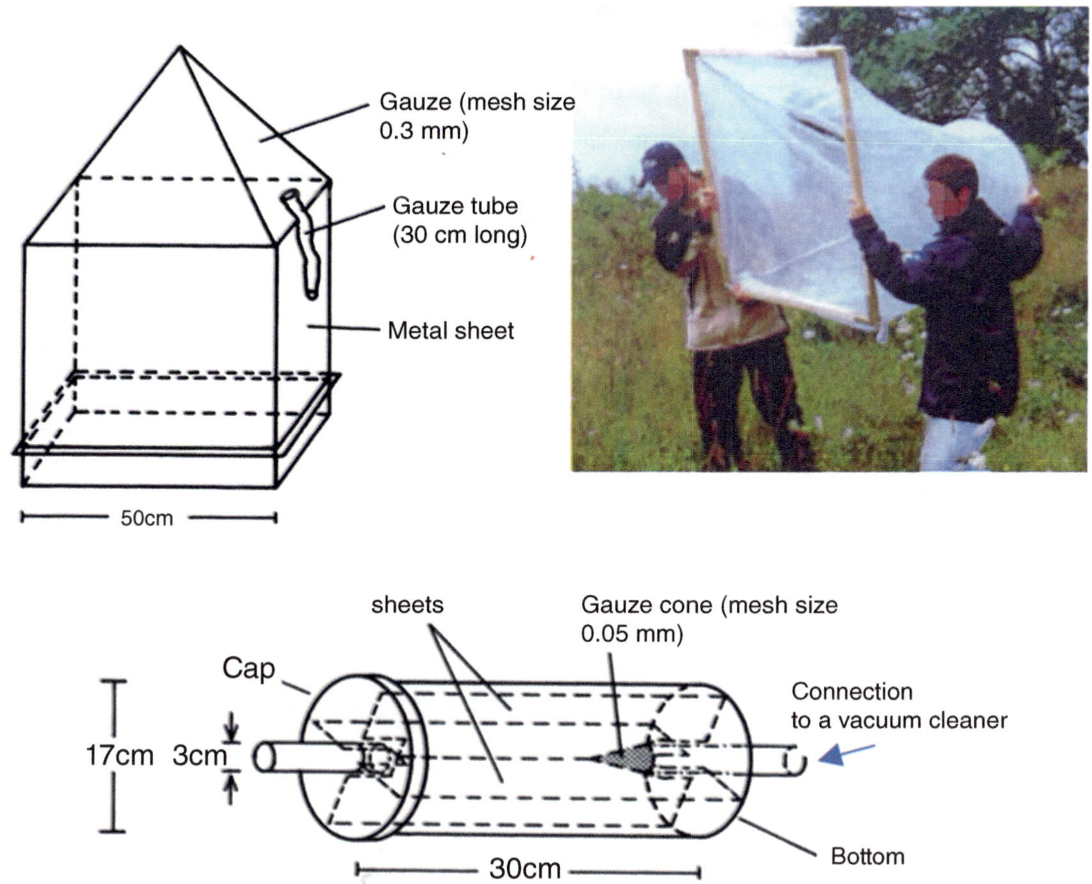

Fig. 2.9 Biocoenometer. Top left: schematic drawing. Top right: drop of an isolation square (1×1 m^2) for estimating abundances of grasshoppers (individuals per unit area). Bottom: custom-made collecting tube (catch can) to connect a vacuum cleaner (Haas 1980)

of the sample points and switched on at the beginning of dusk. Moth species have different flight times (Hitchen et al. 1969). The captured moths were sorted into species and number of individuals. Identifying the species is not always necessary; it is sufficient for our question to separate the moths by size, shape, and pattern, formally documented with working names or numbers ("Morphospecies," see Sect. 2.27.1). If the moths are caught alive and only sorted the following day, they are kept in small mesh cages or net bags in cool and shady places, supplied with cotton soaked in sugar water. The following evening (around 8 pm), the moths are carefully released back into higher vegetation. For population studies, the moths can be marked daily (cf. Baker 1985) using the marking technique for butterflies described in Sect. 5.3.1. For recapture, the light traps are switched on at least 2 h after the release of the moths (within a radius of 50 m). The variation of the light catches (the trap efficiency) between different days can be controlled by comparing the recapture results.

The number of tree species and the occurrence of individual woody plant species are recorded with the Point-Centered-Quarter-Method (see Sect. 4.2.1). To characterize the plant communities of the forest habitats, the herbaceous plants should also be determined (see Sect. 4.1). At least for some moths identified according to specialist literature (with illustrations, e.g., Waring et al. 2017; Skinner and Wilson 2009; Leckie and

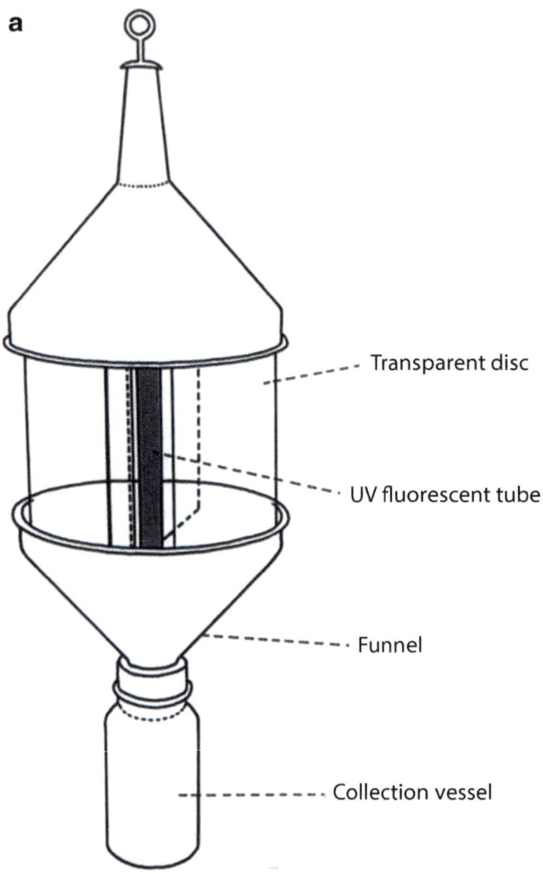

Fig. 2.10 (**a**, **b**) WALZ light trap (BEHRE). Moths respond best to ultraviolet-rich light sources. A car battery provides the electric power for the bulb. A window cross surrounding the bulb increases the efficiency of the trap. The attracted insects bump against the acrylic and fall into the killing jar

Beadle 2018; Beadle and Leckie 2012), a connection between their presence and the occurrence of certain food plants (determined from the literature) has been documented.

The catch efficiency of light traps depends substantially on temperature; it is exceptionally high during warm nights (Kurtze 1974; Persson 1976; Muirhead-Thomson 1991). If you want to compare light traps over the long term, for example, for the study of population fluctuations, you must always set the traps at the same location. Even if a few meters away, the new trap location can lead to altered catch results (Askew and Yalden 1985). The effective range of a light trap is unclear. It must be assumed that flying moths first fly into the influence area of the light trap and then interact with the trap. Most species are pretty mobile, and the individuals caught in the light trap are more likely a fraction of migrating moths than a subpopulation that lives near the light trap (Baker 1985). Experiments with several light traps (equipped with 125-W mercury vapor lamps) have shown an effective catch radius for moths of 10–25 m around the trap (Muirhead-Thomson 1991).

Night-flying insects are attracted by light, as they orient themselves by the (almost) parallel light of stars or planets for directional flight. Diverging light rays of our artificial light sources lead the animals inevitably into a spiral flight to the light. UV fluorescent tubes appear to most insects as brighter light sources than ordinary light bulbs. The flying insects approach the light trap in a ± spiral bump against one of the four cross-standing Plexiglas panes (= 8 catch surfaces), and

Fig. 2.11 Moths and other insects attracted by the light land on a cloth that is hung up. The photo shows a non-systematic capture of moths as a snapshot, taken in the primary rainforest of Mamberamo, West Papua. Each individual belongs to a different species, which is evidence of an exceptionally high level of biodiversity (see Sect. 15.11, diversity indices)

they usually fall into the funnel and the detachable collection container. The traps are hung freely. For live capture under constant control during the lighting time, it would also be sufficient if simple fluorescent tubes or UV tubes were hung at the top end of a table-like stretched white linen cloth, and the moths or other insects settling or fluttering on the linen cloth can be directly checked for marking (Fig. 2.11). A Petromax lamp (https://petromax.com) can be used as a light source if no power source is available, not even a car battery for 12-V fluorescent tubes. However, it has the disadvantage that the heat damages or burns insects that touch the Petromax.

A light trap primarily determines species spectra with the frequency distribution of individual species and also for diversity values of taxocoenoses in different habitats and years. In addition, light traps are suitable for phenological studies (flight times) of various night-flying species, especially moths.

Their main advantage is standardized catching that can be carried out continuously and simultaneously in several habitats. A light trap that is controlled by a timer can restrict catching to specific nighttime intervals. The trapping efficiency of light traps depends on the weather conditions, with warm, calm nights being the most productive.

2.10 Baits Traps for Butterflies

Many tropical butterfly species, predominantly from the families of Nymphalidae and Papilionidae, can be attracted to fruits, especially when the fruits spread a strong smell while fermenting. This can be exploited in bait traps, which can be freely hung for live capture of butterflies. They also enable the capture of butterflies that fly higher in the forest strata or only in the forest canopy (Figs. 2.12 and 2.13).

The following describes a study design for comparing the butterfly fauna in different habitats. You can use bait traps, for example, See Fig. 2.14 below.

The calculation of the sampling effort: three traps per plot (radius of 50 m), six plots per habitat (as replicates), four habitats (gradient) = 72 traps.

Traps must be checked every day. Specimens collected on one sampling day in three traps of one single plot are pooled = 24 sampling stations.

The six plots of one type of habitat (containing 18 traps in total) are visited for 1 day, corresponding to a total sample from one habitat. For four habitats, 4 days are necessary. The processing of the material requires extra days, so 1 week is estimated for one total catch from all four habitats. Five repetitions are planned. This results in 5 weeks for the trapping period. Sites are visited alternately.

Six sampling stations per habitat, four habitats, and five repetitions give 120 sampling stations (= 120 insect point counts).

Fig. 2.12 (**a**) Bait trap for butterflies. (From De Vries 1987). (**b**) Bait trap hanging in the middle stratum in a tropical forest. (**c**) Bait traps for butterflies at different heights for comparison of habitat use in a tropical forest, used in Fermon et al. (2005)

Fig. 2.13 To identify captured butterflies, keys with color plates are helpful as a first step; in complex cases, one can consult with museum collections. The photos show an example from Papua, Indonesia. (**a**) Living butterflies in a bait trap moving upwards to escape. (**b**) Butterflies that are difficult to identify were preserved for later identification. (**c**) Local staff is using color plates from keys to identify specimens. (**d**) Parts of the butterfly reference collection at the Universitas Cenderawasih, Jayapura, Papua

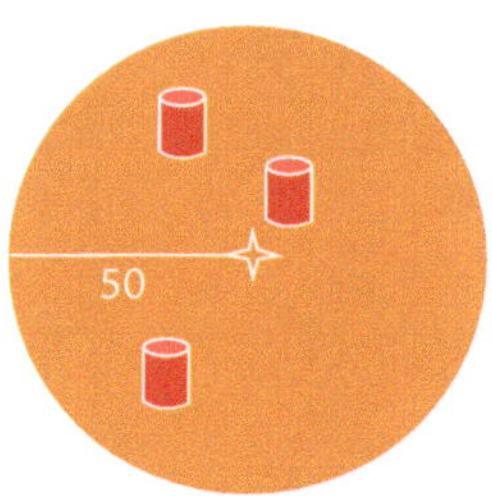

Fig. 2.14 Three traps placed within a sample plot

2.11 Flight Interception Traps

2.11.1 Window Traps

Window traps are particularly useful for collecting beetles (Fig. 2.15a, b). In a forest environment, they are often used to collect insects associated with understory resources such as dead wood. The catch container contains a catch liquid (see Table 2.1). Figure 2.15c shows another type of window trap. For catching beetles, the container can hold salt water as a preservative (see also Table 2.1). A small amount of detergent should be added to the salt water to reduce the surface tension.

2.11.2 Air Eclector

A free-hanging air eclector (Fig. 2.16) captures insects that, like the window traps, fly against a gauze and then try to escape upwards or downwards.

Fig. 2.15 Window traps. (**a**, **b**) Window traps are flight interception traps. In (**a**), it is a self-standing model; in (**b**), it is a model hanging between trees. A drop tray (collecting trough) is attached below a transparent vertical panel in both designs. (**c**) A trap with two perpendicular transparent plastic panes (40 × 60 cm) can be used as a window trap. Insects are captured in a funnel leading down into a 1-L sample bottle

Table 2.1 Advantages and disadvantages of different catch liquids for ground and other traps like window traps (see above)

Catch liquid	Handling when transferring the animals into 70% alcohol	Preservation properties	Attraction properties	Catch result in a controlled comparison	Costs
4% formalin solution	Unpleasant smell and harmful fumes	Very good—suitable for reference collections	Different attractiveness was observed in carabids, not in spiders	The highest species and individual numbers, more single specimens of species	Cheap
Picric acid solution (one part saturated picric acid solution to three parts water)	Little odor nuisance; splashes cause difficulty in removing yellow spots on skin and clothing	Not very good—animal bodies are greatly lightened	Apparently none		Cheap
Ethylene glycol, diluted	Little/no odor nuisance; no harmful effects known so far	Not very good—small/delicate animals may dissolve, can be recognized again by an expert of a group	Apparently none, but can attract some invertebrates	Especially suitable for ground and rove beetles fewer individuals	Expensive

(continued)

Table 2.1 (continued)

Catch liquid	Handling when transferring the animals into 70% alcohol	Preservation properties	Attraction properties	Catch result in a controlled comparison	Costs
Water with an odorless relaxant (e.g., Agepon)	No problems at all	None—therefore emptying at intervals not greater than 2 days; animals die slowly	Apparently none		None
Mixture of 40% alcohol, 30% water, 20% glycerin, 10% acetic acid + detergent (solution according to Renner 1980)	Hardly any	Very good; optimal preparation ability of heavily chitinized insects. Makes samples useless for DNA analyses	Acetic acid can be attractive to some beetle species	Good	Cheap, glycerin is somewhat expensive; the mixture can be reduced to ethanol, distilled water, and glycerin in a ratio of 4:5:1 (Schuch et al. 2020)
Saturated salt solution (Teichmann 1994)	No problems at all	Good; however, the fixation is relatively weak, which can be problematic for delicate animals	None		None

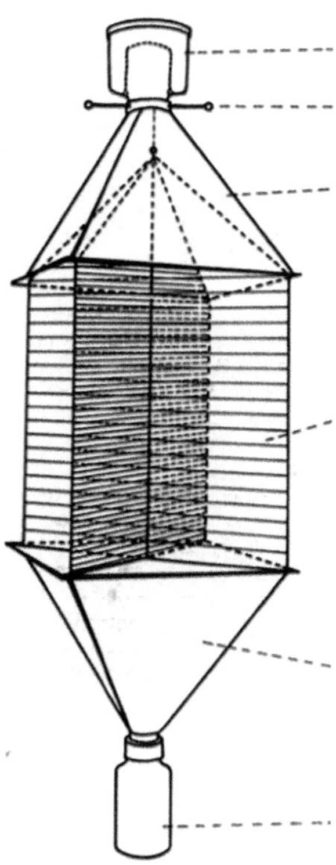

Fig. 2.16 Freely hanging **air eclector** (Behre 1989) complements the determination of species occurrences. It is also suitable for the recapture of marked flying insects

Suppose we want to record flying insects from different heights and compare them between habitats. In that case, we propose the following study design (Fig. 9.3). We will also calculate the time required for such a study (see Case Study in Sect. 9.4). The trap types could be window traps, bait traps for butterflies (see Fig. 2.12), or air eclectors.

2.11.3 Catch Funnel on the Trunk

Arthropods running directly upwards or downwards on the trunk (Fig. 2.17) can be collected with trapping funnels (Fig. 2.18). They are constructed similarly to the eclectors described in Sect. 2.20.

2.11.4 Malaise Traps

The principle of a **Malaise trap** is that flying insects encounter the fine mesh tulle walls set up like a tent, and, in an attempt to escape or evade upwards, they end up in a collection jar over the

Fig. 2.17 Funnels on the trunk catch arthropods running up and down the trunk in catch cans

tent's peak. The side walls are made of black, barely visible mosquito net-like fabrics, and the roof is made of white cloth (Fig. 2.19a). The trap is about 1.5–3 m long and needs to be stabilized with tension ropes (tent cords) on six stakes for a solid setup. The collection container in the roof peak is divided into two parts. The lower part is detachable as a bottle and is usually filled with collection fluid. Instead of this collection container, an eclector head box can also be used. Malaise traps are available in the USA from Forestry Suppliers, Inc. (forestry-suppliers.com) and can be obtained in Germany from ecoTech (ecotech.de) and range between $200 and $400.

Malaise traps are used to determine the species spectrum of flying insects such as flies, beetles, wasps, or other Hymenoptera and the distribution flights of spiders. However, they are less suitable for butterflies and moths. Diurnal flight activities of certain insect groups can be documented, preferably for Diptera and Hymenoptera, and especially for parasitic wasps in the broader sense (Proctotrupidea, Ceraphronoidea, Chalcidoidea,

Fig. 2.18 (**a**) Longhorn beetles (Cerambycidae), like *Monochamus* sp., depend on dead wood for their development. North Mongolia. (**b**) Buprestidae, the "metallic wood borers," are another xylobiontic beetle family, which window traps and catch funnels can capture

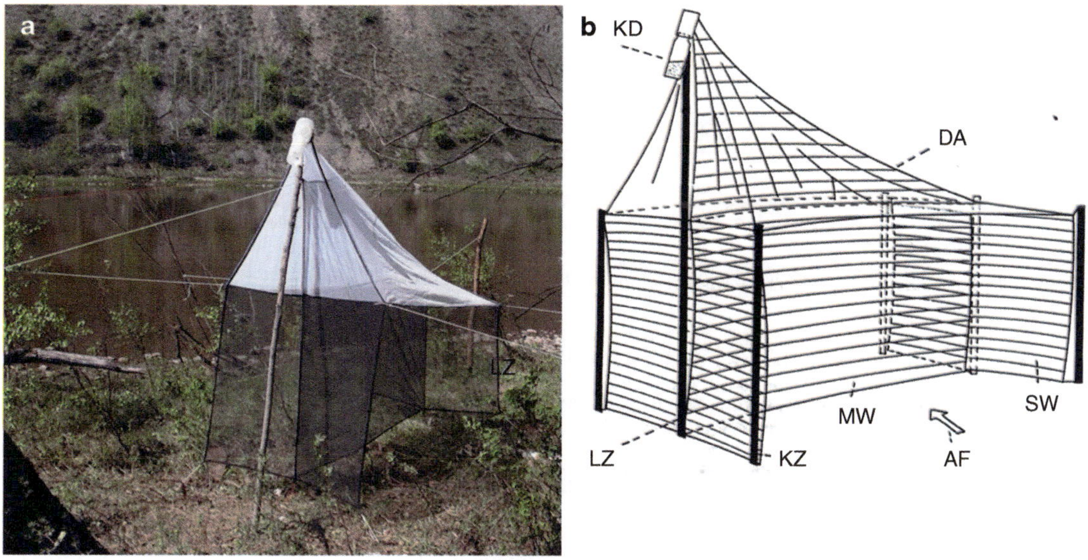

Fig. 2.19 (**a**) Malaise trap at a riverbank in North Mongolia, where they can be used to record the time period and peak of emerging water insects like Plecoptera and Ephemeroptera. (**b**) Malaise trap. KD—head box (collecting chamber), LZ—long tent pole, KZ—short tent pole, MW—center wall, SW—side wall, DA—tent roof, AF—front approach direction

Ichneumonoidea, and Aculeata). Noyes (1989) compared several collection methods for Hymenoptera in a tropical rainforest and found that in addition to catching with an insect net, Malaise traps are most effective.

Advantages include the low amount of work required for trap setup and emptying. The animals can also be read from the tent with an exhaustor during frequent checks (2–3-h intervals). The catch can be substantial, making it necessary to change the collection jar at least every 2 days. The trap tent is not always easy to set up, is more straightforward with at least two people, and requires supervision for the first few times. Catching becomes problematic in locations exposed to strong wind. Large and robust fly species often escape or circumvent Malaise traps (Kentner and Schrade 1991). Flying beetles are usually better captured using window traps. For example, using window traps for ground beetles provides evidence of the respective species' flight ability and indicates their dispersal behavior (Haeck 1971; den Boer 1971; Meijer 1971; van Huizen 1979); see Sect. 11.4. Malaise traps can also be installed in the tree canopy layer (Skvaria et al. 2020).

Case Study: Long-Term Monitoring of Insect Numbers in North America

In a groundbreaking study from Germany, published in 2017, total aerial insect biomass in several of Germany's protected areas declined by 76% between 1989 and 2016 (Hallmann et al. 2017). What makes things even more worrisome is that this decline was detected in areas minimally influenced by human activity and showed a steady decline over almost 30 years. Similar trends were reported from many other areas of the world and were reviewed by Sánchez-Bayo and Wyckhuys (2019) (Fig. 2.20).

As insects are the primary consumers in most terrestrial food chains, parallel declines in higher-level consumers such as mammals, birds, reptiles, and amphibians are expected due to these food shortages (Lister and Garcia 2018). As insects are the most important pollinators, their decline also has troublesome consequences for human food security (Uwingabire and Gallai 2024).

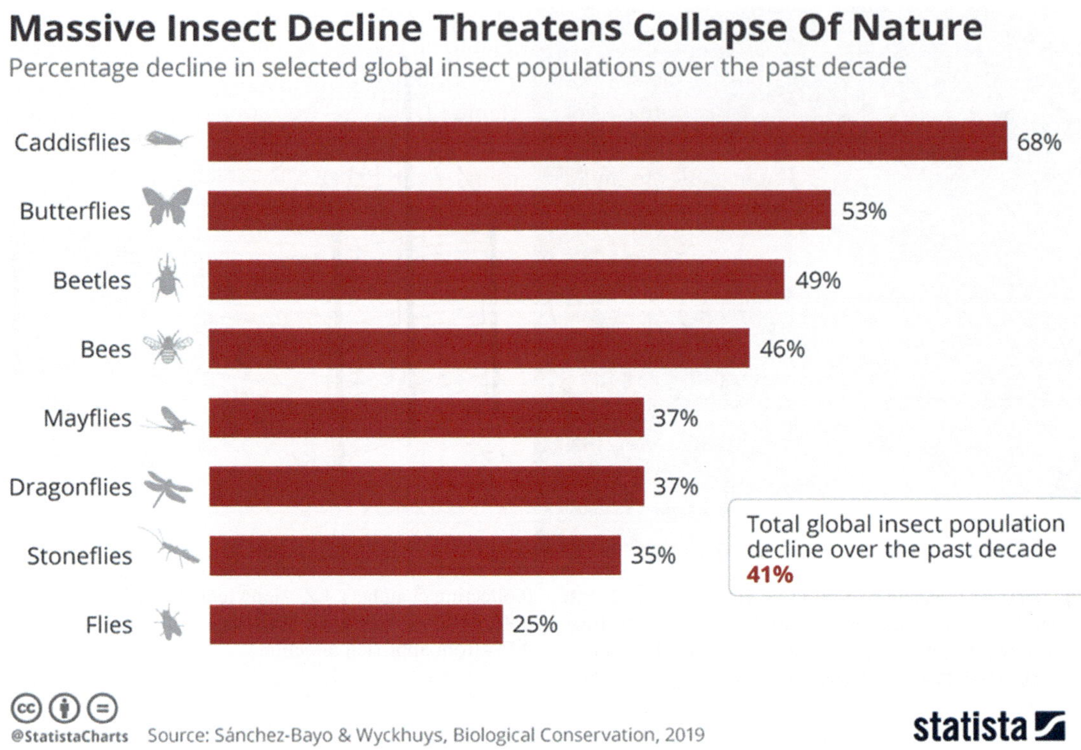

Fig. 2.20 Percentage decline of global insect populations between 2010 and 2019 for major insect orders. (Picture credit: https://cdn.statcdn.com/Infographic/images/normal/16960.jpeg using data from Sánchez-Bayo and Wyckhuys (2019). CC BY-ND)

In 2019, more than 90 research and citizen science groups (North American Insect Abundance Network, NAIAN, https://www.insectabundance.org) initiated a large-scale study to monitor aerial insect biomass all over the North American continent (Dunn et al. 2023). The study follows the same protocol as the German study to facilitate comparability (with some reservations, see below). We capture insects using standard Malaise traps placed in open areas (>20 m from trees) in mostly rural landscapes with predominantly grassy vegetation (Fig. 2.21b). The traps are made from black polyester fabric (96 × 26 mesh/square inch, 680 μm aperture) with an interception area (center panel) of 1.82 m^2 (1.65 × 1.10 m). Insects were collected in a 500 mL bottle partly filled with 70% ethanol.

In order to maximize participation in this study, we developed a protocol that requires only 3 yearly sampling periods at phenologically defined time periods rather than fixed dates to reduce sampling error from seasonal differences between locations. At most locations, samples were collected between April and July. Each sampling period lasted 3 days (72 h) and was based on the breeding biology of tree swallows (*Tachycineta bicolor*), which depend heavily on flying insects to raise their offspring (Winkler et al. 2014). The first sampling period follows the timing of egg laying (typically in late May), the second the hatching date of the eggs (typically

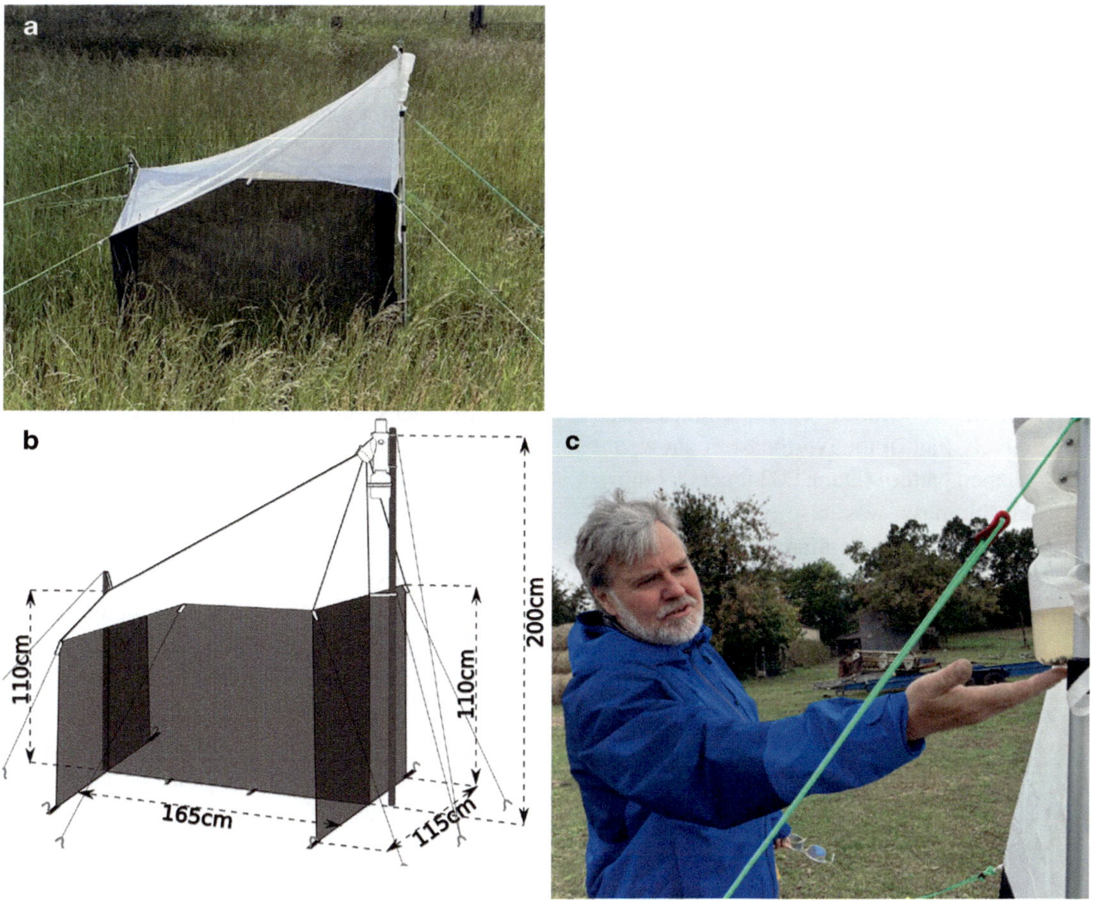

Fig. 2.21 (**a**) Malaise trap, Townes style, BT1011, MegaView Science Co., Ltd., Taiwan. (**b**) One of the Malaise trap locations in a pasture near Adrian, Michigan, USA. (**c**) One of the authors (TW) points to insects preserved in the trap's sample bottle filled with 75% ethanol

13–14 days after clutch completion in early June), and the third period when nestlings reach 12 days of age, reaching a plateau in growth in late June. As we did not have long-term breeding data on tree swallows at all locations, we used a regression equation that used latitude, longitude, and elevation as predictors and explained 52% of the variation in laying date (Dunn et al. 2023).

To open up this project to researchers who are not entomologists and citizen scientists, we simplified post-collection processing by sorting insects only to order rather than finer taxonomic classifications. The only exception was the dipteran suborder Nematocera (including mosquitoes, midges, and blackflies), as they are a significant food source for tree swallows and other insectivorous birds (Twining et al. 2016). After 1 h of sorting the samples into orders and drying them at room temperature, we weighed them to 1 mg on electronic balances. Biomass was measured separately for each order (suborder for Nematocera) and then summed to calculate total biomass.

Our preliminary data for the first 3 years of this study show an overall decline in insect biomass. However, this result was statistically not significant, most likely due to having only 3 years of data with substantial regional variation (Dunn et al. 2023). After adding a fourth year of data in 2024, a preliminary analysis showed significant changes, but the trends seemed to vary depending on the region. Total biomass went down in the Southeast and up in the West. In the Central region, biomass was relatively flat; in the Northeast, it showed a U-shaped pattern (Dunn 2024, pers. comm). Total biomass was also lower in North America than in Germany, where biomass declined 75% over 35 years (data from Hallmann et al. 2017), indicating that North America already has a lower insect abundance than Germany (Fig. 2.22). We intend to continue the study for at least 2024, which will be our sixth year.

A recent study (Uhler et al. 2022) found a 20–55% variation in biomass depending on the type of Malaise trap used in the sampling. The trap size (size of the capture area) did not make a significant difference, but the color and type did. However, a quick reanalysis of the data shared by Uhler et al. (2022) showed that their two most relevant types of traps, the Townes style with a white top (used in our study) and the Long-Term Ecological Research (LTER) trap, which is all black and smaller (1.75 m^2 vs. 1.2 m^2 capture area), showed significant differences in their capture efficiency (4.99 g/day vs. 2.98 g/day, $t = 3.3$, df = 138, $p = 0.0012$) (Dunn 2024, pers. comm.). So standardization is essential, especially in large-scale cooperations where many institutions cover a large geographic area and contribute data. Due to the significant differences between the Townes-style Malaise trap used in the NAIAN study and the German Long-Term Ecological Research (LTER) style, the comparative results between the two studies must be considered cautiously.

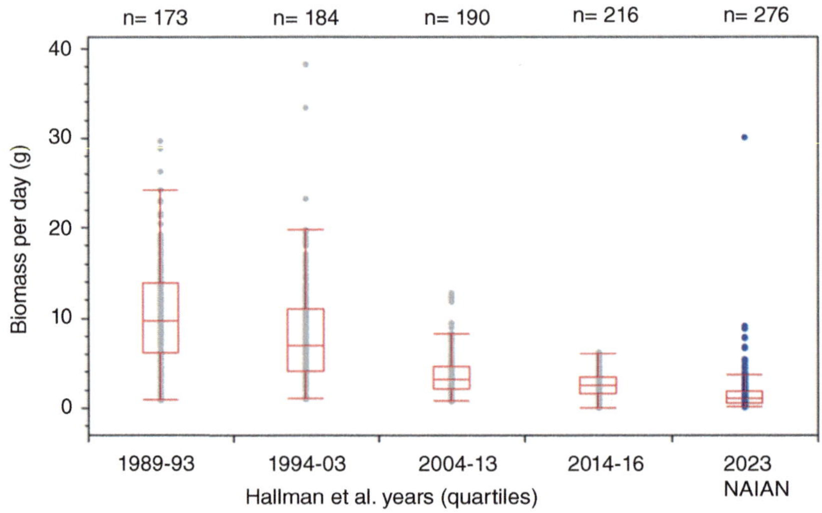

Fig. 2.22 Biomass per day (g) after drying for 10 s (wet weight) from Germany (Hallmann et al. 2017) and North America (NAIAN 2023). Only data from May to July are included. Mean biomass declined significantly between each time period (Tukey-Kramer HSD test $q = 2.73$). Data from Germany covered 1989 to 2016. Boxplots show median (horizontal line), 25th and 75th percentiles (bottom and top of the box, respectively, i.e., the Interquartile range, IQR) and 25th percentile-1.5*IQR and 75th percentile + 1.5*IQR (bottom and top whiskers, respectively)

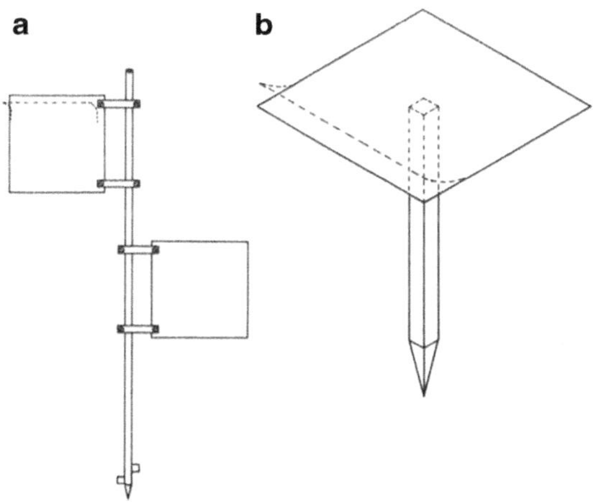

Fig. 2.23 Sticky traps, (**a**) vertical, (**b**) horizontal. (Models after Kühn)

2.12 Killing and Preservative Agents for Traps

Although fewer individuals are caught with ethylene glycol than with formalin, the dominance spectrum of an animal group in the survey remains unchanged.

We recommend formalin traps for short-term use and establishing a reference collection. For a long-term continuous program, in which only significant shifts in species frequency are to be observed, we recommend ethylene glycol traps (Table 2.1).

2.13 Sticky Traps

Sticky traps catch flying or drifting arthropods using Plexiglas plates, which are covered with plastic film and coated with a special glue. The animals must be removed from the adhesive film and brought into the laboratory with a special solvent. The adhesive "Tangle-Trap" from the Tanglefoot Company (available on ARBICO Organics, arbico-organics.com, or Amazon) is an established film coating method. You can remove the coating using purified gasoline (drugstore, pharmacy). Doing this outdoors or under a fume hood is highly suggested. The vertical sticky traps (Fig. 2.23a) are adjustable in height. The horizontal surface traps (Fig. 2.23b) are specifically helpful in answering questions about which animals land in a particular habitat. Adhesive films laid out on the ground (on stones) can also be used to estimate the food supply for lizards (Dunham 1983).

Sticky traps are used to determine activity densities in the air and distribution flights of arthropods (spiders, for example, Greenstone et al. 1987). They also capture a potential food supply of epigaeic predators. Sticky traps can also be installed in inaccessible places, such as in the canopy of tropical rainforests (Grootaert et al. 2010)

Although the construction is simple, animal removal is very labor-intensive, and identifying small arthropods is often tricky. Only light and small arthropods are usually captured with a thin layer of glue. Sticky traps are less selective and more dependent on individual animals' behavior than window traps. Unlike window traps, sticky traps can be exposed for extended periods without control (up to a maximum of 4 weeks).

Grid capture surfaces for drifting spiders (after Kühn) (Fig. 2.24):

Mainly on warm late summer days, many spiders let themselves be carried through the air for dispersal by a thread from their silk gland (*ballooning*). When the flight thread touches these structures, the spiders land on horizontally stretched cords, wires, or vertically erected, simple metal grids. Usually, the spiders stay longer (1 h or longer) before they fly on.

With the grid capture surfaces, one can determine the species and age classes of the spiders spreading through the air.

This capture method is only suitable for specific questions and is only effective for a short time during the year. It can be helpful in combination with sticky traps, with which more extended checks can be carried out. Since the animals get into the net alive, determination of the juvenile stages is more accessible than with sticky trap catches.

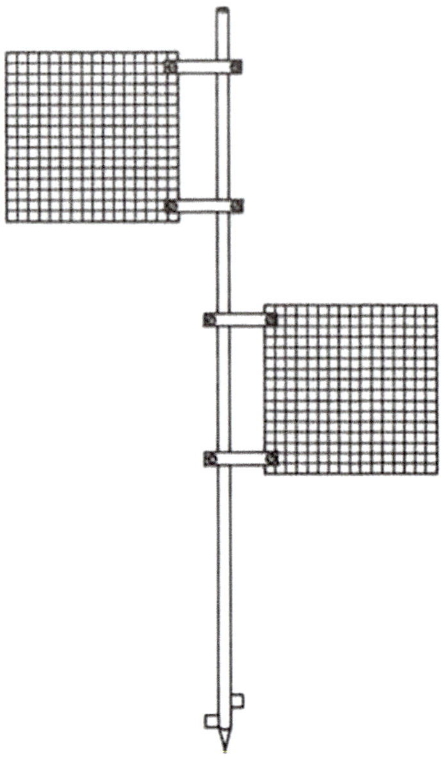

Fig. 2.24 Grid capture surfaces for drifting spiders (after Kühn)

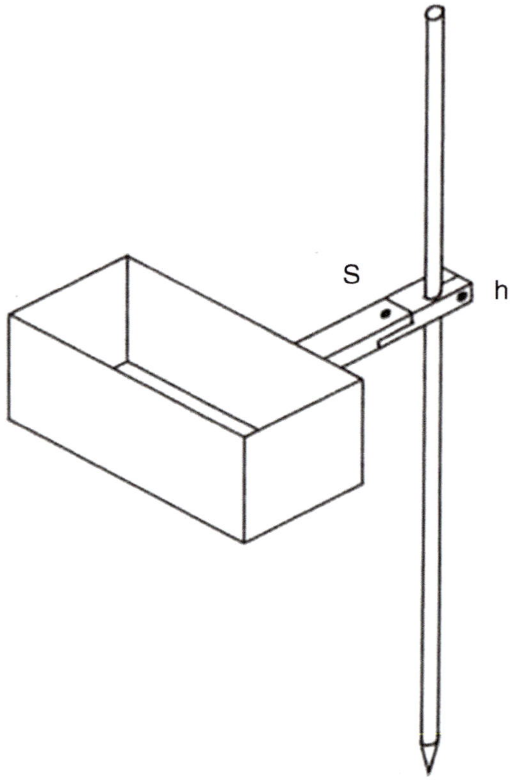

Fig. 2.26 Catch bowl (after I. Kühn)

Fig. 2.25 Yellow bowl in a beet field for controlling the aphid flight and other insect pests

2.14 Pan Traps

Wild bees, hoverflies (Syrphidae), and aphids are attracted by color bowls (Fig. 2.25) set up in the vegetation, best in a meadow at the height of the flowering plants. The color bowls are filled about 2 cm high with collection fluid or water with detergent. Kratochwil (1984) and Klein (1981) were able to catch 83 wild bee and 16 hoverfly species in a semidry grassland in southwestern Germany in 6 months with 8 color bowls using this method; selective catch at flowers yielded 104 wild bees and 57 hoverfly species in 260 h of operation time, however. Ssymank (1991) found that the catch of a yellow bowl for syrphids only accounts for about 5% of the approach but must still be considered one of the most effective hoverfly-catching methods. Yellow water traps also monitor pest insects in oilseed crops (Fig. 2.25) (Metspalu et al. 2015). Parasitic Hymenoptera could also be captured well with yellow bowls in the tropical rainforest of Sulawesi (Noyes 1989).

Color bowls catch mobile arthropods mainly in the herb layer and thus also capture flower visitors. They are filled with catching fluid (see Table 2.1) or, with more frequent control visits, simply with relaxed water (with a few drops of detergent, e.g., Agepon). The bowls are height-adjustable (Fig. 2.26) and can be adapted to the most common flower height. The bowls are often used as color bowls (yellow or blue), with the color attracting certain insects (flower visitors and aphids). The bowls are detached from their stem (Fig. 2.26, s) to empty the catch and shaken through a sieve. The sieve is then rinsed out from behind over a correspondingly large funnel for animal selection.

Color bowls can be used for phenological control of aphid swarms (attracted especially by yellow bowls, Fig. 2.25). In addition, animals running around in the vegetation are also captured, especially if the vertical and horizontal elements of the installation are located amid the vegetation and act like natural running structures. To a limited extent, information about preferences for a habitat structure and vertical migration movements can also be derived.

In contrast to the netting method, color bowls allow for continuous, standardized catching. Emptying traps may disturb the vegetation structure of the immediate surroundings. The time interval to check open bowl traps depends on solar radiation (evaporation) or rainfall. The amount of catch is strongly dependent on weather conditions.

2.15 Carrion Trap and Dung Traps

Carrion traps are used to catch insects attracted by scent. They are mainly used to determine carrion visitors and their successive appearances.

The carrion is placed on sand or sawdust so that fly maggots and other larvae can migrate downwards from the carrion for pupation. If you pile up soil in front of the entry slot, beetles that walk on the ground can also find their way into the trap. Catching many animals scavenging on carrion does not significantly affect the decomposition of the carrion. A 4% formalin solution mixed with detergents is suitable as trapping fluid in the trap channel (drawers). The recommended size for the construction of the trap is 30 cm × 30 cm × 30 cm, a drawer 5 cm × 4 cm × 29 cm, and a glass tube with a barrier, 12 cm long, 4 cm in diameter (∅) (Fig. 2.27).

Carrion traps provide comparable, quantitative trapping results. They are versatile in the type of bait used in scent attraction, e.g., carrion or excrement of various animal species. However, the microclimate is influenced by the trap. Only a portion of the animals reach the carrion or excrement. Environmental parameters also influence the abundance and diversity of carrion beetles visiting carcasses (von Hoermann et al. 2020; Weithmann et al. 2020). If carrion beetles attracted to freely laid-out carcasses are individually marked, the dispersion distances can be determined by rediscoveries at other carcasses laid out at distances of up to 5 km (from von Hoermann et al. 2020, Video of the Bavarian Broadcasting Corporation (in German): https://www.youtube.com/watch?v=syiOZAIczwI). The influence of habitat factors on the colonization of carcasses can be tested using carcasses placed in different habitats (Feng et al. 2022).

Case Study: Phenology and Succession of Dung Beetles

Dung beetles dwell, feed, and reproduce within or close to the feces of terrestrial vertebrates and have coevolved with them since vertebrates conquered the land (Ebert et al. 2021; Hanski and Cambefort 1991; Philips et al. 2004; Stack 2015). Taxonomically, they belong to several families within the superfamily Scarabaeoidea (Krell and Moon 2019) (Fig. 2.28).

Dung beetle communities are especially rich in landscape mosaics that include forest and grassland patches (Meierhofer 1995; Ortega-Martínez et al. 2020; Wassmer 1994) and are inhabited by large mixed herds of megaherbivores such as sheep, goats, cows, and horses (Galetti et al. 2018; Pryke et al. 2016). Dung beetles provide several essential ecosystem services, such as dung removal (Doube 1990; Noriega et al. 2023; Sladecek et al. 2017), reduction of dung-dwelling livestock parasites (Nichols et al. 2008; Ridsdill-Smith

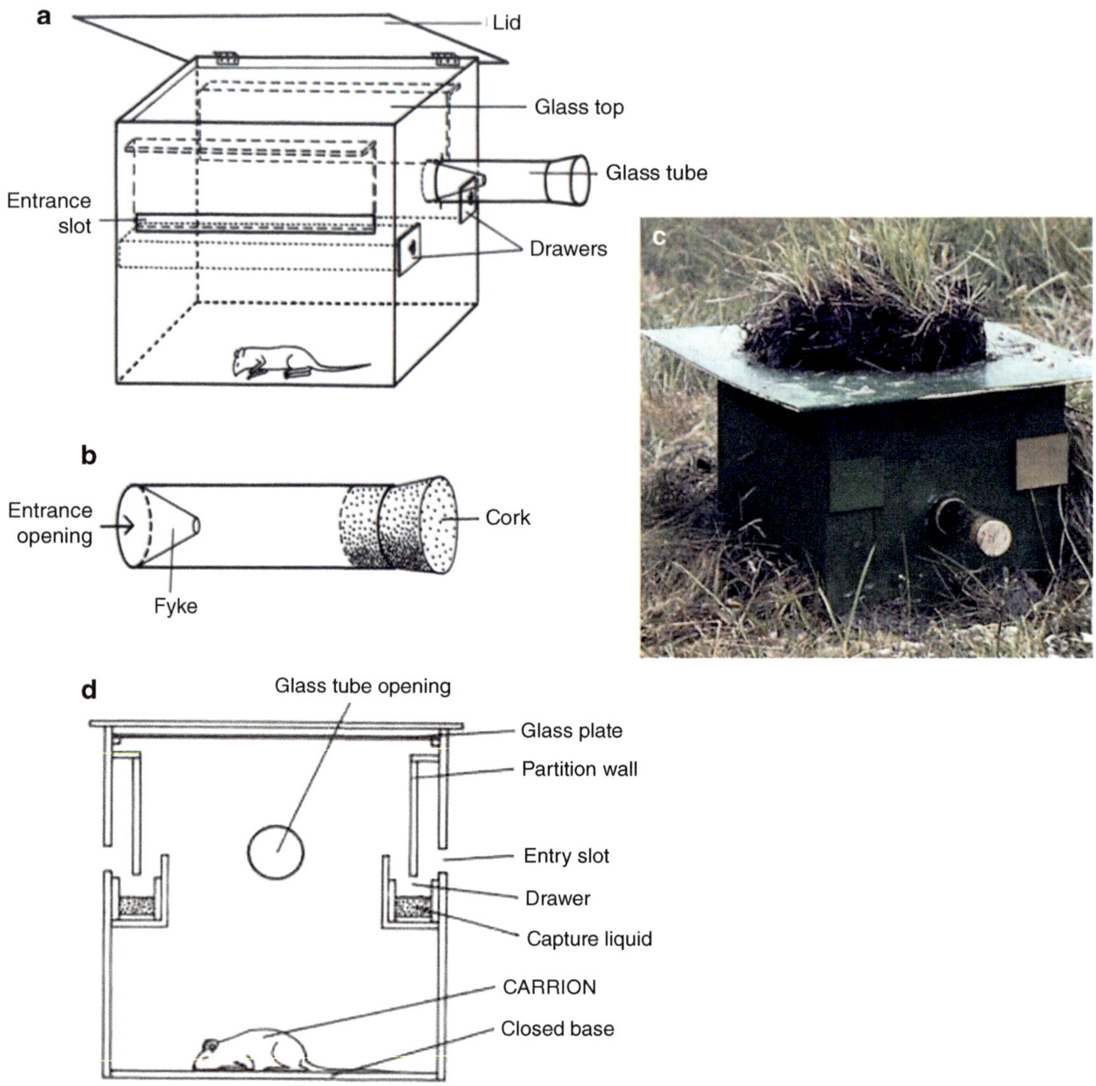

Fig. 2.27 (**a**) Construction of a carrion trap. (**b**) Glass tube. (**c**) Carrion trap in the field. (**d**) Cross-section of the carrion trap with a laid-out carcass

and Hayles 1990; Sands and Wall 2017), soil fertilization (Haynes and Williams 1993; Maldonado et al. 2019; Yamada et al. 2007), and carbon sequestration (Slade et al. 2016; Verdú et al. 2020). They are excellent bioindicators for the environmental health of pasture systems (Carvalho et al. 2020; McGeoch et al. 2002; Waßmer 1995). More information is in the case study in Sect. 9.8 (see Fig. 2.29).

Dung beetle assemblages can be assessed using pitfall traps (Lobo et al. 1988; Wassmer 2020a), flight interceptors (Matthews and Matthews 1972; Bach et al. 2023), and hand collection of dung pads and the soil underneath (Wassmer 1994, 1995) or hand collection of beetles in and underneath dung pads using a pair of feather-steel tweezers (Wassmer 2014), each of which has advantages and disadvan-

Fig. 2.28 Beetles of superfamily Scarabaeoidea present at the University of Sucre campus (Caribbean Columbia) showing the overall form and taxonomic and functional variety of dung beetles. (**a**) *Ataenius* gr. *carinator*. (**b**) *Canthidium aurifex*. (**c**) *Canthon cyanellus*. (**d**) *Canthon juvencus*. (**e**) *Canthon mutabilis*. (**f**) *Deltochilum guildingii*. (**g**) *Diabroctis cadmus*. (**h**) *Dichotomius* aff. *agenor*. (**i**) *Digitonthophagus gazella*. (**j**) *Eurysternus impressicollis*. (**k**) *Eurysternus mexicanus*. (**l**) *Onthophagus marginicollis*. (**m**) *Onthophagus bidentatus*. (**n**) *Pseudocanthon* aff. *perplexus*. (**o**) *Uroxys* cf. *deavilai*. (**p**) *Omorgus suberosus*. Scale bars = 1 cm. (Source: https://doi.org/10.15560/15.4.579, open access article distributed under the terms of the Creative Commons Attribution License (CC BY 4.0))

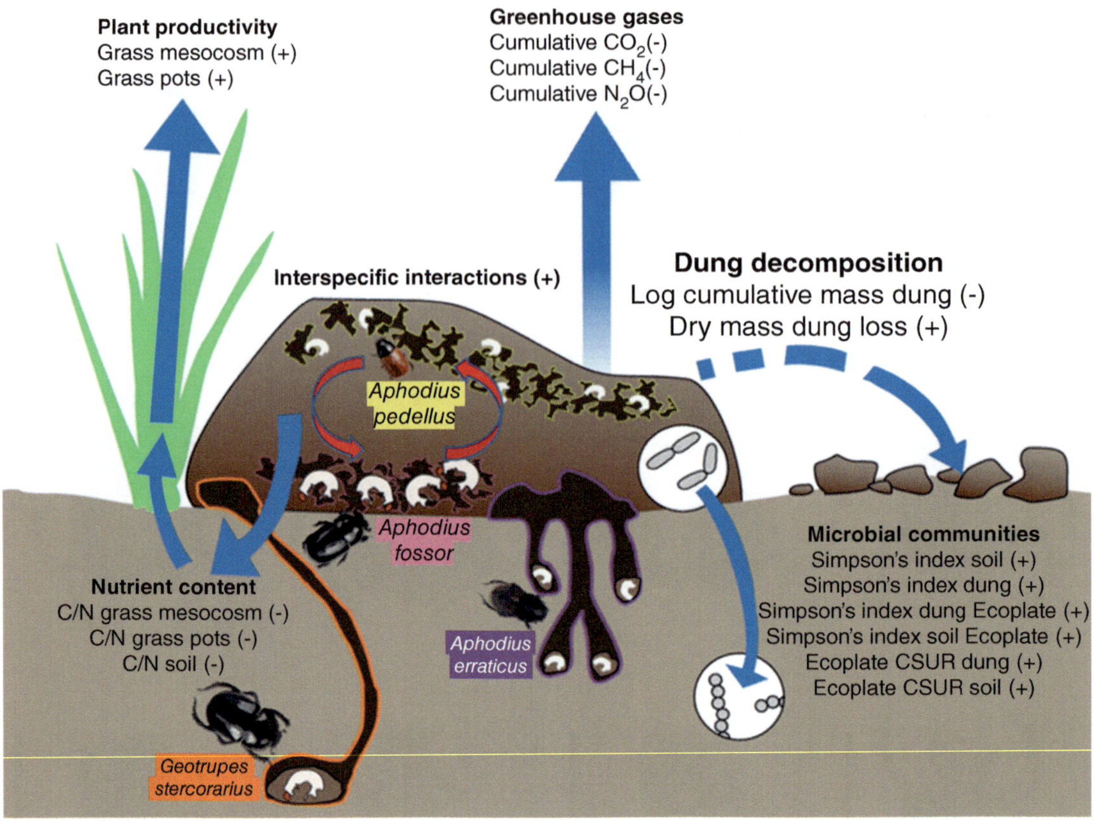

Fig. 2.29 Ecosystem services performed by different functional classes of dung beetles based on the relocation patterns of dung for larval provisioning. (Picture credit: Slade et al. 2017. Reproduced by permission)

tages (Gotelli et al. 2011; De Clerck-Floate et al. 2012; Price and Feer 2012; Silva et al. 2012). Dung is a main waste product of heterotrophic animals and an important food source in the detritus food chain (Wu et al. 2011). As with many limited resources, the use of dung is partitioned in space and time. The temporal niches include daily (Tocco et al. 2021), seasonal (Wassmer 2020a), and successional partitions (Wassmer 2020b). Diel and seasonal segregations can be studied by flight interception traps (Boonrotpong et al. 2012; Lobo and Cuesta 2021), pitfall traps (Wassmer 2020a; Tocco et al. 2021), or the collection of dung pads and the adjacent soil underneath (Wassmer 1994) (Fig. 2.30).

Even hand collection can be used to provide data on occurrence vs. absence (Gotelli et al. 2011; Wassmer 2014). The advantage of this method is that all available microhabitats can be sampled intentionally; the disadvantage is that the biased collection method does not allow the analysis of the quantitative species list (Fig. 2.31). The time course of heterotrophic succession in a dung pad can either be studied by removing some dung pads of an array at specific times (Sladecek et al. 2017; Sullivan et al. 2017) or by removing the insects from a pitfall trap at certain times without altering the dung bait on top of the trap (Wassmer 2020).

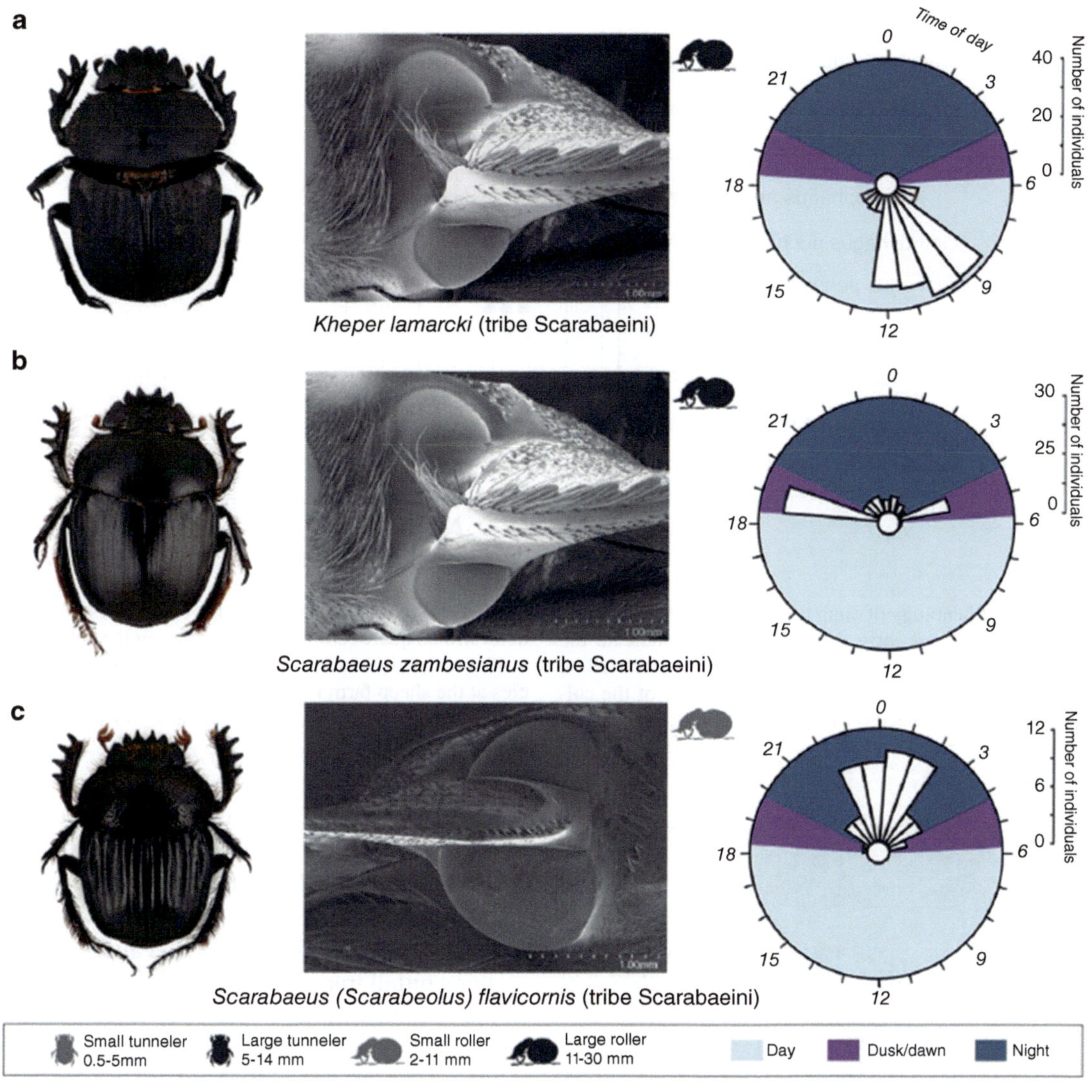

Fig. 2.30 Activity plots and eye morphology of a diurnal species *Kheper lamarcki* (**a**), active during daylight, a crepuscular species active during twilight hours *Scarabaeus zambesianus* (**b**), and a nocturnal species, active during the night *Scarabaeus* (*Scarabaeolus*) *flavicornis* (**c**). The left column shows the habitus (general appearance of the species in dorsal view), the middle column shows a scanning electron microscopy picture of the right eye to depict the eye type, and the right column shows a circular plot of the diel activity period, depicted as the mean abundance of a species in the different sampling times throughout 24 h. (Picture credit: Tocco et al. 2021—selected and reassembled from two figures. CC BY 4.0)

2.16 Pitfall Traps (Barber Traps)

Pitfalls are best to sample surface-dwelling invertebrates like epigaeic arthropods, from which spiders (Araneae) and ground beetles (Carabidae) are dominant taxa, well known as suitable bioindicators (Pfiffner and Niggli 1996; Rainio & Niemelä 2003; Avgin and Luff 2010). Araneae and Carabidae are accepted as taxa that react distinctly to habitat feature changes in the boreal forests following natural disturbances (Hammond 1997). Woodcock (2005) gives a detailed overview of the advantages and limitations of using pitfall traps. Pitfall traps (Figs. 2.32 and 2.33)

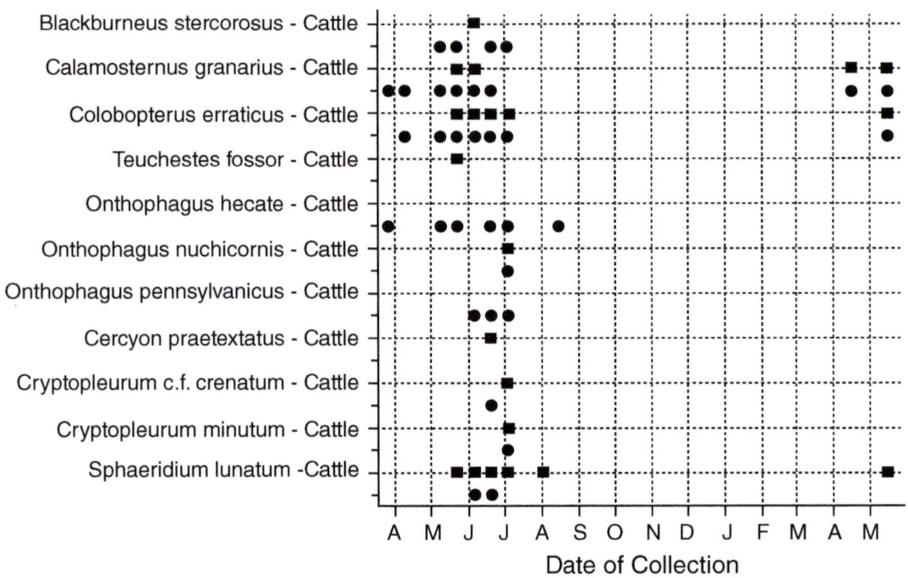

Fig. 2.31 Phenology of dung beetle species that occurred only in spring and early summer. Symbols indicate the presence of a species in each sample. Sampling started on 26 March 2012 (before the first A on the date of the collection axis) and ended on 16 May 2013 (after the last M on the same axis). Circles show the presence of a beetle species on the sheep farm on the respective collection date, while squares show beetle species on the cattle farm on the respective collection date. The presence of a species at the sheep farm is indicated below the labeled presence of the species at the cattle farm. (Picture credit: Wassmer 2014)

Fig. 2.32 Simple ground trap (**pitfall trap**) in the field

depend on the activity of the individual species. As activity can vary between species, the different trapping frequencies do not indicate actual abundance differences between species. Since species not belonging to the epigeal guild can also be caught in the pitfalls, species with fewer individuals (e.g., accounting for only 1–5% of the animal catches) can be cautiously disregarded.

Comparing different recording methods for larger ground-dwelling arthropods (pitfall traps with tracking tunnels, visual surveys, and trail cameras), Bertoia et al. (2023) showed that pitfalls are the best method in New Zealand.

For a pitfall trap, a glass jar or bucket is buried in the ground with its upper edge level with the ground. Epigaeic small animals running around on the earth's surface fall into this container and cannot escape. To prevent the earth from collapsing and keep the hole open when the sample container needs to be changed, a hollow cylinder (plastic tube) can be anchored in the ground. The opening diameter of the standard pitfall glass is about 7 cm (jam jars with 750 mL content, which can be closed and transported with the lid). The gap between the glass opening edge and the inserted cylinder can be covered with a collar sleeve that can be screwed onto the lid thread for easy replacement. The flat ring of the sleeve (approx. 2 cm wide) is made of ± neutral plastic and is roughened up or covered by glued-on sand so that the surface does not repel the animals. The

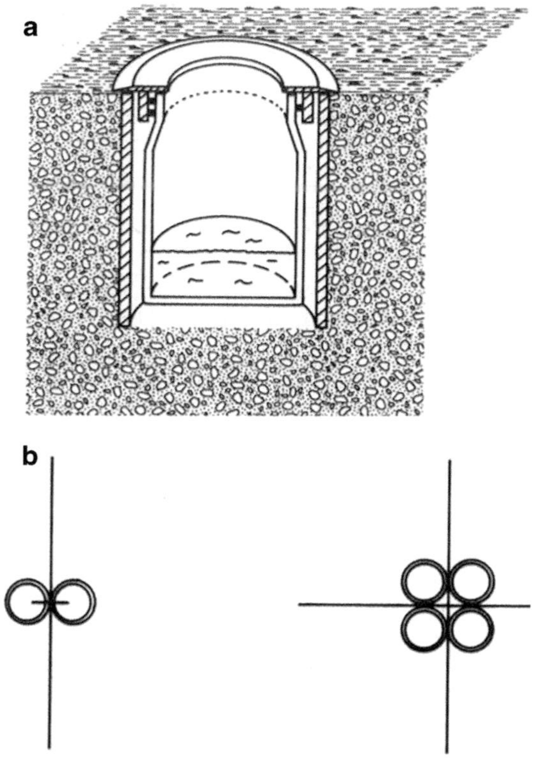

Fig. 2.33 (**a**) Diagram of a pitfall trap with a "collar" to prevent soil material or water from trickling in. (**b**) Simple partitions (compared to Fig. 13.6) made of metal or better plastic (acrylic) increase the trapping efficiency of pitfall traps as epigaeic animals bump into them and usually run along these guidelines. With the appropriate arrangement, the direction from which the animals have come can also be inferred (direction traps, see under Sect. 11.4, Smith 1976)

slight neck constriction typical for jam jars largely prevents animals that have fallen in from escaping. Plastic cups, which often widen toward the opening like an inverted cone, are less efficient. The catch glass can be filled about 1/3 with catching liquid to kill and preserve the animals (see Table 2.1). Pitfall traps are usually emptied or changed every other week. A rain cover from metal sheets changes the microhabitat and possibly attracts nontarget animals to artificial hiding places. Glass rain covers are also not recommended in the field due to the risk of breakage, whereas transparent plastic plates can be used effectively (Wassmer 2020a, b; Wassmer and Armstrong 2023). Even without a rain guard, the traps usually do not fill up with rain, even with a fortnightly change. However, water can flow into the trap if it is inappropriately located in ground grooves or depressions.

Using pitfall traps is a simple, fast, and suitable method to capture many target animals in almost every location, even in large arrays, and for continuous monitoring. The catch can be fractionated into specific time periods within the 24-h cycle by simply closing and opening the lid or using more elaborate devices (e.g., motor-driven turntables with several catch glasses) (Ayre and Trueman 1974). If formalin is used as a preservative, certain animal groups might be more attracted to the traps than others. The catch results also depend on the animals' mobility and activity cycles and may, therefore, be biased.

Heap (1988) expanded pitfall traps by combining them with a light trap and captured more individuals and a wider range of species without including many flying insects, which are often irrelevant when using conventional light trapping.

2.16.1 What Is Better: Dry Traps or Traps with Catching Fluid?

Since the captured animals usually must be identified to the species level using a binocular magnifying glass (stereomicroscope), killing the animals with a catching fluid is usually unavoidable. Even if to the layman, the numbers of caught arthropods seem very high (e.g., in three traps in a German two-furrow meadow within 130 days between May and September, a total of 1114 spiders (Araneae) and 252 ground beetle individuals (Carabidae) were caught (Mühlenberg and Werres 1983)), a loss in this order of magnitude will most likely not harm the arthropod populations. For example, 1 m^2 of meadow can harbor about 130 spiders (Turnbull 1973; Nyffeler 2000). The total number of individuals usually consists of more than 80% dominant and currently non-endangered species, and the natural population fluctuations are orders of magnitude higher. Baars and van Dijk (1984) estimated the catchment area for the ground beetle *Pterostichus versicolor* to be about 200 m and for the less mobile ground beetle *Calathus melanocephalus* at about 80 m. They calculated that 50% of the individuals of *Pterostichus versicolor* caught in a standard series of pitfall traps (8 traps per site) 3–10 m apart from each other come from an area within a radius of 100 m around the traps, and 90% of the individuals caught from an area within a radius of 200 m. The corresponding values for *Calathus melanocephalus* are 50% within 40 m and 90% within 80 m radius from the traps. With a density of 0.5 individuals/m^2, approximately 60,000 individuals of *Pterostichus versicolor* live in an area of 1/8 km^2 ($r = 200$ m) under the assumption of a uniform distribution. With the three pitfall traps described above, 400 individuals of *Pterostichus versicolor* were recorded over a year, less than 1% of this subpopulation. However, Larochelle and Bousque (1977) notice a significant decline in the populations within only a 2-year use of 100 pitfall traps compared to a hand-collected control in a neighboring area.

The captured animals initially remain alive in pitfall traps without preservative fluid (dry traps). Dry traps must be emptied daily and contain leaf litter, moss, grass, or dung as hiding places for the captured animals (e.g., Wassmer and Armstrong 2023). The escape of captured animals can be prevented by inserting a funnel into the opening (see Fig. 2.34). Dry traps are especially suitable for the control of marked animals, whether in the use of recapture methods (Sect. 5.3) or the study of migratory movements (Sect. 11.4). Traps with catching fluid and a few drops of detergent (e.g., unscented dish soap or Agepon from the photo trade) catch spiders more effectively than dry traps (Curtis 1980).

2.16.2 On the Arrangement of Traps in the Field

Although for several statistical test procedures, the initial data should be based on randomly obtained samples, it is usually more practical to arrange pitfall traps uniformly according to a square grid. Regarding the probability of capturing active soil arthropods, traps that are too close together are not equivalent to far-apart traps. According to Westerberg (1977), a minimum distance of 1 m between traps should be sufficient so that the catch of one trap does not reduce the potential catch of the neighboring trap. As removing the caught arthropods also affects the immediate surroundings of a trap, we recommend a more considerable minimum distance of 2 m or more. These distances, however, depend on the taxon, mobility, and habitat. For dung beetles that arrive at baited traps while flying, a minimum trap distance of 50 m is recommended (Silva and Hernández 2015). Generally, the distance between traps should be smaller than the average distance that an individual in the population moves during the catch period. A grid design provides a more accessible location for the traps in the field. It also allows researchers to randomly select unit areas within a grid for equipping with a trap (*stratified random sampling*, Sect. 15.3).

2.16.3 To What Extent Is the Result of a Species Inventory Limited by the Method Used?

With pitfall traps, we discussed the problem of species' detectability depending on the sample size. In addition, we must consider the chosen method (e.g., pitfall traps, handpicking, nets) and can only assess detection deficiencies by comparing different methods. Bährmann (1976) compared net catches of Diptera to pitfall traps, biocoenometers, color dishes, and hand collection and found about 20% additional species using these methods. The efficiency of a suction device in grassland is species-specific (Henderson and Whitaker 1977), with the highest for Diptera (79–98%) and Hymenoptera (60–83%). A method comparison for evaluating catch results is also given by Noyes (1989).

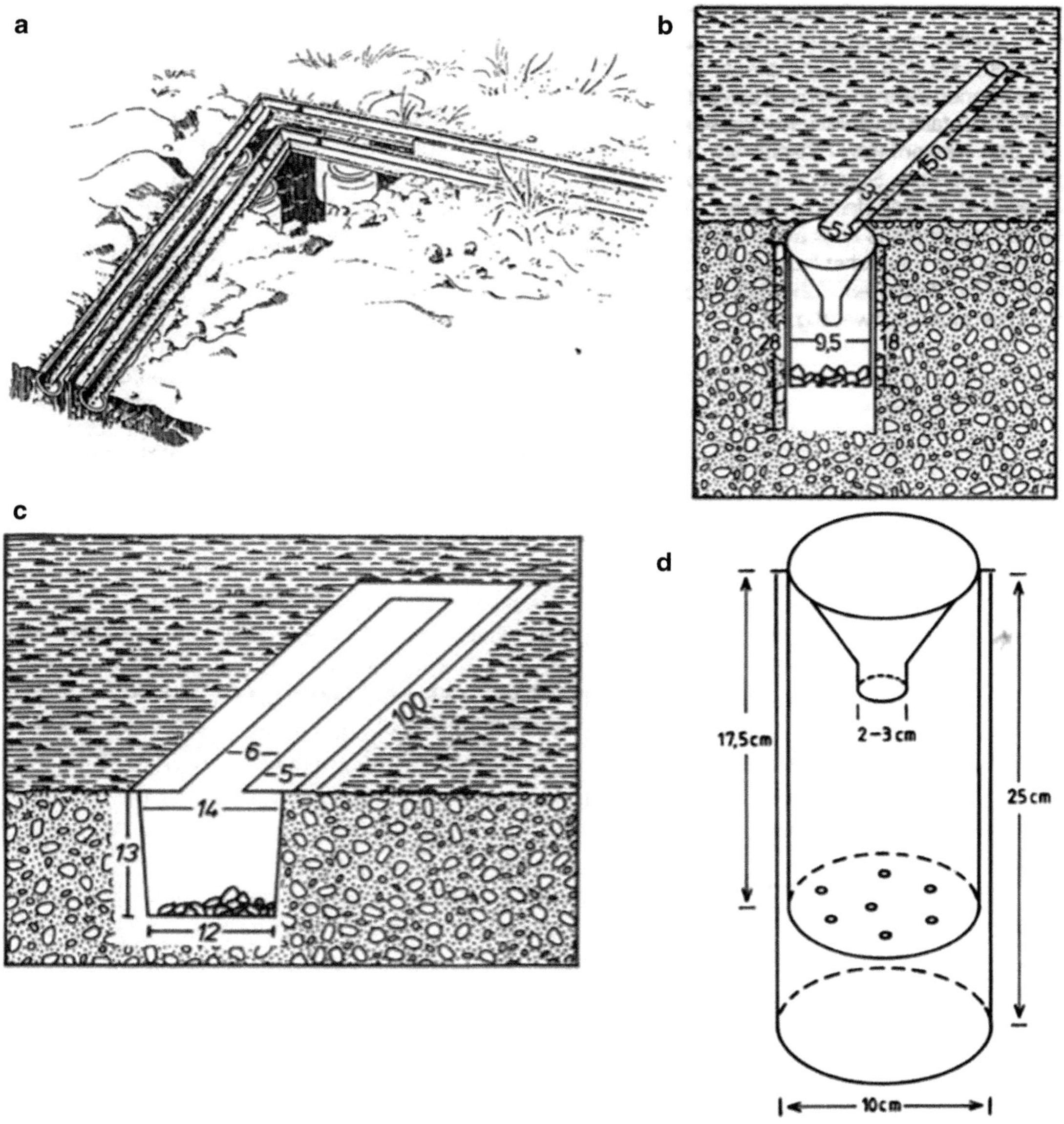

Fig. 2.34 Capture trenches and pits: (**a**) capture trenches according to Schjøtz-Christensen (1965), (**b**) capture trench with pitfall trap as a live trap (dry trap), (**c**) pitfall trap with buried flower box, (**d**) construction of a pitfall trap equipped for live capture: the funnel prevents the fallen animals from escaping; the catch container must have small holes in the ground so that no rainwater accumulates

Case Study: Comparing Arthropod Communities in Forest Stands with Different Disturbance Histories

The most common disturbances in boreal forests are fires and, on a smaller scale, windthrows. In some regions, forest disturbances are caused by pest insects. A chronosequence approach helps study the succession of forests after wildfire or other disturbances because true long-term studies are too expensive in most cases. The questions are whether the fauna shows recovery toward a condition representative of a pre-disturbance state and whether distinct species assemblages are typical of an early successional state or mature timber forest, respectively. A similar approach is the comparison of the fauna among different forest management treatments, like timber harvesting with increasing levels of green-tree retention (0, 10, and 50 m^3/ha), including uncut controls (cf. Lange et al. 2014).

As an example, we select four age classes of *Betula-Larix* forest stands originating from the following wildfire ("pyrogenic" stands), e.g., 1–2 years, 10–15 years, 25–40 years, and >70 years. For replicates, we studied three stands within each age class. We placed six pitfall traps in one stand along one random transect line, with 10 m separating traps (Table 2.2). We can estimate the recovery state for classifying age classes or ask relevant local people.

A row of 6–10 pitfalls (indicated with symbol ¤), each 10 m apart, is established for one vegetation type to compare the species assemblages between different habitat types. That facilitates the statistical analysis of the data.

Data analyses: It is doubtful that such a study design can be performed accurately, and often, the trapping effort varies between stands. Therefore, trap data are standardized to 1000 trap days to compensate for differences in trapping effort among stands. Because estimates of raw species richness are sensitive to sample size, rarefaction analyses are conducted (see Waltert et al. 2004; Collwell et al. 2004; Hsieh et al. 2016).

2.17 Catch Trenches and Pits with Live Traps

By improving the design of trap arrays, the capture efficiency of simple pitfall traps can be significantly increased, e.g., with attached 10 cm high plastic films (Smith 1976) or by digging catch trenches (see below, also Schjøtz-Christensen 1965). In both cases, the added structures guide epigaeic animals into the trap.

The catching principle is similar to simple pitfall traps. The catch trenches and pits shown in Fig. 2.34 are suitable for live capture. Figure 2.34d shows a pitfall trap especially equipped for live capture (compare to Fig. 5.14 in chapter 5). Recapture methods (see Sect. 5.3) rely on high capture numbers, which can be achieved by extending the ground hole into trenches and pits.

Cut plastic pipes (e.g., 5–7 cm in diameter) can be used as capture trenches. In this design (Fig. 2.34), plastic flower boxes serve well as pitfall traps. The flower boxes must be covered a few cm inward at their edges (similar to the collar sleeve of the simple pitfall traps described above) and overlaid with soil material. The epigaeic animals caught alive can usually not fly away. Capture trenches connected to a dry trap are easier to install and empty in the field. Animals in pitfall traps must be retrieved by hand or with an exhaustor. Styrofoam or moss cushions in the live traps shelter the fallen animals. Live traps must be emptied daily, but at least every other day.

Capture trenches have very high capture efficiency. However, their installation is more complex, and frequent emptying is necessary.

It is also possible to **catch small mammals in pitfall traps**. For catching small mammals from a larger area, pitfall traps are set in a long ditch where funnels are placed, making escape difficult (Figs. 2.35 and 2.36). Russian experts use this traditional method for small mammal surveys. The ditches are about 20–30 m long, at minimum 30 cm deep, and have two funnels as pitfalls (Sheftel 1989, 2018; Sheftel and Yakushov 2022). Kaufman et al. (1978) and Shchipanov et al. (2008) compare different methods for detecting small mammals.

While box traps have proven effective for long-tailed mice (Muridae), pit traps are more effective for voles (Arvicolidae). According to some authors, the best box and most expensive box traps are the Longworth live traps (Gurnell and Flowerdew 1982; Clark 1983), whereas Jung

Table 2.2 Comparison of 4 different stages of forest development with pitfalls

	1–2 years	10–15 years	25–40 years	> 70 years
Three study plots as replicates, each with a row of pitfalls along a random transect line			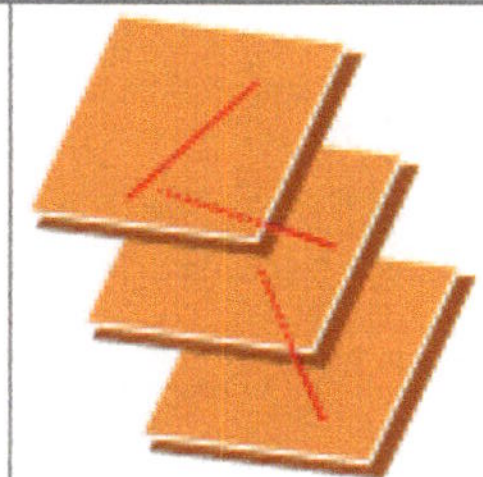	
Six pitfalls per one transect, 20 m separating traps	¤¤¤¤¤¤	¤¤¤¤¤¤	¤¤¤¤¤¤	¤¤¤¤¤¤
Number of traps per succession stage	18	18	18	18
Total sum of traps	72			
Sampling period: 6 weeks in spring, 8 weeks in late summer; traps are emptied every 2 weeks	7 emptying	7 emptying	7 emptying	7 emptying
Sum of all samples	504			

Fig. 2.35 (**a**) Ditch with a pitfall trap (see yellow arrow). (**b**) View into a trap channel dug into the ground with a pitfall trap

Fig. 2.36 The photo shows an old ditch (see yellow arrow) at the margin of the light taiga, the *Betula-Larix* forest. Especially shrews are captured best in ditches because they are difficult to attract to baited life traps. Northern Mongolia

(2016) concludes that Longworth and Ugglan traps are comparable and superior to Sherman traps when capturing small mammals in the Nearctic boreal forest. Simpler traps can be easily made yourself, with wooden boxes protecting the captured animals better from moisture and cold than simple wire mesh traps.

2.18 Life Trapping of Small Mammals and Other Small Vertebrates

Life trapping of small mammals is usually conducted by Sherman traps (see Fig. 2.37), which are placed in a grid of 100 traps, each 10 m from the neighboring trap. The traps are baited with peanut butter or "muesli" (cereal). Each mouse caught is sexed, aged, and measured (Fig. 2.38) and individually marked with a small ear clip. The home range of captured specimens can be estimated through several recaptures of the same individual. In Australia, Sherman traps are called Elliott traps and are available, e.g., from Treetec Environmental Consultancy Services (https://www.treetec.net.au). Tasker and Dickman (2001) provide an overview of small mammal trapping with Elliott traps.

Fig. 2.37 (**a**) Left: Sherman trap unfolded. As many traps are usually required for a study (often 100 or more), these traps can be folded for transport; see (**b**)

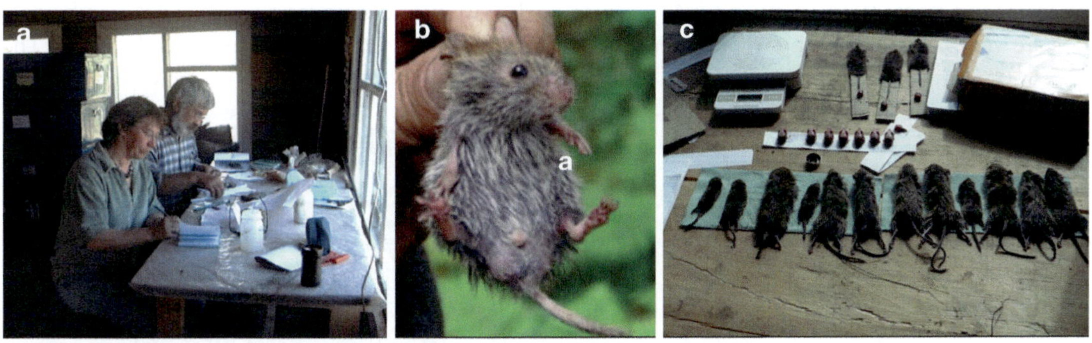

Fig. 2.38 (**a**) Dr. Boris Sheftel, Russian Academy of Science, is taking measures from captured mice with assistance from his wife Dr. Tonya Alexandrova. Mongolia. (**b**) Handling of a vole for sexing, measuring, and weighing. (**c**) For final identification and voucher specimens, some mice were prepared as skins and, together with their skulls, brought to the museum

Case Study: Pitfall Trapping for Estimating Population Fluctuations of Mice

For population estimates, Gurnell and Flowerdew (1982) recommend a trap deployment of no longer than 3 days and nights. Montgomery (1985) recommends 4–5 days per trapping period for short- and long-term studies. The results of more extended trapping periods are influenced by immigrating and emigrating animals and increasingly less likely catchability. Intervals of at least 2–5 weeks are recommended for intensive field studies (see Tserendavaa et al. 2023). The traps should be set up in a grid with a trap distance of

5 m in grassland, 10 or 15 m in the forest, and 20 m in farmland. Twenty traps would be the minimum for quantitative investigations, with two traps per trap point to overcome the disadvantage that an animal passes by an already occupied trap.

If the traps are arranged in a "web," then the density of the mice populations can be determined by the method of distance sampling (see Sect. 5.1). More recaptures should be met in the inner area of the web. If the animals tend to move in home ranges that are small relative to the size of the web and the trap spacing, then the trapping web will perform well (Buckland et al. 2001). The application of trapping webs is described in Anderson et al. (1983), Buckland et al. (2001, 2004) (Fig. 2.39).

Whereas life traps can easily catch squirrels, weasels, and stoats (Figs. 2.40 and 2.41), mustelids are very hard to catch by trapping during the vegetation season. Although pine martens (*Martes martes*) and polecats (*Mustela putorius*) are widespread, little is known about the status of their populations. Lang and Simon (2010) have compared methods for recording the two species and provide information for nationwide monitoring.

Since stoats (*Mustela erminea)* are often caught in hunting districts (particularly in the UK, but also as unwanted invasive species in *conservation reserves* in New Zealand), one could use the trapping and hunting data to determine their abundances. However, animals quickly repopulate after being caught, and the result highly depends on the effort to set the traps (McDonald and Harris 1999, 2002).

The detection of **dormice** (Fam. Gliridae) is challenging and can often only be provided qualitatively as presence evidence, as detections in nest boxes are unreliable. Langer et al. (2020) and Melcore et al. (2020) presented so-called trace tunnels, which seem to be effective for the detection of dormouse species. They can be easily built and attached to bushes and shrubs. The

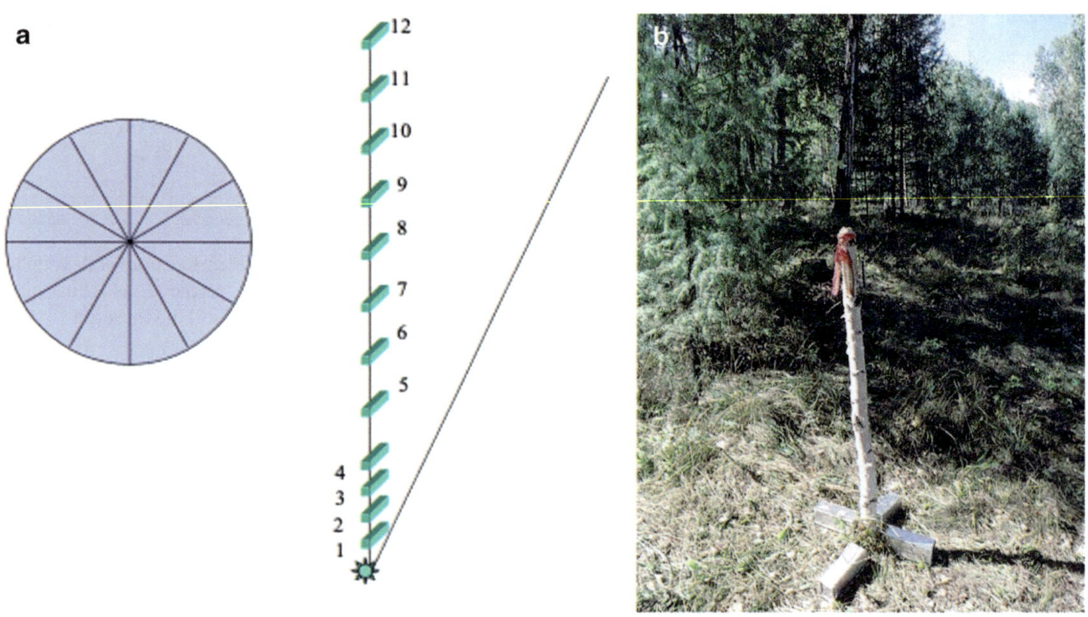

Fig. 2.39 (**a**) Diagram of a trapping web with 148 Sherman life traps. The circular area is divided into 12 radii. Along each radius of 100 m, 12 Sherman traps are set up in a straight line away from the center. On the periphery, the traps are spaced 10 m apart. (**b**) Center of the trapping web in the field, marked with a stick. There are only four traps in the center, pointing in different directions

species can be well determined by their footprints.

2.19 Capturing Reptiles

There are a number of *cost-effective methods* to capture **reptiles** in previously not well-researched areas (Ribeiro-Júnior et al. 2008), such as pitfall traps, funnel traps, and glue traps.

Fig. 2.40 Fox squirrel, *Sciurus niger*, trapped alive in a Tomahawk two-door squirrel trap mounted on a cedar platform attached to the trunk of a tree in Michigan, USA. The second trap door to the right allows for the convenient retrieval of the animal for measurements, attachments of telemetry transmitters or data loggers, and animal release

Glue (or sticky) traps come in a variety of brands, some of which have the glue infused with cockroach odor or banana essence, to lure reptiles in (for an overview, see Norton et al. 2016; Pettit 2018). A bucket dug into the soil as a pitfall trap is also helpful. The success of capturing reptiles depends greatly on the method used depending on the target group (e.g., snakes, various lizard taxa). Ribeiro-Júnior et al. (2008) showed that the best results are achieved when several methods are applied simultaneously. Pitfall traps are suitable and more effective with larger opening diameters. However, especially in tropical areas, many reptiles live arboreally (e.g., skinks and geckos). Henle and Veith (1997) comprehensively describe detection, recording, and marking methods for reptiles, see also National Reptile Survey in reference, compare Fig. 2.42.

Pitfall traps must be checked regularly to ensure the captured reptiles' safety, especially in the tropics, as ants, land crabs, or other predatory reptiles can harm the reptiles in the traps (e.g., Jenkins et al. 2003). For possible reidentifica-

Fig. 2.41 (**a**) Life trap on a fallen tree trunk. The target mammal is attracted by the bait and triggers the door by touching a seesaw mechanism. (**b**) *Tamias sibiricus* often runs along fallen tree trunks. Northern Mongolia. (**c**) Life trap as seen from behind

Fig. 2.42 **Reptiles** such as snakes and **migrating frogs** follow along walls that form barriers to their way. At a greater distance (e.g., 20 m), a funnel-like structure, similar to a fish trap (or lobster pot), makes escape difficult. The construction shown was used to catch snakes. Tropical rainforest of South Mexico

tions, the caught reptiles should be photographed before being released.

In Central Europe, many reptile species are protected, even if they are widespread, such as the sand lizard (*Lacerta agilis*) or the green lizard (*Lacerta viridis*). They are difficult to capture. For conservation, walking a certain distance in a certain time (e.g., 2 h at a speed of about 200–300 m/h) and recording all visual observations is recommended. The route should be recorded cartographically (topographic map) and walked six to ten times a year: three to five times in early summer (May–June, reproduction time) and three to five times in late summer (August–September, when also juveniles are free-roaming) (Bosbach and Weddeling 2005). The standardized setup of a survey for reptiles is described in the "National Reptile Survey" of the UK (http://www.narrs.org.uk/ and https://reptile-survey.arc-trust.org).

For an estimate of the population size, one must apply mark-release-recapture (MMR), also called capture-mark-release (CMR) methods (see Sect. 5.3), which should only be carried out by professional staff (experts). One must temporarily mark individual animals, e.g., with nail polish or similar dyes, to recognize the individuals. Lizards can be best caught with a noose (ImChipBrown, Public domain, via Wikimedia Commons: https://commons.wikimedia.org/wiki/File:Blubelly.JPG) (Elbing 2001; Schedl and Klepsch 2001). The video at https://www.youtube.com/watch?v=CWJbYjJliaw shows how to catch lizards with a lasso snare without harming the animals. The lasso snare with a slipknot on a long stick is slipped over the lizard's nose from a few meters away. They can then be quickly released from the snare. In Germany, an exemption permit according to § 45 Abs. 7 BNatSchG is required for the catch. For construction projects, in nature conservation projects, relocation of caught sand lizards is also possible using foil catch fences similar to those used in amphibian protection, with a catch bucket buried about every 10 m (e.g., Blanke 2020, see Fig. 13.1). The catch buckets must be checked at least twice a day.

Weddeling et al. (2005) summarize survey methods and the effort of estimating population trends in amphibians and reptiles. In addition, McComb et al. (2010) provide literature on inventory and monitoring studies on reptiles and amphibians. Biggs et al. (2015) recommend samples of eDNA for detecting crested newts (*Triturus cristatus*). However, this method cannot be evidence for a current occurrence. Wildlife detection dogs for amphibians and reptiles are superior to other search methods (see Sect. 1.4). Amber et al. (2021) have developed camera traps to record and monitor small mammals, reptiles and amphibians, all of which are difficult to detect in denser grass vegetation. The costs and effort involved in their use are described.

2.20 Recording Arthropods (Macrofauna) of the Forest Floor

For a first qualitative inventory, so-called catch stones can be distributed on the forest floor.

The arthropods of the forest floor (weevils, rove beetles and ground beetles, dipteran larvae and adults, spiders, harvestmen, woodlice, centipedes, and millipedes) can be recorded using various methods (Ellenberg et al. 1986). High species numbers are expected here: Müller (1985)

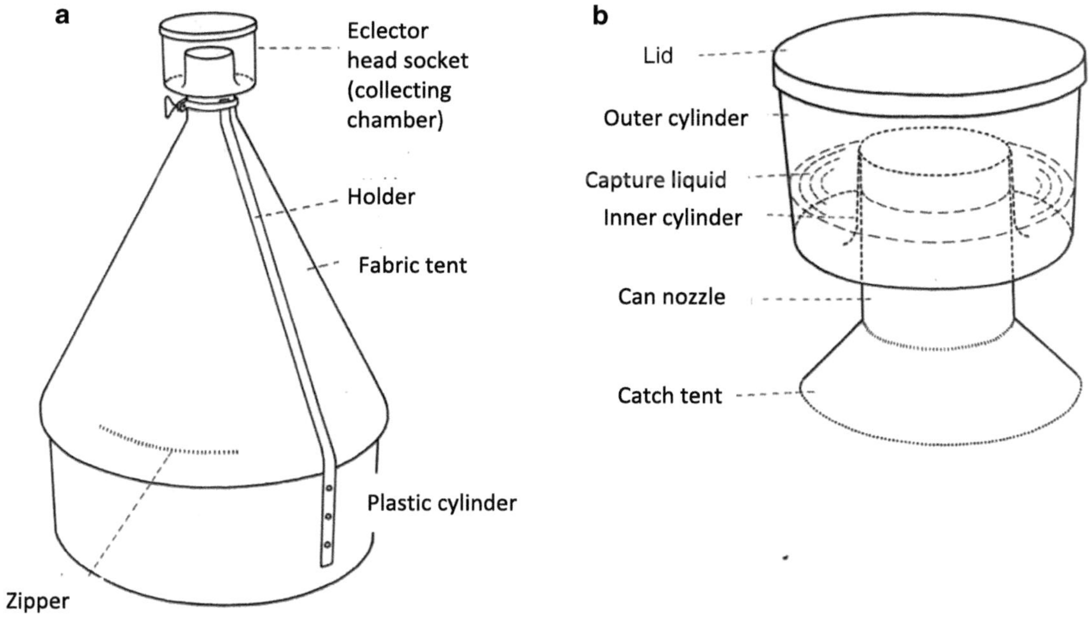

Fig. 2.43 (**a**) Ground eclector, (**b**) eclector head box (Behre 1989)

found 72 carabid species over 3 years in an oak-hornbeam forest in southwestern Germany using pitfall traps. Albert (1982) recorded 38 spider species in an old beech stand in Solling, Germany, using 12 pitfall traps and 67 spider species with tree photoeclectors in one catch period. The arthropods emerging from the ground in spring and summer are quantitatively caught in ground eclectors (Figs. 2.43 and 2.44) (Funke 1971), allowing the evaluation of density (number of individuals per unit area) as well as—for questions of the life cycle—in terms of their temporal occurrence (phenology).

2.20.1 Catch Stones and Natural Cover Objects

Placed on the ground or hidden under leaf litter, catch stones and other natural cover objects offer small animals hiding places. Catch stones are made by creating a wooden "negative," which is greased and filled with plaster or concrete. Catch stones help to identify individual species and gain insight into their habits.

Catch stones can be produced in various variations at low financial cost. They are not suitable for quantitative studies (Fig. 2.45).

2.20.2 Ground Eclectors

(a) Ground Eclector after (Behre 1989) (Fig. 2.43a):

Ground eclectors or emergence traps capture all animals from the confined area that are present or successively emerge from the ground or the overlying substrate and tend to migrate upwards toward the light. The removable catch box (eclectors head box, see below) works on the principle of a trap and is usually filled with a catch or preservation liquid (approx. 4% formalin, picric acid, Table 2.1). The fabric tent should be black for ground eclectors used in the forest. The plastic cylinder is buried 20 cm deep in the ground. Within the eclector, a ground trap is attached to the wall to catch epigaeic predators (e.g., ground beetles and spiders), and the zipper in the fabric tent is used to check

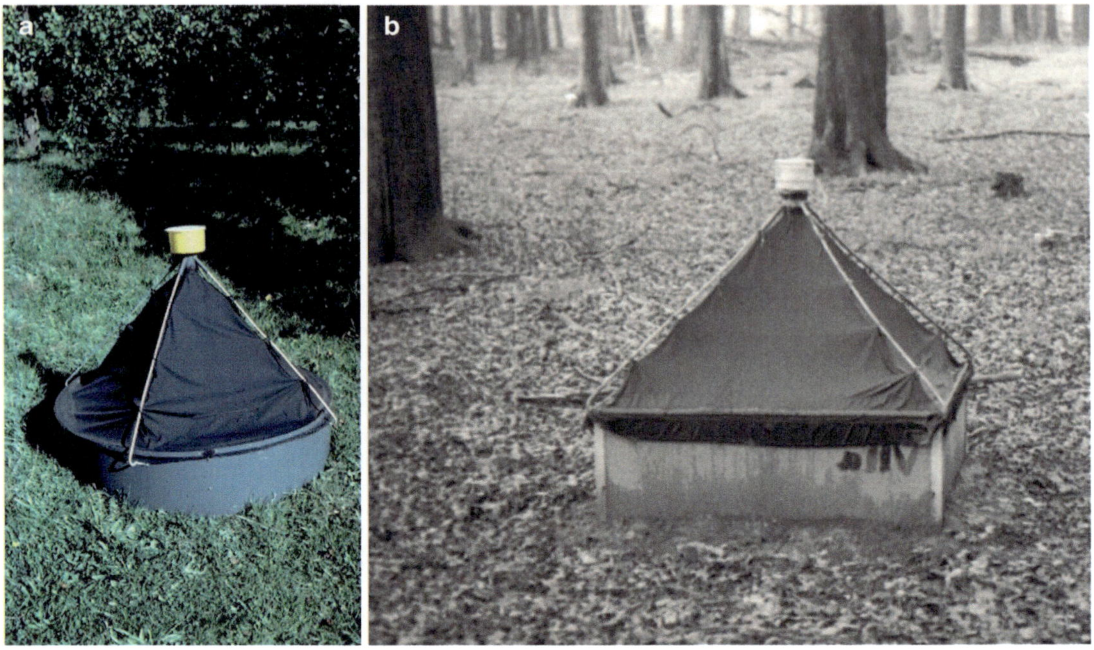

Fig. 2.44 (**a**) Ground eclector. (**b**) Ground eclectors distributed in a beech forest

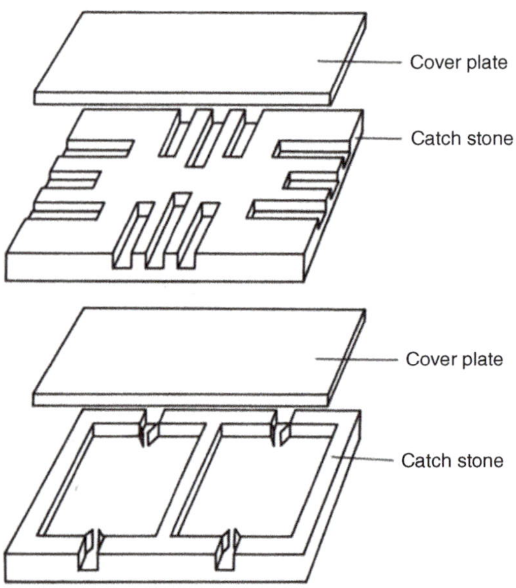

Fig. 2.45 Catch stones. (Modified after Steiner 1963)

the ground trap. If exposed to sunlight over grassland or other vegetation, ground eclectors should have light fabric tents so that vegetation growth is not too impaired. While in the forest, the eclectors can remain in place without problems for a year; they must be moved about every month in grassland due to the fast-growing vegetation.

Ground eclectors determine emergence abundances (animals per unit area, especially for strata-changing arthropods), productivity, and phenological data. An extrapolation to the total emergence abundance is possible with enough samples. Once set up, they are easy to operate by replacing the catch boxes. Their installation is somewhat complex, so they are mainly valuable for stationary operation. Under open sky, the instrument substantially influences the microclimate, which can be reduced using gauze holes in the plastic cylinder. Also, the material cycle of the ecosystem is changed under the eclector.

(b) Eclector Head Box (Fig. 2.43b)

A head box consists of one piece and has a lid. The catch liquid is between a shorter inner cylinder (slightly conical) and a longer outer cylinder. The box is put on as a "head" on the can nozzle, which forms the upper end of cone- or pyramid-shaped catch

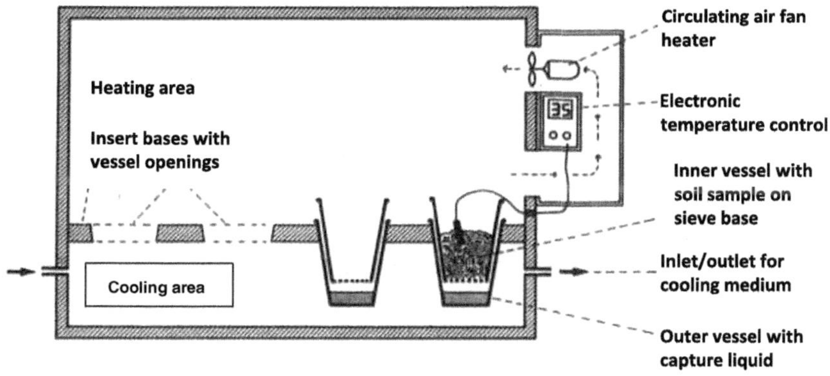

Fig. 2.46 Extraction apparatus according to Kempson (Behre 1987). Diagram

tents. A clearance width of 75 mm for the inner cylinder is suggested so that large insects can pass. The dimensions are diameter (∅) approx. 150 mm and height approx. 110 mm. The catch liquid is only filled as high as the liquid does not overflow the inner cylinder when pouring the content of the head box into a transport container (Fig. 2.44).

Spiders, harvestmen, centipedes, and woodlice are only incompletely captured in ground eclectors, as they are not constantly changing strata. Their absolute density can be determined in two ways: hand selection (square stitch samples, see Sect. 2.3) and heat extraction with a selection apparatus according to Kempson et al. 1963 (see Sect. 2.20.3). For a hand selection (see below), the forest floor is staked out with a square frame (quadrat, e.g., 25 cm × 25 cm), and all loose material is collected up to the humus layer and brought into a 50 cm × 50 cm white tub (edge height not over 10 cm) for systematic selection of the desired animal groups. The frame must be pressed onto the forest floor for sampling. The top layer of soil and twigs along the frame is cut with a knife. Litter and soil material can also be sieved over the selection tub beforehand selection (mesh size 6–10 mm). For the heat extraction, all material collected in the square sample is brought into the Kempson apparatus (Kempson et al. 1963, see Fig. 2.46) and extracted for 14 days with slowly increasing temperature (Schauermann 1982). The extracted animals are first caught in picric acid (two parts saturated picric acid and one part water with a small quantity of detergent) and then transferred to 10% formalin for about 1 day for post-hardening and finally stored in 70% alcohol.

Animal selection in the laboratory with series extraction devices:

2.20.3 Kempson Apparatus

Samples taken in the field (e.g., 25 cm × 25 cm area, 5–10 cm deep) are placed in prepared tubs in the laboratory and gradually warmed and thus dried out from above in daily intervals in a Kempson apparatus (Kempson et al. 1963), see Fig. 2.46. The animals in the sample (dug-out meadow samples, leaf litter samples) migrate downwards into the moister part, which is kept cool with flowing water. Eventually, the animals fall through the wire mesh bottom into the collection tray with preservative liquid (diluted picric acid). The extraction takes about 14 days.

This is the best method for forest soil to extrapolate the abundances of the animals recorded due to their high efficiency (over 90%). It is a simple method but requires high labor and energy expenditure. The study areas are affected, which can be critical for narrow-edge biotopes. The animal-free soil samples dug out by heat extraction should be returned to the holes punched in the study area.

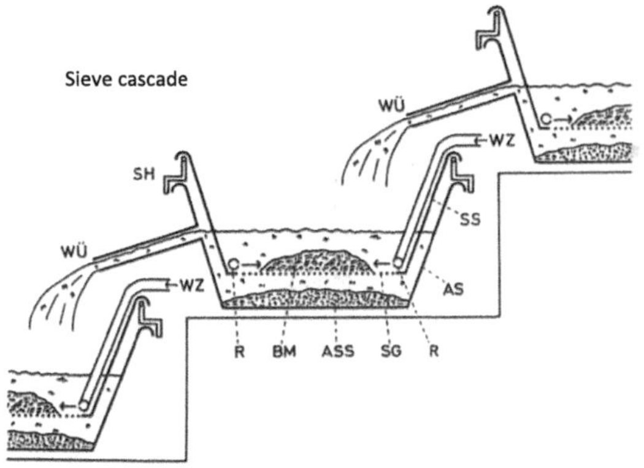

Fig. 2.47 Sieve flotation method. Functional diagram of the sieve cascade (from Behre 1987). AS—collection tray, ASS—collection tray sediment, BM—soil material, R—water inlet ring, SG—sieve mesh, SH—sieve holder, SS—sieve tray, WÜ—water overflow, WZ—water inlet

2.20.4 Sieve Flotation Method

The sieve flotation method (Fig. 2.47) is another method of extracting soil arthropods developed by Behre (1983). Soil samples are rinsed with water through five collection trays arranged in a cascade, each connected by an overflow, in which there is a sieve tray, and the animals in the sieved sample material are then separated from soil and plant material by flotation with a glycerin–water mixture (specific weight 1.14). This flotation separates the animals from the heavier remaining soil and plant material. The method is particularly suitable for Diptera larvae, Collembola, Acarina, and Chilopoda, and it has an efficiency of approximately 100%. Another means of driving arthropods out of the soil is 1,2 dichlorobenzene, which does not kill but impairs the respiration of invertebrates.

Besides the laborious manual selection, it is the only possibility to determine the density of less mobile animals and their developmental stages. However, the procedure is complex. For details of the procedure, see Behre (1987).

2.20.5 Nematode Extraction

Nematodes are also driven out of the soil samples by a generated temperature and humidity gradient while the soil material is constantly moistened from below. They are collected by letting the lower fraction of water in the hose with the sedimented nematodes flow into a catch container by loosening a clamp (Fig. 2.48). Suitable sample containers with gauze bottoms (or metal sieve with 1 mm mesh size) include converted plastic flowerpots; for the funnel, use a powder funnel (∅ 15 cm), and as a heat source, a 15-W light bulb. The water in the funnel should be cool (about 13 °C), and the soil sample should be not too large (about 1–4 cm^3). The extraction takes about 12 h. The nematodes are counted in a glass dish using a stereomicroscope. Place 10–20 animals with some water on a slide to identify species and induce a thermal torpor by applying gentle heat (Dunger 1964). Under a closed plant cover, you can expect over 20 million individuals per m^2.

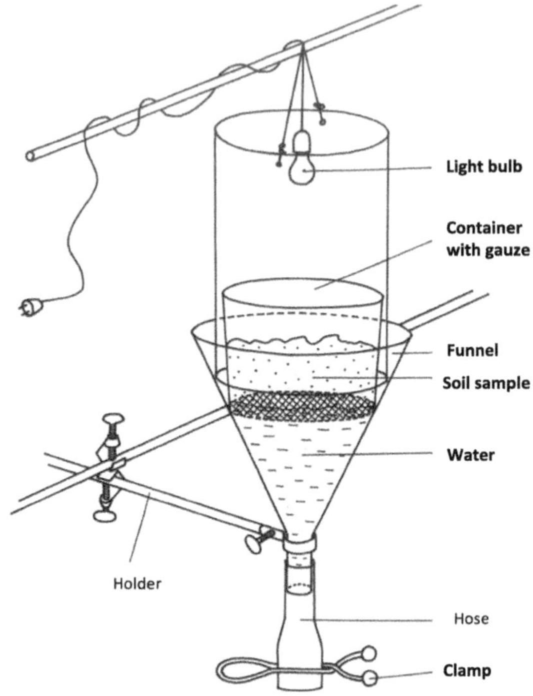

Fig. 2.48 Simple BAERMANN-funnel for nematode extraction

The procedure determines the relative frequencies of nematodes in soil samples. Although the method is easy to use in series, the results only allow for relative statements. Large soil nematodes go mainly undetected, especially since they can stick to the substrate remaining in the sample container.

2.20.6 Hand Selection

In order to overlook as little as possible during the hand selection, an extensive and time-consuming selection with sieving is necessary. A significant advantage is that animals are primarily obtained alive. Heat extraction runs automatically, but it requires 14 days to complete the selection process. Heat extraction has a higher selection efficiency in the end result, which is between 95% and 100% for many animal groups (spiders, chilopods) (Schauermann 1982; Fründ 1983). Usually, the animal densities are underestimated. Schaefer (1983) provides the following values for the macrofauna of predatory epigaeic arthropods for a limestone beech forest near Göttingen (average annual settlement densities per m^2): Staphylinids (larvae and imagines) 103, Carabids (larvae and imagines) 5, spiders 166, harvestmen 19, pseudoscorpions 35, chilopods 241 individuals per square meter. Hockmann et al. (1989) reported from northern Germany that 0.1–0.24 animals/m^2 of the ground beetle species *Carabus auronitens* were found in spring. Roof beetles (Staphylinidae) and spiders were identified with species numbers 85 and 92, respectively. Fründ (1983) found in various forest areas of the northern Steigerwald (eastern Lower Franconia) settlement densities of the Chilopoda (Lithobiidae) between 20 individuals per m^2 in spruce plantations and 170 individuals per m^2 in beech forests with up to 8 syntopic species. The average abundance of all chilopods was 260 animals per m^2 in a red beech forest (Fründ 1987). Schaefer (1996) reported that the densities of Chilopoda are 74–187 individuals per m^2 in beech forests.

The phytophagous insects reach hatching densities of up to 4700 individuals/m^2 in beech forests (Weidemann 1978; Ellenberg et al. 1986). Suppose one separates the upper and lower litter layers during sampling and assesses them separately. In that case, analyses of the vertical distribution of the forest soil fauna can be made. In a deciduous forest, the separation of the topmost layer with the largely undecomposed leaves (litter and upper O-horizon) from the lower layer (O- and upper A-horizon, e.g., Blume et al. 2016) is easy to carry out. Two adjacent square samples (2 × upper, 2 × lower layer) are combined to ensure the sample size is approximately the same as above.

Fig. 2.49 Emergence trap for hatching water insects. Schlitz Limnological Station, Central Germany

2.21 Emergence Traps for Water Insects

Insects of the orders Odonata, Plecoptera, Ephemeroptera, Trichoptera, Megaloptera, and many families of Diptera develop as larvae in water and then hatch into adults that live on land. Their larvae comprise a significant proportion of the macrobenthos in rivers and streams (see Sect. 6.2.6, Thorp and Rogers 2015, 2016).

When they hatch out of the water, they can be caught in small streams with emergence traps (Fig. 2.49) in a similar way to the ground eclectors (see above) (Mundie 1956). The hatching insects fly upwards toward the light in the glasshouse and are intercepted in the same way as in the eclectors with head boxes or catch channels in the roof of the emergence glasshouse. The result is approximate density data for a specific section of the stream, e.g., a 2 m stretch. Aquatic emergence traps must be emptied regularly, e.g., once a week. The quantity of insects can be astonishing.

For stagnant water, floating boxes can be used the same way as terrestrial ground electors (Morgan et al. 1963; Kimerle and Anderson 1967). Floating emergence traps are available, for example, from NHBS (https://www.nhbs.com/aquatic-emergence-trap). Cadmus et al. (2016) provide instructions for two aquatic emergence traps for self-assembly. A significant advantage of emergence traps is that hatching results inform us about the phenology of the species.

2.22 Density Estimation of Earthworms

Depending on the moisture and lime content of the soil, earthworms can be the dominant soil animal group in the forest soil, e.g., with about 10 g dry mass per m^2 in a lime beech forest (Schaefer 1986). An average of 100 individuals/m^2 can be found in the mull of an old oak stand, corresponding to a fresh weight of 0.8–1 t/0.01 km^2 or dry weight (without intestinal content) of 13 g/m^2 (Kloft 1978). Only earthworms can decompose the approximately 4 t/0.01 km^2 leaf litter that falls annually. Their function significantly influences microbial activity in leaf litter decomposition by shredding the substrate and transforming leaf material into fecal particles (Herlitzius and Herlitzius 1977; Beck 1984). In doing so, they also chemically alter the substrate by excreting nitrogenous excreta and contribute to soil reorganization (transport of leaf litter into the soil interior and fecal excretion on the soil surface or in soil cavities) (Schaefer 1986; Edwards and Lofty 1977). In this way, earthworms have a decisive influence on soil fertility (Drake and Horn 2007). In German beech forests, about six species of earthworms (Lumbricidae) live sympatrically in the soil. Schaefer (1996) found up to 11 species in lime beech forests.

The best method to capture as many earthworm species as possible is a combination of hand sorting and expulsion with formalin (Edwards and Lofty 1977). For high efficiency in population estimation of medium-sized earthworms using hand sorting, the excavation of 16 soil samples, each with a 25 cm × 25 cm area and 20 cm depth, is recommended. With hand sorting, earthworms are captured regardless of their activity state. Hand sorting can reveal 95% of the worm biomass and 80% of the total abundance (Axelsson et al. 1971). The extraction with formalin is less time-consuming but less effective (lower abundances, Bouché 1969). For a square sample, the soil is leveled, cleared of leaf litter and other surface plant material, and marked out with a wooden frame measuring 60 cm × 60 cm. Dilute 25 mL of 40% formalin in 4.5 L of water and pour it onto the soil surface of the square sample using a small watering can or a perforated plastic bag (this corresponds to a 0.22% formalin solution, e.g., Raw 1959). With this method, burrowing species are driven to the soil surface, preferentially capturing the species *Lumbricus terrestris* (Lewis and Taylor 1967). Slightly more effective is using more formalin solution through 3 doses of 4 L each successively (Satchell 1969). After the first dose of formalin, wait about 20 min until the first earthworms appear on the surface. After the second dose, you must wait about 20 min before collecting earthworms. Many earthworm species (especially from the genus *Allolobophora*) aestivate during prolonged drought and cannot be brought to the soil surface with formalin (Askew and Yalden 1985).

When collecting the earthworms that appear on the soil surface after the formalin application, care must be taken that the worms have entirely emerged from their burrows. Otherwise, they can retreat into their earth tunnels and not be driven out even with further formalin application. Not all earthworms in the forest soil are leaf litter eaters. Some eat topsoil, humus, animal excreta, and softened wood remains. Even with leaf litter, earthworms select certain leaf types (tree species) and qualities (Satchell and Lowe 1967).

Instead of the toxic formalin, earthworms can also be driven out with a solution of mustard meal, best combined with handpicking of the earthworms (Högger 1993; Fründ and Jordan 2004).

The mustard seed must be ground, and the meal is soaked overnight in water bottles. Table mustard can be stirred into water immediately before being used as a repellent. The mustard meal can also be bought directly in the spice trade.

Schaefer (2017) presents another method for extracting earthworms: the octet method. It is based on driving earthworms out with electricity, a method developed by Thielemann (1986a, b). Weyers et al. (2008) described the construction of the electric device. Eight electrodes, 65 cm in length and 6 mm in diameter, made of stainless steel with handles made of aluminum, are placed in the ground in a circle of 52 cm in diameter. Between the eight electrodes, current flow directions are changed at different voltage strengths, thus generating electromagnetic fields and driving earthworms out of the ground without damaging the soil. Although the efficiency of this method mainly depends on soil moisture and thus on the electrical conductivity in the soil, similar values in terms of the total number of earthworms and the total number of juvenile animals are achieved with handpicking (Čoja et al. 2008). The extraction dates should be adapted to the weather conditions (Schmidt 2001).

Suppose one is more interested in the biology of earthworms and wants to capture all stages of development, including the cocoons. In that case, handpicking and the other methods mentioned above are insufficient. Walther and Snider (1984) have developed a washing method for soil samples with different mesh sizes (5, 1.5, and 1 mm in successive steps) and achieved 97.7% efficiency for all animals and 96.7% for all cocoons as well as 99.9% for the Lumbricidae biomass without cocoons.

2.23 Microhabitat Areas on the Forest Floor

Specific habitat requirements on the forest floor can best be analyzed in microhabitat areas. For the area of soil microarthropods (mites and springtails), soil blocks of 10 cm × 10 cm and a depth of about 6 cm are punched out, frozen in powdered CO_2, and later embedded in 20% gelatine and fixated in 10% formaldehyde (Anderson 1978). Later, 2.5 cm thick sections are counted under the stereomicroscope along a 0.5 mm division according to different microhabitat structures (e.g., leaf fragments, wood fragments, cavities of various sizes, root parts, feces) (Anderson 1978). In the size range of small mammals and macroarthropods, the habitat structure data are counted directly in the field. Dueser and Shugart (1978) separated sympatric rodent species in this way in terms of their microhabitat requirements. Uetz (1975) and Fründ (1987) were able to experimentally demonstrate the dependence of forest floor spiders and centipedes (Chilopoda: Lithobiidae) on vegetation structures. For a description of the vegetation structure, see Sect. 4.2.

Possible questions are:

- How does the spatial distribution of arthropod species on the forest floor differ?
- On which habitat structures does the occurrence of individual species or groups depend?
- Does the species diversity of the forest floor fauna correlate with the vegetation structure's heterogeneity and habitat diversity?

The relevance of a habitat structure is checked by correlating its presence with the occurrence of certain animal species. Since many habitat structures are not independent of each other, partial correlations and multivariate analyses are needed. In particular habitats, the diversity of structures can also be assumed as a resource itself. It can be compared with the help of a diversity index (see Sect. 15.11). The numbers of respective categories are included in the calculation in the formula as p_i-values (see below). To roughly evaluate categories equally—an arbitrary measure that needs to be discussed depending on the research question—one can assign the same maximum point value to all category groups and equalize the category values using conversion factors. For "Habitat Diversity," various structural parameters such as areal proportions of cover types (bare ground, leaf litter, wet spots), vertical zoning, contact points with different plant parts, and ultimately also species numbers for trees, shrubs, or important undergrowth species can be summarized in an index (Mühlenberg et al. 1977).

2.23.1 Tree Stumps as Ecological Microsystems

Many different microsystems stand out as concentration points or action centers within their small subcommunity embedded within a more extensive biotope ("Biochoria," Schaefer and Tischler 1983). Examples are tree stumps, carrion, feces (for detecting dung beetles, e.g., Doube and Giller 1990, Sect. 2.15), fungi, and plant galls.

Dung and carrion are classic examples of patchily distributed and ephemeral resources with a tremendous range of variation in size, density, and duration of availability (see Sect. 2.15). As these resources are easily manipulated, they allow researchers to study the phenomena of competition and predator pressure particularly well (overview by Hanski 1990).

The comparison of tree stumps of different ages can demonstrate the development of a saprophagous community with its predators. You can obtain the age classes of various tree stumps from the local forestry administration. In the field, we record the animals encountered and their respective location. In the advanced stages of decomposition, dissect the tree stumps with a strong knife and a small shovel. Use an exhaustor and long tweezers to collect the arthropods. Part of the fauna can also be caught with a soil photoeclector placed over the stump (see Sect. 2.20.2).

The data can be used to, e.g., study the microsuccession (Fig. 2.50). Five overlapping stages can be distinguished and characterize a development of at least 10 years (compiled according to Szujecki 1987):

I. Degradation by fungi and bacteria and Scolytidae (bark beetles) phase in the first year after felling.
II. Cerambycidae and Pyrochroidae (longhorn and fire beetles) phase starts in the second year.
III. Lucanidae (stag beetles) phase begins in the fourth year.
IV. Formicidae (ants) phase from about the sixth year.
V. Lumbricidae (earthworms) phase from about the seventh year.

Szujecki (1987) provides illustrations of the habitats of various lamellicorn larvae and the form and size of their excrement (important in stages III and IV). Due to their colonization by ants and other arthropods, decayed stumps also offer a food source to several species of woodpeckers. Therefore, the stumps also fulfill a positive function in nature conservation. Stumps can compensate for the food deficit in overgroomed deciduous forests with little dead wood (Schmidt et al. 2016).

Possible research questions:

- How do the small animal faunas (arthropods, earthworms, and snails) of different age classes of tree stumps and dead wood pieces differ?
- Do the decomposition processes in stumps of moist locations proceed faster? Which animal species attack the tree stump from the bark, wood, or root?
- From where do the tree stumps decompose the fastest?
- How do the faunas of coniferous and deciduous tree stumps differ?
- How do ants colonizing a tree stump influence the other fauna taxa?

An overview of the insects on tree stumps and their spatial distribution can be found in Brauns (1991), who also provides information on the biology of the individual species in the different age stages.

Many species found on tree stumps have become rare and were included in the Red List. An example is *Osmoderma eremita* (Coleoptera, Lamellicornia), which develops in the mulm of old deciduous wood and is now included in Annex II and IV of the FFH Directive.[1]

[1]FFH-Directive is known as European Council Directive 92/43/EEC on the Conservation of Natural Habitats and of Wild Fauna and Flora, adopted by the European Community in 1992.

Fig. 2.50 (**a–e**) Tree stumps in a beech forest in various stages of decomposition. (**e**) (Bottom right) shows ants of the genus *Formica,* which have colonized a tree stump. Such an ant nest also serves as a food resource for woodpeckers

2.23.2 Lying Tree Trunks

Just as tree stumps slowly decompose, one can also study lying tree logs at various stages of decomposition. For a classification of the various stages, see Sect. 4.2.2. Deadwood fungi and xylobiont insects depend on substrate quality. For example, tropical deadwood beetles have not been studied very well. An air eclector (the apparatus in Fig. 2.51, see also Fig. 2.16 for a different design) can be hung in logging areas to catch attracted deadwood beetles (Fig. 2.52). A single 6-week catch period in Indonesian Papua yielded more than 200 beetles, which belonged to more than 20 families (Warikar 2011). A similar trap design can be purchased from https://www.e-econex.net/en/insect-traps/crosstrap-with-wet-collection-cup-with-slider-46.html. Such investigations can also have economic significance, as longer-stored logs quickly become infested by fungi and wood insects.

To investigate the influence of differences in the exposure of decaying lying logs on insect flight's temporal sequence (succession), one can lay out test trunks in a forest with and without cover (Fig. 2.53).

Section 4.2.3 addresses the importance of deadwood for nature conservation. How to determine the amount of deadwood in the forest is also discussed in Sect. 4.2.3.

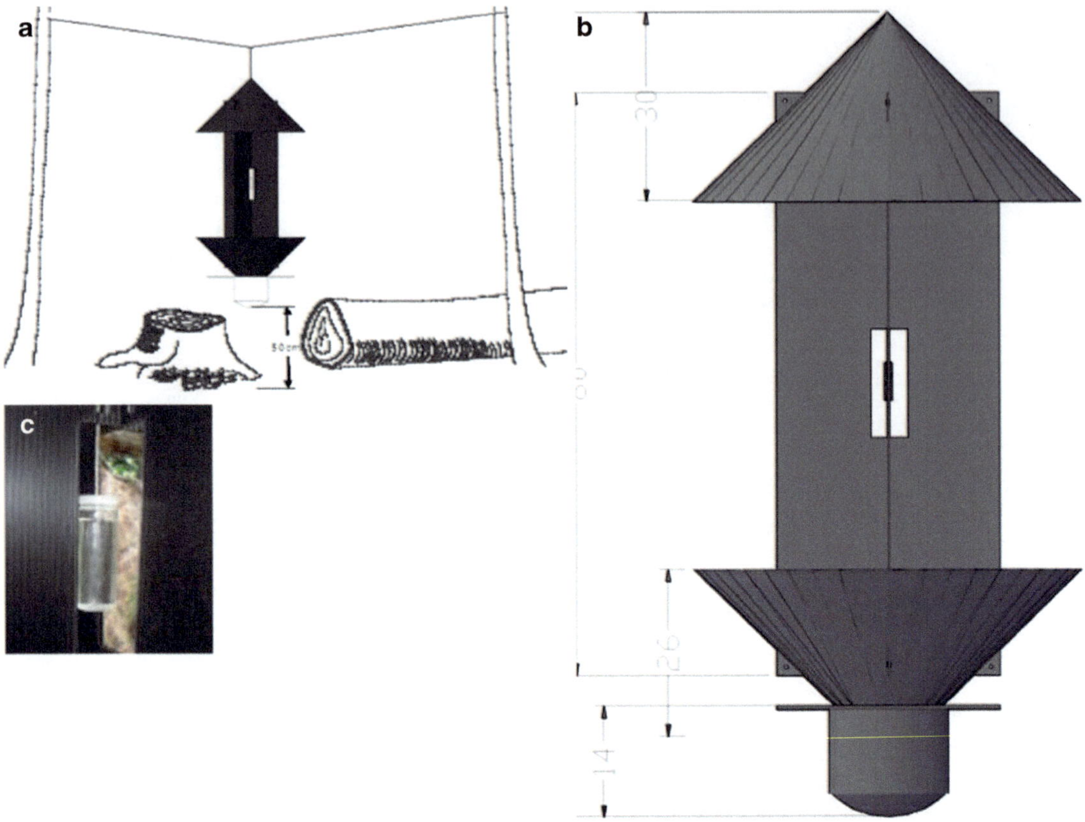

Fig. 2.51 Air eclector trapping design: (**a**) "Cross trap" suspended above deadwood. (**b**) Schematic of the trap with a plastic container at the bottom. (**c**) Transparent container for collecting insects in trap center containing 70% ethanol. After Warikar (2011).

Fig. 2.52 The cross trap in the logged area. (Photo credit E.L. Warikar, Papua-Indonesia)

2.24 Stem Eclectors

Tree photo eclectors were first systematically used by Funke (1971). They primarily capture the so-called "trunk run-up" and provide information on small habitat dwellers, transients, and immigrants from other biotopes. Of the spider fauna of the Solling beech forest in Germany (97 species), 69% of the species inventory was only detected with 2 tree eclector catches from 1 year, whereas 12 ground traps yielded only 54% of the known species spectrum in 5 years. Per trunk and year, 10–20,000 animals were caught in a woodrush beech forest (*Luzulo-Fagetum*) of the Solling. In a beech wood on limestone with a rich herb layer and in an oak mixed forest with deep branching near Ulm, Germany, the numbers were 6–10,000 and 40,000 animals, respectively (Funke 1977). The most abundant taxon was Collembola, which contributed 25–90% of the total catch.

With minor remodeling, tree trunk eclectors can also be attached to lying tree trunks (see below).

Fig. 2.53 (**a**) Shaded tree trunks laid out, Steigerwald, North Bavaria. (**b**) Freely laid-out tree trunks, Steigerwald, North Bavaria

In the open type of the modified eclector, epitruncal insects (insects moving along the log's surface) are caught. In contrast, the closed type captures hatching endotruncal insects (insects that live as larvae inside the log and hatch into the adult form at the log's surface). Compared to the catch using a ground eclector, six pitfall traps, and a window trap in the same area, and over the same sampling period, 39% and 65% of the total individuals were caught with the horizontal tree trunk eclectors. Coleoptera were represented by 25 species from 7 families in the open and 15 species from 9 families in the closed eclector type in a Bavarian Forest (Otte 1987). Hymenoptera and Araneae were also recorded in many individuals and species.

The principle of the trap is that arthropods running up the tree trunk get into the eclector-like tents and are collected in the catch cans. The catch cans correspond to the eclector head cans (see ground eclector, Fig. 2.44b). Stem eclectors can also be constructed so that the arthropods running down the trunk are caught.

A frame is attached to the tree trunk with holding straps and arms, carrying the roof of four fabric pyramids with eclector head cans and four fabric walls (Fig. 2.54). The roof is attached arthropod-tight to the tree trunk. The base is a clear plastic disc with four catch funnels and catch bottles and is approximately 40–50 mm away from the trunk. A smooth ridge about 40 mm high prevents the animals from falling out.

Trunk eclectors are used to determine the inhabitants of the trunk region and the stratum switchers in the forest (animals that hatch in the ground and migrate to the canopy layer during the vegetation period). The eclectors provide essential data on the phenology and activity periods of arthropods in the trunk and canopy layer (cf. Funke 1971). Although high catch numbers with a large species spectrum are obtained, quantification is difficult. Another disadvantage is the complex attachment of the trunk eclector, which should be at 2–4 m height, thus requiring a ladder.

Trunk eclector for lying tree trunks (Behre 1989, Figs. 2.55 and 2.56):

(a) Open type (Fig. 2.55a)

    The open type is essentially similar to the vertical tree trunk eclector and is accessible from both sides. The eclector head cans are supplemented by catch vessels attached below, which are removable and therefore replaceable (e.g., 1 L wide-mouth polyethylene bottles). The cylindrical eclector mantle, consisting of two halves, is adjusted to tree trunks by four spacers. A partition is in the middle; immediately next to it, on each side,

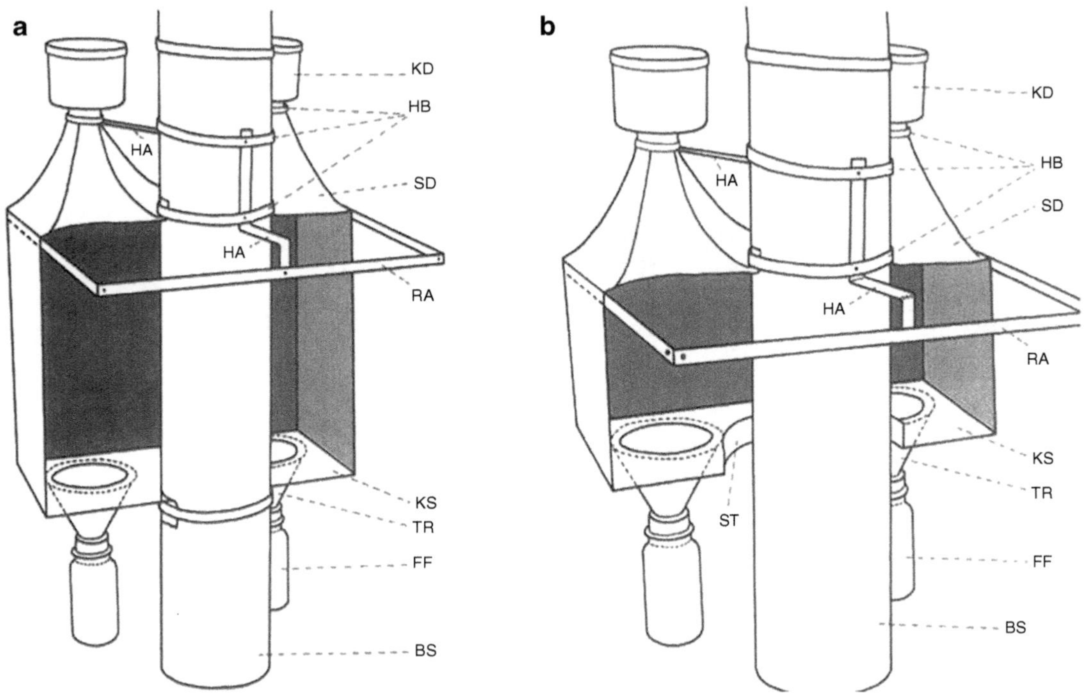

Fig. 2.54 Trunk eclectors for standing trees (Behre 1989). (**a**) Open type, (**b**) closed type. HB—holding straps, HA—holding arms, BS—tree trunk, RA—frame, SD—fabric pyramid roofs, KD—eclector head cans, KS—plastic clear disc, TR—catch funnels, FF—catch bottles, ST—ridge for delimiting the inlet gap

is a funnel-shaped nozzle on which an eclector head can be seated. Correspondingly, two liquid-filled catch vessels are attached below.

(b) Closed Type (Fig. 2.55b)

In this design, the base consists of an opaque plastic plate or fabric with four catch funnels and catch bottles. Both the roof and the base are arthropod-tight and attached to the tree trunk. This type is only suitable for catching bark and wood-dwelling arthropods, as there is no spacing for ectotruncal species.

The insects (bark and wood-dwelling arthropods) emerging from a tree section are intercepted with this type. These eclectors provide a complete recording of the wood-dwelling arthropods. Comparisons between different tree species and age stages of the lying logs are possible. The catch numbers are absolute values related to the volume of the enclosed tree trunk.

The cylindrical eclector mantle, consisting of two halves, is adjusted by four spacers on tree trunks—the end walls on both front sides close tightly to the tree trunk. At the top, near the end walls, there is a nozzle with a head box and a catch vessel at the bottom in the middle attached to a catch funnel (Fig. 2.56).

Depending on their running habits, the arthropods running on the trunk are caught in the eclector head cans or the catch vessels.

These eclectors are used to determine arthropods that visit tree trunks or live on the bark and provide species spectra, phenology data of the species, and, to a limited extent, relative frequencies.

In windthrows or natural forest cells, tree logs left lying are particularly attractive to many arthropods. Although the installation is laborious, the checks are simple.

2.25 Tree Bark Structure and Arthropod Coenoses

Tree barks can be structured differently depending on the tree species, from smooth to corky and torn. Accordingly, different species of arthropods

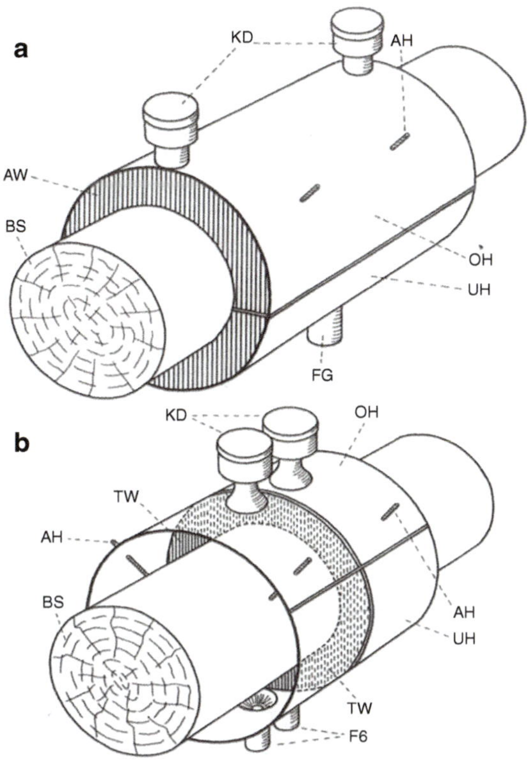

Fig. 2.55 Trunk eclector for lying tree trunks (Behre 1989). (**a**) Open type, (**b**) closed type. Lower half of the eclector mantle, OH—upper half of the eclector mantle, AH—spacers, BS—tree trunk, TW—partition, KD—eclector head can, FG—catch vessels, AW—end wall

Fig. 2.56 Trunk eclectors for lying tree trunks with open (left) and closed type (right). Ecological Research Station Fabrikschleichach, University Würzburg, Bavaria

are found at different frequencies in the bark types (Fig. 2.57). While phytophagous and xylobiont insects seek the trunk structures for feeding or egg laying, predators lurk or actively search for arthropods on the tree trunks. Arthropods in bark crevices form an important food resource for insectivorous birds, especially in winter (Fig. 2.58). Woodpeckers and tree creepers are a well-known group that depend on the arthropods on tree trunks. Since a more fissured bark harbors more arthropods than smooth bark, a forest's composition of tree species is essential for its conservation value. For example, beech trees have a smooth bark in their early stages but show many bark cracks in their old stages (see Fig. 2.57c) and thus become more valuable for conservation. Beech stands with interspersed old beeches (breast height diameter >60 cm, or >100 years old) are populated by significantly more species of woodpeckers (Wesolowski 1995; Mikusinski et al. 2001; Ćiković et al. 2008; Müller et al. 2009; Nationalpark Kalkalpen 2011; Basile et al. 2020). In addition, the settlement density of bird species increases with increasing deadwood share in a forest due to the birds' dependency on deadwood for cavity construction and food acquisition (Hertel 2003; Winter et al. 2005). Dominant species groups among bark dwellers or visitors are Coleoptera, Heteroptera, Neuroptera, and spiders (Araneae). It is reasonable to predict that upon closer species analyses in diverse, mixed-age forests with deadwood, many species listed as endangered on the Red Lists will be found and that such forests are thus highly relevant to conservation.

A comparison of the arthropod coenosis seems easiest when you directly search the trunks and best when you also use an exhaustor (see Fig. 2.3). However, for a comparison between different tree species or between locations, you should consider a few things:

Although you can choose the exact area sizes for direct collection, individual species are not evenly distributed over the tree trunk. They can, for example, accumulate at different heights. As alternatives to handpicking, you can also use *hand/eye searching, trap-logging, log dissection, bark removal, emergence trapping or caging, and external trapping for flying adults* (Leather 2005) (Fig. 2.58).

Fig. 2.57 Examples of arthropods on tree trunks. (**a**) A bark bug (Aradidae). (**b**) A hunting spider of the Sparassidae family on a tree trunk in North Mongolia. (**c**) The orb-weaver spider (*Araneus marmoreus)* hiding on a birch trunk in North Mongolia. (**d**) A praying mantis (Order Mantodea) lurking on a tree trunk for prey, West Papua (Indonesia)

Case Study: Plant Galls in Flower Heads of the Asteraceae (Daisies, Fam. Compositae)

The galls in the flower heads of thistles (*Carduus* and *Cirsium*) and knapweeds (*Centaurea*) represent ecological microsystems ("miniature ecosystems," Zwölfer 1979, 1988), in which researchers can not only study food chains with several trophic levels but also mortality factors of a phytophagous insect population and existing resources and production efficiency of different trophic levels. The informational value of samples of dead flower heads of these composite flowers collected in autumn or winter (from October on) is immense because phytophages, predators, and parasites can be found in the gall or leave specific traces. Therefore, the growth and life history of the gall inhabitants can be reconstructed (Varley 1947; Varley et al. 1980; Redfern 1968; Cameron and Redfern 1974).

The flower heads are inhabited by (1) gall midges, (2) larvae of fungus gnats (Sciaridae) and other scavenging or car-

Fig. 2.58 (**a**) A northern treecreeper (*Certhia familiaris*) searching for arthropod food on an old ash tree (*Fraxinus excelsior*). (**b**) Trunk of the black alder (*Alnus glutinosa*), on which a Treecreeper (*Certhia familiaris*) is searching for arthropods in early spring. (**c**) Older beech (*Fagus sylvatica*), whose trunk is getting cracks. Harz, Germany

nivorous Diptera, and (3) the conspicuous phytophagous larvae of fruit flies (*Urophora* spec., Tephritidae), weevils (*Larinus* spec. and *Rhinocyllus* spec., Curculionidae), as well as moths (*Homoeosoma* spec., Pyralidae) (Zwölfer 1979, 1988). As the most critical guild of this system, group (3) will be the focus of our attention here. Their natural enemies are the parasitic wasps in the broad sense (species of Chalcidoidea, Braconidae, and Ichneumonidae). The galls themselves may contain several cells, in each of which a fruit fly larva must have developed initially. Thirty to 200 flower heads form a local population. While on average, three coexisting phytophagous insect species can be found in a *Centaurea*-gall population, only two live simultaneously in the thistle group (Carduinae) (Zwölfer 1979). The females of the weevil genus *Larinus* differ in their rostrum length, and this is an adaptation to using flower heads of different sizes (Zwölfer and Brandl 1989).

The galls are best collected before the onset of winter, as they later fall off the stems and can lie on the ground. Hold the gall vertically upwards when dissecting and cut off horizontal slices with a scalpel (or a razor blade). The phytophagous insect larvae lie at the bottom of their chambers and are easily recognizable from above and easy to remove (Cameron and Redfern 1974). The following data should be recorded in a table: (a) total number of galls of a composite species examined; (b) empty cells, gall tissue not consumed, phytophage died as a young larva; (c) number of fruit flies, weevils and moth individuals, from which larvae, pupae or cocoons were found; (d) parasitization by primary, secondary or parasitic wasps; (e) predatory gall midge larvae or other predators. To distinguish between insect larvae or their traces, see illustrations in Varley (1947), Varley et al. (1980), and Redfern (1983). The number of animals found, or their traces, is expressed as percentages of all galls examined.

If a phytophagous larva is parasitized and both host and parasite are alive, then the parasite is already recorded as the cause of death for the host. From the compilation of these data, a life table can be constructed for the period from gall formation to the beginning of winter (Varley et al. 1980).

Mortality factors such as early deceased larvae, parasitized larvae, and larvae eaten by caterpillars are listed separately. At other times of the year, other mortality factors play a role in the life of the wood-boring flies or other phytophages. In winter, primarily wood mice bite off some galls and eat their contents. In addition, stormy weather can destroy some galls. In spring, birds destroy some galls, or fallen galls can be flooded during heavy rainfall, and larvae and pupae in the galls drown (Varley et al. 1980). Since the thistles are heterogeneous and patchily distributed, their phytophages are divided into small, isolated subpopulations. The phytophages often comprise less than 100 individuals (*Urophora cardui* and *Larinus* spp.), have minimal dispersal capabilities, and locally frequently die out (Zwölfer 1979).

For production measurements, undamaged larvae are dried at 60–70 °C and weighed; several larvae of the same species can be weighed simultaneously (15–20 animals). Cameron and Redfern (1974) provide averaged caloric values per dry weight for some larvae. The production efficiency of secondary consumers is calculated from the ratio of the total production of parasites in grams (or calories) to the total production of the host animals in grams (or calories). All host chambers (of phytophages) are included (average dry weight of a host larva) since, at some point, every host larva is likely to be eaten. According to Cameron and Redfern (1974), the production efficiency of parasites (secondary consumers) lies between 3.1% and 18.6% of the dry weight of the host animals.

Abrahamson and Weis (1997) presented a comparable study about the biocoenosis on goldenrods encompassing three trophic layers.

2.26 Monitoring Programs for Terrestrial Habitats

Protected areas should fulfill their function. Therefore, researchers and conservation professionals should gather evidence that the goals set for protected areas were achieved compared to unprotected areas. In most cases, the goals relate to many species, especially when the goal is to preserve specific habitats with their species communities. If a single, particularly endangered species is to be protected, its population development is monitored in a specific way. If a "target species" (see Sect. 10.8) is selected, it can be assumed that many other species benefit from their protection without studying their populations. In nature conservation, there is also the so-called "process protection," which usually involves simply refraining from all human interventions in the system and observing how a particular habitat develops under this protection. In all of these cases, one would like to study the changes in the species communities at various levels.

Different habitats and conservation goals cannot be observed using the same universally valid monitoring program. Examples of broader monitoring programs using various recording methods are provided by Bässler et al. (2009), Gaiser et al. (2020), Hausmann et al. (2020), Montgomery et al. (2021), Gärtner et al. (2022), and Müller et al. (2023). Del Vecchio et al. (2019) point out the efforts required by different recording methods.

2.27 Rapid Biodiversity Assessments (BIORAP)

The number of species in the tropics is usually significantly higher than in temperate latitudes. In addition, there is generally much less knowledge about the lifestyle of individual species than in more northern or southern latitudes, where, for example, the density of universities is higher. Unfortunately, land use in the tropics significantly changes the originally species-rich habitats through progressive deforestation alone. A simple estimation of tropical forest loss is equivalent to a rate of 11 football fields per minute (Weisse et al. 2023). For these reasons, quick documentation is often required in locations with an exceptionally high number of species. This knowledge must be used to prioritize protected areas and management options. Areas from which no good data on the occurrence of species are available can be assessed for their biodiversity value in a time- and cost-saving manner using a "Rapid Biodiversity Assessment" (Kerr et al. 2000; Alonso et al. 2011; Patrick et al. 2014).

An example is the MacKinnon list technique: while an observer walks a transect route, (s)he notes all observed and heard bird species, starting with the first species, then the second species, until (s)he has completed a list of ten species. If (s)he sees a species for the second time, this observation is discarded. After completing the list, (s)he starts with a second list (species 11–20). This protocol is continued until enough lists are completed. The suitability of this method was reviewed by Herzog et al. (2002) and statistically analyzed for its quantitative statement value by MacLeod et al. (2011). The advantages of this data collection are:

- Can be standardized and even statistically evaluated.
- Measures the magnitude of species richness.
- Allows the calculation of species accumulation curves, which helps determine when a site is adequately surveyed.
- Since the observations are recorded chronologically, the data can be pooled from many observers.
- The surveys can be conducted throughout the day.
- The method is not particularly sensitive to different ornithological experiences of individual observers or even concerning the probability of discovering species at different times.
- The technique is probably also applicable to the recording of other taxa, such as mammals and some insect groups.

However, with this method, quantified statements can only be cautiously adopted (O'Dea et al. 2004). The different probabilities of discovery of different species complicate comparing species.

In preparation and implementation of a BIORAP, local communities should be involved (see also Sect. 2.28). Local people usually know the locations of certain animals and can, therefore, be helpful in the direct survey, saving time and effort.

2.27.1 Morphospecies

In principle, the BIORAP method can be applied to other taxonomic groups. For butterflies, for example, through visual discovery along transects or a combination of visual identification and net swings (see Sect. 2.5) along transects. Also, trap techniques can be used with this rapid assessment if randomized. Obrist and Duelli (2010) provide examples of the application to arthropods. The collection method must be strictly standardized, and then developments (monitoring) can be tracked over the years. For a BIORAP, it is always advantageous to use several recording methods or traps simultaneously (e.g.,

Table 2.3 Classification of insect groups into different guilds. Excerpt from Table 2 in Obrist and Duelli (2010)

	Carnivores	Herbivores	Pollinators
Lepidoptera		X	X
Carabidae	X		
Cerambycidae		X	X
Buprestidae		X	X
Aculeata			X
Other Hymenoptera	X		
Heteroptera	X	X	
Homoptera		X	
Arachnoidea	X		
Neuroptera	X		

Fig. 2.59 (**a**) An assassin bug (Heteroptera, Reduviidae) is easily recognizable by its forward-stretched beak, West Papua. (**b**) Another predatory bug (Heteroptera, Reduviidae) impaled a small beetle as prey with its proboscis, West Papua

hand capture with insect net, Malaise traps, and light traps; see Sects. 2.9–2.11). The identification of species in BIORAP is treated as "parataxonomic units" = morphospecies instead of a factual determination of species (Oliver and Beattie 1996a; Kerr et al. 2000; Krell 2004; Derraik et al. 2002, 2010). The effectiveness of trapping the epigean fauna and analyzing it with morphospecies was demonstrated by Oliver and Beattie (1996b) using selected beetle families. The differentiation into recognizable taxonomic units has the advantage that nonspecialists can be involved in the process. Morphospecies can be further differentiated depending on the question, e.g., classified according to guilds such as taxa that pollinate flowers or a ratio between carnivorous and herbivorous guilds (McGeoch 1998). Obrist and Duelli assign separately counted taxonomic groups to the three functional guilds as follows (Table 2.3).

In many species, the larvae are, e.g., herbivorous, while the adults are nectar feeders. Among the terrestrial Heteroptera, one could sort out the carnivorous assassin bugs (Reduviidae, Fig. 2.59), while most species belong to the herbivores.

In addition to the advantages mentioned above, the focus on morphospecies has a decisive disadvantage: rare or endangered species, e.g., those on the IUCN Red List, are overlooked or not recognized, and therefore the value for nature conservation is minimal. When conducting a

BIORAP, one should also connect with taxonomic and local experts to increase the credibility of the results and to provide opportunities to train laypeople for BIORAP. Voucher specimens should be collected to document evidence for the species list in areas that have been little explored.

Jones and Eggleton (2000) examined the BIORAP results of **termite** communities in tropical forests (Fig. 2.60). Termites play a crucial role in the decomposition processes in tropical forests. The survey was designed so that termites from dead wood, the ground, termite nests, and other microhabitats up to 2 m above the ground were collected from 20 sections (each 5 × 2 m) along a 100 m transect in one person-hour (according to the principle of "catch per unit effort"), which corresponds to a stratified sampling regime. BIORAP achieved 31–36% of the well-known local termite fauna (species pool) in three tropical forests. Species accumulation curves showed that with the BIORAP method over the collected sections along a transect, the effort (collecting effort in days) is much lower than with the population recording (quantifying the density and spatial distribution of termite populations in a plot, e.g., 2500 m^2), averaging 4 versus 40 person-days. Thus, it could be shown that BIORAP provided a reliable protocol. Termites react sensitively to habitat disturbances and are good indicators for the functional decomposition processes.

Nevertheless, the species range and abundance of termite species in a tropical forest are difficult to determine (Jones et al. 2005). Only 12 % of the termite fauna in a southern Cameroon forest build epigeal or arboreal nests (Eggleton et al. 1996). Some mounds are subterranean, or nests are connected via arboreal runways (polycalic species), so the delimitation between colonies is not straightforward.

In other work, the ant and termite fauna were recorded along line transects (Difouo et al. 2023). Pitfall traps were used to record the ant species, supplemented by manual sampling on placed baits (Bestelmeyer et al. 2000).

In a hitherto unknown area, **orb-weaving spiders** can also be quantitatively recorded. Since the webs are usually species-specific, they can be directly photographed or recorded in the field (Fig. 2.61). The corresponding spiders are collected once and given a working name. Later, a specialist can make definite species determinations. Mühlenberg et al. (1977) provide an example of such fieldwork.

2.28 Citizen Science

Many factors influence the population dynamics of animal and plant species. While on a local scale, it may be possible to identify and experimentally test specific causes of population change, this is usually not possible on a large scale. Climate change is one such cause that leads to changes on a statewide or supra-regional scale (e.g., Heller and Zavaleta 2009). A monitoring system is needed, which records given targets nationwide or internationally simultaneously to exclude that the observed change in biodiversity is a local-specific phenomenon. Such a large-scale monitoring program requires the participation of many nonscientists and nature lovers. It is commonly called "citizen science" and broadly refers to the active engagement of the general public in scientific research tasks (Vohland et al. 2021). Through citizen science, people share and contribute to data monitoring and collection programs (National Geographic Encyclopedia 2022). The benefits are mutual: volunteers gain knowledge about the objects they observe, and the analysis of nationwide data can be used to make population management decisions and inform policy about necessary conservation measures (e.g., Roy et al. 2016).

The value of data gained from citizen science is debated. Cohn (2008) questions whether scientific results can be achieved with the participation of volunteers, whereas Cooper et al. (2008) consider it possible to introduce citizens to science. In any case, it would be beneficial if citizens developed a closer understanding of nature when dealing with data.

The discipline of citizen science is as manifold as its objectives and definitions vary accordingly (Haklay et al. 2021). Kamp et al. (2016) point out that the data collected via citizen sci-

Fig. 2.60 (**a–d**) Termite nests. (**a**) Predominantly carton nests (masticated wood) from the rainforest in West Papua. (**b**) Many termites lay earthy tunnels (arboreal runways) to protect themselves against enemies, especially ants. (**c**) Mushroom-shaped termite nests built from earth material with saliva by *Cubitermes* sp. in the rainforest of Côte d'Ivoire. (**d**) Termite construction from the rainforest of the D.R. Congo. The building material does not let water in. (**e**) Free-roaming termites of the *Nasutitermes* group. Rainforest in West Papua

Fig. 2.61 Different types of spider webs. (**a**) The special light at dawn with the early morning humidity reveals numerous spider webs in a savanna without human interference in Northern Mongolia, suggesting an impact of predatory spiders on the entomofauna, see Chap. 8. Webs of (**b**) Araneidae, (**c**) Linyphiidae, *Neriene* sp., and (**d**) Nephilidae, with a large orb web (diameter >2 m). b-d from West Papua

ence are generally not structured and are therefore subject to bias. In comparison with structured data, unstructured databases can often deliver contradictory results. For example, the observation period is standardized in structured monitoring programs, and a random distribution is required. Brown and Williams (2019) provide an overview of the application of citizen science. They analyze the positive and negative effects on the results, depending on how and according to which scheme the data was collected. Clare et al. (2019) analyze data obtained from citizen science and discuss how the data can be improved. Monitoring with citizen science can also contribute significantly to the knowledge of rare and endangered species, especially if the monitoring is carried out over a more extended period (Surdo et al. 2022; Fontaine et al. 2022). Considering omnipresent budget limitations for ecological research and monitoring, paired with various unfolding challenges the natural world faces, citizen science's role in monitoring biodiversity trends is globally gaining importance. While citizen science in some regions, such as Europe and North America, is already established in ecological research and environmental policy development, there is a particular potential for citizen science in other regions, such as Asia and Africa (Pocock et al. 2019).

2.28.1 Could Citizen Science Be Beneficial for Your Research?

The specific characteristics of your research question may determine if citizen science provides viable methods that will allow you to collect suitable data to address them. Typically, scientists in ecological research turn to citizen science to increase the amount of gathered data to address their research question and expand the spatial and temporal scope of their data collection (Chandler et al. 2017a). Empirical observations rather than complex experiments are more common and more suitable as a methodological approach to collecting data through the participation of citizens (Hecker et al. 2018). The less specialized skills and training required to collect data, the easier it will be to involve and encourage large numbers of participants.

Known data analyses on species occurrence and abundance have already been applied in the eBird, iNaturalist, and iSpot studies. eBird is probably the world's largest environmental citizen science project. More than 50% of the Global Biodiversity Information Facility (GBIF) comes from monitoring via citizen sciences (Chandler et al. 2017b). Smartphone apps and access to mobile networks empower volunteers to be involved in recording observations and environmental monitoring in multiple ways (Luna et al. 2018). Various software programs are available as smartphone apps that can be used to identify plants or certain animal groups. In the USA, phenological data are collected using "Nature's Notebook" https://www.usanpn.org . iNaturalist uses identification programs for crowdsourcing monitoring activities. More localized programs such as "BioBlitzes" are run by local nature observation societies, universities, or museums. Here, volunteers from the public are, for example, sent to different parts of the neighborhood in one day to record given taxa (amphibians, certain easily identified insect groups, bird observations, and others). The data is then compiled by experts and announced to the community. The results also serve as an incentive for the population to pay attention to the existing biodiversity.

Zooniverse is a worldwide platform for volunteers to participate in citizen science projects: https://www.zooniverse.org. By collaborating via this platform, it is also possible for anyone to design a project themselves and carry it out in such a way that it ultimately results in publications. You do not have to be an expert yourself.

References

Abrahamson, W. G., & Weis, A. E. (1997). *Evolutionary ecology across three trophic levels : goldenrods, gallmakers, and natural enemies*. Princeton, N.J.: Princeton University Press. http://archive.org/details/evolutionaryecol0000abra. Accessed 13 August 2024

Albert R. 1982: Studies on the structure and dynamics of spider communities of different vegetation types in the High-Solling. – *University Collection Natural Science, Biology* , Vol. 16 , University Publishing House Freiburg, 147 p.

Allison A., Samuelson G.A. & Miller S.E. 1993: Patterns of beetle species diversity in New Guinea rain forest as revealed by canopy fogging: Preliminary findings. – *Selbyana* 14, 16-20

Allison A., Samuelson G.A. & Miller S.E. 1997: Patterns of beetle species diversity in *Castanopsis acuminatissima* (Fagaceae) trees studied with canopy fogging in mid-montane New Guinea rainforest, 224-236. – In: Stork N.E., Adis J. & Didham R.K. (eds.): *Canopy Arthropods*. – Chapman & Hall.

Alonso L.E., Deichmann J.L., McKenna S.A., Naskrecki P. & Richards S.J. 2011: *Still Counting. Biodiversity Exploration for Conservation.* The First 20 Years of Rapid Assessment Program. – Conservation International, Arlington, Virginia, United States. 315 p.

Amber, A. D., Lipps, G. J. & Peterman, W. E. (2021). Evaluation of the AHDriFT Camera Trap System to Survey for Small Mammals and Herpetofauna. – *Journal of Fish and Wildlife Management 12*(1), 197–207. https://doi.org/10.3996/JFWM-20-016

Anderson J.M. 1978: A Method to Quantify Soil – Microhabitat Complexity and its Application to a Study of Soil Animal Species Diversity. – *Soil Biol. Biochem* . 10, 77–78

Anderson, D.R., Burnham, K.P., White, G.C. and Otis, D.L. 1983: Density estimation of small-mammal populations using a trapping web and distance sampling methods. – *Ecology* 64, 674-80.

Askew R.R. & Yalden D.W. 1985: The Woodchester Park valley. – In: Cook L.M. (ed.): *Case studies in population biology*, 1-26. - Manchester Univ. Press, Manchester, 218 p.

Avgin S.S. & Luff M.L. 2010: Ground Beetles (Coleoptera: Carabidae) as Bioindicators of Human Impact. – *Mun. Ent.Zool* . 5 (1), 209-215

Axelsson G.D., Lohm U., Persson T. & Terow D. 1971: Reliability of estimating Standing crop of earthworms by handsorting – *Pedobiologia* 11, 338–340

Ayre G.L. & Trueman D.K. 1974: A battery operated time-sort pitfall trap. – *The Manitoba Entomologist* 8, 37-40

Baars M.A. & van Dijk Th.S. 1984: Population dynamics of two carabid beetles at a Dutch heathland, I. Subpopulation fluctuations in relation to weather and dispersal. – *J. Anim. Ecol* . 53, 375–388

Baars M.A. 1979: Catches in pitfall traps in relation to mean densities of Carabid beetles. – *Oecologia* 41, 25–46

Bährmann R. 1976: Comparative studies of the results of different trapping methods on brachyceran Diptera from the nature reserve "Leutratal" near Jena (Thür.) – *Entom. Abh. (Dresden)* 41 (2), 19–47

Baker R.R. 1985: Moths: Population estimates, light-traps and migration. – In: Cook L.M. (ed.): *Case studies in population biology* , 188–211. - Manchester Univ. Press, Manchester, 218 p.

Basile M., Asbeck T., Pacioni C., Mikusiński G. & Storch I. 2020: Woodpecker cavity establishment in managed forests: relative rather than absolute tree size matters. – *Wildlife Biology* 2020(1). https://doi.org/10.2981/wlb.00564

Basset Y., Springate N.D., Aberlenc H.P. & Delvare G. 1997: A review of methods for sampling arthropods in tree canopies. 27-52. – In: Stork N.E., Adis J. & Didham R.K. (eds.): *Canopy Arthropods*. – Chapman & Hall.

Bässler C., Förster B., Moning C. & Müller J. (2009): The BIOKLIM Project: Biodiversity Research between Climate Change and Wilding in a temperate montane forest – The conceptual framework. *Forest Ecology, Landscape Research and Nature Conservation* 7: 21-34.

Beadle D. & Leckie S. 2012: *Peterson Field Guide to Moths of Northeastern North America.* - Houghton Mifflin Harcourt, 624 p.

Beck L. 1984: Soil Animals in the Leaves of the Beech Forest. Decisive Role in the Natural Household. – *Research - Communications of the DFG 2/84* , 15–18

Begon M., Townsend C.R. & Harper J.L. 2006 (4th ed.): *Ecology. From Individuals to Ecosystems* . – Blackwell Publishing Ltd, 738 p.

Behre G.F. 1983: The Sieve-Flotation Method. Construction and testing of an ecological work device for the mechanical selection of soil arthropods. – State examination work, unpublished, Univ. Bonn, 100

Behre G.F. 1987: Die Sieb-Flotations-Methode. Bau und Erprobung eines ökologischen Arbeitsgerätes zur mechanischen Auslese von Bodenarthropoden. – über, *naturw. Ver. Wuppertal* 40**,** 52–55

Behre G.F. 1989: Field ecological method for recording the entomofauna (further and new development of devices). – *Jber. naturwiss. Ver. Wuppertal* 42, Fig. 1–6

Bellmann, H. 2004: *Grasshoppers* (CD). The voices of 61 native species. – Edition Ample.

Bertoia A., Murray T., Robertson B.C. & Monks J.M. 2023: Pitfall trapping outperforms other methods for surveying ground-dwelling large-bodied alpine invertebrates. – *J. Insect Conserv.* 27, 679–692

Bestelmeyer B., Agosti D., Alonso L.E., Brandao C.R.F., Brown J.W.L., Delabie J.H.C. & Silvestre R. 2000: Field techniques for the study of ground dwelling ants. p. 122-154. – In Agosti J.D.M.D., Majer J.D., Alonso L.E. & Schultz T. (eds.): *Ants: Standard methods for measuring and monitoring biodiversity*. Smithsonian Institution Press. 280 p.

Biggs J., Ewald N., Valentini A., Gaboriaud C., Dejean T., Griffiths R.A., Foster J., Wilkinson J.W., Arnell A., Brotherton P., Williams P. & Dunn F. 2015: Using eDNA to develop a national citizen science-based monitoring programme for the great crested newt (*Triturus cristatus*). – *Biological Conservation* 183, 19–28.

Blanke I. 2020: Berenbusch Harbor: Relocation of lizards in 2019. Report, https://www.bueckeburg.de › ceasy › resource

Blume, H.-P., Brümmer, G. W., Fleige, H., Horn, R., Kandeler, E., Kögel-Knabner, I., et al. (2016). *Scheffer/Schachtschabel Soil Science*. Berlin, Heidelberg: Springer. https://doi.org/10.1007/978-3-642-30942-7

Bonkowska T. & Ryszkowski L. 1975: Methods of Density Estimation of Carabids (Carabidae, Coleoptera) in Cultivated Fields. – *Pol. Ecol. Stud* . 1, 155–171

Bosbach G. & Weddeling K. 2005: Sand lizard and green lizard, p. 285-294 in Doerpinghaus A. et al.: Methods for recording species of Annexes IV and V of the Fauna-Flora-Habitat Directive. – Nature Conservation and Biological Diversity Issue 20, 449p.

Bouché M.B. 1969: Comparison critique de methodes d'evaluation des populations de lumbricides. – *Pedobiologia* 9, 26–34

Bradley J.S. 1985: Comparative demography of four species of grasshopper on a common site. – In: Cook L.M. (ed.): *Case studies in population biology*, 61–100. – Manchester University Press, Manchester/USA, 218 p.

Brauns A. 1991: *Pocketbook of Forest Insects* . Fourth, revised edition. – Gustav Fischer Verlag, Stuttgart, Jena, 860 p.

Brown E.D. & Williams B.K. 2019: The potential for citizen science to produce reliable and useful information in ecology. – *Conservation Biology* 33 (3), 561-569

Buckland, S.T., Anderson, D.R., Burnham K.P., Laake, J.L., Borchers, D.L. and Thomas, L. 2001: *Introduction to Distance Sampling* . Estimating abundance of biological populations. – Oxford University Press, Oxford, NY, 432p.

Buckland, S.T., Anderson, D.R., Burnham, K.P., Laake, J.L., Borchers, D.L. and Thomas, L. 2004: *Advanced Distance Sampling* . – Oxford University Press, 416p.

Buddle, C.M. 2001: Spiders (Araneae) associated with downed woody material in deciduous forest in central Alberta, Canada. – *Agricultural and Forest Entomology* 3, 241-251

Cadmus, P., Pomeranz, J. P. F., & Kraus, J. M. (2016). Low-cost floating emergence net and bottle trap: comparison of two designs. *Journal of Freshwater Ecology, 31*(4), 653–658. https://doi.org/10.1080/02705060.2016.1217944

Cain, S. A. (1938). The Species-Area Curve. *The American Midland Naturalist, 19*(3), 573–581. https://doi.org/10.2307/2420468

Cameron R.A.D. & Redfern M. 1974: A Simple Study in Ecological Energetics using a Gall-fly and its Insect Parasites. – *Journal of Biological Education* 8 (2), 75–82

Carvalho, R. L., Andersen, A. N., Anjos, D. V., Pacheco, R., Chagas, L., & Vasconcelos, H. L. (2020). Understanding what bioindicators are actually indicating: Linking disturbance responses to ecological traits of dung beetles and ants. *Ecological Indicators, 108*, 105764. https://doi.org/10.1016/j.ecolind.2019.105764

Caughley, G. (2004). *Analysis of Vertebrate Populations*. Caldwell, NJ: The Blackburn Press.

Chandler M., See L., Buesching C.D., Cousins J.A., Gillies C., Kays R.W. et al. 2017a: Involving citizen scientists in biodiversity observation. – In M. Walters & R.J. Scholes (Eds.): *The GEO handbook on biodiversity observation networks,* p.211–237. Cham: Springer International Publishing

Chandler M., See L., Copas K., Bonde A.M.Z., López B.C., Danielsen F., Legind J.K., Masind S., Miller-Rushing A.J., Newman G., Rosemartin A. & Turak E. 2017b: Contribution of citizen science towards international biodiversity monitoring. – *Biological Conservation* 213, 280–294. 10.1016/j.biocon.2016.09.004

Ćiković D., Barišić S., Vesna Tutiš V. & Jelena Kralj J. 2008: Woodpeckers in the Croatian Karst Mountains. – *Bird Census News* 21/1, 2 – 15

Clare J., Townsend P., Anhalt-Depies C., Locke C., Stenglein J., Frett S., Martin K., Singh A., Deelen T. & Zuckerberg B. 2019: Making inference with messy (citizen science) data: when are data accurate enough and how can they be improved? – *Ecological Applications* 29, e01849. https://doi.org/10.1002/eap.1849

Clark M. 1983: *Mammal Watching* . – Hamlyn Paperbacks edition, Feltham, Middlesex, England, 175 p.

Cohn J. 2008: Citizen Science: Can volunteers do real research? – *BioScience* 58(3), 192-197

Čoja, T., Zehetner K., Bruckner A., Watzinger A. & Meyer E. 2008: Efficacy and side effects of five sampling methods for soil earthworms (Annelida, Lumbricidae). – *Ecotoxicology and Environmental Safety* 71(2), 552-565. https://doi.org/10.1016/j.ecoenv.2007.08.002.

Collwell R.K., Mao C.X. and Chang J. 2004: Interpolating, extrapolating, and comparing incidence-based species accumulation curves. - *Ecology* 85, 2717-2727

Cooper C.B., Dickinson J.L., Phillips T. & Bonney R. 2008: Science Explicitly for Nonscientists. – *Ecology and Society 13*(2). http://www.jstor.org/stable/26268008

Curtis D. 1980: Pitfalls in spider community studies. – *Journal of Arachnology* 8, 271-280

Davies N.B. 1977: Prey selection and the search strategy of the Spotted Flycatcher *(Muscicapa striata):* a field study on optimal foraging. – *Anim. Behav*. 25, 1016–1033

De Vries, Ph.J. 1987: *The Butterflies of Costa Rica and their Natural History* (Papilionidae, Pieridae, Nymphalidae). – Princeton University Press, Princeton, New Jersey, 327 p.

Del Vecchio S., Fantinato E., Silan G. & Buffa G. 2019: Trade-offs between sampling effort and data quality in habitat monitoring. – *Biodiversity and Conservation* 28(1):1-19

Dempster, J. P., & McLean, I. F. G. (Eds.). (1999). *Insect Populations: in Theory and in Practice* (1st edition.). Dordrecht: Springer.

Den Boer P.J. (ed.) 1971: *Dispersal and Dispersal Power of Carabid Beetles*. – H.Veenman & Zonen N. V., Wageningen, 151 p.

Derraik J.G.B., Closs G.P., Dickinson K.J.M., Sirvid P., Barratt B.I.P. & Patrick B.H. 2002: Arthropod Morphospecies versus Taxonomic Species: A Case Study with Araneae, Coleoptera, and Lepidoptera. – *Conservation Biology* 16 (4), 1015–23

Derraik J.G.B., Early J.W., Closs G.P. & Dickinson K.J.M. 2010: Morphospecies and Taxonomic Species Comparison for Hymenoptera. – *Journal of Insect Science* 10, 108. doi: https://doi.org/10.1673/031.010.10801. PMID: 20874395; PMCID: PMC3016916.

Difouo G.F., Simo F.T., Kekeunou S., Ebangue G.T., Fokou O.R., Ndoh L.G., Ichu I.G. & Olson D. 2023: Diversity patterns of ants and termites in forest-savanna mosaic habitats in two protected areas of Cameroon. – *African Journal of Ecology* 00, 1-20. https://doi.org/10.1111/aje.13183

Doube, B. M. (1990). A Functional Classification for Analysis of the Structure of Dung Beetle Assemblages. *Ecological Entomology, 15*(4), 371–383. https://doi.org/10.1111/j.1365-2311.1990.tb00820.x

Doube B.M. & Giller P.S. 1990: A comparison of two types of trap for sampling dung beetle populations (Coleoptera: Scarabaeidae). – *Bulletin of Entomological Research* 80, 259–263

Drake H.L. and Horn M.A. 2007: As the Worm Turns: The Earthworm Gut as a Transient Habitat for Soil Microbial Biomes. - *Annual Review of Microbiology* 61, 169-189. https://doi.org/10.1146/annurev.micro.61.080706.093139

Dubois, F., & Binning, S. A. (2022). Predation and parasitism as determinants of animal personalities. *Journal of Animal Ecology, 91*(9), 1918–1928. https://doi.org/10.1111/1365-2656.13781

Du Merle P. 1978: The ant and locust populations of Mont Vetoux. II. The ant populations. – *The Earth and Life*, Suppl. 1, 161–218

Dueser R.D. & Shugart H.H. Jr. 1978: Microhabitats in a forest floor small mammal fauna. – *Ecology* (Durham, N. C.) 59, 89–98

Duffey E. 1974: Comparative sampling methods for grassland spiders. – *Bull. Brit. Arach. Soc* . 3 (2), 34–37

Dunger W. 1964: *Animals in the Soil* . – Ziemsen-Verl., Wittenberg Lutherstadt, 265 p.

Dunger W. 2008: *Animals in the Soil* . – Militzke Publishing GmbH, 290 p.

Dunham A.E. 1983: Realized Niche Overlap, Resource Abundance, and Intensity of Interspecific Competition. – In: Huey R.B., Pianka E.R. & Schoener T.W. (eds.): *Lizard Ecology* , 261–280. – Harvard University Press, Cambridge, Mass. and London, 501 p.

Dunn, P. O., Ahmed, I., Armstrong, E., Barlow, N., Barnard, M. A., Bélisle, M., et al. (2023). Extensive regional variation in the phenology of insects and their response to temperature across North America. *Ecology*, *104*(5), e4036. https://doi.org/10.1002/ecy.4036

Ebert, K. M., Arnold, W. G., Ebert, P. R., & Merritt, D. J. (2021). Hindgut Microbiota Reflects Different Digestive Strategies in Dung Beetles (Coleoptera: Scarabaeidae: Scarabaeinae). *Applied and Environmental Microbiology*, *87*(5), e02100-20. https://doi.org/10.1128/AEM.02100-20

Edwards C A. & Lofty J.R. 1977: *Biology of Earthworms*. – Chapman & Hall, London, 333 p.

Eggleton P., Bignell D.E., Sands W.A., Mawdsley N.A., Lawton J.H., Wood T.G. & Bignell N.C. 1996: The Diversity, Abundance and Biomass of Termites under Differing Levels of Disturbance in the Mbalmayo Forest Reserve, Southern Cameroon. – *Philosophical Transactions of the Royal Society B Biological Sciences* 351(1335), 51-68. DOI:https://doi.org/10.1098/rstb.1996.0004

Elbing K. 2001: The green lizards – two unequal sisters. – *Journal of Field Herpetology* , Supplement 3: 1-143

Ellenberg H. (sen.), Mayer R. & Schauermann J. 1986: *Ökosystemforschung – Ergebnisse des Sollingprojekts: 1966–1986*. – Ulmer, Stuttgart, 507 p.

Ellenberg H. 1987: Floristic changes due to eutrophication. – In: *Ammonia and Acidification* 6, 301–308

Ellenberg H. 1988: Eutrophication – Changes in forest vegetation – Consequences for roe deer browsing and its feedback effects on the vegetation. – *Schweiz. Ztschr. f. Forstwesen* 139 (4), 261–282

Elliott N.C., Tao F.L., Fuentes-Granados R., Giles K.L., Elliott D.T., Greenstone M.H., Shufran K.A. & Royer T.A. 2006: D-vac sampling for predatory arthropods in winter wheat. – *Biological Control* 38, 325-330

Erdelen M. 1978: Quantitative relationships between avifauna and vegetation structure. – Dissertation, University of Cologne, 133 p.

Farag, M. R., Alagawany, M., Bilal, R. M., Gewida, A. G. A., Dhama, K., Abdel-Latif, H. M. R., et al. (2021). An Overview on the Potential Hazards of Pyrethroid Insecticides in Fish, with Special Emphasis on Cypermethrin Toxicity. *Animals*, *11*(7), 1880. https://doi.org/10.3390/ani11071880

Fechter, H. 1977: On the functional relationship between population density, dispersal ability and catch quantity in pitfall traps. – *Spixiana, Munich* , 1, 3–15

Federal Species Protection Ordinance of February 16, 2005 (Federal Law Gazette I p. 258, 896), which was last amended by Article 10 of the Law of January 21, 2013 (Federal Law Gazette I p. 95)

Feng, L., Lysakowski, B., Eisenschmidt, J. and Birkhofer, K. 2022: The impact of wildfire and mammal carcasses on beetle emergence from heathland soils. - *Ecological Entomology* 47(5), 883–894. https://doi.org/10.1111/een.13179

Fermon H., Waltert M., Vane-Wright R.I. & Mühlenberg M. 2005: Forest use and vertical stratification in fruit-feeding butterflies of Sulawesi, Indonesia: Impacts for conservation. – *Biodiversity and Conservation* 14: 333-350

Floren A. 2010: Sampling arthropods from the canopy by insecticidal knockdown. – In *Manual on Field Recording Techniques and Protocols for All Taxa Biodiversity Inventories and Monitoring*; Eymann, J., Degreff, J., Häuser, C., Eds.; Abc Taxa: Brussels, Belgium, 2010; Volume 1, 158–172

Floren A., Linsenmair K.E. & Müller T. 2022: Diversity and Functional Relevance of Canopy Arthropods in Central Europe. – *Diversity* 14(8):660. https://doi.org/10.3390/d14080660

Fontaine A., Simard A., Brunet N. & Elliott K.H. 2022: Scientific contributions of citizen science applied to rare or threatened animals. – *Conservation Biology* 36 (6), https://doi.org/10.1111/cobi.13976

Frazer I.F.D. 1973: Estimating butterfly numbers. – *Biol. Conserv* . 5 (4), 271–276

Fründ H.-Ch. & Jordan B. 2004: Suitability of various mustard preparations as an alternative to formalin for the expulsion of earthworms. – *Communications of the German Soil Science Society* Vol. 103, 25-26

Fründ H.-Ch. 1983: Untersuchungen zur Koexistenz verschiedener Chilopodenarten im Waldboden. – Dissertation, Universität Würzburg, 164 p.

Fründ H.-Ch. 1987: Spatial distribution and coexistence of the Chilopoda in an old beech stand. – *Pedobiologia* 30, 19–29

Funke W. 1971: Food and energy turnover of leaf-eating insects and their influence on primary production. – *Ecol. Studies* 2, 81–93

Funke W. 1977: *The Trunk Region of Forests – Habitat and Transit Zone of Arthropods* . – Verh. Dtsch. Zool. Ges., Erlangen. Gustav Fischer Verlag, Stuttgart, 244 p.

Gaiser E.E., Bell D.M., Castorani M.C.N., Childers D.L., Groffman P.M., Jackson C.R., Kominoski J.S., Peters D.P.C., Pickett S.T.A., Ripplinger J. & Zinnert J.C. 2020: Long-Term Ecological Research and Evolving

Frameworks of Disturbance Ecology. – *BioScience* 70, 141–156

Galetti, M., Moleón, M., Jordano, P., Pires, M. M., Guimarães, P. R. J., Pape, T., et al. (2018). Ecological and evolutionary legacy of megafauna extinctions. *Biological reviews of the Cambridge Philosophical Society*, *93*(2), 845–862. https://doi.org/10.1111/brv.12374

Gärtner S.M., Birk S., del Val E. & Bernauer T. 2022: Process protection in the Black Forest National Park – integrative long-term monitoring of biodiversity development. – *Forest Ecology, Landscape Research and Nature Conservation* 21, 15-30

Gillespie R.G. & Caraco Th. 1987: Risk-Sensitive Foraging Strategies of Two Spider Populations. – *Ecology* 68 (4), 887–899

Gist C.S. & Crossley D.A. 1973: A Method for Quantifying Pitfall Trapping. – *Environ. Entomol* . 2, 951–952

Gossow H. 1976: *Wildlife Ecology*. – BLV-Verlagsgeseilschaft, Munich, 316 p.

Greenstone M.H., Morgan C.E., Hultsh A.-L., Farrow R.A. & Dowse J.E. 1987: Ballooning Spiders in Missouri, USA, and New South Wales, Australia: Family and Mass Distributions. – *The Journal of Arachnology* 15 (2), 163–170

Gromadzka J. & Trojan P. 1967: Comparison of the usefulness of an entomological net, photo-eclector and biocoenometer for investigation of entomocoenoses. – *Ekol. Polska* A XV, 505–529

Grootaert P., Pollet M., Dekoninck W. & van Achterberg C. 2010: *Sampling insects: general techniques, strategies, and remarks.* Chap. 15, 377-399. – In: Eymann J., Degrees J., Häuser Ch., Monte J.C., Samyn Y. & VanvenSpiegel D. (eds.): Manual on field recording techniques and protocols for All Taxa Biodiversity Inventories (ATBIs), Volume 8, part 2. - Abc Tax - The Series of Manuals Dedicated to Capacity Building in Taxonomy and Collection Management. The Belgian Development Cooperation.be. http://www.abctaxa.be/volumes/volume-8-manual-atbi/chapter-15/Chapter_15.pdf/download/en/1/Chapter%2015.pdf

Gurnell J. & Flowerdew J.R. 1982: *Live Trapping Small Mammals - a Practical Guide*. – The Mammal Society, 24 p.

Haas V. 1980: Methoden zur Erfassung der Arthropodenfauna in der Vegetationsschicht von Grasland-Ökosystemen. – *Zool. Anz. Jena* 204, 319–330

Haeck J. 1971: The immigration and settlement of carabids in the new Ijsselmeer-polders. – In: Den Boer P.J. (ed.): Dispersal and Dispersal Power of Carabid Beetles. – *Miscell. Papers, L. H. Wageningen*, 8, 33–52

Haklay M. (Muki), Dörler D., Heigl F., Manzoni M., Hecker S. & Vohland K. 2021: What Is Citizen Science? The Challenges of Definition. – In: Vohland K., Land-Zandstra A., Ceccaroni L., Lemmens R., Perellб J., Ponti M., Samson R. & Wagenknecht K. (eds.): *The Science of Citizen Science* (pp. 13–33). Springer International Publishing, 529 p. 10.1007/978-3-030-58278-4_2

Hallmann, C. A., Sorg, M., Jongejans, E., Siepel, H., Hofland, N., Schwan, H., et al. (2017). More than 75 percent decline over 27 years in total flying insect biomass in protected areas. *PLoS ONE, 12*(10), e0185809. https://doi.org/10.1371/journal.pone.0185809

Hammond H.E.J. 1997: Arthropod biodiversity from *Populus* coarse woody material in north-central Alberta: a review of taxa and collection methods. – *Canadian Entomologist* 129, 1009-1033

Hanski I. 1990: 8. Dung and carrion insects. – In: Shorrocks, B. & Swingland I. R. (eds.): *Living in a patchy environment* , 127–145. – Oxford University Press, Oxford, New York, Tokyo, 246 p.

Hanski, I., & Cambefort, Y. (1991). *Dung beetle ecology*. Princeton, N.J.: Princeton University Press.

Hao, L., Qingdong, S., Imin, B., & Kasim, N. (2020). Methodology for optimizing quadrat size in sparse vegetation surveys: A desert case study from the Tarim Basin. *PLoS ONE, 15*(8), e0235469. https://doi.org/10.1371/journal.pone.0235469

Harper C.A. & Guynn D.C.. 1998: A Terrestrial Vacuum Sampler for Macroinvertebrates. – *Wildlife Society Bulletin (1973-2006)* 26 (2), 302–306

Hausmann A., Segerer A.H., Greifenstein T., Knubben J., Morínièr J., Bozicevic V., Günter A., Ulrich W. & Habel J.C. 2020: Toward a standardized quantitative and qualitative insect monitoring scheme. – *Ecology and Evolution* 10 (9), 4009-4020

Haynes, R. J., & Williams, P. H. (1993). Nutrient Cycling and Soil Fertility in the Grazed Pasture Ecosystem. In D. L. Sparks (Ed.), *Advances in Agronomy* (Vol. 49, pp. 119–199). Academic Press. https://doi.org/10.1016/S0065-2113(08)60794-4

Heap M.A. 1988: The Pit-Light, a new trap for soil-dwelling insects. – *J. Aust. ent. Soc* . 27, 239–240

Hecker S., Garbe L. & Bonn A. 2018: The European citizen science landscape–a snapshot. p. 190-200 – In: Hecker S., Hackley M., Bowser A., Makuch Z., Vogel J. & Bonn A. (eds.): *Citizen Science: Innovation in Open Science, Society and Policy*. – UCL Press, London, 580 p. 10.14324/111.9781787352339

Henderson I.F. & Whitaker T.M. 1977: The efficiency of an insect suction sampler in grassland. – *Ecological Entomology* 2, 57–60

Heller N.E. & Zavaleta E.S. 2009: Biodiversity management in the face of climate change: A review of 22 years of recommendations. – *Biological Conservation* 124(1), 14–32. https://doi.org/10.1016/j.biocon.2008.10.006

Henle K. & Veith M. (eds.) 1997: *Naturschutzrelevante Methoden der Feldherpetologie*. – Mertensiella 7, Bonn, 400 p.

Herlitzius R. & Herlitzius H. 1977: Litter Decay in Deciduous Forests. Studies in Lime and Acid Humus Beech Forests. – *Oecologia* (Berl.) 30, 147–171

Hertel F. (2003): Habitatnutzung und Nahrungserwerb von Buntspecht *Picoides major*, Mittelspecht *Picoides medius* und Kleiber *Sitta europaea* in bewirtschafteten und unbewirtschafteten Buchenwäldern des nordostdeutschen Tieflandes. – *Vogelwelt* 124: 111 – 132.

Herzog S.K., Kessler M. & Cahill T.M. 2002: Estimating species richness of tropical bird communities from rapid assessment data. – *The Auk* 119: 749-769

Hess R. & Ritschel-Kandel G. 1988: Erfassung der Heuschrecken im Grenzbereich zwischen Biotop-Flächen und extensivierten landwirtschaftlichen Flächen des Schweinfurter Beckens. Untersuchungen auf Vertragsflächen des Acker- und Wiesenrandstreifenprogrammes und des Pufferzonenprogremmes. – *Regierung von Unterfranken, Höh. Naturschutzbehörde*. Manuscript. 124 p.

Hitchen J.M., Hitchen E.T., Jackson C.M. & Meers G.H.A. 1969: The flight times of some nocturnal moths. – *Entomologist* **102** , 80–85

Högger C.H. 1993: Mustard flour instead of formalin for the extraction of earthworms in the field. – *Bulletin of the Swiss Soil Science Society* 17, 5-8

Hopkins, B. (1957). The Concept of Minimal Area. *Journal of Ecology*, *45*(2), 441–449. https://doi.org/10.2307/2256927

Hsieh T.C., Ma K.H. and Chao A. iNEXT 2016: an R package for rarefaction and extrapolation of species diversity (Hill numbers). – *Methods Ecol. Evol.* 7, 1451–1456

Ingrisch S. 1977: Beitrag zur Kenntnis der Larvenstadien mitteleuropäischer Laubheuschrecken (Orthoptera: Tettigoniidae). – *Z. ang. Zool.* 64, 459–501

Isaac N.J.B., Cruickshanks K.L., Weddle A.M., Rowcliffe J.M., Brereton T.M., Dennis R.L.H., Shuker D.M. & Thomas C.D. 2011: Distance sampling and the challenge of monitoring butterfly populations. – *Methods in Ecology and Evolution* 2, 585-594. https://doi.org/10.1111/j.2041-210X.2011.00109.x

Jenkins C.L., McGarigal K. & Gamble L.R. 2003: Comparative effectiveness of two trapping techniques for surveying the abundance and diversity of reptiles and amphibians along drift fence arrays. – *Herpetological Review* 34(1), 39-42

Johnson C.G., Southwood T.R.E. & Entwistle H.M. 1957: A new method of extracting arthropods and molluscs from grassland and herbage with a suction apparatus. – *Bull. ent. Res* . 48, 211–18

Johnson, R. M., Wen, Z., Schuler, M. A., & Berenbaum, M. R. (2006). Mediation of Pyrethroid Insecticide Toxicity to Honey Bees (Hymenoptera: Apidae) by Cytochrome P450 Monooxygenases. *Journal of Economic Entomology*, *99*(4), 1046–1050. https://doi.org/10.1093/jee/99.4.1046

Jones D.T. & Eggleton P. 2000: Sampling termite assemblages in tropical forests: testing a rapid biodiversity assessment protocol. – *Journal of Applied Ecology* 37: 191-203

Jones D.T., Verkerk R.H.J. & Eggleton P. 2005: Methods for sampling termites. – In: Leather S. (ed.): *Insect Sampling in Forest Ecosystems*, p. 221-253. - Blackwell Publishing, 303 p.

Jung, T. S. (2016). Comparative efficacy of Longworth, Sherman, and Ugglan live-traps for capturing small mammals in the Nearctic boreal forest. *Mammal Research*, *61*(1), 57–64. https://doi.org/10.1007/s13364-015-0251-z

Kamp J., Oppel S., Heldbjerg H., Nyegaard T. & Donald P.F. 2016: Unstructured citizen science data fail to detect long-term population declines of common birds in Denmark. – *Diversity and Distributions* 22, 1024-1035. https://doi.org/10.1111/ddi.12463

Kaufman D.W., Gentry J.B., Kaufman G.A., Smith M.H. & Wiener J.G. 1978: Density estimation of small mammals, comparison of techniques utilizing removal trapping. – *Acta Theriol.* 23, 147–171

Kempson D., Lloyd M. & Ghelardij R. 1963: A new extractor for woodland litter. – *Pedobiologia* 3, 1–21

Kentner E. & Schrade H. 1991: Die Malaisefalle – Geeignete Methode zur Erfassung von Dipteren? – *Beihefte zu den Verhandlungen der Gesellschaft für Ökologie* 2, 171–176

Kerr J.T., Sugar A. & Packer L. 2000: Indicator Taxa, Rapid Biodiversity Assessment, and Nestedness in an Endangered Ecosystem. – *Conservation Biology* 14 (6), 1726-1734

Kimerle R.A. & Anderson N.H. 1967: Evaluation of Aquatic Insect Emergence Traps. – *Journal of Economic Entomology* 60 (5), 1255–1259, https://doi.org/10.1093/jee/60.5.1255

Klein W. 1981: Quantitative Untersuchungen zur Farbpräferenz blütenbesuchender Insekten (Apoidea, Syrphidae) in einem Halbtrockenrasen im Kaiserstuhl. – Staatsexamensarbeit University of Freiburg, 90 p.

Kloft W J. 1978: *Ökologie der Tiere*. – Ulmer Publishing, Stuttgart, 303 p.

Kratochwil A. 1984: Pflanzengesellschaften und Blütenbesucher-Gemeinschaften: biozönologische Untersuchungen in einem nicht mehr bewirtschafteten Halbtrockenrasen (Mesobrometum) im Kaiserstuhl (Southwest Germany). – *Phytocoenologia* 11 (4), 455–669

Kratochwil A. 1989: Biozönotische Umschichtungen im Grünland durch Düngung. – *NNA-Reports* 2 (1), 46–58

Krebs J.R. 1974: Colonial Nesting and Social Feeding as Strategies for Exploiting Food Resources in the Great Blue Heron *(Ardea herodias)*. – *Behaviour* 51, 99–134

Krell F.T. 2004: Parataxonomy vs. taxonomy in biodiversity studies – pitfalls and applicability of 'morphospecies' sorting. – *Biodivers. Conserv.* 13: 795-812

Krell, F.-T., & Moon, A. R. (2019). Dung beetles. *Current Biology*, *29*(12), R554–R555. https://doi.org/10.1016/j.cub.2019.05.027

Kühnelt W. 1950: *Bodenbiologie*. – Verl. Herold, Wien, 368 p.

Kurtze W. 1974: Synökologische und experimentelle Untersuchungen zur Nachtaktivität von Insekten. – *Zool. Jb. Syst.* 101, 297–344

Lang J. & Simon O. 2010: *Konzept zur Erfassung von Baummarder (Martes martes) und Iltis (Mustela putorius) in Hessen* (Arten des Anhanges V der FFH-Richtlinie). – Institut für Tierökologie und

Naturbildung. Artgutachten Version 2, Hessen Forst. 69 p.

Lange M., Türke M., Pašalić E., Boch S., Hessenmöller D., Müller J.C., Prati D., Socher S., Fischer M., Weisser W. & Gossner M. 2014: Effects of forest management on ground-dwelling beetles (Coleoptera; Carabidae, Staphylinidae) in Central Europe are mainly mediated by changes in forest structure. – Forest Ecology and Management 329, 166-176

Langer F., Erhardt S. & Fietz J. 2020: Erfahrungsbericht: Spurentunnel als Nachweismethode für Bilche. – Maus – *Mitteilungen aus unserer Säugetierwelt* (Zoologisch-Botanische Datenbank) 21, 15-19

Larochelle A. & Bousque Y. (1977): Diminution de populations de Carabidae par récolte intense. – *Cordulia* (Canada) 3 (3), 105–106

Leather S.R. ed. (2005): *Insect sampling in forest ecosystems.* – Methods in Ecology. Blackwell Publishing, 303p.

Leckie S.. & Beadle D. 2018: *Peterson Field Guide to Moths of Southeastern North America.* – Harper Collins, 664 p.

Lévieux J. 1969: The sampling of terrestrial ant populations. – In: Lamotte, M. & Bourliére, F.: Problems of ecology: The sampling of animal populations in terrestrial environments, 289–300. – Masson & Cie. Paris, 320 p.

Lewis T. & Taylor L.R. 1964: Diurnal Periodicity of Flight by Insects. – *Trans. R. ent. Soc. Lond* . 116, 393–476

Lewis T. & Taylor L.R. 1967: *Introduction to Experimental Ecology.* – Academic Press, London, New York, 401 p.

Liebhold, A. M., & Tobin, P. C. (2008). Population ecology of insect invasions and their management. *Annual Review of Entomology, 53*, 387–408. https://doi.org/10.1146/annurev.ento.52.110405.091401

Lister, B. C., & Garcia, A. (2018). Climate-driven declines in arthropod abundance restructure a rainforest food web. *Proceedings of the National Academy of Sciences, 115*(44), E10397–E10406. https://doi.org/10.1073/pnas.1722477115

Lowrie D.C. 1971: Effects of Time of Day and Weather on Spider Catches with a Sweep Net. – *Ecology* 52 (2), 348–350

Luczak J. 1958: Methods of investigating the spiders living on the plants of a pine forest. Influence on the unevenness of the strokes of the scop on the picture of the quantitative dynamics of spiders. – *Ekol. Polska* Ser. B. 4, 283–292

Luna S., Gold M., Albert A., Ceccaroni L., Claramunt B., Danylo O. et al. 2018: Developing mobile applications for environmental and biodiversity citizen science: Considerations and recommendations. – In: Joly A., Vrochidis S., Karatzas K., Karppinen A. & Bonnet P. (eds.): *Multimedia tools and applications for environmental & biodiversity informatics*, p.9–30. Cham: Springer

MacLeod R., Herzog S.K., Maccormick A., Ewing S.R., Bryce R. & Evans K.L. 2011: Rapid monitoring of species abundance for biodiversity conservation: Consistency and reliability of the MacKinnon lists technique. – *Biological Conservation* 144: 1374-1381

Maier Ch.T. & Waldbauer G.P. 1979: Diurnal Activity Patterns of Flower Flies (Diptera: Syrphidae) in an Illinois Sand Area. – *Ann. Entomol. Soc. of America* 72 (2), 237–245

Maldonado, M. B., Aranibar, J. N., Serrano, A. M., Chacoff, N. P., & Vázquez, D. P. (2019). Dung beetles and nutrient cycling in a dryland environment. *CATENA, 179*, 66–73. https://doi.org/10.1016/j.catena.2019.03.035

Marques M.I., Adis J., dos Santos G.B. & Battirola L.D. 2006: Terrestrial arthropods from tree canopies in the Pantanal of Mato Grosso, Brazil. – *Revista Brasileira de Entomologia* 50(2), 257-267

Martin K. 2002: *Ökologie der Biozönosen.* – Springer, 325 p.

May R.M. 1986: The search for Patterns in the Balance of Nature: Advances and Retreats. – *Ecology* 67 (5), 1115–1126

McComb B., Zuckerberg B., Vesely D. & Jordan C. 2010: Monitoring Animal Populations and their Habitats. A Practitioner's Guide. – CRC Press, Taylor & Francis Group, 277 p.

McDonald R.A. & Harris S. 1999: The use of trapping records to monitor populations of stoats *Mustela erminea* and weasels *M. nivalis*: the importance of trapping effort. – *Journal of Applied Ecology* 36: 679–688.

McDonald R.A. and Harris, S. 2002: Population Biology of Stoats *Mustela erminea* and Weasels *Mustela nivalis* on Game Estates in Great Britain. – *Journal of Applied Ecology* 39: 793-805.

McGeoch M.A. 1998: The selection, testing and application of terrestrial insects as bioindicators. – *Biol. Rev.* 73: 181-201

McGeoch, M. A., Rensburg, B. J. V., & Botes, A. (2002). The verification and application of bioindicators: a case study of dung beetles in a savanna ecosystem. *Journal of Applied Ecology, 39*(4), 661–672. https://doi.org/10.1046/j.1365-2664.2002.00743.x

Meierhofer, I. (1995). *Étude phénologique des coléoptères coprophages* (Travaux No. 42) (p. 30). Banyuls sur Mer: Laboratoire Arago. http://www.rnnmassane.fr/1995/12/18/4/. Accessed 20 August 2023

Meijer, J. (1971): Immigration of arthropods into the new Lauwerszee Polder. – In: Den Boer, P. J. (ed.): Dispersal and Dispersal Power of Carabid Beetles. – *Miscell. Papers, L. H. Wageningen*, 8, 53–64

Melcore I., Ferrari G. and Bertolino S. 2020: Footprint tunnels are effective for detecting dormouse species. – *Mammal Review* 50, 226-230.

Metspalu L., Veromann E., Kaasik R., Kovacs G., Helvi I. and Mänd W.u.M. 2015: Comparison of sampling methods for estimating the abundance of *Meligethes aeneus* on oilseed crops. – *International Journal of Pest Management* 61(4), 312-319, https://doi.org/10.1080/09670874.2015.1058992

Mikusinski G., Gromadzki M. and Chylarecki P. 2001: Woodpeckers as Indicators of Forest Bird Diversity. – *Conservation Biology* 15 (1): 208-217

Modlmeier, A. P., Keiser, C. N., Wright, C. M., Lichtenstein, J. L., & Pruitt, J. N. (2015). Integrating animal personality into insect population and community ecology. *Current Opinion in Insect Science*, *9*, 77–85. https://doi.org/10.1016/j.cois.2015.03.008

Montgomery G.A., Belitz M.W., Guralnick R.P. & Tingley M.W. 2021: Standards and Best Practices for Monitoring and Benchmarking Insects. – *Frontiers in Ecology and Evolution* 8, 1-18

Montgomery W.I. 1985: Interspecific competition and the comparative ecology of two congeneric species of mice. – In: Cook L.M. (ed.): *Case studies in Population biology*, 126–187. Manchester University Press, Manchester, 218 p.

Moore N.W. 1975: Butterfly transects in a linear habitat. 1964–1973. – *Entomol. Gaz. (London)* 26, 71–78

Morgan N.C., Waddell A.B. and Hall W.B. 1963: A Comparison of the Catches of Emerging Aquatic Insects in Floating Box and Submerged Funnel Traps. – *Journal of Animal Ecology* 32 (2), 203-219

Morris, R. F. (1960). Sampling Insect Populations. *Annual Review of Entomology*, *5*(Volume 5, 1960), 243–264. 10.1146/annurev.en.05.010160.001331

Mühlenberg M. & Werres W. 1983: Lebensraumverkleinerung und ihre Folgen für einzelne Tiergemeinschaften. Experimentelle Untersuchung auf einer Wiesenfläche. – *Natur und Landschaft* 58 (2), 43–50

Mühlenberg M., Leipold D., Mader H.-J. and Steinhauer B. 1977: Island Ecology of Arthropods. I. Diversity, Niches and Resources on Some Seychelles Islands. – *Oecologia* (Berlin) 29, 117–134

Muirhead-Thomson R.C. 1991: *Trap Responses of Flying Insects. The Influence of Trap Design on Capture Efficiency*. – Academic Press, Harcourt Brace Jovanovich, Publishers, London, San Diego, New York, Sydney, Tokyo, Toronto, 287 p.

Müller H.J. 1978: Strukturanalyse der Zikadenfauna (Homoptera Auchenorrhyncha) einer Rasenkatena Thüringens (Leutratal bei Jena). – *Zool. Jb. Syst* . 105, 258–334

Müller J., Hothorn T., Yuan Y., Seibold S., Mitesser O., Rothacher J., Freund J., Wild C., Wolz M. and Menzel A. 2023: Weather explains the decline and rise of insect biomass over 34 years. – *Nature* (27.09.2023). https://doi.org/10.1038/s41586-023-06402-z

Müller J., Pöllath J., Moshammer R. and Schröder B. 2009: Predicting the occurrence of Middle Spotted Woodpecker *Dendrocopos medius* on a regional scale, using forest inventory data. – *Forest Ecology and Management* 257, 502–509

Müller J.K. 1985: Konkurrenzvermeidung und Einnischung bei Carabiden (Coleoptera). – *Z. zool. Syst. Evolut.-forsch.* 23, 299–314

Mundie J.H. 1956: Emergence traps for aquatic insects. – *Internationale Vereinigung für Theoretische und Angewandte Limnologie: Mitteilungen* 7 (1), 1-13. Published online: 01 Dec 2017: https://doi.org/10.1080/05384680.1956.11904089

Murray, D. L., & Sandercock, B. K. (Eds.). (2020). *Population Ecology in Practice* (1st edition.). Hoboken: Wiley-Blackwell.

National Geographic Encyclopedia. (n.d.). 2022: Citizen science. – In *National Geographic*. Retrieved March 22, 2022, from https://www.nationalgeographic.org/encyclopedia/citizen-science/

National Park O.ö. Kalkalpen Ges.m.b.H. (2011): Bericht „Jahrestagung der Arbeitsgruppe Spechte der Deutschen Ornithologen-Gesellschaft 2009" – Band 10; Schriftenreihe Nationalpark Kalkalpen, Molln, 102 p.

National Reptile Survey: http://www.narrs.org.uk/ and https://reptile-survey.arc-trust.org

Nentwig W. 1982 a: Epigeic Spiders, Their Potential Prey and Competitors: Relationship Between Size and Frequency. – *Oecologia* (Berl.) 55, 130–136

Newton I. 1980: The role of food in limiting bird numbers. – *Ardea* 68, 11–30

Nichols, E., Spector, S., Louzada, J., Larsen, T., Amezquita, S., & Favila, M. E. (2008). Ecological functions and ecosystem services provided by Scarabaeinae dung beetles. *Biological Conservation*, *141*(6), 1461–1474. https://doi.org/10.1016/j.biocon.2008.04.011

Noriega, J. A., Hortal, J., deCastro-Arrazola, I., Alves-Martins, F., Ortega, J. C. G., Bini, L. M., et al. (2023). Dung removal increases under higher dung beetle functional diversity regardless of grazing intensification. *Nature Communications*, *14*(1), 8070. https://doi.org/10.1038/s41467-023-43760-8

Norton T.M., Andrews K.M. & Smith L.L. 2016: Techniques for Working with Wild Reptiles. – In: Veterian Key, Chap. 29. https://veteriankey.com/techniques-for-working-with-wild-reptiles/

Noyes J.S. 1989: A study of five methods of sampling Hymenoptera (Insecta) in a tropical rainforest, with special reference to the Parasitica. – *Journal of Natural History* 23(2), 285-298, https://doi.org/10.1080/00222938900770181

Nyffeler M. 2000: Ecological impact of spider predation: a critical assessment of Bristowe's and Turnbull's estimates. - Bull Br Arachnol Soc. 11, 367–373

O'Dea N., Watson J.E.M. & Whittaker R.J. 2004: Rapid assessment in conservation research: a critique of avifaunal assessment techniques illustrated by Ecuadorian and Madagascan case study data. – *Diversity and Distribution* 2010: 55-63

Obrist M.K. & Duelli P. 2010: Rapid biodiversity assessment of arthropods for monitoring average local species richness and related ecosystem services. – *Biodiversity and Conservation* 19: 2201-2220

Oelke H. 1977: Vogelsiedlungsdichten und ornitho-ökologische Differenzierungen der Laubwälder und Laubholzanlagen des Westharzes. – *Ber. Int. Symp. Int. Vergg. Veg.kunde Rinteln* , 353–396

Ohnesorge B. 1976: *Animals as Plant Pests*. – Thieme-Verl. Stuttgart, 288 p.

Olechowicz E. 1970: Evaluation of Number of Insects Emerging in Meadow Environment. – *Bull. Acad. Pol. Sci* ., II, 18, 389–395

Oliver I., & Beattie A.J. 1996a: Invertebrate morphospecies as surrogates for species: a case study. – *Conservation Biology* 10, 99-109

Oliver I. and Beattie A.J. 1996b: Designing a Cost-Effective Invertebrate Survey: A Test of Methods for Rapid Assessment of Biodiversity. - *Ecological Applications* 6(2), 594-607

Onsager J.A. and Henry J.E. 1977: A method for estimating the density of Rangeland grasshoppers (Orthoptera, Acrididae) in experimental plots. – *Acrida,* 231–237

Ortega-Martínez, I. J., Moreno, C. E., Rios-Díaz, C. L., Arellano, L., Rosas, F., & Castellanos, I. (2020). Assembly mechanisms of dung beetles in temperate forests and grazing pastures. *Scientific Reports, 10*(1), 391. https://doi.org/10.1038/s41598-019-57278-x

Oschmann M. 1969: Bestimmungstabellen für die Larven mitteldeutscher Orthopteren. – *Dtsch. Ent. Z* . 16 (1/3), 277–291

Otte J. 1987: Begünstigen nicht geräumte Windwurfflächen die Insektenfauna? – Diplomarbeit an der Julius-Maximilians-Universität Würzburg, 240 p.

Owen D.F. 1975: Estimating the abundance and diversity of butterflies. – *Biological Conservation* 8 (3), 173-183

Paarmann W. & Stork N.E. 1987: Canopy fogging, a method of collecting living insects for investigations of life history strategies. – *Journal of Natural History* 21, 563-566

Patrick B., McCllelan R., Martin T., Tocher M., Borkin K., McKoy, J., & Smith, D. 2014: *Guidelines for undertaking rapid biodiversity assessments in terrestrial and marine environments in the Pacific*. – Apia, Samoa: SPREP, Wildlands. 54 p.

Persson B. 1976: Influence of weather and nocturnal illumination on the activity and abundance of populations of noctuids (Lep.) in south coastal Queensland (Austr.). – *Bull. ent. Res* . 66 (1), 33–63

Pettit Rebekah. 2018: 5 Popular Ways To Find Reptiles And Amphibians In The Wild. – In :*Reptiles Magazine*. September 18, 2018 7:53 pm https://reptilesmagazine.com/5-popular-ways-to-find-reptiles-and-amphibians-in-the-wild/

Pfiffner L. and Niggli U. 1996: Effects of Biodynamic, Organic and Conventional Farming on Ground Beetles (Col. Carabidae) and Other Epigaeic Arthropods in Winter Wheat. – *Biological Agriculture & Horticulture*, 12:4, 353-364, 10.1080/01448765.1996.9754758

Philips, T. K., Pretorius, E., & Scholtz, C. H. (2004). A phylogenetic analysis of dung beetles (Scarabaeinae : Scarabaeidae): unrolling an evolutionary history. *Invertebrate Systematics, 18*(1), 53–88.

Piechocki R. 1985: Makroskopische Präparationstechnik. Teil II Wirbellose. – Fischer-Verl., Stuttgart, 308 p.

Pimm S.L. & Redfearn A. 1988: The variability of population densities. – *Nature* 334, 613–614

Pocock M.J.O., Roy H.E., Augus, T., Kuria A., Barasa F., Bett J., Githiru M., Kairo J., Kimani J., Kinuthia W., Kissui B., Madindou I., Mbogo K., Mirembe J., Mugo P., Muniale F.M., Njoroge P., Njuguna E.G., Olendo M.I., Opige M., Otieno T.O., Ng'weno C.C., Pallangyo E., Thenya T., Wanjiru A. & Trevelyan R. 2019: Developing the global potential of citizen science: Assessing opportunities that benefit people, society and the environment in East Africa. – *Journal of Applied Ecology 56*(2), 274–281. https://doi.org/10.1111/1365-2664.13279

Pollard E. & Yates T.J. 1993: *Monitoring Butterflies for Ecology and Conservation*. – Chapman & Hall, 292 p.

Pollard E. 1977: A method for assessing the abundance of butterflies. – *Biol. Conserv*. 12, 115–134

Pollard E. 1982: Monitoring butterfly abundance in relation to the management of Nature Reserves. – *Biol. Conserv* . 24, 317–328

Pollard E. 1988: Temperature, rainfall and butterfly numbers. – *Journal of Applied Ecology* 25, 819–828

Pollard E. 1991: 6. Monitoring butterfly numbers. – In: Goldsmith, F. B. (ed.): *Monitoring for Conservation and Ecology*, 87–111.– Chapman and Hall, London, New York, Tokyo, Melbourne, Madras, 275 p.

Pollierer M.M., Drescher J., Potapov A., Kasmiatun K., Mawan A., Mutiari M., Nazarreta R., Hidayat P., Buchori D. & Scheu S. 2023: Rainforest conversion to plantations fundamentally alters energy fluxes and functions in canopy arthropod food webs. – *Ecology Letters* (06). https://doi.org/10.1111/ele.14276.

Pryke, J. S., Roets, F., & Samways, M. J. (2016). Wild Herbivore Grazing Enhances Insect Diversity over Livestock Grazing in an African Grassland System. *PloS one, 11*(10), e0164198. https://doi.org/10.1371/journal.pone.0164198

Rainio, J. and Niemelä, J. 2003: Ground beetles (Coleoptera: Carabidae) as bioindicators. – *Biodiversity and Conservation* 12, 487-506

Raw F. 1959: Estimating earthworm populations by using formalin. – *Nature* 184 (4699), 1661-1662

Redfern M. 1968: *The natural history of spear thistle-heads*. – In: Carkty J.D. (ed.): Field Studies, London, 2 (5), 669–717

Redfern M. 1983: *Insects and thistles*. – Cambridge University Press, Cambridge, 64 p.

Regensburg J.T. 1977: A Modified Sweep Net for Quantitative Sampling. – *Entomol. News* 88, 141–142

Renner, K., 1980. Faunistisch-ökologische Untersuchungen der Käferfauna pflanzensoziologisch unterschiedlicher Biotope im Evessel-Bruch bei Bielefeld-Sennstedt. – *Berichte des Naturwissenschaftlichen Vereins für Bielefeld* Sonderheft 2, 145-176.

Ribeiro-Júnior M.A., Gardner T.A. & Ávila-Pires T.C. 2008: Evaluating the effectiveness of herpetofaunal sampling techniques across a gradient of habitat change in a tropical forest landscape. – *Journal of Herpetology* 42(4): 733-749

Richards O.W. and Waloff N. 1954: Studies on the biology and population dynamics of British grasshoppers. – *Anti-Locust Bulletin* 17, 1–186

Ridsdill-Smith, T. J., & Hayles, L. (1990). Stages of bush fly, Musca vetustissima (Diptera: Muscidae), killed by scarabaeine dung beetles (Coleoptera: Scarabaeidae) in unfavourable cattle dung. *Bulletin of Entomological Research*, *80*(4), 473–478. https://doi.org/10.1017/S0007485300050744

Robinson J.V. 1981: The Effect of Architectural Variation in Habitat on a Spider Community: An Experimental Field Study. – *Ecology* 62 (1), 73–80

Rockwood, L. L. (2015). *Introduction to Population Ecology* (2nd edition.). Chichester, West Sussex, UK ; Hoboken, NJ, USA: Wiley-Blackwell.

Roy H.E., Baxter E., Saunders A. and Pocock M.J.O. 2016: Focal plant observations as a standardized method for pollinator monitoring: Opportunities and limitations for mass participation citizen science. *PLOS ONE* 11: e0150794

Rümer H. & Mühlenberg M. 1988: Kritische Überprüfung von „Minimalprogrammen" zur zoologischen Bestandserfassung. – *Schriftenreihe Bayer. Landesamt für Umweltschutz, München,* 83, 151–157

Sánchez-Bayo, F., & Wyckhuys, K. A. G. (2019). Worldwide decline of the entomofauna: A review of its drivers. *Biological Conservation*, *232*, 8–27. https://doi.org/10.1016/j.biocon.2019.01.020

Sands, B., & Wall, R. (2017). Dung beetles reduce livestock gastrointestinal parasite availability on pasture. *Journal of Applied Ecology*, *54*(4), 1180–1189. https://doi.org/10.1111/1365-2664.12821

Satchell J.E. & Lowe DG. 1967: Selection of leaf litter by *Lumbricus terrestris*. – In: Graff O. & Satchell J. E. (eds.): *Progress in Soil Biology* , 102–119. – North Holland Publishing Co. Amsterdam, 656 p.

Satchell J.E. 1969: Studies on methodical and taxonomical questions. – *Pedobiologia* 9, 20–25

Schaefer C. 2017: *Untersuchung zu einem neuen elektrischen Gerät zur Regenwurmextraktion*. – München, GRIN Publishing, 47 p., https://www.grin.com/document/378284

Schaefer M. & Tischler W. 1983: *Wörterbücher der Biologie: Ökologie*. – Fischer-Verl. Stuttgart, 354 p.

Schaefer M. 1975: Experimental studies on the Importance of Interspecies Competition for the Lycosid Spiders in a Salt Marsh. – *Proc. 6th Arachn. Cong* . 1974, 86–90

Schaefer M. 1978 a: Some Experiments on the Regulation of Population Density in the Spider *Florinia bucculenta* (Araneida: Linyphiidae). – *Symp. Zool. Soc. Lond* . 42, 203–210

Schaefer M. 1983: Räuberische Arthropoden in der Streuschicht eines Kalkbuchenwaldes: Biomasse, Energiebilanz, „Feinddruck" und Aufteilung der Ressourcen. – *Verh. Dtsch. Zool. Ges* ., 206 p.

Schaefer M. 1986: Ökosystemforschung in Wäldern: Zur Funktion der Tiere. Forschungsberichte – In: *Georgia Augusta (Göttingen)* 44, 29–38

Schaefer M. 1996: *Die Bodenfauna von Wäldern: Biodiversität in einem ökologischen System.* – Akademie der Wissenschaften und der Literatur, Abhandlungen der Mathematisch-Naturwissenschaftliche Klasse Jg. 1996 (2), Mainz, 81 p..

Schauermann J. 1982: Verbesserte Extraktion der terrestrischen Bodenfauna im Vielfachgerät modifiziert nach Kempson und Macfayden. – *Mitt. SFB* 135 (1), 47–50

Schedl H. & Klepsch R. 2001: Population structure of the green lizard (*Lacerta viridis* (Laurenti 1768) – A comparison of two study areas in Vienna (Austria). – In: Elbing K. & Nettmann H-K. (eds.): Contributions to the Natural History and Conservation of the Green Lizards (*Lacerta* s. str.). – *Mertensiella* 13: 215-228

Schjøtz-Christensen B. 1965: Biology and population studies of Carabidae. II. Technique, 1. Trapping. – *Natura Jutlandica* 11, 1–173

Schmidt M., Meyer P., Mölder A. & Hondong H. 2016: New or re-colonization? The late 19th century range expansion of the black woodpecker (*Dryocopus martius*) in northwest Germany and its underlying causes. – *Forstarchiv* 87 (5), 135-151

Schmidt O. 2001: Appraisal of the electrical octet method for estimating earthworm populations in arable land. – *Annals of applied biology* 138 (2), 231-241

Schmucki R., Pe'er G., Roy D.B., Stefanescu C., Van Swaay C.A.M., Oliver, T.H., Kuussaari, M., Van Strien A.J., Ries L., Settele J., Musche M., Carnicer J., Schweiger O., Brereton T.M., Harpke A., Heliölä J., Kühn E. & Julliard R. 2016: A regionally informed abundance index for supporting integrative analyses across butterfly monitoring schemes. – *Journal of Applied Ecolology* 53, 501-510. https://doi.org/10.1111/1365-2664.12561

Schoener T.W. 1983: Field Experiments on Interspecific Competition. – *The American Naturalist* 122 (2), 240–285

Schuch S., Ludwig H. and Wesche K. 2020: Recording methods for insect monitoring. A collection of materials. – BfN-Scripts 565, Federal Agency for Nature Conservation, Bonn. 83p.

Shchipanov N.A., Litvinov Y.N. and Sheftel B.I. 2008: Rapid method for estimating local biodiversity of a community of small mammals. – *Contemp. Probl. Ecol* . 1, 596–602

Sheftel B.I. & Yakushov V.D. 2022: Impacts of Climate Warming on Terrestrial Species in the Middle Yenisei Taiga. – *Contemporary Problems of Ecology* 15(1), 1-10

Sheftel B.I. 1989: Long-term and seasonal dynamics of shrews in Central Siberia. – *Ann. Zool. Fennici* 26, 357-369

Sheftel B.I. 2018: Methods of accounting for small mammals. – *Russ. J. Ecosyst. Ecol* . 3 (3), 1-21

Silva, P. G. da, & Hernández, M. I. M. (2015). Spatial Patterns of Movement of Dung Beetle Species in a Tropical Forest Suggest a New Trap Spacing for Dung Beetle Biodiversity Studies. *PLOS ONE*, *10*(5), e0126112. https://doi.org/10.1371/journal.pone.0126112

Simberloff, D. S., & Wilson, E. O. (1970). Experimental Zoogeography of Islands. A Two-Year Record of Colonization. *Ecology*, *51*(5), 934–937. https://doi.org/10.2307/1933995

Skinner B. & Wilson D. 2009: *Colour Identification Guide to the Moths of the British Isles*: Macrolepidoptera. (3rd rev. ed.) – Brill Publisher, 325 p.

Skuhravy V. 1957: The Pitfall Trap Method (Metoda zemnich pasti). – *Acta societatis entomologicae cechoslovenìae* Praha 54 (1), 27–40

Skvaria M., Larson J.L., Fisher R. & Dowling A.P.G. 2020: A Review of Terrestrial and Canopy Malaise Traps. - Annals of the Entomological Society of America 114(1), 1-21. DOI: https://doi.org/10.1093/aesa/saaa044

Slade, E. M., Kirwan, L., Bell, T., Philipson, C. D., Lewis, O. T., & Roslin, T. (2017). The importance of species identity and interactions for multifunctionality depends on how ecosystem functions are valued. *Ecology*, *98*(10), 2626–2639. https://doi.org/10.1002/ecy.1954

Slade, E. M., Riutta, T., Roslin, T., & Tuomisto, H. L. (2016). The role of dung beetles in reducing greenhouse gas emissions from cattle farming. *Scientific Reports*, *6*(1), 18140. https://doi.org/10.1038/srep18140

Sladecek, F. X. J., Segar, S. T., Lee, C., Wall, R., & Konvicka, M. (2017). Temporal Segregation between Dung-Inhabiting Beetle and Fly Species. PLOS ONE, 12(1), e0170426. https://doi.org/10.1371/journal.pone.0170426

Smith B.J. 1976: A new application in the pitfall trapping of insects. – *Transaction Kentucky Academy Science* 37 (3–4), 94–97

Southwood T.R.E. & Henderson P.A. 2000: *Ecological Methods* (3 rd ed.). – Blackwell Science, 575 p.

Ssymank A. 1991: The application of color bowls in biozoonology using the example of the Syrphidae. – *Supplements to the Negotiations of the Society for Ecology* 2, 119–128

Stack, C. (Ed.). (2015). *Beetles: Biodiversity, Ecology and Role in the Environment* (UK ed. edition.). Hauppauge, New York: Nova Science Pub Inc.

Steiner G. 1963: *Das Zoologische Laboratorium*. – E. Schweizerbart'sche Verlagsbuchhandlung, Stuttgart, 557 p.

Surdo S., Galasso P., Cusimano C., Reale M., & Zafarana M.A. 2022: Citizen science project to monitor wildlife: a first census of wintering Booted Eagle *Hieraaetus pennatus* in Sicily. – *Rivista Italiana Di Ornitologia* 92(2), 33–38. https://doi.org/10.4081/rio.2022.583

Szujecki A. 1987: *Ecology of Forest Insects*. – Series Entomologica, 26. – Polish Scientific Publishers, Warsaw, 600 p.

Tanier M. 1977: Empty catch with pitfall traps. – *Cordula* 3, 105–106

Tasker E.M. & Dickman C.R. 2001: A review of Elliott trapping methods for small mammals in Australia. - *Australian Mammalogy* 23(2) 77 - 87

Teichmann B., 1994: Eine wenig bekannte Konservierungsflüssigkeit für Bodenfallen. – *Entomologische Nachrichten und Berichte* 38, 25-29.

Thielemann U. 1986a: Elektrischer Regenwurmfang mit der Oktett-Methode. – *Pedobiologia* 29 (4), 296-302

Thielemann U. 1986b: Untersuchungen zur Erfassung der Lumbricidenfauna mittels einer neu entwickelten Elektro-Methode (Oktett-Methode) in verschiedenen Waldökosystemen. – In: *Bodenökologie. Akademie für Naturschutz und Landschaftspflege, Laufener Seminarbeiträge* 7/86, 42-50

Thomas D.B. and Sleeper E.L. 1977: The Use of Pitfall Traps for Estimating the Abundance of Arthropods, with Special Reference to the Tenebrionidae (Coleoptera). – *Ann. Entomol. Soc. Amer* . 70 (2), 242–248

Thomas J.A. 1983: A quick method of estimating butterfly numbers during surveys. – *Biol. Conserv* . 27, 195–211

Thompson D. 2019: Sampling Rangeland Grasshoppers, 219-233. – In: Capinera J.L. (ed.): *Integrated Pest Management on Rangeland* . A Shortgrass Prairie Perspective. 450 p. https://doi.org/10.1201/9780429043673-15.

Thorp J.H. & Rogers D.C. (eds.) 2015: *Thorp and Covich's Freshwater Invertebrates* (4th ed.). Vol. I: *Ecology and General Biology*. – Academic Press, Elsevier Inc., 1148 p.

Thorp J.H. & Rogers D.C. (eds.) 2016: *Thorp and Covich's Freshwater Invertebrates* (4th ed.). Vol. II: *Keys to Nearctic Fauna*. – Academic Press, Elsevier Inc., 740 p.

Tocco, C., Dacke, M., & Byrne, M. (2021). The finely defined shift work schedule of dung beetles and their eye morphology. *Ecology and Evolution*, *11*(22), 15947–15960. https://doi.org/10.1002/ece3.8264

Tolbert W.W. 1977: The use of a water sprayer in web building spider density determinations. – *J. Arachnol* . 4, 211–214

Tserendavaa Purevjal, E. Erdenechimeg, K. Hackländer, Samiya Ravchig, Boris I. Sheftel, Undrakhbayar Enkhbat, A. Nomin, M. Muehlenberg 2023: A 12-year population study of the reed vole (Alexandromys fortis) (Rodentia: Cricetidae) in Khonin Nuga, West Khentei, Mongolia. - *Russian J. Theriol. 22* (2): 142–149

Turnbull AL. 1973: Ecology of the true spiders. (Araneomorphae). - Annu. Rev. Entomol. 18, 305–348. doi: https://doi.org/10.1146/annurev.en.18.010173.001513.

Twining, C. W., Brenna, J. T., Lawrence, P., Shipley, J. R., Tollefson, T. N., & Winkler, D. W. (2016). Omega-3 long-chain polyunsaturated fatty acids support aerial insectivore performance more than food quantity. *Proceedings of the National Academy of Sciences*,

113(39), 10920–10925. https://doi.org/10.1073/pnas.1603998113

Uetz G.W. & Unzicker J.D. 1976: Pitfall trapping in ecological studies of wandering spiders. – *J. Arachnology* 3, 101–111

Uetz G.W. 1975: Temporal and spatial variation in species diversity of wandering spiders (Araneae) in deciduous forest litter. – *Environmental Entomology* 4, 719–724

Uetz G.W. 1979: The influence of variation in litter habitats on spider communities. – *Oecologia* 40, 29–42

Uhler, J., Haase, P., Hoffmann, L., Hothorn, T., Schmidl, J., Stoll, S., et al. (2022). A comparison of different Malaise trap types. *Insect Conservation and Diversity*, *15*(6), 666–672. https://doi.org/10.1111/icad.12604

Uwingabire, Z., & Gallai, N. (2024). Impacts of degraded pollination ecosystem services on global food security and nutrition. *Ecological Economics*, *217*, 108068. https://doi.org/10.1016/j.ecolecon.2023.108068

van Huizen T.H.P. 1979: Individual and environmental factors determining flight in carabid beetles. – *Miscell. Papers, L. H. Wageningen* 18, 199–211

van Swaay C., Nowicki P., Settele J. & van Strien A.J. 2008: Butterfly monitoring in Europe: Methods, applications and perspectives. – *Biodiversity and Conservation* 17(14), 3455-3469. DOI:https://doi.org/10.1007/s10531-008-9491-4

Varley G.C. 1947: The natural control of population balance in the knapweed gallfly *(Urophora jaceana)*. – *J. Anim. Ecol.* 16, 139–187

Varley, G. C., Gradwell, G. R., & Hassell, M. P. (1974). *Insect Population Ecology: An Analytical Approach*. Berkeley: University of California Press.

Varley G.C., Gradwell G.R. & Hassel M P. 1980: *Populationsökologie der Insekten. Analyse und Theorie*. – Thieme-Verl. Stuttgart, 211 p.

Verdú, J. R., Sánchez-Piñero, F., Lobo, J. M., & Cortez, V. (2020). Evaluating long-term ivermectin use and the role of dung beetles in reducing short-term CH4 and CO2 emissions from livestock faeces: a mesocosm design under Mediterranean conditions. *Ecological Entomology*, *45*(1), 109–120. https://doi.org/10.1111/een.12777

Vohland K., Land-Zandstra A., Ceccaroni L., Lemmens R., Perelló J., Ponti M., Samson R. & Wagenknecht K. (eds.) 2021: *The Science of Citizen Science*. – Springer, 529 p., open access https://doi.org/10.1007/978-3-030-58278-4

von Hoermann C., Weithmann S., Deißler M., Ayasse M., Steiger S. (2020): Forest habitat parameters influence abundance and diversity of cadaver-visiting dung beetles in Central Europe. – *Royal Society Open Science* 7 (3): 191722. https://doi.org/10.1098/rsos.191722

Waltert, M., Mardiastuti, A., Mühlenberg, M. 2004: Effects of land use on bird species richness in Sulawesi, Indonesia. – *Conservation Biology* 18: 1339-1346

Walther P.B. & Snider R.M. 1984: Techniques for sampling earthworms and cocoons from leaf litter, humus and soil. – *Pedobiologia* 27, 293–297

Warikar E.L. 2011: Study Diversity of Log Dependent Beetles in District of Bonggo, Regency of Sarmi, Province of Papua, Indonesia. – Master Thesis at the Georg-August University Göttingen, 62 p.

Waring P., Townsend M. & Lewington R. 2017: *Field Guide to the Moths of Great Britain and Ireland*. (3rd ed.) – Bloomsbury, 464 p.

Wassmer, T. (1994). Seasonality of Coprophagous Beetles in the Kaiserstuhl Area near Freiburg (Sw Germany) Including the Winter Months. *Acta Oecologica-International Journal of Ecology*, *15*(5), 607–631.

Waßmer, T. (1995). Mistkäfer (Scarabaeidae und Hydrophilidae) als Bioindikatoren für die naturschützerische Bewertung von Weidebiotopen. *Zeitschrift für Ökologie und Naturschutz*, *4*, 135–142.

Wassmer, T. (2020a). Phenological Patterns and Seasonal Segregation of Coprophilous Beetles (Coleoptera: Scarabaeoidea and Hydrophilidae) on a Cattle Farm in SE-Michigan, United States Throughout the Year. *Frontiers in Ecology and Evolution*, *8*, 293. https://doi.org/10.3389/fevo.2020.563532

Wassmer, T. (2020b). Attractiveness of Cattle Dung to Coprophilous Beetles (Coleoptera: Scarabaeoidea and Sphaeridiinae) and Their Segregation During the Initial Stages of the Heterotrophic Succession on a Pasture in Southeast Michigan. *Journal of Insect Science*, *20*(3). https://doi.org/10.1093/jisesa/ieaa040

Wassmer, T. (2014). Seasonal Occurrence (Phenology) of Coprophilous Beetles (Coleoptera: Scarabaeidae and Hydrophilidae) from Cattle and Sheep Farms in Southeastern Michigan, USA. *The Coleopterists Bulletin*, *68*(3), 603–618. https://doi.org/10.1649/072.068.0327

Wassmer, T., & Armstrong, E. (2023). Population structure of Phanaeus vindex (Coleoptera: Scarabaeidae) in SE Michigan. *Journal of Insect Science*, *23*(4), 2. https://doi.org/10.1093/jisesa/iead050

Weddeling K., Hachtel M., Schmidt P., Ortmann D. & Bosbach G. 2005: Determining population trends in species of the Habitats Directive: Methodological suggestions for monitoring using the example of the amphibian and reptile species of Annexes IV and V. – In: Doerpinghaus, A., Eichen C., Gunnemann H., Leopold P., Neukirchen M., Petermann J. & Schröder E. 2005: Methods for recording species of Annexes IV and V of the Fauna-Flora-Habitat Directive. – Nature Conservation and Biological Diversity 20, Federal Agency for Nature Conservation Bonn-Bad Godesberg, 422-449

Weidemann G. 1971: D. Food and Energy Turnover of Predatory Arthropods of the Soil Surface. – *Ecol. St. Anat. & Synth.* 2, 111–118

Weidemann G. 1978: Über die Bedeutung von Insekten im Ökosystem Laubwald. – *Mitt. Deut. Ges. Allg. Angew. Ent.* 1, 196–204, (Soll.-Nr. 222).

Weisse M., Goldman L. and Carter S. 2023: Tropical Primary Forest Loss Worsened in 2022, Despite International Commitments to End Deforestation. - Global Forest Watch, https://

www.globalforestwatch.org/blog/forest-insights/global-tree-cover-loss-data-2022/

Weithmann S., Kuppler J., Degasperi G., Steiger S., Ayasse M., von Hoermann C. (2020): Local and Landscape Effects on Carrion-Associated Rove Beetle (Coleoptera: Staphylinidae) Communities in German Forests. – *Insects* 11 (12), 828. https://doi.org/10.3390/insects11120828

Wesolowski T. 1995: Value of Bialowieza Forest for the Conservation of White-Backed Woodpecker *Dendrocopus leucotos* in Poland. – *Biological Conservation* 71, 69-75

Westerberg D. 1977: Evaluation of the pitfall method in inventorying the field – and ground layer's lower fauna. – Swedish Environmental Protection Agency PM 844, VINA Rapp. 5, 1–72. – Cited in: Hågvar, S., Østbye, E. & Melåen, J. (1978): Pit-fall catches of surface-active arthropods in some high mountain habitats at Finse, south Norway. II. General results at group level, with emphasis on Opiliones, Araneida and Coleoptera. – *Norw. J. Ent*. 2, 195–205.)

Weyers S. L., Schomberg H. H. and Hendrix P. F. 2008: Construction of an electrical device for sampling earthworm populations in the field. – *Applied Engineering in Agriculture*, 24 (3), 391

Winkler, D. W., Ringelman, K. M., Dunn, P. O., Whittingham, L., Hussell, D. J. T., Clark, R. G., et al. (2014). Latitudinal variation in clutch size–lay date regressions in Tachycineta swallows: effects of food supply or demography? *Ecography*, *37*(7), 670–678.

Winter S., Flade M., Schumacher H., Kerstan E. and Moeller G. (2005): The importance of near-natural stand structures for the biocoenosis of lowland beech forests. – *Forest Snow and Landscape Research* 79 (1 – 2): 127 – 144.

Wise D.H. 1981: Inter-and Intraspecific Effects of Density Manipulations upon Females of Two Orb-Weaving Spiders (Araneae: Araneidae). – *Oecologia* (Ber.) 48, 252–256

Witsack W. 1975: Eine quantitative Keschermethode zur Erfassung der epigäischen Arthropoden-Fauna. – *Entomol. Nachrichten* 8, 123–128

Woodcock B.A. 2005: Pitfall trapping in ecological studies. – In: Leather S. (ed.): *Insect Sampling in Forest Ecosystems*, p. 37-57. - Blackwell Publishing, 303 p.

Wolf, M., & Weissing, F. J. (2012). Animal personalities: consequences for ecology and evolution. *Trends in Ecology & Evolution*, *27*(8), 452–461. https://doi.org/10.1016/j.tree.2012.05.001

Workman C. 1978: Life cycle and population dynamics of *Trochosa terricola* Thorell. – *Ecological Entomology* 3, 329–340

Yamada, D., Imura, O., Shi, K., & Shibuya, T. (2007). Effect of tunneler dung beetles on cattle dung decomposition, soil nutrients and herbage growth. *Grassland Science*, *53*(2), 121–129. https://doi.org/10.1111/j.1744-697X.2007.00082.x

Zou Y., Feng J., Xue D., Sang W & Axmacher J.C. 2012: A Comparison of Terrestrial Arthropod Sampling Methods. – *J. Resour. Ecol*. 3 (2), 174-182

Zwölfer H. and Brandl R. 1989: Niches and size relationships in Coleoptera associated with Carduae host plants: adaptations to resource gradients. – *Oecologia* 78, 60–68

Zwölfer H. 1979: Strategies and Counterstrategies in Insect Population Systems Competing for Space and Food in Flower Heads and Plant Galls. – *Fortschr. Zool*. 25 (2/3), 331–333

Zwölfer H. 1980: Artenschutz für unscheinbare Tierarten? – *Schriftenreihe Naturschutz und Landschaftspflege (München)* 12, 81–88

Zwölfer H. 1988: Evolutionary and ecological relationships of the insect fauna of thistles. – *Ann. Rev. Entomol.* 33, 103–122

3 Recording Abiotic Parameters

Abstract

In all ecological studies, knowing the physico-chemical conditions of specific locations is as important as knowing what species occur and how extensive their populations are. Abiotic parameters are measured with devices, and the measured values can now be recorded continuously using data loggers. Modern devices can measure and record many parameters simultaneously. We also present devices and techniques that can be used to collect data if no electricity is available. Ecologically, sums of parameter values, such as temperature sums and sunshine duration, are often more significant. Detailed instructions for precise measurements are provided.

3.1 Data Loggers

Measuring devices with an electrical signal output can be combined in a data logger and scanned in a specific sequence and at a selectable time interval (see Fig. 3.1). The advantages of such devices for fieldwork are apparent. Until recently, environmental conditions could only be acquired as a finalized series of several measured variables that lasted days or weeks using fixed measuring stations with mechanical or electric data writers. However, the situation is now completely different. Small, robust, portable, and battery-operated devices with a long service life of 1 year or longer can collect environmental data in almost any conceivable biocoenosis without any significant intervention from the researcher. These devices will give new impetus to field research and should, therefore, be presented and tested in practice, e.g., in practical field courses if possible. Bingham and Long (1988), Beddows and Mallon (2018), Cannon et al. (2022), and Gandra et al. (2015) provide overviews of the devices and their manufacturers.

Temperature and humidity are the most frequently recorded of all environmental conditions, but other environmental parameters can also be measured continuously or at predetermined intervals using appropriate sensors. You can learn about suitable devices from manufacturer sites on the internet, e.g., Gemini Data Loggers (2024) and Onset Computer Corporation (2024).

Appropriate software programs are also available that can be used to read the stored data from the data logger to a PC. The data can be displayed on the computer in tabular form or as diagrams, e.g., in Microsoft Excel, if required. The Excel tables can also be exported into graphical analysis programs such as OriginLab's OriginPro (www.originlab.com) and statistics programs such as Statistica (statista.com), SPSS (ibm.com/spss), SAS (sas.com), Stata (stata.com), MiniTab (minitab.com), or used in R (r-project.org). There are also data logging systems available for self-sufficient outdoor operation, which can be connected to various power supplies and upload data via the internet or satellite (e.g., myDatalo-

M. Mühlenberg, T. Waßmer, *Field Ecology*, https://doi.org/10.1007/978-3-031-87547-2_3

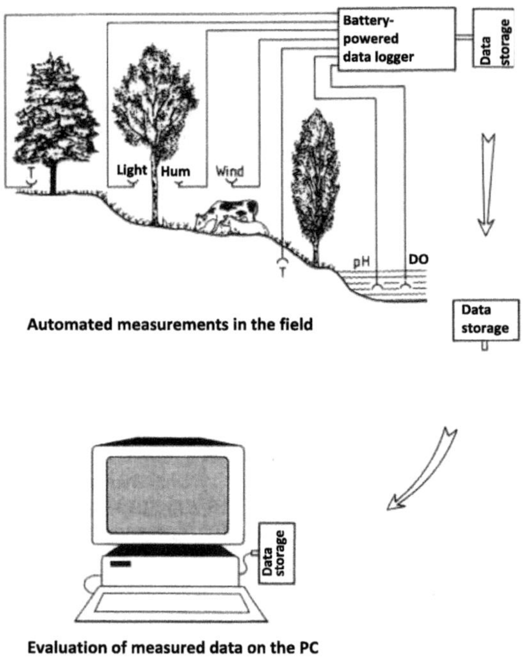

Fig. 3.1 Principle of a data logger. Top panel: automatic measurement in the field. Bottom panel: evaluation of the measurement data on the PC. *T* temperature, *Hum* humidity, *DO* dissolved oxygen

gEASY IoT by Microtronics Engineering GmbH, Microtronics Engineering GmbH 2024, YouTube video: https://www.youtube.com/watch?v=Ta8iXurCASA).

The technology of data loggers has improved tremendously during the last 10 years, so biological signals from animals, such as data on ecophysiology and behavior, can also be received and recorded. See Sect. 4.6 ("bio-logging") for more information. Good sources for information are also "Ecology and Behaviour of Free-Ranging Animals Studied by Advanced Data-Logging and Tracking Techniques" (Wassmer et al. 2020) as well as (Börger et al. 2020; Morrison 2021) and "An overview of behavioral, physiological, and environmental sensors used in animal biotelemetry and biologging studies" (Whitford and Klimley 2019). Finally, we recommend the upcoming proceedings of the International Eighth Bio-Logging Science Symposium 2024 (https://www.biomedcentral.com/collections/bls8). Data loggers for underwater recording are described in Chap. 5 (Limnology, water quality). Data loggers are especially useful when combined with LiDAR and GIS (Stickley and Fraterrigo 2021).

3.2 Radiation Measurements

In ecological studies on meadow areas, the recording of incident radiation is particularly interesting if the test areas have different inclinations and exposures or if mountains, tall trees, or buildings partially obscure the horizon. In contrast to the comparison with the local climate of a forest, for example, instantaneous values are of secondary importance, as these can be pretty similar over more extended periods of the day. The sum of the daily irradiation and its dependence on the terrain is primarily important for the comparison.

Determining the shortening of the theoretically possible duration of sunshine is straightforward by narrowing the horizon with a horizontoscope. The simple device consists of a Plexiglas dome reflecting the surrounding horizon. This shadow outline is projected onto a supplied base with a fixed hour division and the seasonal sun paths by looking exactly vertically from above. This makes it easy to read off the theoretically possible duration of sunshine and the effects of horizon elevations, shadow-casting obstacles, and terrain inclination (Fig. 3.2). The device is available from the Institut für Tageslichttechnik Stuttgart at http://ift-stuttgart.de upon contacting the owner. The same principle but with often inferior quality is used in instruments known as the Solar Pathfinder (solarpathfinder.com), Solmetric SunEye (solmetric.com), and Horicatcher (meteonorm.com), which perform shading analyses, mainly used for photovoltaic installations. Examples of recent ecological research using the horizontoscope include Scherer et al. (2021) and Schwarz and Fartmann (2021, 2022).

While the maximum possible sunshine duration for a location can be determined with the horizontoscope, recording the actual hours of sunshine during a certain period of time, depending on the weather conditions, requires recording sunshine detectors.

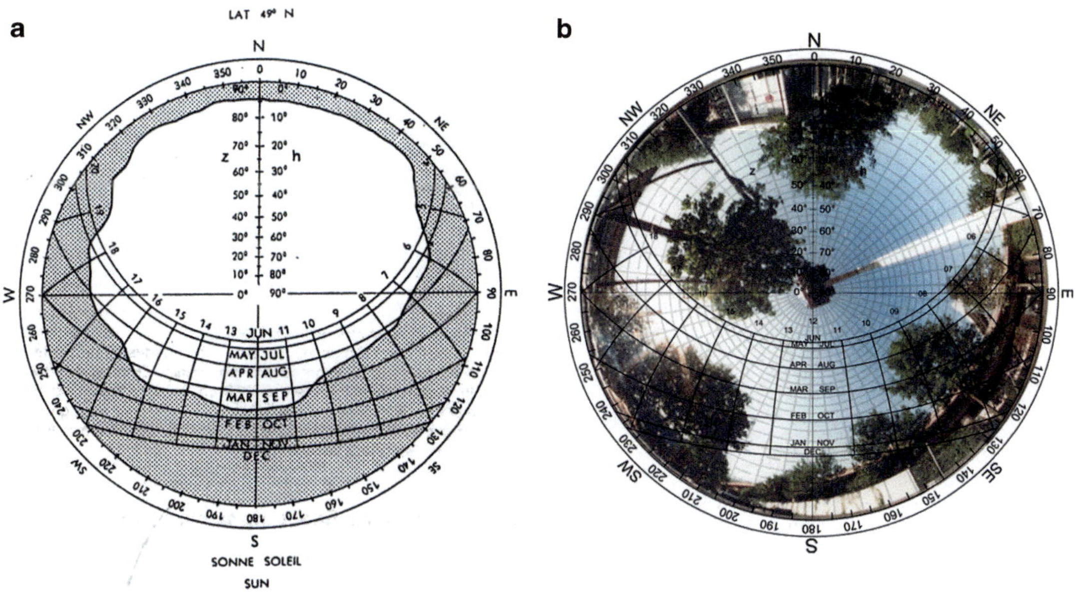

Fig. 3.2 Determination of the maximum possible sunshine duration for a location using a horizontoscope (**a**) View of a horizontoscope onto which the surrounding horizon is reflected as a silhouette. Further explanations are in the text body. (Available from Institut für Tageslichttechnik, Osterbronnstr. 30 D-70565 Stuttgart, ift-stuttgart.de). (**b**) Projection of structures around the horizon onto the horizontoscope. The black lines show the path of the sun throughout the months of the year and allow an estimation of the overall received solar radiation. See text for further details. (Picture credit: Institut für Tageslichttechnik Stuttgart, ift-stuttgart.de with permission)

The Campbell–Stokes sunshine recorder (heliograph) (Fig. 3.3) is a simple, not exactly cheap, device that works without any power supply.

It consists of a glass sphere that concentrates the solar radiation regardless of the sun's position and combines it in a (moving) focal point behind the sphere. In the concentric focal plane behind the sphere, there is a strip of paper that can be changed daily, and the path of the moving sun is visible as a focal trace that disappears when the sky is overcast. Sanchez-Lorenzo et al. (2013) provide new insights into the history of the Campbell–Stokes sunshine recorder.

In addition to this old mechanical sunshine recorder, there are numerous electronic sunshine duration sensors, such as the Kipp and Zonen CSD3 Sunshine Duration Sensor (kippzonen.com, https://www.youtube.com/watch?v=c_bJ2JBWyl8).

Direct solar radiation is not the only form of radiation important for a location's climate but also diffuse skylight and reflected radiation from the surroundings. The sum, known as global radiation, can be measured most easily with so-called (solar) radiation thermometers. These are two precision thermometers melted into airless glass sleeves whose spherical mercury vessels are reflective (white) or absorbent (black) (Fig. 3.4a, c).

The radiation converted into heat on the black surface creates a temperature difference between the two thermometers that is proportional to the energy of the incident radiation. The two thermometer bodies occupy a relatively small solid angle, so the radiation from all directions is approximately equal. It is, therefore, a space-integrating, energy-dependent (caloric) radiation measurement, whereby it must be considered that the effective spectral range (approx. 0.3–3.0 µm) extends from long-wave UV to far infrared.

Thermopile pyranometers (also called thermoelectric pyranometers), offered by several companies in various designs, also work according to the principle of caloric radiation measure-

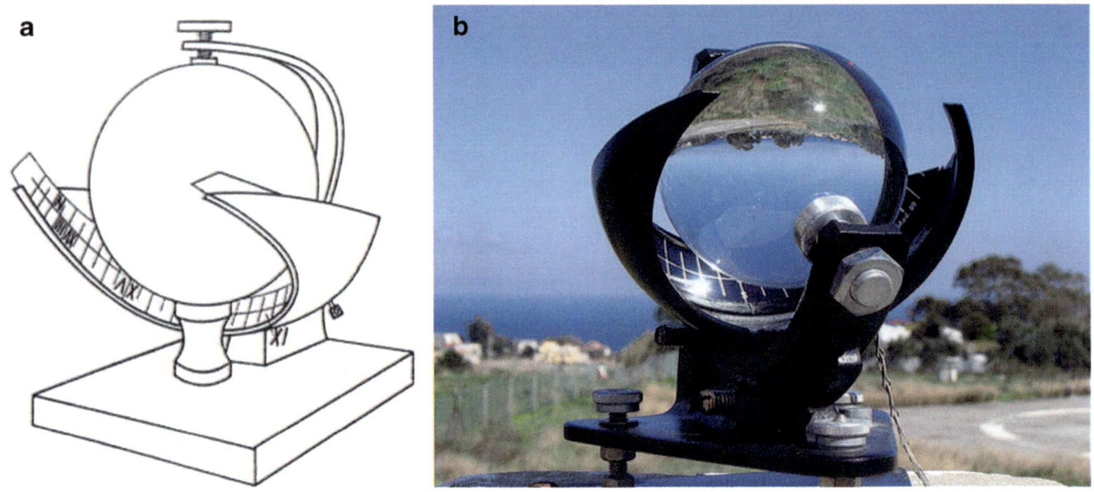

Fig. 3.3 Recording of the actual hours of sunshine during a certain period of time (**a**) Sun recorder (heliograph) according to Campbell–Stokes with writing strip in place. See text for further details. (**b**) Image of a heliograph, an instrument used in meteorology to measure the sunshine hours during a day. (Image credit: Mª Victoria Fernández Arboleya on Wikimedia. CC BY-SA 3.0)

ment (Fig. 3.4b). These are almost always so-called black-and-white pyranometers, in which black and white receiver surfaces are arranged in a specific pattern under a glass dome. When radiation is incident, a temperature difference is formed between the fields due to the strongly differing radiation absorption, which built-in thermocouples can measure (Ruitenbeek and Wely 2016).

Photovoltaic pyranometers, also known as photoelectric pyranometers, are usually offered with a grid-dependent display or recording devices, initially making them appear less suitable for fieldwork. However, the output voltages of most pyranometers are of a magnitude that they can be easily measured with a modern 3½-digit (better still, 4½-digit) battery-operated digital voltmeter.

Like many other radiation-measuring instruments, pyranometers typically have flat receiver surfaces. Such receivers should satisfy the cosine law, i.e., their output signal must decrease according to the cosine of the angle of incidence. Edge shading and reflection angles caused by the construction and material can result in significant deviations from the cosine law, which must be compensated for by the manufacturer using suitable corrective measures. If a device with correct cosine correction is available, its display is a good measure of the total energy irradiated on a horizontal surface in the spectral range between approx. 300 and 2000 nm (depending on the dome material)—regardless of the spectral composition. Such devices also allow for quickly determining inclined surfaces' radiation levels by aligning them parallel to the slope.

At this point, however, it should be pointed out that although a calorimetric measurement of the type just described can be used to make statements about the energy irradiated (e.g., during the day), no overly far-reaching conclusions can be drawn from it regarding the effectiveness of this radiation regarding photosynthesis. For one thing, plants are generally not exclusively horizontally oriented radiation receivers, and for another, the photosynthetically active radiation (PAR) always represents only a certain proportion of the total radiation, which varies with the spectral composition. This objection is particularly important when (e.g., for financial reasons) light measuring devices that exhibit pronounced spectral sensitivity differences have to be used. This objection applies first and foremost to the well-known lux meter, whose sensitivity characteristics are adapted to the brightness impression of the human eye. Such devices are, at best, use-

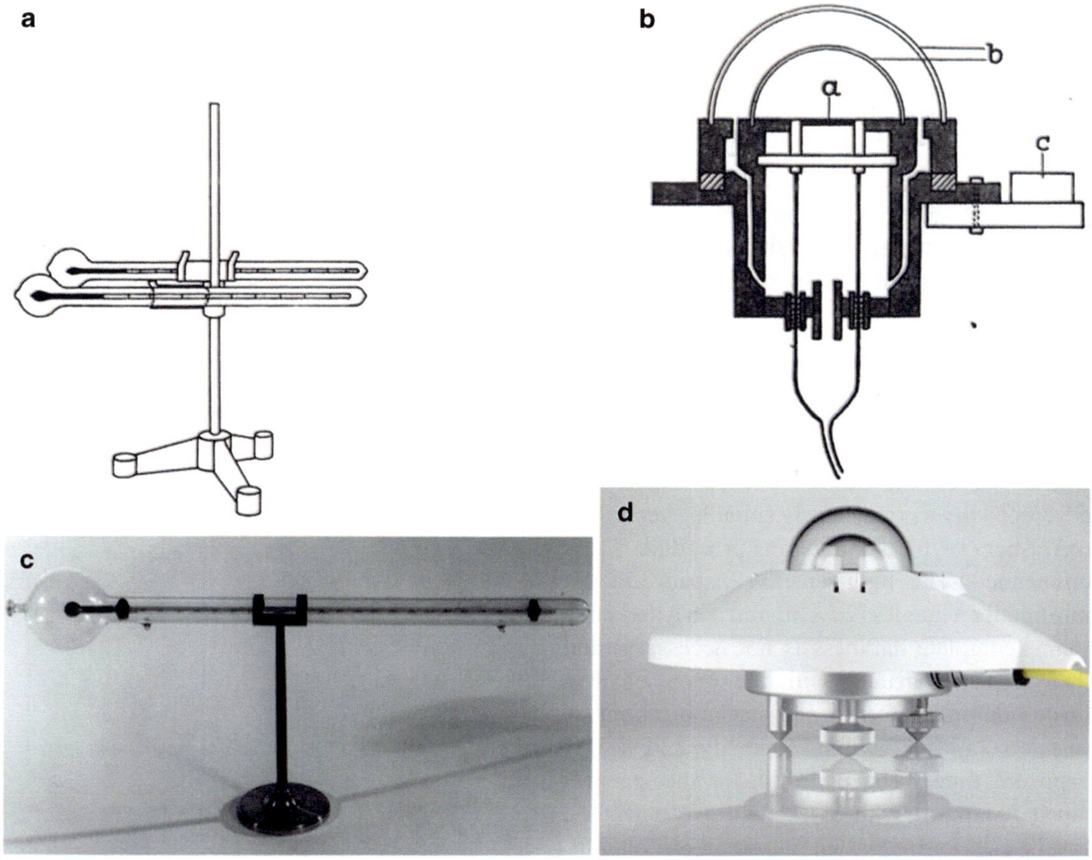

Fig. 3.4 (**a**) Radiation measuring devices. See the text for more detailed explanations. Radiation thermometer. (**b**) Electrical black-and-white pyranometer (solarimeter) according to Moll–Gorczynski. Abbreviations: a measuring sensor, b glass hemispheres, c spirit level. (**c**) Hicks solar radiation thermometer. (Picture credit: John William Christopher Draper and James Christopher Draper, CC0 1.0 UNIVERSAL). (**d**) The Kipp and Zonen SMP22 is a spectrally flat Class A pyranometer that combines an intelligent interface with quartz domes and sensor technology. The SMP22 has a wide spectral range, excellent directional response, and reduced thermal offsets. The directional error is below 5 W/m². (Picture credit: kippzonen.com)

ful for orientation measurements. The quality and price of PAR sensors have recently improved tremendously, allowing them to be used more frequently (Comella et al. 2022; Scordo et al. 2024).

Another difficulty with light measurements concerns the averaging of rapidly changing light conditions. Although there are electronic options for integration, they are costly and can usually only be used in conjunction with a particular type of device. Most pyranometers have time constants of 10 s to 1 min; this means that the device only reaches 98% of the final display value after this time. This apparent inertia already ensures a certain smoothing of the light curve. Nevertheless, there will often be situations where it is no longer possible to calculate a reasonable average value even if the reading intervals are considerably shortened. Such strongly fluctuating measured variables are not all too rare in ecological studies in the field (air temperature, wind speed); however, this fact is often masked by the inertia of the measuring instruments. This phenomenon can be observed clearly if an electrical black-and-white pyranometer replaces the "radiation thermometer" already mentioned (Fig. 3.4), which reacts much more slowly to radiation changes than most pyranometers; its relatively large inertia makes it unsuitable for detecting rapid radiation changes over time, but on the other hand, this design-related property results in a certain degree of temporal integration and thus a significant smoothing of the light curve.

3.3 Integrated Light Measurement with Light-Sensitive Paper of Film

Light-sensitive papers or films offer a simple, in most cases sufficiently accurate, and inexpensive way of measuring light. Light-sensitive blueprint paper, also called diazo photosensitive paper, which can be exposed as platelet stacks and developed dry, is particularly suitable. According to Wasner (1976), the following procedure is recommended: The light-sensitive papers are cut into small rectangles (20 × 30 mm^2) in a darkened room (complete darkness is not necessary) and then stapled together (with the photosensitive layer side up) to form small stacks of about 14 platelets (at full outdoor intensity and 1 day exposure time). The underside of the stack is taped lightproof with black adhesive tape. The finished stacks are twisted into small rolls and fitted into 5 mL rolling rim vials with the layer side facing outwards. After fitting the snap-on lid and light-tight sealing in film canisters or similar, the light-sensitive cartridges are ready for exposure on the test surface (Fig. 3.5).

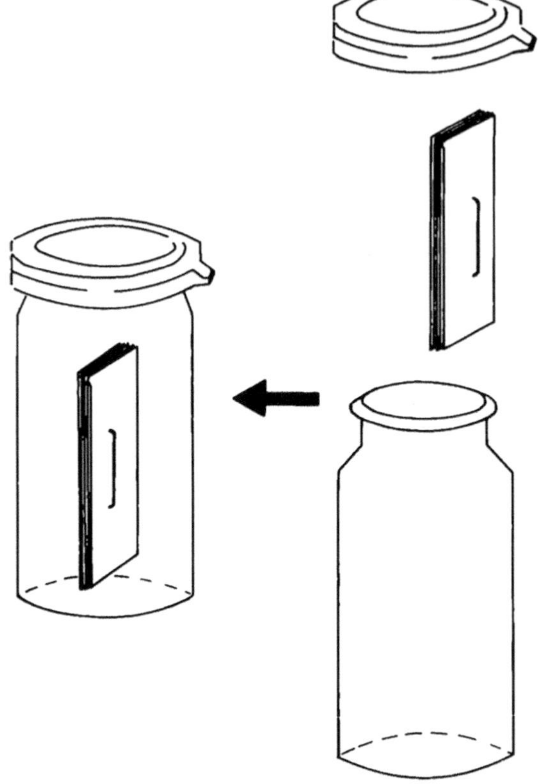

Fig. 3.5 Preparation of a measurement capsule containing light-sensitive paper. (Picture credit: Glück (1981). Reproduced by permission)

In order to achieve comparable conditions, the cartridges should be exposed strictly horizontally or parallel to the slope and in the same orientation. In our experience, this is best achieved by fixing the vials to a thin wooden stick with a prepared wire loop (staple upwards) and then pushing them into the vegetation from the side.

The cartridges are collected and developed at the end of the exposure time. To do this, disassemble the stacks of platelets in a darkened room and attach the platelets (in the correct order!) to a suitable base (polystyrene foam). The paper strips are then developed by placing them into a sealable container with ammonia. It takes a maximum of 20 min and results in a deep brown coloration in unexposed papers; exposed or partially exposed papers do not color at all or only slightly. During the evaluation, all plates on which the shadow of the staples holding them together can still be seen are counted as exposed. The average number of exposed plates provides a measure of the relative light intensity at the place of exposure. The (experimentally determined) correlation between the mean illuminance (in lux) and the number of bleached sheets is presented by Wasner (1976). A similar method is based on photosensitive film (Bontempo et al. 2018).

3.4 Integrated Temperature Measurement with the Sugar Inversion Method

The sugar inversion method offers an exciting possibility for determining long-term average values of the temperature in the soil (Jones and Court 1980; Mills and Macdonald 2005; Pallmann et al. 1940). It can be designed for different time periods (1 week to several months), and due to its comparatively low costs, it can be used for large numbers of samples; however, it requires an accurate circular polarimeter for evaluation. Reactions of organic molecules are time reactions, and their speed is greatly influenced by temperature. Increasing the temperature by 10 °C accelerates such a reaction by a factor of 2–3. Therefore, the rate of a suitable reaction can be used to infer the temperature during the course of the reaction. A suitable reaction is the acid-catalyzed hydrolysis of sucrose.

$$C_{12}H_{22}O_{12} + H_2O \rightarrow C_6H_{12}O_6 + C_6H_{12}O_6$$

$$\text{Saccharose}\left(\text{Sucrose}\right) + \text{Water} \rightarrow \text{Glucose} + \text{Fructose}$$

Sucrose is optically active; it turns polarized light to the right, i.e., positive. The mixture of glucose and fructose (inverted sugar) produced during the reaction, on the other hand, rotates polarized light to the left (glucose is dextrorotatory, whereas fructose is levorotatory). Therefore, the respective rotation angle between the initial and final value measures how far the reaction has progressed at the measurement time.

The relationship between reaction rate and temperature is not linear. Therefore, the average value determined is not an arithmetic mean but a so-called exponential mean value, in which higher temperatures are considered disproportionately. Strictly speaking, such average values ("effective temperatures") may not be compared with those measured in other ways. However, the deviations are minimal if the temperature range measured during the measurement period is not too wide. This requirement is usually met when measuring soil temperatures—at least in deeper soil layers.

In practice, we carry out the following measurements: Prepare vials with approx. 20 mL sucrose solution (1 L finished solution contains 387.5 g sucrose, 11.88 g Na-citrate dihydrate (for buffering), and 3.5 mL formaldehyde (to prevent bacterial degradation)). Adjust to the desired pH value using concentrated HCl immediately before the planned application. Attach vials to a wooden stick with adhesive tape at a predetermined distance and lower them into a borehole (soil auger). At the same time, the initial angle of rotation of the solution (R_O) must be determined. The pH value of the sucrose solution to be adjusted beforehand with hydrochloric acid depends on the intended exposure time and the expected temperatures. A pH of 1.7 has been shown to be favorable for a measuring period of 1 week; at average temperatures below 10 °C, it should be lowered to pH 1.5 or 1.3. Before measuring with the polarimeter, the sucrose solution must be brought to room temperature (equilibrium adjustment of α- and β-forms).

The following formula is used to calculate the effective temperature (in Kelvin):

$$T_{\text{eff}} = \frac{-5856.6}{\text{pH} - 20.1998 - \log t + \log\left(\log\left(R_O - R_{OO}\right)\right) - \log\left(R_1 - R_{OO}\right)}$$

pH = adjusted pH value
t = exposure time in days
R_O = angle of rotation of the sugar solution at the beginning
R_t = angle of rotation after t days
R_{OO} = angle of rotation at the end of the reaction

3.5 Humidity and Evaporation

3.5.1 Humidity

When measuring relative humidity, difficulties similar to temperature measurement can arise. In contrast to absolute humidity, this variable strongly depends on temperature, so significant radiation errors can occur without appropriate precautions. This applies to the frequently encountered hair hygrometers because, unlike psychrometers, the radiation error is not mitigated by an actively generated airflow. On the other hand, hair hygrometers have the great advantage of working without any external power supply. This takes advantage of the fact that most hair shortens as it dries out. When the hairs are bundled together (or combined into "hair harps"), the resulting actuating forces are large enough to move a pointer or a writing arm. Despite certain disadvantages, hair hygrometers are still widely used as an essential part of self-writing thermo-hygrographs.

One of these disadvantages is the low long-term stability of hairs, which is particularly noticeable in the case of repeated sharp dehydration. This error can be kept within certain limits by regularly regenerating the measuring elements at high humidity (>90%) and recalibrating the display. In our context, it is of minor importance but should be a reason to regenerate thermo-hygrographs before starting practical work and to check their synchronization. Equally important, however, is the strictly radiation-protected installation of the thermo-hygrographs in the field, as this is the only way to obtain comparable daily curves. This requirement is difficult to fulfill in open meadow terrain; even weather huts do not entirely eliminate the radiation error.

Actively aspirated psychrometers, in which the air is moved past two temperature sensors by a spring mechanism or electric motor and thus constantly exchanged, are much less sensitive to the effects of radiation. One of the two temperature sensors is kept moist by a water-soaked coating and shows a certain under-temperature when air passes by due to evaporative cooling. Above a certain wind speed (approx. 1.2 m/s), the size of this under-temperature depends only on the relative humidity of the air drawn past. With the psychrometric difference and the temperature of the dry thermometer (air temperature), the relative humidity can be calculated or taken from special psychrometer charts. Figure 3.6 shows the classic ASSMANN psychrometer and the portable Bacharach sling psychrometer with two conventional thermometers (Fig. 3.6). There are now direct-reading digital devices with electrical temperature sensors. They respond quickly, are accurate, and are suitable for continuous measurements. However, they also have the potential disadvantage of all aspirated psychrometers: their powerful airflow can disturb the air stratification of the crop and falsify the measured value. The measurement with psychrometers is, therefore, not free of repercussions. Although this objection applies in principle to almost all measuring sensors, the influence can be neglected in most cases—but not necessarily when using psychrometers.

3.5.2 Evaporation

Although wind speed and relative humidity are not directly essential to the plant, they determine the extent of water loss when the stomata are open, influencing the point at which stomatal regulation begins and the amount of produced substances. This applies above all to periods of difficult water supply, in which transpiration and gas exchange are already clearly limited by the beginning closure of the stomata. A good measure of the actual or potential water losses of a plant stand is provided by the amount of evaporation, which, in contrast to the actual transpiration of the plant, is relatively easy to measure, e.g., with evaporation scales, ceramic plates (evaporation meter according to CZERATZKI), or water-soaked paper disks (PICHE evaporimeter, Fig. 3.7).

In our context, the latter method is exciting because PICHE evaporimeters are very simple, inexpensive (if necessary, you can make them yourself), and easy to install. Therefore, they are suitable for recording profiles and transects or for simultaneously measuring several sample points. However, it should be noted that they (like all evaporimeters) must be exposed to radiation. The evaporating paper sheets should be pierced in the middle with a needle to allow air to penetrate the water-filled measuring tube.

3.6 pH Measurement of Soil

The pH value of the soil plays a decisive role in the processes in the plant's root zone. As a directly acting site factor, it is of no great importance for most plants (within certain limits). However, it influences the soil structure, nutrient balance, soil life, and other soil factors and,

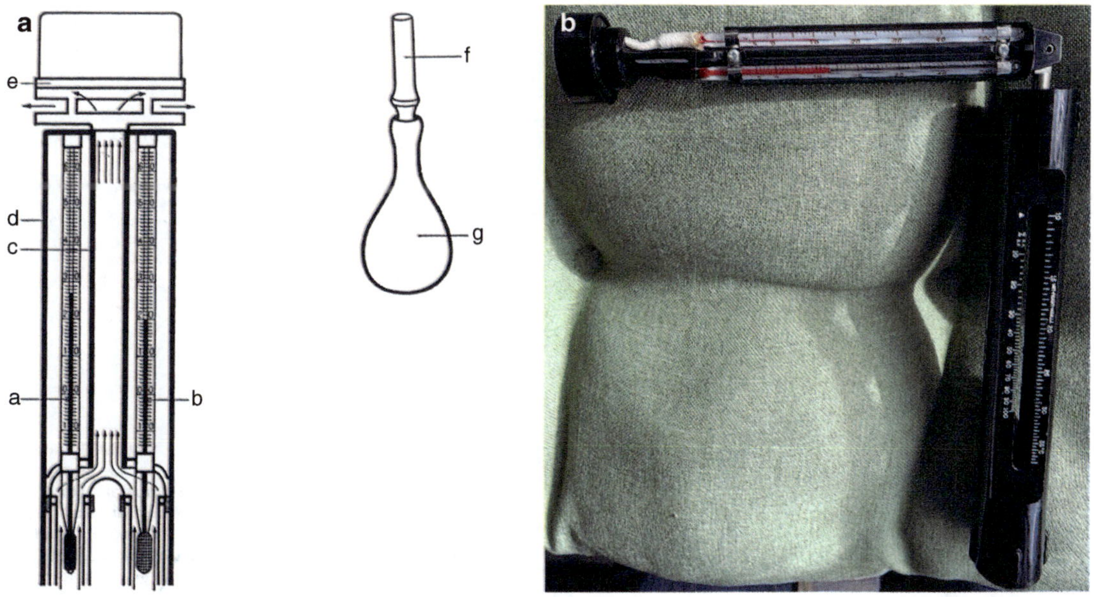

Fig. 3.6 (**a**) Aspiration psychrometer according to ASSMANN. a—dry thermometer, b—wet thermometer, c—inner sheath tube, d—outer sheath tube, e—head section with aspirator, f—humidification tube with g—rubber bulb. (**b**) A Bacharach sling psychrometer. The horizontal top part contains the wet bulb thermometer (with a wick for wetting) on top and the dry bulb thermometer below. The hand holds the tube that serves as a handle to spin it around and as a sheath for the instrument. It also aligns the two temperatures and reads the relative humidity

therefore, represents one of the critical factors of the edaphic situation (see also Scheffer & Schachtschabel 1989).

The cation exchangers of the soil (mainly clay minerals and organic matter) are more or less firmly occupied by exchangeable protons, Al, Fe, and Mn ions. They can be exchanged by alkali or alkaline earth ions of the surrounding soil solution, whereby hydrogen ions are released directly or indirectly. A measure of the activity of the released hydrogen ions is the pH value, which can be measured relatively easily and accurately in a soil suspension using color indicators or glass electrodes (accuracy with glass electrodes approx. 0.05 pH).

Most soils are too dry for a direct pH measurement with a glass electrode. Therefore, suspending the soil samples in excess distilled water or a dilute salt solution ($CaCl_2$ or KCl) is usual. While a 0.01 M $CaCl_2$ suspension solution (occasionally also 0.1 M KCl or 1 M KCl) is generally used for agricultural purposes, an aqueous suspension has often proved more favorable for ecological studies.

As a rule, it is not advisable to push the accuracy of the pH measurement to the equipment's very limits. The pH values of the soil fluctuate quite enormously in space and time (up to one pH unit), and it makes more sense to invest additional effort in repeat or parallel measurements.

The seasonal variation of the pH value is closely correlated with the water content of the soil. It primarily reflects the ion content of the soil water, which causes an increasing exchange of protons from the cation exchangers of the soil with increasing ion concentration and, thus, a lowering of the pH value. This effect can be easily observed by comparing the values measured in an aqueous suspension and a diluted salt solution (e.g., $CaCl_2$). The soil samples suspended in a salt solution are generally approx. 0.5 pH units below the values in an aqueous sample and thus represent the condition of very dry soil with a moderately concentrated external solution. In

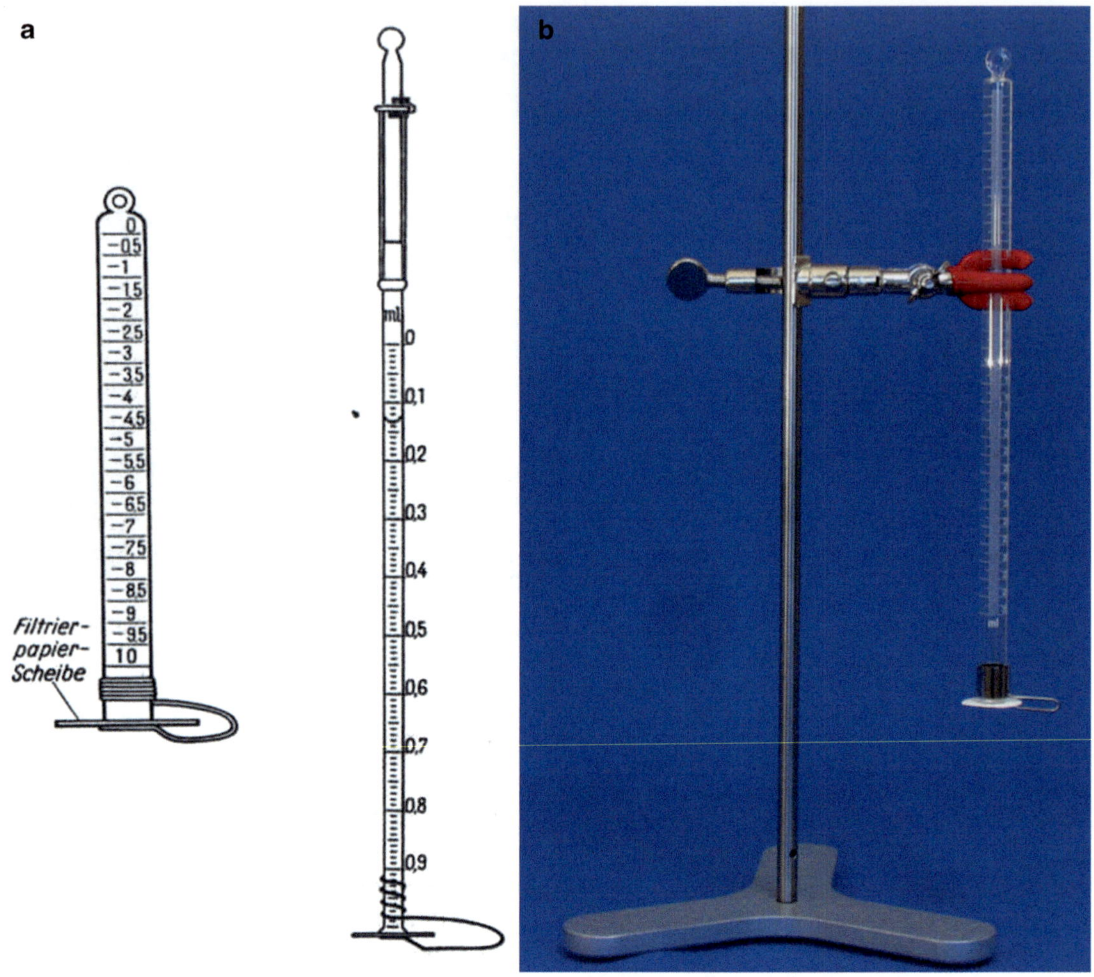

Fig. 3.7 Left: (a) Evaporimeter according to PICHE, (b) short-term PICHE evaporimeter for measuring evaporation. (c) Evaporimeter according to PICHE: glass measuring tube with etched graduation, porous paper disc Ø30 mm serving as measuring element and disc support. Measuring range: 0...25 mm, evaporation height graduation of measuring tube: 0...30 mL, dimensions of measuring tube: Ø14 mm × 310 mm length. (Picture credit: Meteorologische Instrumente K, rfuess-mueller.de) See text for further details

contrast, the substantial dilution of an aqueous suspension tends to reflect the conditions of water-saturated soil.

Small animals and earthworms usually tolerate a range of pH values. Earthworms can partially neutralize the carbonic acid produced during respiration in the soil. Earthworms have calcareous glands in their gullet. Specialized oribatids can be found on the surface in highly acidic soils, such as in bogs. Certain species of ground beetles have been found in strongly alkaline soils (Kühnelt 1950).

3.6.1 Practical Procedure

For each soil sample, fill three screw- or rolled-rim vials (25 mL) about half full with fresh soil and then fill them with approximately 2.5 times the weight or volume of aqua dest. (1), 0.01 M $CaCl_2$ (2), and 0.1 M KCl (3). Close the vials tightly and shake them for approximately 15 min (approx. 2 h if the soil has dried out), then measure by immersing the glass electrode in the suspension. If the samples are allowed to stand for some time, the supernatant and sediment will separate. In this case, it must be stated in which of the two phases the measurement was made, as the pH value of the sediment tends to be a few tenths of a pH unit lower than that of the supernatant (suspension effect). Furthermore, the soil depth, the suspension medium, and the electrode type must be specified in the protocol. Before each measurement and after completion of the measurement series, the calibration of the pH electrode must be checked with two calibration buffers (pH 4 and pH 7). Please refer to the accompanying leaflets for information on adequately handling glass electrodes.

3.7 Weather Stations

Weather stations record various environmental factors, such as temperature (in air, water, and soil), humidity, air pressure, wind speed and direction, precipitation, solar radiation, UV index, and air quality. Typical providers are Davis Instruments (davisnet.com), Ambient Weather (https://ambientweather.com/), Ecowitt (ecowitt.com) (Fig. 3.8), and other Fine Offset clones (e.g., DNT, Froggit, GARNI, Huygens, Misol, PanTech, ProTech, Steinberg, Tycon, Waldbeck, Watson) and AcuRite Weather Stations (acurite.com).

These weather stations upload automatically to several weather networks (requires internet connection), such as Weather Underground (wunderground.com), Weathercloud https://weathercloud.net/en, Met Office Weather Observation Website (wow.metoffice.gov.uk), AWEKAS (awekas.at), CWOP (Citizen Weather Observer Program, wxqa.com), Ambient Weather Network (ambientweather.net) (Fig. 3.9).

3.8 Smartphone Apps

Millions of people worldwide now use smartphones and apps (application software), constituting a capable combination of sensors, information transfer, and computing power with immense potential for ecological research. Usage fields are data collection, evaluation, and distribution (Fig. 3.10). Especially promising is open-source software, which nonexperts can customize to their needs. Smartphones and their apps might

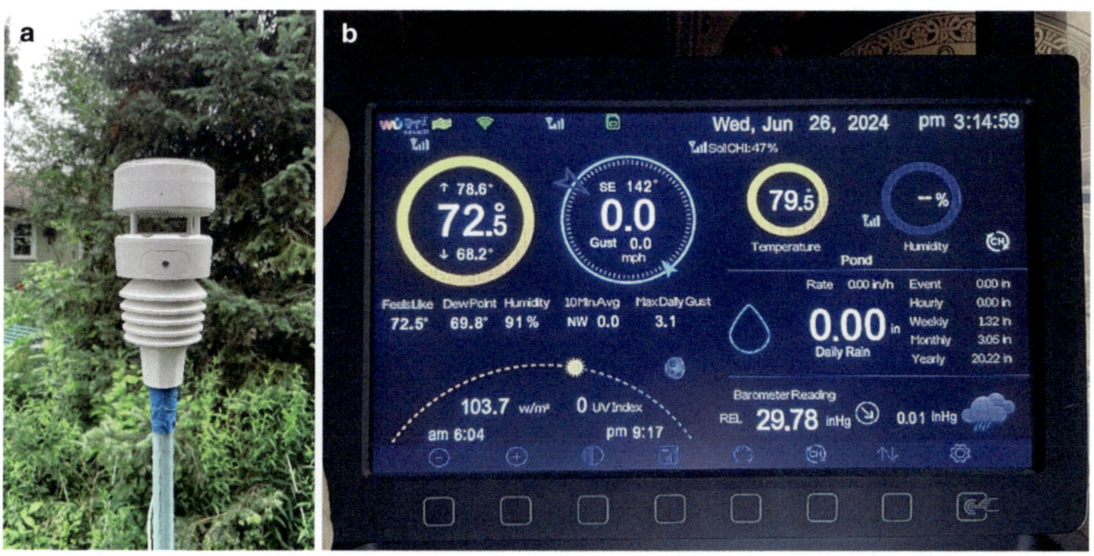

Fig. 3.8 The Ecowitt WS90 (Wittboy) weather station (**a**) with the HP2560_C display console (**b**). Also attached are several external sensors, e.g., for soil and water temperature

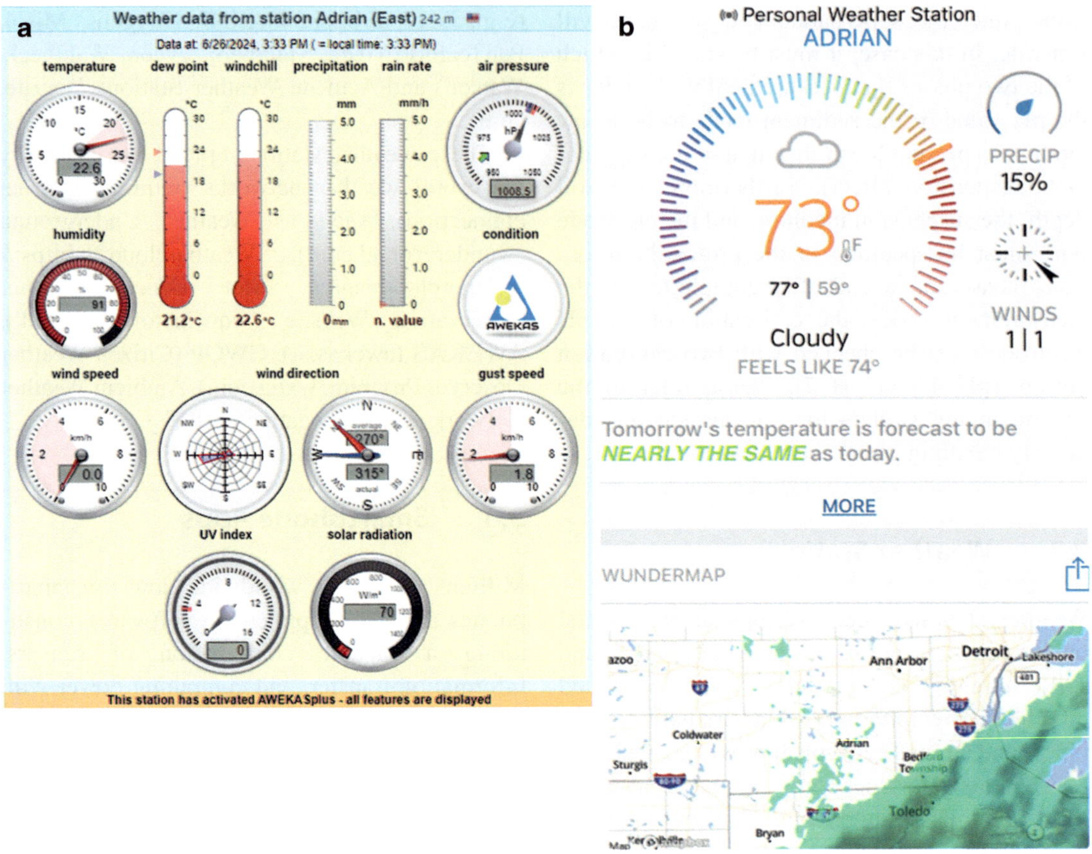

Fig. 3.9 Uploads from the above-shown weather station to two weather networks: (**a**) AWEKAS (https://www.awekas.at/wp/?lang=en) and (**b**) Weather Underground (wunderground.com)

be able to replace many traditional sensors, calculators, and data storage devices in ecological research and revolutionize high-throughput collection, analysis, and storage of complex ecological information (ATREE Young Researchers Network 2017; Teacher et al. 2013).

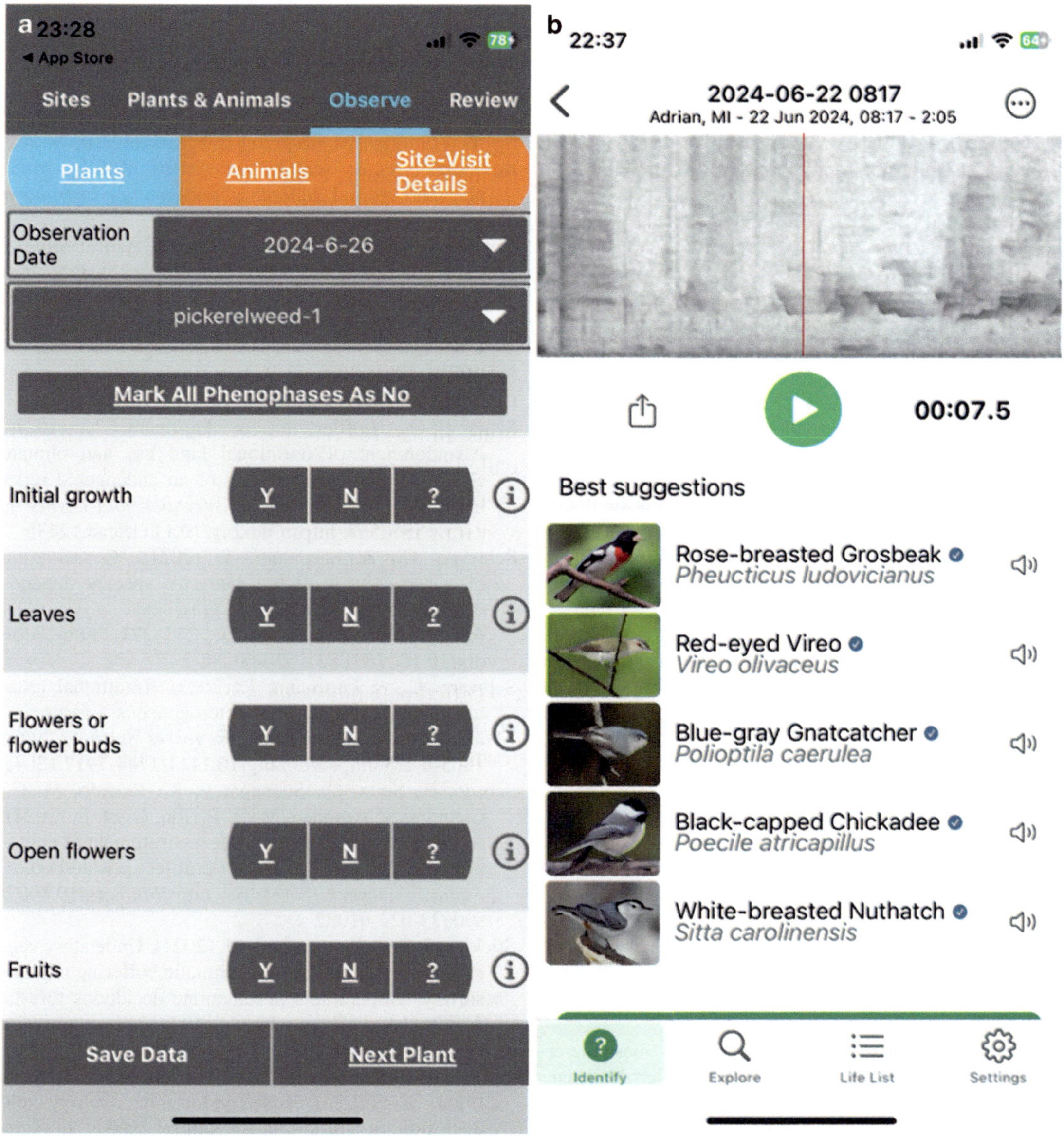

Fig. 3.10 Two iPhone apps that can be used for ecological research and citizen science. (**a**) Nature's Notebook from the USA National Phenology Network, usanpn.org, allows users to enter phenological data of followed plants and animals. (**b**) Cornell Laboratory of Ornithology's Merlin, merlin.allaboutbirds.org, allows users to identify birds by features, pictures, and audio recordings. Pictures and audio recordings are auto-identified using a powerful artificial intelligence engine that improves through every use

References

ATREE Young Researchers Network. (2017, August 9). Use Your Smartphone Smartly: Apps in Ecological Research. *ATREE Young Researchers Network*. https://ayrnstories.wordpress.com/2017/08/09/apps-in-ecological-research/. Accessed 25 June 2024

Beddows, P. A., & Mallon, E. K. (2018). Cave Pearl Data Logger: A Flexible Arduino-Based Logging Platform for Long-Term Monitoring in Harsh Environments. *Sensors*, *18*(2), 530. https://doi.org/10.3390/s18020530

Bingham M.J. & Long S.P. 1988: *Equipment for Crop and Environmental physiology*: Specifications, sources, costs. – Colchester, 43 p.

Bontempo, A., Smith, J., Brandt, M., Peterlunger, E., & Stoll, M. (2018). Evaluation of photosensitive films for light measurements in the fruiting zone of grapevine canopies. *VITIS - Journal of Grapevine Research*, *57*(4), 159–165. https://doi.org/10.5073/vitis.2018.57.159-165

Börger, L., Bijleveld, A. I., Fayet, A. L., Machovsky-Capuska, G. E., Patrick, S. C., Street, G. M., & Vander Wal, E. (2020). Biologging Special Feature. *Journal of Animal Ecology*, *89*(1), 6–15. https://doi.org/10.1111/1365-2656.13163

Cannon, J., Warren, L., Ohlson, G., Hiers, J., Shrestha, M., Mitra, C., et al. (2022). Applications of low-cost environmental monitoring systems for fine-scale abiotic measurements in forest ecology. *Agricultural and Forest Meteorology*, *321*, 108973. https://doi.org/10.1016/j.agrformet.2022.108973

Comella, L. M., Goldschmidtboeing, F., Kluppel, J., Hager, E., & Woias, P. (2022). An innovative sensor for the simultaneous measurement of Photosynthetic Active Radiation (PAR) and Leaf Area Index (LAI). In *2022 IEEE Sensors* (pp. 1–4). Presented at the 2022 IEEE Sensors. https://doi.org/10.1109/SENSORS52175.2022.9967119

Gandra, M., Seabra, R., & Lima, F. P. (2015). A low-cost, versatile data logging system for ecological applications. *Limnology and Oceanography: Methods*, *13*(3), e10012. https://doi.org/10.1002/lom3.10012

Gemini Data Loggers. (2024). Data loggers for ecological applications | Tinytag. *geminidataloggers.com*. https://www.geminidataloggers.com/info/outdoor-environmental-monitoring. Accessed 25 June 2024

Glück, E. (1981). Beleuchtungsstärke der Neststandorte bei einigen Finken (Fringillidae) und beim Grauschnäpper Muscicapa striata. *Anzeiger der ornithologischen Gesellschaft in Bayern*, *20*, 35–44.

Jones, R. J. A., & Court, M. N. (1980). The Measurement of Mean Temperatures in Plant and Soil Studies by the Sucrose Inversion Method. *Plant and Soil*, *54*(1), 15–31.

Kühnelt W. 1950: *Bodenbiologie*. Mit besonderer Berücksichtigung der Tierwelt. – Verlag Herold, Wien. 368 p.

Microtronics Engineering GmbH. (2024). myDatalogEASY IoT. *Microtronics*. https://microtronics.com/en/produkt/mydatalogeasy-iot/. Accessed 25 June 2024

Mills, S. E., & Macdonald, S. E. (2005). Factors Influencing Bryophyte Assemblage at Different Scales in the Western Canadian Boreal Forest. *The Bryologist*, *108*(1), 86–100.

Morrison, G. (2021). PeerJ Award Winners: 7th Bio-Logging Symposium | PeerJ Blog. https://peerj.com/blog/post/115284885148/peerj-award-winners-7th-bio-logging-symposium/. Accessed 14 August 2024

Onset Computer Corporation. (2024). Data Loggers & Sensor Networks for Ecological Monitoring. https://www.onsetcomp.com/environmental-research-and-protection/ecology. Accessed 25 June 2024

Pallmann, H., Eichenberger, E., & Hasler, A. (1940). Eine neue Methode der Temperaturmessung bei ökologischen oder bodenkundlichen Untersuchungen. *Berichte der Schweizerischen Botanischen Gesellschaft = Bulletin de la Société Botanique Suisse*, *50*, 337. https://doi.org/10.5169/seals-34260

Ruitenbeek, K., & Wely, L. van. (2016). Moll's thermopiles and beyond. *Bulletin of the Scientific Instrument Society*, *130*, 11–16.

Sanchez-Lorenzo, A., Calbó, J., Wild, M., Azorin-Molina, C., & Sanchez-Romero, A. (2013). New insights into the history of the Campbell-Stokes sunshine recorder. *Weather*, *68*(12), 327–331. https://doi.org/10.1002/wea.2130

Scheffer F. & Schachtschabel P. 1989: *Textbook of Soil Science*. 12.ed. - Enke-Verl., Stutgart, 442 p.

Scherer, G., Löffler, F., & Fartmann, T. (2021). Abandonment of traditional land use and climate change threaten the survival of an endangered relict butterfly species. *Insect Conservation and Diversity*, *14*(5), 556–567. https://doi.org/10.1111/icad.12485

Schwarz, C., & Fartmann, T. (2021). Conservation of a strongly declining butterfly species depends on traditionally managed grasslands. *Journal of Insect Conservation*, *25*(2), 255–271. https://doi.org/10.1007/s10841-020-00288-2

Schwarz, C., & Fartmann, T. (2022). Traditional grazing management creates heterogeneous swards and fosters grasshopper densities. *Insect Science*, *29*(6), 1805–1818. https://doi.org/10.1111/1744-7917.13041

Scordo, F., Seitz, C., Suenaga, E. K., Piccolo, M. C., Chandra, S., Amodeo, M., & Perillo, G. M. E. (2024). An inexpensive method for the measurement of photosynthetically active radiation profiles in waterbodies. *Aquatic Sciences*, *86*(3), 68. https://doi.org/10.1007/s00027-024-01082-x

Stickley, S. F., & Fraterrigo, J. M. (2021). Understory vegetation contributes to microclimatic buffering of near-surface temperatures in temperate deciduous forests. *Landscape Ecology*, *36*(4), 1197–1213. https://doi.org/10.1007/s10980-021-01195-w

Teacher, A. G. F., Griffiths, D. J., Hodgson, D. J., & Inger, R. (2013). Smartphones in ecology and evolution: a guide for the apprehensive. *Ecology and Evolution*, *3*(16), 5268–5278. https://doi.org/10.1002/ece3.888

Wassmer, T., Murray, D., Boutin, S., Fahlman, A., & Jensen, F. H. (Eds.). (2020). *Ecology and behaviour of free-ranging animals studied by advanced data-logging and tracking techniques*. Lausanne: Frontiers Media SA. https://doi.org/10.3389/978-2-88963-792-8

Wasner U. 1976: Eine Methode zur Mikroklimamessung im Freiland. Eichtabellen zur integrierten Lichtmengenmessung nach FRIEND. – *Zool. Jb. Syst.* 103, 355–360.

Whitford, M., & Klimley, A. P. (2019). An overview of behavioral, physiological, and environmental sensors used in animal biotelemetry and biologging studies. *Animal Biotelemetry*, *7*(1), 26. https://doi.org/10.1186/s40317-019-0189-z

4 Mapping Landscape and Locating Individuals

Abstract

A description of the vegetation is essential to characterize study areas. The vegetation can be described floristically by the composition of the plant community. However, the structural composition of the vegetation is also essential for the occurrence of animals. Various methods can be used to describe the structure, enabling a quantitative comparison of study sites. Locations are specified using the global positioning system (GPS). Recently, unmanned aerial vehicles (UAVs) have been increasingly used as conservation drones to survey wildlife in an area and for the spatial and phenological recording of the vegetation cover. The locations and movements of tagged animals can be tracked using radio telemetry and other tracking methods. Various applications are described in detail.

4.1 Description of Vegetation Types

In order to work in a specific area, this area must be categorized geographically and in terms of the prevailing vegetation. This is, first and foremost, a question of scale. Global maps and descriptions of the world's ecoregions are available for terrestrial, limnic, and marine areas (The Nature Conservancy 2023). Geographically, one can describe vegetation types from satellite images (e.g., Landsat MSS) to aerial photographs (scale 1:5000 to 1:25,000) and topographic maps (usually 1:25,000) to terrestrial habitat surveys (e.g., 100 × 100 m). Faber-Langendoen et al. (2014) provide an overview of methods for describing and classifying vegetation types. The authors also developed a new vegetation description called the EcoVeg Approach. In addition to the floristic components, criteria such as growth form and structure should also be considered. EcoVeg can be applied to global, regional, and local vegetation classification.

In the following, we present a typical floristic characterization of a site's vegetation. Floristic methods take into account the ecological relationships and plant associations like the vegetation survey, according to the *Braun-Blanquet method*, also called a *relevé* (Braun-Blanquet et al. 1932; Moore 1955; Westhof and van der Maarel 1978; Minnesota Department of Natural Resources 2007; Wikum and Shanholtzer 1978). A Braun-Blanquet record consists of a complete list of species and their estimated frequencies/coverage of the surveyed area following a specific and intuitive scale (Table 4.1). This list is supplemented by topographical, geological, pedological, and historical information. With some practice and experience, this method provides results of sufficient accuracy in a moderate amount of time. It has become widely accepted in Central Europe (e.g., Mueller-Dombois and Ellenberg 1974; Wilmanns 1993; Kent and Coker 1992; Kratochwil and Schwabe 2001) but also,

M. Mühlenberg, T. Waßmer, *Field Ecology*, https://doi.org/10.1007/978-3-031-87547-2_4

Table 4.1 The Braun-Blanquet abundance-cover scale

Level	Abundance	Coverage
r	1 specimen	
+	2–5 specimen	
1	6–50 specimen	
2m	>50 specimen	
2a	Numbers irrelevant	Cover 5–15%
2b	Numbers irrelevant	Cover 16–25%
3	Numbers irrelevant	Cover 26–50%
4	Numbers irrelevant	Cover 51–75%
5	Numbers irrelevant	Cover 76–100%

e.g., in Southern and Eastern Europe (Filis et al. n.d.; Penksza et al. 2024; Tavilla et al. 2023), Africa (Aardt et al. 2020; Daemane et al. 2024), North America (Minnesota Department of Natural Resources 2007; Peinado et al. 2008, 2014; Talbot et al. 2005), South America (Arbeláez and Duivenvoorden 2004; Zeballos et al. 2023)), the Middle East (El-Sheikh et al. 2017), Central Asia (Gafurov et al. 2024) and the Arctic (Walker et al. 2018).

A problematic and frequently discussed problem is the correct choice of sample area. It should be uniform, i.e., not contain any disturbances, marginal structures, or small areas that differ in location, and it should have a more or less even distribution of the plant species involved. These preconditions do not allow for a random selection of the plot area and, therefore, need to be selected by the surveyor, which some scientists consider a biased sampling method (Lájer 2007; Lepš and Šmilauer 2007). However, a study comparing the Braun-Blanquet method to randomized sampling methods found comparably accurate results when analyzing simulated plant cover data (Damgaard 2014). The method based on the dominant vegetation is more suitable for large, more uniform vegetation units, such as those found in the boreal zone (Aleksandrova 1969; Walter and Breckle 1991). In this Scandinavian and Russian method, the vegetation is classified according to its dominant species, regardless of whether these species also occur in other vegetation units.

In contrast to the Central European classification, species that contribute only a tiny proportion of the vegetation are not considered (Dulamsuren 2004). The Braun-Blanquet method is too small-scale orientated in these regions and requires extensive recording material (Dierschke 1994). Dulamsuren (2004) compared both methods for northern Mongolia. She also showed that the plant-sociological relationship between the ground vegetation and the tree layer, which is constant in Central Europe (Wilmanns 1993), is disturbed by the frequent occurrence of fire.

4.1.1 How Large Should the Survey Area Be?

On the one hand, it should contain as many species of the plant community in question as possible; on the other hand, its minimum size should not be set too high because otherwise, it is very easy to come into conflict with the requirements mentioned above for uniformity and quasi-homogeneity of the plant distribution. Experience values that have proven themselves in practice are:

- Forests 200–500 m^2.
- Meadows and dry grassland 50–100 m^2.
- Arable weed communities 25–100 m^2.
- Heaths 10–25 m^2.
- Spring meadows 1–5 m^2.

Based on 41,174 relevés with an indication of plot size, published in six major European journals focusing on phytosociology from 1970 to 2000 and an additional data set of 27,365 relevés from the Czech National Phytosociological Database, Chytrý and Otýpková (2003) recommend four plot sizes as possible standards for the European vegetation:

- Aquatic vegetation and low-grown herbaceous vegetation: 4 m^2.
- For most grassland, heathland, and other herbaceous or low-scrub vegetation types or low scrub: 16 m^2.
- For scrubland: 50 m^2.
- For woodlands: 200 m^2.

The shape of the survey area is irrelevant. In the case of joint surveys, it is advisable to mark out the corners of the area so that each participant can see the boundaries at all times and avoid entering the stand prematurely.

Before a species list is drawn up, general information on the sample area should be provided, such as:

1. Identification number.
2. Date (and time).
3. Preliminary community name (e.g., meadow, dry grassland, deciduous forest).
4. Location (exact name of the site, geographic coordinates, from GPS and map).
5. Altitude above sea level (from altimeter or topographic map).
6. Terrain shape (slope, ridge, valley), exposure (compass direction of downhill slope), inclination (slope, determined with inclinometer or estimated).
7. Geological subsoil (particular geological map or own exploration).
8. Cultivation/usage/history (e.g., three-cut meadow).
9. Notable observations (e.g., not mowed, old stalks, drainage ditch).
10. Size of the sample area (and as far as can be determined of the entire stand).
11. Vertical stratification, height, and cover of the layers (tree layers, shrub layers, herb layers, moss layer).

The ecological information provided by the presence or absence of a plant can be made more precise by specifying the quantity of the individual species. Braun-Blanquet uses a scale that considers the number of individuals (abundance) and the cover percentage of the total area occupied by each species (dominance). Both variables are combined in a seven-level class classification. Cover (or degree of cover) refers to the area covered when projected vertically onto the ground surface. In contrast, the number of individuals refers to the number of shoots or individual plants. Table 4.1 shows the slightly modified nine-level classification that is most commonly used.

As the vegetation descriptions above are qualitative—or at most semiquantitative—they do not allow many methods of statistical evaluations (Aubry et al. 2024; Lájer 2007). To measure vegetation structures more quantitatively, we need to consider different approaches. Refer to the textbooks, e.g., by Avery and Burkhart (2015), Kirby and Hall (2019), and research articles, e.g., Atkins et al. (2023), for methods to describe forests and to determine tree volumes, area distribution, and forest inventories. We will present a few simple field methods without statistics in the following.

4.2 Structural Parameters of Terrestrial Habitats

4.2.1 PCQ-Method for Woody Plants

The Point-Centered-Quarter-Method (PCQ-Method) (Cottam and Curtis 1956; Stearns 1949; Mueller-Dombois and Ellenberg 1974) is suitable for characterizing forest sites and various microhabitats on the forest floor.

A list of variables for forest floor structures is compiled in Table 4.2, along with other vegetation structure variables. Characterizing a study area of 60 m × 100 m by 40 samples for the crown profile measurement at ten measurement points according to the PCQ method takes about 2–5 h when recording all variables listed in Table 4.2.

For example, data collection using the square method is carried out with a cross that can be assembled from 1 m long rods. At the respective measurement point for the *PCQ-Method*, a quadrat is marked with the cross (see Fig. 4.1).

Numerical values of habitat variables (No. 12 in Table 4.2) or the proportional area (Variable No. 15, 16) are estimated within this square sample. The habitat components of interest are listed in advance according to the terrain and prepared in tabular form for the protocol.

4.2.2 Decay Stages of Lying Tree Trunks and Fire Marks in Tree Trunks

Mäkinen et al. (2006) provide a classification scheme for the decay stages (Fig. 4.2):

The decomposition of deadwood depends on both the tree species and exogenous factors. A long-term study (10 years) on the mass loss of tree trunks of 13 tree species under different con-

Table 4.2 List of habitat variables and their recording methods, helpful in describing the vegetation structure and other resources

S. no	Habitat variable	Measurement method and calculation	Statement value
Shrub and tree layer			
1	Average tree species number	PCQ 10–20	G
2	Tree density	PCQ 10–20	G, P
3	Horizontal tree heterogeneity	PCQ 10–20, calculation acc. ROTH	G
4	Average tree circumference	PCQ 10–20	G
5	Percent canopy closure	PROFM 40–50, 100% = vegetation in all ten height layers	G, Average
6	Foliage height diversity	PROFM 40–50, diversity index acc. SHANNON (p_i = relative proportion of vegetation amount in the *i*-th layer)	G
7	Heterogeneity of foliage	PROFM 40–50, coefficient of variation of canopy closure percentage values	G
8	Combined leaf density index	PCQ 10–20, calculated as the relative value from the variables (2), (4), (5), and (17), see text	G
Herb layer and forest floor			
9	Stump density	PCQ 10–20	G, P
10	Average stump age	PCQ 10–20, estimation according to a three-stage scale (1 = fresh, 2 = half-rotted, 3 = rotten)	G, P
11	Number of cavities at trunk and stump base	PCQ counts on PCQ trees and stumps of all cavities with at least mouse hole width	P
12	Number of cavities in 1 m radius	QM, count at PCQ points	P
13	Density of fallen wood	PCQ 10–20, calculated like tree density (2) for deadwood >7.5 cm diameter	G, P
14	Number of herbaceous plant species	QM, at PCQ, points 1 m² plant recording	G
15	Degree of coverage with herbaceous plants	QM, estimated according to phytosociological methods for 1 m² at PCQ points	G
16	Percentage of bare ground (or stone) without leaf litter	QM, estimated for 1 m² at PCQ points	P
17	Litter layer thickness	PCQ, four measurements at the endpoints of a 1 m cross; see text—Insertion of a metal rod	G, P
18	Leaf litter interstitial volume	PCQ, four measurements as in (17). Compression of the leaf litter with standard weight; see text	G, P

The statement value G refers to characterizing the entire study area, and P to the individual samples. *PCQ* "Point-Centered-Quarter-Method," *PROFM* profile measurement according to Fig. 4.4, *QM* Quadrate method, recording on a sample area. (Modified after Fründ 1983)

ditions was carried out by Edelmann et al. (2023). Organismic diversity and microbial activity have a strong influence on the decay process. Mass loss was higher in regions with warmer, drier, and nutrient-poor soil conditions. Gossner et al. (2016) showed that about 30% of the existing fauna of saproxylic beetles colonize tree trunks in Germany in the first 3 years.

For the history of tree stands, age can be determined by drilling into the tree trunks (increment core samples, Fig 4.3), which also reveals past traces of fire. The question of past fire events is particularly important in forests that go through natural fire cycles (cf. Sect. 13.7). A more reliable determination of past fires can be achieved by studying tree discs (Fig. 4.3) or partial cross-sectioning large-diameter live trees (Arno and Sneck 1977; Dietrich and Swetnam 1984; Byambasuren 2011; Bakker et al. 2019). A non-invasive method for a rough age estimate is based on the diameter of the tree (diameter at breast height, DBH, 4.5 ft/1.37 m above the ground) and a growth factor available for many North American trees (Walas 2024): Growth Factor × DBH (in in.) = Age in years (International Society of Arboriculture, ISA).

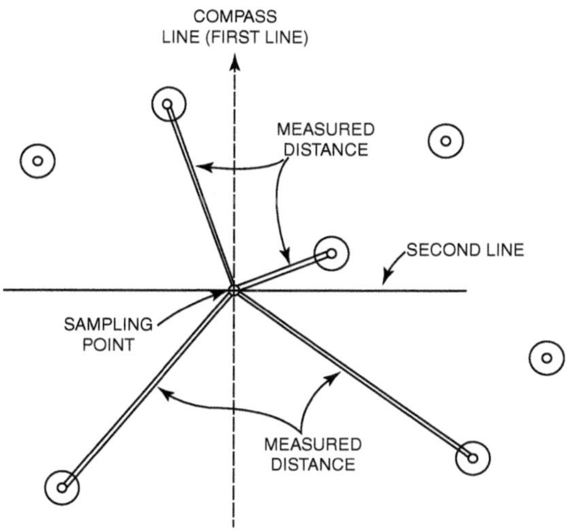

Fig. 4.1 Measurement of the density of woody plants and other resources according to the "***Point-Centered-Quarter-Method***." The cross for determining the four quadrants is placed at random points in the field. (From Mueller-Dombois and Ellenberg 1974)

4.2.3 Deadwood (Coarse Woody Debris)

Deadwood, regarded as coarse woody debris (CWD) in forests, is emphasized in nature conservation (e.g., Kirby 1993; Peterken 1996; Humphrey et al. 2005; Bunnell and Houde 2010; USDA 2020; Grove et al. 2024). A diverse fauna of invertebrates depends on it (Kirby and Drake 1993). Vertebrates benefit from standing deadwood (snags) and the availability of tree cavities (e.g., Casey et al. 2015; Anderson and Wecker 2011; Thiemann 2016; Kozma on YouTube 2020; Stowe 2024). The quantitative recording of CWD is described in Kirby (1993), Kirby et al. (1998), and Kirby and Hall (2019).

An 8-year study by Edelmann et al. (2022) showed that the diversity of deadwood-dependent beetles is richer in more extensively managed forests of Central Europe. The saproxylic beetle assemblages are more influenced by the proportion of conifers and the degree of canopy cover. Beetle specialists for broadleaf tree species were hardly affected by canopy cover. The composition of the tree species and the canopy cover are the most critical factors for the saproxylic beetle community. Management intensity was described with an index based on the proportion of locally nonnative tree species, the ratio of natural to man-made deadwood, and the ratio of woody biomass removed to woody biomass left standing (Kahl and Bauhus 2014). When recording the saproxylic beetles, it must be considered that the beetle assemblages are also distributed vertically in the forest. This is particularly true in tropical forests (Weiss et al. 2019). Beetles can be recorded in upper strata using flight-intercept traps (see Sect. 2.11).

Lying deadwood can be quantitatively estimated in line transects (Kirby and Hall 2019). On 50 m long transects, the number of CWD intersecting the transect line is counted (line-intercept sampling). At the same time, the diameter of the deadwood lying on the transect is measured. The total length of deadwood per hectare is then calculated as follows: $L = \pi \times 10^4 \times N \div 2\,l$.

The volume is calculated using the following formula:

$$V = \pi^2 \times 10^4 \div 8l \times \sum nd^2,$$

where n is the number of intersections for logs of diameter d, and l is the total length of the transect.

An overview of different sampling methods for assessing coarse woody debris (CWD) was provided by Ståhl et al. (2001). When measuring CWD, it must also be noted that the decay rate can significantly vary in different forests. Therefore, CWD's currently measured volume value cannot be easily compared between different forests. When selecting random samples, it must also be considered that, for example, many trunks fall in the same direction during storm events and that there can easily be a bias when counting along line transects. This bias can be avoided by counting from a transect line at different points in a randomly selected direction.

In seminatural forest sites, the amount of deadwood is between approx. 15 and 20 m^3/ha; in more natural forests, over 50 m^3 (Kirby and Hall 2019). In European old-growth forests, up to 200 m^3 deadwood/0.01 km^2 can be reached (MELFF 2002); in managed forests, 10 m^3/0.01 km^2 is rarely exceeded. Nature conservation in Germany requires 10% deadwood in a forest (Scherzinger 1996). In primeval forests of the temperate zone, the proportion of deadwood can reach over 30%.

Fig. 4.2 (**a**–**e**) Different decay stages of lying tree trunks are classified as 1–5. **a**–**c** are birch trunks, and **d** and **e** are larch trunks. The photo examples of the five classes are from Mongolia

4.2.4 Angle Count Sampling with Dendrometer

Several methods describe woodland (e.g., Kirby and Hall 2019). Here, we present an easy-to-perform angle count sampling method that can be used to quantitatively describe tree species composition and woodland structure over large areas. The angle count sample (Kramer and Akça 1995) is carried out from individual sample points, which are either placed in a grid (e.g., with a mesh size of 100 × 100 m) over the survey area or distributed over random points in the field (Fig. 4.4a). The sample points are determined using a compass and GPS (see Sect. 4.4).

The angle count sample is primarily used to record the basal area of the individual tree species in m^2/ha. All trees are checked at 1.3 m height (diameter at breast height, DBH), and if they are wider than the window gap of the dendrometer (Fig. 4.4b), they are counted as "1." The sum of counted trees corresponds to the basal area (in m^2) per 0.01 km^2. The probability that the image captures a tree is proportional to its diameter.

The species, diameter class, and tree type are noted for each tree recorded during the angle count sample. Parameters for the standing stock are compiled in a worksheet (Table 4.3). This also differentiates the tree types.

4.2.5 Measurement of Vertical Structure and Canopy Closure in Forests

Often, "macro features" such as canopy closure and the presence or absence of a vegetation layer on the ground are cumbersome on a micro-scale.

Fig. 4.3 (**a**) Students drill a tree trunk with a professional trunk corer used in forestry operations. (**b**) The annual rings are visible in the drill core, possibly also traces of fire. (**c**) Tree disc with visible age rings. The age of past fire marks (gray lines) can be determined by counting the annual rings

Many macro features can be collected more economically through aerial image evaluation than on-site. Vertical photography can easily quantify canopy closure (Fig. 4.5a). Fig. 4.5b demonstrates the quantification of such measurements.

To determine the canopy closure, the eyepiece (Fig. 4.5c) is held vertically upwards (the plumb line then falls parallel to the edge of the tube), and it is read in the crosshairs (f) whether vegetation is in the field of view or not.

The GRS Densitometer (Geographic Resource Solutions, Arcata, California) uses a similar but more elegant approach (Fig. 4.6). To determine an area's elemental canopy coverage percentage, equally distribute sampling points along linear transects covering the area (Fig. 4.6d, e). Level up the densitometer at each point (Fig. 4.6c) and peer through the mirror to determine if the canopy covers the point (Fig. 4.6b). After completing the transects, divide the number of points with canopy coverage by the total number of points sampled. The resulting figure is the percentage of canopy coverage for that area.

To measure the vertical structure in a forest, you can first document the conditions with photographs (Fig. 4.7a). In order to evaluate the structures quantitatively, the coverage of horizontal strips by vegetation can be measured (Fig. 4.7b).

Modern techniques open up new methods for quantifying the complex forest structure. Light detection and ranging (LIDAR) is one such laser scanning system (Dassot et al. 2011; Vaglio Laurin et al. 2020). Among other things, LIDAR can be used to measure canopy space and the connectedness of tree canopies, on which forest growth, light conditions, and biodiversity depend (e.g. Atkins et al. 2023). Ehbrecht et al. (2021) used this method to compare the structural complexity of primary forests in different biomes (from boreal to tropical). Higher structural complexity is associated with higher plant diversity.

Fig. 4.4 (**a**) Angle count sampling using a dendrometer. The observer is looking through the opening of the dendrometer (see photo right), which is held exactly 50 cm in front of the eye and turns around. A second person assists in keeping the records. (**b**) Dendrometer

Climate change impacts forest structure, but the extent of the consequences for biodiversity is not yet fully understood (Ehbrecht et al. 2021).

4.2.6 Density Estimation in Grassland Vegetation

A "vegetation stratimeter" developed by Oppermann (1989) works according to the light barrier principle (example for a simple device, Fig. 4.8). The battery-powered device takes objective samples of vegetation density at different heights in fodder and litter meadows, perennial flower beds, lean grasslands, and fallow stages. The density measurements correlate significantly with biomass values. You should expect a time requirement of 20–30 min for the measurement in ten rows with up to 6 height layers per test area and date. An electronic recording of the measured values allows for direct evaluation with the computer.

Another method to measure the vertical structure in grasslands using photography is offered by Zehm et al. (2003).

4.3 Terrain Mapping

A terrain map representing trap locations and sampling sites is necessary for almost all field trials. It also forms the prerequisite for the reproducibility of the experiments. Topographic maps at a scale of 1:25,000 serve as the starting material for terrain orientation. Aerial photos allow for more precise work and detailed enlargements of specific areas. For areas in the size range below 100 × 500 m, we recommend conducting your own surveys. For

Table 4.3 Working sheet with the variable list describing forest stand features

forest structure survey

area:	
tract no.:	
study site:	
coordinates:	
slope inclination (°):	
date:	
observer:	

tree spec	abbr.	tree spec	abbr.
		Betula fusca	BiG
birch B.p.	Bi	hawthorn	Cra
larch	Lä		
aspen	Zpa	poplar2	Pos
spruce	Fi	poplar3	Pol
fir	Ta	great sallow	SWei
Scots pine	Kie	willow	Wei
Cedar pine	Zki	rowan	Esch

BHD cm	BHD-class
<5	1
6-12,5	2
12,5-25	3
26-50	4
51-75	5
76-100	6

tree types	class
alive, only little decay	1
tree with dead parts	2
conspicuous fire scars	3
large parts decayed/hollowed out	4
wind broken	5
staying dead, snag	6

dead wood-type	class
decomposition	
starting	1
medium	2
old	3

stand of trees ("Winkelzählprobe", window gap 1, measure of basal area)

tree-No.	tree-spec	BHD-class	tree type	tree No.	tree spec	BHD-class	tree type	tree No.	tree spec	BHD-class	tree type
1				26				51			
2				27				52			
3				28				53			
4				29				54			
5				30				55			
6				31				56			
25				50				75			

stand of lying wood > 25 cm diameter (sample plot with 15 m radius)

tree-No.	tree spec	BHD-class	dead wood type	length % o. m	tree No.	tree spec	BHD-class	dead wood type	length % o. m	remarks
1					6					
2					7					
5					10					

Drafted by Dr. Hermann Hondong, Georg-August-Universität Göttingen. An example of evaluating such data from different forest types can be found in Mühlenberg et al. 2004

this, you will need ranging rods (or wooden stakes), approximately 500 m of string, a tape measure (possibly a measuring rod), a compass, and, in steeply rising terrain, an altimeter and pocket inclinometer. The iPhone app Theodolite by Craig Hunter (Hunter 2023) might also be helpful, combining a compass, two-axis inclinometer, rangefinder, GPS, map, nav calculator, and geo-overlay photo/movie camera into one app.

To measure the study area (Fig. 4.9), person A directs persons B and C, holding the tape measure, along the line of sight marked with stakes (1–2, 2–3, 3–4, 4–1). If there are fewer than three people, it is possible to mark the lines of sight with a string and then measure them separately. Suppose a terrain slope or a general height difference needs to be specified. In that case, the terrain angle α (e.g., between points 4 and 5, or 5 and 3, Fig. 4.9) is determined with a pocket inclinometer, and the distance (c) between these terrain points is measured. The elevation (a) of point 5 compared to 4 is $a = c \sin \alpha$.

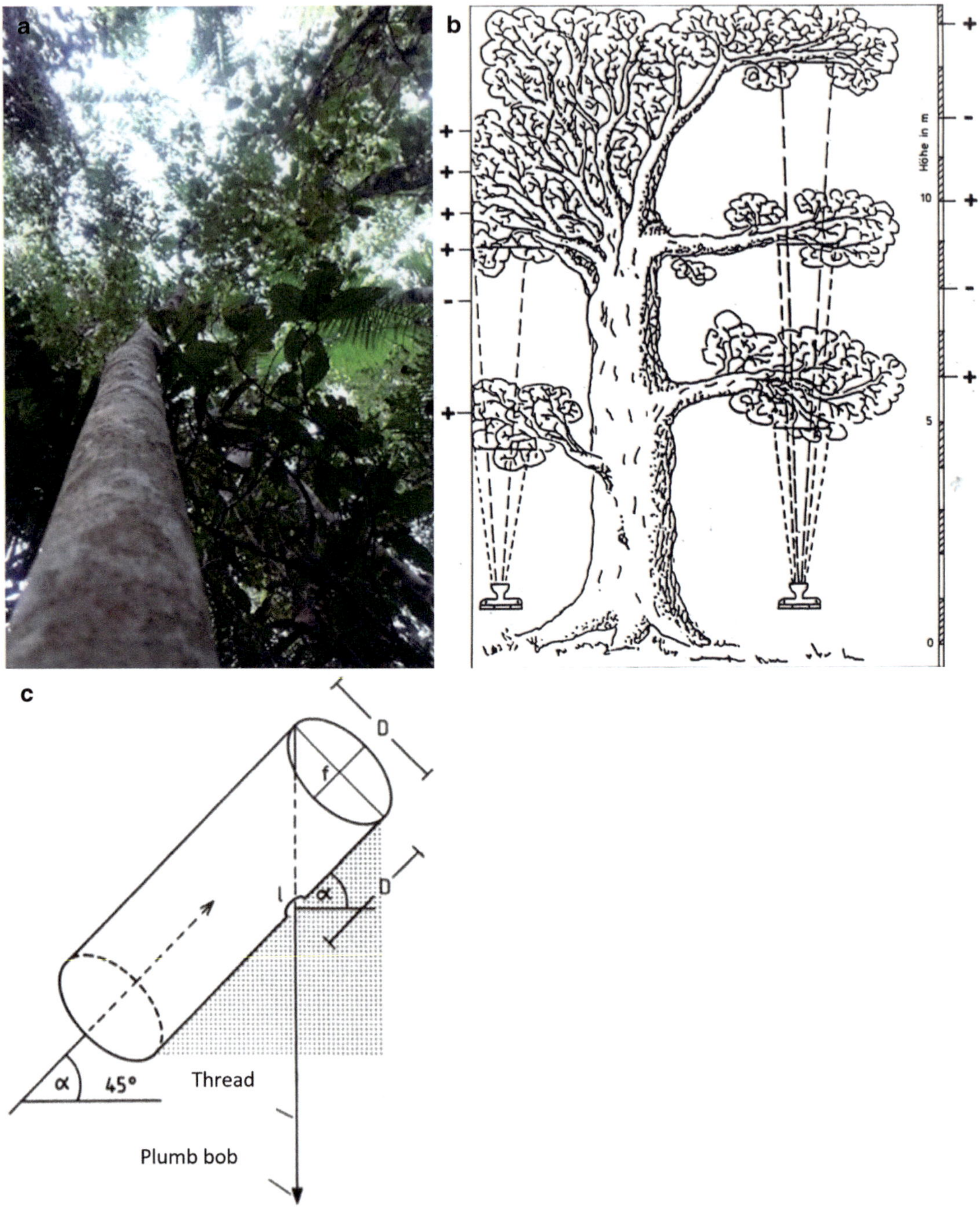

Fig. 4.5 (**a**) Vertical shot into the canopy of a rainforest. The open space in the canopy can be roughly estimated using percentage values or BRAUN-BLANQUET classes (see Sect. 4.1) West Papua. (**b**) Measurement of leaf distribution in the trunk zone and forest canopy layer. The registrations with + and – at a certain distance from the camera indicate whether leaf-bearing branches or leaves can be observed or not detected with the telephoto lens (focus). (**c**) Eyepiece for estimating canopy closure and ground cover according to Cyr and Oelke (1976)

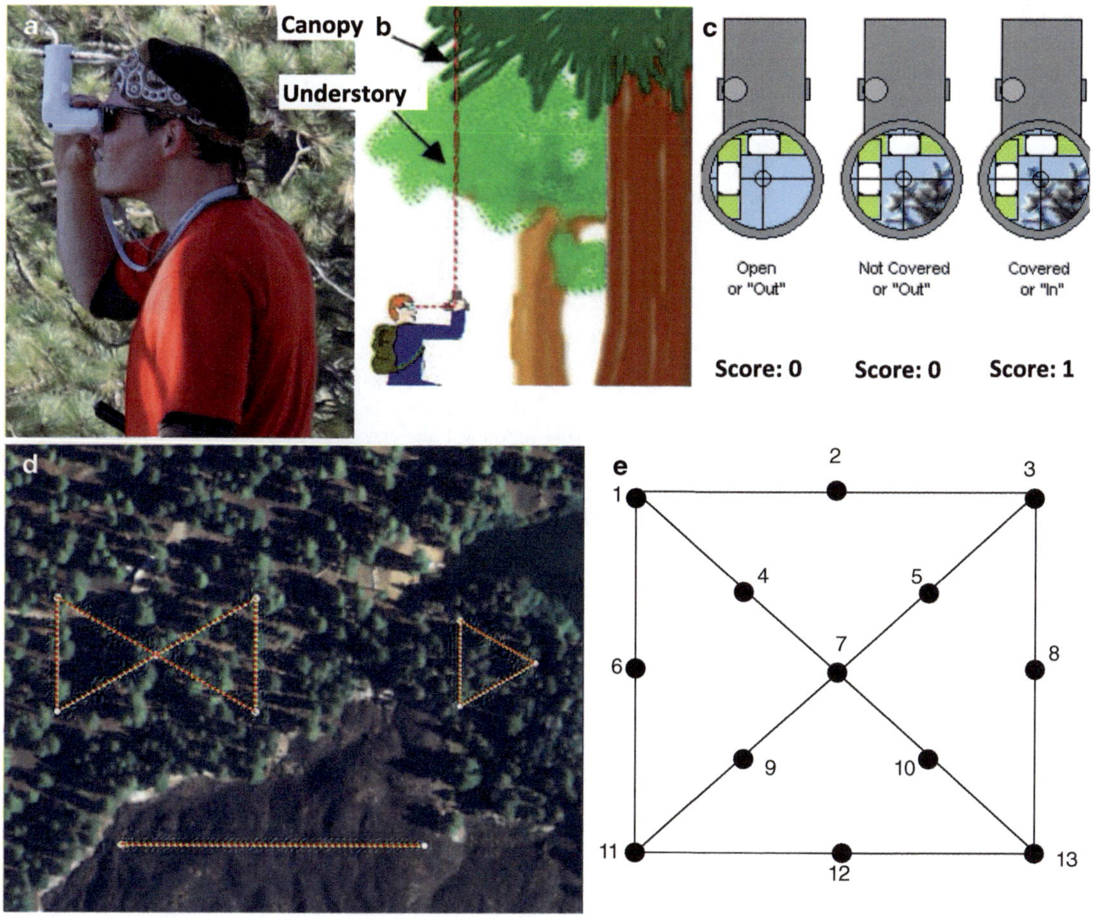

Fig. 4.6 (**a**) Use of the densitometer. (**b**) Projected line of vision of the observer. (**c**) Position the densitometer strictly vertically by centering both level bubbles. Scoring: If canopy/understory leaves are in the center of the cross-hair, score 1—if not, score 0. (**d**) Transect types used in the Lassen Volcanic National Park Project. (**e**) Transect points in a 10 × 10 m grid used in a forest Braun-Blanquet relevé. (Picture credits: (**a**) through (**d**) Geographic Resource Solutions, grsgis.com, reproduced by permission)

The study area is divided into a coordinate grid to measure the location and extent of individual formations on the map. This task is often greatly facilitated by using suitable aerial photos. Primarily, contact prints or digital images from image flights carried out for official survey offices are considered because they are relatively inexpensive, and with a scale of approximately 1:30,000, they come quite close to the base topographic maps (TK 25 in Germany). However, it should be noted that these aerial photographs only represent exact vertical shots in the center of the image and show increasing area distortions toward the edge. Distortion-free enlargements (orthophotos) are often also offered, which do not have this disadvantage.

If aerial photos are to be used as mapping aids, the season of the shot is of some importance. While some features such as stand boundaries, bush strips, individual trees, and paths are almost always easy to recognize, some structures and boundaries relevant to mapping in the vegetation only stand out clearly at certain times of the year and with suitable lighting. Comparing the structures found in the aerial photo with the situation in the field will allow you, over time, to extract the information contained in the aerial photo by intuition and sensitivity. With more experience, you may be able to deduce some vegetation types of the area directly from the aerial photo. However, this should always be checked in the field, at least initially.

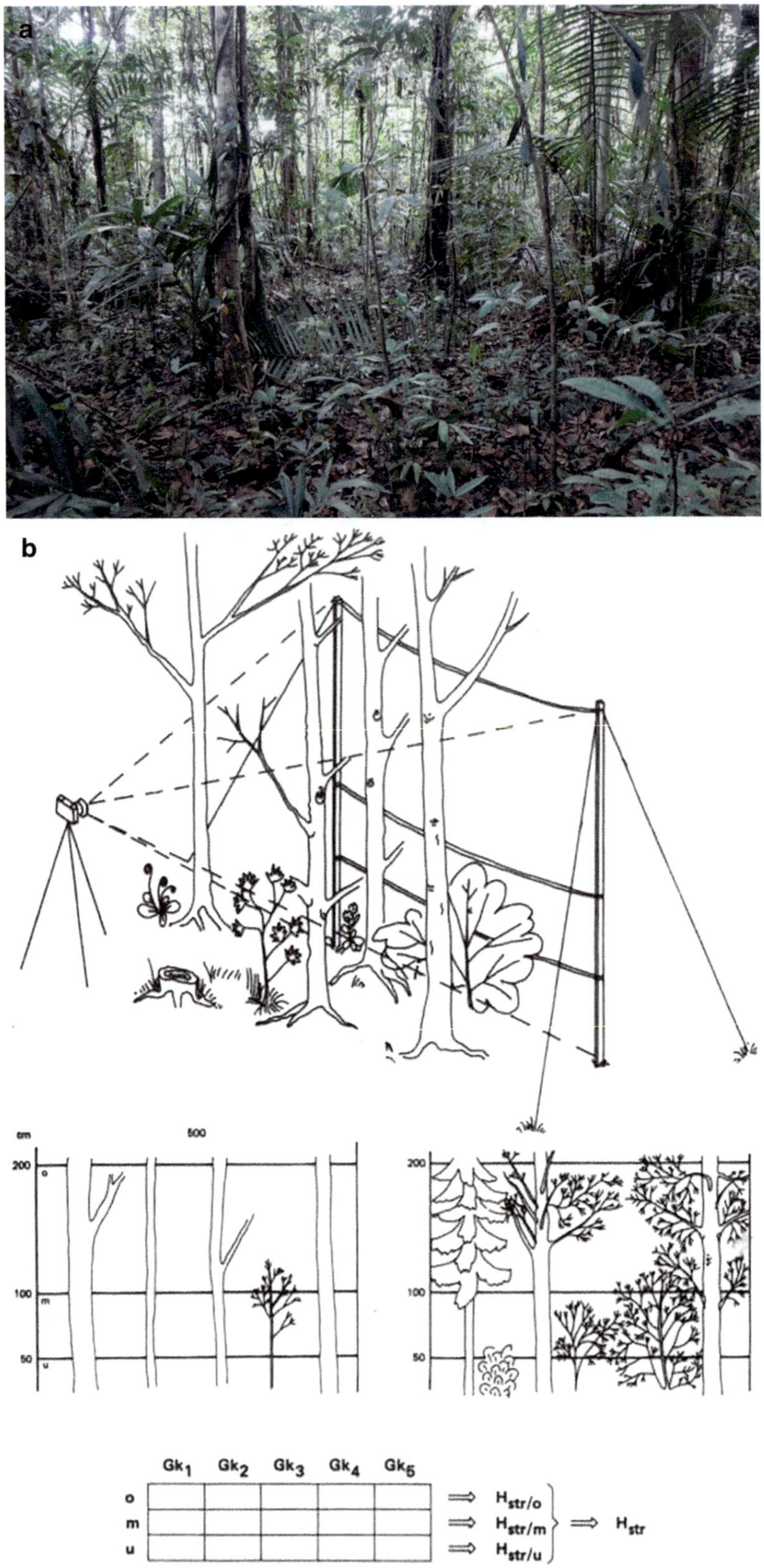

Fig. 4.7 (**a**) Photographs can also document the vertical structuring in a forest. West Papua. (**b**) Photographic spatial structure recording. Measurement strips at 0.5, 1, and 2 m in height. Width of the image and the measurement strip arrangement five: m.– Center: 2 examples of a photograph of various diverse habitats.— Below: Evaluation scheme: **o**—top, **m**—middle, **u**—bottom. **Gk**—Size class 1,2,... of the interruptions of the measurement strips on the photo. $\mathbf{H}_{\mathbf{str}}$—a measure of diversity calculated from the individual rows or the structure's overall distribution of size classes. (From Mader and Mühlenberg 1981)

Special catch and measurement points in the field are entered into the map according to their coordinates. The map should contain the scale, north direction, recording date, and symbols with a legend for a more detailed terrain description. In addition, photos are suitable for characterizing habitats, which should ideally be taken from an elevated point.

Habitats are mapped after possible identification from satellite or drone images; field inspections with rough classification of plant formations would then be necessary for particular questions. If too many signatures exist in one representation, producing several maps with different themes covering the same area is advisable.

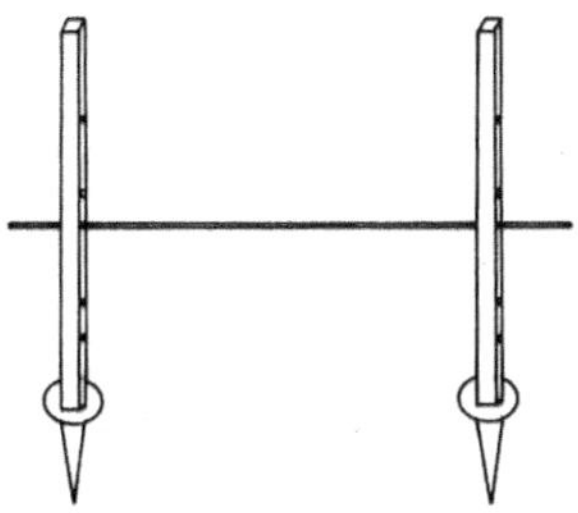

Fig. 4.8 Simple device for measuring vegetation density in grassland. The vertical rods are spaced 30 cm apart until the round slice is driven into the ground, and then a horizontal wire piece is inserted into the corresponding heights on the side. For further explanations, see the text

For a uniform representation, the following map legends are suggested (see Fig. 4.10). For Central Europe, we suggest the classification of plant formations according to Wilmanns (1993) and Ellenberg (1973).

Using digital evaluation systems facilitates data integration into a *geographic information system* (GIS). These information systems are now widespread and represent an essential source of information. They are often cloud-based, allowing data entry from mobile devices and decentralized processing (Nowak et al. 2020). An overview is given by Schilcher and Fritsch (1989), Bajjali (2018), Brebbia and Pascolo (1998), Haines-Young et al. (1993), Haslett (1990), Johnston (1998), Lopez and Frohn (2018), Lovett and Appleton (2020), Wadsworth and Treweek (1998), and Zhu (2016), and applications are described by Göpfert (1991), Göpfert

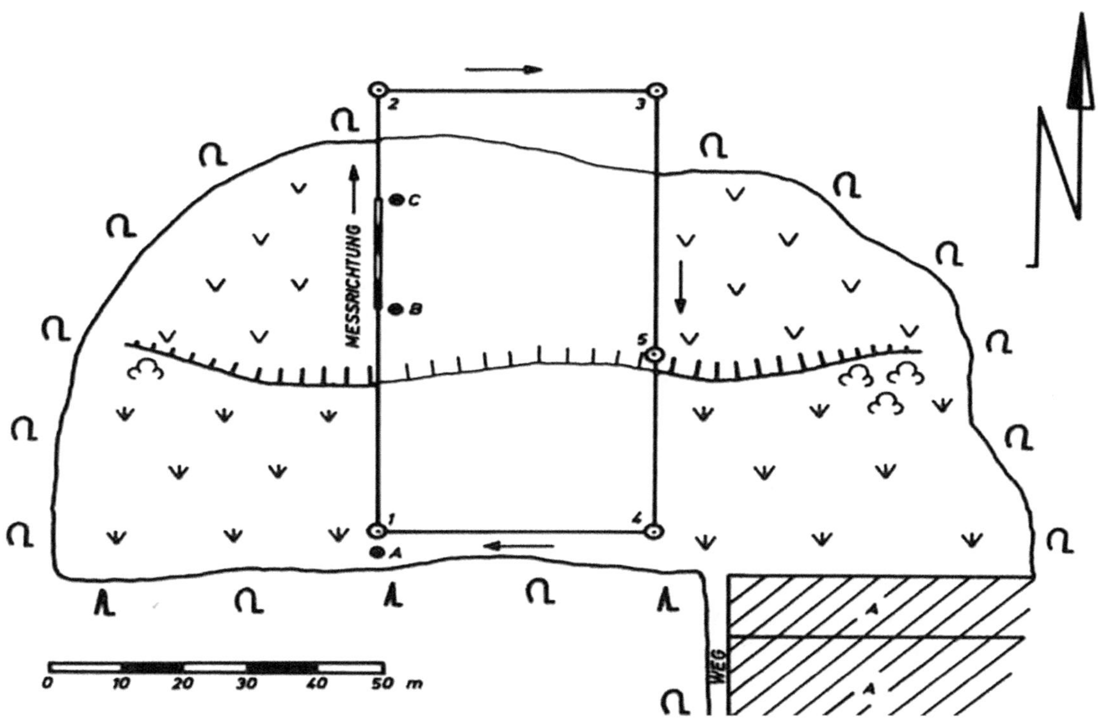

Fig. 4.9 Measuring a study area. Explanation: see text

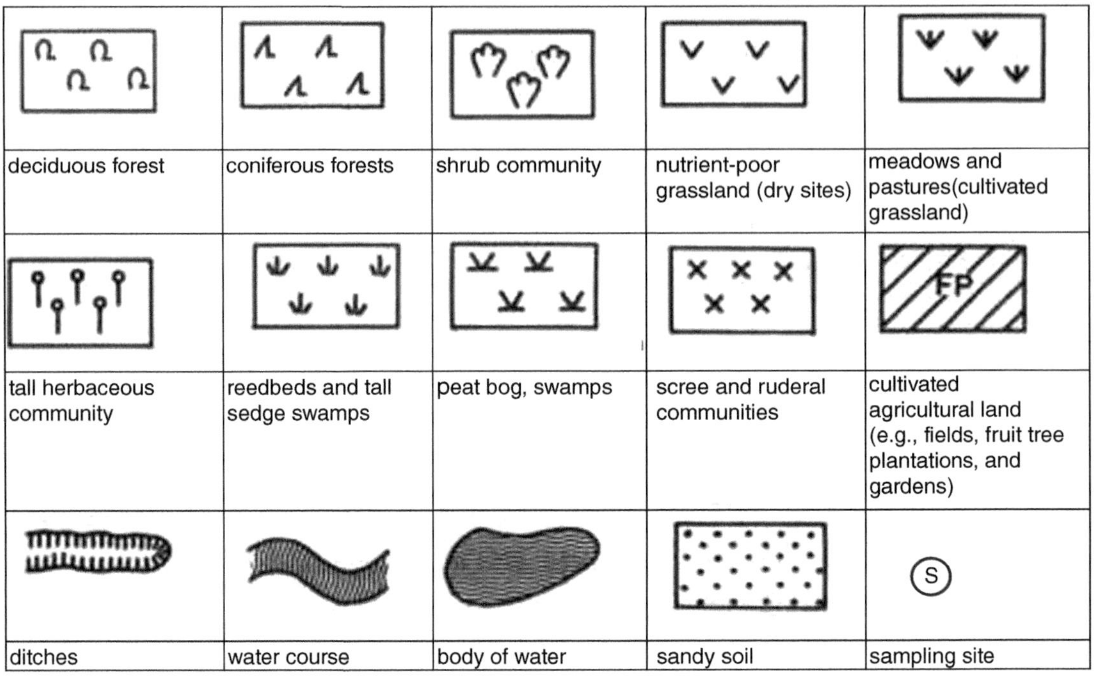

Fig. 4.10 Suggestions for the legend for field mapping

et al. (1989), Petropoulos and Chalkias (2024), Singh and Vyas (2022), Wang et al. (2010), and Yu et al. (2019). Observation points and other features in the field are transferred into the geographic information system for visualization and further data analysis (overview in Longley et al. 2015; Xia et al. 2024; Zhu 2024).

4.4 Use of Global Positioning System (GPS)

A global positioning system (GPS) can determine your positions or routes in the field.

A handheld GPS device is almost indispensable for observers in the field. Most of the products on offer are from Garmin (https://www.garmin.com)—the simplest units, e.g., the eTrex 10, cost about US$ 140. With a handheld GPS device, you can determine your position, the route you took, and the distance to specific destinations. You can record the route you have traveled as a track without needing an internet connection. These options are essential for setting up monitoring transects and determining the randomly selected starting position. GPS devices also offer an electronic compass. You can quickly learn how to operate your GPS device using its operating instructions or tutorials on blogs or YouTube by searching the internet for your device's exact name.

If you have planned transects yourself, the corresponding waypoints can also be communicated to other observers so that they can continue monitoring the same transects. An example is provided in the case study on deer counts in Sect. 5.2. For foot transects, it makes sense to lay out a triangular transect so that you return to the starting position and do not walk the same route twice. The route (= the sequence of waypoints or locations that leads you to your final destination) is also displayed on the GPS device in a current map section and can be edited. It is also saved on your device as a track (= a recording of your path). The data in the GPS device can be easily transferred to a computer, e.g., with a usually provided cable between the mini-USB port on your device and the USB port on the computer. Some devices can also do this wirelessly using Bluetooth, allowing data to be transferred to a smartphone. The instructions for this can also be found in the corresponding manual for the GPS device. You may need to install a corresponding app on your smartphone to transfer some features. A large-scale application of GPS for foot transects in the African savannah can be found in Waltert et al. (2008).

If you use the handheld GPS device to scan a specific area in the orbit (perimeter), you can also use the GPS to calculate the area of the selected area. However, all manipulations of recorded waypoints, tracks, or calculations can be more accessible when importing your GPS data into a GPS or GIS software, such as Google Earth, ArcGIS, or EasyGPS. Suppose you are unfamiliar with a specific GIS software. In that case, we recommend the freely available Google Earth Pro program, which plots your waypoints and tracks, allows you to export them as HR images, and allows for length and area measurement.

GPS devices can also be attached to animals by a collar or harness or glued to their bodies. In that case, all data on an animal's movements over a specific time can be obtained by downloading the recorded positions (see Sect. 4.6).

4.5 Use of Conservation Drones

Conservation drones are low-cost, autonomous, and operator-friendly unmanned aerial vehicles (UAVs) that can be used for surveying, mapping, and monitoring habitat and biodiversity. Wich and Koh (2018) and Wich et al. (2016) describe the use and advantages of drones. By the use of drone technology, you can obtain data on:

Case Study: Drone-Based Assessments of Invasive Shrubs in Ramsdell Park, Lenawee County, Michigan

Researchers: Thomas Bakoway and Thomas Wassmer, Siena Heights University

Prairie environments are essential ecosystems in many parts of the North American Midwest. They harbor astonishing biodiversity (Allart et al. 2024; Kemmerling et al. 2022, 2023) and provide refuge for many pollinators (Coleman and Wait 2022; Drobney et al. 2020; Schulte et al. 2017). Invasive autumn olive (*Elaeagnus umbellata*), Russian olive (*E. angustifolia*), and bush honeysuckle (*Lonicera tatarica and L. maackii*) shrubs (Fig. 4.11) overrun prairies, lowering biodiversity and reducing ecosystem services (Bach and Kleiman 2021; Grant et al. 2020; Rook et al. 2011). Remote sensing provides a cost- and labor-effective way to assess the invasion without surveying every square foot on the ground (Huelsman et al. 2023; Kedia et al. 2021; Klemas 2015; Sun et al. 2021), especially as drones, camera technology, and mapping software have become more sophisticated and affordable (Fricke and Olden 2023). This ongoing study aims to develop a drone-based map of the park in high resolution, compare the accuracy of drone-based assessment to on-site identification, and quantify the extent of invasive shrubs present.

The study occurred at Ramsdell Park, Lenawee County, Michigan, USA, which encompasses approximately 728,434 m^2 and is located 20 km west of Adrian, Michigan, at 41.909 N°, 84.258°W. The park's landscape mosaic consists of old field vegetation, prairies, deciduous forests, and shallow ponds. The photo acquisition was completed using a DJI Phantom 4 Pro drone equipped with an RGB camera (Fig. 4.12a) controlled by the iPad app Pix4Dcapture Pro version 1.59.0 (Pix4D SA, Prilly, Switzerland, Fig. 4.12b).

Orthomosaic maps were generated using an educational license of the software Agisoft Metashape Professional version 2.1.1 (Agisoft LLC, St. Petersburg, Russia). Maps were geo-referenced in Google Earth Pro and added as an image overlay to existing satellite records (Fig. 4.13a).

Fig. 4.11 (**a**) Ground image of an invasive bush honeysuckle (*Lonicera tartarica*, *L. maackii*, or *L. x bella*) in Ramsdell Park. (**b**) Invasive autumn olive (*Elaeagnus umbellata*) with a noticeable blue tint compared to the surrounding native vegetation and bush honeysuckle. (Picture credit: Thomas Bakoway)

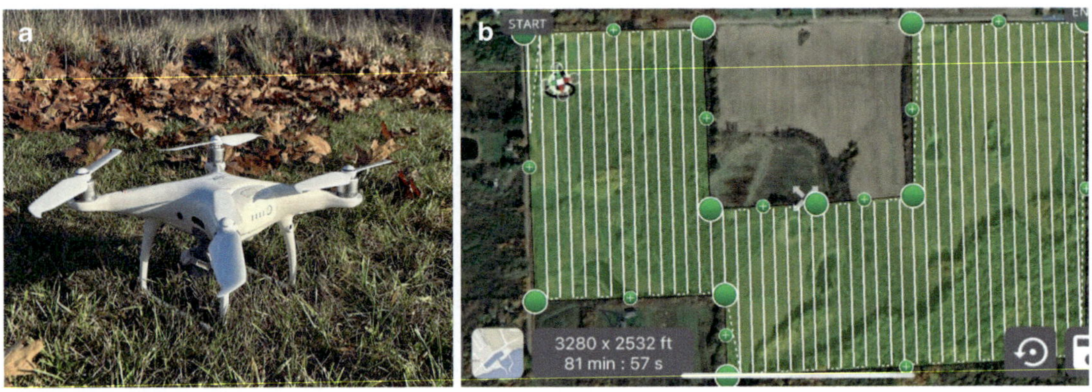

Fig. 4.12 (**a**) DJI Phantom 4 Pro Drone used in the study at Ramsdell Park ready to take off to complete the imaging cycle. (**b**) Example of flight planning using the Pix4DCapture software showing the corner points, flight path, and estimated flight time with an 80% overlay of pictures and a flight altitude of 60 m. (Picture credit: Thomas Bakoway)

A ground-based picture set of 20 invasive and 20 native shrubs geo-referenced using a handheld Garmin eTrex 10 was compared to the drone-based assessment (Fig. 4.13b). The overall accuracy distinguishing invasive vs. native shrubs using drone-based ortho-maps was 92.5% (Chi-square value: 29.00, p-value <0.0001). The estimated coverage of invasive shrubs in prairie restoration plots was 15–20% and varied highly between different prairie plots.

These results show that drone-based mapping in late fall or early spring when only invasive shrubs have leaves (Fridley 2012) is a cheap and fast assessment method to accurately identify and quantify invasive shrubs. The method is less effi-

Fig. 4.13 Orthomosaic maps created from drone images using the Agisoft Metashape software. (**a**) Overlay of a Google Earth map with two ortho-maps from flyovers in early October (top part) and late October (bottom part). Note the loss of green tint. (**b**) Detail from the lower ortho-map in **a** with invasive and native shrubs circled for verification on site. Red ovals indicate suspected invasive shrubs, while blue ovals indicate native shrubs

cient in wooded areas (Krasylenko et al. 2023) and at the margins of flight patterns due to Agisoft Metashape's inability to distinguish image overlaps correctly. Ongoing efforts focus on comparing early spring mapping to late autumn mapping, adding full-spectrum imaging using a DJI Mavic 3 Multispectral drone, and training an AI to detect invasive shrubs automatically (Rominger and Meyer 2021). Extensions of this study include distinguishing between autumn olives and bush honeysuckles and categorizing invasive shrubs into different age classes that require different management methods.

- Land-cover classification.
- Animal counts (Hodgson et al. 2018, Hodgson and Koh 2016, Chrétien et al. 2016).
- Detection of ground nests (Israel and Reinhard 2017; Scholten et al. 2019; compare also Shewring and Vafidis 2021).

You can find guidelines for responsible drone use in biodiversity conservation online (https://conservationdrones.org/).

4.6 Tracking the Movements of Animals Using Radiotelemetry and Satellite Telemetry

Migratory movements of large animals can be tracked with the help of radiotelemetric studies or satellite telemetry (Hofman et al. 2019). This type of study is also a prerequisite for landscape corridors to be identified and evaluated (Sect. 11.5). GPS use to analyze animal movements is now widespread. Nathan et al. (2022) is cited as a comprehensive publication[1]. Tracking individuals by visual observation is only possible in a few cases. A transmitter must be attached to animals to locate them and study their movements. However, the animals must first be captured. Regarding the techniques to catch vertebrates, refer to Bateman (1982) and Stubbe et al. (1995). An example of catching deer is shown in Fig. 4.14. The capture of songbirds for population counts is addressed in Sect. 1.14, and extensively covered in Bub (1977, 1991).

Attaching a transmitter to an animal and the wireless transmission of signals falls into the area of *radiotelemetry* (*radio tracking*). The methods of accurately locating an animal with a radio transmitter are discussed, for example, by

[1] https://huji.move-ecol.com/users/ran-nathan

Fig. 4.14 An example of catching deer: Wooden Gate in the Hustai National Park, Mongolia. The animals are lured with food within the enclosure (seen behind) during the transition from winter to spring. Due to the sluice-like facility, the animals cannot find their way out alone. There is a wooden chamber on the enclosure wall where deer can be grabbed by the head through a window-like opening. You can then fit a collar with a transmitter to the deer without any problems with the antlers. In the foreground, you can see wild deer grazing

MacDonald et al. (1980), Kenward (2001), and Millspaugh and Marzluff (2001). The location may be based on triangulation (see also Sect. 1.9). Special, portable receivers suitable for fieldwork are used to receive a fixed frequency. Most miniature transmitters are usually glued to the animal's skin; medium-sized transmitters are attached as a "backpack," and larger transmitters are attached to collars on the animal. The production of the transmitters should be planned for a few weeks. Advanced modern transmitters for almost all animal groups are offered, for example, by *Advanced Telemetry Systems, Inc. (ATS)*, with headquarters in Isanti, Minnesota, USA. This company cooperates with *Wildlife Drones* (https://wildlifedrones.net). The combination of radiotelemetry and UAVs allows the simultaneous tracking of several animals, even in rugged terrain. Radio tags transmitting at a very high frequency (VHF radio tags) are needed, and the drones must be equipped with the wildlife drone system. The drone pilot receives the signals on the flight controller and can directly follow the animals. The company offers a short video (https://vimeo.com/744461578) and free demos upon request.

The necessary number of transmitters for a study should not be underestimated, as high losses must be expected. Their application requires special skills and adaptation to the selected animal group. As a necessary introduction, we recommend the book by Kenward (1987, 2001) and the articles by Altobelli et al. (2022), Balmori (2016), Gould et al. (2024), Hebblewhite and Haydon (2010), Javed et al. (2003), Keefe et al. (2019), Manville II et al. (2024), Matley et al. (2022), McKinnon and Love (2018), Ruzickova and Elek (2023), Seidel et al. (2018), and Walker et al. (2012). Transmitters of various sizes, adapted to different taxa, are, for example, also offered by Holohil Systems (https://www.holohil.com/transmitters, 2020). For the evaluation of the data, refer to White and Garrott (1990), Frair et al. (2010), Kie et al. (2010), Seidel et al. (2018), and Whoriskey et al. (2019). Radiotelemetry collects an enormous amount of data about individual animals. However, they do not permit drawing conclusions from individual animals to the behavioral patterns of an entire population.

The time required for radiotelemetric studies is often even more underestimated than the financial effort. Therefore, one must critically reconsider the research goals and one's possibilities. Radiotelemetric methods serve to locate an animal without having to recapture it. The location determinations can record the local movements of an animal and calculate its home range. Locating an animal repeatedly over many days by radio tracking can also help to observe it directly throughout a more prolonged study (*radio-assisted surveillance*). Radiotelemetry can also be used to record physiological data (respiration rate, heartbeat, body temperature) or activity phases (resting, moving, feeding). This is often addressed as "biotelemetry." Here, we will focus on location determinations. Before using radiotelemetry, the following questions should be answered (refer to Kenward 2001):

- Is the marking of animals with transmitters the best method?
- Isn't it more manageable, cheaper, and less time-consuming to mark ten times as many

individuals optically (color signs, collars, colored leg rings)?
- Isn't it better to analyze the stomach contents of multiple animals than to watch a few eat?
- Can indigestible, harmless, colored fibers be fed for rediscovery or tracking of their migration paths, which are recognized again in the feces (Randolph 1977; Kruuk 1978) or attach coils with unrolling, thin threads (Miles et al. 1981)?
- Do I have enough time to prepare and implement radiotelemetry? Can I catch enough animals?
- Will the animals not be shocked or injured during the capture and attachment of the transmitters?
- Do you only catch animals with certain characteristics from the population (systematic error)?
- Are marked animals representative of the population or species at large?
- Can the animals be equipped with transmitters without affecting their behavior?
- Will I receive enough data to carry out a proper statistical analysis?

Optical marking of birds with color is also possible in some cases (Ellenberg and Dreifke 1988). Fox et al. (2012) describe a method and the materials used. The individual marking of birds with color ring combinations is described in Sect. 1.14.

Kenward (1987) provides information on the selection and attachment of transmitters. For a tiny transmitter, a crystal-controlled VHF tag, Kenward indicates a weight of about 0.8 g. Such a transmitter can be used for an animal weighing at least 8 g. As a rule of thumb, payloads that hardly impair movement are 10–15% of the body weight for small animals under 50 g, but only 4–6% for animals over 50 g—and even only 2–3% for large animals (Kenward 1987). However, a payload of just 5% of the body mass in tiny animals such as bats might alter the tagged animal's behavior, mobility, and health (Meierhofer et al. 2024). The weight of the transmitter represents a compromise between payload and lifespan: the longer the lifetime of a transmitter, the more batteries are required. According to Holohil Systems (holohil.com), a transmitter of 7–22 g (collar attachment) can transmit for 26–250 weeks, while miniature transmitters (around 0.3 g) can have a lifespan of up to 3 weeks. The transmission duration also depends on the transmission power, which influences the range.

Primarily, the antenna dimension of a transmitter determines the maximum distance from which the transmitter can still be received. Secondarily, the habitat structure plays a role in the "discovery range": While the same transmitter with the same equipment can be received up to 20 km away in an open landscape, it is only received from about 1 km away on a forest floor!

Kenward (1982) located squirrels with collar transmitters within a forest from 1 km when the animal was on a tree and he, as an observer, was on the ground, but only from 400 m distance when the animal was on the ground. From a hilltop with a clear view, he could locate squirrels from remote locations up to 7 km away.

With the most miniature transmitters, terrestrial arthropods such as insects can be tracked telemetrically (Kissling et al. 2014; Klink et al. 2022; Rhodes et al. 2022). The Martin Marietta Corp., Bethesda/Virginia, USA, developed one of the first transmitters. It was a solar-powered semiconductor chip that could be glued on the thorax of bees. The prototype transmitters weighed 35 mg and transmitted infrared signals (of about 800 nm), which a particular receiver could perceive up to 1½ km away. Low weight is only possible by using passive tags on animals that do not transmit signals and, therefore, do not need a battery. Instead, they reflect an external HF radio (Ye et al. 2023) or radar signal (Colpitts and Boiteau 2004; Psychoudakis et al. 2008). Holmes (2022) reports about a research team in Tasmania that fitted bumblebees with passive harmonic laser tags (Brebner et al. 2021). A video on YouTube[2] shows how bumblebees and bees can be fitted with transmitters and then tracked in the field. This way, various behavioral components and environmental influences on the bees can be studied (Wikelski et al. 2010; Crall et al.

[2]Bumblebees given radio backpacks in insect-tracking trial | ABC News: https://youtu.be/acg3Y0WdESk

2018). Besides Hymenoptera, larger beetles were also tracked successfully (Růžičková and Elek 2023). Despite the apparent potential of these passive tags, this tracking method might severely impact an insect's behavior and mobility, so the pros and cons must be carefully weighed (Batsleer et al. 2020).

Another method for individual tracking of ground beetles is "radar tracing" (Wallin and Ekbom 1988; Hockmann et al. 1989). A special reflector (Schottky diode) is glued to the beetle's elytra with superglue and sealed with a two-component adhesive. With the help of a radar device, the beetle can be located from a distance of 5–6 m if a 5 cm long antenna is soldered to the reflector diode. The trail can be followed day and night, and the beetle's movement pattern in different microhabitat structures can be individually and continuously analyzed.

Besides radio tracking, satellite navigation systems are used to track the movement patterns of individual animals. Attached GPS locators are getting small enough even to track smaller mammals and birds (e.g., Conenna et al. 2019; Harris et al. 2024; Williamson and Witt 2021). The animal's position can be determined from afar, over large distances, and in difficult-to-access terrain through satellite telemetry. The accuracy reaches about 5 m, so home ranges and habitat analyses are possible. Compared to radiotelemetry, you get much more data with higher quality and cost efficiency. You also no longer need to triangulate to get the location, and the data are typically uploaded to the satellite in certain time intervals, thus minimizing the danger of data loss. For more information on satellite telemetry, consult, e.g., with Tomkiewicz et al. (2010), Hofman et al. (2019), Nield et al. (2015), and Renshaw et al. (2023).

The further development of telemetry devices with data logging capacities enables the continuous recording of biological data from individual animals on land, in water, and in the air (Wassmer et al. 2020). Bio-logging tools provide continuous data on movement patterns and the selected animals' ecophysiology, energy expenditure, and behavior. Bio-logging also opens opportunities to study the reaction and performance of animals to changes in the environment. Relevant topics may include reactions to disturbances and climate change. However, applying bio-logging tools requires validation studies and system stress tests under controlled conditions (Wassmer et al. 2020).

Various animal species can now be tagged with customized tracking devices. Suppose many individuals of the same species are tagged. In that case, behavioral responses to environmental disturbances and migrations can be continuously tracked via satellite networks interlinking small data loggers attached to individual animals on a global scale. Such an initiative is International Cooperation for Animal Research Using Space (ICARUS), which was initiated in 2002. Its tracking system was installed on the International Space Station (ISS) in 2018, switched on in 2019, and began operations in 2020 (Wikelski 2024; Jetz et al. 2022).

4.7 Calculation of Home Ranges of Animals

Many methods have been developed to calculate the home range. Here we introduce a simple one that can also be evaluated for quantitative habitat use analyses. It goes back to the occupation of grid squares, into which the study area is divided (Southwood 1978; Kenward 1987). In Germany, the German base map 1:5000 (2 cm = 100 m) serves as the basis for the division of the study area. The grid should consist of squares, the side length of which is not smaller than the resolution of the location technology, e.g., 50 m × 50 m squares when tracking magpies (Ellenberg and Dreifke 1988). Reeve (1982) used a grid of 25 m × 25 m to study the home range of the hedgehog. The series of rediscoveries of a marked animal (localization) is entered into the square grid of the study area map with time stamps. Corresponding to the grid, x and y coordinates are applied to the study area, and all fixed positions of the animal are assigned x and y values. On this basis, the arithmetic mean, first the "action center," $(\overline{x},\overline{y})$ is calculated by adding up all x values and then dividing the sum by the sam-

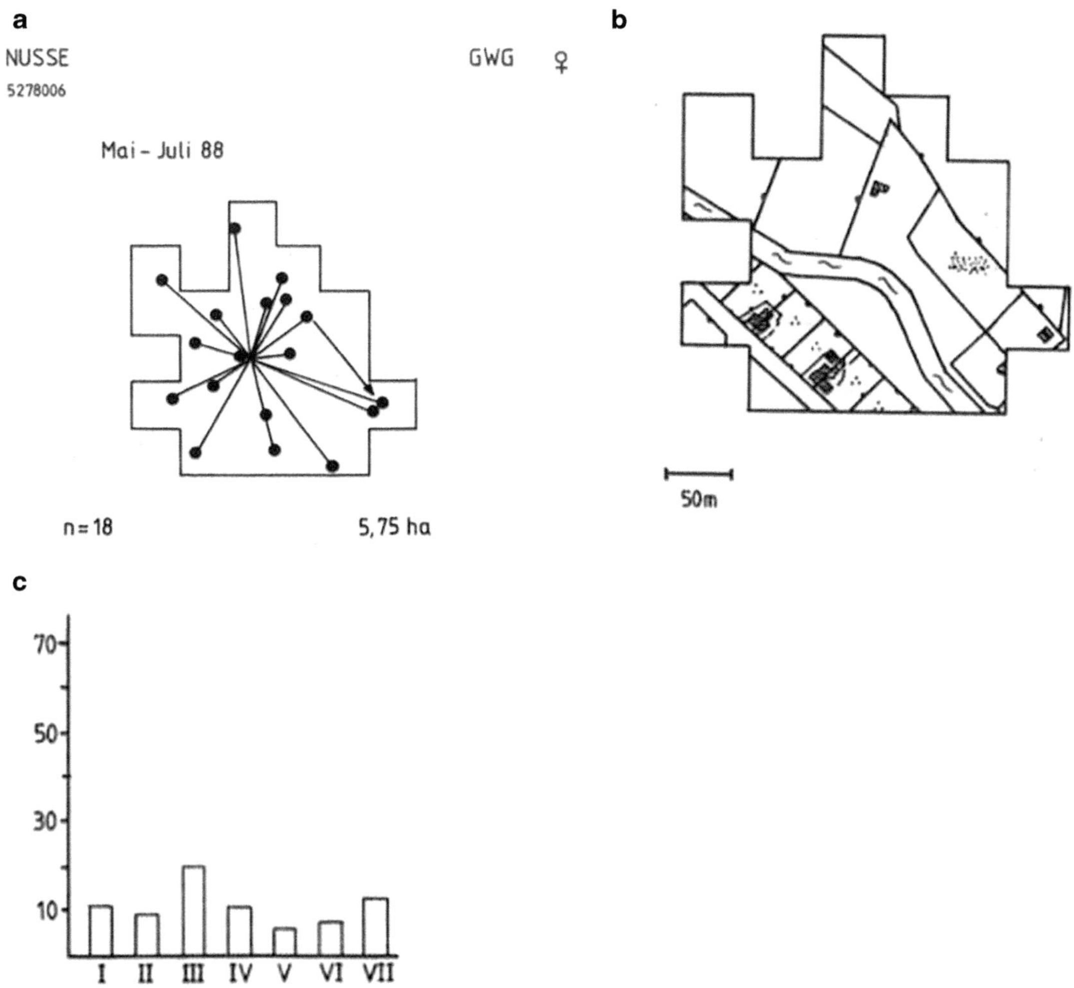

Fig. 4.15 Determination of the home range and habitat use using the example of a marked magpie. (**a**) Rediscoveries of the marked magpie (points), represented as the path from the calculated action center. All grid squares touched by the paths together make up the home range. (**b**) Area usage map of the determined residential area of a magpie in the same section as **a**). (**c**) Percentage information on the area distribution of the terrain structures. I = fields, II = grassland, III = buildings and gardens, IV = bushes, V = roads, VI = bodies of water, VII = fallow land. (From Ellenberg and Dreifke 1988)

ple numbers (= $\overline{x}$), then the summed y values are divided by n (= $\overline{y}$). From the action center $(\overline{x},\overline{y})$ entered into the map, straight lines are drawn to all location fixes (Fig. 4.15). All squares that are touched or crossed by these lines contribute to the estimated home range: the sum of their areas is considered the total area of the action space for the controlled period.

The uneven distribution of the determined animal positions within the entire action space (many localizations multiple times in the same squares) shows that the animals prefer certain parts of their residential area. One speaks of a "core area," which often only makes up 10% of the determined residential area but where the animal spends more than 50% of its activity time (Reeve 1982). Continuous tracking of the marked animal is only possible with great effort. If you aim at short intervals, one after the other, you can continuously record the exact location movement ("trackogram"), and much more information about the animal's behavior can be obtained. The movement speeds can also be calculated from the measured distances between two fixed points and the corresponding time. From their average, we can deduce the time interval required for bearings, according to the resolution of the tracking technology, for the most accurate tracking. Reeve (1982) recommends, for example, two bearings per hour for the nocturnal tracking of hedgehogs.

The number of locations necessary to determine the habitat depends on the movement pattern of the animal to be studied. The following examples are mentioned: Kenward (1982) conducted three location fixings per day on the gray squirrel and, after ten days, could no longer observe any expansion of the determined home range. Reeve (1982) used 20 time-scattered bearings on the hedgehog during its active season as the lower limit for determining the home range. Birks and Linn (1982) considered 10 days (one bearing in the morning and evening each) to be a sufficient minimum for the American mink. Neaf-Daenzer (1993) used 20–40 bearings per hour to locate small birds and to tune their home range.

Powell (2000) provides a recent overview of the importance of determining the home ranges of animals and the approaches that exist for this. Animal radio tracking techniques (see Sect. 4.6) significantly expand the data that can be used to estimate home ranges. Noonan et al. (2019) analyzed different models for calculating home ranges based on the abundance of data. The size of the action area of individual animals also allows conclusions to be drawn about the area requirement of a population if the necessary population size for the survival of a species can be estimated (Hovestadt 1990).

References

Allart, L., Dumont, B., Joly, F., Mosnier, C., Alvarez, G., Galliot, J.-N., et al. (2024). Species richness: A pivotal factor mediating the effects of land use intensification and climate on grassland multifunctionality. *Journal of Applied Ecology*, *61*(5), 1053–1066. https://doi.org/10.1111/1365-2664.14627

Aleksandrova V.D. 1969: Klassifikacija rastitel'nosti (The Classification of Vegetation). - St. Petersburg

Altobelli, J. T., Dickinson, K. J. M., Godfrey, S. S., & Bishop, P. J. (2022). Methods in amphibian biotelemetry: Two decades in review. *Austral Ecology*, *47*(7), 1382–1395. https://doi.org/10.1111/aec.13227

Anderson P. & Wecker M. 2011: *Snags – The Wildlife Tree.* The Importance of Snags in Your Neighborhood. – Washington Department of Fish and Wildlife, 7 p. https://wdfw.wa.gov/sites/default/files/2020-07/snags-the_wildlife_tree-1.pdf

Arbeláez, M. V., & Duivenvoorden, J. F. (2004). Patterns of plant species composition on Amazonian sandstone outcrops in Colombia. *Journal of Vegetation Science*, *15*(2), 181–188. https://doi.org/10.1111/j.1654-1103.2004.tb02253.x

Arno S.F. and Sneck K.M. 1977: *A method for determining fire history in coniferous forests in the Mountain West.* - Gen. Tech. Rep. INT-GTR-42. Ogden, UT: U.S. Department of Agriculture, Forest Service, Intermountain Forest and Range Experiment Station. 28 p.

Atkins, Jeff W., Parth Bhatt, Luis Carrasco, Emily Francis, James E. Garabedian, Christopher R. Hakkenberg, Brady S. Hardiman, et al. (2023). Integrating Forest Structural Diversity Measurement into Ecological Research. *Ecosphere* *14*(9), e4633. https://doi.org/10.1002/ecs2.4633

Atkins, J. W., Bhatt, P., Carrasco, L., Francis, E., Garabedian, J. E., Hakkenberg, C. R., Hardiman, B. S., Jung, J., Koirala, A., LaRue, A. A., Oh, S., Shao Gang, Shao Guofan, Shugart, H. H., Spiers, A., Stovall, A. E. L., Surasinghe, T. D., Tai X., Zhai, L., Zhang, T. & Krause, K. (2023). Integrating forest structural diversity measurement into ecological research. – *Ecosphere* 14 (9), 17, https://doi.org/10.1002/ecs2.4633

Aubry, P., Francesiaz, C., & Guillemain, M. (2024). On the impact of preferential sampling on ecological status and trend assessment. *Ecological Modelling*, *492*, 110707. https://doi.org/10.1016/j.ecolmodel.2024.110707

Avery, T. E. & Burkhart, H. E. (2015). *Forest Measurements.* 5th ed. - Waveland Press, 456 p.

Bach, E. M., & Kleiman, B. P. (2021). Twenty years of tallgrass prairie restoration in northern Illinois, USA. *Ecological Solutions and Evidence*, *2*(4), e12101. https://doi.org/10.1002/2688-8319.12101

Bakker J.D., Jones E. and Sprenger C.B. 2019: Evidence of a historical frequent, low-severity fire regime in western Washington, USA *Canadian Journal of Forest Research* 49(2), 575-585. DOI:https://doi.org/10.1139/cjfr-2018-0354

Bateman J. 1982: *Animal Traps and Trapping.* (2nd ed.) Biddles Limited for David & Charles (Publishers) Limited, Newton Abbot, London, 286.

Bajjali, W. (2018). *ArcGIS for Environmental and Water Issues* (1st ed. 2018.). Cham: Springer International Publishing: Imprint: Springer. https://doi.org/10.1007/978-3-319-61158-7

Balmori, A. (2016). Radiotelemetry and wildlife: Highlighting a gap in the knowledge on radiofrequency radiation effects. *Science of the Total Environment*, *543*, 662–669. https://doi.org/10.1016/j.scitotenv.2015.11.073

Batsleer, F., Bonte, D., Dekeukeleire, D., Goossens, S., Poelmans, W., Van der Cruyssen, E., et al. (2020). The neglected impact of tracking devices on terrestrial arthropods. *Methods in Ecology and Evolution*, *11*(3), 350–361. https://doi.org/10.1111/2041-210X.13356

Braun-Blanquet, J., Fuller, G. D., & Conard, H. S. (1932). *Plant sociology: the study of plant communities* (1st ed.). New York, London: McGraw-Hill Book Company, Inc.

Brebbia, C. A., & Pascolo, P. (Eds.). (1998). *GIS technologies and their environmental applications*. Presented at the Conference on Geographical Information Systems in the next Millennium, Southampton; Boston: WIT Press.

Brebner, J. S., Makinson, J. C., Bates, O. K., Rossi, N., Lim, K. S., Dubois, T., et al. (2021). Bumble bees strategically use ground level linear features in navigation. *Animal Behaviour*, *179*, 147–160. https://doi.org/10.1016/j.anbehav.2021.07.003

Bub H. 1977: Vogelfang und Vogelberingung, Teil. II. – A. Ziemsen-Verlag, Wittenberg Lutherstadt, 182 p., Teil I (1978), 169 p.

Bunnell F.L. & Houdel I. 2010. Down wood and biodiversity – implications to forest practices. – *Environmental Reviews*. 18(NA): 397-421. https://doi.org/10.1139/A10-019

Birks, J. D. S. & Linn, I. J. (1982). Studies of Home Range of the Feral Mink, *Mustela vison*. - *Symp.zool. Soc*. Lond. 49, 231–257.

Bub, H. (1991). *Bird trapping and bird banding: a handbook for trapping methods all over the world*. Ithaca, NY: Cornell University Press.

Byambasuren O. 2011: Fire and stand dynamics in different forest types of the West Khentey Mountains, Mongolia. - Dissertation University Göttingen, 118 p.

Casey D., Altman B. & Stringer D. 2015: Snags, Bark Beetles, and Cavity-Nesting Birds: Conservation and Management in Ponderosa Pine Forests of the Pacific Northwest. – *American Bird Conservancy*. https://abcbirds.org/wp-content/uploads/2015/05/Cavity-nester-booklet.pdf

Chrétien L.P., Théau J. & Ménard P. 2016: Visible and thermal infrared remote sensing for the detection of white-tailed deer using an unmanned aerial system. – *Wildlife Society Bulletin* 40, 181–191

Crall J.D., Switzer C.M., Oppenheimer R.L., Versypt A.N.F., Dey B., , Brown A., Mackay Eyster M., Guérin C., Pierce N.E., Combes S.A. & de Bivor B.J. 2018: Neonicotinoid exposure disrupts bumblebee nest behavior, social networks, and thermoregulation. – *Science* 362 (6415), 683–686. https://doi.org/10.1126/science.aat1598

Chytrý, M., & Otýpková, Z. (2003). Plot sizes used for phytosociological sampling of European vegetation. *Journal of Vegetation Science*, *14*(4), 563–570. https://doi.org/10.1111/j.1654-1103.2003.tb02183.x

Coleman, A. L., & Wait, D. A. (2022). Urban Prairie Plots and Gardens Can Sustain Plant-Pollinator Interactions Similar to Established Rural Prairies. *The American Midland Naturalist*, *188*(1), 102–118. https://doi.org/10.1674/0003-0031-188.1.102

Colpitts, B. G., & Boiteau, G. (2004). Harmonic radar transceiver design: miniature tags for insect tracking. *IEEE Transactions on Antennas and Propagation*, *52*(11), 2825–2832. Presented at the IEEE Transactions on Antennas and Propagation. https://doi.org/10.1109/TAP.2004.835166

Conenna, I., López-Baucells, A., Rocha, R., Ripperger, S., & Cabeza, M. (2019). Movement seasonality in a desert-dwelling bat revealed by miniature GPS loggers. *Movement Ecology*, *7*(1), 27. https://doi.org/10.1186/s40462-019-0170-8

Cottam, G., & Curtis, J. T. (1956). The Use of Distance Measures in Phytosociological Sampling. *Ecology*, *37*(3), 451–460. https://doi.org/10.2307/1930167

Cyr, A. & Oelke, H. (1976). Vorschläge zur Standardisierung von Biotopbeschreibungen bei Vogelbestandsaufnahmen im Waldland. - *Die Vogelwelt* 97, 161–175.

Daemane, M. E., Adelabu, S., & Ramoelo, A. (2024). Assessing species richness, diversity and assemblage of forest patches within a grassland matrix in the Afrotemperate ecosystems. *Phytocoenologia*. https://doi.org/10.1127/phyto/2024/0442

Damgaard, C. (2014). Estimating mean plant cover from different types of cover data: a coherent statistical framework. *Ecosphere*, *5*(2), art20. https://doi.org/10.1890/ES13-00300.1

Dassot M., Constant T. & Fournier M. 2011: The use of terrestrial LiDAR technology in forest science: application fields, benefits and challenges. – *Ann.Sci.* 68, 959–974

Dierschke H. 1994: Pflanzensoziologie. Eugen Ulmer, Stuttgart, 683.

Dietrich, J.H. and T.W. Swetnam. 1984. Dendrochronology of a fire-scarred ponderosa pine *For. Sci.* 30(1): 238-247

Dulamsuren Ch. 2004: *Floristische Diversität, Vegetation und Standortbedingungen in der Gebirgstaiga des Westkhentej, Nordmongolei*. - Dissertation Universität Göttingen, Berichte des Forschungszentrums Waldökosysteme Reihe A, Bd. 191, 290 p.

Drobney, P., Larson, D. L., Larson, J. L., & Viste-Sparkman, K. (2020). Toward Improving Pollinator Habitat: Reconstructing Prairies with High Forb Diversity. *Natural Areas Journal*, *40*(3), 252–261. https://doi.org/10.3375/043.040.0322

El-Sheikh, M. A., Thomas, J., Alfarhan, A. H., Alatar, A. A., Mayandy, S., Hennekens, S. M., et al. (2017). SaudiVeg ecoinformatics: Aims, current status and perspectives. *Saudi Journal of Biological Sciences*, *24*(2), 389–398. https://doi.org/10.1016/j.sjbs.2016.02.012

Edelmann P., Ambarlı D., Gossner M.M., Schall P., Ammer C., Wende B., Schulze E-D., Weisser W.W.

& Seibold S. 2022: Forest management affects saproxylic beetles through tree species composition and canopy cover. – *Forest Ecology and Management* 524:120532. DOI:https://doi.org/10.1016/j.foreco.2022.120532

Edelmann P., Weisser W.W., Ambarlı D., Bässler C., Buscot F., Hofrichter M., Hoppe B., Kellner H., Minnich, C., Moll J., Persoh D., Seibold S., Seilwinder C., Schulze E.D., Wöllauer S. & Borken W. 2023: Regional variation in deadwood decay of 13 tree species: Effects of climate, soil and forest structure. – *Forest Ecology and Management* 541(S1): 121094. https://doi.org/10.1016/j.foreco.2023.121094

Ehbrecht M., Seidel D., Annighöfer P., Kreft H., Michael Köhler M., Zemp D.C., Puettmann K., Nilus R., Babweteera F., Willim K., Stiers M., Soto D., Boehmer H.J., Fisichelli N., Burnett M., Juday G., Stephens S.L. & Ammer C. 2021: Global patterns and climatic controls of forest structural complexity. – *Nature Communications* https://doi.org/10.1038/s41467-020-20767-z

Ellenberg H. (sen.) 1973: *Ökosystemforschung*. – Springer-Verlag, Berlin – Heidelberg – New York, 280.

Ellenberg H. & Dreifke R. 1988: Ursachen für Verteilung und Häufigkeit, einschließlich des Fehlens, von Elstern auf zwei Probeflächen in Schleswig-Holstein. 2. Jahresbericht zu einem Forschungsauftrag im Rahmen des BMFT-Projekts (Mühlenberg, Kaule, Ellenberg) über Minimalareale für Tierpopulationen aus Naturschutzsicht, Manuskript, 35 p.

Faber-Langendoen D., Keeler-Wolf T., Meidinger D., Tart D., Hoagland B., Josse C., Navarro G., Ponomarenko S., Saucier J-P., Alan Weakley A. & Comer P. 2014: EcoVeg: a new approach to vegetation description and classification. – *Ecological Monographs* 84(4), 533–561

Fox A.G., Hill C.E. and Young R.F. 2012: A New Method to Dye-Mark Wild Birds. – *North American Bird Bander* 37 (2), 49-53

Filis, E., Kyrkas, D., Mantzos, N., Dimitrakopoulos, P. G., Fotiadis, G., & Konstantinou, M. (n.d.). Grassland flora of ultramafic areas in Northern Pindus (Greece). *Ecological Research*, *n/a*(n/a). https://doi.org/10.1111/1440-1703.12469

Frair, J. L., Fieberg, J., Hebblewhite, M., Cagnacci, F., DeCesare, N. J., & Pedrotti, L. (2010). Resolving issues of imprecise and habitat-biased locations in ecological analyses using GPS telemetry data. *Philosophical Transactions of the Royal Society B-Biological Sciences*, *365*(1550), 2187–2200. https://doi.org/10.1098/rstb.2010.0084

Fricke, R. M., & Olden, J. D. (2023). Technological innovations enhance invasive species management in the anthropocene. *BioScience*, *73*(4), 261–279. https://doi.org/10.1093/biosci/biad018

Fridley, J. D. (2012). Extended leaf phenology and the autumn niche in deciduous forest invasions. *Nature*, *485*(7398), 359–362. https://doi.org/10.1038/nature11056

Fründ, H.-C. (1983). *Untersuchungen zur Koexistenz verschiedener Chilopodenarten im Waldboden*. Diss. Uni. Würzburg, 164 p.

Gafurov, A., Prokhorov, V., Kozhevnikova, M., & Usmanov, B. (2024). Forest Community Spatial Modeling Using Machine Learning and Remote Sensing Data. *Remote Sensing*, *16*(8), 1371. https://doi.org/10.3390/rs16081371

Gould, L. A., Manning, A. D., Mcginness, H. M., & Hansen, B. D. (2024). A review of electronic devices for tracking small and medium migratory shorebirds. *Animal Biotelemetry*, *12*(1). https://doi.org/10.1186/s40317-024-00368-z

Grant, T. A., Shaffer, T. L., & Flanders, B. (2020). Resiliency of Native Prairies to Invasion by Kentucky Bluegrass, Smooth Brome, and Woody Vegetation. *Rangeland Ecology & Management*, *73*(2), 321–328. https://doi.org/10.1016/j.rama.2019.10.013

Göpfert W. 1991: Ein raumbezogenes Informationssystem für Raumordnung und Umweltplanung auf der Grundlage rechnergestützter Kartographie und Fernerkundung. – *Kartographische Nachrichten* (KN) 41, Bonn, 1–9.

Göpfert W., Wei L. & Weisensee M. 1989: *Quantitative Landschaftsanalysen durch Integration von Biotopkartierungen*. IRIS 1. – In: Fachgebiet Kartographie, TH Darmstadt, Darmstadt, 29 p.

Gossner M.M., Wende B., Levick S.R., Schall P., Floren A., Linsenmair K.E., Steffan-Dewenter I., Schulze E.D. & Weisser W. 2016: Deadwood enrichment in European forests – Which tree species should be used to promote saproxylic beetle diversity? – Biological Conservation 201, 92-102. DOI:https://doi.org/10.1016/j.biocon.2016.06.032

Grove S., Hopkins A., Harrison C., Yee M., Stamm L., Wardlaw T. & Mohammed C. 2024: *Coarse Woody Debris, Old Trees and Biodiversity Conservation in Production Forests*. – Forestry Tasmania. 36. https://www.academia.edu/56050733/Coarse_Woody_Debris_Old_Trees_and_Biodiversity_Conservation_in_Production_Forests?uc-sb-sw=54460780

Hockmann P., Schlomberg P., Wallin H. & Weber,F. 1989: Bewegungsmuster und Orientierung des Laufkäfers *Carabus auronitens* in einem westfälischen Eichen-Hainbuchen-Wald (Radarbeobachtungen und Rückfangexperimente). – *Abhandlungen aus dem Westfälischen Museum für Naturkunde* 51 (1), 73. https://doi.org/10.3390/d5030479

Hodgson J.C. & Koh L.P. 2016: Best practice for minimising unmanned aerial vehicle disturbance to wildlife in biological field research. – *Current Biology* 26, R404–R405

Hodgson J.C., Mott R., Baylis S.M., Pham T.T., Wotherspoon S., Kilpatrick A.D., Segaran R.R., Reid I., Terauds A. & Koh L.P. 2018: Drones count wildlife more accurately and precisely than humans. – *Methods in Ecology and Evolution* 9, 1160–1167

Hofman M.P.G., Hayward M.W., Heim, M., Marchand P., Rolandsen C.M., Mattisson J., Urbano F., Heurich M., Mysterud A., Melzheimer J., Morellet N., Voigt

U., Allen B.L. …, Balkenhol N. 2019: Right on track? Performance of satellite telemetry in terrestrial wildlife research. – *PLoS One* 14(5), e0216223. https://doi.org/10.1371/journal.pone.0216223

Holmes A. 2022: Researchers put radio backpacks on bumblebees in Tasmania in insect-tracking trial. – https://www.abc.net.au/news/2022-10-10/bumblebees-radio-transmitters-research-insect-tracking/101515280

Holohil (2020). Holohil Transmitters List. Available at: https://www.holohil.com/transmitters/#tableall

Hovestadt T. 1990: Possibilities and criteria for the determination of "minimal areas" of animal populations and ecosystem stocks. – *Series of Publications for Landscape Conservation and Nature Protection* 32, 175-186, Bonn-Bad Godesberg

Humphrey, J.W., Sippola, A., Lempérière, G., Dodelin, B., Alexander, K. and Butler, J. 2005: Deadwood as an indicator of biodiversity in European forests: from theory to operational guidance. – In: M. Marchetti (ed.) *Monitoring and Indicators of Forest Biodiversity in Europe* – From ideas to operationality. Joensuu: European Forest Institute. pp. 193–206.

Haines-Young, R. H., Green, D. R., & Cousins, S. (Eds.). (1993). *Landscape ecology and geographic information systems*. London; New York: Taylor & Francis.

Harris, M. E., Hobson, K. A., & Morrissey, C. A. (2024). Hirundo rustica (Barn Swallows) and Tachycineta bicolor (Tree Swallows) select wetlands in agriculturally intensive landscapes, as revealed by GPS tracking. *Ornithological Applications*, duae012. https://doi.org/10.1093/ornithapp/duae012

Haslett, J. R. (1990). Geographic information systems. A new approach to habitat definition and the study of distributions. *Trends in Ecology & Evolution*, *5*(7), 214–218. https://doi.org/10.1016/0169-5347(90)90134-Y

Hebblewhite, M., & Haydon, D. T. (2010). Distinguishing technology from biology: a critical review of the use of GPS telemetry data in ecology. *Philosophical Transactions of the Royal Society B-Biological Sciences*, *365*(1550), 2303–2312. https://doi.org/10.1098/rstb.2010.0087

Huelsman, K., Epstein, H., Yang, X., Mullori, L., Červená, L., & Walker, R. (2023). Spectral variability in fine-scale drone-based imaging spectroscopy does not impede detection of target invasive plant species. *Frontiers in Remote Sensing*, *3*. https://www.frontiersin.org/articles/10.3389/frsen.2022.1085808. Accessed 11 July 2023

Hunter, C. (2023, September 23). Theodolite. *App Store*. https://apps.apple.com/us/app/theodolite/id339393884. Accessed 28 June 2024

Israel, M. & Reinhard, A. (2017). Detecting nests of lapwing birds with the aid of a small unmanned aerial vehicle with thermal camera. - *Conference: International Conference on Unmanned Aircraft Systems June 2017*. https://doi.org/10.1109/ICUAS.2017.7991393

Javed, S., Higuchi, H., Nagendran, M., & Takekawa, J. (2003). Satellite telemetry and wildlife studies in India: Advantages, options and challenges. *Current Science*, *85*(10), 1439–1443.

Jetz, W., Tertitski, G., Kays, R., Mueller, U., Wikelski, M., Åkesson, S., et al. (2022). Biological Earth observation with animal sensors. *Trends in Ecology & Evolution*, *37*(4), 293–298. https://doi.org/10.1016/j.tree.2021.11.011

Johnston, C. A. (1998). *Geographic information systems in ecology*. Oxford; Malden, MA: Blackwell Science.

Kedia, A. C., Kapos, B., Liao, S., Draper, J., Eddinger, J., Updike, C., & Frazier, A. E. (2021). An Integrated Spectral–Structural Workflow for Invasive Vegetation Mapping in an Arid Region Using Drones Drones, *5*(1), 19. https://doi.org/10.3390/drones5010019

Keefe, R. F., Wempe, A. M., Becker, R. M., Zimbelman, E. G., Nagler, E. S., Gilbert, S. L., & Caudill, C. C. (2019). Positioning Methods and the Use of Location and Activity Data in Forests. *Forests*, *10*(5). https://doi.org/10.3390/f10050458

Kemmerling, L. R., McCarthy, A. C., Brown, C. S., & Haddad, N. M. (2023). Butterfly biodiversity increases with prairie strips and conservation management in row crop agriculture. *Insect Conservation and Diversity*, *16*(6), 828–837. https://doi.org/10.1111/icad.12675

Kemmerling, L. R., Rutkoski, C. E., Evans, S. E., Helms, J. A. I., Cordova-Ortiz, E. S., Smith, J. D., et al. (2022). Prairie Strips and Lower Land Use Intensity Increase Biodiversity and Ecosystem Services. *Frontiers in Ecology and Evolution*, *10*. https://doi.org/10.3389/fevo.2022.833170

Kie, J. G., Matthiopoulos, J., Fieberg, J., Powell, R. A., Cagnacci, F., Mitchell, M. S., et al. (2010). The home-range concept: are traditional estimators still relevant with modern telemetry technology? *Philosophical Transactions of the Royal Society B-Biological Sciences*, *365*(1550), 2221–2231. https://doi.org/10.1098/rstb.2010.0093

Kissling, W. D., Pattemore, D. E., & Hagen, M. (2014). Challenges and prospects in the telemetry of insects. *Biological Reviews*, *89*(3), 511–530. https://doi.org/10.1111/brv.12065

Klemas, V. V. (2015). Coastal and Environmental Remote Sensing from Unmanned Aerial Vehicles: An Overview. *Journal of Coastal Research*, *31*(5), 1260–1267. https://doi.org/10.2112/JCOASTRES-D-15-00005.1

Krasylenko, Y., Rydlo, K., Atamas, N., Sosnovsky, Y., Horielov, O., Maceček, I., et al. (2023). Druid Drone—A portable unmanned aerial vehicle with a multifunctional manipulator for forest canopy and mistletoe research and management. *Methods in Ecology and Evolution*, *14*(6), 1416–1423. https://doi.org/10.1111/2041-210X.14058

Kahl T. & Bauhus J. 2014: An index of forest management intensity based on assessment of harvested tree volume, tree species composition and dead wood origin. – *Nature Conservation* 7(2), 15-27. DOI:https://doi.org/10.3897/natureconservation.7.7281

Kent M. & Coker P. 1992: *Vegetation Description and Analysis. A Practical Approach.* – Boca Raton (CRC), 363 p.

Kenward R. 1987: *Wildlife Radio Tagging, Equipment, Field Techniques and Data Analysis.* – Academic Press, Inc. (London) Ltd. (Publ.), Cambridge University Press, Cambridge, 222 p.

Kenward R. 2001: *A Manual for Wildlife Radio Tagging*, 2nd ed. – New York: Academic Press. 350 p.

Kenward R.E. 1982: Techniques for Monitoring the Behaviour of Grey Squirrels by Radio. – *Symp. zool. Soc. Lond.* 49, 175–196

Kirby K.J. & Drake C.M. 1993: Dead wood matters: the ecology and conservation of saproxylic invertebrates in Britain. – *English Nature Science* No. 7, 105.

Kirby K.J. & Hall J.E. 2019: *Woodland Survey Handbook. Collecting Data for Conservation in British Woodland.* – Pelagic Publishing, 253.

Kirby K.J. 1993: Assessing nature conservation values in British woodland. – *Arboricultural Journal* 17, 253–276.

Kirby K.J., Reid C.M., Thomas R.C. & Goldsmith F.B. 1998: Preliminary estimates of fallen dead wood and standing dead trees in managed and unmanaged forests in Britain. – *Journal of Applied Ecology* 35: 148–155.

Kozma J. 2020: *Identifying Snags with Cavities: Protecting a Critical Wildlife Resource.* – YouTube Video., The Washington Department of Fish and Wildlife. https://www.youtube.com/watch?v=EvilMA3ihN0

Kramer H. and Akça A. 1995: *Leitfaden zur Waldmesslehre* (3rd ed.). – Frankfurt, J.D.Sauerländer's Verlag, 266 p.

Kratochwil A. & Schwabe A. 2001: *Ökologie der Lebensgemeinschaften.* – UTB für Wissenschaft, Eugen Ulmer Verlag,756.

Kruuk H. 1978: Spatial Organisation and territorial behaviour of the European badger *Meles meles.* – *Journal of Zoology*, London, 184, 1–19

Lájer, K. (2007). Statistical tests as inappropriate tools for data analysis performed on non-random samples of plant communities. *Folia Geobotanica*, *42*(2), 115–122. https://doi.org/10.1007/BF02893878

Lepš, J., & Šmilauer, P. (2007). Subjectively sampled vegetation data: Don't throw out the baby with the bath water. *Folia Geobotanica*, *42*(2), 169–178. https://doi.org/10.1007/BF02893883

Longley P.A., Goodchild M.F., Maguire D.J. & Rhind D.W. 2015: *Geographic Information Science and Systems*, 4th ed. – Wiley, New York, NY, 496 p.

Lopez, R. D., & Frohn, R. C. (2018). *Remote sensing for landscape ecology: monitoring, modeling, and assessment of ecosystems* (Second edition.). Boca Raton: CRC Press, Taylor & Francis Group.

Lovett, A. A., & Appleton, K. (Eds.). (2020). *GIS for Environmental Decision-Making* (1st edition.). CRC Press.

Mader, H.-J. & Mühlenberg, (1981). Artenzusammensetzung und Ressourcenangebot einer kleinflächigen Habitatinsel, untersucht am Beispiel der Carabidenfauna. - *Pedobiologia* 21, 46–59.

Manville II, A. M., Levitt, B. B., & Lai, H. C. (2024). Health and environmental effects to wildlife from radio telemetry and tracking devices-state of the science and best management practices. *Frontiers in Veterinary Science*, *11*. https://doi.org/10.3389/fvets.2024.1283709

Matley, J. K., Klinard, N., V., Martins, A. P. B., Aarestrup, K., Aspillaga, E., Cooke, S. J., et al. (2022). Global trends in aquatic animal tracking with acoustic telemetry. *Trends in Ecology & Evolution*, *37*(1), 79–94. https://doi.org/10.1016/j.tree.2021.09.001

McKinnon, E. A., & Love, O. P. (2018). Ten years tracking the migrations of small landbirds: Lessons learned in the golden age of bio-logging. *AUK*, *135*(4), 834–856. https://doi.org/10.1642/AUK-17-202.1

Meierhofer, M. B., Tena, E., Lilley, T. M., Dechmann, D. K. N., Voigt, C. C., Troitsky, T. S., et al. (2024). Re-weighing the 5% tagging recommendation: assessing the potential impacts of tags on the behaviour and body condition of bats. *Mammal Review*, *n/a*(n/a). https://doi.org/10.1111/mam.12369

Minnesota Department of Natural Resources. (2007). *A handbook for collecting vegetation plot data in Minnesota: The relevé method.* St. Paul, MN: Minnesota County Biological Survey, Minnesota Natural Heritage and Nongame Research Program, and Ecological Land Classification Program.

Mühlenberg, M. (1993). Freilandökologie. - UTB für Wissenschaft 595, 3rd ed., Quelle & Meyer, Heidelberg, Wiesbaden, 512 p.

Müller, J. & Bütler, R. (2010). A review of habitat thresholds for dead wood: a baseline for management recommendations. – *European Journal of Forest research* 129, 981–992.

MacDonald D.W., Ball F.G. & Hough N.G. 1980: The evaluation of home range size and configuration using radio tracking data. – In: Amlaner C.J & MacDonald D.W. (eds.) 2013: *A handbook on biotelemetry and radio tracking.* Proceedings of an International Conference on Telemetry and Radio Tracking in Biology and Medicine, Oxford, 20-22 March 1979, 405–424. Pergamon Press, Oxford, 824 p.

Mäkinen H., Hynynen J., Siitonen J. and Sievänen R. 2006: Predicting the decomposition of Scots Pine, Norway spruce, and Birch stems in Finland. - *Ecological Applications* 16. 1865-79. https://doi.org/10.1890/1051-0761(2006)016[1865:PTDOSP]2.0.CO;2.

MELFF 2002: *Richtlinie zur Sicherung von Alt- und Totholzanteilen im Wirtschaftswald.* – Heft G1, Ministerium für Ernährung, Landwirtschaft, Forsten und Fischerei Mecklenburg-Vorpommern, 16 p.

Miles M.A., de Souza A.A. & Povoa M.M. 1981: Mammal tracking and nest location in Brazilian forest with an improved spool-and-line device. – *Journal of Zoology*, London, 195, 331–347

Millspaugh J. & Marzluff J. 2001: *Radio Tracking and Animal Populations.* – New York, Harcourt Publishers. 400 p.

Moore M.E.D. 1955: The Use of Phytosociological Methods in Ecological Investigations: I. The Braun-Blanquet System. – *Journal of Ecology* 43 (1), 226-244

Mueller-Dombois L.D. & Ellenberg H. (sen.) 1974: *Aims and Methods of Vegetation Ecology*. – John Wiley & Sons, New York, London, Sydney, Toronto, 547.

Mühlenberg, M. (1993). *Freilandökologie*. – UTB für Wissenschaft 595, 3rd ed., Quelle & Meyer, Heidelberg, Wiesbaden, 512.

Mühlenberg, M., Hondong, H., Dulamsuren, Ch. & K.v. Gadow (2004). Large-scale biodiversity research in the southern taiga, Northern Mongolia. – In: Szaro, R.C., Peterson, C.E., Gadow, K.v., Kräuchi, N. and Broggi, M.F. (eds.): Creating a Legacy for Sustainable Science-based Forest Management: Lessons Learned from Field Experiments. – *Journal of Forest, Snow and Landscape Research* 78, 1/2: 93–118.

Müller J. & Bütler R. 2010: A review of habitat thresholds for dead wood: a baseline for management recommendations. – *European Journal of Forest research* 129, 981-992

Naef-Daenzer B. 1993: A New Transmitter for Small Animals and Enhanced Methods of Home-Range Analysis. – *Journal of Wildlife Management* 57(4), 680-689

Nathan R., Monk C.T., Arlinghaus R., Adam T., Alós J., Assaf M., Baktoft H., Beardsworth C.E., Bertra M.G., Bijleveld A.I., Brodin T., Brooks J.L., Campos-Candela A., Cooke S.J., Gjelland K.Ø, Gupte P.R., Harel R., Hellström G., Jeltsch F., Killen S.S., Klefoth T., Langrock R., Lennox R.J., Lourie E., Madden J.R., Orchan Y., Pauwels I.S., Říha M., Roeleke M., Schlägel U.E., Shohami D., Signer J., Toledo S., Vilk O., Westrelin S., Whiteside M.A. & Jarić I. 2022: Big-data approaches lead to an increased understanding of the ecology of animal movement *Science* 375, eabg1780. Download: https://www.science.org on February 14, 2024

Nield A.P., Enright N.J. & Ladd P.G. 2015: Study of seed dispersal by emu (*Dromaius novaehollandiae*) in the Jarrah (*Eucalyptus marginata*) forests of south-western Australia through satellite telemetry. – *Emu 115*, 29–34

Noonan M.J., Tucker M.A., Fleming C.H., Akre T., Albertz S.C., Ali A.H., Altmann J., Antunes P.C., Belant J.L., Berens D., Beyer D., Blaum N., Böhning-Gaese K., Paula L.C. Jr, Dekker J., Drescher-Lehman J., Farwig N., Fichtel C., Fischer C., Ford A., Goheen J.R., Janssen R., Jeltsch F., Kauffman M., Kappeler P., Koch F., LaPoint S., Markham A.C., Medici E.P., Morato R.G., Nathan R., Oliveira-Santos L.G.R., Patterson B.D., Paviolo A., Ramalho E.E., Roesner S., Selva N., Sergiel A., Silva M.X., Spiegel O., Ullmann W., Zieba F., Zwijacz-Kozica T., Fagan W.F., Mueller T. & Calabrese J.M. 2019: A comprehensive analysis of autocorrelation and bias in home range estimation. – *Ecological Monographs* 89(2), 1-21. e01344 : 89

Nowak, M. M., Dziób, K., Ludwisiak, Ł., & Chmiel, J. (2020). Mobile GIS applications for environmental field surveys: A state of the art. *Global Ecology and Conservation, 23*, e01089. https://doi.org/10.1016/j.gecco.2020.e01089

Oppermann, R. (1989). Ein Meßinstrument zur Ermittlung der Vegetationsdichte in grasig-krautigen Pflanzenbeständen. - *Natur und Landschaft* 64(7/8), 332–338.

Peinado, M., Delgadillo, J., Aparicio, A., Aguirre, J. L., & Macías, M. Á. (2014). Major plant communities of the Carson Desert (Nevada), North America's coldest and driest desert. *Plant Biosystems - An International Journal Dealing with all Aspects of Plant Biology, 148*(5), 945–955. https://doi.org/10.1080/11263504.2013.845267

Peinado, Manuel, Aguirre, J. L., Delgadillo, J., & Macías, M. Á. (2008). A phytosociological and phytogeographical survey of the coastal vegetation of western North America. Part I: plant communities of Baja California, Mexico. *Plant Ecology, 196*(1), 27–60. https://doi.org/10.1007/s11258-007-9334-5

Penksza, K., Saláta, D., Fűrész, A., Penksza, P., Fuchs, M., Pajor, F., et al. (2024). Are Hungarian Grey Cattle or Hungarian Racka Sheep the Best Choice for the Conservation of Wood-Pasture Habitats in the Pannonian Region? *Agronomy, 14*(4), 846. https://doi.org/10.3390/agronomy14040846

Petropoulos, G. P., & Chalkias, C. (Eds.). (2024). *Geographical Information Science: Case Studies in Earth and Environmental Monitoring* (1st edition.). Elsevier.

Psychoudakis, D., Moulder, W., Chen, C.-C., Zhu, H., & Volakis, J. L. (2008). A Portable Low-Power Harmonic Radar System and Conformal Tag for Insect Tracking. *IEEE Antennas and Wireless Propagation Letters, 7*, 444–447. Presented at the IEEE Antennas and Wireless Propagation Letters. https://doi.org/10.1109/LAWP.2008.2004512

Peterken G.F. 1996: *Natural Woodland: Ecology and conservation in northern temperate regions*. – Cambridge University Press, 540.

Powell R.A. 2000: Animal home ranges and territories and home range estimators. Pages 65–110 – In: L. Boitani and T. F. Fuller, editors. Research techniques in animal ecology (2nd ed.). Columbia University Press, New York, New York, USA, 464 p.

Randolph S.E. 1977: Changing spatial relationships in a population of *Apodemus sylvaticus* with onset of breeding. – *Journal of Animal Ecology* 46, 653–676

Reeve N.J. 1982: The home range of the Hedgehog as revealed by a radio tracking study. – *Symp. zool. Soc. Lond.* 49, 207–230

Renshaw, S., Hammerschlag, N., Gallagher, A. J., Lubitz, N., & Sims, D. W. (2023). Global tracking of shark movements, behaviour and ecology: A review of the renaissance years of satellite tagging studies, 2010–2020. *Journal of Experimental Marine Biology and Ecology, 560*, 151841. https://doi.org/10.1016/j.jembe.2022.151841

Rhodes, M. W., Bennie, J. J., Spalding, A., French-Constant, R. H., & Maclean, I. M. D. (2022). Recent advances in the remote sensing of insects. *Biological Reviews*, *97*(1), 343–360. https://doi.org/10.1111/brv.12802

Rominger, K. R., & Meyer, S. E. (2021). Drones, Deep Learning, and Endangered Plants: A Method for Population-Level Census Using Image Analysis. *Drones*, *5*(4), 126. https://doi.org/10.3390/drones5040126

Rook, E. J., Fischer, D. G., Seyferth, R. D., Kirsch, J. L., LeRoy, C. J., & Hamman, S. (2011). Responses of Prairie Vegetation to Fire, Herbicide, and Invasive Species Legacy. *Northwest Science*, *85*(2), 288–302. https://doi.org/10.3955/046.085.0216

Ruzickova, J., & Elek, Z. (2023). Beetles on the move: Not-just-a-technical review of beetles' radio-tracking. *Entomologia Experimentalis et Applicata*, *171*(2), 82–93. https://doi.org/10.1111/eea.13260

Růžičková, J., & Elek, Z. (2023). Beetles on the move: Not-just-a-technical review of beetles' radio-tracking. *Entomologia Experimentalis et Applicata*, *171*(2), 82–93. https://doi.org/10.1111/eea.13260

Scholten, H, Scheres, A, de Water, E, Graf, U, Granic, I, Luijten, M. Behavioral trainings and manipulations to reduce delay discounting: A systematic review. Psychon Bull Rev. 2019 Dec;26(6):1803–1849. https://doi.org/10.3758/s13423-019-01629-2. PMID: 31270766; PMCID: PMC6863952.

Schwabe, A. & Mann, P. (1990). Eine Methode zur Beschreibung und Typisierung von Vogelhabitaten, gezeigt am Beispiel der Zippammer (Emberiza cia). – Ökol. Vögel 12, 127–157.

Schulte, L. A., Niemi, J., Helmers, M. J., Liebman, M., Arbuckle, J. G., James, D. E., et al. (2017). Prairie strips improve biodiversity and the delivery of multiple ecosystem services from corn–soybean croplands. *Proceedings of the National Academy of Sciences*, *114*(42), 11247–11252. https://doi.org/10.1073/pnas.1620229114

Seidel, D. P., Dougherty, E., Carlson, C., & Getz, W. M. (2018). Ecological metrics and methods for GPS movement data. *International Journal of geographical Information Science*, *32*(11), 2272–2293. https://doi.org/10.1080/13658816.2018.1498097

Singh, A., & Vyas, V. (2022). A Review on remote sensing application in river ecosystem evaluation. *Spatial Information Research*, *30*(6), 759–772. https://doi.org/10.1007/s41324-022-00470-5

Sloan, M. (2020). A Review of the Biological Significance of Coarse Woody Debris in Forest Ecosystems and its Positive Impact on Forest Characteristics. – Honours Bachelor of Environmental Management Thesis, Faculty of Natural Resources Management, Lakehead University, Thunder Bay ON. 54 p.

Stearns, F. W. (1949). Ninety Years Change in a Northern Hardwood Forest in Wisconsin. *Ecology*, *30*(3), 350–358. https://doi.org/10.2307/1932615

Sun, Z., Wang, X., Wang, Z., Yang, L., Xie, Y., & Huang, Y. (2021). UAVs as remote sensing platforms in plant ecology: review of applications and challenges. *Journal of Plant Ecology*, *14*(6), 1003–1023. https://doi.org/10.1093/jpe/rtab089

Scherzinger W. 1996: *Naturschutz im Wald. Qualitätsziele einer dynamischen Waldentwicklung*. – Verlag Eugen Ulmer, 447.

Schilcher M. & Fritsch D. 1989: *Geoinformationssysteme*. – Wichmann Verlag, Karlsruhe, 340.

Schwabe A. & Mann P. 1990: Eine Methode zur Beschreibung und Typisierung von Vogelhabitaten, gezeigt am Beispiel der Zippammer (*Emberiza cia*). – *Ökol. Vögel* 12, 127–157.

Shewring, M. P. & Vafidis, J. O. 2021: Using UAV-mounted thermal cameras to detect the presence of nesting nightjar in upland clear-fell: A case study in South Wales, UK. – *Ecological Solutions and Evidence* 2(1), 1–10.

Sloan M. 2020. A Review of the Biological Significance of Coarse Woody Debris in Forest Ecosystems and its Positive Impact on Forest Characteristics. – Honours Bachelor of Environmental Management Thesis, Faculty of Natural Resources Management, Lakehead University, Thunder Bay ON. 54.

Southwood T.R.E. 1978: *Ecological Methods – with particular reference to the study of Insect Populations*. – Chapman and Hall, London, John Wiley & Sons, New York, 524 p.

Ståhl G., Ringvall A. & Friedmann J. 2001: Assessment of coarse woody debris - a methodical overview. – *Ecological Bulletins* 49, 57 - 70

Stowe J. 2024: *Many Animal Species Benefit From Snags*. – Spartanburg.com News MARCH 26, 2024

Stubbe C., Ahrens M., Stubbe M. and Goretzki J. 1995: *Lebendfang von Wildtieren:* Fangtechniken - Methoden – Erfahrungen. – Deutscher Landwirtschaftsverlag, Berlin, GmbH, (In German), 208 p

Talbot, S. S., Talbot, S. L., & Daniëls, F. J. A. (2005). Comparative phytosociological investigation of subalpine alder thickets in southwestern Alaska and the North Pacific. *Phytocoenologia*, 727–759. https://doi.org/10.1127/0340-269X/2005/0035-0727

Tavilla, G., Lamoliere, A., Gabarretta, J., Attard, V., Henwood, J., Stevens, D. T., et al. (2023). Climate Change and Wetland Ecosystems: The Effects on Halophilous Vegetation Diversity in Il-Ballut ta' Marsaxlokk Natura 2000 Site (Malta). *Land*, *12*(9), 1679. https://doi.org/10.3390/land12091679

The Nature Conservancy 2023: *Geospatial Conservation Atlas*. Authoritative TNC conservation data at global, regional and national scales. – Terrestrial Ecoregions: https://geospatial.tnc.org/datasets/b1636d640ede4d6ca8f5e369f2dc368b/about. Freshwater Ecoregions of the world: https://geospatial.tnc.org/datasets/TNC::freshwater-ecoregions-of-the-world-feow/about

Thiemann P.J. 2016: *Woodpecker holes and owls*. https://applegater.org/pdf/2016/v09n04/v09n04p11.pdf

Tomkiewicz S.M., Fuller M.R., Kie J.G., & Bates K.K. 2010: Global positioning system and associated technologies in animal behaviour and ecological research. – *Philosophical Transactions of the Royal Society of London B: Biological Sciences* 365, 2163–2176

USDA 2020: *Snags, den trees, and coarse woody debris for wildlife habitat.* – Conservation Stewardship Program E6660. 6 p. https://www.nrcs.usda.gov/sites/default/files/2022-11/E666O%20May%202020.pdf

Vaglio Laurin G., Puletti N., Grotti M. et al. 2020: Species dominance and above ground biomass in the Białowie˙za Forest, Poland, described by airborne hyperspectral and lidar data. – *International Journal of Applied Earth Observation and Geoinformation* 92, 102178.

van Aardt, A. C., Codron, D., Theron, E. J., & du Preez, P. J. (2020). Plant community structure and possible vegetation changes after drought on a granite catena in the Kruger National Park, South Africa, *Koedoe*, *62*(2), 11. https://doi.org/10.4102/koedoe.v62i2.1585

van Klink, R., August, T., Bas, Y., Bodesheim, P., Bonn, A., Fossøy, F., et al. (2022). Emerging technologies revolutionise insect ecology and monitoring. *Trends in Ecology & Evolution*, *37*(10), 872–885. https://doi.org/10.1016/j.tree.2022.06.001

Wadsworth, R., & Treweek, J. (1998). *Gis for Ecology: An Introduction* (1st edition.). Harlow: Addison-Wesley.

Walas, M. (2024). How to Determine the Age of a Tree. *The Living Urn*. https://www.thelivingurn.com/blogs/news/79236289-how-to-determine-the-age-of-a-tree. Accessed 14 August 2024

Walker, D. A., Daniëls, F. J. A., Matveyeva, N. V., Šibík, J., Walker, M. D., Breen, A. L., et al. (2018). Circumpolar Arctic Vegetation Classification. *Phytocoenologia*, 181–201. https://doi.org/10.1127/phyto/2017/0192

Walker, K. A., Trites, A. W., Haulena, M., & Weary, D. M. (2012). A review of the effects of different marking and tagging techniques on marine mammals. *Wildlife Research*, *39*(1), 15–30. https://doi.org/10.1071/WR10177

Wang, K., Franklin, S. E., Guo, X., & Cattet, M. (2010). Remote Sensing of Ecology, Biodiversity and Conservation: A Review from the Perspective of Remote Sensing Specialists. *Sensors*, *10*(11), 9647–9667. https://doi.org/10.3390/s101109647

Whoriskey, K., Martins, E. G., Auger-Methe, M., Gutowsky, L. F. G., Lennox, R. J., Cooke, S. J., et al. (2019). Current and emerging statistical techniques for aquatic telemetry data: A guide to analysing spatially discrete animal detections. *Methods in Ecology and Evolution*, *10*(7), 935–948. https://doi.org/10.1111/2041-210X.13188

Wikum, D. A., & Shanholtzer, G. F. (1978). Application of the Braun-Blanquet cover-abundance scale for vegetation analysis in land development studies. *Environmental Management*, *2*(4), 323–329. https://doi.org/10.1007/BF01866672

Williamson, J. L., & Witt, C. C. (2021). A lightweight backpack harness for tracking hummingbirds. *Journal of Avian Biology*, *52*(9). https://doi.org/10.1111/jav.02802

Wallin H. & Ekbom B.S. 1988: Movements of carabid beetles (Coleoptera: Carabidae) inhabiting cereal fields: a field tracing study. – *Oecologia* (Berlin) 77, 39–43

Walter H. and Breckle S.-W. 1991: Ökologische Grundlagen in globaler Sicht. - In: Ökologie der Erde Bd. I, Stuttgart, 238 p.

Waltert, M., Meyer, B., Shanyangi, M.W., Balozi, J.J., Kitwara, O., Qolli, S., Krischke, H. and Mühlenberg, M. (2008): Foot surveys of large mammals in the woodlands of western Tanzania. – *Journal of Wildlife Management* 72: 603-610

Wassmer T., Jensen F.H., Fahlman A. & Murray D.L. 2020: Editorial: Ecology and Behaviour of Free-Ranging Animals Studied by Advanced Data-Logging and Tracking Techniques. – *Front. Ecol. Evol.* 8, Sec. Behavioral and Evolutionary Ecology. https://doi.org/10.3389/fevo.2020.00113

Weiss M., Didham R.K., Procházka J., Schlaghamerský J., Basset Y., Odegaard F., Tichechkin A., Schmidl J., Floren A., Curletti G., Aberlenc H.-P., Bail J., Barrios H. , Leponce M., Medianero E., Fagan L.L., Corbara B., Cizek L. 2019: Saproxylic beetles in tropical and temperate forests – a standardized comparison of vertical stratification patterns. – *For. Ecol. Manag.* 444 (2019), 50-58. https://doi.org/10.1016/j.foreco.2019.04.021

Westhoff V. and van der Maarel E. 1978: The Braun-Blanquet approach. p. 287-399. – In: Whittaker R.H. (ed.): *Handbook of vegetation science.* Part V. Classification of plant communities. - Springer, The Hague, The Netherlands. 408 p.

White, G.C. & Garrott R.A. 1990: *Analysis of Wildlife Radio-Tracking Data.* – Academic Press Inc., San Diego, California, 383.

Wich S., Scott L.. & Koh L.P. 2016: Wings for wildlife: the use of conservation drones, challenges and opportunities. p. 153–167. – In Sandvik K.B. & Jumbert M.G. (eds.): *The Good Drone*. Routledge, Oxon, UK, 213 p.

Wich S.A. & Koh L.P. 2018: *Conservation Drones: Mapping and Monitoring Biodiversity.* – Oxford University Press, Oxford, UK, 132 p.

Wikelski M. 2024: *The Internet of Animals: Discovering the Collective Intelligence of Life on Earth.* – Greystone Books, 264 p.

Wikelski M., Moxley J., Eaton-Mordas A., López-Uribe M.M., Holland R., Moskowitz D., Roubik D.W. & Kays R. 2010: Large-Range Movements of Neotropical Orchid Bees Observed via Radio Telemetry. – PLoS One 5(5): e10738. https://doi.org/10.1371/journal.pone.0010738

Wilmanns O. 1993: *Ökologische Pflanzensoziologie* (5. Aufl.). – Quelle & Meyer, Heidelberg, Wiesbaden, 479 p.

Xia, J., Huang, Q., Gui, Z., & Tu, W. (2024). *Open GIS*. Springer.

Ye, Z., Yang, M., Ren, Y., Hung, C.-H. J., Wu, C.-T. M., & Chen, P.-Y. (2023). Review on Recent Advances and Applications of Passive Harmonic RFID Systems. *IEEE Journal of Radio Frequency Identification*, *7*, 118–133. Presented at the IEEE Journal of Radio

Frequency Identification. https://doi.org/10.1109/JRFID.2023.3276310

Yu, H., Liu, X., Kong, B., Li, R., & Wang, G. (2019). Landscape ecology development supported by geospatial technologies: A review. *Ecological Informatics*, *51*, 185–192. https://doi.org/10.1016/j.ecoinf.2019.03.006

Zeballos, S. R., Acosta, A. T. R., Agüero, W. D., Ahumada, R. J., Almirón, M. G., Argibay, D. S., et al. (2023). Vegetation types of the Arid Chaco in Central-Western Argentina. *Vegetation Classification and Survey*, *4*, 167–188. https://doi.org/10.3897/VCS.100532

Zehm A., Nobis M. & Schwabe A. (2003). Multiparameter analysis of vertical vegetation structure based on digital image processing.—*Flora - Morphology, Distribution, Functional Ecology of Plants* 198(2), 142–160. https://doi.org/10.1078/0367-2530-00086

Zhu, X. (2016). *GIS for Environmental Applications: A practical approach* (1st edition.). London; New York, NY: Routledge.

Zhu, X. (2024). *Geographical Information Systems: A Practical Approach* (2nd edition.). New York: Routledge.

5 Procedures for Estimating Populations

Abstract

Many ecological studies on animal populations require an estimate of the population size or density of individuals. Distance sampling (DS) is the most commonly used method today. Distance sampling can also be used with indirect counting methods. Another group of methods for determining population size are mark-recapture methods. For these, the animals must be individually recognizable or marked. We discuss how to mark animals of different taxa so that hundreds of individuals can be identified using an adapted numbering key. Many animal species are individually identifiable by facial or color patterns. Photo databases help to archive these patterns.

5.1 Estimation of Abundance by Distance Sampling

Many studies of biological populations require estimates of population density, size, or rate of population change. Population size is closely related to population viability (depending on disturbances) and habitat capacity (dependent on habitat quality) and is, therefore, a concern for conservation issues (e.g. Groom et al. 2006, Cardinale et al. 2020).

Chapter 1 described various methods for capturing vertebrates. The most comprehensive and modern introduction to the estimation of population size using distance sampling (DS) was provided by Buckland et al. (2001, 2004); Schmidt et al. (2022), and it includes free software (Thomas et al. 2002, 2010)—for more recent updates, see also Miller et al. (2013). In some cases, distance sampling cannot be applied. Alternative methods are, for example, mark-release-recapture (MMR, also known as capture-mark-recapture, CMR) methods (see below) and, e.g., Feldmann et al. (1995), Settele et al. (1998), Kendall (1999), Silver et al. (2004), Sutherland (2006), Olsen (2009), and Seber and Schofield (2019).

Density and population size are related as:

$$N = D * A$$

(where A is an area, N is the number of individuals as population size, and D is the number per unit area.)

Distance sampling (DS) uses the perpendicular distances between objects seen from the transect line and the frequency of each distance class (Figs. 5.1 and 5.4).

The density is estimated by a function of objects seen within a certain distance from the transect. The detection function is central to the DS concept (Fig. 5.2).

Distance affects the detection: detectability usually decreases with increasing distance from the random line or point. Only a proportion of the objects of interest is detected in field surveys. Density estimation with DS requires a minimum

M. Mühlenberg, T. Waßmer, *Field Ecology*, https://doi.org/10.1007/978-3-031-87547-2_5

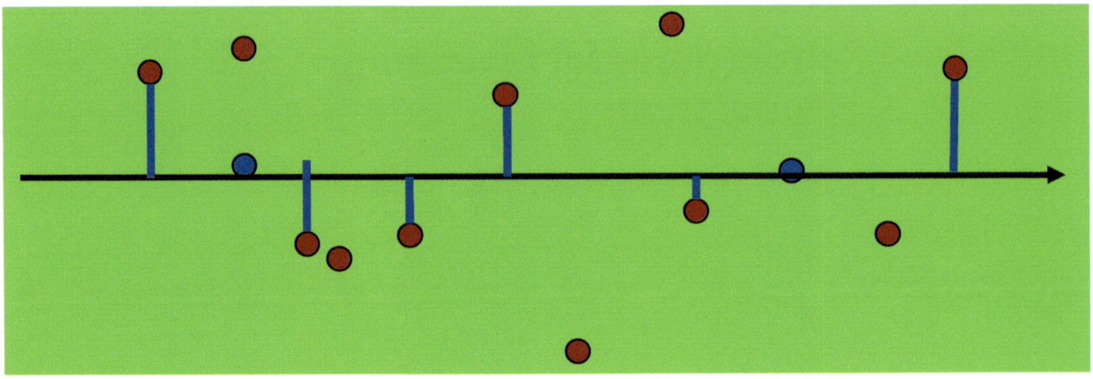

Fig. 5.1 Scheme for counting objects along a transect. The points connected by a blue line are detected and estimated based on their perpendicular distance to the transect; unconnected red points are overlooked objects not included in the count

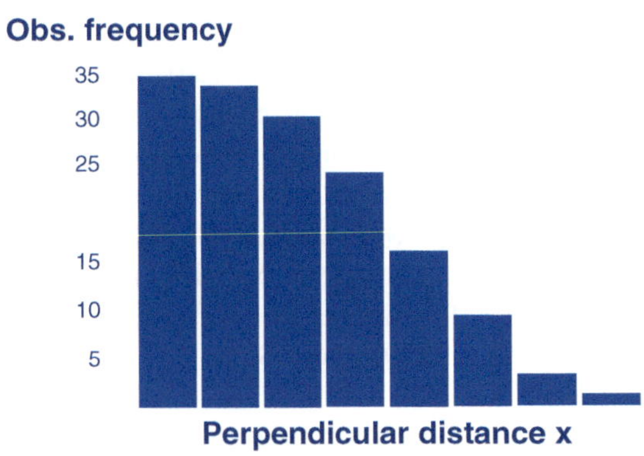

Fig. 5.2 An example of the distribution of recorded distances corresponding to the detected objects (detection function)

of about 40 observations (detections) to fulfill the statistical requirements (Buckland et al. 2001).

If an animal species lives in social groups, the detection of individuals may be replaced by clusters (distance measurement between transect line and center of the group, to calculate Indiv./km^2, the average group size must be recorded) (Figs. 5.3 and 5.4).

In larger areas like steppes or savannas, it may be easier to drive along tracks with a suitable vehicle and search the right and left sides for wildlife with two trained observers ("road strip counts," see Fig. 5.5). In this way, larger areas can be covered (Olson et al. 2005). It has been found that more animals are discovered when the car drives evenly and slower.

A prepared table for recording the observations could look like Table 5.1.

In expansive, easily visible areas, larger animals can also be counted using airplanes (Norton-Griffiths 1978; Steinhauer 1981; Caro 1999). To count a specific area from the air, the pilot must maintain the same altitude above the land surface. Two light rods are mounted on each side of the aircraft to define a counting strip. An aircraft similar to a Cessna (Fig. 5.6) is best suited, as it can fly relatively slowly, and the wings are located above the viewing windows, allowing an unobstructed view downwards. A four-seater machine is sufficient for the pilot and counters on each side. Counters must be tested before deployment for both recognition of animal species and their ability to withstand fluctuating flight without problems.

Two differences can be noted in comparing DS theory with classical finite population sampling theory:

- The size of the sample area is sometimes unknown.
- Many objects may not be detected for whatever reason.

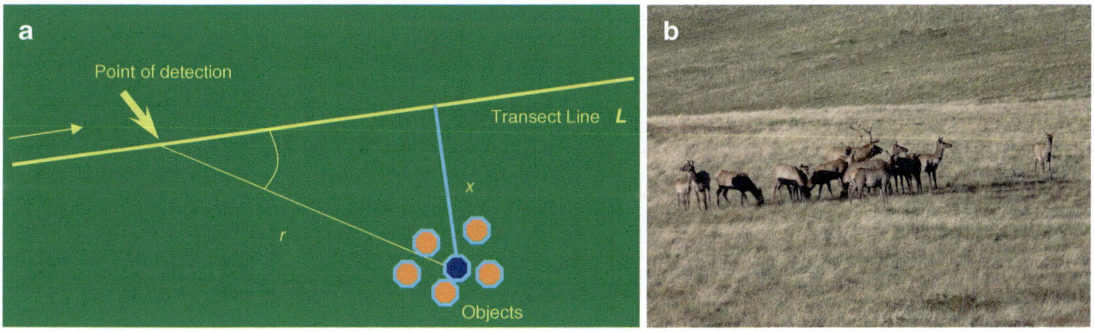

Fig. 5.3 (**a**) Recording animals in social groups. (**b**) Deer as an example of a social group. Hustai-NP, Central Mongolia

Fig. 5.4 Example of how a transect walker in a tropical forest should record the distances of birds discovered in the treetops. All objects detected are counted, and their perpendicular distance to the transect line (red) is noted. The hornbill is only 1 m beside the transect line, although it sits 30 m high in the canopy. For the software DISTANCE, use the distance projected on the ground. For example, for the butcherbird shown on the left side of the picture, record a 10 m perpendicular distance. Estimating distances in the forests requires trained observers, as the view is often too obstructed to use optical rangefinders that can be useful in more open habitats. (Quesada et al. 2010; Ransom 2011; Ransom et al. 2012)

Fig. 5.5 A pick-up model vehicle for road strip counts. When counting wildlife, the two observers stand in the loading area and can hold on to the high bar at the driver's cabin. Example from Comoé National Park in the north of the Ivory Coast, West Africa

While it seems that in large, primarily open landscapes like African savannas, *aerial surveys* of larger herbivores would be the most effective method for population estimates, Kiffner and Waltert (2009) have shown that *foot surveys* using the distance sampling method usually provide more precise data (Waltert et al. 2008; Greene et al. 2017; Petso et al. 2022).

Gray et al. (2012) provide examples of ungulate counts using the distance sampling method in Southeast Asian forests. Other recent examples of the use of distance sampling for different animal groups and also combinations with camera

Table 5.1 Example of a worksheet for recording wildlife observations on a road trip

Observer's name: ——— Date: ——— Weather conditions: ——— Examined region: ——

ID	Transect	Transect length (m)	Time	Distance along transect (m)	Species	No. of individuals	Distance (m)	Angle	Sex	Age class	Remarks
.											.

Fig. 5.6 A four-seater Cessna for aerial wildlife counts in the Comoé National Park, Ivory Coast

trapping can be found in Bellier et al. (2023), Crego et al. (2024), Cripps et al. (2021), Li et al. (2023), and Nolan et al. (2023).

Many mammal species are either shy or difficult to detect in denser vegetation. Then, it makes sense to record indirect signs of these species according to distance sampling.

We want to demonstrate an indirect method (pellet counts) combined with distance sampling using the example of deer, which are difficult to count everywhere (e.g., Smart et al. 2004; Krause 2008).

5.2 Population Estimates for Deer (*Cervus elaphus*)

Counting deer in a forest is challenging but would be the prerequisite for adaptive management of the deer population regarding hunting quotas and forest regeneration. Deer populations in Mongolia (*Cervus elaphus canadensis,* "Maral") are so severely threatened by hunting that they are legally protected in the country and are listed in the "Red List" category as "*critically endangered.*" Within 15 years, their population decreased by 92% due to illegal and uncontrolled hunting (Zahler 2004).

In many countries hosting populations of *Cervus elaphus,* like Canada, the United States, and Germany, the species is a highly regarded game animal with solid economic importance (e.g., Skonhoft et al. 2013; Gordon et al. 2004). Most management efforts in these countries have been directed toward producing harvestable populations of *Cervus elaphus* on a sustained yield basis (Mohler and Toweill 1982). The exact analysis of the existing deer population in quantity, structure, and quality is essential for a management plan regulating and controlling the deer harvest. However, determining these population parameters in the field is hard, and all known methods for counting deer in forested habitats have been somewhat inaccurate (Wagenknecht 2000; D'Eon et al. 2006). As the visibility in habitats with deer populations varies greatly, the accuracy of detecting animals during distance sampling can be considerably improved using thermal imaging (Gill et al. 1997; compare also Hughes et al. 2021). O'Connel & Ilse (1999) provide a compilation of the bibliography of population size estimation methods for deer populations in America.

Many camera traps are needed for abundance estimations by camera trapping (see Fig. 5.8b).

Fig. 5.7 (**a**) Mountain slope with low ground cover. (**b**) Forest habitat with high ground cover, Mongolia

Case Study: Pellet Counts for Deer Surveys

Line transect counts are often applied as a census technique for monitoring *Cervus elaphus* (O'Connell and Ilse 1999). When sampling the animals directly, they must be detected before they are disturbed and flee from the observer, which is a significant drawback of this method, especially when applied to Mongolia, as the assumed low population density and the elusive behavior of Maral will likely result in sampling bias (Marques et al. 2001). Counting dung of these species is often the best remaining option as an index of abundance. *Pellet counts* were used to inventory Maral and other hoofed big game animals (Davis et al. 2022; Franceschi et al. 2022). Mark-recapture methods can also provide density estimates if the collected pellet groups can be genetically analyzed. The DNA extracted from the pellets resembles an individual marker (Goode et al. 2014; see also Sect. 1.3).

Smart et al. (2004) compared three methods of estimating deer populations in forests and found that *fecal standing crop* was the most accurate estimation method, surpassed only by *thermal imaging by distance sampling*.

Considerable research has been conducted on pellet group counts for deer. The method was applied as a standard census technique for different deer species in many states of the United States (Lewis 1970; Sample et al. 2022), Britain (Marques et al. 2001), and New Zealand (Forsyth 2005). Generally, the counts were made in winter or spring, and the last snowfall incident was used to estimate the period during which the pellets were deposited. In other studies, the sample plots were cleared from old pellets (Marques et al. 2001) to estimate the age of the samples. If plots or transects cannot be cleared before the study, all pellets must be counted, and the local decomposition rate must be determined. Ideally, the decomposition time of dung should be monitored by locating random, or at least representative, fresh pellet groups throughout the study area (Marques et al. 2001). The defecation rates should be estimated for the specific population. This can be achieved best by observing captive animals (Fig. 5.8).

The decomposition rate was estimated by experimenting with fresh dung on two sites, open land and forest (Fig. 5.7).

The vast study area was subdivided into strata according to vegetation types (forest,

Fig. 5.8 Maral (*Cervus e. canadensis*) in a pen in Northern Mongolia, object for estimating the defecation rate

steppe) and altitude. If possible, the stratification should be done in advance using satellite imagery or vegetation maps. Within each stratum, implement randomly distributed line transects using random numbers. Plan to establish 15 to 20 transects within a stratum (Marques et al. 2001). At least 40 to 80 sightings per stratum must be recorded to obtain statistically sound data. The minimum transect length is determined by each stratum's pellet density encountered in the field. Before sampling, the pellet defecation and decay rates are evaluated by observing fenced Maral (Fig. 5.8).

Sampling Procedure

For the data collection, two random points were generated on a Google Earth map of the Khentei region, Northern Mongolia, from which two quadrats were drawn up (5 × 5 km) in a northwest direction, covering an area of 50 km². This map was then overlaid by a 1 × 1 km grid, resulting in 25 intersections, which are the starting points of the transects. The transects were surveyed triangularly to return to the starting point after one survey (Fig. 5.9). The triangle had a side length of 400 m, resulting in a transect length of 1.2 km. The 50 starting points were located in the field using GPS (see Sect. 4.4). In total, 60 km of transect lines were surveyed by this field design.

A recording sheet could look like Table 5.2

The density N of the Maral can be calculated as follows:

$$N = \text{Dung density} / \text{decomposition rate} / \text{daily defecation rate} = 100 \times N \text{ per km}^2.$$

As it is impossible to distinguish individuals with certainty, this method is of little use, see Sect. 1.2.

Distance sampling also records birds on line transects and point-stop counts, see Sects. 1.7 and 1.8. In Chap. 2, methods for recording invertebrates and small vertebrates were presented, which also provided information on the abundance of the animals. If possible, it should always be checked whether the distance sampling method can also be used for invertebrates.

5.3 Population Size Estimates from Recapture Results

The principle of population size estimation using mark-recapture methods is that the ratio of the frequencies of known to unknown individuals in a population can be used to extrapolate the total number of individuals in a population (summary in Begon 1979).

Animals are caught in a specific area, marked with a suitable technique, and released. After a customized sampling interval, individuals from the same population are again captured, marked, and released. Ideally, this procedure should be repeated for at least ten sampling intervals (minimum three intervals). At each sampling date, the number of animals captured in total, the number of marked and released unharmed animals, and the number of marked and recaptured animals with a note of their marking date (see example below). Be careful when applying calculation methods, as the biological prerequisites for the

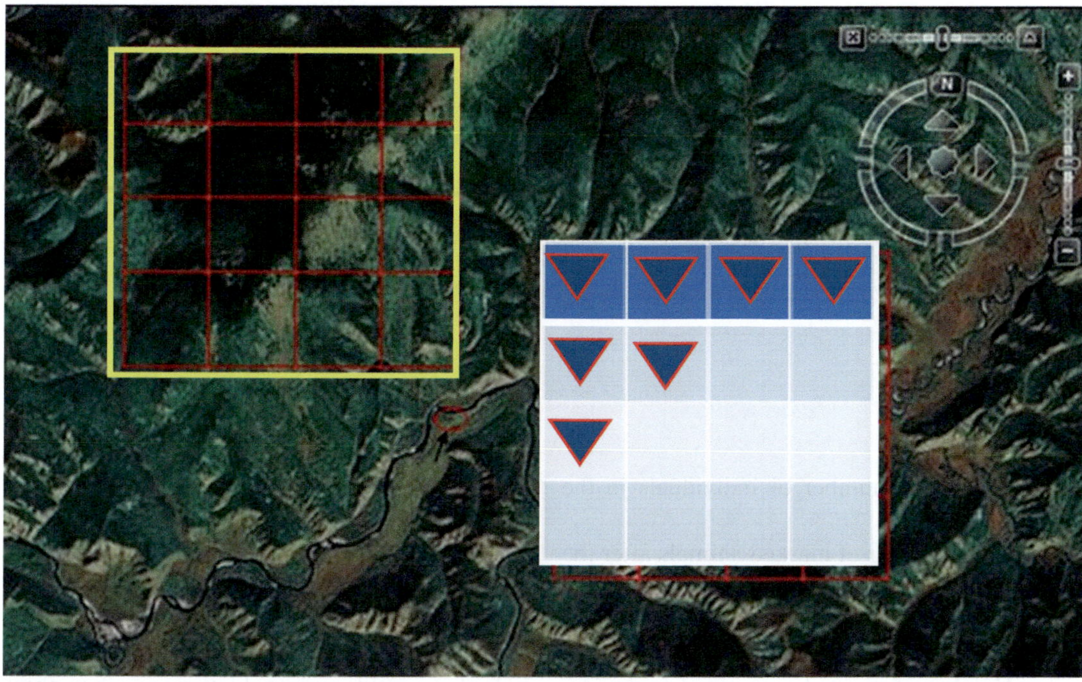

Fig. 5.9 Satellite picture of the study area and the location of the quadrats. They were drawn up with Google Earth in 2008. (From Krause 2008)

Table 5.2 Possible recording sheet for pallet counts

Transect no., Date/ Time	Droppings/Degree of decomposition	Distance from transect line in m	Type of vegetation	Coordinate's location
.				.

selected calculation rule must be matched (Begon 1983). Often, more straightforward calculations require stricter assumptions. One of the advantages of the Jolly–Seber method (see below) is that it allows for documenting population fluctuations between individual days.

As a manageable, summarizing book on recapture analyses, we recommend Begon (1979). Refer to Otis et al. (1978) for more extensive statistical analyses. Buckland (1980) extends the computational analysis and gives an example of a bird study.

The following is intended to provide only a brief overview of standard calculation methods:

Mark-recapture methods can be divided into three stages in field practice:

1. Mark once and recapture once.
2. Mark once and recapture multiple times (or reobserve, mark, and resight).
3. Mark multiple times and recapture multiple times.

1. Mark once and recapture once.

With only one recapture (e.g., Lincoln-Index), the population must be "closed," meaning immigration/emigration must not occur. During the collection period, there should be no unknown births or deaths. If the marking refers to parts of the population (e.g., certain age classes), then mortality must not differ between these classes.

(1) The Petersen- or Lincoln-Index:

$$\frac{N}{c} = \frac{m}{r} N = \frac{m \cdot c}{r}$$

Table 5.3 Compilation of recapture data following the days the animals were last captured for the analysis according to the Jolly–Seber method. R_i = Sum of all animals from a_i that were marked and released on day i and last recaptured after day i

Total number captured on day i, n_i	Total number released on day i, a_i	Last capture day j (i – 1) Number of marked animals recaptured					
n_{i1}	a_{i1}	1					
n_{i2}	a_{i2}	r_{22}	2				
n_{i3}	a_{i3}	r_{23}	r_{33}	3			
n_{i4}	a_{i4}	r_{24}	r_{34}	r_{44}	4		
n_{i5}	a_{i5}	r_{25}	r_{35}	r_{45}	r_{55}	5	
							6
R_i =							

N = Estimated number of individuals in the total population.

m = Total number of marked animals after the first catch.

c = Total number of captured (observed) animals in the second sample.

r = Number of marked recaptured (rediscovered) animals in the second sample.

2. If it is difficult to catch and mark the animals multiple times, the easiest way is to repeat the recapture or reobservation multiple times after the first catch and marking. An average value for N can then be formed from the respective estimates after calculation (1).
3. If you can catch and mark your study animals multiple times in a consecutive series, you can estimate the population size and the loss of marked animals (deaths and emigrations) and, from this, calculate their survival rate. Of several possible calculation methods, we highly recommend the Jolly–Seber method. It is still relatively easy to calculate "by hand" and reliable if you can capture 9% or more of the population (Table 5.3). Specifically for small samples (<10%), the so-called Fischer–Ford method would be preferable, but its calculation is quite complicated (see Begon 1979). For the Jolly–Seber method, several pieces of software are available for evaluation on PCs and other platforms.

Further suggestions for calculating the Jolly–Seber capture–recapture models are given by, for example, Buckland (1980) and Pledger et al. (2003). The capture–recapture method is also extensively treated in Henderson (2021).

The free software MARK (White and Burnham 1999; White and Cooch 2019) analyzes mark-recapture data based on almost any scenario or mathematical model developed. Its tutorials also provide an introduction to each method.

When capturing animals for marking, additional data that cannot be determined by observation, such as sex and size measurements (body length, wing length, tail length), infestation with parasites, or temporarily transported animals (e.g., *Triungulinus* larvae, mites), are usually collected. Such data are part of the standard protocol for bird ringing.

Sweep-net captures are also used for mark-recapture studies to estimate the population size. For example, the caught insect is marked, released, and recaptured in another sample. From the recapture rate of marked specimens, it is possible to calculate the relative density of populations according to different models. For further explanations, see, e.g., Ehrlich and Davidson (1960), Feldmann et al. (1995), Henderson (2021), Settele et al. (1998), and Krebs (2014).

Prerequisites for applying mark-recapture methods are easy detectability of the individuals and a marking method that allows individual marks for each individual, cannot be lost, and does not increase or decrease the carrier's fitness. All animals must be equally catchable, and there should be a good mixing of (marked) individuals within the population. The lifespan of the individuals must exceed the study time frame. It is advisable to test the durability and legibility of the markings over

Fig. 5.10 An individual number can be written on the underside of a butterfly's wing with a felt-tip pen

the entire study period in the lab—or even better, in a pilot in the field. Some simple procedures also depend on closed populations without births, deaths, immigration, or emigration.

These conditions are well met in ground beetles (Carabidae), tiger beetles (Cicindelidae), wolf spiders (Lycosidae), cicadas (Cicadina), field grasshoppers (Caelifera), damselflies (Zygoptera) in wet or riverside meadows, selected fly groups (Brachycera) and butterflies (Lepidoptera), small mammals (esp. Arvicolidae and Muridae), and birds (Passeriformes), in exceptional cases also for snails with shells.

May (1978) found surprisingly well-matching results between the mark-recapture and the D-Vac capture method for Auchenorrhyncha (cicadas). He collected the Delphacidae with an exhauster for the recapture, as the studied species primarily lives at the base of grass tufts and is challenging to catch with sweep nets.

5.3.1 Marking of Individuals

The marking of individuals depends on the species. An overview of various marking methods is provided by Stonehouse (ed. 1978), specifically for birds (Lockley and Russell 1953; Bub 1995; Bub and Oelke 1980). Options for different animal groups are presented below.

Butterflies are caught for marking with the net, taken directly from the net with the fingers, and the wings are labeled with a waterproof fiber pen (see Fig. 5.10). The animals must not be pressed, especially on the thorax. Staedtler pens are reliable, such as the Lumocolor 318F, available in various colors. Depending on how the butterfly usually holds its wings at rest, label either the top or bottom for later observation. You can easily write/paint a two- to three-digit number on day butterflies. To minimize the loss of scales, press the wing onto a small metal base for labeling. Another pattern of individual marking of up to 1000 animals is shown in Fig. 5.11. The marked butterflies are then released back into the center of their territory (Ehrlich and Davidson 1960). Gall (1985) provides an overview of applying and evaluating various mark-recapture methods for butterflies, and Schtickzelle et al. (2002) and Turlure et al. (2018) provide recent examples of MRR studies of butterflies.

One way to apply long-lasting markings on the wing is by using small, self-adhesive plates. However, the wing must be pierced at its leading edge for a small hole so that the self-adhesive plates stick together from both sides. Such a marking is standard for migratory butterflies. Suppose the animals need to be recaptured for identification within a study area. In that case, the method is disadvantageous as butterflies do not get caught as easily during the second, third, and subsequent catches. They evade more quickly. In such a case, evaluating the recapture results will be quite complicated (Singer and Wedlake 1981; Mallet et al. 1987).

The study of larval biology (mortality, dispersal, weight gain) is indispensable in butterfly population analyses. It is usually the last larval stage, at which the increase in weight determines the reproductive capacity of the adult female (White and Singer 1987). Butterfly caterpillars are marked with color dots, preferably on lateral bristles or protrusions, so an easily recognizable code with 240 variations for three marks per individual using six colors at six positions is achieved (Fig. 5.11i). Usually, the last larval stage lives for 2–3 weeks. When studying butterfly caterpillars, it must be considered that many species only climb into the vegetation to search for food at night and can retreat during the day. In this case, nighttime checks become essential.

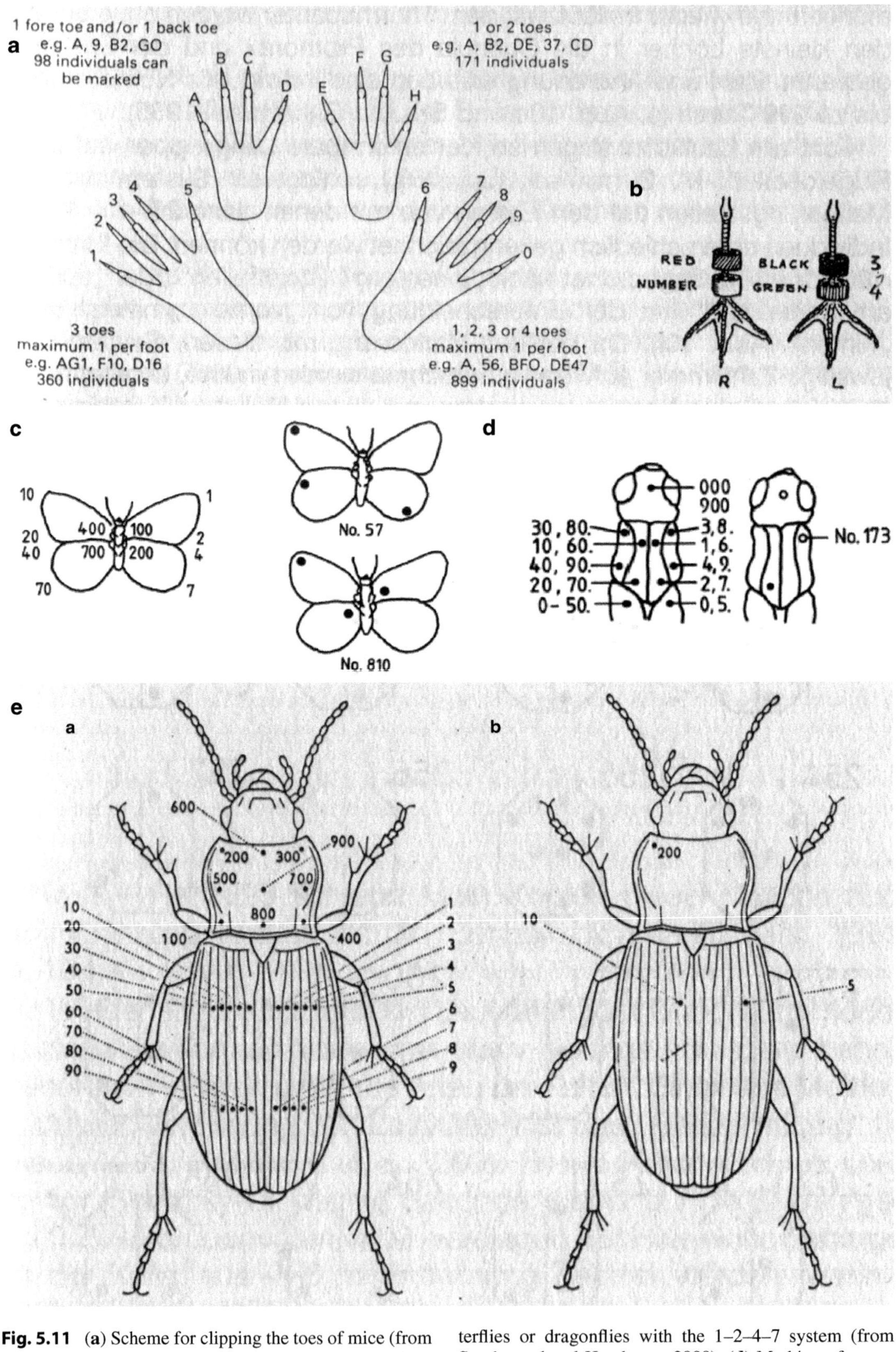

Fig. 5.11 (**a**) Scheme for clipping the toes of mice (from Begon 1979), (**b**) Color banding of birds (from Lockley and Russel 1953), see also Sect. 1.14. (**c**) Marking of butterflies or dragonflies with the 1–2–4–7 system (from Southwood and Henderson 2000). (**d**) Marking of grasshoppers. (From Southwood and Henderson 2000).

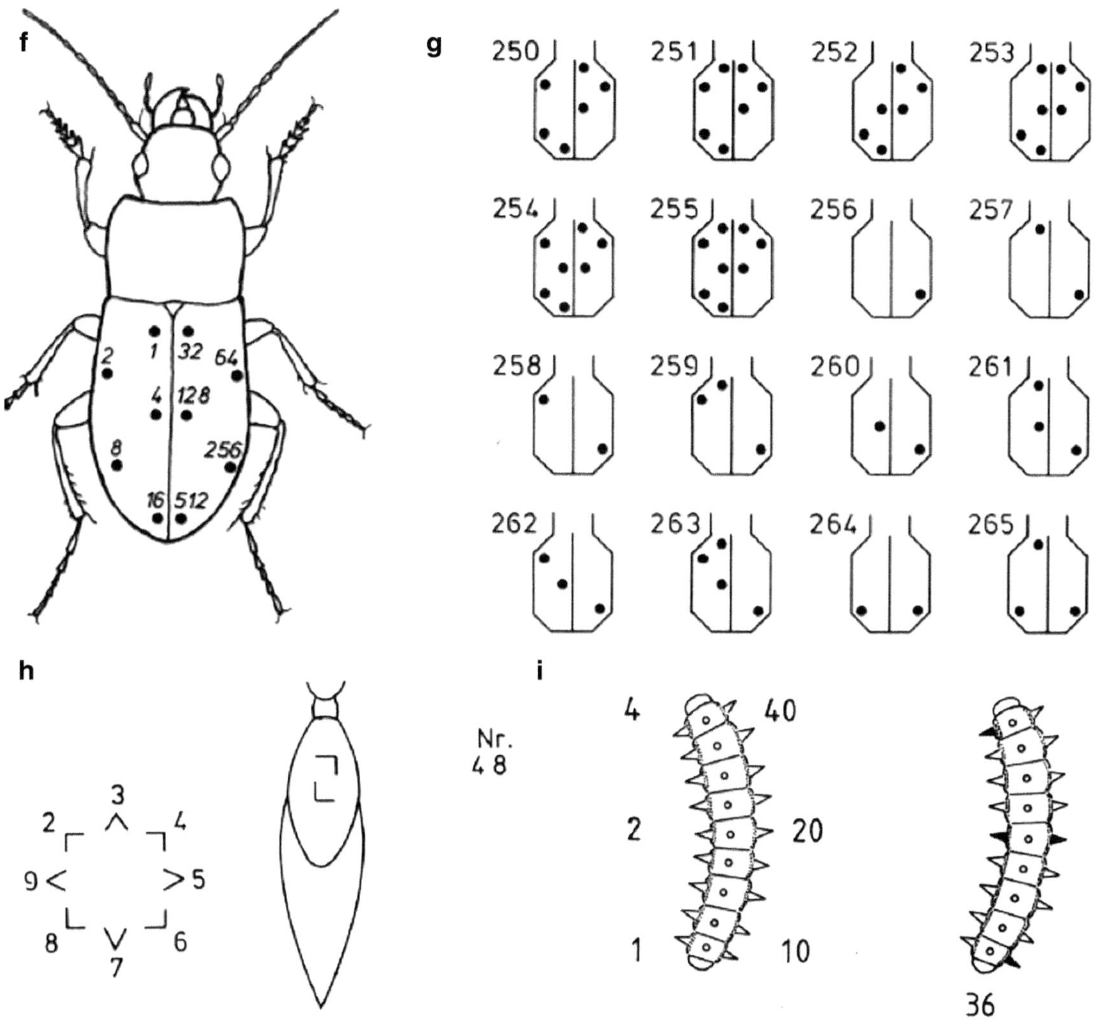

Fig. 5.11 (continued)

A *ground beetle* population can be studied by catching the animals alive with a grid net at pitfall traps (see Sect. 2.16), marking them in the laboratory or field with a thermocauter (available from medical products distributors at prices of about $/€ 20, modern thermocauters are more expensive) on the wing covers (elytra), and rereleasing them. With the thermocauter, tiny holes are burned into the cuticle of the prothorax and the elytra without damage. Their number and arrangement allow an individual numbering up to 999 animals (see Fig. 5.11e and Schjøtz-Christensen 1965).

Not all ground beetles carry such recognizable striae (longitudinal ribs) on the wing covers. M. Butterweck (unpublished) proposed a system with ten marking points on the elytra, with which $2^{10}-1 = 1023$ specimens can be marked individu-

Fig. 5.11 (continued) (**e**) Marking of ground beetles with the thermocauter, (**f**) Alternative to (**e**) for marking ground beetles when the striae on the wing covers are not recognizable, (**g**) Marking template according to (**f**). (**h**) Marking of slugs. (**i**) Marking of butterfly caterpillars ally. The marking locations are arranged according to "inside on the suture" or "outside on the margin" with the distinction of "front," "middle," and "back" (Fig. 5.11f). Since marking with this system would always require the respective number to be calculated laboriously, it is advisable to prepare a marking template with all possible combinations from No. 1 to No. 1023 and use copies of this template for record keeping (M. Butterweck, written comm. 1991, cf. Fig. 5.11g).

Generally, it is advantageous always to use the same number of marking points in a point system, e.g., three. Should a marking point fail, this error is immediately recognized, and the numbering is not misinterpreted. In the above example (d), two colors are needed for 1–99 and 10 for the hundreds. In the 1–2–4–7 system (example c above), up to 1000 animals can be marked individually with only one color: 1 = 1, 2 = 2, 3 = 1 + 2, 4 = 4, 5 = 4 + 1, 6 = 4 + 2, 7 = 7, 8 = 7 + 1, 9 = 7 + 2, 10 = 7 + 4. Proceed accordingly with the coding for the tens and hundreds. In example (e) above, the point pattern provides the possibility to distinguish up to 999 animals individually. B shows a ground beetle with the marking number 215. The points are burned onto the cuticle with a thermocauter (e.g., available at https://shop.bioskin.it/gb/81-thermocautery) (Table 5.4).

The mark-recapture method is preferable to the quadrat method for ground beetles when the species have a relatively low abundance, particularly for larger species.

For field crickets, the quadrat method (see above) is superior to the mark-recapture method for multi-week studies, as the juvenile stages molt more often and thereby lose their markings. For example, in England, the juvenile stages of *Chorthippus brunneus* last between 10 and 20 days, and the adults live on average 50 days (Bradley 1985). Overall, the biocoenometer method has also been shown to be reliable for abundance determinations in grasshoppers. 87% of all grasshoppers in a meadow were found in the upper, 13% in the lower vegetation layer, or on the ground when the animals are sucked out of the biocoenometer (Köhler 1987). The relative catchability is most balanced throughout the vegetation period using quantitative netting ("catch per unit effort," Sect. 2.5). However, this method only catches 9–14% of the grasshoppers compared to the biocoenometer.

Running spiders are more difficult to mark. To prevent the animals from losing legs when being handled after catching them in ground traps or by hand in the field, spray them briefly with a freezing agent (available in drugstores, sports stores, or electronics retail). Anesthetizing the spiders with carbon dioxide is more cumbersome, but it also works in dense vegetation like in grass tufts. A steel cylinder is placed over the chosen habitat structure, into which the anesthetic gas is introduced from a CO_2 bottle. As the gas is denser than air, it fills the tube from the bottom up. The spiders move down in the tube and are anesthetized. The duration of the gassing is tested beforehand in the laboratory and adapted to the cylinder volume. Experience values for a cylinder with a 10 cm diameter and 20 cm height are around 40 s (Schmid 1991). Nail polish, model paint, and craft paint, which become water-insoluble like dispersion paint or small cans of car paint, are suitable for color marking. The color point is applied with a needle head (dissecting needle) on the prosoma (do not touch the legs with their trichobothria!). Non-damaging are colorless shellacs, dissolved in undenatured, absolute ethanol and mixed with various powder colors. To

Table 5.4 Number of possible combinations when marking birds with color rings (Fig. 5.11b)

Number of rings (incl. Numbered rings)	Colors (and their numbers)				
1	Yellow (1)	Red or Pink (2)	Green (3)	Black (4)	White (5)
2	6	12	18	24	30
3	12	8	108	192	300
4	10	80	270	640	1250
5	6	90	486	1536	3750

ensure the marking does not harm the animal, keep marked animals temporarily in a terrarium for the same duration as the planned field investigation.

Cicadas are marked with color points on the prothorax and *small dragonflies* on the wings (Parr 1965). Although grasshoppers can also be easily marked with color points on the prothorax, a line transect is generally more suitable for direct density determination than recapture methods. Note the objective: estimation of density/area or population size (see above).

Chilopods can be fed with Sudan Black. This harmless dye makes the fats green and visible. Adult chilopods, grasshoppers, and woodlice can be provided with a piece of reflective foil. The animals can then be easily found again in the dark with a flashlight, significantly increasing the overall recapture rate. For field grasshoppers, it is recommended to wrap small self-adhesive foil strips (about 2 × 4 mm) around the middle femur with tweezers and stick them together at the open ends (Herold 1990). Since the foil reflects most strongly in the direction of the light beam (flashlight at eye level, headlamp), animals can be located up to 50 m or even further by the flashing of the reflection point. Both marking methods do not allow recognition of individual animals but identify them as previously captured specimens, increasing the chances of a recapture.

Scorpions fluoresce UV light. It is, therefore, easy to spot them at night with a UV torch, e.g., in open country (desert, semi-desert) (e.g., Kloock et al. 2010; Lourenço 2012; Gaffin et al. 2012). If you walk a transect through the area, the fluorescent scorpions can be easily detected and counted. Begon (1978) and Begon et al. (1975) dusted *Drosophila flies* with a UV-fluorescent, finely dispersed agent for density determination (see Crumpacker 1974) and then identified animals recaptured at bait traps with UV light. The flies were stunned in a small jar with CO_2 and briefly rolled in the fluorescent dust.

Since some *snails* adjust their activity to the microclimate and retreat when the weather conditions are unfavorable for them, they are often very clumped and, therefore, poorly suited for counting using the quadrat method. A mark-recapture method is recommended for shell snails: the snails are marked with small holes drilled into the shell near the mouth and randomly distributed in the field in the evening when the daytime temperatures drop again. The control catches are made at dawn during their highest activity (Greenwood 1974). To allow sufficient mixing of marked animals with unmarked ones, the intervals between the sampling samples should be longer than one week, preferably three weeks. For studies under one year, numbers painted on the shell with weatherproof paint are sufficient for marking. The recapture method has also been successfully used to monitor migration movements in snails (e.g., *Cepaea nemoralis*), which provide information about immigration and emigration in open populations and about the dispersal rate (Cameron and Williamson 1977).

Slugs can be individually marked by "cold branding" (Richter 1976). A right-angled silver bar with an edge length of 4 mm and a thickness of 1 mm in a tool holder is used. This marking rod is dipped in liquid nitrogen for about 1 min, and then the angle is pressed onto the mantle shield of the snail for 2–4 s. Numbers can now be assigned to the orientation of the angled burn wound (Fig. 5.11h). If you first burn in one, then two numbers, you get 72 marking possibilities, which last at least six months in the large adult road and leech snails. For more extended checks, the snails must be rebranded.

Many *tail amphibians* (Urodela) have individual drawing patterns, which can be recognized in subsequent catches when recorded in a photo database. In fire salamanders, it is the dorsal drawing pattern of the three- and multi-year-old animals (Feldmann 1981; Schulte et al. 2024; see Fig. 5.12c,d); in pond and crested newts, it is the belly spot pattern (Hagstrom 1973). A multi-year capture–recapture study on the movement patterns of fire salamanders showed that the mean movement distances in disturbed habitats with interspersed residential areas were greater than in a more natural forest habitat (Kiss et al. 2022). For the importance of stepping stones, see also Sects. 11.3 and 11.4.

Some toad species have an individually recognizable dorsal pattern, see Fig. 5.12a, b.

Fig. 5.12 (**a**) Mongolian toad, *Strauchbufo raddei*, Southern Mongolia, Gobi. The dorsal pattern is individually different. (**b**) *Strauchbufo raddei*, Northern Mongolia. (Photo credit: Gantulga Bayandonoi). (**c**) Fire salamander (*Salamandra salamandra)* in a German forest. The drawing pattern can be used to distinguish individual animals. (Photo credit: Christoph Parsch). (**d**) Fire salamander (*Salamandra salamandra)* in the National Park Harz, Central Germany. You can see the different pattern compared to the individual in Fig. **c**. (Photo credit: Eva Mühlenberg-Horn)

Photographs of individual toads have also been used for citizen science monitoring (Smith et al. 2018) (Sect. 2.28).

A comprehensive overview of the marking of amphibians and reptiles is given by Ferner (1979), Henle et al. (1997), Odum and Sonntag (2010), and Woodbury et al. (1956). Alternative but controversial methods are toe-clipping (May 2004; McCarthy and Parris 2004) and visible implant elastomer (VIE) tags (Heard et al. 2008; Brannelly et al. 2013; Petit et al. 2012; Waudby and Petit 2020). The movement patterns of juveniles of spotted salamanders (*Ambystoma maculatum*) were also tracked using fluorescent powder (Pittman and Semlitsch 2013).

A new technique in the individual marking of vertebrates consists of the use of microchips (passive integrated transponder (PIT) or radiofrequency identification (RFID) tags). Tiny microchip tags (e.g., 11 mm long, 2 mm diameter) are injected subcutaneously into individual animals without surgery or are attached using bands or neck collars (Fig. 5.13a) and allow a unique, long-term identification with a scanner without touching the animal. However, as the scanner/reader and the tag must be in close proximity to each other, the researcher either has to be physically close to the (recaptured) animal using a handheld scanner, or a battery-powered and weatherproofed field scanner needs to be close to a location that animals frequently visit, such as a feeding station (Fig. 5.13). The technology is extensively used for amphibia (Ribas et al. 2022; Testud et al. 2019), birds (Bonter and Bridge 2011; Hughes et al. 2021; Sandoval et al. 2022), and mammals (Rose et al. 2020, 2022). The equipment for the electronic animal identification system (handheld or field reader, micro transpon-

Fig. 5.13 Use of an RFID tag to track the activity of individual squirrels. (**a**) A fox squirrel collared with a passive integrated transponder (PIT) in the dorsal portion of the neck collar is visiting the feeding station in a Tomahawk life trap. To get to the nut, the animal has to pass under the receiver (**b**), which consists of the Generation 3 RFID receiver (Bridge and Wilhelm 2021) and a two-loop antenna

der in implantation cannula, implant device, and battery charger) is available, for example, from https://www.telinject.de for Europe and North America from https://www.uidevices.com and https://cyntag.com.

Much has been written about the capture and marking of mice. Summaries can be found in Delany (1974), Kaufman et al. (1978), Gurnell and Flowerdew (1982), and Clark (1983). The systematic recording of mice occurrences and their population densities is described in Sect. 2.18.

The marking of rodents is done, for example, by punching out a slight semicircle on the outer edge of the ears, applying commercially available ear tagging (Wassmer and Refinetti 2016, 2019) or ear tattoos (Petit et al. 2012), or toe-clipping according to a particular pattern (see Fig. 5.11a). When clipping toes, the selected toe is anesthetized with a drop of ethyl chloride, and

Case Study: Analyzing the Population Structure of the Rainbow Scarab by Mark-Release-Recapture Studies

A study by Thomas Wassmer and Elise Armstrong, Siena Heights University, published as an open-access article (Wassmer and Armstrong 2023).

Until now, little is known about the population structure and mobility of temperate dung beetles, including the rainbow scarab, *Phanaeus vindex* (MacLeay 1819), although this knowledge is essential for their conservation as pastures become increasingly rare and the landscape fragmented by monocultures and urban sprawl.

Populations of animals can be studied by marking or tagging animals reliably to conduct mark-release-recapture (MRR) studies, also known as capture-mark-recapture (CMR) studies (Henderson 2021; Krebs 1999; Seber 2002). For MRR studies, dung beetles can be caught alive using a modified pitfall trap with a funnel inserted into the pitfall bucket (Fig. 5.14a–c), allowing beetles to enter the trap easily while minimizing the chance of escape (Wassmer and Armstrong 2023). Captured dung beetles can then be marked using an oil paint marker (Fig. 5.14d) or by applying an individual tattoo using a Dremel tool (Fig. 5.14e, f).

Tattoo markings are more time-consuming to apply and require a delicate balance of enough force to apply a shallow scraping into the elytron cuticle while using enough restraints not to penetrate the hemolymph and injure or kill the beetle (Fig. 5.14e and this video: https://youtu.be/wxSU3c-TOy6M). However, due to the burying

Fig. 5.14 (**a–c**) Live trap design. (**a**) 1-gallon bucket in the ground inserted funnel. (**b**) 19 mm wire mesh placed over the funnel and staked down with dung bait. (**c**) Transparent rain guard. Rainbow scarabs were marked using an oil paint marker (**d**) or a Dremel tool (**e**) to apply a tattoo pattern of 1–6 dots per elytron (**f**)

activity of dung beetles, paint marks did not last long enough to allow the identification of recaptures.

Of 470 rainbow scarabs, only 14 were recaptured once, and two were recaptured twice. One beetle was recaptured after 338 days of surviving the cold Michigan winter. Population estimates ranged between 217 and 491 beetles per pasture and were calculated using the MARK 9.0 and 8.2 software (White and Burnham 1999; White and Cooch 2019). This software allows for estimating populations under many different scenarios and models. We decided on the Jolly–Seber model with the POPAN parametrization (Schwarz and Arnason 1996) as it does not assume a closed population (which most natural populations are not) and adjusts for births, immigration, deaths, and emigration. It requires repeated sampling and marking and assumes that the likelihood of capturing marked and unmarked individuals is the same (Fig. 5.15).

This approach allowed us to estimate population size N and three primary parameters: φ_i—apparent survival probability, p_i—recapture probability (catchability), and $pent_i$—the probability of entering the population (recruitment rate: combining birth and immigration). Given the low recapture rate in this study, the only model we considered a priori was $\varphi(.)\,p(.)\,pent(t)$, in which we kept survival φ and catchability p constant $\varphi(.)\,p(.)$ because we could not estimate a time variation in these variables. In contrast, we could use the recruitment rate $pent(t)$ as we captured many new individuals in each sampling event. Survival and catchability may not be constant over time, leading to biased population size estimates (N. Schtickzelle, personal communication).

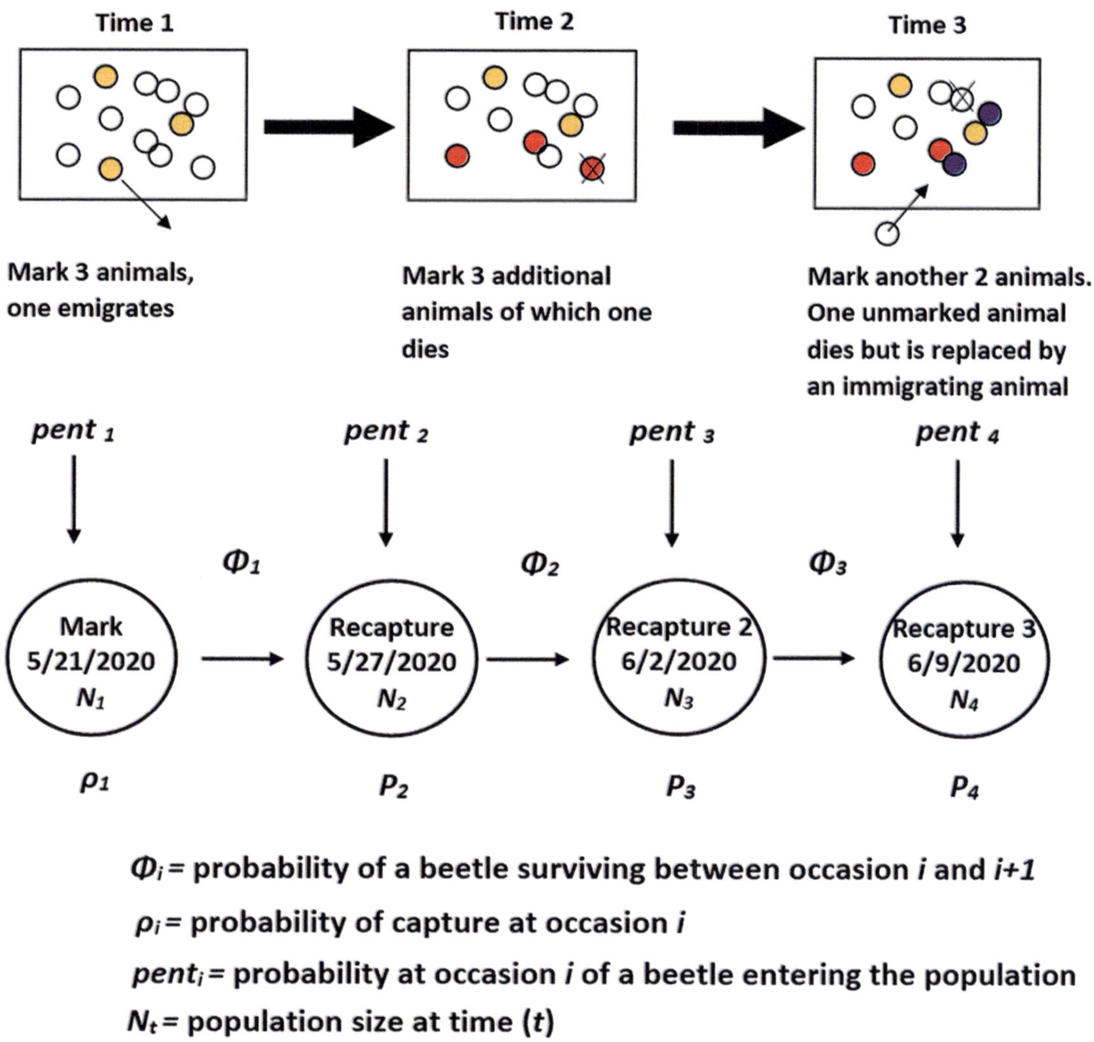

Fig. 5.15 Upper panel: Principle of the Jolly–Seber population estimation model. Lower panel: Conceptual Jolly–Seber model (POPAN formulation) used to estimate population size

after a few seconds, the nail and first joint are quickly cut off with scissors (Begon 1979). The choice of method depends on the duration of the study, the question at hand (Giles 1969; Twigg 1975; Mader and Pauritsch 1981), as well as the avoidance of artifacts and ethical considerations (Jung et al. 2020; Perry et al. 2011).

For the evaluation of recapture data, Nichols and Pollock (1983) recommend calculations according to the Jolly–Seber model. Smith (1968) compared the results of a population estimate of mice (genera *Mus* and *Peromyscus*) based on recapture data with a direct count of the animals by excavating all mouse burrows in the field. In short-term studies on the population structure using live capture, systematic errors arise because the sexes' movement activity is unequal and depends on the reproductive cycle. Pregnant field mouse females cover daily distances of several 100 m, although field mice have been attributed to an action range of only 300–400 m^2 (H. M. Steiner, written comm. 1990). More extended collection periods can mitigate this error (bias). The recapture calculations overestimated one part of the population while underestimating another. Baker (1990) provides a comprehensive overview of the systematic errors that can occur when trapping mice (trap biases). Heise and Wieland (1991) have tested the hole-treading method for simple and time-saving density estimation in field mouse populations (*Microtus arvalis*). All field mouse holes are plugged into an area of 16 m × 16 m, and the number of holes reopened after 24 h is determined. According to experience, one individual opens, on average, 2.5 holes. The relationship of (number of holes reopened)/2.5 can be used to estimate the number of animals living on 1000 m^2. The factor 2.5 is valid for spring and autumn population controls. In early summer, one should instead use a factor of 1.34. Dickman (1986) describes the use of hair collection tubes for population estimates of small mammals in urban areas. The tubes contain a tape to which visiting mice lose loose hairs. The hairs are later collected and identified (Hair Atlas in Sect. 7.3). The tubes are cut to 10 cm from a PVC pipe with an inner diameter of 3 cm and given a double-sided adhesive tape on the inside ceiling. A few oat flakes, fly pupae, or apple pieces are scattered on the bottom of the tube as bait. Dickman found that the hair tubes effectively capture the species in an area compared to conventional live trapping but that they cannot provide solid abundance determinations. At best, one obtains a frequency index in comparison.

With individually marked animals, recapture or reobservation opens up various possibilities for conducting residence and migration controls. Studies of movement patterns are particularly interesting for animals in the human-altered cultural landscape (e.g., Mader 1979, see Sect. 11.5, Animal Dispersal-Corridors).

5.4 Differentiation of Photo Recordings

If individuals of an elusive species can be distinguished from each other, *photo traps* are optimal in estimating the population size. Some relevant new papers are Karanth and Nichols (1998), Karanth et al. (2004), Jennelle et al. (2002), Carbone et al. (2002), and Bian et al. (2023).

Upon closer analysis, many more mammal species can be individually identified photographically (Karczmarski et al. 2022). In addition

Fig. 5.16 African Painted Dogs (*Lycaon pictus*) in Selous NP, Tanzania. Note the different spotting of the individuals

Fig. 5.17 Masai Giraffe (*Giraffa tippelskirchi*) in Saadani NP, Tanzania. Note the individual patterns of the reticulated spots

Fig. 5.18 Grevy's Zebra (*Equus grevyi*) in Ruaha NP, Tanzania. The stripe pattern differs significantly between individuals

Fig. 5.19 African Savanna Elephant (*Loxodonta africana*) in Mkomazi NP, Tanzania. The head and folds of the ear show individual variation

to the striking examples of African wild dogs, giraffes, and zebras, individual structures can

Case Study: Photo-ID Using Camera Trapping

Based on: Using camera-trap photographs to identify individual fox squirrels (*Sciurus niger*) in the Southeastern United States (Tye et al. 2015).

In the SE of the United States, it has been challenging to study fox squirrels (*Sciurus niger*) due to their rarity, extended home ranges, and low capture rates. Remote cameras efficiently investigate larger mammals, including identifying individual animals based on their unique body patterns. To determine whether cameras help identify individual fox squirrels, the authors (1) examined digital photographs of fox squirrels, (2) identified individual variations in facial and pelage features suitable to identify individual squirrels, (3) tested if students of wildlife biology and professionals could identify individual squirrels using these features, and (4) assessed whether guidance could improve students' and professionals' skills to identify individual squirrels. The authors established that individual fox squirrels can be identified by applying a blend of sex and ten variable body patterns, including the color morph, ring around the eye, extent of white on the snout, and facial patterns. In total, they tested ≥25,920 possible feature combinations to identify southeastern fox squirrels and found that wildlife students and professionals involved in the survey could identify fox squirrels without training (Fig. 5.20). However, their average performance increased notably from 73.6% to 80.3% after training ($t[\mathrm{df} = 106] = 5.068$, $p < 0.001$). These results reveal that camera trapping is a valuable tool for identifying individual squirrels, especially if combined with training that teaches participants

Fig. 5.20 Ten highly variable features most efficiently identified individual southeastern fox squirrels from camera-trapping photographs in the study by Tye et al. (2015): (**a**) sex, (**b**) ventral color, (**c**) dorsal color, (**d**) eye ring, (**e**) facial pattern, (**f**) toe color, (**g**) dark spot on foot, (**h**) distribution of white on the rostrum, (**i**) forearm patterns, and (**j**) melanistic dorsal stripe. (Source: (Tye et al. 2015) licensed under CC-BY 4.0)

which features vary between individuals. Photo-ID via camera trapping thus provides an alternative to live trapping for population assessment studies, at least in southeastern fox squirrels.

Recent studies reporting photo-ID using camera traps in various species include, e.g., puma (Alexander and Craighead 2022), Saimaa ringed seal (Koivuniemi et al. 2016), Mediterranean monk seal (Kurt and Gücü 2021), European polecats (Hofmeester et al. 2024), snow leopard (Bohnett et al. 2023), wolf (Jiménez et al. 2023), European beaver (Dytkowicz et al. 2023), brown bear (Clapham et al. 2020), giant anteaters (Möcklinghoff et al. 2018), and gorillas (Brust et al. 2017).

also be found, for example, on the heads of African elephants (Chui and Karczmarski 2022) or lions (Searle et al. 2025) or individual spot patterns in spotted hyenas (Dheer et al. 2022) (Figs. 5.16, 5.17, 5.18 and 5.19). For the evaluation, software programs, including R-packages, have been developed for photo identification (Keen et al. 2022).

5.5 Estimation of Population Density by Removal Trapping

The principle of population density estimation by removal trapping consists of the successive removal of known animal numbers, which makes the trapping result progressively smaller. The true

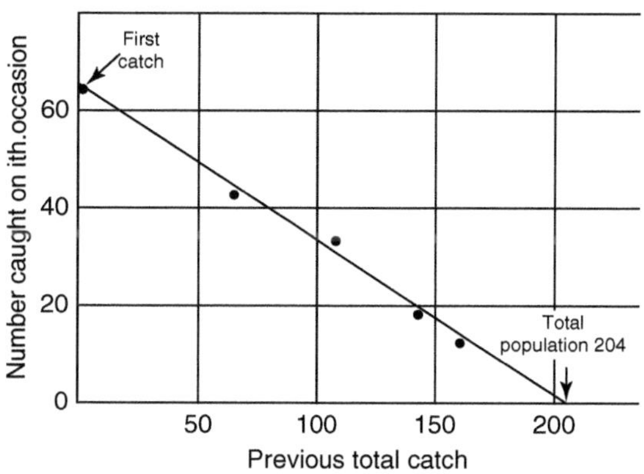

Fig. 5.21 Example of a graphical extrapolation for population estimation using the removal method. (After Southwood 1978)

population is then estimated by extrapolating the trapping curve to the value 0. The representation of an empty catch attempt is exemplified in Fig. 5.21.

The slope of the trapping curve is determined by the unknown population size and the known number of animals trapped each time. The curve can then be assumed to be a straight line (and drawn by the eye) if the following conditions are met: The trapping procedure must not itself influence the subsequent trapping effectiveness. The population must remain stable during the trapping period. As many animals as possible should be able to be captured at each trapping; all animals should have the same probability of being captured.

References

Albert C.H., Rayfield B., Dumitru M. & Gonzalez A. 2017: Applying network theory to prioritize multispecies habitat networks that are robust to climate and land-use change. – *Conservation Biology* 31, 1383-1396

Alexander, P. D., & Craighead, D. J. (2022). A novel camera trapping method for individually identifying pumas by facial features. *Ecology and Evolution, 12*(1), e8536. https://doi.org/10.1002/ece3.8536

Baker A.E.M. 1990: 12. Does group selection occur in commensal house mice (*Mus domesticus*)? – In: Shorrocks B. & Swingland I.R. (eds.): *Living in a patchy environment* , 197–218. – Oxford University Press, Oxford, New York, Tokyo, 246 p.

Begon M. 1978: Population densities in *Drosophila obscura* Fallen and *D. obscura* Collin. – *Ecol Ent* 3, 1–12

Begon M. 1979: *Investigating Animal Abundance: capture-recapture for biologists*. – E. Arnold (Publishers) Limited, London, 97 p.

Begon M. 1983: Abuses of mathematical techniques in ecology: applications of Jolly's capture-recapture method. – *Oikos* 40 (1), 155–158

Begon M., Milburn O. & Turner D. 1975: Density estimates of Drosophila in Southern England. – *J. Nat. Hist*. 9, 315–20

Bellier, E., Rectenwald, J., Sisson, C., Terhune II, T. M., & Martin, J. A. (2023). Estimating population dynamics trajectories of raptors from a multi-species hierarchical distance sampling model. *Ecological Informatics, 75*, 102024. https://doi.org/10.1016/j.ecoinf.2023.102024

Bian X., Liang X., Weckworth B., Jyal D., Hull V. & Yang L. 2023: Spatial density estimate of the snow leopard, Panthera uncia, in the Central Tibetan Plateau, China, – *Integrative Zoology* 18 (4), 677-687

Bohnett, E., Holmberg, J., Faryabi, S. P., An, L., Ahmad, B., Rashid, W., & Ostrowski, S. (2023). Comparison of two individual identification algorithms for snow leopards (*Panthera uncia*) after automated detection. *Ecological Informatics, 77*, 102214. https://doi.org/10.1016/j.ecoinf.2023.102214

Bonter, D. N., & Bridge, E. S. (2011). Applications of radio frequency identification (RFID) in ornithological research: a review. *Journal of Field Ornithology, 82*(1), 1–10. https://doi.org/10.1111/j.1557-9263.2010.00302.x

Bradley J.S. 1985: Comparative demography of four species of grasshopper on a common site. In: Cook L.M. (ed.): *Case studies in population biology*, 61–100. – Manchester University Press, Manchester/USA, 218 p.

Brannelly, L. A., Chatfield, M. W. H., & Richards-Zawacki, C. L. (2013). Visual implant elastomer (VIE) tags are an unreliable method of identification in adult anurans. *Herpetological Journal, 23*(3), 125–129.

Brust, C.-A., Burghardt, T., Groenenberg, M., Kading, C., Kühl, H. S., Manguette, M. L., & Denzler, J. (2017). Towards Automated Visual Monitoring of Individual Gorillas in the Wild. In *2017 IEEE International Conference on Computer Vision Workshops (ICCVW)* (pp. 2820–2830). Presented at the 2017 IEEE International Conference on Computer Vision Workshops (ICCVW). https://doi.org/10.1109/ICCVW.2017.333

Bub H. 1995: *Bird Trapping and Bird Banding: A Handbook for Trapping Methods All over the World.* Translated by Hamerstrom F. & Wuertz-Schaefer K. – Cornell University Press, 448 p.

Bub H. & Oelke H. 1980: *Marking methods for birds*. – A. Ziemsen-Verlag Wittenberg Lutherstadt, 152 p.

Buckland S.T. 1980: A modified analysis of the Jolly-Seber capture-recapture model. – *Biometrics* 36, 419-435

Buckland S.T., Anderson D.R., Burnham K.P., Laake J.L., Borchers D.L. & Thomas L. 2001: *Introduction to Distance Sampling* . Estimating Abundance of Biological Populations. – Oxford University Press, Oxford, New York, 432p.

Buckland S.T., Anderson D.R., Burnham K.P., Laake J.L., Borchers D.L. & Thomas L. 2004: *Advanced Distance Sampling*. – Oxford University Press, 416.

Cameron R.A.D. & Williamson P. 1977: Estimating Migration and the Effects of Disturbance in Mark-Recapture Studies on the Snail *Cepaea nemoralis* L. – *J. Anim. Ecol* . 46, 173–179

Carbone C., Christie S., Conforti K., Coulson T., Franklin N., Ginsberg J.R., Griffiths M., Holden J., Kinnaird M., Laidlaw R., Lynam A., MacDonald D.W., Martyr D., McDougal C., Nath L., O'Brien T., Seidensticker J., Smith J.L.D., Tilson R. & Wan Shahruddin 2002: The use of photographic rates to estimate densities of cryptic mammals: response to Jennelle et al. – *Animal Conservation* 5, 121-124

Cardinale B.J., Primack R.B. & Murdoch J.D. 2020: Conservation Biology. – Sinauer Associates Inc., 584 p. (+ references and index)

Caro T.M. 1999: Conservation monitoring: estimating mammal densities in woodland habitats. – *Animal Conservation* 2, 305-315

Carpenter S.R. Cole J.J., Hodgson J.R., Kitchell J.F., Pace M.L., Bade D., Cottingham K.L., Essington T.E., Houser J.N. & Schindler D.E. 2001: Trophic cascades, nutrients, and lake productivity: whole-lake experiments. – *Ecological Monographs* 71, 163-186

Chui S.Y.S. & Karczmarski L. 2022: Everyone matters: identification with facial wrinkles allows more accurate inference of elephant social dynamics. – *Mamm Biol.* 102, 645–666

Clapham, M., Miller, E., Nguyen, M., & Darimont, C. T. (2020). Automated facial recognition for wildlife that lack unique markings: A deep learning approach for brown bears. *Ecology and Evolution, 10*(23), 12883–12892. https://doi.org/10.1002/ece3.6840

Clark M. 1983: *Mammal Watching*. – Hamlyn Paperbacks edition, Feltham, Middlesex, England, 175 p.

Crego, R. D., Wells, H. B. M., Connette, G., Stabach, J. A., Soit, N., & Thompson, S. (2024). Monitoring spatiotemporal dynamics of large herbivores across an African rangeland using hierarchical multi-species distance sampling. *African Journal of Ecology, 62*(1), e13219. https://doi.org/10.1111/aje.13219

Cripps, J. K., Nelson, J. L., Scroggie, M. P., Durkin, L. K., Ramsey, D. S. L., & Lumsden, L. F. (2021). Double-observer distance sampling improves the accuracy of density estimates for a threatened arboreal mammal. *Wildlife Research, 48*(8), 756–768. https://doi.org/10.1071/WR19136

Crumpacker D.W. 1974: The use of micronized fluorescent dusts to mark adult *Drosophila pseudoobscura*. – *Am. Midi. Nat.* 91, 118–129

Cummings S. 2024: How ants twarted lions on the African savanna. – *Science News* 25 Jan 2024.

D'Eon R.G., Wilson S.F. & Hamilton D. 2006: *Ground-based Inventory Methods for Ungulate Snow-track Surveys*. Standards for Components of British Columbia's Biodiversity No. 33a. – Published by the Resources Information Standards Committee, The Province of British Columbia. 19 p.

Davis N.E., Di Stefano J., Whelan J. Wright J., Taylor L., Coulson G. & Sitters H. 2022: Power of faecal pellet count and camera trapping indices to monitor mammalian herbivore activity. – *Wildlife Research*, https://doi.org/10.1071/WR21135, 49(8), 686-697

Delany M.J. 1974: *The ecology of small mammals* . – Studies in Biology no. 51. Institute of Biology, London, 60 p.

Dheer A., Samarasinghe D., Dloniak S.M. & Braczkowski A. 2022: Using camera traps to study hyenas: challenges, opportunities, and outlook. *Mamm Biol.* 102, 847–854

Dickman C.R. 1986: A method of censusing small mammals in urban habitats. – *Journal of Zoology (London)* 210A, 631–636

Dyer L.A. & Letourneau D.K. 2013: Can Climate Change Trigger Massive Diversity Cascades in Terrestrial Ecosystems? – *Diversity 5*(3), 479-504

Dytkowicz, M., Hinds, R., Megill, W. M., Buttschardt, T. K., & Rosell, F. (2023). A camera trapping method for the targeted capture of Eurasian beaver (Castor fiber) tails for individual scale pattern recognition. *European Journal of Wildlife Research, 69*(2), 39. https://doi.org/10.1007/s10344-023-01654-6

Ehrlich P.R. & Davidson S.E. 1960: Techniques for capture-recapture studies of Lepidoptera populations. – *Journal of the Lepidopterists' Society* 14 (4), 227–229

Feldmann R. (ed.) 1981: The Amphibians and Reptiles of Westphalia. – *Treatises from the State Museum for Natural History in Münster in Westphalia* 43 (4), 161 p.

Feldmann T., Henle K. & Settele J. 1995: Applicability of Mark-Recapture Methods to Butterfly Conservation Research: A Pilot Study, 624-626. – In: Bissonette J.A. & Krausmann P.R. (eds). Integrating people and wildlife for a sustainable future. *Proceedings of the first International Wildlife Management Congress* . The Wildlife Society, Bethesda

Ferner J. 1979: A review of marking techniques for amphibians and reptiles. – *Herpetolog. Circular (Lawrence, Kan.)* 9, 1–40

Forsyth D.M. 2005: *Protocol for estimating changes in the relative abundance of deer in New Zealand forests using the Faecal Pellet Index (FPI)*. – Manager, Science and Research Conservation Sciences Centre Department of Conservation P.O. Box 10-420, Wellington, 22 p.

Forsyth D.M., Comte S., Davis N.E., Bengsen A.J., Côté S.D., Hewitt D.G., Morellet N. & Mysterud A. 2022: Methodology matters when estimating deer abundance: a global systematic review and recommendations for improvements. – Wildlife Management and Research 86 (4). e22207. https://doi.org/10.1002/jwmg.22207

Franceschi S., Bongi P., Del Frate M., Fattorini L. & Apollonio M. 2022: A sampling strategy for habitat selection, mapping, and abundance estimation of deer by pellet counts. – *The Journal of Wildlife Management*, https://doi.org/10.1002/jwmg.22345, 87, 2

Gaffin D.D., Bumm L.A., Taylor M.S., Popokina N.V. & Mann S. 2012: Scorpion fluorescence and reaction to light. *Animal Behaviour* 83 (2), 429-436

Gall L.F. 1985: Measuring the size of lepidopteran populations. – *Journal of Research on the Lepidoptera* 24 (2), 97–116

Giles R.H. (ed.) 1969: *Wildlife Management Techniques* (3rd ed, revised). – Wildlife Society, Washington D. C., 623 p.

Gill R.M.A., Thomas M.L. and Stocker D. 1997: The Use of Portable Thermal Imaging for Estimating Deer Population Density in Forest Habitats. – *Journal of Applied Ecology* 34(5), 1273-1286

Goode M.J. , Beaver J.T., Muller L.I., Clark J.D., van Manen F.T. , Harper C.A. & Basinger P.S. 2014: Capture – recapture of white-tailed deer using DNA from fecal pellet groups. – *Wildlife Biology* 20, 270–278

Gordon I.J., Hester A.J. & Festa-Bianchet M. 2004: REVIEW: The management of wild large herbivores to meet economic, conservation and environmental objectives. – *Journal of Applied Ecology* 41 (6), 1021-1031

Gray T.N.E., Phan C., Pin C. & Prum S. 2012: Establishing a monitoring baseline for threatened large ungulates in eastern Cambodia. – *Wildlife Biology* 18 (4), 406-413

Greene K., Bell D., Kioko J. & Kiffner C. 2017: Performance of ground-based and aerial survey methods for monitoring wildlife assemblages in a conservation area of northern Tanzania. – *European Journal of Wildlife Research* 63 (5): 77

Greenwood J.J.D. 1974: Effective population numbers in the snail *(Cepaea nemoralis)*. – *Evolution* 28 (4), 513–526

Groom M.J., Meffe G.K. & Carroll C.R. 2006: *Principles of Conservation Biology* (3rd ed.). – Sinauer Associates Inc., 793 p.

Gurnell J. & Flowerdew J.R. 1982: *Live Trapping Small Mammals - a Practical Guide*. The Mammal Society, 24 p.

Hagstrom T. 1973: Identification of new specimens (Urodela, *Triturus)* by recording the belly pattern and a description of photographic equipment for such registrations. – *Brit. J. Herpetol.* 4, 321–326

Heard G.W., Scroggie M.P. and Malone B. 2008: Visible Implant Alphanumeric tags as an alternative to toe-clipping for marking amphibians - A case study. – *Wildlife Research* 35(8), https://doi.org/10.1071/WR08060

Heise S. & Wieland H. 1991: Zu den Methoden der Abundanzbestimmung bei Feldmauspopulationen als Grundlage eines umweltgerechten Pflanzenschutzes. – *Nachrichtenbl. Deut. Pflanzenschutzd. (Braunschweig)* 43 (2), 30–33

Heleno R., Devoto M. & Pocock M. 2012: Connectance of species interaction networks and conservation value: Is it any good to be well connected? – *Ecologicale Indicators* 14, 7-10

Henderson P.A. 2021: *Southwood's Ecological Methods*. 5th ed. – Oxford University Press, 528 p.

Henle K., Kuhn J., Podloucky R., Schmidt-Loske K. & Bender C. 1997: Individual recognition and marking of Central European amphibians and reptiles: Overview and evaluation of methods; Recommendations from a nature and animal protection perspective. – In: Henle K. & Veith M.: Methods of Field Herpetology Relevant to Nature Conservation. – *Mertensiella* 7, 133-184

Herold D. 1990: Das Wiesenrandstreifenprogramm: Auswirkungen auf Verteilung, Migration und Populationsentwicklung von Feldheuschrecken (Acrididae: Gomphocerinae). – Diploma Thesis, University of Erlangen-Nuremberg, 127 p.

Hofmeester, T. R., Erath, N., Mos, J., & Thurfjell, H. (2024). Mustelid mugshots: photographing facial masks of European polecats (Mustela putorius) for individual recognition and density estimation using camera traps. *Mammal Research*. https://doi.org/10.1007/s13364-024-00751-4

Hughes, E. J., Mady, R. P., & Bonter, D. N. (2021). Evaluating the accuracy and biological meaning of visits to RFID-enabled bird feeders using video. *Ecology and Evolution*, *11*(23), 17132–17141. https://doi.org/10.1002/ece3.8352

Jennelle C.S., Runge M.C. & MacKenzie D.I. 2002: The use of photographic rates to estimate densities of tigers and other cryptic mammals: a comment on misleading conclusions. – *Animal Conservation* 5 (2), 119-120

Jiménez, J., Cara, D., García-Dominguez, F., & Barasona, J. A. (2023). Estimating wolf (Canis lupus) densities using video camera traps and spatial capture–recapture analysis. *Ecosphere*, *14*(7), e4604. https://doi.org/10.1002/ecs2.4604

Jung, T. S., Boonstra, R., & Krebs, C. J. (2020). Mark my words: experts' choice of marking methods used in capture-mark-recapture studies of small mammals. *Journal of Mammalogy*, *101*(1), 307–317. https://doi.org/10.1093/jmammal/gyz188

Kamaru D.N., Palmer T.M., Riginos C., Ford A.T., … Goheen J.R. 2024: Disruption of an ant-plant mutualism shapes interactions between lions and their primary prey. – *Science*. 2024;383(6681):433-438. doi: https://doi.org/10.1126/science.adg1464

Karanth K.U. & Nichols J.D. 1998: Estimation of tiger densities in India using photographic captures and recaptures. – *Ecology* 79 (8), 2852-2862

Karanth K.U., Chundawat R.S., Nichols J.D. & Kumar N.S. 2004: Estimation of tiger densities in the tropical dry forests of Panna, Central India, using photographic capture-recapture sampling. – *Animal Conservation* 7, 285-290

Karczmarski L., Chan S.C.Y., Rubenstein D.I., Chui S.Y.S. & Cameron E.Z. 2022: Individual identification and photographic techniques in mammalian ecological and behavioural research – Part 1: Methods and concepts. – *Mammalian Biology* 102, 545–549

Kaufman D.W., Gentry J.B., Kaufman G.A., Smith M.H. & Wiener J.G. 1978: Density estimation of small mammals, comparison of techniques utilizing removal trapping. – *Acta Theriol.* 23, 147–171.

Keen E.M., Wren J., O'Mahony É. & Wray J. 2022: catRlog: a photo-identification project management system based in R. – *Mamm Biol.* 102, 995–1004

Kendall, W.L. 1999: Robustness of closed capture-recapture methods to violations of the closure assumption. – *Ecology* 80 (8), 2517-2525

Kiffner C. & Waltert M. 2009: Wildlife Population Assessments in the Katavi Ecosystem. – *Miombo Newsletter* (Wildlife Conservation Society of Tanzania) 34: 1-8

Kiss I., Vörös, J. and Hamer A.J. 2022: Movement patterns within an urban population of fire salamanders highlight the importance of conserving small habitat patches. – *Journal of Zoology* 316, 240–249. https://doi.org/10.1111/jzo.12949

Kloock C.T., Kubli A. & Reynolds R. 2010: Ultraviolet light detection: a function of scorpion fluorescence. – *The J. of Arachnology* 38(3), 441-445. 10.1636/B09-111.1, and 10.1038/468734c

Köhler G. 1987: Die quantitative Erfassung von Feldheuschrecken (Saltatoria: Acrididae) in zentraleuropäischen Halbtrockenrasen – ein Methodenvergleich. – *Wiss. Z. Jena, Naturwiss. Reihe* 36, 375–390

Koivuniemi, M., Auttila, M., Niemi, M., Levänen, R., & Kunnasranta, M. (2016). Photo-ID as a tool for studying and monitoring the endangered Saimaa ringed seal. *Endangered Species Research*, *30*, 29–36. https://doi.org/10.3354/esr00723

Krause J. 2008: Density estimation of a Maral (*Cervus e. canadensis*) Population in the Khentej-Region, Mongolia. – Master thesis at the University Goettingen, 69p. (in German)

Krebs C.J. 2013: *Ecology: The Experimental Analysis of Distribution and Abundance* (6th ed.). – Pearson, 652.

Krebs, C. J. (2014). *Ecological methodology* (3rd ed.). Vanvouver: Charles Krebs. https://www.zoology.ubc.ca/~krebs/books.html

Kurt, M. and Gücü, A. C. (2021). Demography and population structure of Northeastern mediterranean monk seal population. *Mediterranean Marine Science*, *22*(1), 79–87. https://doi.org/10.12681/mms.21913

Lewis C. 1970: Wildlife Census methods. A resume for the Oklahoma Cooperative Wildlife Research Unit, Oklahoma State University. – *Journal of Wildlife Diseases* 6

Li, Z., Shi, X., Lu, J., Fu, X., Fu, Y., Cui, Y., et al. (2023). Assessing mammal population densities in response to urbanization using camera trap distance sampling. *Ecology and Evolution*, *13*(10), e10634. https://doi.org/10.1002/ece3.10634

Liere H., Kim T.N., Werling B.P., Meehan T.D., Landis D.A. & Gratton C. 2015: Trophic cascades in agricultural landscapes: indirect effects of landscape composition on crop yield. – *Ecological Applications* 25(3), 652-661. https://doi.org/10.1890/14-0570.1

Lockley R.M. & Russell R. 1953: *Bird-Ringing. The Art of Bird Study by Individual Marking*. – Crosby Lockwood & Son Ltd., London, 119 p.

Lourenço W.R. 2012: Fluorescence in scorpions under UV light; can chaerilids be a possible exception? – *ScienceDirect* 335 (12), 731-734

Mader H.-J. 1979: *Die Isolationswirkung von Verkehrsstraßen auf Tierpopulationen untersucht am Beispiel von Arthropoden und Kleinsäugern der Waldbiozönosen*. – Sch.-R. Landschaftspflege u. Natursch. 19, 131

Mader H.-J. & Pauritsch G. 1981: Nachweis des Barriereeffektes von verkehrsarmen Straßen und Forstwegen auf Kleinsäuger der Waldbiozönose durch Markierungs- und Umwetzungsversuche. – *Natur und Landschaft* 56 (12), 451–454

Mallet J., Longino J.T., Murawski D. & Simpson de Gamboa A. 1987: Handling effects in *Heliconius:* Where do all the butterflies go?. – *J. Anim. Ecol.* 56 (2), 377–386

Marques F.F.C., Stephen T. Buckland S.T., Goffin D., Dixon C.E., Borchers D.L., Mayle B.A. & Peace A.J. 2001: Estimating deer abundance from line transect surveys of dung: sika deer in southern Scotland. – *Journal of Applied Ecology* 38, 349-363

May Y.Y. 1978: A Population Study of *Stenocranus minutus* (Fab.) (Hemiptera: Delphacidae). – *Res. Popul. Ecol.* 20 (1), 61–78

May, R. M. (2004). Ecology: ethics and amphibians. *Nature*, *431*(7007), 403. https://doi.org/10.1038/431403a

McCarthy, M. A., & Parris, K. M. (2004). Clarifying the effect of toe clipping on frogs with Bayesian statistics. *Journal of Applied Ecology*, *41*(4), 780–786. https://doi.org/10.1111/j.0021-8901.2004.00919.x

Miller, D. L., Burt, M. L., Rexstad, E. A., & Thomas, L. (2013). Spatial models for distance sampling data: recent developments and future directions. *Methods in Ecology and Evolution*, *4*(11), 1001–1010. https://doi.org/10.1111/2041-210X.12105

Möcklinghoff, L., Schuchmann, K.-L., & Marques, M. I. (2018). New non-invasive photo-identification technique for free-ranging giant anteaters (Myrmecophaga tridactyla) facilitates urgently needed field studies. *Journal of Natural History*, *52*(37–38), 2397–2411. https://doi.org/10.1080/00222933.2018.1537407

Mohler L.L. & Toweill D.E. 1982: Regulated elk populations and hunter harvests. – In: Thomas J. W. and Toweill, D. E. (Eds.) *Elk of North America: ecology and management* , 561-597. Stackpole, Harrisburg, U.S.A.

National Audubon Society 2013: Woodpeckers as Keystone Species. – https://www.audubon.org/news/woodpeckers-keystone-species, *BirdNote* by M. McCann, August 20, 2013.

Nichols J.D. and Pollock K.H. 1983: Estimation methodology in contemporary small mammal capture-recapture studies. – *J. Mamm.* 64 (2), 253–260

Nolan, V., Wilhite, N., Howell, P. E., Chandler, R. B., Ingram, D., Yeiser, J. M., et al. (2023). Distance sampling and spatial capture-recapture for estimating density of Northern Bobwhite. *Ecological Informatics*, *78*, 102330. https://doi.org/10.1016/j.ecoinf.2023.102330

Norton-Griffiths M. 1978: *Counting animals*. Handbook No. 1: Serengeti ecological monitoring programme. 2 nd ed. – African Wildlife Foundation, Afropress, Nairobi, Kenya.

O'Connell A.F. & Ilse L. 1999: Annotated Bibliography of Methodologies to Census, Estimate and Monitor the Size of White-Tailed Deer Populations. Department of the Interior. National Park Service Boston Support Office. Online available at: https://pubs.usgs.gov/publication/5200224

Odum, R. A. O., & Sonntag, E. (2010). Amphibian Ark: Methods to individually identify the amphibian specimen. https://www.academia.edu/74029276/Methods_to_Individually_Identify_the_Amphibian_Specimen

Olsen, I. C. (2009). *The Analysis of Continuous Mark-Recapture Data: With examples*. VDM Verlag, 124 p.

Olson K.A., Fuller T.K., Schaller G.B., Odonkhuu D. & Murray M.G. 2005: Estimating the population density of Mongolian gazelles *Procapra gutturosa* by driving long-distance transects. – *Oryx* 39(2), 164-169

Otis D.L., Burnham K.P., White G.C. & Anderson D.R. 1978: Statistical inference from capture data on closed animal populations. – *Wildlife Monographs* 43 (62), 1–135

Parr M.J. 1965: A population study of a colony of imaginal *Ischnura elegans* (van der Linden) (Odonata, Coenagridae) at Dale, Pembrokeshire. – *Field studies, London*, 2, 237–282

Pakkala, T., Tiainen, J., Piha, M., & Kouki, J. (2018). Three-toed Woodpecker cavities in trees: A keystone structural feature in forests shows decadal persistence but only short-term benefit for secondary cavity-breeders. Forest Ecology and Management, 413, 70–75. https://doi.org/10.1016/j.foreco.2018.01.043

Perry, G., Wallace, M. C., Perry, D., Curzer, H., & Muhlberger, P. (2011). Toe Clipping of Amphibians and Reptiles: Science, Ethics, and the Law1. *Journal of Herpetology*, *45*(4), 547–555. https://doi.org/10.1670/11-037.1

Petit, S., Waudby, H. P., Walker, A. T., Zanker, R., & Rau, G. (2012). A non-mutilating method for marking small wild mammals and reptiles. *Australian Journal of Zoology*, *60*(1), 64–71. https://doi.org/10.1071/ZO11088

Petso T., Jamisola Jr. R.S. & Mpoeleng D. 2022: Review on methods used for wildlife species and individual identification. – *Eur. J. Wildl. Res.* 68 (1)

Pittman S.E. and Semlitsch R.D. 2013: Habitat type and distance to edge affect movement behavior of juvenile pond-breeding salamanders. – *Journal of Zoology* 291 (2), 154-162

Pledger S., Pollock K.H. & Norris J.L. 2003: Open capture-recapture models with heterogeneity: I. Cormack-Jolly-Seber model. – *Biometrics* 59(4), 786-794

Quesada, J., Guallar, S., Perez-Ruiz, N. J., Estrada, J., & Herrando, S. (2010). Observer Error Associated with Band Allocation Is Negligible in Large Scale Bird Monitoring Schemes, but How Precise Is the Use of Bands at All? *ARDEOLA-INTERNATIONAL JOURNAL OF ORNITHOLOGY*, *57*, 23–32.

Ransom, J. I. (2011). Customizing a rangefinder for community-based wildlife conservation initiatives. *Biodiversity and Conservation*, *20*(7), 1603–1609. https://doi.org/10.1007/s10531-011-0040-1

Ransom, J. I., Kaczensky, P., Lubow, B. C., Ganbaatar, O., & Altansukh, N. (2012). A collaborative approach for estimating terrestrial wildlife abundance. *Biological Conservation*, *153*, 219–226. https://doi.org/10.1016/j.biocon.2012.05.006

Ribas, M. P., Alonso-Almorox, P., Espunyes, J., Martinez-Silvestre, A., & Cabezon, O. (2022). Evaluation of passive integrated transponder tags for marking urodeles. *Ecological Indicators*, *145*, 109690. https://doi.org/10.1016/j.ecolind.2022.109690

Richter K.O. 1976: A method for individually marking slugs. – *J. moll. Stud.* 42, 146–151

Rose, A., Tschapka, M., & Knörnschild, M. (2020). Visits at artificial RFID flowers demonstrate that juvenile flower-visiting bats perform foraging flights apart from their mothers. *Mammalian Biology*, *100*(5), 463–471. https://doi.org/10.1007/s42991-020-00048-4

Ross, R., Anderson, B., Bienvenu, B., Scicluna, E. L., & Robert, K. A. (2022). WildTrack: An IoT System for Tracking Passive-RFID Microchipped Wildlife for Ecology Research. *Automation*, *3*(3), 426–438. https://doi.org/10.3390/automation3030022

Sample R.D., Tomey R.D., Delisle Z.J., Trumbower A.R., Habeck P.J., Brooke J.M. & Jenkins M.A. 2022: Comparing Methods of Estimating Fecal-Pellet-Group Density in Woodlots of the Midwestern United States. – *Journal of Fish and Wildlife Management* 13 (2), 483–493. https://doi.org/10.3996/JFWM-21-098

Sandoval, B. A., Bridge, E. S., Fulton, C. F., & Ruyle, J. E. (2022). Phantom Design and Verification Enabling Effective Antennas for Bird Tracking Technology. *IEEE Antennas and Wireless Propagation Letters*, *21*(8), 1703–1707. Presented at the IEEE Antennas and Wireless Propagation Letters. https://doi.org/10.1109/LAWP.2022.3178292

Schjøtz-Christensen B. 1965: Biology and population studies of Carabidae. II. Technique, 1. Trapping. – *Natura Jutlandica* 11, 1–17

Schmid B. 1991: Die Bedeutung von Stufenrainen als Überwinterungshabitat für Spinnen (Araneae) der Agrarlandschaft. – Diplomarbeit, Universität Würzburg, 69 p.

Schmidt, J. H., Thompson, W. L., Wilson, T. L., & Reynolds, J. H. (2022). Distance sampling surveys: using components of detection and total error to select among approaches. Wildlife Monographs, 210(1), e1070. https://doi.org/10.1002/wmon.1070

Schtickzelle, N., Le Boulengé, E., & Baguette, M. (2002). Metapopulation Dynamics of the Bog Fritillary Butterfly: Demographic Processes in a Patchy Population. *Oikos*, *97*(3), 349–360.

Schulte, L., Faul, C., Oswald, P., Preißler, K., Steinfartz, S., Veith, M., & Caspers, B. A. (2024). Performance of different automatic photographic identification software for larvae and adults of the European fire salamander. *PLOS One*, *19*(4), e0298285. https://doi.org/10.1371/journal.pone.0298285

Searle C.E., Strampelli P., Parsais S.N. and Dickman,A. 2025: Spatially explicit camera trap-based lion monitoring in Tanzania's Selous–Nyerere landscape. - *Journal of Zoology*, p. 12. https://doi.org/10.1111/jzo.70019.

Seber, G. A. F., & Schofield, M. R. (2019). *Capture-Recapture: Parameter Estimation for Open Animal Populations* (1st ed. 2019 edition.). Springer.

Settele J., Feldmann R., Henle K., Kockelke K. & Poethke H.-J. 1998: Populationsgrößenschätzung bei Tieren. Ausgewählte Verfahren für den Einsatz in Populationsökologie und Naturschutz - Kurzfassung einer Internet-Version. – *Naturschutz und Landschaftsplanung* 30 (6), 174-181

Silver, S.C., Ostro, L.E.T.; Marsh, L.K.; Maffei, L.; Noss, A.J.; Kelly, M.J.; Wallance, R.B.; Gómez, H.; Ayala, G. 2004: The use of camera traps for estimating jaguar *Panthera onca* abundance and density using capture/recapture analysis. – *Oryx* 38 (2), 148-154

Singer M.C. & Wedlake P. 1981: Capture does affect probability of recapture in a butterfly species. – *Ecological Entomology* 6, 215–216

Skonhoft A., Veiberg V., Gauteplass A., Olaussen J.O., Meisingset E. & Mysterud A. 2013: Balancing income and cost in red deer management. – *Journal of Environmental Management* 115, 179-188

Smart J.C.R., Ward A.I. & White P.C.L. 2004: Monitoring woodland deer populations in the UK: an imprecise science. – *Mammal Review* 34 (1-2), 99–114

Smith M.H. 1968: A Comparison of Different Methods of Capturing and Estimating Numbers of Mice. – *Journal of Mammalogy* 49 (3), 455–461

Smith S., Young T. and Skydmore D. 2018: Effectiveness of the field identification of individual natterjack toads (*Epidalea calamita*) using comparisons of dorsal features through citizen science. – *Herpetological Journal* 28, 31-38

Southwood T.R.E. 1978: *Ecological Methods – with particular reference to the study of Insect Populations.* – Chapman and Hall, London, John Wiley & Sons, New York, 524 p.

Steinhauer B. 1981: Wildlife assessment in the Comoé Nationalpark, Ivory Coast. – Dissertation Univ. Würzburg

Stonehouse B. (ed.) 1978: *Animal Marking: Recognition Marking of Animals in Research.* – Macmillan, London, 257 p.

Sutherland W.J. 2006: *Ecological Census Techniques,* a handbook. 2nd ed. – Cambridge University Press, 432 p.

Testud, G., Vergnes, A., Cordier, P., Labarraque, D., & Miaud, C. (2019). Automatic detection of small PIT-tagged animals using wildlife crossings. *Animal Biotelemetry*, *7*(1), 21. https://doi.org/10.1186/s40317-019-0183-5

Thomas, L., Buckland, S. T., Rexstad, E. A., Laake, J. L., Strindberg, S., Hedley, S. L., et al. (2010). Distance software: design and analysis of distance sampling surveys for estimating population size. *Journal of Applied Ecology*, *47*(1), 5–14. https://doi.org/10.1111/j.1365-2664.2009.01737.x

Turlure, C., Pe'er, G., Baguette, M., & Schtickzelle, N. (2018). A simplified mark–release–recapture protocol to improve the cost effectiveness of repeated population size quantification. *Methods in Ecology and Evolution*, *9*(3), 645–656. https://doi.org/10.1111/2041-210X.12900

Twigg G.I. 1975: Marking Mammals. – *Mammal Review* 5, 110–116

Tye, C. A., Greene, D. U., Giuliano, W. M., & McCleery, R. A. (2015). Using camera-trap photographs to identify individual fox squirrels (Sciurus Niger) in the Southeastern United States. *Wildlife Society Bulletin*, *39*(3), 645–650. https://doi.org/10.1002/wsb.573

Vanbergen A.J., Woodcock B.A., Heard M.S. & Chapman D.S. 2017: Network size, structure, and mutualism dependence affect the propensity for plant-pollinator extinction cascades. – *Functional Ecology* doi:https://doi.org/10.1111/1365-2435.12823

Wagenknecht E. 2000: *Red Deer* (5th ed.). – Nimrod Publishing, Suderburg, 584 p.

Waltert M., Meyer B., Shanyangi M.W., Balozi J.J., Kitwara O., Qolli S., Krischke H. & Mühlenberg M. 2008: Foot surveys of large mammals in the woodlands of western Tanzania. – *Journal of Wildlife Management* 72, 603-610

Wassmer, T., & Armstrong, E. (2023). Population structure of Phanaeus vindex (Coleoptera: Scarabaeidae) in SE Michigan. *Journal of Insect Science*, *23*(4), 2. https://doi.org/10.1093/jisesa/iead050

Wassmer, T., & Refinetti, R. (2016). Daily Activity and Nest Occupation Patterns of Fox Squirrels (Sciurus Niger) throughout the Year. *PLOS One*, *11*(3), e0151249. https://doi.org/10.1371/journal.pone.0151249

Wassmer, T., & Refinetti, R. (2019). Individual daily and seasonal activity patterns in fox squirrels (Sciurus Niger) quantified by temperature-sensitive data loggers. *Frontiers in Ecology and Evolution*, *7*, 179. https://doi.org/10.3389/fevo.2019.00179

Waudby, H. P., & Petit, S. (2020). Comments on the Efficacy and Use of Visible Implant Elastomer (VIE) for Marking Lizards. *The South Australian Naturalist*, *85*(1), 7–13. https://doi.org/10.3316/informit.251857167920976

White, G. C., & Burnham, K. P. (1999). Program MARK: survival estimation from populations of marked animals. *Bird Study*, *46*(sup1), S120–S139. https://doi.org/10.1080/00063659909477239

White, G. C., & Cooch, E. G. (2019). *Program MARK: a gentle introduction* (19th ed.). http://www.phidot.org/software/mark/docs/book/

White R.R. & Singer M.C. 1987: Marking technique for larvae. – *Pan-Pacific Entomologist* 63 (4), 341–345

Woodbury, A. M., Ricker, W. E., Cottam, C., Taber, R. D., & Pendleton, R. C. (1956). Symposium: Uses of Marking Animals in Ecological Studies. *Ecology*, *37*(4), 665–689. https://doi.org/10.2307/1933058

Zahler, P., Lhagvasuren, B., Reading, R. P., Wingard, J. R., Amgalanbaatar, S., Gombobaatar, S., et al. (2004). Illegal and Unsustainable Wildlife Hunting and Trade in Mongolia. *Mongolian Journal of Biological Sciences*, *2*(2), 23–31

Freshwater Ecology: Limnology 6

Abstract

The study of freshwater ecosystems is also known as limnology. Compared to terrestrial studies, limnic systems can be better delineated as their shorelines present clear boundaries between fundamentally different environments. Freshwater systems can be distinguished into lotic or flowing water and lentic or standing water systems. River habitats can be characterized by flow velocity and discharge, streambed, bank and riparian conditions, and habitat quality for organisms. Detailed instructions for appropriate measurements and collection methods for macrobenthos and fish are provided. We present simple quality indices for habitat, water quality, and macrobenthos that can be compared to other rivers. Detailed descriptions and instructions for studying lakes, including the vertical stratification of light, temperature, and dissolved oxygen (DO), are also provided. Methods for plankton sampling and calculating density estimates and biodiversity indices are presented. In addition, we describe how macrophytes at the shoreline of lakes can be surveyed and how these data can be presented in a density map. Waters that only carry water intermittently throughout the year are classified as temporary. These include small bodies of water known as temporary ponds with their specific fauna.

6.1 Introduction to Lotic and Lentic Freshwater Systems

Only about 3% of Earth's water is freshwater. Freshwater is one of the most essential resources for all terrestrial life on Earth—second only to air. Only 0.3% of this freshwater occurs on the Earth's surface as lakes, rivers, and wetlands, covering approximately 2% of the Earth's surface. Despite this scarcity, freshwater habitats harbor and nourish an astounding biodiversity of at least 140,000 of the approximately 1.8 million described species—almost 12% of the entire biodiversity (Dudgeon et al. 2006) concentrated in only 2% of Planet Earth's surface area (Reid et al. 2019). This disproportionately high biodiversity would already be enough reason to study and preserve freshwater habitats, but their importance goes beyond their system limits. Due to their extensive ecosystem services, freshwater habitats are vital to many terrestrial and marine organisms, warranting that they are studied, preserved, and restored even more. Despite their importance, freshwater habitats are highly endangered by anthropogenic overexploitation, pollution, fragmentation, habitat destruction or degradation, invasive species, and climate change (Geist 2011; Woodward et al. 2010). Freshwater ecosystems are central to five of the tipping points that threaten life on Earth as we know it: declining freshwater availability, forest loss,

M. Mühlenberg, T. Waßmer, *Field Ecology*, https://doi.org/10.1007/978-3-031-87547-2_6

dwindling biodiversity, climate change, and human population growth, but are often missing from the discussion (e.g., Lenton et al. (2019)).

Freshwater habitats are commonly divided into lotic and lentic systems (Dodson 2004). Lotic systems are based on (fast) flowing waters, including rivers and streams, springs, intermittent streams, canals, backwaters, and estuaries. Lentic systems are based on standing or slow-flowing water and comprise lakes, ponds, reservoirs, and several types of wetlands, such as marshes, swamps, fens, and bogs. In this chapter, we will focus on rivers as the most studied lotic systems and lakes as the most studied lentic systems.

6.2 Lotic Systems: Rivers

6.2.1 Research Questions and Methods for Lotic Systems (Rivers)

- How do you determine the flow and discharge of a river?
- What types of habitats exist along a stretch of river?
  - How can these be characterized?
  - What organisms can be found in each habitat?
- How clean/healthy is a river?

6.2.2 Methods to Approach These Research Questions

The research methods are pretty different for slow-flowing, relatively shallow rivers that are safe for researchers to enter and move around (wadeable rivers) and rivers that are fast-flowing or relatively deep and, therefore, do not provide researchers safe access (non-wadeable rivers/streams). We will focus almost entirely on wadeable rivers, which are easily accessible to students, volunteers, and researchers. In contrast, non-wadeable streams can only be studied from bridges or using specialized, pricey equipment. For more information about the study of non-wadeable rivers, see the bibliography for this chapter (Blocksom and Johnson 2009; Bo et al. 2023; Collins and Flotemersch 2014; Di Sabatino et al. 2016; Gibson-Reinemer et al. 2016; Kislik et al. 2020; Pan et al. 2012; Parker et al. 2020; Pingram et al. 2020; Reid et al. 2021; U.S. Environmental Protection Agency, Office of Water 2023; Wessell et al. 2008; Wilbanks et al. 2020; Wilhelm et al. 2005).

Safety is paramount! Always follow the precautions below when entering a river (Hoosier Riverwatch 2019; RHS Team et al. 2018).

- If the water is too deep or flows too fast, select another site or postpone the study. Do not monitor the stream after heavy rainfall when it is at the flood stage. Any stream can be dangerous when flooded. Check the most recent weather forecast. Do not wade into streams if severe weather is predicted or during storms.
- Never venture out into the stream alone without another person available. Always take a buddy along! Always survey with at least one partner in sight to assist you in an emergency.
- Do not wade into areas where the water depth exceeds hip depth.
- Be careful when walking in and around the river.
  - Wear functional waders that provide traction.
  - Rivers with a rocky bottom may be slippery and contain deep pools.
  - Streams with a muddy bottom and accumulations of clay, silt, and sand might get you stuck or sink below the safe water depth (hip-high).
  - If you must cross the river, use a stick or your net pole to help you steady yourself and probe the bottom in front of you.
  - Watch out for barbed wire or sharp objects that may penetrate your waders.
  - Do not walk on the edge of riverbanks. Disturbing these banks and their vegetation

may accelerate erosion and lead to their collapse.

- Be aware that many streams are known to be unsafe to handle. If in doubt, assume the river is polluted and wear waders, latex gloves, and protective eyewear. Do not drink the water from the stream, and when you are done monitoring, wash your hands with antibacterial soap or use a hand sanitizer, especially before eating and drinking.
- Accidents happen; always take a first aid kit and a cell phone with you. Develop a safety plan, including location and directions to the nearest medical facility.
- Watch for unsupervised dogs, livestock, wildlife, insects, and ticks. Know what to do if you are bitten or stung. Also, be aware of skin-irritating vegetation such as poison ivy and poison sumac.

As it is too time-consuming, most rivers cannot be studied over their entire length. You must focus on a particular part of its reach called a stretch, typically approximately 50–100 m long. River stretches with a steep gradient (slope), as in the mountainous headwaters tend to flow rapidly in a straight channel, while rivers flowing through the flatter areas of floodplains will meander slowly back and forth across the landscape following a snake-like path unless they were channeled (Fig. 6.1a, b). Slow-moving rivers often have multiple or braided channels. As flowing water is the defining characteristic of lotic systems, the first thing you may want to study is the velocity of the flow, and by factoring in the cross-sectional area of the river stretch, you can determine the flow rate or discharge (volume of water flowing through the river per unit time). The flow rate determines if a river stretch is more prone to erosion or sediment deposition and shapes habitats that favor certain plants and animals.

6.2.3 How Do You Determine the Velocity and Discharge of a River?

The velocity of the stream is measured at regular intervals across the stream using a flow rate sensor. At each interval, the depth of the streambed must also be determined. The distance of these intervals from the shoreline and their depths determine the cross-sectional area of the stream. Finally, calculate the stream's discharge by multiplying the cross-sectional area by the average velocity of the cross-section.

First, select a suitable site within your 100 m stretch of the river. Two or three sites that are as far apart as possible and represent the overall conditions of the stretch will give you a reasonable estimate. Unless you want to characterize the discharge of stream disturbances, such as

Fig. 6.1 (**a**) Natural river in a mountainous region, flowing rapidly over riffles. Northern Mongolia. (**b**) Natural river meandering. Slow-flowing and flat, therefore easy to walk on. Northern Mongolia

bends or breaks in the channel caused by rocks or sandbars, it is better to exclude such sites. However, to estimate the average discharge of a stretch, choose one site with a swift flow found near a riffle and a second site with a more moderate or slow flow like that found after a pool.

Measure the width and depth at several points across the river section at each site to create a cross-sectional channel profile. This can be done low-tech using a measuring tape and a meter stick (described below) or high-tech using a drone and a laser profiler (Mandlburger et al. 2016; Bandini et al. 2023).

Using the measuring tape, determine and record the width of the river cross-section in meters. Depending on the width of your river, divide the cross-section into several equally spaced sections, each about 1–2 m wide (Fig. 6.2a). Using a meter stick, measure the stream's depth in meters at each equally spaced point along the cross-section. Record the depth and the distance out from one shore edge. Always measure from the same shore. Be sure to include both the initial distance and depth and the final distance and depth (Fig. 6.2b). Use software to plot your stream width against your depth data and calculate the integral below the curve. Vernier Graphical Analysis is free, available on Windows, Mac, Chrome, iOS, and Android, and allows CSV data import (Vernier Science Education 2021) (Fig. 6.2c). Commercial alternatives are OriginLab Origin or Synergy KaleidaGraph. Alternatively, you can use Excel to determine and sum up the area of each trapezoid or resort to a less accurate estimate, reducing each trapezoid to a rectangle and averaging out the number of rectangles over- and underestimating the trapezoid areas between two section points (Fig. 6.2d).

At each transect point used to calculate the cross-sectional area (besides the two shore points), you also measure the velocity of the flow using a flowmeter (Fig. 6.3a). For consistent measurements, submerge the flow meter's impeller (or other sensor) to about 40% of the depth measured at each section point. Some instruments include risers (Fig. 6.3b) that make keeping the impeller in the same spot and oriented in the same direction easier. Point the impeller upstream against the flow (Fig. 6.3c). To calculate the flow rate or discharge of your river segment, first determine the average velocity of all transect points you measured, then multiply this velocity with the cross-sectional area you determined before. If you use square meters for the area and meters per second for the velocity, you arrive at the correct unit of cubic meters per second for the flow rate.

High-tech instruments to measure flow rate/velocity are Doppler sensors and velocity radar, which can be placed into or above rivers for prolonged periods of time and allow continuous measurements of flow rates. Non-wadeable rivers can be measured from a bridge or, if no bridges exist, using a cableway system (e.g., ott.com).

We provide a simple calculation sheet for the calculation of discharge at https://tinyurl.com/2ax46jwr.

6.2.4 Habitat Assessment of Rivers

What types of habitats exist along a stretch of river?

How can these habitats be characterized?

Especially in the meanders of floodplains, we find different habitats for plants and animals characterized by different water depths, flow rates, and substrates. Depending on the landscape's slope, natural stream channels seldom flow in a straight line but typically meander. Along any stretch of a natural river, we find a variety of habitats (Fig. 6.4) such as pools, runs, riffles (Fig. 6.5), undercut banks, and snags (fallen limbs or small log piles, Fig. 6.6), each with various plant and animal species. The more types of habitats present in a stretch of a river, the greater the possibility for diverse plant and animal communities. A consistently straight or deep channel provides less diverse habitats than river stretches with variable depths and flow rates. In addition, a river's habitat diversity depends on the areas immediately adjacent to the

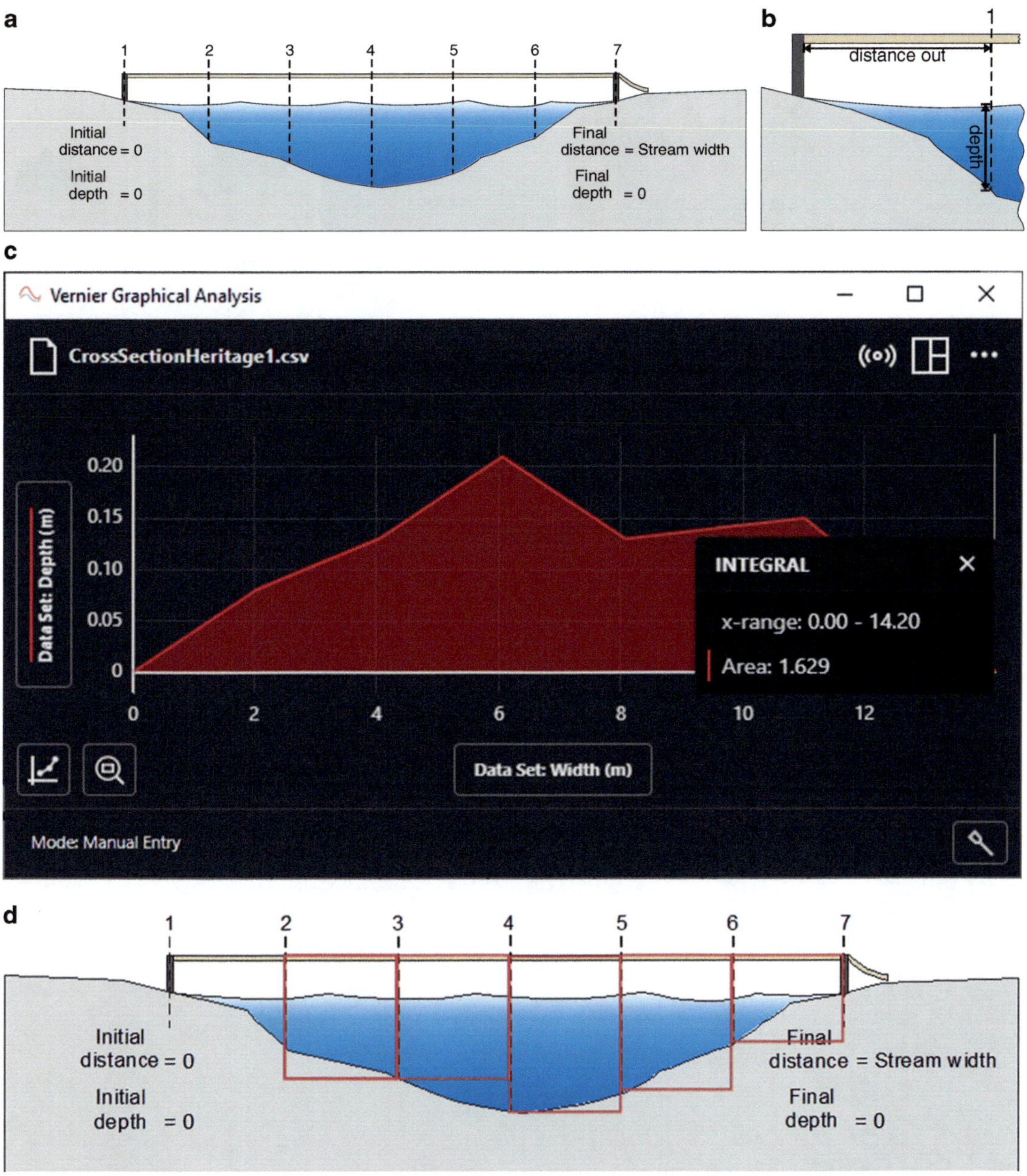

Fig. 6.2 Determination of a river's cross-sectional area. (**a**) Distribution of measuring points across a section of a river. (**b**) Measurement of distance and depth at each point. (**c**) Use of Vernier Graphical Analysis to determine the integral below the river's profile (here plotted as elevation). (**d**) Rough estimate of the cross-sectional area. (Picture credits for **a**, **b**, and **d** (Holmquist et al. 2015) with permission from Vernier Science Education)

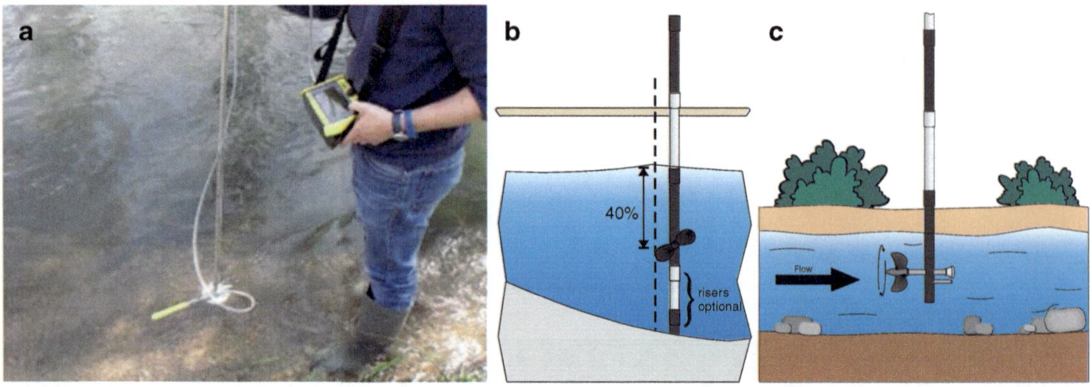

Fig. 6.3 Measurement of a river's flow. (**a**) Flow meter (FLUVIA RC3—Current Metering System), (**b**) Submerging the flow meter's impeller to about 40% of the measured depth, (**c**) Directing the impeller upstream directly into the flow. (Picture credit for **a**. with permission from Aqua-Data.com, **b** and **c**. (Holmquist et al. 2015) with permission from Vernier Science Education)

Fig. 6.4 Onon-Balj River in the north of Mongolia. Almost all habitat types of a river system can be found in this natural river section

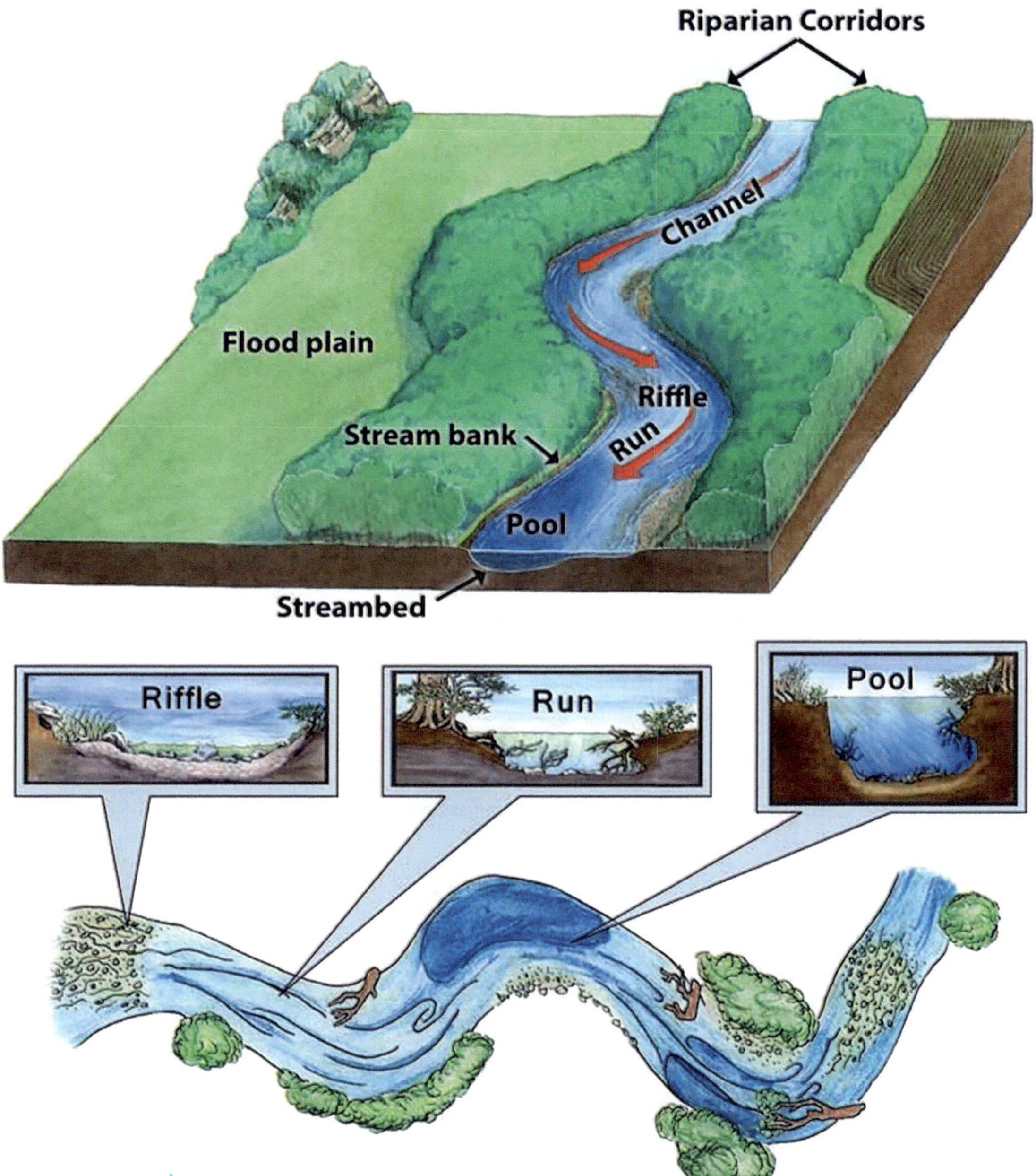

Fig. 6.5 Parts of a stream. (Artwork courtesy Texas Parks and Wildlife Department © 2004, texasaquaticscience.org reproduced with permission)

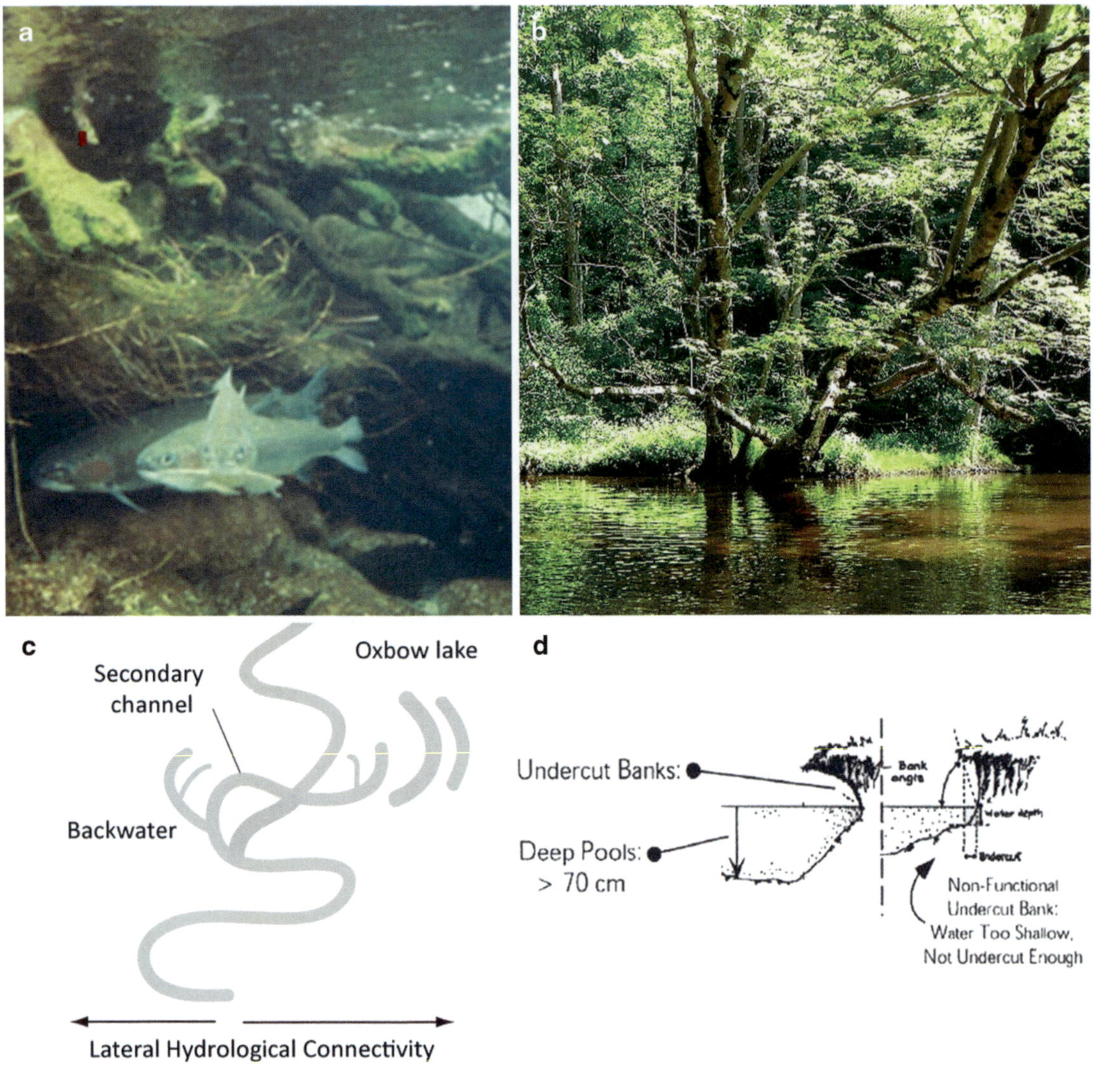

Fig. 6.6 Fish cover types. (**a**) Underwater tree roots and rootlets. (**b**) Overhanging shrubs and trees. (**c**) Backwater, oxbows, or side channels (here called secondary channels). (**d**) Undercut riverbanks. (Picture credits: **a**. Pat Higgins, Eel River Recovery Project https://www.eelriverrecovery.org, **b**. kathrynwarmstrong.wordpress.com, **c**. (Gallardo 2009), **d**. (Hoosier Riverwatch 2019). All reproduced with permission)

river, called the riparian zone, especially the type and structure of vegetation. Riparian vegetation determines the stability of a river's banks and controls erosion, flooding, and habitat for fish and wildlife. In many countries, riparian strips are either protected, or their exclusion from agricultural use is financially compensated (see Sect. 10.17).

There are many different assessment systems for river habitats (e.g., Barbour et al. (1999), Bain and Stevenson (1999), RHS Team et al. (2018), and Vaughan (2010)). In the following, we will present an amazingly fast and simple, but at the same time meaningful scoring system developed by the Ohio Environmental Protection Agency in the 1990s as a "Citizens" companion to the Qualitative Habitat Evaluation Index (CQHEI) used by Ohio's professional staff. We follow the excellent compilation of the CQHEI by the Hoosier Riverwatch, a program of the Indiana Department of Environmental Management (Hoosier Riverwatch 2019).

6.2.5 The CQHEI Habitat Evaluation

The index measures habitat integrity and diversity, which generally corresponds to the diversity of fish and macroinvertebrates. The CQHEI score can be used to compare changes at one site over time or to compare two different sites.

A sampling site is a 200-foot river segment (exact conversion is 61 m; however, it can be anything between 50 and 100 m). Use local landmarks (bridges, trees) or survey tape to mark the boundaries of your sampling site. Take photos and supplement with sketches to record site elements not visible on your images (Fig. 6.7). When choosing and working on the site, consider safety regarding water depth and bank accessibility and respect private ownership.

When completing the assessment, evaluate your entire 50–100 m stretch. For each survey category, choose the most prevalent answer option. If sections of your channel or the riverbanks show entirely different characteristics, you may check two boxes, average their point values, and record a score for the subsections (a), (b), or (c).

The maximum total points for the CQHEI is 114. You have an exceptionally high-quality stream if your score is over 100 (as in 100%). A set of ranges for excellent, medium, poor, and extremely poor has not yet been developed for this index. However, scores above 60 are acceptable (generally conducive to warm-water fauna).

I. **Substrate (Bottom Type): Max 24 pts**

(*Note:* "smothering" is the same as "embeddedness.") See Fig. 6.7. Check "yes" for smothering if the stream bottom is more than 50% embedded.

II. **Fish Cover (Hiding Places): Max 20 pts**

Select all the cover types you see using Fig. 6.6 as a guide. Add up the points.

III. **Stream Shape and Human Alterations: Max 20 pts**

For a) "Curviness" or "sinuosity" of the channel, refer to Fig. 6.8.

IV. **Stream Forests and Wetlands (Riparian Areas) and Erosion: Max 20 pts.**

(a) Width of the Riparian Forest or Wetland—*This is not the stream's width!* Estimate the width of the area containing trees or wetlands on each side of the stream by answering: "Can you throw a rock to the other side?"

(b) Conservation tillage (any tillage and planting system that covers 30% or more of the soil surface with crop residue after planting).

Figure 6.9a, b show natural rivers that meander widely in the lowlands to be assessed with 8 pts. according to Fig. 6.8.

For (a) Riparian width, (c) bank erosion, and (d) stream shading refer to Fig. 6.10.

Figure 6.11a shows a river that has eroded a large bank wall in the outer bend. This bank wall is a suitable habitat for sand martin swallows, which form colonies there. In the tropics, bee-eaters (Meropidae) inhabit this particular habitat for their breeding tubes.

V. Depth & Velocity: Max 15 pts

(a) Deepest Pool: Select the maximum depth if your stream has a consistent depth (Fig. 6.5).

(b) Select all the flow types that you see and add the points.

VI. Riffles/Runs (where the current is turbulent): Max 15 pts

Use the lower diagrams (Fig. 6.5) as a guide.

A blank form, a filled-in example form, and a digitized form packed in a ZIP archive are available at https://tinyurl.com/42ay5m2z.

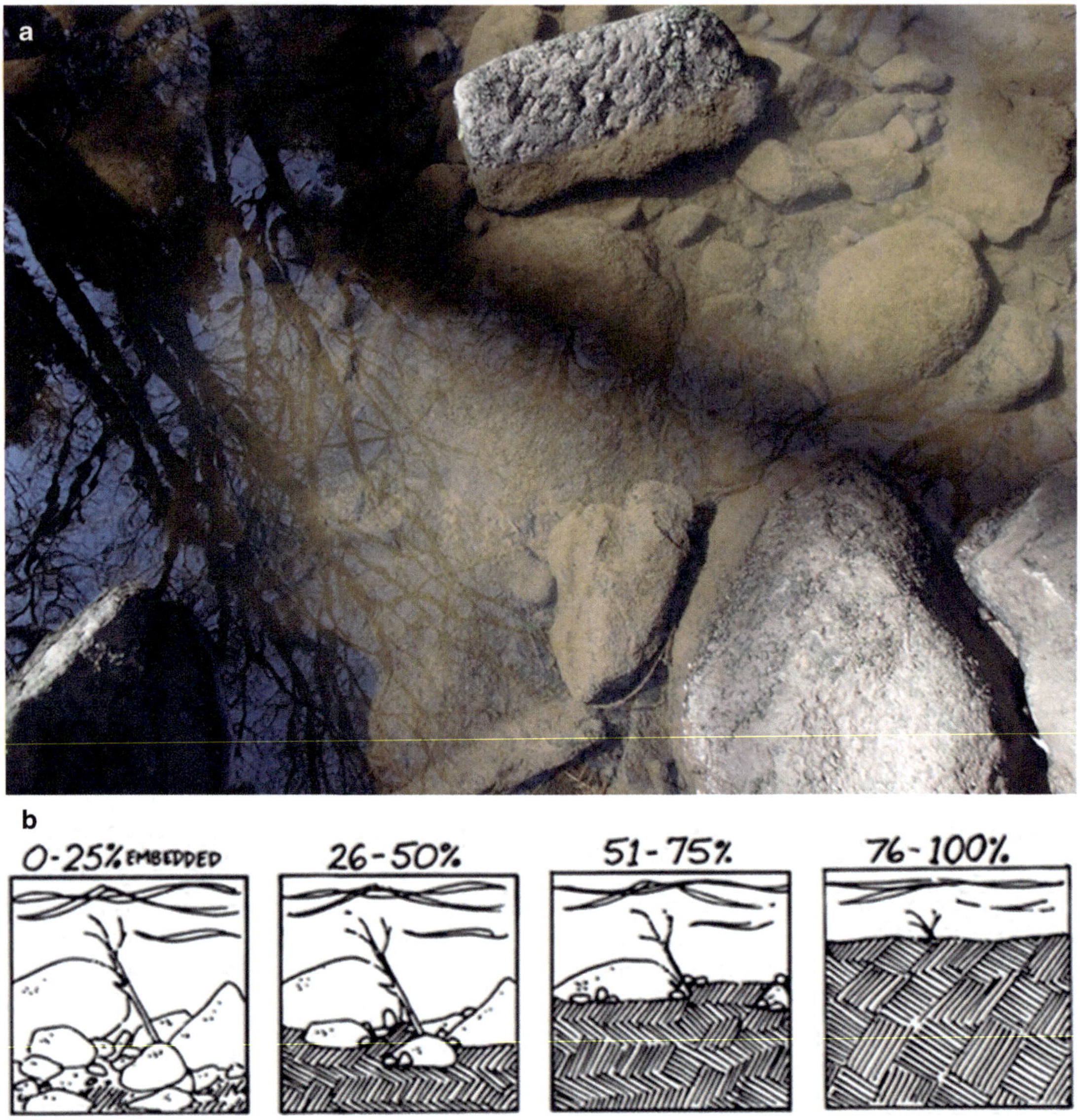

Fig. 6.7 Smothering or embeddedness of a river by fine sediment. (Picture credits: **a**. North York Moors National Park Authority, http://www.northyorkmoors.org.uk/ **b**. (Hoosier Riverwatch 2019). Both are reproduced with permission)

6.2.6 Sampling Benthic Macroinvertebrates from Rivers

6.2.6.1 What Are Benthic Macroinvertebrates?

Benthic macroinvertebrates (macros) are invertebrates that live at the bottom of rivers and lakes and are large enough to be seen with the unaided eye. They consist of freshwater clams, mussels, snails, crustaceans such as crayfish, and aquatic insects like the larvae of mayflies, stoneflies, caddisflies, midges, and adult and larval beetles.

6.2.6.2 Why Monitor Benthic Macroinvertebrates?

Macros are significant contributors to the freshwater food chain. Due to their abundance and intermediate position in the aquatic food chain, macros play a crucial role in the natural flow of

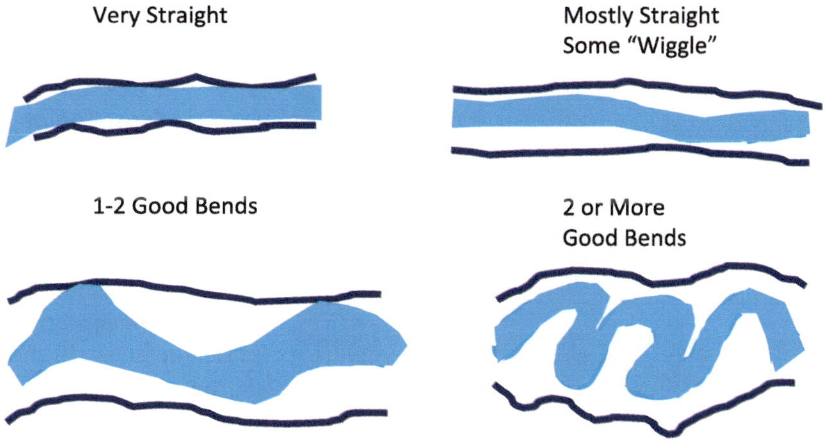

Fig. 6.8 Sinuosity of a channel

Fig. 6.9 (**a**) Meandering river in the Biebza protected area, northeast Poland. (**b**) Meandering river Onon, which flows downstream into the Amur, North Mongolia

Fig. 6.10 Riparian vegetation and erosion. (**a**) Show a narrow riparian width park vegetation (grass) and raw collapsing banks lacking shading, whereas (**b**) shows an expansive riparian forest with well-vegetated banks and partial shading. (Picture credits: **a**. butlerswcd.org with permission, **b**. AnjaIgnjat on Wikimedia, CC BY-SA 4.0)

Fig. 6.11 (**a**) River Tuul with an eroded bank wall in the river bend. The red arrow depicts the location of the sand martin colony shown below. North Mongolia. (**b**) The eroded bank wall is ideal for settling a sand martin colony (*Riparia riparia*)

energy and nutrients. This makes them well-suited as indicator species for the biological integrity of aquatic ecosystems. Changes in their biodiversity and populations over time can indicate the impact of human activities on ecosystems, particularly habitat change and pollution. Furthermore, macros can be sampled and identified using simple and inexpensive equipment and are the indicator organisms of choice for scientists, natural resource officers, and volunteer citizen scientists (Hoosier Riverwatch 2019).

6.2.6.3 Sampling Methods

It would be best to collect a maximum of three samples per year. If you gather more frequently, you can impact populations and communities. Set a limit to your collection time—in most cases, 30 min should be sufficient. The best time for collecting specimens in cold-temperate climates is mid-April to mid-November. The season you sample will affect what macros you will find due to their life cycles. Barbour et al. (1999) describe two main sampling techniques: the single-habitat riffle-run method using a 1 m kick net and the multi-habitat approach using a D-frame dip net. Here, we will only describe the dip net method.

Dip Net Sampling Method

Choose a 100 m reach, which is representative of the characteristics of the river. The sampling area should be at least 100 m upstream of road or bridge crossings whenever possible to minimize impact on flow rate, depth, and overall habitat quality. No major tributaries should flow into the stream in the study area. Before sampling, assess the habitat (Sect. 6.2.5), document the weather conditions, and take water samples if you plan to document the water quality (Sect. 6.3.5). Take at least 20 jabs in various habitats (Figs. 6.5, 6.12, and 6.14 and Sect. 6.2.5). Sample different habitat types in proportion to their representation of the total surface area (Fig. 6.13).

For example, tree roots, snags, and submerged logs comprise 40% of the habitat, and riffles contribute 30%. Take eight of your 20 jabs in the woody debris and six in riffle areas. Unless time allows for more jabs, do not sample habitat types that make up less than 5%. Record the number of jabs taken in each habitat type. For each jab, thrust the net forcefully over the river bottom for about 0.5 m, starting close to your body and finishing with arms fully outstretched (Fig. 6.14 and this YouTube video: https://youtu.be/28j4D5o2RqQ). However, the best sampling technique depends on the habitat type (Table 6.1).

6.2.6.4 Collecting the Catch and Preparing It for Identification Later in the Lab

After two to three jabs, dump the gathered materials into a little white dishpan (Fig. 6.15). This figure 6.15 shows how to transfer the catch from a dip net into a white plastic or enamel pan where the macroinvertebrates are well visible.

Remove the debris as well as possible without accidentally removing macros. Keep the white tub out of the sun and safe from accidental spilling. If you find macros, put them into a sampling jar containing 75–80% ethanol for later identification in the lab.

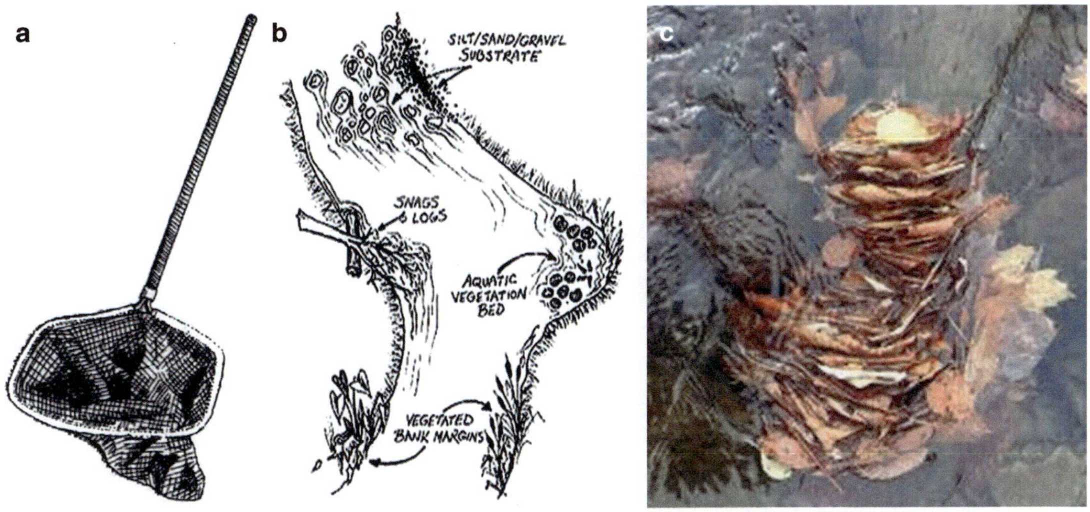

Fig. 6.12 (**a**) A dip net or D-net. (**b**) Some of the habitats sampled by D-nets. (**c**) Leaf packs at a riffle. (Picture credit: **a** and **b** (Hoosier Riverwatch 2019), (**c**) Stroud Water Research Center: https://stroudcenter.org/research/landmarks/watershed-tea/. All reproduced with permission)

Fig. 6.13 Numerous aquatic macroinvertebrate (macrobenthos) taxa collected from a stream. (Picture credit: Kelly O Maloney, Research Ecologist https://www.usgs.gov/, public domain)

6.2.6.5 Calculating the Benthic Index of Biotic Integrity as a Measure of Water Quality

Alone in the EU, at least 300 different metrics are used to assess the status of surface water using aquatic organisms (Birk et al. 2012). Here, we describe the usage of a particularly useful and fast bioassessment method, Hilsenhoff's family-level biotic index (FBI) (Haney et al. 2013a; Hilsenhoff 1988), developed for the cold-temperate United States but adopted in more or less modified form worldwide.

1. Take about one teaspoon of specimen from your sample and spread it over the bottom of a petri dish.
2. Sort specimens alike into ice cube trays (Fig. 6.16). Use the online resource https://www.macroinvertebrates.org/key/ (Louw et al. 2020) to assist you. This key was developed for the USA, but the general form variety should work for many geographic areas.
3. Go through each ice cube tray using a suitable key—for North America, e.g., (Bouchard 2004; Haney et al. 2013b; Louw et al. 2020; Valley City State University Macro-Invertebrate Lab 2014)—and determine each specimen to family level (some taxa might only need to be determined to order or class).
4. Calculate Hilsenhoff's HBI based on family-level identification (Hilsenhoff 1988) and their tolerance values (Soil & Water Conservation Society of Metro Halifax 2015)

Fig. 6.14 Use of the dip net to collect macrobenthos in the river. North Mongolia

Table 6.1 Optimal sampling techniques depend on habitat type

Riffles	Place the net on the bottom downstream of the riffle you want to sample. Use 1 foot to disturb the sediment right upstream of the net
Sediments (sand/mud)	At low flow rates, jab the net into the sediment and sweep it away from you. Refer to riffles for higher flow rates. To rinse, keep the net opening well above the water surface and wash small particles out of the net by moving back and forth
Tree roots, snags (accumulations of debris), and submerged logs	Scrub the surface of roots, logs, or snags directly with the net, or use a stick or your hand or foot while positioning the net downstream to capture any dislocated organisms
Undercut Banks	Place the net below any overhanging vegetation and repeatedly move it over the bank in a bottom-up motion to loosen organisms
Rocks/cobbles as sediment	Pick up rocks and cobbles and inspect them. There are often specialized macrobenthos on stones (Fig. 6.13a) that can be picked up with featherweight forceps (Fig. 6.13b)
Leaf packs	Shake the leaf pack in the water to release organisms, scooping them up with your net or picking them up with your hands. Separate leaves and inspect for organisms picked up with forceps

Modified after Larson (2000)

and rate the water quality accordingly. Tolerance values seem to vary between different regions of a country (Barbour et al. 1999; Carlisle et al. 2007) and even more between continents (Chang et al. 2014). Use the most appropriate tolerance values, as your water quality rating depends on this (McGauley et al. 2018)!

Fig. 6.15 Transfer from the D-net to a white plastic tub. In this way, the organisms can be separated from soft soil particles

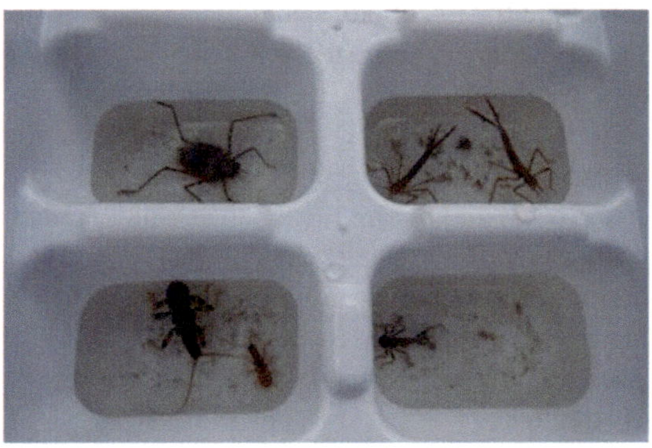

Fig. 6.16 Separating similar benthic macroinvertebrates using ice cube trays. Upper left: dragonfly naiad (Corduliidae), upper right: damselfly larvae (Calopterygidae), lower left: stonefly larvae (Perlidae), lower right: mayfly larva (cf. Ephemerellidae)

$$HBI = \frac{\sum n_i \times a_i}{N}$$

n = number of specimens in taxon i,

a = tolerance value of taxon i,

N = total number of specimens in all taxa of the sample.

6.2.7 Detection of Fish in Streams

There are three different approaches to collecting fish in a stream: catching with traps such as fish traps, catching with nets, and catching with electrofishing. For example, various fish traps are presented in Southwood and Henderson (2000). Fishing with nets is not permitted in many countries. Most fish die in the nets.

Fig. 6.17 Electrofishing in a small stream. The sampling team of two persons must wear rubber boots to avoid being shocked by the electric current. The electric power is supplied by batteries carried in a backpack. In front, Peter Melchert, Universität Göttingen, behind Mrs. Mendsaikhan from the Mongolian Academy of Sciences, North Mongolia, collecting the stunned fishes. Both are walking upstream

6.2.7.1 Electrofishing

Electrofishing is the most effective method for catching fish in a small stream (Cowx 1990; Price and Peterson 2010; Benejam et al. 2012; Pregler et al. 2015). Electrofishing requires special training and an official license. Two people are always required. The first person carries the device with pulsed direct current (via anode and cathode electrodes), and the second person collects the paralyzed fish (Fig. 6.17). If the electric current is suitably adjusted to the conditions in the stream, the fish only suffer a shock and remain alive. The conductivity of the water body influences the efficiency of electrofishing. The gear should be modified according to the environmental conditions. Repeated electrofishing using the same set period of time can be used to get closer to an absolute estimate of population sizes, following the principle of "empty trapping" (see removal trapping in Sect. 5.5).

6.3 Lakes

6.3.1 Research Questions for Lentic Systems (Lakes)

As a comprehensive methodology of the limnology of lentic systems would fill a book by itself, we will focus on a few research questions that we have used in our research and teaching of lentic systems and methods to answer them.

- How can we characterize, map out, and evaluate lake shorelines? How healthy is your lakeshore?
- How can we determine the vertical stratification of a lake?
- How can we identify and determine the density of plankton in a lake?

Some of these questions can be partially addressed from the shore. However, most need either a second vantage point from out on the water looking or sampling toward the shoreline (evaluation of lakeshores) or being entirely conducted on the water (vertical stratification and plankton analyses). These studies require kayaks, canoes, or a boat—and anchors and ropes to fix the vessel into a location and prevent drifting.

6.3.2 Vertical Stratification

6.3.2.1 Introduction

Depending on the turbidity of the water, light may penetrate lakes only for a few meters or up to 100 m deep, separating a well-lit top layer that allows plant and algae growth (euphotic zone) from a dark bottom layer without plant growth (aphotic zone). In addition, lakes in many regions of the planet, especially in the temperate climates, show a pronounced stratification in temperature, dissolved oxygen (DO), and other parameters (deFischer-Ooster and Brunner 1849; Dodds and Whiles 2019; Dodson 2005; Hutchinson 1957; Schwoerbel and Brendelberger 2022). These vertical patterns identify lake strata of high and low productivity and limits of animal life due to hypoxia. We will determine the turbidity/murkiness of the water using a Secchi disk and determine the depth profile of water temperature, dissolved oxygen, and light using sensors and data loggers.

6.3.2.2 Turbidity and Light Penetration

The easiest way to measure turbidity and how deep light can enter a lake is by using a Secchi disk invented by the Italian priest Angelo Secchi in 1865 (Secchi 1864) (Fig. 6.18).

To avoid tipping the boat over, stay low and submerge the Secchi disk into the water over **the shaded side** of the boat for a better view of the disk. Continue lowering the Secchi disk along the calibrated line until it becomes invisible, and record the depth. Then, slowly raise the disk until it reappears and record the depth at which it reappears. The average between the depth at which the disk disappeared and the depth at which it reappeared is the Secchi depth (Fig. 6.18b). If you can see the Secchi disk reaching the bottom of the lake without disappearing from sight, record this depth and add "on bottom" in parentheses. Do not use sunglasses or an aquascope/bathyscope for accurate and comparable measurements when taking the disk reading. You may want to watch the following video to better understand the procedure: https://youtu.be/xiRT2j54Y2U (3:33 min).

The Secchi (disk) depth can be used to estimate the depth of the euphotic zone, the upper portion of a lake that receives enough light for plants and algae to produce energy for growth and reproduction (primary productivity). The bottom of the euphotic zone receives approximately 1% of the surface light levels and corresponds to about three times the Secchi depth (Dodson 2005).

The most accurate determination of the euphotic depth is achieved by using quantum sensors for photosynthetically active radiation (PAR), like the LI-193 (LI-COR Biosciences, Lincoln,

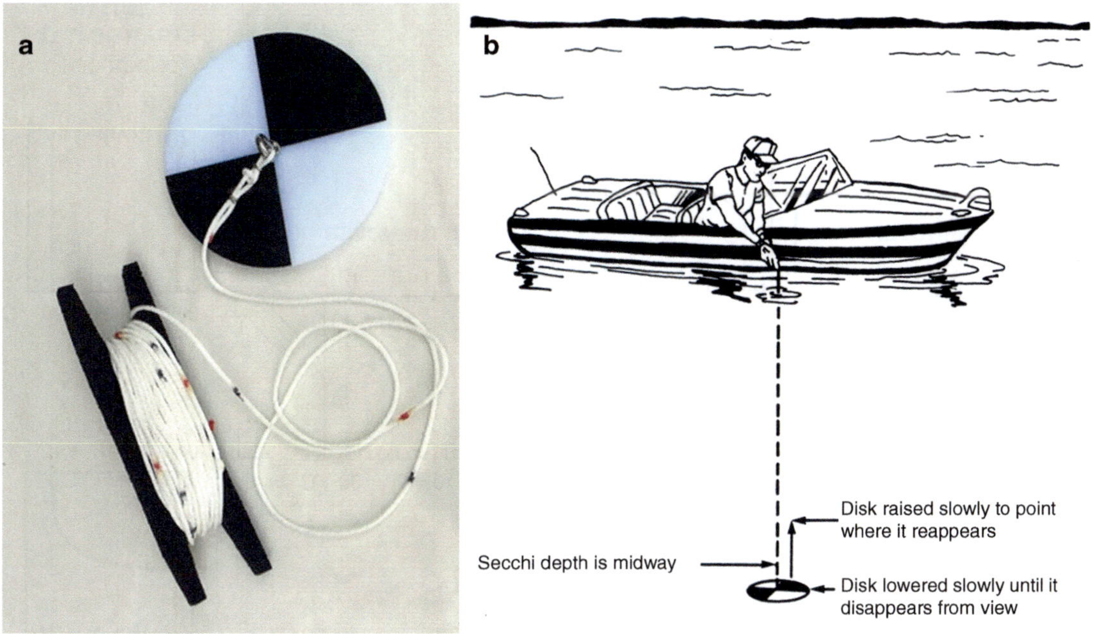

Fig. 6.18 (**a**) Secchi disk with calibrated line (calibration markings on every foot/0.5 m). (**b**) The Secchi depth is the mean between the depth when the disk disappears when it is lowered into the water column and the depth when it reappears when it is raised back. (Picture credit for (**b**) US EPA (2001), CC0 1.0)

NE, USA), but these sensors are pretty expensive. The Secchi disk method can be significantly improved using scalable and locally calibrated methods instead of a single fixed coefficient (Luhtala and Tolvanen 2013).

6.3.2.3 Dissolved Oxygen/Temperature Profile Measurements

The parameters most accessed when a lake's depth changes are temperature and dissolved oxygen (DO). Both variables can be used to define the depth (and thickness) of the upper lake portion (epilimnion) and lower part of the water column (hypolimnion) and their dividing thermocline (rapid decline in temperature per meter depth) in the middle portion of the lake (metalimnion). In lakes with a summer stratification, the thermocline is typically located at a depth of 5–15 m (Fig. 6.19).

Both variables can be determined by gradually lowering a probe with the sensors attached to a calibrated cable (e.g., in meter intervals) through the water column. A small weight attached at the end of the cable near the probe aids the cable/probe to sink straight down through the water column. At specific depths determined using the markings on the calibrated cable, readings can be made with a handheld data monitor. One of the most widely used instruments that can record temperature and DO is the YSI Pro20 (YSI Inc., Yellow Springs, OH, USA), shown in Fig. 6.20. Newer devices have an integrated pressure gauge that determines the depth of the measurements, offer up to four sensors, and communicate/export data to computers and USB drives. There are sensors available, e.g., for DO, temperature, conductivity, turbidity, oxidation/reduction potential (ORP), pH, ammonium, nitrate, chloride, total dissolved solids (TDS), depth, total algae, and chlorophyll. It is recommended that the measurements be performed on a calm day, as the cable/probe may drift away from a straight line and distort the readings. This video shows the use of a similar device, the HACH HQ30d flexi: https://youtu.be/HLMtB87YWlQ?t=27.

As these professional devices are costly, we want to introduce you to a low-cost alternative for temperature measurements. The Fish Hawk

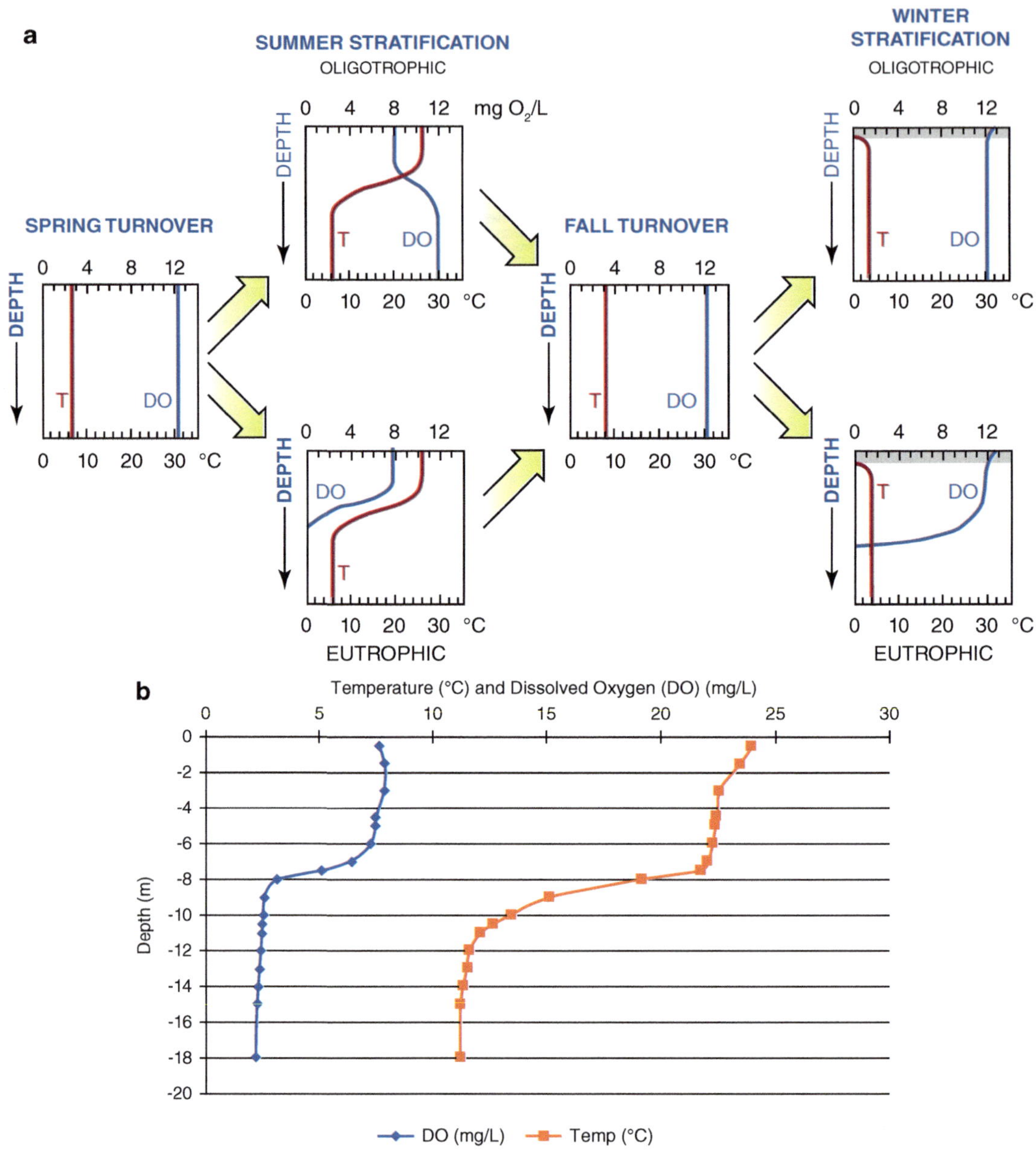

Fig. 6.19 (**a**) Typical seasonal stratification of temperature (T) and dissolved oxygen (DO) in cold-temperate lakes. (Picture credit: WOW. 2004. Water on the Web—Monitoring Minnesota Lakes on the Internet and Training Water Science Technicians for the Future—A National On-line Curriculum using Advanced Technologies and Real-Time Data (http://WaterOntheWeb.org). Reproduced by permission. (**b**) DO and water temperature in Devil's Lake, a mesotrophic inland lake in Michigan, on August 15, 2012, measured by a YSI 55Pro)

TD (Fish Hawk Electronics, Brainerd, MN, USA) combines temperature and depth measurements every 5 ft (1.5 m) and stores the values in its internal memory. Depth-temperature values then have to be manually recalled on the screen and can be copied into a spreadsheet.

Vertical lake profiles can also be determined by taking water samples at specific depths using Niskin, Nansen, Ruttner, Van Dorn, or Kammerer bottles, which can be triggered to close the bottles by a trigger mechanism called a messenger (Figs. 6.21 and 6.22). You can see a Niskin bottle

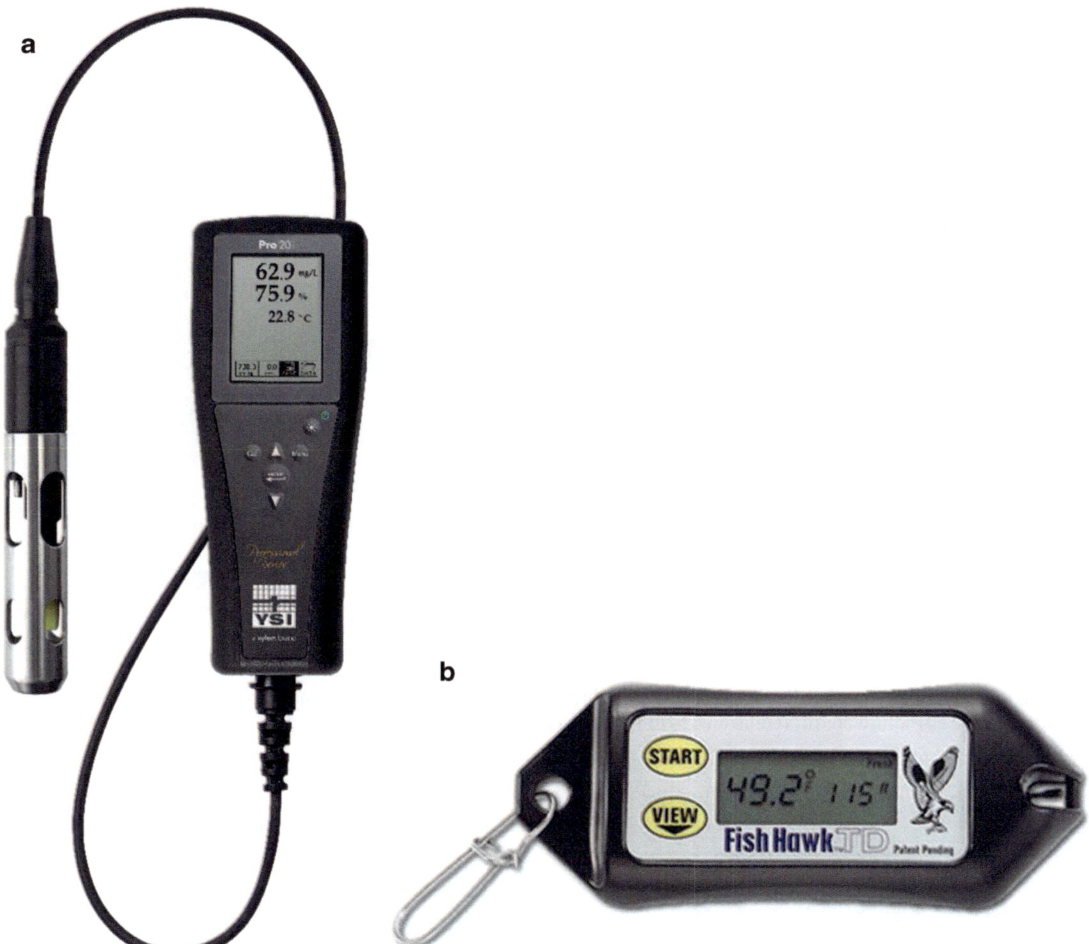

Fig. 6.20 (**a**) YSI Model Pro20. Picture credit: YSI, a Xylem brand, ysi.com. (**b**) Fish Hawk Electronic TD Digital Water Temperature & Depth Sensor. (Picture credit: fishhawkelectronics.com. Both pictures reproduced by permission)

in action here: https://www.youtube.com/watch?v=YvJBI9nO6Yg and a Van Dorn bottle here: https://youtu.be/DvxIpe_JxTc. For temperature, the sample water needs to be analyzed immediately after retrieving the sampler. Other tests can be performed on chilled samples in the laboratory. Niskin bottles are often combined with CTDs (conductivity, temperature, and depth meters), quantum sensors, and transmissometers and provide a depth profile of temperature, salinity, density, dissolved oxygen, fluorescence, and transmittance while being gradually submerged, collecting water samples at defined depths to be processed later (Fig. 6.23).

The most complete and convenient recordings of the changes in the vertical stratification of lakes are automatic measurements using stationary buoys equipped with temperature (and light) sensors/data loggers and a cable array transversing the entire water column from the surface to the bottom. This relatively cost-efficient method is used by the Lakes Environmental Association (LEA) in the US state of Maine (Fig. 6.24). Sensors are attached to a floating line held in place by a regulatory-style buoy and an anchor. The sensors are attached at 2-m intervals, beginning 1 m from the bottom and ending approximately 1 m from the top (Fig. 6.24a). Each buoy

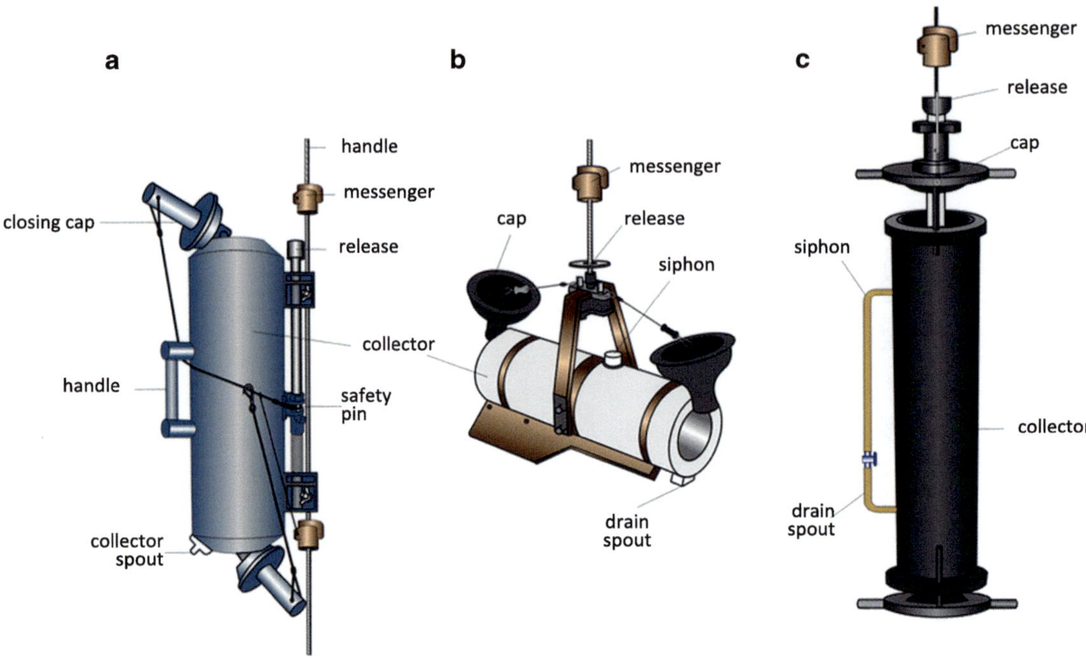

Fig. 6.21 Water collection bottles: **a**, Niskin; **b**, Van Dorn; and **c**, Kammerer (**a**, adapted from the Hydro-Bios catalog at hydrobios.de). (Picture credit: (Calazans 2011) with permission from the author)

Fig. 6.22 A water scoop after Ruttner in use (see Ruttner 1962)

apparatus is deployed at the deepest point of the basin it monitors. HOBO combined temperature and light sensors/data loggers (Fig. 6.24b) are programmed to record temperature readings every 15 min during the entire ice-free time of the year.

For temperature profiles, LEA switched from the handheld YSI meter to the stationary logger array, allowing them to monitor more lakes throughout the ice-free season. They still use handheld YSI meters when performing biweekly field sampling to measure dissolved oxygen data. However, the sampling of DO data is much more streamlined as LEA uses the profiles received from their stationary logger arrays to determine the depths at which to take water samples. This offsets the time investment needed to deploy and retrieve the loggers. In addition, the data provides much greater detail, including daily temperature fluctuations, brief mixing events caused by storms, the date and time of stratification setup and breakdown, and the timing of seasonal high temperatures (Fig. 6.25).

The most complete and convenient recording of the changes in the vertical stratification of lakes but the most costly solution is the automatic

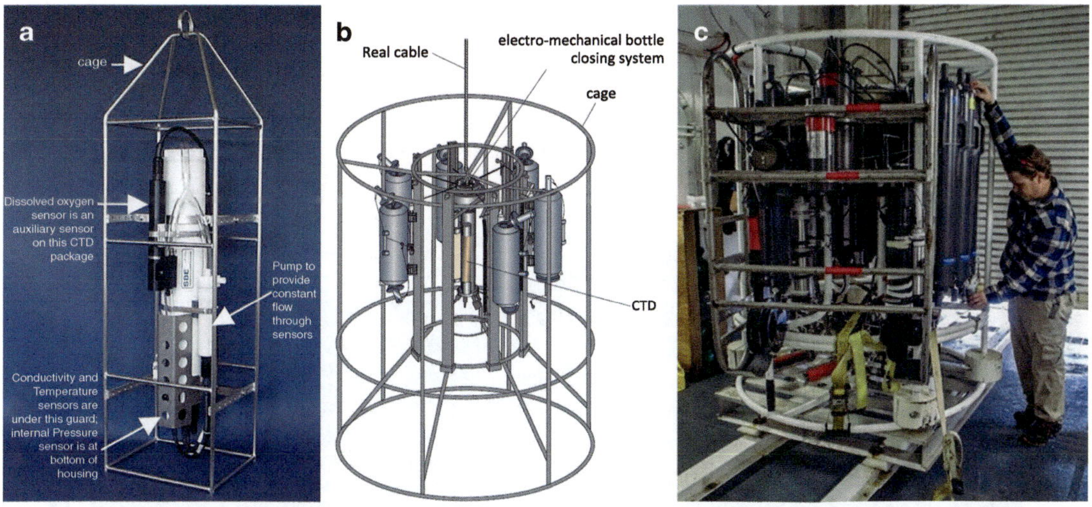

Fig. 6.23 (**a**) SBE 19plus V2 SeaCAT CTD (conductivity, temperature, and depth meter) can provide power to and integrate data from up to seven auxiliary sensors. (Picture credit: With permission from Sea-Bird Scientific (seabird.com)) (**b**) Rosette with CTD and multiple Niskin bottles used in Brazil. (Picture credit: (Calazans 2011) with permission from the author) (**c**) Peter Shipton collects water samples from the 20 Niskin bottles on this CTD that took water samples from various depths, starting near the seafloor and ending close to the surface. (Image courtesy of Caitlin Bailey, GFOE, The Hidden Ocean 2016: Chukchi Borderlands)

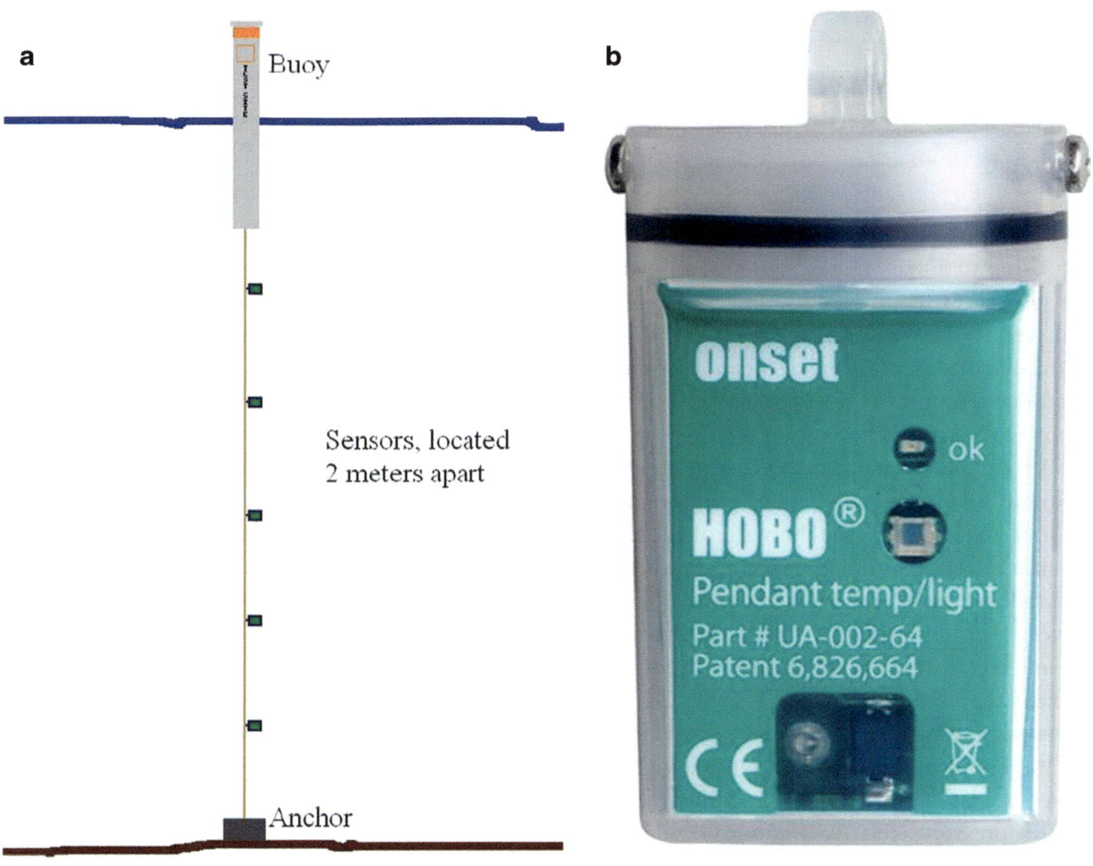

Fig. 6.24 (**a**) HOBO sensor setup. (Picture credit: with permission by the Maine Lakes Environmental Association (Lakes Environmental Association 2023) (**b**) HOBO temperature/light sensor used for the array in (**a**))

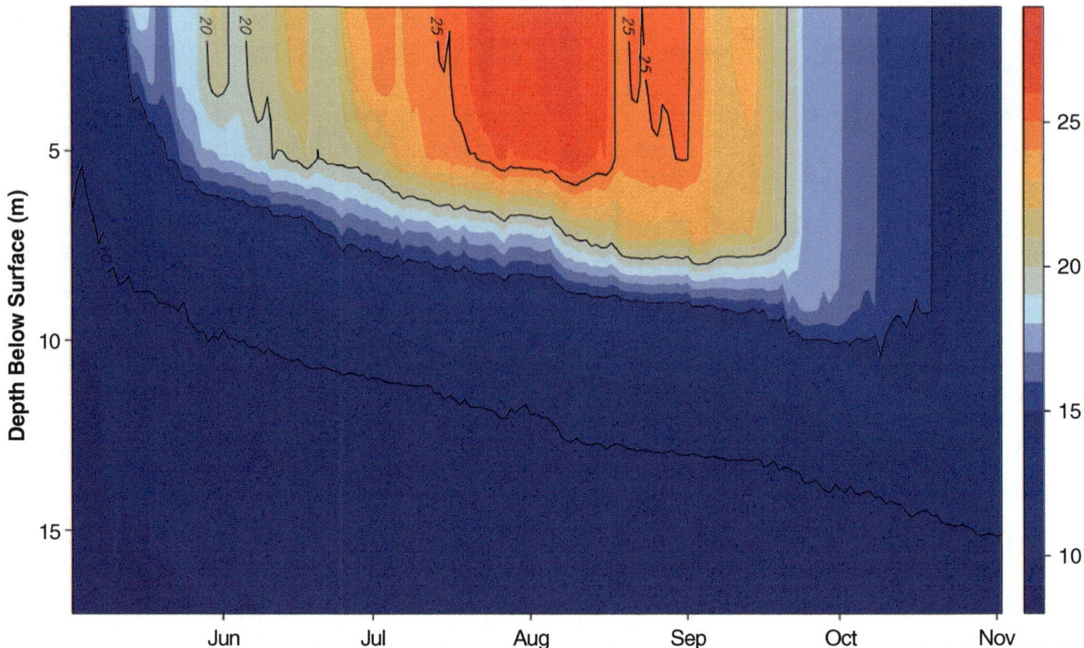

Deployment Date	Peak Temperature (°C/ °F)	Full Mixing	Retrieval Date
5/2/2022	28.6 / 83.5	After Retrieval	11/2/2022

Fig. 6.25 Temperature isopleth heatmap based on a sensor array in Hancock Pond, Maine, USA. The same colors and vertical lines indicate isothermal temperature periods, and the curved lines indicate summer stratification. (Picture credit: Lakes Environmental Association 2023 with permission)

simultaneous measurement of multiple parameters using stationary buoys equipped with various sensors at different depths and a cable array transversing the entire water column from the surface to the bottom. One of the most commonly used buoy systems is the NexSens CB-450 Data Buoy (NexSens Technology, Inc., Fairborn, OH, USA). Again, we will portray the usage of data buoys at LEA in Maine, which operates buoys in two of their lakes. The buoys collect data at 15-min intervals throughout the spring, summer, and fall and transmit them in real time. The buoys have seven temperature and oxygen sensors mounted at 2-m intervals from the lake's surface to near the bottom (Fig. 6.26). Two solar radiation sensors and a chlorophyll sensor are mounted on the buoy cage underneath the buoy. A small weather station measuring precipitation, barometric pressure, relative humidity, wind speed and direction, and air temperature is located on top of the buoy above the water surface. The buoys use a power supply of three 10-watt solar panels and a rechargeable battery. In addition to the sensors/data loggers used by LEA, the buoys can be equipped with the LI-COR LI-192 Underwater PAR Sensor (LI-COR Biosciences, Lincoln, NE, USA) and the YSI EXO2 multiparameter water quality sonde (YSI Inc., Yellow Springs, OH, USA) that records temperature, conductivity, depth, dissolved oxygen, pH, oxidation/reduction potential (ORP), total algae (phycocyanin or phycoerythrin along with chlorophyll), turbidity, fluorescent dissolved organic matter (fDOM), rhodamine WT, UV nitrate, ISE ammonium, ISE nitrate, and ISE chloride and calculates the derived parameters salinity, specific conductance, total dissolved solids (TDS), and total suspended solids (TSS).

The simultaneous measurements of water temperature, dissolved oxygen, algae (chloro-

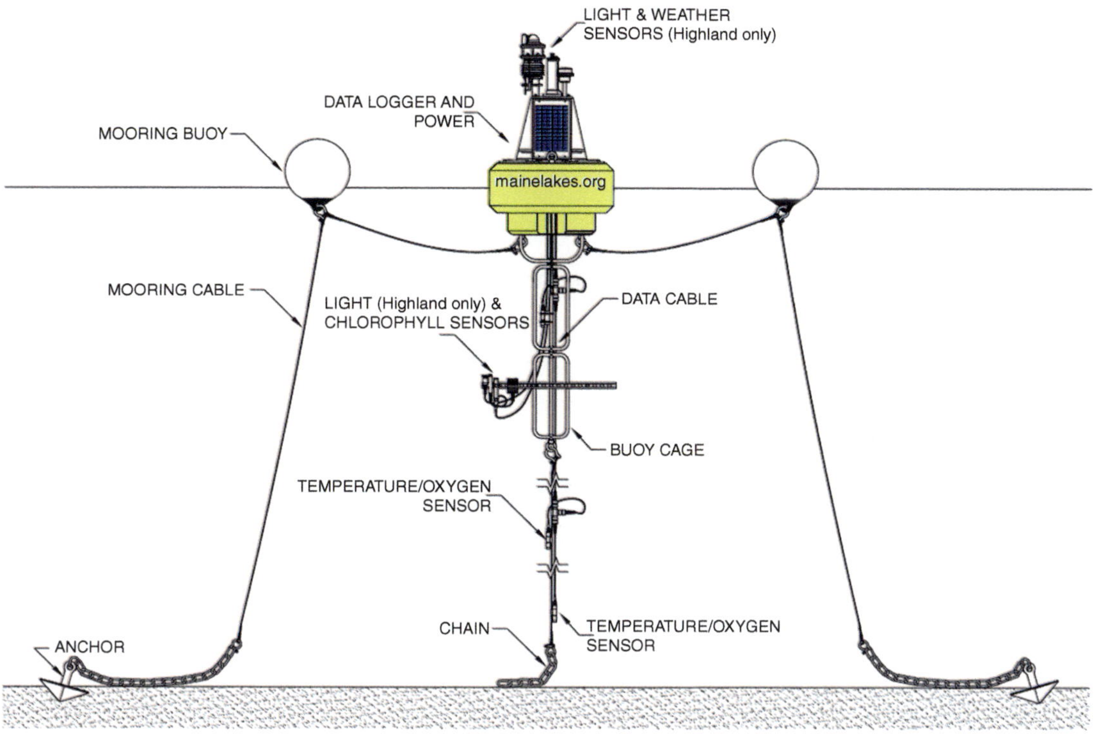

Fig. 6.26 Schematic of the automated buoy system used by the Lakes Environmental Association in Maine, USA. (Picture credit: Lakes Environmental Association 2023 with permission)

phyll), water clarity, and weather conditions allow LEA to see the effects of air temperature, wind, and precipitation on the water quality of their lakes in real time. Furthermore, LEA collaborates with researchers deploying data buoys throughout New England and the world by joining the Global Lake Ecological Observatory Network (GLEON). Below is an example of the simultaneously recorded temperature and DO levels in Highland Lake, Maine, USA (Fig. 6.27).

6.3.3 Plankton

6.3.3.1 Introduction

Plankton consists of tiny organisms drifting or floating in the sea or freshwater. It is commonly divided into autotrophic phytoplankton, providing the primary productivity of the open water of lakes and the oceans, and heterotrophic zooplankton, which are the primary consumers in the food chain of the open waters and feed on phytoplankton. As zoologists, we will focus on the harvesting, density estimation, and biodiversity analysis of zooplankton in lakes. Most zooplankton species are filter feeders, extracting millions of phytoplanktonic algae by passing the entire volume of lakes through their bodies in just a few days. In return, zooplankton becomes the most critical food source for smaller planktivorous fish and other organisms, which larger piscivorous fish consume. Therefore, the population density of zooplankton determines the composition and diversity of the entire lake community and the success of recreational and commercial fisheries. Suppose we want to protect higher trophic levels of the lake food web, including larger fish, turtles, otters, and humans. Consequently, the monitoring and safeguarding of viable populations (compare Sect. 10.8) and biodiverse zooplankton communities are vitally important.

Large mesozooplankton ($\geq$1.0 mm) are suitable indicators of habitat and food conditions for

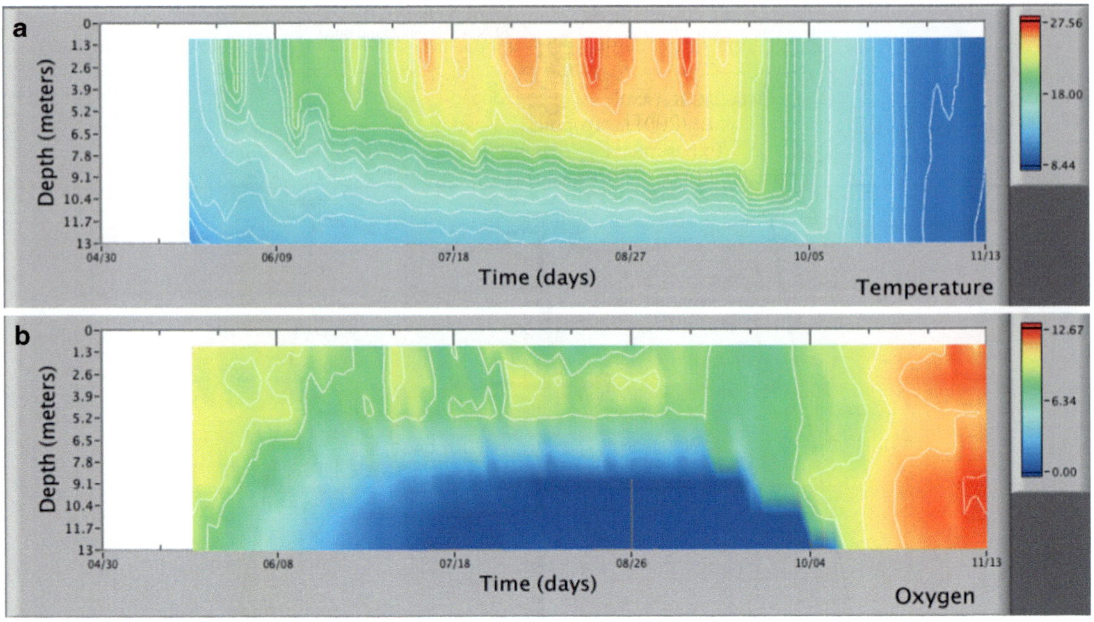

Fig. 6.27 Isopleth heatmaps for (**a**) temperature and (**b**) oxygen concentration for Highland Lake (Maine) from April through November. The vertical axis is depth, and the horizontal axis is time. The numbers on the upper graph indicate the temperature in °C, and those on the lower graph indicate oxygen concentration in mg/L (PPM). The same colors and vertical lines indicate isothermal periods of temperature. The blue color from the top to the bottom in (**a**) indicates lake mixing in October. The lower panel (**b**) shows oxygen levels over the same time period. The dark blue on the bottom shows the level and extent of anoxia (lack of oxygen). (Picture credit: mainelakes.org with permission)

commercial and recreational fish (Fig. 6.28). For example, lakes with larger zooplankton have, on average, faster-growing bluegills (Theiling 1990) and larger adult bluegills (Schneider 1990). Galbraith (1975) found that poor trout lakes had fewer large zooplankters per net haul than good ones. Furthermore, the mean length of zooplankter may indicate general fishing conditions (Mills et al. 1987). Zooplankton sampling during spring and summer in many US states is a standard element of routine lake surveys. These surveys provide information to determine young fish's relative year class strength, the relative abundance of older planktivores, and predator–prey ratios.

6.3.3.2 Plankton Harvesting

The easiest and most efficient way to sample plankton is by using a plankton net (Fig. 6.29).

To sample zooplankton in temperate regions, the mesh size of the plankton net should be at least 150–200 mm, while the mesh size should be 70–100 mm in the tropics due to the smaller size of tropical zooplankton. However, smaller mesh sizes create more resistance to the flow of water through the net and lower efficiencies due to clogging and backflow and require slower towing speeds, which may allow some mobile taxa of zooplankton to avoid the plankton net (Smith et al. 1968; Jacobs and Grant 1978).

Plankton nets can be towed horizontally or vertically (Fig. 6.30), potentially resulting in different populations and communities.

Horizontal plankton tows sample much more of a lake's surface area but only along a single stratum, usually close to the water surface—the epilimnion. If the sampling is done during the day, much of the zooplankton might not be caught due to their daily vertical migration (DVM) into deeper layers of the water column during the day, by which they are avoiding radiation and visual predators (Bandara et al. 2021). To get an idea of

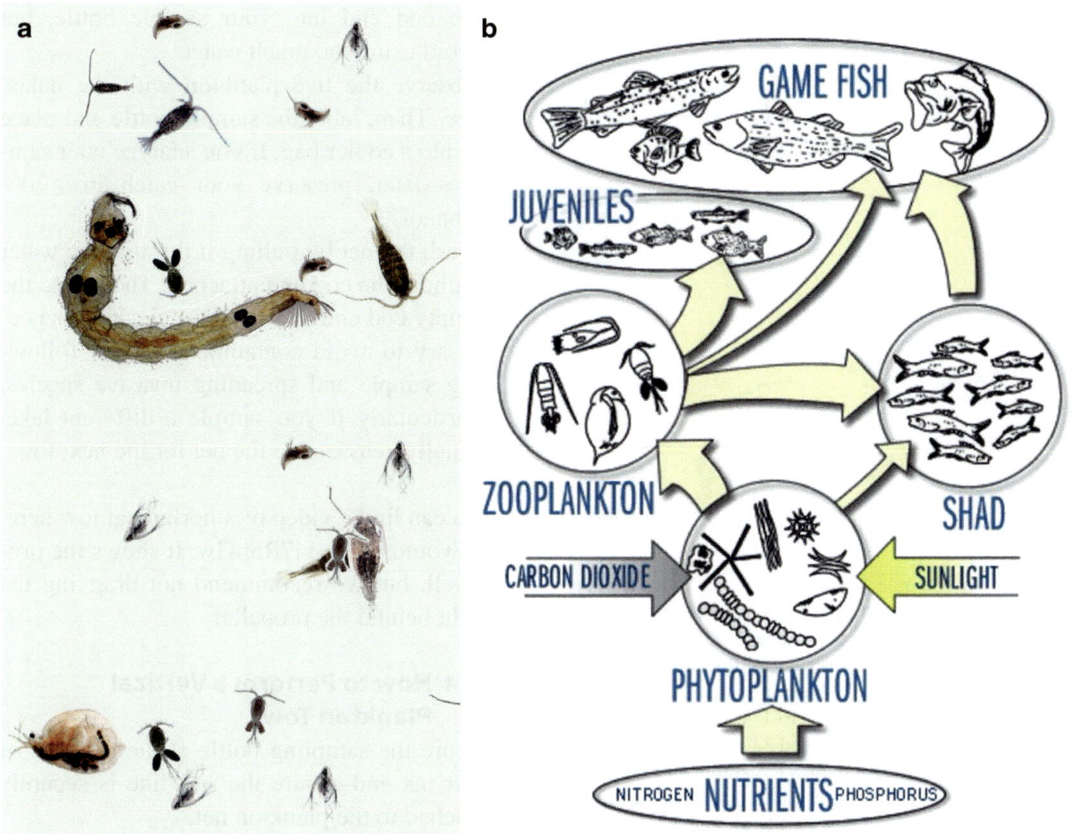

Fig. 6.28 (**a**) Representative zooplankton species from Devil's Lake, Michigan. (**b**) Simplified lake food web. (Picture credit: WOW. 2004. Water on the Web—Monitoring Minnesota Lakes on the Internet and Training Water Science Technicians for the Future—A National On-line Curriculum using Advanced Technologies and Real-Time Data (http://WaterOntheWeb.org). Reproduction by permission)

the local DVM of plankton, you can sample the same area during the night using the same method as during the day.

6.3.3.3 How to Perform a Horizontal Plankton Tow

1. Secure the sampling bottle at the cod end of your net and attach the tow line firmly to the plankton net.
2. Hold the mouth ring using your thumb and forefinger. Loop the tow line and keep it loosely in the same hand as the net. Grasp the other end of the line with your free hand or better secure it to the boat.
3. Lower the net into the water and allow it to sink into the waterbody until the mouth ring is fully submersed, but avoid letting it sink more than a foot. If air bubbles get trapped in the net, retrieve the net and start again.
4. If all is fine, the boat can begin moving at a slow and constant speed (1.5–2 knots = 75–100 cm/s), then allow the line to feed out with the net until it is 10–20 ft away from the boat.
5. Record the time (and take a GPS reading) and tow the net for about 30–100 m (Fig. 6.30a)—make sure you also record the end time of your tow (and a second GPS reading) and the speed of the boat. You will need this information to calculate the pulled distance from the velocity of the ship and the pulling time (GPS: waypoints for the start and end of the tow) that you will need to calculate the plankton density.

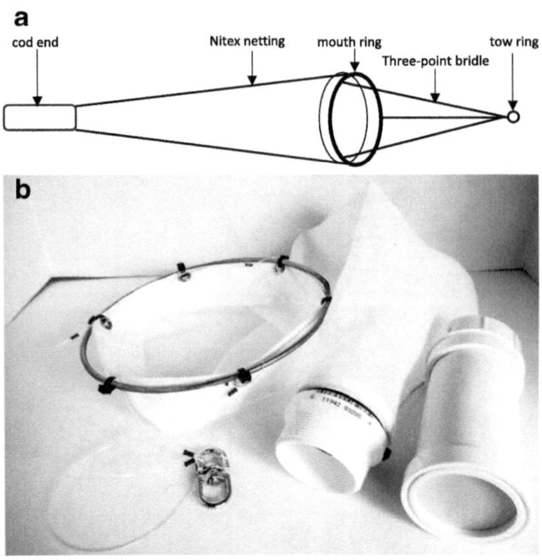

Fig. 6.29 (**a**) Simple plankton net made of a metal (mouth) ring and bridle/harness, a heavy-duty nylon net in different mesh sizes, and a detachable PVC cod end or collecting bucket allowing for accessible sample collection. (**b**) High-quality plankton net made from Nitex micron cloth netting (resistant to UV and salt), with an 8-in. diameter mouth opening and 34 in. long (1:4 mouth-to-length ratio for maximum flow). A stainless steel swivel is attached to the bridle to prevent the twisting of the tow line. The triple-sewn net is securely attached to the solid stainless steel mouth ring with reinforced brass rings. The 2 × 5 in. PVC cog end features a micron mesh to optimize flow and reduce backflow during towing. (Picture credit: Aquaculture Nursery Farms, https://www.aquaculturenurseryfarms.com. Reproduced with permission)

6. Retrieve the net hand-over-hand at approx. 0.5 m/s (1.5 ft/s). The key to a good plankton sample is a slow and steady retrieval.
7. Rinse the net by raising it straight up while keeping the cod end of the net at the water's surface. Flush organisms from the net into the cod end by repeatedly lowering the net back into the water, keeping the plankton net's mouth well above the water surface to avoid losing your catch, and raising the net again. After you have removed all the organisms inside the net into the cod end, quickly pull the net straight out of the water.
8. Unscrew the cod end bottle without spilling the sample. Pour your catch into a wide-mouth sample bottle. Use a wash bottle with lake water to rinse plankton that remained in the cod end into your sample bottle, but avoid using too much water.
9. Observe the live plankton with the naked eye. Then, label the sample bottle and place it into a cooler bag. If you analyze your samples later, preserve your catch in ≥70% ethanol.
10. Wash the net by pulling it through the water without the cod end attached. Then rinse the empty cod end as well. This cleaning is necessary to avoid contaminating your following sample and spreading invasive species, particularly if you sample a different lake. Finally, reassemble the net for the next tow.

You can find a video of a horizontal tow here: https://youtu.be/dx-vj7RnhGw. It shows the process well, but we recommend not dragging the net right behind the propeller.

6.3.3.4 How to Perform a Vertical Plankton Tow

1. Secure the sampling bottle at the cod end of your net and ensure the tow line is securely attached to the plankton net.
2. Lower the net at least 30 m (100 ft) below the water surface but stay 1 m above the lake bottom (Fig. 6.30b). Record the depth to which you lowered the net. You will need this distance if you want to calculate the density of the plankton. If your net is not sinking sufficiently, attach a sounding weight to the cod end.
3. Keep the net at the maximum depth for 30 seconds and then recover it hand-over-hand at a rate of 0.5 m/s (1.5 ft/s). The key to a good plankton sample is a slow and steady retrieval.
4. Follow steps seven to ten described for horizontal tows above.

You can find a video of a vertical tow here: https://youtu.be/fS213qP8pS8.

6.3.3.5 Determining Plankton Density

Determine the volume of the water filtered through your plankton net using Eq. (6.1), assuming an efficiency of 100% without any clogging. If the net clogs up, a pressure wave will cause

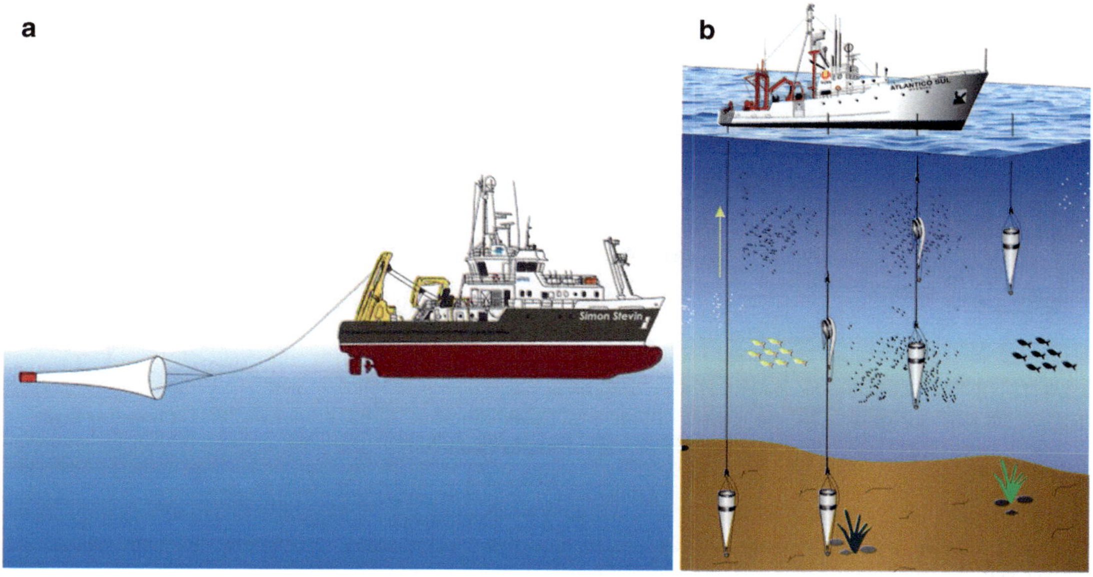

Fig. 6.30 (**a**) Horizontal plankton tow close to the water surface. Picture credit: Vlaams Instituut voor de Zee VLIZ, vliz.be. (**b**) Vertical plankton tow. (Picture credit: (Calazans 2011). Both pictures are reproduced by permission)

backflow out of the plankton net. If clogging occurs, reduce the velocity of the tow/retrieval of the net. If it still happens, estimate the filtering efficiency and adjust the maximum volume of water filtered (e.g., 90% filtering efficiency means $0.9 \times V_m$).

$$V_m = \pi \times r^2 \times d \tag{6.1}$$

where V_m is the maximum volume of filtered water, r is the radius of the net opening (convert the diameter of your net to the radius in meters), and d is the towing distance of a horizontal tow in meters or the depth to which the net was lowered in a vertical tow in meters.

Preparing Your Sample

1. If you want to restrict your analysis of zooplankton to larger species, pour your sample from the sample bottle through a strainer with the chosen mesh size (e.g., small mesozooplankton using a 200 µm mesh or large mesozooplankton using a 1000 µm mesh, Jacobs and Grant (1978)) placed above a large enough container to receive the ethanol and small samples that drain through the mesh size. Set aside the receiving container with the ethanol and smaller zooplankton. Rinse the retained zooplankton (e.g., ≥200 µm) in the strainer thoroughly above the sink using a squirt bottle containing water to get rid of any residue preservative (ethanol).
2. Transfer the retained zooplankton into a 500 mL beaker with marked volumes (in 50 mL increments). Using a squirt bottle containing water, ensure that you transfer all visible zooplankton from the strainer into the beaker.
3. Fill the beaker with tap water to the 50-mL mark. If the sample contains a high density of zooplankton, diluting it further by adding more water to the beaker (in 50 mL increments) is desirable. Record the total volume of the sample.
4. Use a 3 mL disposable pipette, swivel the plankton sample in the beaker to keep plankton floating, then stir the plankton sample in the beaker in a figure-8 pattern, and take a 1 mL subsample and place it carefully into a Sedgewick–Rafter counting cell (Fig. 6.31). Follow the instructions in this YouTube video: https://www.youtube.com/watch?v=df07VG9nQ8M, precisely the sequence between 1:26 and 3:16 min.

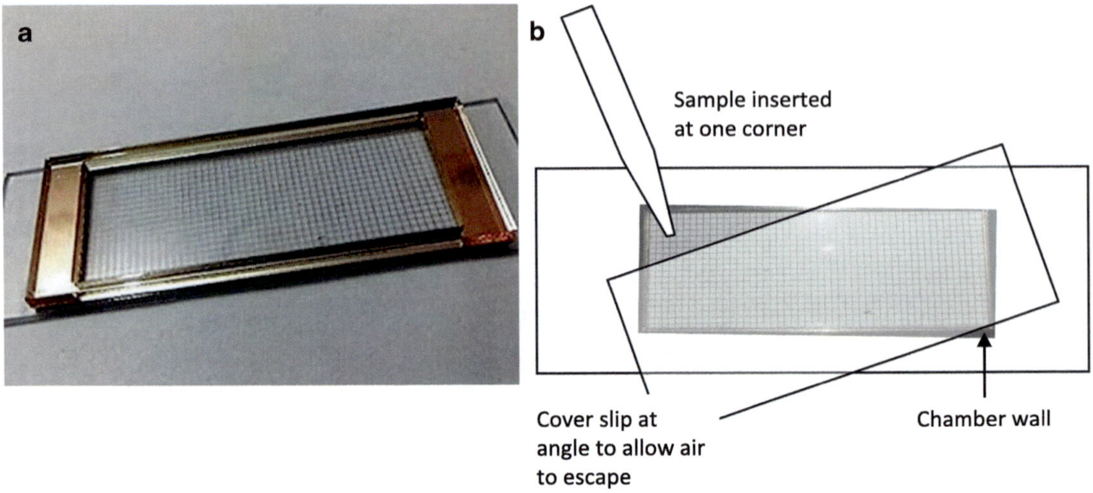

Fig. 6.31 Sedgewick–Rafter counting cell. (**a**) 1 mL high-quality glass cell with an etched grid. (**b**) Sample application into the cell

Counting Your Sample

1. If your counting cell has no mm grid, you can place the counting cell on a transparent mm grid and line it up with your cell.
2. Scan through the cell as shown in Fig. 6.32 and count all planktonic organisms. Be careful to count the entire dish but do not overlap the passes, thus avoiding the same individual being counted twice.
3. After the sample has been counted, pour the content of the counting cell back into the beaker, rinse the counting cell out using the squirt bottle, dispose of the rinse water, and dry the cell using a paper towel. Prepare another 1 mL subsample in the manner described above and count it the same way. Count a total of 4 × 1 mL subsamples per sample.
4. After you have counted four samples, pour the content of the beaker back through the strainer and squirt the plankton in the strainer back into the sample jar using a squirt bottle containing 80% ethanol.

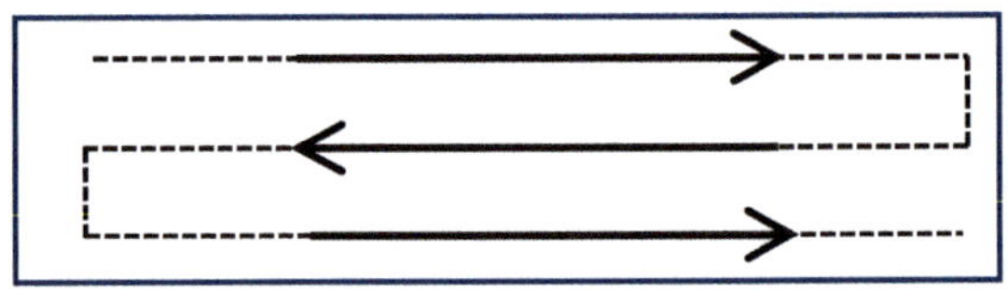

Fig. 6.32 Scanning through a counting cell to avoid double counting. You may need to take more than two turns to cover your sample

Calculating Plankton Densities

1. Calculate the mean value of the 4 × 1 mL counts you performed.
2. Calculate the total count for your sample by multiplying this mean value for 1 mL by the total volume of water in the total sample (Preparing your sample, step 3). Compute the total of zooplankters in the original sample following Eq. (6.2)

$$Q = T \times V / v \quad (6.2)$$

where Q is the number of zooplankton equal to or larger than your sieve's mesh size (e.g., ≥200 μm) in the entire sample, T is the average number of zooplankton in a 1 mL subsample, V is the volume of the original sample (Preparing your sample, step 3), and v is the volume of your subsample (here: 1 mL).

Example: 100 zooplankton specimens are in 1 mL, and the total volume is 250 mL, which results in 25,000 zooplankters in 250 mL.

3. Plankton density can be expressed as the number of organisms per volume of water strained through the plankton net. To calculate this,

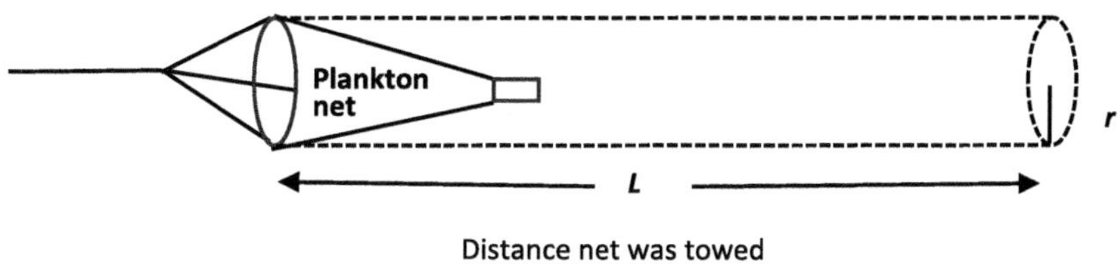

Fig. 6.33 Variables needed to calculate plankton density from the total count in the sample

you will need the area of the mouth of the plankton net and the distance or depth the net was towed or raised (Fig. 6.33).

4. You can calculate the water volume filtered through the plankton net during the tow (V_T) following Eq. (6.3).

$$V_{\mathrm{T}} = \pi \times r^2 \times L \tag{6.3}$$

where $\pi \times r^2$ is the area of the net opening, and L is the distance the net was towed (Fig. 6.33). It is advisable to convert all measures to meters, resulting in a volume of cubic meters (m^3).

5. Express the population density of zooplankton as the number of zooplankters per cubic meter of water filtered. Another standard unit is the number of zooplankters per L of water, which can be derived by dividing the number per cubic meter by 1000.

A problem with this simple computation is that the volume of the filtered water is often less than the calculated volume because the water builds up at the net's mouth, especially if the net is towed or pulled too fast. For this reason, following the recommended towing/pulling speed is vital to receive a consistent estimate of the relative density across samples.

6.3.3.6 Identifying Plankton and Calculating Its Biodiversity

This section will provide introductory advice on identifying zooplankton taxa and using the taxon list to calculate standard biodiversity metrics, including species richness, evenness, and the Shannon–Weaver index. We will also portray a simple procedure to compare the communities deducted from a horizontal tow to the ones gained from a vertical tow. It is widely assumed that high biodiversity is synonymous with ecosystem health. Most experts believe diverse communities have increased stability and productivity, providing better resistance and resilience to invasion and other disturbances (Anonymous 2017; Castro et al. 2016; Launchbaugh 2009; Quinney 2020).

Preparing Your Sample

Follow the instructions under *Determining Plankton Density*.

ID and Count Your Sample

1. Scan through the cell as shown in Fig. 6.32 and ID and count all planktonic organisms. Be careful to count the entire dish, but do not overlap the passes to avoid counting the same individual twice. ID at least to the family level using appropriate keys. Good guides for the Great Lakes region are the online resources of Haney et al. (2013a, b) and McNaught (2014) and the print resource of Balcer et al. (1984). We recommend identifying one species at a time, looking through the entire petri dish, counting how many you have of this species, and then doing the same for your subsequent species.
2. After finishing your ID in one 1 mL subsample, repeat the steps three more times (for a total of four 1-mL aliquots subsamples) to increase the chance of finding rare species.
3. Express your taxon list as numeric abundance in your four subsamples. You can also factor in the amount of water filtered through your plankton net and calculate each species' density per cubic meter or liter. See above under *Determining Plankton Density* for details.

Calculating Biodiversity Indices

Use the quantitative species list to compare your sample to published results—most appropriate to the study area and calculate biodiversity indices. When calculating diversity, the number of species and the distribution of individuals among the species present are important. There are various indices to describe diversity. For calculation and comparison of the indices, see Sect. 15.11. Other ways to describe your species community are rank-abundance diagrams (Avolio et al. 2019) and pie charts of the dominance structure in your samples based on numeric abundance and biomass (Wassmer 2020).

6.3.4 Mapping Lake Macrophytes

6.3.4.1 Introduction

Natural lake shores show several broad belts of macrophytes (macroscopic plants, large enough to be visible to the naked eye). Intact shore ecosystems provide essential services to the other components of the ecosystem. The vegetation above the ground reduces the wind speed, and due to this, shoreline erosion. They trap and store carbon and other nutrients in their bodies. Belowground vegetation (roots, rhizomes, and tubers) stabilizes the sediment and reduces erosion–better than an artificial lake wall. In addition, the oxygen they produce is speeding up nutrient recycling. Finally, the plants inside the water filter out nutrients and sediment from the water, facilitate the uptake of nutrients, and reduce the force of the breach and currents (Kochi et al. 2020).

Different types of macrophytes occur depending on the depth of the water. Emergent plants occur in the shore zone from the shoreline to about 1 m depth and grow upright and tall above the water. Slightly further from the shoreline in the shallow littoral zone in water depths between 1 and 3 m, we find floating-leaved macrophytes like water lilies (family Nymphaeaceae) and some of the pondweeds (family Potamogetonaceae). They are rooted to the bottom of the water body, and their leaves float on the water's surface. Submerged macrophytes grow entirely underwater and occur primarily in the deep littoral zone at water depths above 3 m. They are either rooted to the bottom (e.g., *Myriophyllum spicatum*) or lack root systems (e.g., *Ceratophyllum demersum*). Finally, there are primarily tiny free-floating macrophytes with roots in the water, for example, duckweed (genera *Lemna* and *Spirodela*) and watermeal (*Wolffia*) blown about by the wind.

There are no easy methods to map aquatic plants. The following will portray a general assessment method as a snapshot of the existing conditions (Wandell et al. 2007). It is time-consuming and needs to be performed by a small team over a day on small lakes or several days on large lakes. The sampling method described (rake toss method) is mainly for comparative purposes and not for quantitative statistical analysis (unless using models based on presence or absence). Plant density scores can be directly affected by the procedures used. Even substantial differences in density scores should be cautiously treated (Institute for Environment and Sustainability et al. 2014; Vondracek et al. 2014). However, the outcomes of this kind of sampling can be helpful for the development of a macrophyte management plan if combined with the knowledge of the adjacent shoreline land use, nutrient concentrations, and watershed features (Azzella et al. 2013; Brucet et al. 2013; Capers 2000; Deppe and Lathrop 1992; Jäger et al. 2004; Labat et al. 2022; Mikulyuk et al. 2010; Spears et al. 2009).

6.3.4.2 Methods and Procedures

The most common way to sample a lake's macrophytes is to use transects running perpendicular to the shoreline toward the lake's interior. The number of transects depends on the lake's size and how much the shoreline still has a macrophyte belt. Usually, transects should be at least 100 m apart. Ideally, work in groups of four to five people, with two in each of two canoes/kayaks and one support person on land. The sampling will be done with the help of a special double rake called a grapnel (Figs. 6.34 and 6.35) attached to a rope to retrieve the samples.

Tie the rope's end to your canoe or kayak to ensure you do not accidentally throw the rake with its rope into the water. Practice throwing and retrieval first in the open water, and throw the rake about the same length (about 1–5 m) each

Fig. 6.34 Assembly of a grapnel from two garden rakes. Cut the handles off two garden rakes and bolt the rakes back to back with two "C" bolts. Use a small hose clamp between the rake tines to prevent side-to-side slipping. Drill a hole in the remaining wooden handle core and twist a moderately large eye bolt into the hole. The rope should be about 6 m (20 ft) long. File off any sharp edges. Wear gloves when using the rake to protect the hands from cuts.

time. This "standardizing" will increase the comparability of your results.

Approach your first location at a 1-foot depth from the water. Anchor the boat facing the shore—the 12:00 clock position of the four samples (Fig. 6.36). Use the anchor to stabilize the position of the back of the canoe/kayak. Tie the front end to a long rope. The second kayak can paddle to the shore and tie the rope to a tree to stabilize the front of the sampling canoe or kayak. Tighten that rope so that the canoe is fixed in the direction of the transect. This should minimize drifting by wind and waves. Take the GPS position of your transect at the 1-foot position. Take photos of the site to give a general overview. Take the first photo in the direction of the open water. The second photo should show the view from the open water toward the shore.

Pitch the grapnel toward the shore and retrieve it. Remove all vegetation collected and sort it into piles for each known and unknown species. ID the plants using an appropriate key. Try to ID in the field—it is much easier when the samples are fresh and less damaged than later in the lab. This process will go quite fast once you have seen a handful of the most common species. However, if you are uncertain, place three specimens of each unknown species in a sample bag or bottle for later identification and label it appropriately, e.g.,

Fig. 6.35 Aquatic plant sampling rake (grapnel). (**a**) Tossing the rake for a sampling of lake macrophytes. (**b**) Retrieved grapnel with submersed macrophytes attached. (Credit for both pictures: Hamilton County Soil and Water Conservation District. Reproduced by permission)

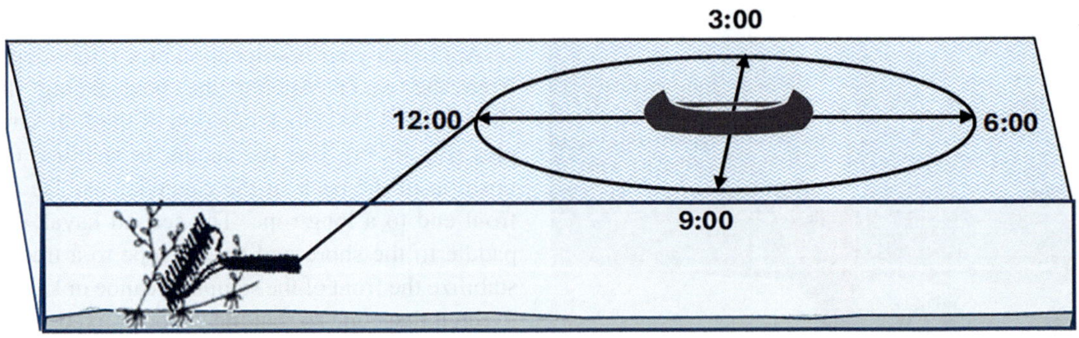

Fig. 6.36 Clock positions of the grapnel sampling throws. (Graphic design by Thomas Wassmer after Simpson (1991))

the transect number, the transect point, and the clock position, followed by a consecutive number of unknowns. Use the same number in your field record sheet. Record the presence of each species on the field recording sheet. Repeat the procedures at the 3:00, 6:00, and 9:00 clock positions (Fig. 6.36).

After completing all four clock positions, provide a density rating for each plant:

- Dense (D): Species fills the rake in all four casts (12, 3, 6, and 9 o'clock).
- Heavy (H): Species was mixed with other plants in all four casts of the rake.
- Moderate (M): Species in three of the four rake casts.
- Sparse (S): Species in two of the four rake casts.
- Found (F): Species in one of the four rake casts.

If you are experienced with identifying macrophytes, you can try to ID plants using the bathyscope/aquascope in the four compass (clock) directions. One look through the bathyscope/aquascope is considered one sample, like one throw with the grapnel (Fig. 6.37).

With the rake, collecting floating plants is nearly impossible, and collecting emergent shoreline plants is difficult. To account for this deficiency, visually check for these plants in each clock position. If they are present in sufficient quantities to appear in the path of an imaginary rake tow, record their presence. Take a sample if you cannot ID a floating or emergent plant from afar.

After finishing the data collection at the 1-foot transect point, paddle out along the transect line and anchor at 4 ft depth (use a calibrated sounding line to measure depth). The support person on land must untie the rope and tie it back once the canoe reaches its destination. Sample and record your data as you did at 1-foot depth and do the same once more at 8 ft depth. Try to have all three sample locations in a line perpendicular to the shoreline. Results can be displayed on a map (Fig. 6.38).

A box sampler can sample plants or macrobenthos at greater depths (see Fig. 6.39). It is lowered with open jaws into the water, sinks to the bottom due to its weight, and is then closed with a drop weight.

6.3.5 Water Quality

Many types of tests can be performed to assess the water quality of rivers and lakes. Some methods are specific for either rivers or lakes, and the same tests can be meaningful and essential for one of the major freshwater habitat types but not the other. In addition, different research questions may require specific tests or measurements. The scope of this handbook on ecological methods is not to provide an exhaustive list of possible methods but rather to describe some essential

Fig. 6.37 (**a**) Principle of looking underwater with a bathyscope with its glass bottom submerged under the water surface. (**b**) View through a bathyscope showing plant animals and sediment clearly. (Picture credit: **a**: Osculati S.r.l. http://www.osculati.com. Reproduced by permission. **b**. NHBS.com wildlife, ecology, and conservation books and equipment. Both pictures are reproduced by permission)

Fig. 6.38 Macrophyte Map using Google Earth Pro resulting from four transects. Labels indicate the three depths for each transect (1, 4, and 8 ft), the plants encountered (code numbers from Wandell et al. (2007)), and their density scaling (capital letters—see above)

Fig. 6.39 A box sampler to take macrophytes from the bottom. (**a**) Opened to plunge into the water, (**b**) closed, brought up again with macrophytes

methods in sufficient detail to allow the reader to jumpstart their own research or data collection. We will, therefore, restrict this section to the nine water quality parameters considered by the US National Sanitation Foundation (NSF) to be most helpful in determining the water quality of rivers (Brown et al. 1970, 1973; Ott 1978). These parameters can be used to calculate a water quality index (WQI), which allows meaningful comparison between rivers and the long-term monitoring of the same river and detects if management practices lead to improved water quality. Due to the abundant availability of budget testing kits for these parameters, allowing for easy and safe field application, they are also often chosen by governmentally endorsed citizen science institutions such as the Hoosier Riverwatch, a program of the Indiana Department of Environmental Management (https://www.hoosierriverwatch.com). Table 6.2 provides these nine tests.

Each water quality measurement is converted into a Q-value (quality value) using a conversion table or graph based on the original NSF publication (Brown et al. 1970, 1973; Ott 1978), translating each measurement into a score between 1 and 100. There are also a few web conversion applications and Excel sheets available: https://www.knowyourh2o.com/outdoor-3/water-quality-index-calculator-for-surface-water

WQI Calculation.xls, developed by Kim Irvine, Thammasat University, provided at https://tinyurl.com/4azewmnt

The *Q*-value scales your results to a metric that expresses the overall health or quality of your river's water. The higher the *Q*-value, the better the test results (100 is the maximum value; 0 is the minimum). After obtaining the *Q*-value, it is multiplied by a "weighting factor" based on each test's importance in water quality. These weighing factors are also based on the original NSF method and sum up to 1. The resulting water quality index (WQI) is a 100-point (percent) scale that summarizes results. A team of 142 water experts assisted in establishing the list of significant parameters, developing a standard ranking scale, and assigning weights to the selected water quality variables (Brown et al. 1970, 1973; Ott 1978). The overall assessment of the river's water quality follows the below point values:

Table 6.2 Water quality parameters of the NSF water quality index (WQI) (Brown et al. 1970, 1973; Ott 1978)

Parameter	Unit of measurement	Weighing factor	Time to complete	Location
Water temperature change (1 mile)	°C	0.10	5 min	In the field
Dissolved oxygen	mg/L and % saturation	0.17	5 min	In the field
Biochemical oxygen demand (BOD)	mg/L	0.11	5 days to incubate, then 5 min to test	In the lab
pH	Standard units	0.11	2 min	In the field
Total dissolved solids (TDS)	mg/L	0.07	5 min	In the field
Total phosphate (TP)[a]	mg/L	0.10	Orthophosphate 5 min, TP 1 h	In the field/ the lab
Nitrate/nitrite	mg/L	0.10	2 min	In the field
Transparency/turbidity	cm (Secchi depth) converted to NTU	0.08	5 min	In the field
E. coli and general coliforms	cfu/100 mL	0.16	24 h to incubate, 15+ min to count	In the lab

[a]TP, which includes orthophosphate, metaphosphate, and organic phosphate, requires 1 h of boiling the water sample in concentrated sulfuric acid

We also provide the excellent scoresheet developed by the Hoosier Riverwatch, a program of the Indiana Department of Environmental Management (Hoosier Riverwatch 2022), and its adaptation into a digital form: https://tinyurl.com/2enzwc8w.

Now, some notes about each of the nine selected parameters:

6.3.5.1 Water Temperature (Change)

Water temperature is critical to water quality, affecting dissolved oxygen, photosynthesis, and the metabolic rate of organisms living in the water. This metric looks at water warming through a 1-mile (1.6 km) long stretch of river. Conduct the measurements like this:

(a) Position a waterproofed digital or the bulb of a simple liquid thermometer below the water's surface at about the same depth at which water samples for the other tests are collected. Ideally, obtain the temperature reading in the main streamflow if this does not violate your safety protocol.

(b) Keep the thermometer submerged, swirling it gently for about 2 min or until the reading stabilizes.

(c) As soon as possible, select a stretch of the river approximately 1 mile upstream with a similar level of shade and flow as in step 1 and conduct the same test using the same thermometer.

(d) Calculate the ΔT difference between the downstream and upstream temperature as:

$$\text{Downstream Temp} - \text{Upstream Temp} = \text{Temperature Change } (+/-).$$

The Utah State University Water Quality Extension provides a helpful video: https://www.youtube.com/watch?v=Pk8UM3SWJdA

Use Fig. 6.40 to determine the temperature change Q-value.

6.3.5.2 Dissolved Oxygen (DO)

Percent (%) Saturation

To determine dissolved oxygen levels as percent saturation, you must measure the DO concentration in parts per million (ppm) or mg/L and the water temperature. Percent saturation specifies how the measured amount of dissolved oxygen

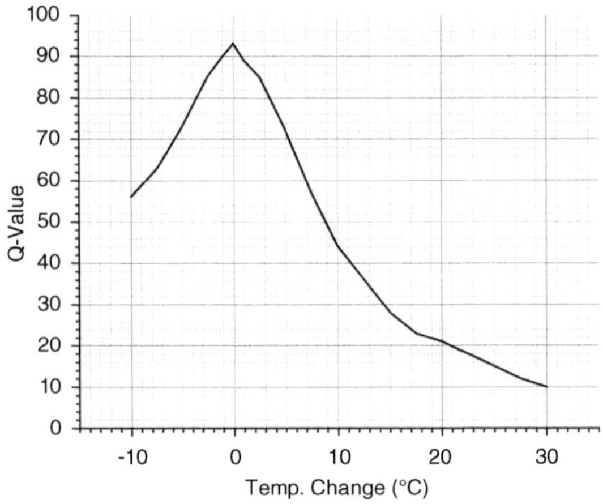

Fig. 6.40 Water quality values (Q-values) for the parameter temperature change

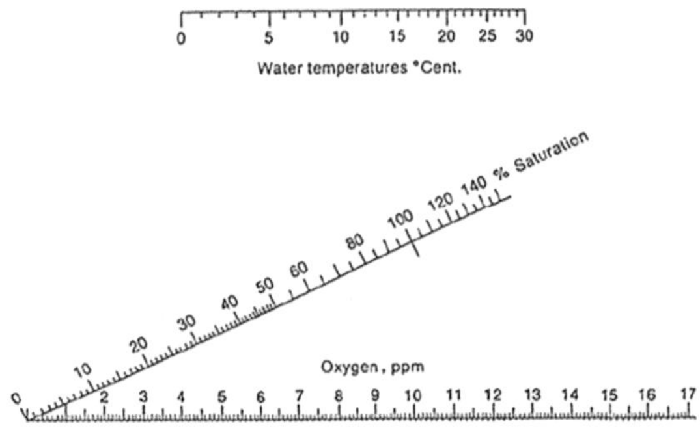

Fig. 6.41 The oxygen saturation nomogram: Run a straight edge from the measured water temperature (top scale) to oxygen, ppm, to determine % saturation along the angled (middle) scale

relates to the maximum amount of oxygen that can dissolve in 1 L of water at the same temperature and pressure. You can use the below oxygen solution nomogram (Fig. 6.41) to determine the approximate % solution without considering salinity (neglectable in freshwater with less than 0.5 ppm) and elevation above sea level.

For a more accurate conversion, you can refer to online calculators such as https://www.waterontheweb.org/under/waterquality/dosatcalc.html or Excel spreadsheets such as the Dissolved Oxygen at Saturation Spreadsheet at the State of Oregon Department of Environmental Quality: https://www.oregon.gov/deq/FilterPermitsDocs/RPADOSaturationEquation.xls—also provided at https://tinyurl.com/mrx4s6db.

For example, water at 20 °C is 100% saturated with 9.07 ppm dissolved oxygen. However, water at 10 °C can hold up to 11.25 ppm DO before it is 100% saturated. Due to this, daily and seasonal temperature changes and thermal pollution, e.g., from industrial wastewater, can significantly affect oxygen levels and the livelihood of aquatic life in rivers.

Supersaturation

Water can become supersaturated, holding more than 100% of oxygen at a given temperature. Supersaturation may occur if a river contains abundant aquatic plants and algae that enrich the water with oxygen from their photosynthesis. Supersaturation may also occur at the base of waterfalls or dams due to increased water pressure downstream. The adaptation of aquatic organisms to supersaturation during daytime hours with its high photosynthetic activity may lead to oxygen depletion during the evening and night hours due to overconsumption and lead to the death of sensitive aquatic organisms. This is why supersaturation is commonly regarded as detrimental to a river's water quality and causes the QA values to decline above 100% (Fig. 6.43).

DO Tests and Sensors

These instructions are for use with the CHEMetrics Dissolved Oxygen Test Kit K-7512 (https://www.chemetrics.com/product-category/test-kits/oxygen-dissolved). Refer also to the videos: https://youtu.be/GCQ9ckIbhbA and https://youtu.be/9Jk0AmmuLSY. This is a well-priced colorimetric test using a color reaction of the sample with the test solution that can be compared to a color scale (Fig. 6.42). Similar test kits are available in many countries.

In addition to these budget colorimetric DO tests, a more generous budget may allow for portable spectrophotometers that detect the wavelength of the color-indicated sample solution instead of requiring the user to compare to a reference chart, e.g., the Hach DR1900 or electric

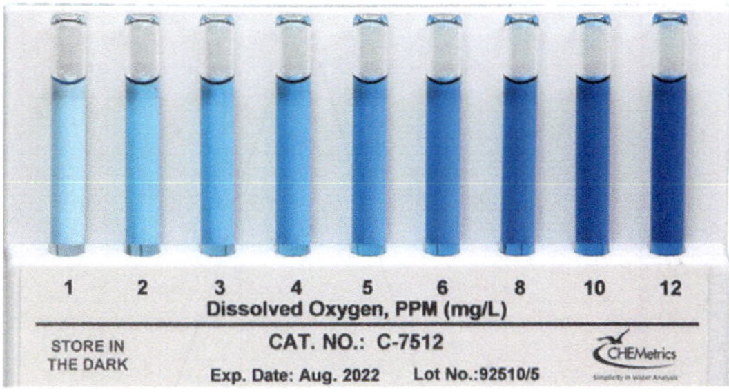

Fig. 6.42 The CHEMetrics Dissolved Oxygen Test Kit K-7512 color scale to which the sample vial can be compared

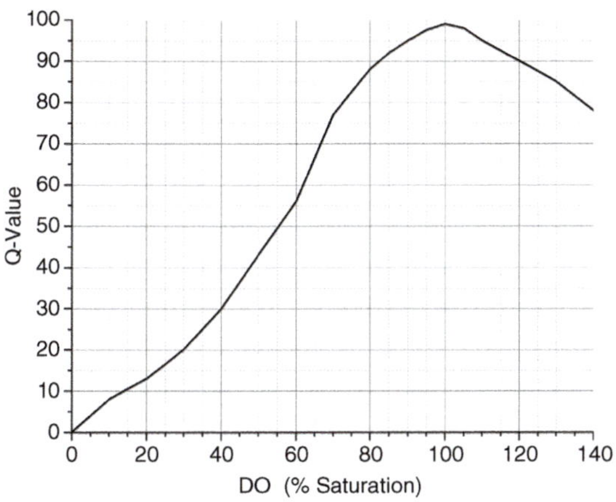

Fig. 6.43 Water quality values (Q-values) for the parameter dissolved oxygen. Please note that if DO % saturation >140, $Q = 50.0$

and optical DO probes. Electric probes usually require the user to calibrate the instrument before use, while the newer optical probes are ready to go. An example of an electrical probe is the classic YSI Pro 20, while the YSI ProSolo is an optical probe. Optical probes can also be data loggers, allowing continuous deployment in rivers and lakes, e.g., the Onset HOBO U26–001.

6.3.5.3 Biochemical Oxygen Demand (5-Day)

The biochemical oxygen demand over 5 days (BOD5) measures the oxygen aerobic microorganisms consume as they break down organic wastes under aerobic conditions over five days. High levels indicate large quantities of organic matter in the river. Polluted rivers, or rivers with abundant plant growth (and consequently decay), are usually characterized by high levels of BOD5, while clean and vegetation-poor rivers usually have low BOD5 levels.

In slow-flowing, warm, and polluted rivers, bacteria consume most of the low DO levels such rivers have, making them unsuitable for most other freshwater organisms. On the contrary, fast-flowing, turbulent, and cold rivers have higher DO levels. They can still be suitable for various aquatic organisms even if polluted with higher concentrations of organic material. Therefore, the significance of BOD5 levels depends upon other river conditions. Nevertheless, you can expect that a healthy river usually has high DO and low BOD5 levels. Sources of BOD include natural sources such as leaves and woody debris, dead plants and animals, and animal manure, but also anthropogenic sources such as wastewater from pulp and paper mills, water treatment plants, feedlots, and food-processing plants, as well as failing septic systems and urban stormwater runoff.

Hoosier Riverwatch (2019) provides the following rough guide to what various BOD5 levels indicate:

- 1–2 mg/L: Clean water with little organic waste.
- 3–5 mg/L: Fairly clean with some organic waste.
- 6–9 mg/L: Poor: Lots of organic material and bacteria.
- 10+ mg/L: Very poor water quality with exceptionally high amounts of organic material.

For this test, you can use the same test kit used for DO, e.g., the CHEMetrics Dissolved Oxygen Test Kit K-7512, but collect your sample in a darkened (light-free) sample bottle. Follow the below instructions.

1. Rinse a dark (light-free) bottle several times with the river water, then fill it by lowering it below the water's surface. Let water flow into the bottle for approximately 2 min, reducing the chances of any air bubbles. Then, close the bottle with the stopper or lid while it is underwater and remove it from the water.

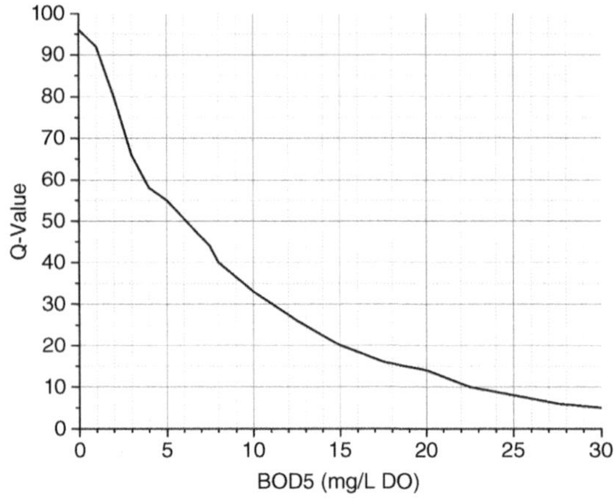

Fig. 6.44 Water quality values (*Q*-values) for the parameter biological oxygen demand after 5 days (BOD5). Please note that if 5-day BOD >30.0, $Q = 2.0$

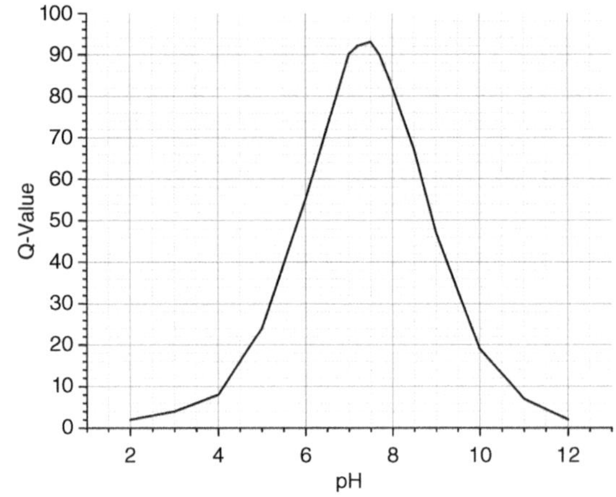

Fig. 6.45 Water quality values (*Q*-values) for the parameter pH. Please note that if pH is <2 or >12, $Q = 0.0$

2. In the lab, store the BOD sample undisturbed in a light-free location (e.g., a drawer or cabinet) at room temperature at approximately 20 °C for 5 days.
3. After 5 days, remove the BOD bottle and perform the DO test using the BOD sample water.
4. Calculate the BOD5 level by *subtracting* the DO reading of the BOD sample (mg/L or ppm) from the original DO sample taken 5 days earlier.
5. Use Fig. 6.44 to calculate the *Q*-value for this parameter.

6.3.5.4 pH

One in ten million water molecules (H_2O) dissociates into hydrogen ions (H^+) and hydroxide ions (OH^-). The concentration of the H^+ ions determines whether a solution is acidic or basic. In pure water at 20 °C, the H^+ concentration [H^+] is 1 in ten million or 10^{-7}. The pH is the negative base-10 logarithm ($-\log_{10}$) of [H^+]—in this case, pH 7. Pure water is neutral as the OH^- concentration [OH^-] is also 10^{-7}. If [H+] is higher than 10^{-7}, e.g., 10^{-3}, the watery solution is an acid—in this example, with a pH of 3.

Many environmental processes affect the pH. The pH is temperature dependent, resulting in a slightly lower pH—thus reacting acidic in water bodies with higher temperatures. One factor that moves the pH toward the basic (alkaline) part of the spectrum is algae blooms that remove too much carbon dioxide (CO_2) from the water during photosynthesis, raising the pH to ≥9.

Anthropogenic factors influencing freshwater pH include runoff from the land of abandoned mines (acid mine drainage), which may lower pH substantially. Even more dangerous is a side effect. Acidic runoff and groundwater increase the solubility of soil metal cations, such as aluminum and copper, which are incredibly toxic to aquatic organisms.

Depending on the geology, most natural water bodies range between pH values of 5.0 and 8.5. Rainwater is usually slightly acidic with a pH of 5.5–6.0 due to CO_2 in the atmosphere—these values became slightly higher due to the anthropogenic increase in atmospheric CO_2 levels. In addition, air pollution, especially nitric and sulfuric oxides from combustion engines and coal-burning power plants, creates acid rain, which is even more acidic, reaching pH values as low as 2.85 (Likens et al. 1976). Alkaline soils and minerals (limestone) can somewhat buffer the effects of acid rain and may raise the pH to 8.0–8.5.

The LaMotte Wide Range pH Test Kit 3353–01 is a simple, low-budget pH test. There is a short video showing this method at https://youtu.be/Y9Bb4G0buC4. Use Fig. 6.45 to convert your pH to a WQI value.

6.3.5.5 Total Dissolved Solids (TDS)

Total dissolved solids (TDS) include inorganic cations (mainly calcium, magnesium, potassium, and sodium), anions like bicarbonates, chlorides, and sulfates, and a small amount of organics dissolved in water. It measures the number of dissolved ions but does not tell us what ions are present and at what concentrations. Unpolluted water typically has a TDS content of <30 mg/L. TDS is usually not a significant problem at such levels unless the concentration exceeds 1500 mg/L, which decreases dissolved oxygen levels and possible oxygen saturation. Drinking water levels (secondary water) must be below 500 mg/L due to reduced palatability, scaling, and corrosion in water pipes (USEPA 2015).

The most common method for measuring TDS is a conductivity meter, which measures the ability of the water to conduct electricity. As most of the TDS are ions, conductivity is directly proportional to the concentration of dissolved solids. A reliable meter is the LaMotte Salt/TDS/Temp Tracer (Code 1749–01). However, this method only estimates TDS levels. For a complete TDS determination, you must filter, weigh, and evaporate a small amount of your sample water in the lab. Use Fig. 6.46 to convert your TDS into a WQI value.

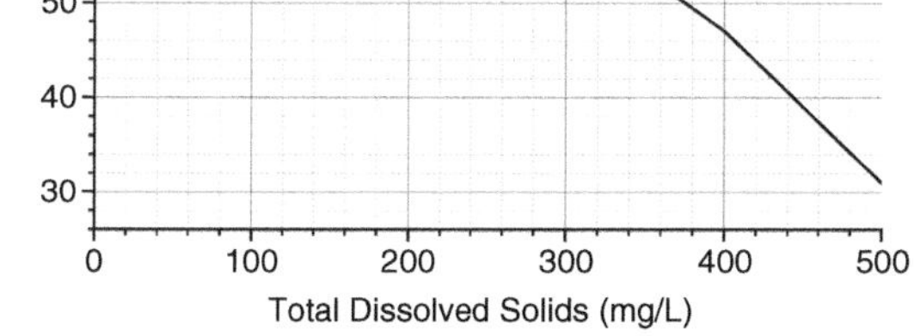

Fig. 6.46 Water quality values (Q-values) for total dissolved solids (TDS). Please note that if TDS is >500 mg/L, $Q = 20.0$

6.3.5.6 Total Phosphate (TP)

Phosphorus is one of the critical elements necessary for the growth of plants and animals. In lake ecosystems, it tends to be the growth-limiting nutrient, meaning that all other nutrients are typically abundant in sufficient quantities, while phosphorus is rare and controls plant growth. Phosphorus occurs in nature in the form of phosphates (PO_4) in three forms: orthophosphate (dissolved inorganic phosphorus or soluble reactive phosphorus), metaphosphate (polyphosphate or condensed phosphates), and organic phosphate—together addressed as total phosphate. Orthophosphate averages about 0.001 mg/L or ppm worldwide in unpolluted rivers, whereas total phosphates average about 0.0025 mg/L (Meybeck 1982, 1993).

Orthophosphate in river water often doubles or even quadruples during and following heavy rainfall or at the beginning of snowmelt. Agricultural runoff may increase orthophosphates to 0.015–0.03 mg/L, while municipal sewage overflows may push orthophosphates to over 0.3 mg/L (Meybeck 1982, 1993). Orthophosphate levels above 0.03 g/L lead to increased plant and algae growth. Their concentration is, therefore, a valuable indicator of the potential for algae blooms and *eutrophication*. Extensive plant and algae growth also leads to more decomposition, depleting the dissolved oxygen in the water. Low oxygen levels are called *hypoxia*, and complete oxygen depletion is called *anoxia* and causes widespread death of all aerobic organisms in a water body. Under hypoxia, phosphates are released from previously insoluble forms in the bottom sediment back into the water, leading to a vicious cycle of constantly increasing nutrients called *eutrophication*. Moderate hypoxia and eutrophication may be natural events in many rivers and lakes during some seasons but were undoubtedly intensified by human interference and pollution.

For rivers, orthophosphate concentration should be <0.03 mg/L, while total phosphates should be below 0.3 mg P/L.

Determining total phosphate concentrations requires boiling the sample for 10 min in concentrated sulfuric acid (H_2SO_4), which is unsafe to do in the field, thus requiring the sample to be brought to the lab. A test kit that includes all reagents needed for TP is the Hach Phosphate Test Kit, Total Ortho–/Meta-, Model PO-24. As orthophosphates can be easily identified in the field using a budget colorimetric test, many citizen science projects are only determining orthophosphates. A good test kit is the CHEMetrics Phosphate Test Kit K-8510. For the procedure, refer to the video at: https://youtu.be/GCQ9ckIbhbA.

Unfortunately, the results of the orthophosphate test cannot be entered into the calculation of a comprehensive WQI as there are currently no national or state water quality standards for orthophosphate. However, in most WQI metrics, an index can be calculated with at least 6 of 9 parameters. Another strategy is entering the orthophosphate level instead of TP. As we know, TP is higher than orthophosphates, so you are using a conservative underestimate that this will make the water quality appear slightly better than it is (Fig. 6.47).

6.3.5.7 Nitrate (& Nitrite)

In water, nitrogen occurs as nitrate (NO_3), nitrite (NO_2), and ammonia (NH_3). It originates from human and animal waste and decomposing organic matter within the water and surrounding land through runoff. Human activities significantly increase the amount of nitrogen in water through fertilizer runoff from lawns and crops. Nitrates are essential for plant growth. However, nitrates are not a growth-limiting nutrient in most cases because they are much more common than phosphates. Blue–green algae can fix nitrogen gas from the atmosphere into bioavailable ammonia, which is further oxidized into nitrate. Phosphates and nitrates from fertilizers enhance the growth of aquatic plants and algae, leading to eutrophication and hypoxia. In marine ecosystems, especially estuaries and coastal waters, nitrogen is a leading cause of dead zones caused by hypoxia. Nitrates are typically measured in mg/L (ppm). These values can be converted into mg/L of nitrogen by dividing the nitrate concentration by 4.43.

The LaMotte Nitrate-Nitrogen Tablet Test Kit 3354–01 is a simple and affordable field test. Take a look at the procedure at https://youtu.be/jmcod9MRflo.

Healthy lakes usually have nitrate concentrations of less than 0.05 mg/L, whereas healthy rivers rarely exceed 1 mg/L. The US EPA recommends 1.5 mg/L as the dividing line between mesotrophic and eutrophic rivers. Use Fig. 6.48 to convert your nitrates into a WQI value.

6.3.5.8 Turbidity and Transparency

Turbidity is the relative clarity of water. Turbid water is cloudier due to suspended particles such

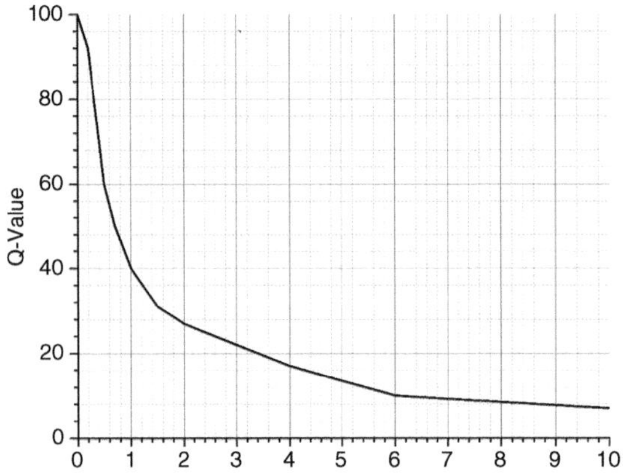

Fig. 6.47 Water quality values (*Q*-values) for total phosphate (TP). Please note that if TP >10, $Q = 2.0$

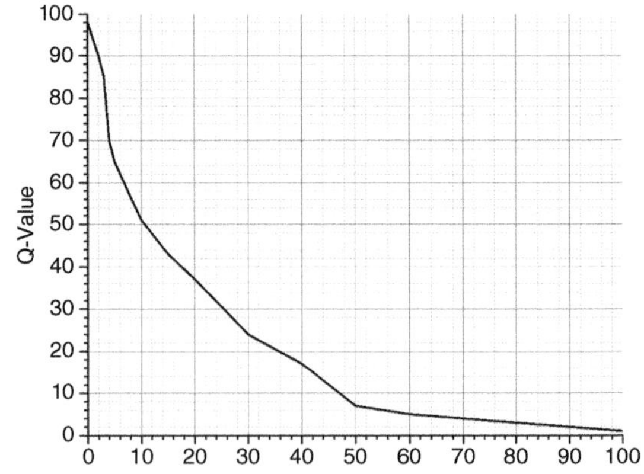

Fig. 6.48 Water quality values (*Q*-values) for the parameter nitrate. Please note that if nitrate >100, $Q = 1.0$

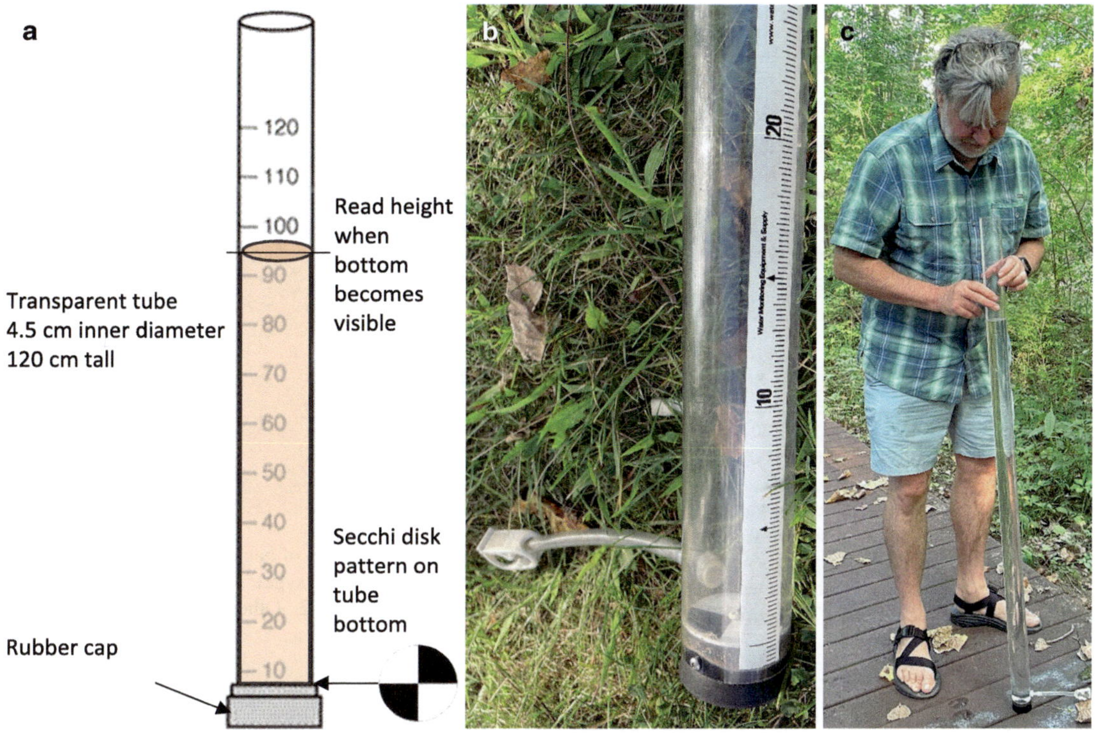

Fig. 6.49 Details and usage of a transparency tube. (**a**) Components of a transparency tube and principle of usage: Measure the height of the water column when the Secchi pattern at the bottom becomes visible. (**b**) Details of the bottom of the tube with the Secchi disk pattern and outflow to reduce the height of the water column. (**c**) Constant monitoring of the tube bottom for the appearance of the Secchi pattern while the water slowly flows out, lowering the water column

as sediment, organic matter, and algae. These particles scatter and absorb light, causing the turbidity. Do not confuse turbidity with color, as colored water can still be clear. The amount of light scattering determines a water sample's transparency and can be measured as the depth at which a pattern can be seen through the water column (Fig. 6.49). The depth at which the pattern becomes invisible can then be converted into the units of turbidity (NTUs).

Water becomes turbid due to soil erosion and runoff from agriculture, lawns, parking lots, construction sites, or the stream bank. Nutrient-rich and warm water facilitates the growth of algae and contributes to turbidity. Turbid water limits light availability, impairing photosynthesis and, thus, the ecosystem's primary productivity. Furthermore, water temperature increases due to increased light absorption at the suspended particles.

Increased turbidity due to sediment influx can also inhibit egg oxygen uptake from fish and benthic insects (Lemly 1982; Strayer and Dudgeon 2010). This can be counteracted by riparian vegetational strips, which reduce sediment influx and eutrophication (Osborne and Kovacic 1993; Urban et al. 2006; Wang et al. 2023).

Transparency can be assessed with a transparency tube (Fig. 6.49). Short video: https://youtu.be/8xzpG9IDUq8. Use Fig. 6.50 to convert your turbidity value into a WQI value.

6.3.5.9 Fecal Coliform (FC) Bacteria

Fecal coliform bacteria, including *Escherichia coli* (*E. coli*), thrive in the digestive tract and feces of humans, livestock, and waterfowl. Unpolluted surface and ground waters are typically low in these bacteria. However, they can be contaminated by runoff from livestock, combined sewer or water treatment plant overflows, poor

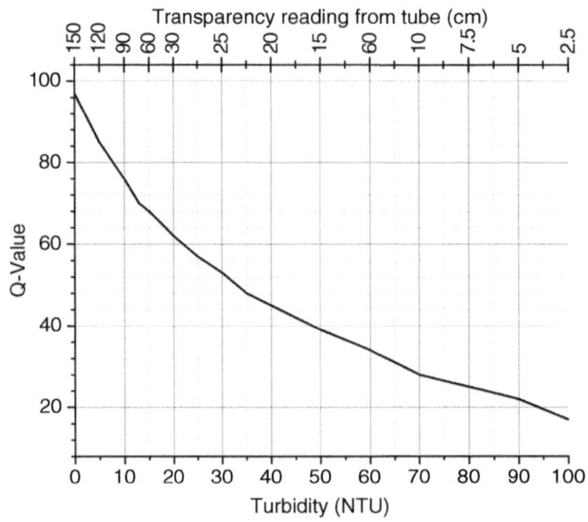

Fig. 6.50 Water quality values (Q-values) for the parameter turbidity. The top x-axis shows the water column's maximum depth, which still allows us to see the Secchi pattern. The bottom x-axes show the Secchi disk depths converted to Nephelometric Turbidity Units (NTUs). If the turbidities >100, $Q = 5$

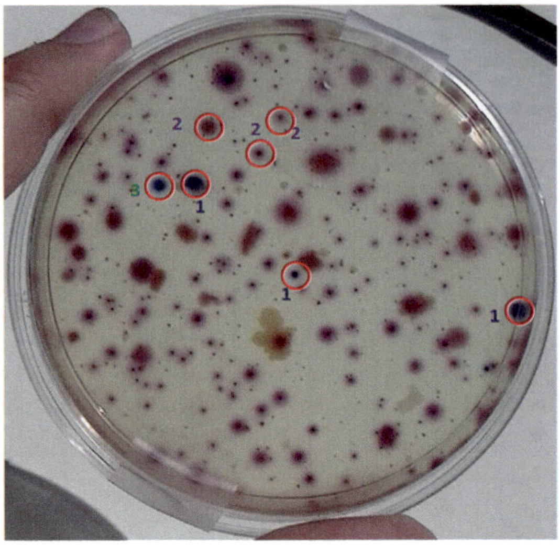

Fig. 6.51 Identification of bacterial colonies by different stains using Coliscan Easygel®. Colony color guide: 1. *E. coli* dark blue/purple, 2. Other coliforms: pink/red, 3. Non-coliforms: teal/green

septic systems, and wildlife and livestock access to streams. FC bacteria can enter the human body during full-body contact when swimming or partial-body contact while fishing, wading through the mouth and other body openings, scratches, and other skin lesions. In many parts of the world, rivers are too contaminated by FC bacteria to allow full-body contact, e.g., 38% (25,793 km) of rivers in the US state of Indiana (IDEM 2012).

Only a few strains of *E. coli* are pathogenic themselves. However, they often occur with other intestinal tract bacteria, viruses, and protozoa that may be dangerous to human health (Fig. 6.51). FC bacteria are, therefore, indicators for the microbiological safety of water (see also Huq et al. 2005). The USEPA has determined that FC bacteria counts above 235 colonies per 100 mL may be unsafe for partial-body exposure (USEPA 2012).

High FC counts usually coincide with higher levels of organic pollutants and water temperatures. However, they might also indicate reduced dissolved oxygen levels, impacting many aquatic organisms and potentially shifting the species' communities. High FC counts may also change the aquatic microbiome, altering the flow of energy and matter through aquatic ecosystems and leading to community changes and biodiversity loss (Paruch et al. 2019). This is especially threatening as climate change, especially increased temperatures, facilitates the survival and spread of FC bacteria (Jeon et al. 2019). Furthermore, the growth of FC bacteria seems to be accelerated by organic and inorganic phosphorus (Chudoba et al. 2013), another agricultural nutrient pollutant showing the synergism between different anthropogenic environmental stressors.

Two easy and inexpensive quantitative methods are Coliscan Easygel® (https://www.micrologylabs.com/coliscan-easygel/) and 3M™ Petrifilm™ (https://www.3m.com/3M/en_US/p/c/b/petrifilm/). Stepenuck et al. (2011) compared the two methods' usability for citizen science water studies and their accuracy compared with specialized laboratory results. A video of the usage of Coliscan Easygel can be found at https://youtu.be/OYXdT4s4RJU, and for 3M™ Petrifilm™, here: https://youtu.be/23tkG4pp-vc.

Use Fig. 6.52 to convert your coliform bacteria count into a WQI value.

6.3.6 Temporary Waters

6.3.6.1 Ephemeral Ponds

Ephemeral ponds are periodic bodies of water, i.e., they dry out occasionally and are usually not large in area (1–10 ha in the tropics, usually much less in less arid places) and are often shallow (less than 2 m deep) (Fig. 6.53). They are easily mixed from top to bottom by light wind. As they can dry out, the organisms that colonize these waters must be adapted to these extreme conditions. Temporary pools are threatened by intensive agriculture and urbanization, which lead to physical destruction, eutrophication, and hydroperiod changes. The fauna in these ephemeral waters, therefore, contains many endangered species. The most important representatives are large branchiopods and amphibians (e.g., Eder et al. 2014; Wisconsin Wetlands Association 2017; Withford and Duval 2020; Craft 2022). Since the time span for the living conditions in the water can be short, some organisms develop quickly under favorable conditions. We want to focus on macroscopic arthropods in small bodies of water, especially crustaceans (class Crustacea) and some water insects.

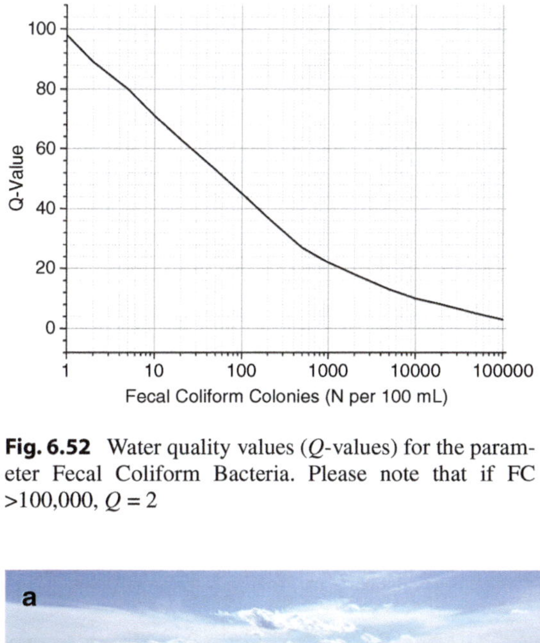

Fig. 6.52 Water quality values (Q-values) for the parameter Fecal Coliform Bacteria. Please note that if FC >100,000, $Q = 2$

Simple nets (D-nets, kick nets) are sufficient to catch the animals (refer to Sect. 6.2.6.3), the contents of which are then rinsed out into white bowls for an overview of the captured animals (Fig. 6.54).

The water quality is measured according to the standard procedures described above (see Sect. 6.3.5). Ecologically relevant factors are also the degree of eutrophication and the turbidity of the pond water. Clear ponds offer advantages for predators, while turbid waters favor particle filterers. Since the animals in the small bodies of water are easy to capture, we can also deal with the ratio of particle eaters to phytophagous and carnivorous arthropods.

6.3.6.2 Specialized Invertebrates of Ephemeral Ponds

Conspicuous representatives of the Crustacea in the temporary waters are the Branchiopods and Phyllopods (Fig. 6.55). These seemingly primitive crustaceans are filter feeders that live on microorganisms. The larger species of

Fig. 6.53 (**a**) A periodic small body of water that is turbid due to sediment. North Mongolia. (**b**) Shallow pond with clear water. North Mongolia

Fig. 6.54 (**a**) Fishing for crustaceans using a small hand net. North Mongolia. (**b**) The contents of the net are rinsed out into a white bowl

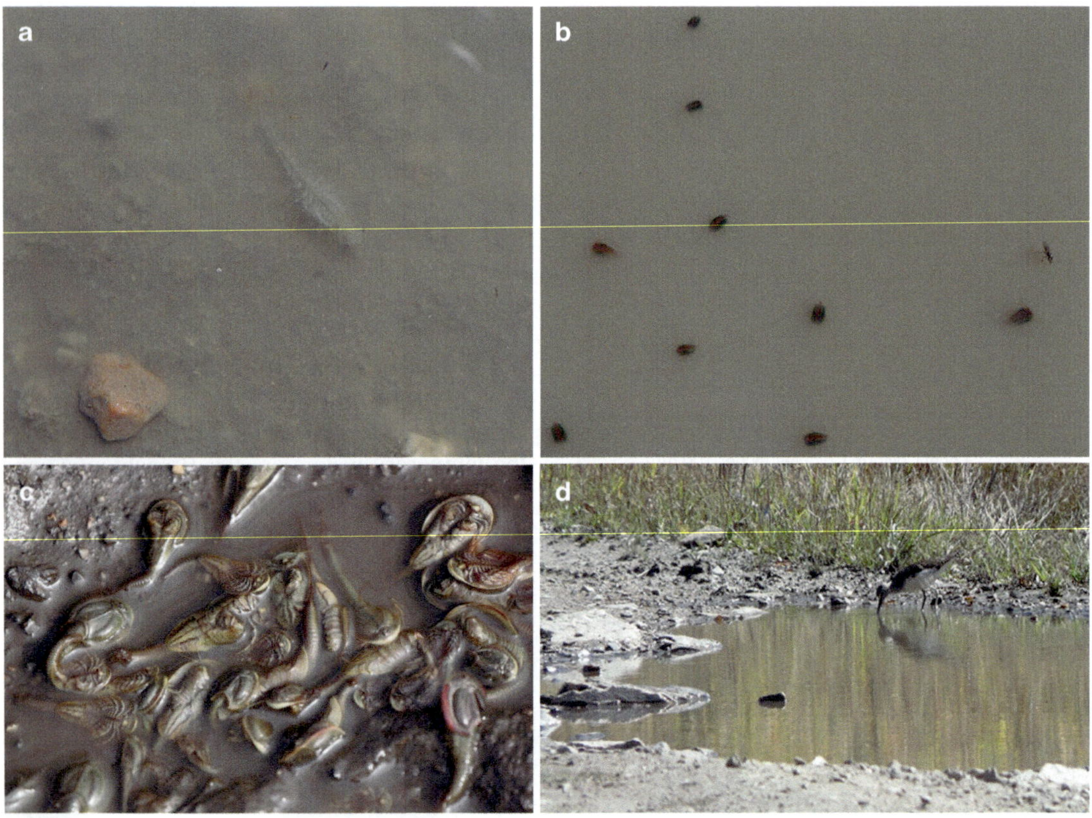

Fig. 6.55 (**a**) A Branchiopod (superorder Anostraca), *Branchipodopsis affinis,* in a turbid small body of water. (**b**) Conchostraca (*Lynceus dauricus*, order Onychura) in a turbid pond. (**c**) Gathering of *Triops numidicus* (Order Notostraca) in a just drying out small body of water. Central Mongolia. (**d**) A Green Sandpiper (Tringa ochropus) as an external predator fishing for Crustacea in a pond. North Mongolia

Phyllopods, such as *Triops* sp. (Fig. 6.55c), also prey on tiny organisms from the bottom. They can also be prey for birds themselves. The life forms of these crustacean groups can survive very long periods of drought or freezing as eggs for 7–9 years (Kaestner 1967). Larger Branchiopods are globally threatened mainly due to the loss of their typical habitats (Angeler et al. 2008; Brendonck et al. 2008).

If you know a pond that is currently dried up, you can take soil samples and hatch them in an aquarium by watering. In this way, large branchiopods can be identified later.

6.3.6.3 Large Temporary Waters in Tropical Regions

In the seasonal tropics, even large bodies of water carry water only periodically due to the change between rainy and dry seasons. Larger fish can occur in these waters, which are adapted to the survival of dry periods. These include, for example, Silurids and Polypterids (Fig. 6.56b, c).

Rivers in the tropics can also change significantly in terms of water flow, from having much water during the rainy season to dropping dry during the dry season. Then, they contain only small ponds (Fig. 6.57).

Fig. 6.56 (**a**) Floodplain (mare, mere) during the rainy season. Comoé National Park, Ivory Coast. (**b**) The temporary body of water (mere) dries up, and the fish of the genus *Polypterus* (gray bichir) becomes visible. In adaptation to low water levels, this fish genus has a pair of primitive lungs instead of a swim bladder. Comoé National Park, Ivory Coast. (**c**) The fish *Polypterus* sp. in the dried-up mere. (**d**) Yellow-billed storks (*Mycteria ibis*) prey on the fish that are easy to catch in the dried-up floodplain. Comoé National Park, Ivory Coast

Fig. 6.57 (**a**) The Comoé River in the Parc National de la Comoé during the rainy season. North Ivory Coast. (**b**) In the dry season, this river is almost completely dry, with individual ponds containing water remaining between the stones at the bottom of the river. (Photo **b** was taken from the same position as **a**)

References

Angeler D.G., Viedma O., Sánchez-Carrillo S. & Alvarez-Cobelas M. 2008: Conservation issues of temporary wetland Branchiopoda (Anostraca, Notostraca: Crustacea) in a semiarid agricultural landscape: What spatial scales are relevant? – *Biological Conservation* 141, 1224-1234

Anonymous. (2017). Why biodiversity matters. *Nature Ecology & Evolution*, *1*(1), 0042. https://doi.org/10.1038/s41559-016-0042

Avolio, M. L., Carroll, I. T., Collins, S. L., Houseman, G. R., Hallett, L. M., Isbell, F., et al. (2019). A comprehensive approach to analyzing community dynamics using rank abundance curves. *Ecosphere*, *10*(10), e02881. https://doi.org/10.1002/ecs2.2881

Azzella, M. M., Ricotta, C., & Blasi, C. (2013). Aquatic macrophyte diversity assessment: Validation of a new sampling method for circular-shaped lakes. *Limnologica*, *43*(6), 492–499. https://doi.org/10.1016/j.limno.2013.04.001

Balcer, M. D., Korda, N. L., & Dodson, S. I. (1984). *Zooplankton of the Great Lakes : a guide to the identification and ecology of the common crustacean species*. Madison, Wis.: University of Wisconsin Press.

Bandara, K., Varpe, Ø., Wijewardene, L., Tverberg, V., & Eiane, K. (2021). Two hundred years of zooplankton vertical migration research. *Biological Reviews*, *96*(4), 1547–1589. https://doi.org/10.1111/brv.12715

Bandini, F., Kooij, L., Mortensen, B. K., Caspersen, M. B., Thomsen, L. G., Olesen, D., & Bauer-Gottwein, P. (2023). Mapping inland water bathymetry with Ground Penetrating Radar (GPR) on board Unmanned Aerial Systems (UASs). Journal of Hydrology, 616, 128789. https://doi.org/10.1016/j.jhydrol.2022.128789

Barbour, M. T., Gerritsen, J., Snyder, B. D., & Stribling, J. B. (1999). *Rapid Bioassessment Protocols for Use in Streams and Wadeable Rivers: Periphyton, Benthic Macroinvertebrates, and Fish - Second Edition | Bioassessment | US EPA* (No. EPA 841-B-99-002). Washington, D.C.: U.S. Environmental Protection Agency, Office of Water. https://archive.epa.gov/water/archive/web/html/index-14.html. Accessed 28 June 2022

Benejam L., Alcaraz C., Benito J., Caiola N., Casals F., Maceda-Veiga A., de Sostoa A. & García-Berthou E. 2012: Fish catchability and comparison of four electrofishing crews in Mediterranean streams. – *Fisheries Research* 123–124, 9–15

Blocksom, K., & Johnson, B. (2009). Development of a regional macroinvertebrate index for large river bioassessment. *Ecological Indicators*, *9*(2), 313–328. https://doi.org/10.1016/j.ecolind.2008.05.005

Bo, T., Doretto, A., Marino, A., Laini, A., & Candiotto, A. (2023). Taxonomic and functional responses of macroinvertebrate communities to dam construction in a non-wadeable river. *Knowledge & Management of Aquatic Ecosystems*, (424), 18. https://doi.org/10.1051/kmae/2023015

Bouchard, R. W. (2004). *Guide to aquatic macroinvertebrates of the Upper Midwest*. St. Paul, MN: Water Resources Center, University of Minnesota.

Brendonck L., Rogers D.C., Olesen J., Weeks S. & Hoeh W.R. 2008: Global diversity of large branchiopods (Crustacea: Branchiopoda) in freshwater. – *Hydrobiologia* 595, 167-176

Brown, R. M., McClelland, N. I., Deininger, R. A., & Tozer, R. G. (1970). A water quality index-do we dare? *Water and Sewage Works*, *October*, 339–343.

Brown, R. M., McClelland, N. I., Deininger, R. A., & Landwehr, J. (1973). Validating the WQI. In

Proceedings of the National Meeting of American Society of Civil Engineers on Water Resources Engineering. Presented at the National Meeting of American Society of Civil Engineers on Water Resources Engineering, Washington, D.C.

Brucet, S., Poikane, S., Lyche-Solheim, A., & Birk, S. (2013). Biological assessment of European lakes: ecological rationale and human impacts. *Freshwater Biology*, *58*(6), 1106–1115. https://doi.org/10.1111/fwb.12111

Calazans, D. (2011). *Estudos Oceanográficos: do instrumental ao prático*. Pelotas, RS, Brazil: Universidade Federal do Rio Grande and Edirora Textos.

Capers, R. S. (2000). A comparison of two sampling techniques in the study of submersed macrophyte richness and abundance. *Aquatic Botany*, *68*(1), 87–92. https://doi.org/10.1016/S0304-3770(00)00102-9

Carlisle, D. M., Meador, M. R., Moulton, S. R., & Ruhl, P. M. (2007). Estimation and application of indicator values for common macroinvertebrate genera and families of the United States. *Ecological Indicators*, *7*(1), 22–33. https://doi.org/10.1016/j.ecolind.2005.09.005

Castro, P., Azul, A. M., & Paiva, J. (2016). Conservation of Biological Resources: Why Does It Matter? In P. Castro, U. M. Azeiteiro, P. Bacelar-Nicolau, W. Leal Filho, & A. M. Azul (Eds.), *Biodiversity and Education for Sustainable Development* (pp. 13–28). Cham: Springer International Publishing. https://doi.org/10.1007/978-3-319-32318-3_2

Chang, F.-H., Lawrence, J. E., Rios-Touma, B., & Resh, V. H. (2014). Tolerance values of benthic macroinvertebrates for stream biomonitoring: assessment of assumptions underlying scoring systems worldwide. *Environmental Monitoring and Assessment*, *186*(4), 2135–2149. https://doi.org/10.1007/s10661-013-3523-6

Chudoba, E. A., Mallin, M. A., Cahoon, L. B., & Skrabal, S. A. (2013). Stimulation of fecal bacteria in ambient waters by experimental inputs of organic and inorganic phosphorus. *Water Research*, *47*(10), 3455–3466. https://doi.org/10.1016/j.watres.2013.03.047

Collins, S. E., & Flotemersch, J. E. (2014). Evaluation of the Sounding Rod Method for Sampling Coarse Riverbed Sediments in Non-Wadeable Streams and Rivers. *River Research and Applications*, *30*(8), 1065–1069. https://doi.org/10.1002/rra.2697

Cowx I.G. 1990 (ed.): *Developments in Electric Fishing*. – Fishing News Books, Blackwell Scientific Publications, Oxford, UK. 358 p.

Craft C. 2022: Inland marshes, 117-161 in: Creating and Restoring Wetlands (2nd ed.) Chap. 5. https://doi.org/10.1016/B978-0-12-823981-0.00014-9.

deFischer-Ooster, M., & Brunner, C. (1849). *Recherches Sur La Température Du Lac De Thoune À Différentes Profondeurs Et Dans Toutes Les Époques De L'année*. Genève, Suisse: Société de Physique et d'Histoire Naturelle de Genève. https://play.google.com/books/reader?id=E_wTAAAAQAAJ&pg=GBS.PA477&hl=en

Deppe, E. R., & Lathrop, R. C. (1992). *A comparison of two rake sampling techniques for sampling aquatic macrophytes* (Findings 32 No. PUBL-RS 732-92) (4). Madison, WI: Wisconsin Department of Natural Resources Research Findings.

Di Sabatino, A., Cristiano, G., Di Sanza, D., Lombardo, P., Giansante, C., Caprioli, R., et al. (2016). Leaf-Nets (LN): A New Quantitative Method for Sampling Macroinvertebrates in Non-Wadeable Streams and Rivers. *River Research and Applications*, *32*(6), 1242–1251. https://doi.org/10.1002/rra.2976

Dodds, W. K., & Whiles, M. R. (2019). *Freshwater Ecology: Concepts and Environmental Applications of Limnology* (3rd edition.). London San Diego Cambridge, MA Kidlington, Oxford: Academic Press.

Dodson, S. (2004). *Introduction to Limnology* (1st edition.). New York: McGraw-Hill Education.

Dodson, S. I. (2005). *Introduction to limnology* (1st ed.). New York: McGraw-Hill.

Dudgeon, D., Arthington, A. H., Gessner, M. O., Kawabata, Z.-I., Knowler, D. J., Lévêque, C., et al. (2006). Freshwater biodiversity: importance, threats, status and conservation challenges. *Biological Reviews of the Cambridge Philosophical Society*, *81*(2), 163–182. https://doi.org/10.1017/S1464793105006950

Eder E., Schernhammer T., Zuna-Kratky T. and Nüsken U. 2014: Temporäre Gewässer und ihre naturschutzfachliche Bedeutung, 251-264. - In: Süßwasserwelten Kap. 33, - Publisher: Linz, Austria. DOI: 10.13140/RG.2.1.3189.7360

Galbraith, M. G. (1975). *The use of large Daphnia as indices of fishing quality for rainbow trout in small lakes.* (No. 1827). Lansing, MI, USA: Michigan Department of Natural Resources, Fisheries Division. http://name.umdl.umich.edu/aag2862.1827.001

Geist, J. (2011). Integrative freshwater ecology and biodiversity conservation. *Ecological Indicators*, *11*(6), 1507–1516. https://doi.org/10.1016/j.ecolind.2011.04.002

Gibson-Reinemer, D. K., Stewart, D. R., Fritts, M. W., DeBoer, J. A., & Casper, A. F. (2016). Estimating the Effects of Environmental Variables and Gear Type on the Detection and Occupancy of Large-River Fishes in a Standardized Sampling Program Using Multiseason Bayesian Mixture Models. *North American Journal of Fisheries Management*, *36*(6), 1445–1456. https://doi.org/10.1080/02755947.2016.1206642

Haney, J. F., Aliberti, M. A., Allan, E., Allard, S., Bauer, D. J., Beagen, W., et al. (2013a). Hilsenhoff Biotic Index. University of New Hampshire Center for Freshwater Biology. http://www.cfb.unh.edu/StreamKey/html/biotic_indicators/indices/Hilsenhoff.html

Haney, J. F., Aliberti, M. A., Allan, E., Allard, S., Bauer, D. J., Beagen, W., et al. (2013b). An Image-Based Key to Stream Insects. University of New Hampshire Center for Freshwater Biology. http://www.cfb.unh.edu/StreamKey/html/index.html

Hilsenhoff, W. L. (1988). Rapid Field Assessment of Organic Pollution with a Family-Level Biotic Index.

Journal of the North American Benthological Society, 7(1), 65–68. https://doi.org/10.2307/1467832

Holmquist, D. D., Redding, K., & McDaniel, C. M. (2015). *Water Quality with Vernier Lab Book*. Beaverton, OR: Vernier Science Education. Accessed 3 August 2020

Hoosier Riverwatch. (2019). *Volunteer stream monitoring training manual 2019*. Indianapolis, IN: Indiana Department of Environmental Management. https://www.in.gov/idem/riverwatch/training-manual. Accessed 13 March 2022

Hoosier Riverwatch. (2022). *Volunteer stream monitoring training manual 2022* (10th ed.). Indianapolis, IN: Indiana Department of Environmental Management. https://www.in.gov/idem/riverwatch/training-manual. Accessed 11 June 2023

Huq, A., Grim, C., Colwell, R. R., & Nair, G. B. (2005). Detection, Isolation, and Identification of Vibrio cholerae from the Environment. In *Current Protocols in Microbiology*. John Wiley & Sons, Inc. https://doi.org/10.1002/9780471729259.mc06a05s02

Hutchinson, G. E. (1957). *A Treatise on Limnology, Volume 1*. New York: Wiley.

Institute for Environment and Sustainability, Mjelde, M., Phillips, G., Ecke, F., Hellsten, S., Poikane, S., et al. (2014). *Water Framework Directive intercalibration technical report: northern lake macrophyte ecological assessment methods*. LU: Publications Office of the European Union. https://data.europa.eu/doi/10.2788/75735. Accessed 24 June 2023

Jacobs, F., & Grant, G. C. (1978). *Guidelines for Zooplankton Sampling in Quantitative Baseline and Monitoring Programs* (No. EPA 600-3-78-026; SPECIAL SCIENTIFIC-83; EPA-R-804147-010) (p. 61). Corvallis, OR,USA: Environmental Protection Agency, Office of Research and Development, Environmental Research Laboratory. Accessed 21 June 2023

Jäger, P., Pall, K., & Dumfarth, E. (2004). A method of mapping macrophytes in large lakes with regard to the requirements of the Water Framework Directive. *Limnologica, 34*(1), 140–146. https://doi.org/10.1016/S0075-9511(04)80033-1

Jeon, D. J., Ligaray, M., Kim, M., Kim, G., Lee, G., Pachepsky, Y. A., et al. (2019). Evaluating the influence of climate change on the fate and transport of fecal coliform bacteria using the modified SWAT model. *Science of The Total Environment, 658*, 753–762. https://doi.org/10.1016/j.scitotenv.2018.12.213

Kaestner A. 1967: *Crustacea*. - In: Lehrbuch der Speziellen Zoologie, Bd. I, Wirbellose, 2. Teil. (2nd ed.), 879-1233. - Gustav Fischer Verlag, Stuttgart

Kislik, C., Genzoli, L., Lyons, A., & Kelly, M. (2020). Application of UAV Imagery to Detect and Quantify Submerged Filamentous Algae and Rooted Macrophytes in a Non-Wadeable River. *Remote Sensing, 12*(20), 3332. https://doi.org/10.3390/rs12203332

Kochi, L. Y., Freitas, P. L., Maranho, L. T., Juneau, P., & Gomes, M. P. (2020). Aquatic Macrophytes in Constructed Wetlands: A Fight against Water Pollution. *Sustainability, 12*(21), 9202. https://doi.org/10.3390/su12219202

Labat, F., Thiébaut, G., & Piscart, C. (2022). A new method for monitoring macrophyte communities in small shallow lakes and ponds. *Biodiversity and Conservation, 31*(5), 1627–1645. https://doi.org/10.1007/s10531-022-02416-7

Lakes Environmental Association. (2023). *Lakes Environmental Association 2022 Water Testing Report. Chapter 3—High-Resolution Temperature Monitoring* (p. 20). Bridgton, ME, USA: Lakes Environmental Association (LEA). https://mainelakes.org/wp-content/uploads/2023/03/Chapter-3-High-Res.-Temp.-Monitoring.pdf

Larson, M. (2000). *Teacher Training Manual*. Rochester Hills, MI: Clinton River Watershed Council.

Launchbaugh, K. (2009). Why Measure Biodiversity? *Principles of Vegetation Measurement & Assessment and Ecological Monitoring & Analysis*. Education. https://www.webpages.uidaho.edu/veg_measure/Modules/Lessons/Module%209(Composition&Diversity)/9_2_Biodiversity.htm. Accessed 8 September 2020

Lemly, A. D. (1982). Modification of benthic insect communities in polluted streams: combined effects of sedimentation and nutrient enrichment. *Hydrobiologia, 87*(3), 229–245. https://doi.org/10.1007/BF00007232

Lenton, T. M., Rockström, J., Gaffney, O., Rahmstorf, S., Richardson, K., Steffen, W., & Schellnhuber, H. J. (2019). Climate tipping points — too risky to bet against. *Nature, 575*(7784), 592–595. https://doi.org/10.1038/d41586-019-03595-0

Likens, G. E., Johnson, N. M., Galloway, J. N., & Bormann, F. H. (1976). Acid Precipitation: Strong and Weak Acids. *Science, 194*(4265), 643–645. https://doi.org/10.1126/science.194.4265.643

Louw, M., Morse, J., & Wenzel, J. W. (2020). Macroinvertebrate Identification Key. *Macroinvertebrates.org*. Education. https://www.macroinvertebrates.org/key/. Accessed 17 August 2020

Luhtala, H., & Tolvanen, H. (2013). Optimizing the Use of Secchi Depth as a Proxy for Euphotic Depth in Coastal Waters: An Empirical Study from the Baltic Sea. *ISPRS International Journal of Geo-Information, 2*(4), 1153–1168. https://doi.org/10.3390/ijgi2041153

Mandlburger, G., Pfennigbauer, M., Wieser, M., Riegl, U., & Pfeifer, N. (2016). Evaluation of a novel UAV-borne topo-bathymetric laser profiler. In The International Archives of the Photogrammetry, Remote Sensing and Spatial Information Sciences (Vol. XLI-B1, pp. 933–939). Presented at the XXIII ISPRS Congress, Commission I (Volume XLI-B1) - 12–19 July 2016, Prague, Czech Republic, Copernicus GmbH. https://doi.org/10.5194/isprs-archives-XLI-B1-933-2016

McGauley, E., Tregunno, B., & Jones, F. C. (2018). Coarse taxonomy (tolerance-value averaging) biases Hilsenhoff's family-level biotic index. *Environmental Monitoring and Assessment, 190*(8), 446. https://doi.org/10.1007/s10661-018-6817-x

McNaught, A. S. (2014). Zooplankton of the Great Lakes. Mount Pleasant, MI: Central Michigan University. http://people.cst.cmich.edu/mcnau1as/zooplankton%20web/

Meybeck, Michel. (1982). Carbon, nitrogen, and phosphorus transport by world rivers. *American Journal of Science, 282*(4), 401–450. https://doi.org/10.2475/ajs.282.4.401

Meybeck, M. (1993). C, N, P and S in Rivers: From Sources to Global Inputs. In R. Wollast, F. T. Mackenzie, & L. Chou (Eds.), *Interactions of C, N, P and S Biogeochemical Cycles and Global Change* (pp. 163–193). Berlin, Heidelberg: Springer. https://doi.org/10.1007/978-3-642-76064-8_6

Mikulyuk, A., Hauxwell, J., Rasmussen, P., Knight, S., Wagner, K. I., Nault, M. E., & Ridgely, D. (2010). Testing a methodology for assessing plant communities in temperate inland lakes. *Lake and Reservoir Management, 26*(1), 54–62. https://doi.org/10.1080/07438141003666848

Mills, E. L., Green, D. M., & Schiavone, A. (1987). Use of Zooplankton Size to Assess the Community Structure of Fish Populations in Freshwater Lakes. *North American Journal of Fisheries Management, 7*(3), 369–378. https://doi.org/10.1577/1548-8659(1987)7<369:UOZSTA>2.0.CO;2

Osborne, L. L., & Kovacic, D. A. (1993). Riparian vegetated buffer strips in water-quality restoration and stream management. *Freshwater Biology, 29*(2), 243–258. https://doi.org/10.1111/j.1365-2427.1993.tb00761.x

Ott, W. (1978). *Environmental indices: Theory and practice*. Ann Arbor, MI: Ann Arbor Science.

Pan, Y., Hughes, R. M., Herlihy, A. T., & Kaufmann, P. R. (2012). Non-wadeable river bioassessment: spatial variation of benthic diatom assemblages in Pacific Northwest rivers, USA. *Hydrobiologia, 684*(1), 241–260. https://doi.org/10.1007/s10750-011-0990-z

Parker, J., Pescitelli, S. M., Epifanio, J., & Cao, Y. (2020). Relationships among side-scan sonar classified substrates and fish-catch rates at multiple spatial scales. *River Research and Applications, 36*(8), 1579–1587. https://doi.org/10.1002/rra.3663

Paruch, L., Paruch, A. M., Eiken, H. G., & Sørheim, R. (2019). Faecal pollution affects abundance and diversity of aquatic microbial community in anthropozoogenically influenced lotic ecosystems. *Scientific Reports, 9*(1), 19469. https://doi.org/10.1038/s41598-019-56058-x

Pingram, M. A., Clapcott, J. E., Hamer, M. P., Atalah, J., & Özkundakci, D. (2020). Exploring temporal and spatial variation in cotton tensile-strength loss to assess the ecosystem health of non-wadeable rivers. *Ecological Indicators, 108*, 105773. https://doi.org/10.1016/j.ecolind.2019.105773

Pregler K.C., Vokoun J.C., Jensen T. & Hagstrom N. 2015: Using Multimethod Occupancy Estimation Models to Quantify Gear Differences in Detection Probabilities: Is Backpack Electrofishing Missing Occurrences for a Species of Concern? – Transactions of the American Fisheries Society 144:1, 89-95. https://doi.org/10.1080/00028487.2014.968291

Price A.L. & Peterson J.T. 2010: Estimation and Modeling of Electrofishing Capture Efficiency for Fishes in Wadeable Warmwater Streams. – *North American Journal of Fisheries Management* 30 (2), 481-498

Quinney, M. (2020, May 22). 5 reasons why biodiversity matters to humans. *World Economic Forum*. https://www.weforum.org/agenda/2020/05/5-reasons-why-biodiversity-matters-human-health-economies-business-wellbeing-coronavirus-covid19-animals-nature-ecosystems/. Accessed 23 June 2023

Reid, A. J., Carlson, A. K., Creed, I. F., Eliason, E. J., Gell, P. A., Johnson, P. T. J., et al. (2019). Emerging threats and persistent conservation challenges for freshwater biodiversity. *Biological Reviews, 94*(3), 849–873. https://doi.org/10.1111/brv.12480

Reid, S. M., Haxton, T., & Gutowsky, L. F. G. (2021). Boat-electrofishing transect location and flow levels: influence on riverine fish monitoring in non-wadeable habitats. *Environmental Monitoring and Assessment, 193*(10), 680. https://doi.org/10.1007/s10661-021-09444-w

RHS Team, UK Environment Agency, & River Restoration Centre. (2018). *River Habitat Survey in Britain and Ireland*. Bristol: RHS Team, UK Environment Agency and River Restoration Centre.

Ruttner, F. (1962). *Grundriß der Limnologie: (Hydrobiologie des Süßwassers)* (3rd ed.). Berlin: De Gruyter. https://doi.org/10.1515/9783111646480

Schneider, J. C. (1990). *Classifying bluegill populations from lake survey data* (No. 90–10). Lansing, MI, USA: Michigan Department of Natural Resources, Fisheries Division.

Schwoerbel, J., & Brendelberger, H. (2022). *Einführung in die Limnologie: Stoffhaushalt - Lebensgemein schaften - Technologie* (11. Aufl. 2022nd edition.). Berlin Heidelberg: Springer Spektrum.

Secchi, A. (1864). Relazione delle esperienze fatte a bordo della pontificia pirocorvetta l'Immacolata concezione per determinare la trasparenza del mare; Memoria del P. A. Secchi. *Il Nuovo Cimento (1855-1868), 20*(1), 205–238. https://doi.org/10.1007/BF02726911

Simpson, J. T. (1991). *Volunteer Lake Monitoring: A Methods Manual*. Washington, D.C.: U.S. Environmental Protection Agency, Office of Wetlands, Oceans, and Watersheds, Assessment & Watershed Protection Division.

Smith, P. E., Counts, R. C., & Clutter, R. I. (1968). Changes in Filtering Efficiency of Plankton Nets Due to Clogging Under Tow. *ICES Journal of Marine Science, 32*(2), 232–248. https://doi.org/10.1093/icesjms/32.2.232

Soil & Water Conservation Society of Metro Halifax. (2015). Taxa Tolerance Values. http://lakes.chebucto.org/ZOOBENTH/BENTHOS/tolerance.html. Accessed 5 June 2023

Southwood, R., & Henderson, P. A. (2000). Ecological methods (3rd ed.). Oxford ; Malden, MA, USA: Blackwell Science

Spears, B. M., Gunn, I. D. M., Carvalho, L., Winfield, I. J., Dudley, B., Murphy, K., & May, L. (2009). An evaluation of methods for sampling macrophyte maximum colonisation depth in Loch Leven, Scotland, *Aquatic Botany*, *91*(2), 75–81. https://doi.org/10.1016/j.aquabot.2009.02.007

Stepenuck, K. F., Wolfson, L. G., Liukkonen, B. W., Iles, J. M., & Grant, T. S. (2011). Volunteer monitoring of E. coli in streams of the upper Midwestern United States: a comparison of methods. *Environmental Monitoring and Assessment*, *174*(1), 625–633. https://doi.org/10.1007/s10661-010-1483-7

Strayer, D. L., & Dudgeon, D. (2010). Freshwater biodiversity conservation: recent progress and future challenges. *Journal of the North American Benthological Society*, *29*(1), 344–358. https://doi.org/10.1899/08-171.1

Theiling, C. H. (1990). *The relationships between several limnological factors and Bluegill growth in Michigan lakes* (Thesis). The University of Michigan, Ann Arbor, MI, USA. Retrieved from http://deepblue.lib.umich.edu/handle/2027.42/114184

U.S. Environmental Protection Agency, Office of Water. (2023). *National Rivers and Streams Assessment 2023/24: Field Operations Manual—Non-Wadeable* (Other Policies and Guidance No. EPA-841-B-22-007) (p. 125). Washington D.C.: U.S. Environmental Protection Agency, Office of Water. https://www.epa.gov/national-aquatic-resource-surveys/national-rivers-streams-assessment-2023-24-field-operations-0. Accessed 3 January 2024

Urban, M. C., Skelly, D. K., Burchsted, D., Price, W., & Lowry, S. (2006). Stream communities across a rural–urban landscape gradient. *Diversity and Distributions*, *12*(4), 337–350. https://doi.org/10.1111/j.1366-9516.2005.00226.x

US EPA, (2001). *Marinas and Boating: National Management Measures* (Overviews and Factsheets). Washington, DC: U.S. Environmental Protection Agency, Office of Water, Nonpoint Source Control Branch. https://www.epa.gov/nps/marinas-and-boating-national-management-measures. Accessed 18 June 2023

USEPA. (2012). *Recreational Water Quality Criteria* (No. 820- F-12–058) (p. 69). Washington, DC: Office of Water, U.S. Environmental Protection Agency. https://www.epa.gov/sites/production/files/2015-10/documents/rwqc2012.pdf

USEPA. (2015). Secondary Drinking Water Standards: Guidance for Nuisance Chemicals. https://www.epa.gov/sdwa/secondary-drinking-water-standards-guidance-nuisance-chemicals. Accessed 11 June 2023

Valley City State University Macro-Invertebrate Lab. (2014). Digital Key to Freshwater Invertebrates of North Dakota. Valley City, ND: Valley City State University. http://www.ndfreshwaterinverts.vcsu.edu/

Vernier Science Education. (2021). Vernier Graphical Analysis™ Pro. https://www.vernier.com/products/graphical-analysis/. Accessed 13 March 2022

Vondracek, B., Koch, J. D., & Beck, M. W. (2014). A comparison of survey methods to evaluate macrophyte index of biotic integrity performance in Minnesota lakes. *Ecological Indicators*, *36*, 178–185. https://doi.org/10.1016/j.ecolind.2013.07.002

Wandell, H. D., Wolfson, L. G., & Michigan Lake and Stream Associations Inc. (2007). *A Citizen's Guide for the Identification, Mapping and Management of the Common Rooted Aquatic Plants of Michigan Lakes*. Lansing, Michigan: Michigan State University Extension.

Wang, X., Li, J., Tan, L., Yao, J., Zheng, Y., Shen, Q., & Tan, X. (2023). The impact of land use on stream macroinvertebrates: a bibliometric analysis for 2010–2021. *Environmental Monitoring and Assessment*, *195*(5), 613. https://doi.org/10.1007/s10661-023-11235-4

Wassmer, T. (2020). Phenological Patterns and Seasonal Segregation of Coprophilous Beetles (Coleoptera: Scarabaeoidea and Hydrophilidae) on a Cattle Farm in SE-Michigan, United States Throughout the Year. *Frontiers in Ecology and Evolution*, *8*, 293. https://doi.org/10.3389/fevo.2020.563532

Wessell, K. J., Merritt, R. W., Wilhelm, J. G. O., Allan, J. D., Cummins, K. W., & Uzarski, D. G. (2008). Biological evaluation of Michigan's non-wadeable rivers using macroinvertebrates. *Aquatic Ecosystem Health & Management*, *11*(3), 335–351. https://doi.org/10.1080/14634980802297729

Wilbanks, K. A., Mullis, D. L., & Colón-Gaud, J. C. (2020). Comparison of a wood sampler for macroinvertebrate bioassessment of non-wadeable streams in the southeastern coastal plain. *Journal of Freshwater Ecology*, *35*(1), 429–448. https://doi.org/10.1080/02705060.2020.1852122

Wilhelm, J. G. O., Allan, J. D., Wessell, K. J., Merritt, R. W., & Cummins, K. W. (2005). Habitat assessment of non-wadeable rivers in Michigan. *Environmental Management*, *36*(4), 592–609. https://doi.org/10.1007/s00267-004-0141-7

Wisconsin Wetlands Association 2017: Ephemeral ponds: Seasonal habitat for wildlife. Apr 14, 2017. https://www.wisconsinwetlands.org/updates/ephemeral-ponds-seasonal-habitat-for-wildlife/

Withford W.G. and Duval B.D. 2020: Wind and Water Processes, 73.107 in: Ecology of Desert Systems (2nd ed.) Chap. 4. - Academic Press. https://doi.org/10.1016/B978-0-12-815055-9.00004-7.

Woodward, G., Perkins, D. M., & Brown, L. E. (2010). Climate change and freshwater ecosystems: impacts across multiple levels of organization. *Philosophical Transactions of the Royal Society of London. Series B, Biological Sciences*, *365*(1549), 2093–2106. https://doi.org/10.1098/rstb.2010.0055

Part II

Focus on Analyses

Investigating Ecological Relationships

Ecological relationships are complex and typically form a network-like structure. The work of Schreiber et al. (2010) shows how food webs are influenced by plant diversity. The effects are more robust at lower trophic levels (herbivores) than at higher trophic levels, such as carnivores and hyperparasites.

Ecological relationships are often characterized by the fact that the disappearance of one species often triggers the local extinction of many other species. This phenomenon is known as the cascade effect (e.g., Dyer and Letourneau 1999, 2013; Groom et al. 2006). The loss of a "keystone species" can also lead to changes in the food web of many species (e.g., compare the food web in a lake, Sect. 6.3, Fig. 6.28 Plankton, and Carpenter et al. 2001). Another example of an endangered species was provided by Vanbergen et al. (2017) in plant-pollinator networks. A keystone species can be a top predator or a habitat modifier (e.g., woodpeckers, beavers, Shukla et al. 2023). Key species are, therefore, also considered "target species" in nature conservation (see Sect. 10.8). The multiple links in a food web can also be described quantitatively with the "linkage density" (McKenney and Rund 2023). Multiple linkages reduce the risk of extinction (Heleno et al. 2012). They should also be the focus of habitat protection (e.g., Albert et al. 2017; Liere et al. 2015). A striking example of the impact of relationships is the altered prey spectrum of lions in the East African savannah, caused by an invasive ant species (Gaynor 2024; Kamaru et al. 2024).

References

Albert, C. H., Rayfield, B., Dumitru, M., & Gonzalez, A. (2017). Applying network theory to prioritize multispecies habitat networks that are robust to climate and land-use change. *Conservation Biology*, *31*(6), 1383–1396. https://doi.org/10.1111/cobi.12943

Carpenter, S. R., Cole, J. J., Hodgson, J. R., Kitchell, J. F., Pace, M. L., Bade, D., et al. (2001). Trophic Cascades, Nutrients, and Lake Productivity: Whole-Lake Experiments. *Ecological Monographs*, *71*(2), 163–186. https://doi.org/10.1890/0012-9615(2001)071[0163:TCNALP]2.0.CO;2

Dyer, L. A., & Letourneau, D. K. (1999). Trophic cascades in a complex terrestrial community. *Proceedings of the National Academy of Sciences*, *96*(9), 5072–5076. https://doi.org/10.1073/pnas.96.9.5072

Dyer, Lee A., & Letourneau, D. K. (2013). Can Climate Change Trigger Massive Diversity Cascades in Terrestrial Ecosystems? *Diversity*, *5*(3), 479–504. https://doi.org/10.3390/d5030479

Gaynor, K. M. (2024). A big-headed problem drives an ecological chain reaction. *Science*, *383*(6681), 370–371. https://doi.org/10.1126/science.adn3484

Groom, M. J., Meffe, G. K., & Carroll, C. R. (2006). *Principles of Conservation Biology*. Sunderland, Mass: Sinauer Associates: Oxford University Press.

Heleno, R., Devoto, M., & Pocock, M. (2012). Connectance of species interaction networks and conservation value: Is it any good to be well connected? *Ecological Indicators*, *14*(1), 7–10. https://doi.org/10.1016/j.ecolind.2011.06.032

Kamaru, D. N., Palmer, T. M., Riginos, C., Ford, A. T., Belnap, J., Chira, R. M., et al. (2024). Disruption of an ant-plant mutualism shapes interactions between lions and their primary prey. *Science*, *383*(6681), 433–438. https://doi.org/10.1126/science.adg1464

Liere, H., Kim, T. N., Werling, B. P., Meehan, T. D., Landis, D. A., & Gratton, C. (2015). Trophic cascades in agricultural landscapes: indirect effects of landscape composition on crop yield. *Ecological Applications*, *25*(3), 652–661. https://doi.org/10.1890/14-0570.1

McKenney, E., & Rund, E. (2023). Chapter 10: Food Webs. In *Applied Ecology*. Montreal, QC: Pressbooks. https://ncstate.pressbooks.pub/appliedecology/chapter/chapter-10-content-food-webs/. Accessed 23 July 2025

Shukla, I., Gaynor, K. M., Worm, B., & Darimont, C. T. (2023). The diversity of animals identified as keystone species. *Ecology and Evolution*, *13*(10), e10561. https://doi.org/10.1002/ece3.10561

Vanbergen, A. J., Woodcock, B. A., Heard, M. S., & Chapman, D. S. (2017). Network size, structure and mutualism dependence affect the propensity for plant–pollinator extinction cascades. *Functional Ecology*, *31*(6), 1285–1293. https://doi.org/10.1111/1365-2435.12823

7 Foraging Ecology

Abstract

Food relationships provide the energetic basis for life. They also determine which species live together. We can rarely directly observe what food an animal eats. Usually, we have to rely on analyzing feces, pellets, and pluckings. Instructions are provided on analyzing these samples for hairs, plant remains, or pollen taken from flower visitors. Artificial food sources can be used to study animals, such as ants and flower visitors, e.g., bumblebees. Many fruit-producing plants are dependent on animals for their propagation. Large animals play a unique role in trees and palms. Analyzing the prey spectrum of carnivorous arthropods also requires laboratory work.

7.1 Problems in Recording an Animal's Diet

Direct observation only allows us to determine a bird's prey in rare cases for insectivorous birds. When young birds are fed in their nest, one can sometimes monitor the food-carrying adult birds with a telescope. However, the insect and other invertebrate prey can usually only be distinguished to the systematic level of orders. Neck ring methods for recording a nestling's food by flushing the stomach content of the living bird are only suitable if done by professionals (Berthold 1980, examples in Kristin 1989; Heiniger 1991; Timmermann and Becker 2003). In most cases, we must rely on an indirect recording method of food by examining fecal samples, pellets, and pluckings. In the feces of flycatchers accumulating under their perches and in the feces of swallows collected under their nests, almost all wing remains of the prey animals are still preserved in a determinable form (Davies 1977; Bryant 1973). Thus, insect orders and many families of flies (e.g., Bibionidae, Empididae, Syrphidae, Calliphoridae) can easily be distinguished. A broad overview of the foraging behavior of birds from different countries and the methodology to study their feeding can be found in Morrison et al. (1990).

Not only the quality but also the quantity of the food consumed or fed to nestlings is of interest. For this question, however, the energy content of the food parts must be estimated. It can be related to the energy requirements of the animals studied. The average caloric value per dry weight can be estimated for individual taxonomic groups based on sample measurements with a micro bomb calorimeter, according to Phillipson (1964). There is a relationship between the ash-free dry weight of a prey animal (insect) and the body size, which can be estimated from the group-specific wing length of the insect. Bryant (1973) gives an example of brachyceran and cyclorrhaphous Diptera. Bernays and Barbehenn (1987) provide an overview of the importance of nitrogen and other chemical components of grass in the food choice of grasshoppers. Southwood

M. Mühlenberg, T. Waßmer, *Field Ecology*, https://doi.org/10.1007/978-3-031-87547-2_7

and Henderson (2000) discussed laboratory studies on energy budgets.

The most critical questions concerning energetic considerations for the field ecologist are the following:

- Do animals choose the most energy-rich prey types when foraging?
- Is the food supply energetically sufficient to guarantee the operational metabolism and the rearing of the young?
- Does the food supply limit the number of broods or the number of successfully fed young animals (see Martin 1987; Lundberg 1987; Nilsson 1987; Newton 1980; Wiens 1989; Awmack and Leather 2002)?

We do not address energy transfer problems within an ecosystem here as they require extensive laboratory investigations on the production at the various trophic levels. For such analyses, e.g., assimilation and metabolic rates, must be quantified (Funke and Weidemann 1971; Albert 1983; Potapov et al. 2021, 2024; Wilder et al. 2013). Glück (1985) worked on the selection of different plant foods based on their energy content, using the example of the seed diet of goldfinches. Newer work includes Marone et al. (2022) and Orłowski et al. (2017).

Fig. 7.1 A researcher has donned in a camouflage suit to observe unnoticed in the middle of a nesting colony of Azur-winged magpies

In the following, we primarily want to focus on how to find out what individual animals have eaten. Using environmental DNA (eDNA), samples that were collected in the field can be analyzed in the laboratory (Taberlet et al. 2018; Brun et al. 2022; Wanniarachchi et al. 2022). We initially limit ourselves to direct evidence that can be observed in the field.

In some cases, it may be possible to observe breeding birds carrying food close to a nest, e.g., at ground nests of larger birds. In order to avoid disturbing the birds, observing them from a camouflage tent would be advantageous. Camouflage tents are available in various designs, with prices between 60 and 150 euros/dollars (internet research 2024) varying according to size, transportability, and opening method. If you want to observe closer to a bird colony or remain mobile, you can use a camouflage suit (Fig. 7.1).

7.2 Pellets and Pluckings

With some certainty, pellets (Fig. 7.2a) can only be collected at owl nests or breeding cavities. To collect pluckings (Fig. 7.2b), one must know the territories and aeries of birds of prey such as sparrowhawks and goshawks. In the forest, their nests can be found most easily in winter when the trees are without leaves (see Sect. 1.13). Detailed descriptions of the preferred nesting sites of raptors can be found in the literature. For example, Link (1986) provides information for the goshawk. In Europe, feathers and bird skulls can be identified with the work by Brown et al. (2003) and Elbroch et al. (2001). Many photos of pellets can be found in Drewitt (2024). Hager and Cosentino (2006) wrote an identification key to rodent prey in owl pellets from the Northwestern and Southeastern United States. A new identifica-

Fig. 7.2 (**a**) Pellets and components from a long-eared owl (*Asio otus*). (Photo credit: Bastien M. 2007, Wikimedia, public domain). (**b**) The plucking of a great spotted woodpecker (*Dendrocopus major*) by a bird of prey. (Photo credit: Eva Mühlenberg-Horn)

tion guide for feathers of birds in Western Europe is provided by Fraigneau (2021) and for 438 North American birds by the U.S. Fish & Wildlife Service, Forensics Laboratory (2023). North American bird feathers can be identified with the Feather Atlas (2023).

Pellets and pluckings are determined according to März (1987), Hansen and Oelke (1978), skull and feather remains according to Schaefer (2018), Husson (1962), Corbet (1969), Peterson et al. (1985), Day (1966), Svensson (1970), Brown et al. (1987, 2003), and Olson (2013). Erfurt (2003) provides a detailed identification table of mammal skulls in prey remains and pellets. Hager and Cosentino (2006) provide an identification key for owl pellets in North America.

At fox dens (*Vulpes vulpes*), many prey remains can be identified in May/June (see Fig. 7.3). Your district forester might be able to give you hints about the location of dens. However, it should be noted that too much disturbance at the den causes the fox to move. If you want to observe foxes, you can find detailed descriptions in Clark (1983) and Burrows (1968). Only in rare cases is it possible to observe a fox (*Vulpes vulpes*) directly carrying its prey (Fig. 7.4). For a detailed food analysis, search the fox's hiding places by following the trail of a fox for a longer time (Sequeira 1980; Lockie 1959; Murie 1936).

Witt (1976, 1980) writes about the evaluation of food analyses. It should be considered that foxes also eat many earthworms, which leave hardly any traces. In feces, however, eaten earthworms can easily be detected. For example, they form the primary food for badgers (Kruuk 1989).

Fig. 7.3 Prey remains in front of a fox den entrance

Fig. 7.4 Red fox (*Vulpes vulpes*) with prey (here a hare, *Lepus europaeus*)

Under the microscope, the characteristic bristles of earthworms ("setae") can easily be identified in the suspended feces. They only serve as qualitative evidence; a single earthworm can have about 1000 of them. However, the transparent remains of the muscular stomach ring, of which each earthworm has only one, can be used to quantify the number of earthworms ingested (Kruuk 1989).

Detailed analyses of the actual food spectrum of birds are obtained through stomach flushings (Berthold et al. 1980; Ford et al. 1982; Major 1990; Hess 1997; Gionfriddo et al. 1995). Due to the danger to the sampled animal, only trained professionals should apply this method. For quantitative statements, you must consider that different food components are digested differently.

A direct assessment of the prey brought in is undoubtedly superior to its subsequent determination via pellets or prey remains (examples in Fig. 7.5 with vertebrates as prey, Fig. 7.6 with arthropods as prey). Kochanek (1990) used a small-format camera at kestrel nests, which was automatically triggered when the falcon approached for landing. By comparing the actual prey input with the nest content, he could evaluate previous food analyses based only on the nest content.

Nicole Wasmund (2013) used cameras for video surveillance to record nests of the Red Kite (*Milvus milvus*) (Fig. 7.7). Photos of small animals feeding at the nest or burrow entrance are not easy to use for determining the prey brought in (Fig. 7.6a and b).

Fig. 7.5 (**a**) Kingfishers of the group Halcyoninae often catch fish for food. Many species in the tropics also take different prey, from larger insects to small vertebrates. *Todiramphus sacer* (Alcedinidae) on the Fiji Islands has caught a lizard. (**b**) A vole was speared by the great gray shrike (*Lanius excubitor*) as food storage in the thorny bush, Thuringia, Germany

7.3 Hair Atlas for Determining the Prey Remains of Carnivorous Mammals

In the feces of carnivorous mammals, the prey species can be determined by the undigested hair of the prey animals, which differ specifically between species. Only feces with particular species affiliation (identification aids, e.g., at Bang and Dahlström 2006) are collected. Air-dry the feces and then completely crumble and disassemble them. Hair samples are separated from the feces and stored dry in collection jars. In the lab, the hairs are washed in a shampoo solution and degreased in 70% alcohol. At the beginning of the analysis, a grouping of similar hairs is carried out by presorting with the help of a magnifying glass (Hoppe-Dominik 1984). To avoid possible contact with the fox tapeworm (*Echinococcus multilocularis*), the collected *scats* should first be frozen for 24 h at −80 °C and then warmed up for 48 h at +60 °C (Reynolds and Aebischer 1991; Tiralla 2013). For species determination of the hairs, it is recommended to assemble a reference collection, e.g., a hair atlas. If rodents are unavailable from your own catches in the area, a hair atlas can be easily self-constructed from any museum collection of prey animals, e.g., rodent species and lagomorphs. To learn about the species specificity of hair, consult with the following literature: Toldt (1935), Williamson (1938), Stoves (1944), Oyer (1946), Benedict (1957), Stains (1958), Mandelli (1961), Day (1966), Dreyer (1966), Kunze (1971a, b, 1972), Dziurdzik (1973), Ryder (1973), Brunner and Coman (1974), Deedrick and Koch (2004), Meyer et al. (2001, 2002), Cavia et al. (2008). For West European mammals, a hair atlas and identification key are available (Teerink 1991). The determination of the hairs is also a prerequisite for population estimates of mice with hair collection tubes (Dickman 1986). It is not easy to quantitatively determine the prey species in the feces. Instead, frequency of occurrence is usually used (Ciucci et al. 1996; Klare et al. 2011; Tiralla 2013).

If one can definitively assign the feces of mammals to species (e.g., Fig. 7.8), the feces found also serve as evidence of the presence of this species (see also Sect. 5.2). Many mammal species are challenging to discover directly. An example in Central Europe is the otter (*Lutra lutra*). Since the otter marks his territory with a feces sample

Fig. 7.6 (**a**) Great spotted woodpecker (*Dendrocopos major*) feeding in a natural cavity in a mixed beech forest, Lower Saxony, Germany. (**b**) Wren (*Troglodytes troglodytes*) feeding arthropods. Southern Sweden. (**c**) The brown dipper (*Cinczlus pallasii*) searches between the cliffs in the water for insects to feed its young. It can also dive briefly underwater to do this. No other birds in this particular habitat pick insects out of the water. Tajikistan. (**d**) The red-rumped wheatear (*Oenanthe moesta*) prefers to collect scorpions as food for the young. Tunisia. (**e**) The collared kingfisher (*Todiramphus chloris*) with a large arthropod prey in the rainforest of West Papua

Fig. 7.7 (**a**) Technical equipment used for video recording. Case with battery, control element for the video camera (Videojet X SN Network Video Server of the company Bosch), external hard drive, and a laptop to check the recordings. (Photo credit: N. Wasmund 2011). (**b**) Red Kite nest with installed camera on the branch. (Photo credit: N. Wasmund 2011)

Fig. 7.8 (**a**) Otter (*Lutra lutra*) feces deposited on a large stone on the bank of a river section. Northern Mongolia. (**b**) Butterflies on wolf feces, predominantly yellows of the genus *Colias* sp. Northern Mongolia. (**c**) These excrements of a brown bear (*Ursus arctos*) in Northern Mongolia show the seeds that the bear has eaten in the *Pinus sibirica* forest. (**d**) Wolf feces. In addition to hair, many bones and teeth of rodents can be seen. Northern Mongolia

(Fig. 7.8a), this can serve as evidence for its occurrence in a particular stream or river section, which is critical to know for nature conservation. Trained wildlife detection dogs are best suited for targeted searching of such feces samples (Grimm-Seyfarth et al. 2017) (see Sect. 1.4).

7.4 Grass-Cuticle Atlas for Herbivorous Mammals and Field Crickets

In the feces of herbivorous mammals, the species of the grasses eaten can be reliably determined based on cuticle residues. The method builds on the fact that parts of the eaten grasses survive digestion and can be detected in the feces (Stewart 1967). Owaga (1977) and Anthony and Smith (1974) showed that the results of feces analyses largely agree with those of rumen content examinations. A comprehensive overview of the methods and their application was provided by Scotcher (1979).

A reference atlas of the cuticular structures of individual grass species is required to identify the grass fragments. Therefore, grass species from the areas in question for the investigation are collected. Stomata formation is species-specific for the monocotyledons and some dicotyledons' families (Dünn et al. 1965; Stewart 1965; Barthlott and Martens 1979). To be able to recognize the identification features of plant fragments in the feces (see Fig. 7.9), one must view these with the microscope at 100–400× magnification.

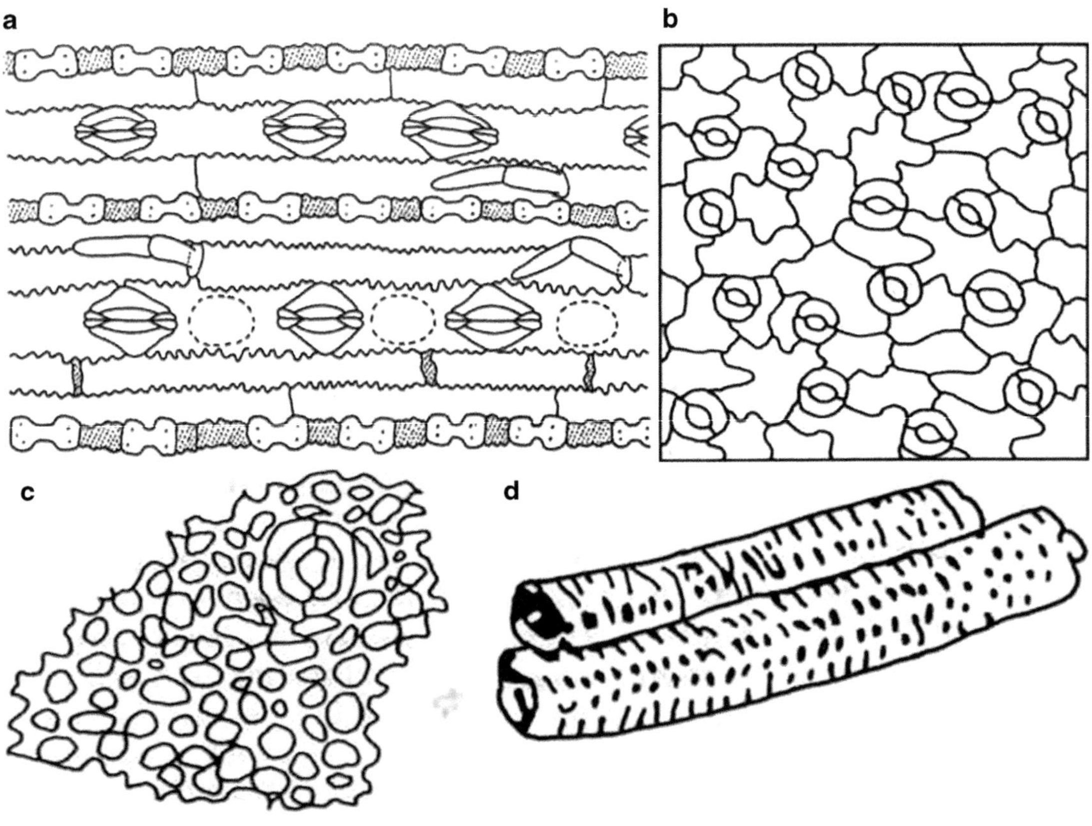

Fig. 7.9 Examples of the epidermis of various plants from fecal samples of herbivorous mammals after preparation in the laboratory. (After Radl 1987). (**a**) Epidermis of a grass. Cells and stomata are arranged in longitudinal rows, and silica cells are recognizable. (**b**) Epidermis of a dicotyledonous leaf, as an example of "non-grass." The stomata are scattered and not of the gramineous type. (**c**) Epidermis of a fruit. The cell walls are thick, and stomata are very rare. (**d**) Conductive elements are often recognizable by the elongated, round-shaped thickening of the cell walls

Case Study: What Did Herbivorous Mammals Feed On?

The fecal samples were prepared in the laboratory in the following steps (according to Radl 1987):

- 1 g (wet weight) of the fecal sample is transferred to a 1000 mL Erlenmeyer flask. The "pellets" are crushed with a flattened glass rod, avoiding excessive pressure so as not to destroy the plant fragments in the feces.
- Add 500 mL of water with a few drops of a highly diluted NaOH solution and stir with a magnetic stirrer until the slurry appears homogeneous.
- After filtering through very coarse filter paper (No. 604), the residue is washed with warm water until the wash water no longer appears cloudy.
- The residue is now taken up again in 100 mL of water. The suspension is now cloudy and light brown.
- 5 mL of Domestos (sanitary cleaner; ingredients include sodium hypochlorite, alkali carbonate, and sodium hydroxide) are added. Directly upon addition, a clearing should be observed in the suspension. If this is not the case, another 5 mL of Domestos are added.
- After about 12 h, the lightening of the plant fragments is complete. This is easily recognizable by the white color of the sedimented particles. The dosage and exposure time of Domestos were shown to be uncritical in experiments. The fragments remain well-evaluable even after 24 h of exposure and at higher Domestos concentrations.
- After filtering again and washing the residue with warm water, it is taken up in 10 mL of FAA (1 L from 100 mL of formaldehyde 40% + 850 mL of ethanol 70% + 50 mL of acetic acid). In this form, the fragments can be kept for some time.

To estimate the proportions of different plant species, 0.1 mL of the prepared samples are dripped onto slides with an etched millimeter grid or an underlying projection film piece with a millimeter print and glued with protein glycerin. After air drying, the cuticles can be stained by overlaying with a few drops of Sudan III-glycerin-red. The staining solution is prepared by adding 0.1 g of Sudan III to 50 mL of ethanol (96%) and heating to a boil while stirring. After cooling, undissolved parts are filtered off. The filtrate is mixed with the same volume of double-distilled glycerin (67%). Such a staining solution remains usable for several months and can be filtered again (Radl 1987). The staining solution is rinsed off, and the preparation can then be embedded, for example, in glycerin gelatin.

The mm grid under the slide serves as a guide to avoid double counting. One searches for all particles along a millimeter strip in a line-like manner and then skips a "line" (compare density estimation of plankton Sect. 6.3.3.5).

The sample size depends on the question. In order to eliminate individual feeding habits of wild animals for the analysis as much as possible, fecal samples from at least seven animals are collected, and an average sample is drawn from them. If the grass fragments are to be determined to species level, 100 fragments per solution sample are determined. Only fragments with at least some silica bodies, long cells, and a stoma are counted. If the research question is about the quantitative proportion of dicotyledons in the total diet, 5 × 100 plant fragments from each sample are classified according to the following criteria (Radl 1987): (1) Grass leaf: stomata of the grass type, silica cells are present. (2) Leaf of "non-grasses": stomata not of the grass type, no silica cells present. This group also includes any sedges found that lack silica cells. (3) Conductive elements: Elongated fragments without recognizable epidermal cells, helical or dotted structures.

(4) Fruit: cell boundaries as in dicotyledons, but not recognizable, with no to few stomata. The epidermis is very thick and colored red differently. Determining the percentage of dicotyledons in the diet provides only relative values for comparing food intake, e.g., between different seasons or a rough comparison between different species (deer and sheep or hares). Eisfeld (1984) gives an overview of the nutritional needs of deer, while Laitat (1983) has written about identifying food items in deer. General summaries about the diet of wild animals can be found in Robbins (1983), Siuda et al. (1969), Ueckermann (1971), Bicudo et al. (2010), Birnie-Gauvin et al. (2017), Hansson (2001), Nagy (2001), Nagy et al. (1999), Simpson and Raubenheimer (2012) and Warne et al. (2019). Ortmann et al. (2006) compare methods for determining primate diet.

While the predominantly grass-eating herbivores' quantitative food supply can be easily determined (see Sect. 7.6), fruit supply is more difficult to estimate in quantity. Fruit falls can be captured using fabric or plastic funnels distributed over the study area, each covering a 1 m^2 area. For fabric funnels, the foil is hung over a simple frame (e.g., a tripod) at about 1 m height, providing pores for the rainwater drainage at the bottom (see Fig. 7.16). The total amount of fruit can be roughly estimated from the determination of the fruit supply from branch samples and the tree crown or shrub volume. To compare the fruit production of a tree with the fruit consumption by birds or squirrels, as a control, matching branches are covered with gauze, or an entire second comparable shrub or small tree is completely covered with a so-called protection net against bird feeding (available in garden stores).

7.5 Food Analysis in Field Locusts

Similar to large herbivores, the feces of *field grasshoppers* can also be examined for food residues of various grass species. Feces are collected from grasshoppers caught alive from the field and kept separately for 12 h in gauze cages (e.g., a caterpillar breeding cage) with a small water dish. During this period, the animals have usually completely defecated and could be released back into the wild or used for other experiments. When catching the animals from the wild, it should be noted that grasshoppers usually do not eat at temperatures below 16 °C (Kaufmann 1965). Food residues in the foregut of the animals, which are collected around noon, usually come from food intake between 6 a.m. and 12 p.m. (Bernays and Chapman 1970). The fecal sample is processed for cuticular analysis in a similar way as described above for large mammals (Sect. 7.4). However, because of the small amount of feces, using a centrifuge (hand centrifuge) instead of filtering is recommended.

Grasshoppers killed with ethyl acetate can be stored in 75% alcohol, and their gut can be prepared at any time. The food residues in the foregut are isolated and determined by their cuticular structures using the same procedures described for fecal samples. If working on large batches, the foregut (esophagus and crop) content can be spread directly in phenol on a slide under the microscope. In the digestive tract, dicotyledon residues are sometimes even better to determine than those from fecal samples. This can be interesting for specific questions, especially for some bush crickets. In this case, a reference atlas must be built, whereby the selection of the dicotyledons should be based on the biology of the grasshopper species and its habitat partitioning. Gandar (1982) recommends pretreating the crop content of the field grasshoppers with gentle stirring in acid for about 15 min at 70 °C. The solution consists of 20 mL of 10% aqueous chromic acid, 10 mL of concentrated nitric acid, 20 mL of distilled water, and 4 mL of saturated potassium chlorate solution. After rinsing, the fragments are stained with ruthenium red and brought to a slide for microscopic examination. Due to the considerable intraspecific variability, an extensive amount of reference material is needed for determination. In *bush crickets*, the analysis of fecal pellets also provides insight into the food spectrum, which often ranges from plant material to residues of invertebrates (Fig. 7.10). For example, Cherril (1989) investigated the diet of the wart-biter (*Decticus verrucivorus*).

Fig. 7.10 (**a**) Male bush cricket (*Gampsocleis sedakovii*) feeding on a dead conspecific, Northern Mongolia. (**b**) Female bush cricket (Tettigonioidea, *Gampsocleis sedakovii*) feeds on plant seeds for protein intake, Northern Mongolia. (**c**) Subadult field grasshopper (Acrididae), feeding on young tree leaves (*Corylus* sp.), Northern Mongolia

7.6 Estimation of the Grazed Phytomass in a Meadow

Plants as food for phytophages are readily determinable in their frequency and distribution in the meadow. On the one hand, we can be interested in what proportion of primary production is eaten by phytophagous insects and, on the other hand, what role the quality of the plant food plays. Quality refers to different plant species, different parts of the same individual plant (e.g., leaves, flowers, stem, root), and the plants' age (stage of development). We want to focus on grasshoppers and caterpillars among the phytophagous insect groups.

To determine the phytomass and plant productivity in a meadow, Bogenrieder (1993) gives the following instructions: Frequency or coverage data for the question of food supply for certain animal groups are often insufficient. For the determination of the aboveground biomass of individual plant species, we recommend using the harvest method. Depending on the question, one must consider the entire aboveground phytomass (including dead plant parts and litter) or only their living parts. In both cases, the plant cover of the test area is harvested as close to the ground as possible, separated by species, and then dried in the laboratory. There are different opinions on whether sorting by species should already be done in the field or later in the laboratory. For both protocols, immediate separation into tussock-growing, multistemmed grasses, dead and readily crumbling plant parts, or falling fruits or seeds is advantageous. Mosses are usually summarized and weighed together.

Despite all care, an undefinable remainder of plant parts will hardly be avoidable. This remainder should not exceed 10% of the total dry weight. The sorted and dried material can be used, e.g., for nitrogen determination and calorific value (see Bogenrieder 1993).

The plant portion of the biomass of a meadow (phytobiomass, stock supply, *standing crop)* is relatively easy and quick to determine by simple harvesting and subsequent drying in a drying cabinet at 80 °C until constant weight. Separation into species groups (herbs, grasses, sedges, legumes) and the separate determination of dead plant parts make sense when later questions about the quality or developmental stage of the feed are to be discussed. For statistical reasons (mean, standard deviation), working on several subareas (approx. 30 × 30 cm^2) is usually more advantageous instead of a single, large test area. This is especially true when comparing your results to another stock or determining productivity. If the biomass determination is carried out immediately before mowing, the harvest quantity per ha can be calculated with sufficient accuracy. The productivity in a certain period can only be determined by two consecutive biomass determinations. The time interval should not be less than 3–4 weeks; otherwise, samples cannot be regarded as independent for statistical purposes, and you will need a higher number of replicas (subareas) to verify your results statistically.

The phytomass's dry weight is not identical to the organic substance. Its proportion of the total weight can be determined as weight loss when glowing in a muffle furnace. This test is needed if you expect a high proportion of inorganic substances (e.g., silica in sedges and horsetails). The residue after incineration can be used as a starting substance to determine the mineral portion of the phytomass. Your analytical possibilities primarily determine the further procedure (Bogenrieder 1993). For an estimate of the locust density, see Sects. 2.3 and 2.8.

You might be interested in the following questions:

- How much phytomass is consumed in a meadow by the locust community?
- Are the individual locust species monophagous or polyphagous?
- Are polyphagous locusts dominant?
- To what percentage do the caterpillars of the butterflies use the plant species of interest in the meadow as food plants?
- What reasons could cause the absence of butterflies on their potential food plants?

Suppose the locust density (see Sect. 2.3) is known in the study area, and you weigh the locusts (for example, random samples of 50 individuals locked in gauze cages). In that case, the food consumption of *locusts* can be quantitatively determined (Sinclair 1975; Gandar 1982). Fresh grass plants with roots are also weighed and offered to the animals with their roots in water. As a control, the plants must be kept and weighed in the same way in a gauze cage but without animals. After every 24 h, the plants with the locusts and the control plants are weighed. The amount of plants eaten should be converted to dry weight by drying the control plants and then weighing them again. You will also need to weigh the grass that was not eaten but wasted by the locusts and remains at the bottom of the cages. During numerous individual observations in the meadow, check whether the locusts have specialized to a food plant or behave polyphagously by noting the type of food plant each time a feeding locust is observed. Polyphagy can be quantitatively described in a calculation of the niche width (Sect. 15.11), representing the different food plants for the individual habitat classes.

If there are not enough prominent encounters of feeding locusts on their forage plants in the field, one can also test mono- or polyphagy through food choice experiments in the gauze cages. However, one cannot easily conclude the actual feeding behavior in the field from only a short observation period of food choice experiments.

Only the comparison between gut content (or feces analysis) and food supply in the habitat answers the question of whether the locusts in the grassland forage selectively. The food supply is qualitatively determined by point samples along a transect (encountered frequency of the individual grass species, measured as contact points with a pointer held vertically into the vegetation). A quantitative estimate is obtained by counting the leaves of each species that root in 10 cm × 10 cm quadrats. The quadrats are randomly selected along the transect strip, dug out, brought to the laboratory in plastic bags, and analyzed within 3 days (cf. Bernays and Chapman 1970).

The gut content of *butterfly caterpillars* can also be analyzed regarding food plant species using similar methods. The food spectrum for individual species is provided in many butterfly books.

7.7 Exploitation of a Food Source by Ants

How a comparable, constant food source is exploited can be studied in-depth on *ants*. The different behavioral patterns, the dominance of individual species, and the temporal sequence in the visit of artificial food sources provide clues about the competition between ants and their foraging efficiency. Diluted honey is placed on slides and distributed across the terrain. You can also offer other types of food on a slide in the field, such as meat-based food sources (insect larvae or pieces of canned tuna) or oatmeal as a plant-based source. The temporal sequence in the appearance of the ant species is recorded. The following questions could arise:

- What ant species come to the feeding stations?
- How long does it take ants to discover the food source?
- In what order do the species appear?
- Is there interference among the species at the feeding stations?
- How do the individual species behave at the feeding stations?
- Which species are solitary hunters, which recruit nest mates by laying trails?
- What differences exist in exploiting the food source depending on whether honey-water or meat food is offered?

A preliminary experiment can test whether or not the ants orient themselves by chemical scent trails by covering the suspected trail with sand. As a further experiment, one can let the finder animal run along a path between the nest and the food source over a cardboard strip (best as a bridge) and mislead nest mates with this strip going nowhere. By marking many individuals using different colors, closely located colonies of the same species can be kept apart. Traveled exit routes can be marked with colored pins.

Exemplary works on the questions addressed come among others from Brian (1952, 1955, 1958), Brian et al. (1965, 1966), Bzuder and Gypta (1972), Cammaerts-Tricot (1974), Dobrzanska (1958), Elmes (1971), Hangartner and Bernstein (1964), Szlep and Jacobi (1967), Wallis (1964), Waloff and Blackith (1962), Hölldobler (1979), Levings and Traniello (1981) and Hölldobler and Wilson (1990).

The time protocol of the appearance of the ants at the food source should be graphically represented. Species differences regarding the discovery time for the food source and the maximum number of individuals reached at the food source should be tested according to the *U*-test (see Sect. 15.9). According to Wilson (1971), the type of food source exploitation leads to the following classification: opportunists, complete exploiters (extirpators), and insinuators.

Actual proof of competition is difficult (den Boer 1980; May 1986; Montgomery 1981; Jacobs 1985; Schoener 1983; Wise 1981; Wirtz et al. 1988). Observations of aggression at the food source alone are not sufficient. Often, one species is displaced by another in their use of a resource or even spatially, without the second seeming to have a visible advantage over the absence of the first species (amensalism). The potential competitors should use a shared resource that is currently or potentially limited. Conspecific individuals (in ants' colonies) should also be in intraspecific competition. Resources and populations should be manipulated separately in the field. Adding or removing one of the competitors should lead to predictable consequences.

Experimental density changes were carried out by Schaefer (1978) on web-building spiders (Linyphiidae): He collected egg sacs from one experimental area and used them to increase the egg density in another experimental area. In addition, adult Linyphiids can also be relocated easily. Schaefer (1978) manipulated the food supply by placing *Drosophila* jars with laboratory cultures of short-winged fungus gnats on the ground. He checked reactions to increased food supply and population densities by weighing collected females (dry weight after 3–4 days at 55 °C) and determining their fertility by breeding them. The number of available catch-net sites was indicated as the population-limiting factor.

Observed differences in the food spectrum do not indicate a causal niche separation (Dickmann 1988). The following additional questions might be investigated:

- What food does species A prefer regardless of species B's presence or absence and vice versa?
- Do the potential competitors also differ in the use of microhabitats?
- Is the food distributed differently across the microhabitats?
- Does competition (evidenced by experimental removal of species B and vice versa) primarily influence the choice of microhabitat or directly influence the selection of food items?

Interspecific interactions also occur in large carnivores that live sympatrically. As large carnivores are mostly endangered species and almost only occur in protected areas, knowledge of the factors that enable them to coexist is necessary for conservation strategies. Translocation experiments can hardly be carried out. We have to rely on observations. Li et al. (2024) have analyzed many papers on the coexistence of tigers and leopards. In prey-rich regions, the overlap in diet is more significant. In addition to prey density, spatial elevation, and habitat structures play a role in the coexistence mechanisms.

7.8 Pollen Analysis in Flower Visitors

Flower visitors (wild bees, bumblebees, hoverflies) cannot always be directly followed during their feeding flights. However, if they collect or eat pollen, the visited flowering plant species (or genus) can be determined by identifying the pollen. While pollen can easily be brushed off the legs of bumblebees (see Fig. 7.11a, "pollen baskets or corbicular pollen," see Kratochwil and Kohl 1988), it usually must be prepared from the crop content in hoverflies and beetles.

Pollen is also collected by field wasps like the European paper wasp *Polistes* sp. (Fig. 7.12a). In their nests, they also store nectar (Fig. 7.12b). The largest, dominant female is often found in the center of the nest (Fig. 7.12b). The larvae are fed with insect meat. Feeding also influences caste formation (Judd et al. 2015). For the nesting ecology of *Polistes*, see Höcherl and Tautz (2015).

It is helpful to create a reference pollen collection from potential flowers. There are also publicly accessible databases of different types of pollen (e.g., Pollen and Spores Image Database n.d.).

For analysis and comparison to a database, the pollen is embedded in glycerin–gelatin or treated in the following way:

Fig. 7.11 (**a**) Bumblebee (*Bombus* sp.) with "corbicular pollen or pollen baskets." Lower Saxony, Germany. (**b**) Longhorn beetles (Cerambycidae) of the genus *Strangalia* feeding on pollen on a composite flower; Lower Saxony

Fig. 7.12 (**a**) Field wasp *Polistes* sp. (European paper wasp) visiting flowers to feed on pollen. Germany. (**b**) Nest of the field wasp *Polistes* sp. Licked-up nectar is also stored in some cells as a reserve (shiny droplets in the lower cells)

1. Prepare the acetolysis mixture (prepare fresh daily!): Mix acetic anhydride and concentrated sulfuric acid in a 9:1 ratio. Approximately 50 mL are needed for eight samples.
2. The pollen mass (or, in the case of herbarium material, the flower ground on a metal sieve) is rinsed with the acetolysis mixture into a centrifuge tube.
3. Place the centrifuge tube in a water bath of about 70 °C, which is quickly brought to a boil, and let it boil for 3–5 min while stirring occasionally.
4. Centrifuge, decant, and wash three times with distilled water.
5. Prepare the unbleached specimens by embedding them in glycerin–gelatin.
6. Bleaching (necessary for all pollen grains with a more substantial wall, not generally recommended for monocotyledons): Fill the centrifuge tube halfway with the remains of the unbleached pollen mass with glacial acetic acid; add two to three drops of hydrochloric acid and concentrated $KClO_3$ solution.
7. Centrifuge after a short time, decant, and wash three times with distilled water.
8. Prepare bleached specimens (as in step 5).
9. Store the remains of the pollen mass with a few drops of pure glycerin in small glass tubes for further sampling.

Note: To determine the size changes during processing and more extended storage of the specimens, adding some hazel pollen (collected once from one bush in late winter/early spring!) to all samples before acetolysis is recommended. For targeted observations, some pollen species can be sprinkled in the field with safranin or fluorescent dyes and later easily identified. You can also collect pollen remains in the brood tubes of solitary bees after hatching. The pollen grain remains are then found in the feces balls left behind. The fecal pellets must first be dissolved in a drop of water (Amiet and Krebs 2019).

7.9 Studies on Flower Ecology in Flower-Visiting Insects

The flowers in the meadow provide natural nectar and pollen. However, they are not only heterogeneously distributed in space but also produce a temporally fluctuating food supply.

Many research directions open up when working on insects as flower visitors (cf. Jones and Little 1983). One research field is about the morphological adaptations of insects as pollinators and the structure of the flower (Schremmer 1949, 1953, 1962a, 1973; Kugler 1970; Proctor and Yeo 1973; Pekkarinen 1976; Gilbert 1981; Barth

Fig. 7.13 (**a**) Hornet (*Vespa crabo*) patrols over a flower stand to search it for bumblebees, bees, and other insects. The captured insects are then chewed up and fed as meatballs to the larvae (see also Schremmer 1962b). (Photo credit: Eva Mühlenberg-Horn). (**b**) The hornet has caught a bee and is processing its prey for transportation home. Northern Germany. (Photo credit: Eva Mühlenberg-Horn)

1982; Kratochwil 1988). Another is about niche separation of closely related species of flower visitors (Brian 1957; Teräs 1976; Inouye 1978; Morse 1977a; Strickler 1979). Sensory physiological questions about pollinators approaching the flowers can only be partially clarified in the field (overview, e.g., in Barth 1982). Another area of concern for nature conservation lies in the relationship between phytocoenosis and the flower visitor community (Kratochwil 1984, 1985; Das and Das 2023; Rader et al. 2020; Rands et al. 2023; Shannon et al. 2024). For identification and information on the lifestyle of solitary bees, see, e.g., Benton (2017), Danforth et al. (2019), Mee and Barribeau (2023) and Raine and Rundlöf (2024). Some works deal with the strategy of foraging from the perspective of energy gain with changing food supply (Pyke 1978, 1980; Heinrich 1979; Real and Caraco 1986; Schmid-Hempel 1987).

Flower visitors can also fall prey to carnivorous insects or spiders. While, for example, paper wasps (Vespidae) of the genus *Dolichovespula* and *Vespula* search for nectar in flowers, hornets (*Vespa crabo*, see Fig. 7.13) catch wasps or bumblebees in flowers to bring them as food for their larvae into the nest.

Beewolves also seize flower visitors to carry them to their nest as prey. They lurk on the flowers (Fig. 7.14a) and then attack honeybees (Fig. 7.14b). The beewolf recognizes honeybees by smell. After paralyzing the honeybee, it carries it to its nest. There, it covers the paralyzed bee with an egg. The developing larva then feeds on the honeybee prey (Else 1997; Witt 1998).

Case Study: Bumblebees as Flower Visitors

We wanted to study questions of flower ecology on *bumblebees*, as they can be represented simultaneously with up to five species on a meadow and are easily markable. Most bumblebee species (up to ten) are found at forest edges and other fringe communities. *Bombus* is the most populous genus among the apoid Hymenoptera of lawn communities next to the genera *Halictus/Lasioglossum* (Kratochwil 1984). For the identification of the species of workers, there are several specialized books: Prys-Jones and Corbet (1987), Free and Butler (1959), Alford (1975), von Hagen (1994), Gokcezade et al. (2018),

Fig. 7.14 (**a**) A beewolf (*Philanthus triangulum*) awaits a honeybee as prey. (Photo credit: Eva Mühlenberg-Horn). (**b**) The beewolf has attacked a honeybee in a flash and paralyzed it with a sting. Northern Germany. (Photo credit: Eva Mühlenberg-Horn)

Williams et al. (2014) and Wilson and Carril (2015).

Possible research questions are:

- Which bumblebee species fly on the meadow under investigation?
- Do individual species or individuals always visit the same types of flowers or the same flowers?
- In what way are flower stands searched?
- Is there a direct relationship between proboscis length and the location of the nectar source in the flower?
- How often and at what intervals is a particular flower visited consecutively by the same individual or species?
- What amounts of nectar do individual flowers offer, and at what times of the day?
- Do flower visitors remember the locations of individual plants?
- In which species is nectar theft found?
- Are flowers with a constant amount of nectar preferred over those with a highly variable amount, even if the difference is small?

Individual animals must be caught for a reference collection to identify the bumblebees. However, one should not catch foraging queens in the spring (they are significantly larger than workers) as they are responsible for the presence of entire colonies. Moreover, all bumblebees in Germany are particularly protected species according to the FEDERAL SPECIES PROTECTION REGULATION of Dec. 1986, and the recommendations outlined in Sect. 2.1 regarding animal trapping apply.

Various flowers visited by bumblebees are morphologically examined, and the location of the nectaries in the flower is determined. You can find information about this approach in Kugler (1970) and Proctor and Yeo (1973). Kugler distinguished the following types of flowers: disk flowers (e.g., *Helianthemum),* funnel flowers (e.g., *Gentiana),* bell flowers (e.g., *Soldanella),* stem plate flowers (e.g., *Primula),* lip flowers (e.g., *Mimulus),* butterfly flowers (e.g., *Lathyrus),* and basket flowers (e.g., *Centaurea).* In Düll and Kutzelnigg (1992), we find information on the flower types of most native plants in Germany, while Mader et al. (2010) provide similar information about the USA.

The proboscis length is measured on captured bumblebee workers. To a limited extent, it can also be estimated from the wing length of some species (Morse 1977b; Pekkarinen 1976). The proboscis length varies interspecifically and intraspecifically (Morse 1978). The combined length of the glossa and prementum is measured on preferably fresh animals (Lewis and Taylor 1967; Prys-Jones and Corbet 1987). Many observation samples collected data on flower species and visiting bumblebee species. Kratochwil (1984) and Kratochwil and Kohl (1988) have, for example, analyzed the food plant spectrum of eight bumblebee species in a semidry grassland (*Mesobrometum*) in southwestern Germany and presented extensive observational material. When studying the flower visitors, it is also important to observe whether and how the behavior of the individual insect at the flower contributes to the flower's pollination by picking up pollen and how the individual animal is guided at the flower (Kugler 1943; Manning 1956). One can test the flower fidelity by color marking the bumblebee individuals (drops of paint on the prothorax). Part of the observation area is sketched with the locations of the flowers visited, and the flower visit of the bumblebees is logged with the help of a dictaphone with time stamps.

Among the bumblebee species are species with short glossae and long glossae. In Central Europe, the short-snouted ones include, for example, the group of so-called earth bumblebees (*Bombus terrestris* s.l.). Their glossa length is about 8–10 mm. The longest-snouted bumblebee species in Central Europe is the monkshood bumblebee (*Bombus gerstaeckeri*), with a snout length of up to 23 mm. This bumblebee species mainly visits the flowers of the monkshood plant *Aconitum* sp. Bees with long glossae visit more different flowering plants than short-nosed ones (Harder 1985). Short-snouted species can also become nectar robbers, i.e., they bite long flower tubes at the base and take the nectar directly from the bite site without contributing to the pollination of the flower (Fig. 7.15b). Loverty and Plowright (1988) compare specialized bumblebees with generalist species. Bumblebees also differ individually when visiting flowers (Heinrich 1976).

Several experiments are suitable for studying the factors determining the type of flower visitors (cf. Barth 1982), e.g., translocating flowers (with flowerpots) and offering flower dummies. To test the flower visitors' UV sensitivity, one can color artificial flowers with lead white and zinc white. Ultraviolet is only reflected by lead white. A pinboard is suitable for choice experiments. To find out whether the optical or chemical properties of a specific flower are responsible for attracting a particular insect to the flowers, the following demonstration is suggested (Schremmer 1941): Place glass tubes open only at the bottom over the flowers and observe whether the approaching insect turns toward the lower opening of the glass tube (scent attraction decisive) or tries to reach the flower directly through the glass tube (optical properties of the flower decisive). The glass tube is held by an iron wire stuck in the ground. Usually, various flower stimuli play a role in the sequence of actions from approach to finding the stigma (compare also to Tsuji et al. 2018; Tsuji and Coe 2014).

To determine the nectar production of a flower, one can exclude individual flowers from being visited by insects using gauze bags and suck up the nectar from the flowers at specific intervals with capillaries (available in laboratory shops) or very fine syringes or glass pipettes and weigh the capillary tubes before and after the filling using a precision scale, or measure the height of the nectar in the capillary in mm (Pleasants and Zimmermann 1979; Corbet 2003). The sugar content of the nectar, which varies between 15% and 88% of the weight (Howe and Westley 1988), can be determined with a refractometer.[1] To avoid

[1]A hand refractometer for determining the sugar content costs about 30 € (2022) and is easy to use.

Fig. 7.15 (**a**) Earth bumblebee (*Bombus terrestris*) collects pollen during flower visits. Germany. (**b**) Earth bumblebee (*Bombus terrestris*) robbing nectar from *Lonicera* sp. flowers without collecting pollen

injuring the nectaries with the glass capillaries, McKenna and Thomson (1988) developed a wick method: They soak up the nectar with punched and pointed pieces of filter paper, let the filter pieces dry in the air, and later determine the carbohydrate content of the nectar sample in the laboratory. Sowig (1991) determined the nectar content of a flower with the glucose test paper "Glucose Test" from Boehringer Mannheim. The paper changes color upon contact with sugar solution from yellow to green. You can cut the paper into small strips and insert them from behind into plucked flowers. The size of the green spot is a measure of the amount of nectar: one uses microcapillaries for calibration and applies known volumes of a glucose solution to the test strips. The area of the green spots must then be measured, e.g., under a stereo microscope. The nectar production is subject to a daily periodicity and usually lies between 1 and 5 mg per flower per day (Barth 1982). The pollen can sometimes be weighed by filling it into small, homemade funnels made of tin foil.

Not every flower has to produce nectar to attract insects. There are also so-called "cheaters" (Gilbert et al. 1991), which save the costs of nectar production and are still visited by insects in association with similar flowers. The proportion of deceptive flowers depends on the insect's recognition time and the time the insect spends at the flower during a regular visit ("handling time"). Flowers can also save nectar through reabsorption when flower visitors have not consumed the produced nectar, or further flower visits would reduce seed production (Búrquez and Corbet 1991).

To what extent certain flower species and flower visitors, or flower colors and flower visitors are related can be checked via the association analysis (see Sect. 15.11, Fauna similarity). Differences in the proboscis length of closely related species are tested with the *t*-test. A correlation between proboscis length and length of the visited flower crown (distance from nectaries to the flower edge) is established through correlation analysis (see Sect. 15.10 and corresponding books).

The pollen collected on the abdomen and legs of bumblebees and wild bees can be stripped off and analyzed for its flower origin (see Sect. 7.8). Pollen from hoverflies can be obtained by preparing the crop in water or alcohol.

Avoidance of competition (regarding interference) can be observed when different-sized bumblebee species visit the same flower stands. Usually, smaller species avoid the larger species, sometimes only to the outermost flower branches, which can no longer bear the weight of a larger bumblebee species (Morse 1977a).

To what extent functional morphology might enforce resource partitioning of flower species among different bumblebee species can be determined by experimental interventions. Possible research questions are:

- Does a bumblebee species visit flowers of another plant species more frequently if an alternative flower in the area is removed or made inaccessible (example in Inouye 1978; Heinrich 1979)?
- Does it expand its flower visits to more plant species under competitive pressure?

There can also be competition between domesticated honeybees and bumblebee species. Evidence can be provided by comparing the bumblebees in flower visitation and their size between regions with and without honeybees (Goulson and Sparrow 2008). Wignall et al. (2020) demonstrated through exclusion experiments on lavender flowers (*Lavandula x intermedia)* that honeybees visited the flowers more often after removing bumblebees. For the experimental design, three areas are compared: plots with the exclusion of honeybees, the exclusion of bumblebees, and control areas without excluding bees and bumblebees. Bees were excluded throughout the day using a light trap with a bamboo cane (Balfour et al. 2015).

A shift in resource use (*niche shift)* can be described by calculating the niche overlap and width values. Flower species or parts of the inflorescences are treated as resource classes (proximity to the stem, distance from the stem), and the frequency of visits by the bumblebee species in each resource class is considered a quantitative measure of their use. A calculation is demonstrated in Sect. 15.11 (niche calculations).

For economic analyses of flower visitation, the selected flowers must be continuously checked for their nectar content, and all visiting insects must be recorded. Usually, a bumblebee's food resources are patchily distributed, and the bumblebees adjust their search flights accordingly (Pleasants and Zimmermann 1979). The mapping of the small-scale distribution of individual plant species serves as the basis for comparing the visits to single-standing flowers and the visits to flowers in small or large groups. If the food search is to be analyzed as a strategy, experiments with artificial flowers and known nectar content are particularly suitable (e.g., Real 1981). We must remember that for rearing their offspring, bumblebees mainly collect pollen for their proteins and, therefore, cannot rely only on the nectar offer. With its sugar content, nectar mainly serves the operational metabolism. Since bumblebees can collect pollen at almost every melittophilous flower, species-specific differences in the food plant spectrum predominantly appear in the nectar plants (Kratochwil 1984). A detailed analysis of the energy intake and energy consumption in bumblebees was presented by Heinrich (1979). A bumblebee with 0.5 g body weight should need 150 calories per hour, for which it needs to collect nectar from, e.g., 500 blueberry flowers (*Vaccinium myrtillus*) (Howe and Westley 1988). Therefore, bumblebee foraging is under selection pressure to be optimized.

7.10 Large Animals as Seed Dispersers

About 85% of the woody plants in tropical forests depend on birds and mammals for dispersal (Ghazoul and Sheil 2017). Some mammal and bird species are the only dispersers for certain tree species. If these vertebrates disappear, the forest can no longer regenerate (Stoner et al. 2007). The role of selected mammals or birds in seed dispersal can be investigated in two ways:

1. Collect the feces of the vertebrate and set up a so-called germination bed.
2. Collect the feces and have the fruits or seeds identified by an expert or through a reference collection afterward.

Method (1) has the advantage that the germination ability of the fecal seeds is actually shown in the germination bed (Fig. 7.17). However, the growing woody plants must then also be determined to species (Theuerkauf et al. 2000). Cui et al. (2023) studied the dispersal of fruit seeds of a particular tree species by various mammals. They found that the dispersal success varied depending on what mammalian species ate the fruit.

Fig. 7.16 (**a**) Checking the presence of seedlings, poles, and young trees beneath a mother tree in a tropical forest of Ivory Coast. (**b**) Fruit trap for estimating the amount of fruit produced and the phenology of their drop. Tropical forest in the Ivory Coast

Especially in the tropical rainforest, the propagules do not grow under their mother tree, which can be shown by checking all seedlings under a mother tree (Fig. 7.16a). McConkey et al. (2018) conducted studies on the effectiveness of megafauna seed dispersal in Thailand's tropical rainforest using direct observation and camera traps. Other studies studied the movement of potential seed dispersers by GPS tracking and then inferred the possible dispersal distance of seeds (e.g., Nield et al. 2015) (Figs. 7.17 and 7.18).

Where large mammals are absent, large birds take on the role of seed dispersers. In Papua, these are the cassowaries (Fig. 7.18).

In montane conifer forests of the temperate zone and boreal needle forests, the seeds from the cones are usually collected by birds such as Eurasian nutcrackers (Fig. 7.19) and Eurasian nuthatches, transported away, and usually hidden (cached) at specific places. Observations on departing nutcrackers showed that the seeds are usually hidden and deposited 1–4 km from the collected tree (Hayashida 2003). Using binoculars and telescopes, the researchers observed nutcrackers collecting tree seeds during seed maturity and also recorded the number of seeds taken up in their sublingual pouch. Several pines sprouted at all discovered food cache sites, providing evidence of successful seed dispersal. The nuthatch carries only single seeds to nearby food caches in the bark of trees or similar structures (Fig. 7.20). Among the smaller mammals, squirrels and chipmunks (Siberian chipmunks (*Tamias sibiricus),* see Fig. 2.41b) also carry away many seeds but do not bury them at a great distance (for an overview of animals as seed dispersers of *Pinus* see Lu 2006).

The observation that birds crack open or crush the pine nuts and then consume them confirms these animals as seed predators. Individual species can be both seed dispersers and seed predators. Van Leeuwen et al. (2022) provide a theoretical overview of the interactions between plants and animals. In the tropics, flying foxes (Megachiroptera) are both fruit eaters and fruit dispersers (Fig. 7.21). In this way, flying foxes can also reforest African woodlands (van Toor et al. 2019, see also video at https://www.youtube.com/watch?v=eq2iesVsIgE).

Methods for estimating seed removal range from direct observation to the use of camera traps

Fig. 7.17 (**a**) Elephant dung, southern Ivory Coast. (**b**) Elephant dung with fruits and plant remains. (**c**) Seedlings growing from a pile of elephant dung. (**d**) Tree planting using collected elephant dung with the included seeds. Bossematié-Forest, southeastern Ivory Coast

Fig. 7.18 Droppings from the cassowary (*Casuarius unappendiculatus*), the largest bird in New Guinea here from the rainforest of North Papua, Indonesia. In this forest, many tree species and numerous palm species depend on cassowaries' distribution of their fruits (Pangau-Adam and Muehlenberg 2014a, b)

(McConkey et al. 2018). Some authors marked seeds to track their removal and dispersal (e.g., Hirsch et al. 2012). The New Zealand pigeon (*Hemiphaga novaeseelandiae,* Fig. 7.22c) feeds on fruits and fresh leaves of at least 70 native tree species. Since many large seed dispersers in New Zealand went extinct, this large pigeon species plays an essential role in the seed dispersal of native woody plants. It is, therefore, also crucial in restoring original forest areas (see Fig. 11.2 in Sect. 11.2, Forest Islands in NZ) (Wilson 2004). The species' feeding behavior was studied through direct observation and radio-tagged pigeons. Fruit doves are both fruit eaters and seed

Fig. 7.19 (**a**) Nutcracker (*Nucifraga caryocatactes*) collects seeds from the pine cones of *Pinus sibirica*, Northern Mongolia. (**b**) The nutcracker (*Nucifraga caryocatactes*) flies away from a pine (*Pinus sibiricus*) with seeds collected in its sublingual pouch. Northern Mongolia

Fig. 7.20 (**a**) The nuthatch (*Sitta europaea asiatica*) carries a collected pine seed in its beak to hide in a tree stump. Northern Mongolia. (**b**) A cached pine nut hidden by the nuthatch. (**c**) A great tit (*Parus major*) steals the pine seeds hidden by the nuthatch in the bark

Fig. 7.21 A flying fox carries a fruit in its mouth. Samoa

dispersers (Fig. 7.22), also Birds of Paradise (Fig. 7.22a). Usually, fruit doves of the same size occur in different habitats, while different-sized species can occur syntopically.

Since pine nuts are also commercially harvested by humans, producing pine cones must be estimated to develop a sustainable harvesting method. Counting the cones in the tree crowns directly is not effective, as only about half of the cones in a tree can be seen. In addition, information is needed on how much of the tree crown production was already consumed by birds. Likewise, fallen cones on the ground are eaten or carried away by mammals (see also Fig. 7.8c, excrements of brown bear). Therefore, one of the few options left is to use funnel traps to catch falling cones under randomly selected trees. These are similar to the fruit traps mentioned above (see also above knocking methods Fig. 2.6 in Sect. 2.6). Alternatively, you can spread plastic foil around the tree (see Fig. 2.6). Forest management also influences seed production. An example is given by Wang et al. (2022).

7.11 Determination of the Prey Spectrum of Carnivorous Arthropods

Food webs within the arthropod community can be studied using stable isotope analyses (Petersen and Fry 1987; Gratton and Forbes 2006; Michener and Lajtha 2007). Stable isotope analyses have often been used to study the food webs of arthropods on the forest floor (Ponsard and Ardith 2000; Scheu and Falca 2000). However, such analyses require extensive laboratory work and expensive instrumentation.

Seifert and Scheu (2012) examined the food web of larval and adult dragonflies through stable isotope analysis by determining the ratio between $^{14}N/^{15}N$ and $^{12}C/^{13}C$ in various tissue parts of larvae and adult dragonflies. In addition, the isotopes in potential prey animals in the aquatic habitat (larval development) and the terrestrial area (adult stage) were compared. In this way, the authors could demonstrate that the food of adult dragonflies differs from that of the larvae and even between males and females.

While it is only rarely possible to spot a carnivorous arthropod with prey (e.g., Fig. 7.23), it is easy to identify the prey remains by regular control (intervals of 1–2 h) of the *capture nets of orb-weaving spiders* (Araneidae) (Fig. 7.24) and cobweb spiders (Theridiidae). Not all web-building spiders create food depots or keep their prey remains long in the net. Some species renew their capture net at least once a day or change the location in case of food shortage (Stowe 1986; Riechert and Harp 1987). Of the insects entangled in the capture net, not every species is consumed as prey (Olive 1980; Riechert and Luczak 1982). On the other hand, there are insects as potential prey that manage to free themselves quickly from a spider web (e.g., some moths, Eisner et al. 1964; Nyffeler and Benz 1981a) or visually recognize the spider web in flight to avoid it (e.g., some hoverflies, Nentwig 1982a; Craig 1988). Therefore, the prey spectrum of web-building spiders was shown to be selective, and native web spiders predominantly catch less flight-active insects such as mosquitoes, aphids, mayflies, and thrips (Nentwig 1982b, see also Rypstra 1982; Eberhard 1986; Stowe 1986,

Fig. 7.22 (**a**) A male of the red bird of paradise (*Paradisaea rubra*) is extracting seeds out of fruits. Gam Island, West Papua. (**b**) A fruit dove (mountain fruit-dove, *Ptinilopus bellus*). Arfak Mountains, Vogelkop, New Guinea. (**c**) New Zealand pigeon, kereru (*Hemiphaga novaeseelandiae*), endemic and endangered in New Zealand. (**d**) The western crowned pigeon (*Goura cristata*) is one of the few larger birds (body length up to 75 cm) in New Guinea that disperses large plant seeds. Like the cassowary, it is endemic in West Papua and rare, foraging on the ground for fallen fruit and seeds in undisturbed lowland forests and is threatened by hunting (Keiluhu et al. 2019)

Fig. 7.24b). A comparison of the insects in the spider web with the potential prey spectrum can be made by setting up sticky traps (see Sect. 2.13) and direct observation controls of insects approaching the nets. There are also predators that steal prey from the spider web as kleptoparasites (Fig. 7.25).

It is relatively easy to determine the prey spectrum *of brood-caring digger wasps* (Sphecidae) or *spider wasps* (Pompilidae, see Fig. 7.26). Once the brood tube entrances have been found, you can intercept the digger wasps arriving at the nest with prey and identify their prey. Detailed presentations on the provision of larvae with food in solitary wasps can be found in Clausen (1940), Olberg (1959), Evans (1966), Malyshev (1968), Evans and West-Eberhardt (1970), and Yeo and Corbert (1983). For illustration, Blösch (2012) can serve. Spider wasps usually transport their spider prey to a suitable place, lay the spider down, and dig a brood tube, into which they transport the spider with the attached egg. Some species also use existing cavities (e.g., Jacobs and Renner 1988).

Even at nests of larger *ant* species, one can identify prey brought in. In temperate latitudes, species of the *Formica* group (e.g., *Formica rufa* and *polyctena*, the "red wood ant") can be sea-

Fig. 7.23 (**a**) Robber-fly (Asilidae) has captured a butterfly. It sucks the prey with its strong proboscis, similar to many assassin bugs (Reduviidae). See Fig. 2.59b in Sect. 2.27. (**b**) A Huntsman spider of the Sparassidae family has preyed on a lizard in West Papua

sonal pest controllers (see Fig. 7.27). However, ants do not only feed on insects or their larvae but also on the honeydew of aphids and scale insects, which belong to the "pest insects," but are cared for by the ants. To what extent ants are beneficial, i.e., increase the expected harvest, cannot be generally answered (Anjos et al. 2022). On a small scale, however, studying the effects of ants is possible. One should choose an annual plant, e.g., a grain plant or bean infested with aphids. On one plant, one counts the subpopulation of aphids visited by ants. At the same time, one counts the individuals of the aphids on another plant, where ants are prevented from reaching the aphids. For such a comparison, one needs, e.g., ten replicates. Then, after several weeks, one observes the increase or decrease of the aphid individuals in the respective subpopulations with or without ant visit. Comparative works were carried out by Banks (1958), Bristow (1984), and Breton and Addicott (1992). Also, the host plant can benefit from the presence of the ants (Skinner and Whittaker 1981).

In the tropics, many species of ants search for prey, inform each other, and then gather to collectively overpower and transport the prey ("*conventional solitary scouting and group recruitment*," Fig. 7.28) (e.g., Maschwitz and Mühlenberg 1975). The "*true army ants*" perform group foraging and raiding (overview by Hölldobler and Wilson 1990). Many tropical ant species primarily hunt termites (compilation by Hölldobler and Wilson 1990). In the tropics, some species of jumping spiders (Salticidae) imitate their prey (mimicry) (Fig. 7.29a, b).

Among the *hunting spiders***,** which seize their prey without using capture threads, the primarily diurnal wolf spiders (Lycosidae, e.g., *Pardosa* spp., see Fig. 7.30a) are most commonly encountered on the ground surface. To assess their prey spectrum, many individuals must be caught alive in the field (best by a turned-over glass tube) and immediately checked for prey pieces between the chelicerae (Nyffeler 1982). Lycosids without prey are released (record!), and spiders with prey are killed and preserved in alcohol for later determination. Since the prey-capture rate with one prey item per wolf spider per day is relatively low and the duration of prey consumption is 30–60 min, the probability of encountering spiders with prey is only about 5% (Nyffeler 1982).

Fig. 7.24 (**a**) Orb-weaver spider (Araneidae, *Gasteracantha* sp.) with a butterfly (Satyridae) as prey in the catch net, West Papua. (**b**) A spider's web (family of orb-web spiders) with the catch of many midges (Nematoceridae). Northern Mongolia. (**c**) *Araneus diadematus* with a fly (Tabanidae) as prey in the catch net, Germany. (**d**) A crab spider (Thomisidae, *Misumena* sp.) lurks, colorfully camouflaged, in a flower and catches a hoverfly (Syrphidae) as prey, Germany

The food spectrum of wolf spiders is broad. Springtails (Collembola) always form a significant part of the prey (Nyffeler and Benz 1981b). Through their juvenile stages and various species, wolf spiders act as predators with a relatively broad range of body sizes; accordingly, their prey spectrum also varies (Nentwig 1982a).

The control of prey also has implications for applied ecology: it is crucial to estimate to what extent, for example, spiders and ground beetles play a role as pest exterminators in the cultivated land (cf. Nyffeler and Benz 1979; Nyffeler 1982; Scherney 1959). How to quantitatively estimate predator pressure is discussed in Sect. 8.1.

Within the species-rich family of rove beetles (Staphylinidae), the species of the genus *Stenus* (Fig. 7.31) possess a unique adhesive prey-capture apparatus that enables them to catch elusive prey such as springtails over a distance of several millimeters (Koerner et al. 2012; Betz et al. 2018). This prey-capture apparatus consists of a protruding elongated labium with the paraglossae modified into adhesive pads covered by an adhesive secretion (Fig. 7.32, Koerner et al.

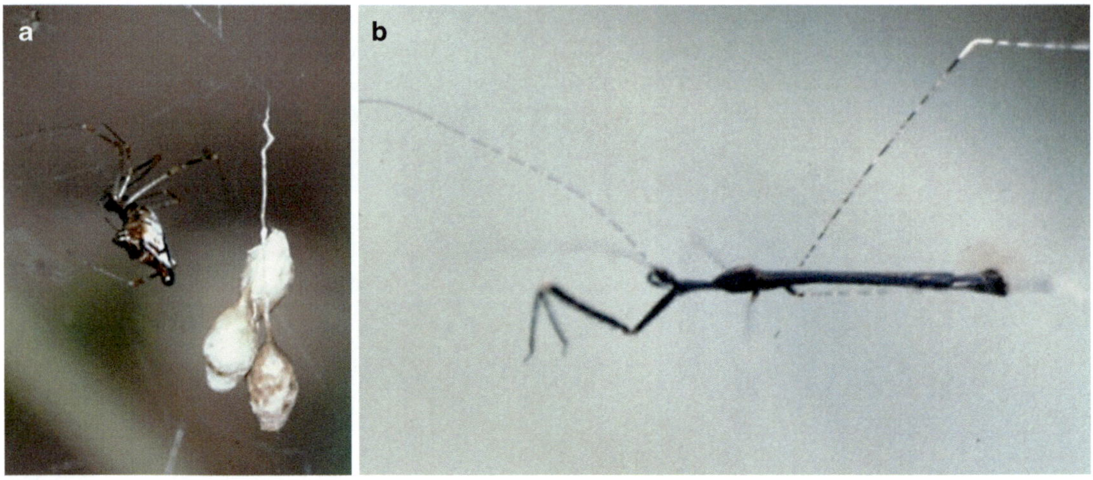

Fig. 7.25 (**a**) A dewdrop spider (*Argyrodes* sp., Theriidae) as a thief in the web of a Pisauride (Nursery web spider). Thief spiders are food thieves (kleptoparasites). The Gambia, West Africa. (**b**) A thread-legged bug (Reduviidae, Emesinae) sneaks into a spider web to hunt for small insects caught in the web. Yunnan, SW China

Fig. 7.26 (**a**) A spider wasp (Pompilidae) has paralyzed a spider and is towing it into its brood tube in the ground. Mongolia. (**b**) A digger wasp (sand wasp *Ammophila* sp., Sphecidae) carries a paralyzed insect caterpillar as food for its larva into its brood tube, Mongolia. (**c**) *Ammophila* has laid down the caterpillar and opens its sealed brood tube. (**d**) Now *Ammophila* pulls the caterpillar (arrow) backward into the brood tube (red arrow)

Fig. 7.27 (**a**) Nest mound of *Formica* sp. ants, Germany. (**b**) An entrance into the anthill of the wood ant (*Formica* sp.), where, e.g., prey is brought in. (**c**) A group of *Formica* sp. ants has overpowered a Lepidoptera larva (gypsy moth, *Lymantria dispar*) in North Mongolia. The gypsy moth can appear massively as a pest on woody plants (e.g., Elkinton and Liebhold 1990)

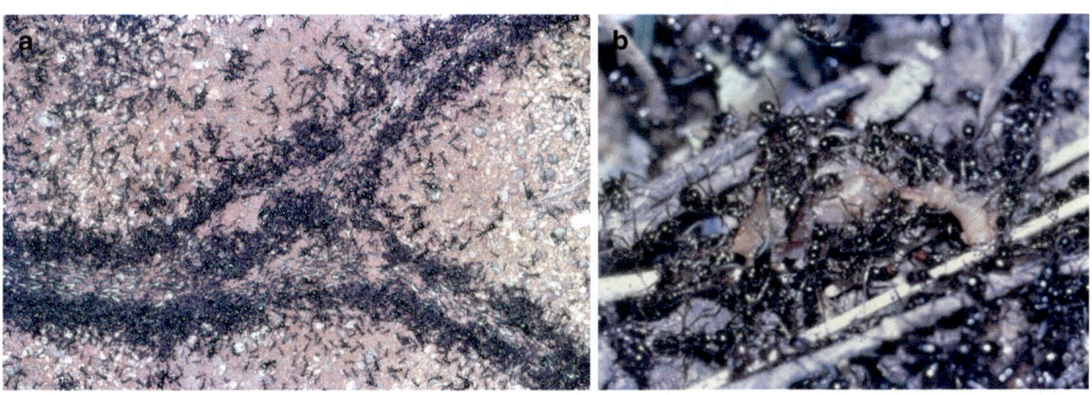

Fig. 7.28 (**a**) African driver ants (*Dorylus* sp.) return home after a raid on termites. The larger-headed individuals are the soldier caste and flank the street. These driver ants form raids in broad fronts. (**b**) Ants of *Leptogenys* sp. overwhelm an earthworm after group recruitment (Maschwitz and Mühlenberg 1975), Sri Lanka

Fig. 7.29 (**a**) A spider of the family Salticidae imitates its possible prey, the *Oecophylla*—ants. West Papua. (**b**) Weaver ants (*Oecophylla smaragdina*) during nest building

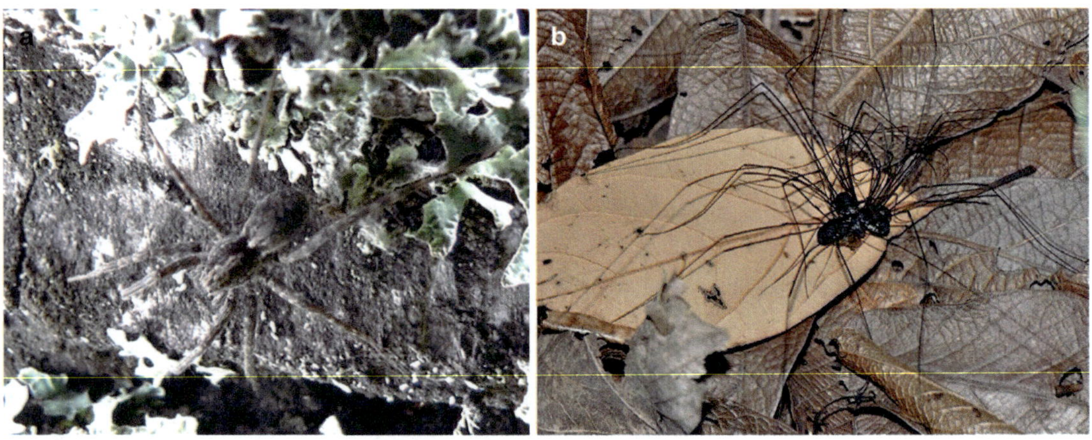

Fig. 7.30 (**a**) Wolf spider Lycosidae, *Pardosa* sp. (**b**) Harvestmen (or daddy longlegs, Order Opiliones) belong to the arachnids but have no venom and spinning glands. They capture small insects or dead arthropods with their small mandibles. Mexico

2012, 2017; Betz et al. 2018). During predation, the labium can be rapidly ejected toward potential prey by increased hemolymph pressure. *Stenus* beetles are usually small (less than 1 cm) and live in waterside environments such as reeds or sparsely vegetated sites on a river or lake shore or in the litter and humus layer of tropical forests; in the tropics, they may also be arboreal.

Remains of the prey of harvestmen (Opiliones, Fig. 7.30b), stone runners (centipedes, Chilopoda-Lithobiidae), and many species of ground beetles (Carabidae, e.g., *Pterostichus* sp.) can be identified after preparation of the digestive tract, especially the crop (Fründ 1983; Koehler 1977; Loreau 1983, 1986). For ***intestinal content analyses***, animals that were freshly

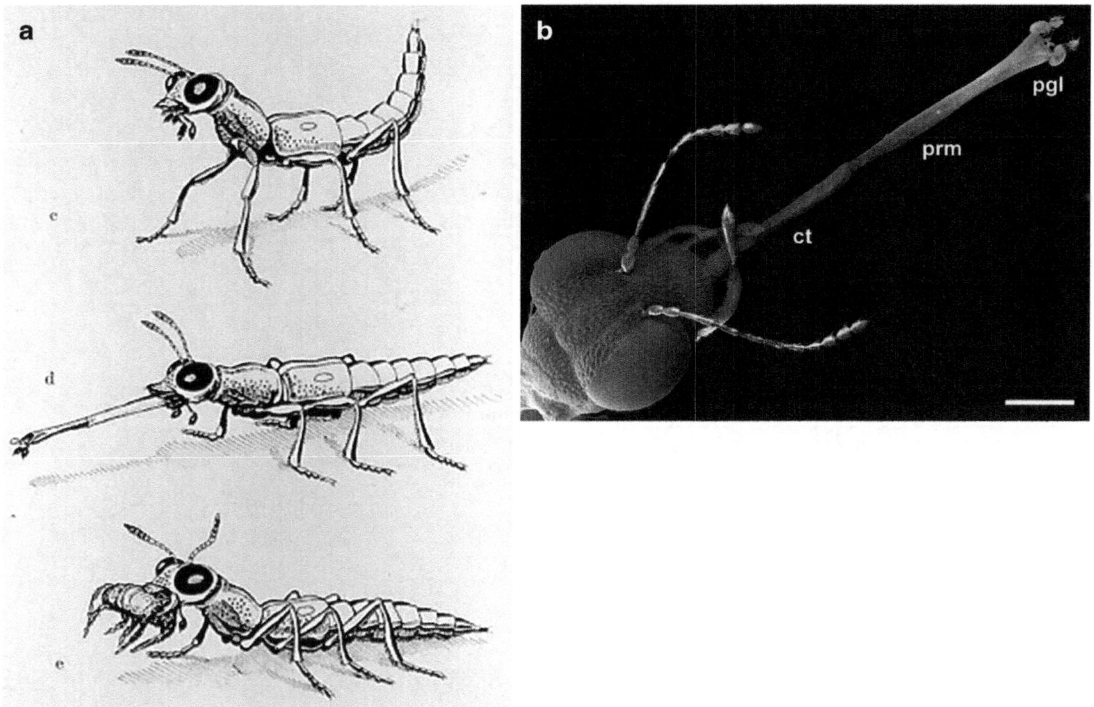

Fig. 7.31 (**a**) Prey-capture sequence in *Stenus comma*. (From Weinreich 1968, used with kind permission from Springer-Verlag, Heidelberg). (**b**) The adhesive trapping apparatus in *Stenus* spp., *S. bimaculatus*. Scanning electron microscope (SEM) image. Head with extended labium. Scale bar: 0.5 mm. *pgl* paraglossae (modified into adhesive pads), *prm* prementum, *ct* connecting tube. (Koerner et al. 2017, used with kind permission from Oxford University Press)

caught from the wild or preserved in alcohol or 4% formalin are suitable (e.g., from pitfall traps; see Sect. 2.16). The animals' intestines are dissected and opened, and the contents are transferred to slides in Faure's solution. For Faure's solution, 50 mL of distilled water, 30 g of powdered gum arabic, 20 mL of glycerin, and 50 g of chloral hydrate are mixed under heat. The solution is then filtered through glass wool (after Weidemann 1971).

Loreau (1983) isolated the crop content of carabids in water and added a drop of detergent to the suspension so that the prey's hydrophobic cuticular pieces remained in solution. The suspension was rinsed with distilled water into a bacteriological filter apparatus (Millipore Ltd.) with a glass filter and grid-marked cellulose acetate filter (see Anderson 1972). The individual food fragments then lie ± evenly distributed on the filter surface. The filter is then dried at about 50 °C and placed on a slide with Canada balsam for examination and more extended storage, making the filter transparent. Chiverton (1984) cut the four intestinal sections of carabids (crop, proventriculus, midgut, and hindgut) under a binocular microscope on a slide in 50% glycerol to identify solid food residues. Although most ground beetle (Carabidae) species feed carnivorously, some carabids, e.g., from the *Amara* sp. group, eat plant seeds (Fig. 7.33).

For a *prey-reference collection*, larger chitinous objects can also be cleared as a whole in warm lactic acid. The chitin fragments and other indigestible components are scanned with a microscope at 100× and 400× magnification in transmitted light and phase contrast. Identification is most reliably done with the help of a photo database of reference specimens, for which

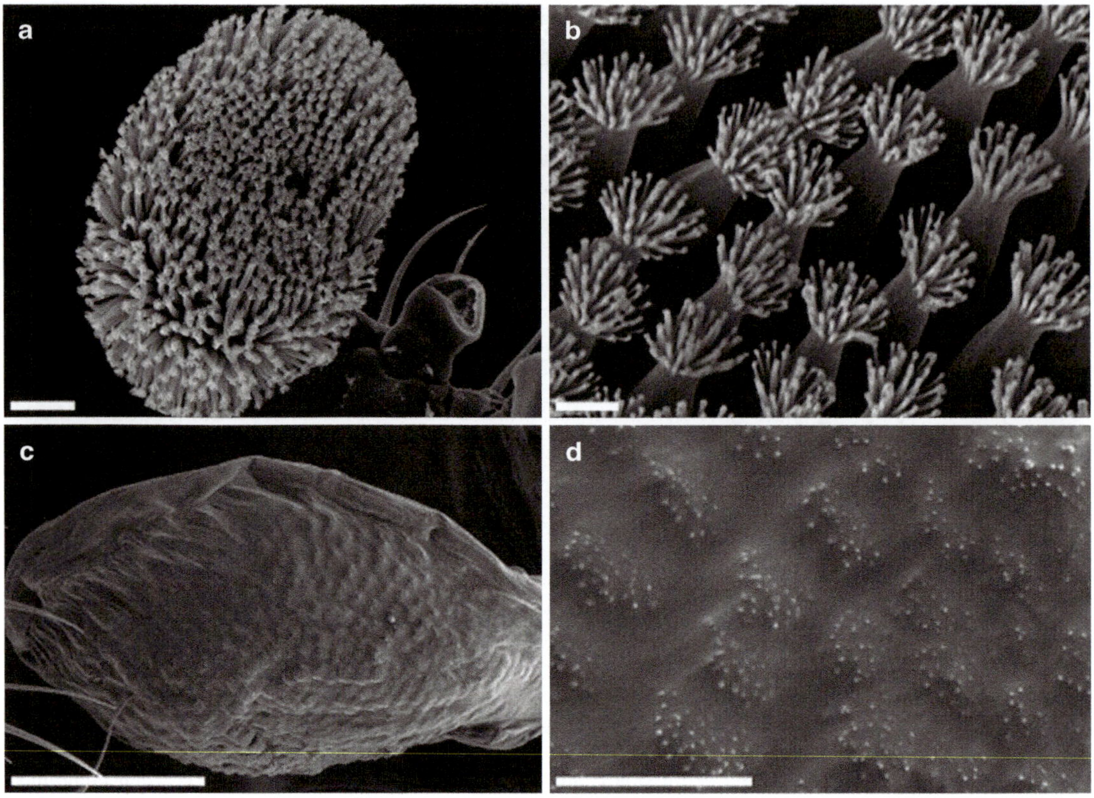

Fig. 7.32 Images of the adhesive pads of *S. juno*. (**a–b**) SEM images, (**c–d**) cryo-SEM images. (**a**) Ventral view (scale = 20 µm). (**b**) Adhesive hairs with terminal branches (scale = 2 µm). (**c**) During prey capture, the adhesive hairs are entirely wetted with the adhesive secretion. (scale = 30 µm). (**d**) Only the terminal branches of the guard hairs protrude from the adhesive secretion (scale = 2 µm). (From Koerner et al. 2012, used with kind permission from Elsevier)

Fig. 7.33 A ground beetle (*Amara* sp.) feeds on grass seeds. Austria

potential prey from various prey groups were prepared as squash preparations. It makes sense to stain the chitin, e.g., with Direct Black. A solution of Direct Black (Chlorazol Black, dissolved in 96% alcohol and filtered) is added drop by drop to the object initially in water until a blue coloration of the chitin is just visible under the magnifying glass. The next steps to the permanent preparation are 70% alcohol (5 min)—96% alcohol (5 min)—100% Alcohol (5 min)—Xylene (10 min)—Balsam in Xylene (15 min). Before the balsam solidifies, the cover glass is applied under slight pressure (cf. Steiner 1963).

The assignment of fragments from the intestine to certain prey groups is made possible by the various, very characteristic structures of the prey animals. For example, scales are good identifiers of Lepidoptera, entomobryid Collembola (e.g., *Lepidocyrtus* sp., *Tomocerus* sp.), and Curculionidae (e.g., *Phyllobius* sp., *Polydrosus* sp., *Strophosomus* sp.). Mouthparts are typical for, e.g., Staphylinidae, beetle larvae, Aphidinae, and fungus gnat larvae. Antenna fragments can identify Diptera, and the typically shaped bristles (setae) can identify Enchytraeids. Limb fragments characterize mites, while Diptera (e.g., Lithobiids) are characterized by their spines. Also recognized can be the sickle-shaped chitin hooks of the crown feet of caterpillars, the toothed or combed tarsal claws of the Araneae, the chelicerae of the Araneae, and the mucrones of the Collembola. Many prey animals carry very characteristic surface structures, such as the pebble pattern of *Fannia* larvae and pupae (Diptera), the net-like pattern of *Lonchoptera* larvae (Diptera), the tile-like feathered cuticle of the Lithobiids, the chagrin of the front edges of the tergites of Staphylinidae, or the ring-shaped structures of the Aphidinae (compilation according to Fründ 1983 and Koehler 1977). Soft-bodied prey animals like snails are usually not identifiable. Plant tissue can always be distinguished from animal remains.

Araneae and some ground beetle species (Carabini, Agonini) digest their prey extraorally so that no solid prey remains can be found in the digestive tract. In these cases, prey proteins in the body fluid of the predators can be detected immunologically by *precipitin tests* (Boreham and Ohiagu 1978; Ohiagu and Boreham 1978; Davies 1969; Miller et al. 1979; Fründ 1983; Allen and Hagley 1982; Denison and Hodkinson 1983). It is recommended that precipitin tests be performed in collaboration with a corresponding laboratory (e.g., in virology). Southwood and Henderson (2000) describe various serological methods. The principle of this serological test on predator-prey relationships is that an animal (usually a rabbit) is stimulated to form antibodies against the proteins of the prey species. The prey proteins present in the predator's digestive tract after consumption of the prey now react with the rabbit serum in an antigen-antibody reaction, which can be made visible as a precipitate under appropriate conditions.

For the quantitative determination of predator pressure, you must check how long a consumed prey animal remains detectable in the predator's intestine (compared to Dempster 1960).

References

Albert, A. M. (1983). Energy Budgets for Populations of Long-Lived Arthropod Predators (Chilopoda: Lithobiidae) in an Old Beech Forest. *Oecologia, 56*(2/3), 292–305.

Albert C.H., Rayfield B., Dumitru M. & Gonzales A. 2017: Applying network theory to prioritize multispecies habitat networks that are robust to climate and land-use change. – *Conservation Biology* 31(6), 1383–1396

Alford D.V. 1975: *Bumblebees*. – Davis-Poynter Lim., London, 352.

Allen W.R. & Hagley E.A.C. 1982: Evaluation of Immunoelectroosmophoresis on Cellulose Polyacetate for Assessing Predation of Lepidoptera (Tortricidae) by Coleoptera (Carabidae) Species. – *The Canadian Entomologist* 114 (11), 1047 – 1054. DOI: https://doi.org/10.4039/Ent1141047-11

Amiet F. & Krebs A. 2019: *Bienen Mitteleuropas. Gattungen, Lebensweise, Beobachtung*, (3rd ed.). – Haupt Verlag, 423.

Anderson J.M. 1972: Food and feeding on *Notiophilus biguttatus* F. (Coleoptera: Carabidae). *Rev. Ecol. Biol. Sol, Paris* IX (2), 177–184

Anjos DV, Tena A, Viana-Junior AB, Carvalho RL, Torezan-Silingardi H, Del-Claro K, Perfecto I. 2022: The effects of ants on pest control: a meta-analysis. – *Proc. R. Soc. B* 289: 2022.1316. https://doi.org/10.1098/rspb.2022.1316

Anthony R.G. & Smith N.S. 1974: Comparison of rumen and fecal analysis to describe Deer diets. – *J. Wildl. Management* 38 (3), 535–540

Awmack C.S. & Leather S.R. 2002: Host plant quality and fecundity in herbivorous insects. – *Annual Review of Entomology* 47, 817-844

Balfour N.J., Gandy S. & Ratnieks F.L.W. 2015: Exploitative competition alters bee foraging and flower choice. – *Behav Ecol Sociobiol* 69, 1731–1738. https://doi.org/10.1007/s00265-015-1985-y

Bang P. & Dahlström P. 2006: *Animal Tracks and Signs*. – Oxford University Press, 264.

Banks C.J. 1958: Effects of the ant *Lasius Niger* (L.) on the behaviour and reproduction of the black bean aphid *Aphis fabae* Scop. – *Bull. Ent. Res* . 49, 701-714

Barth F.G. 1982: *Biologie einer Begegnung*. – Deutsche Verlags-Anstalt, Stuttgart, 304.

Barthlott W. & Martens B. 1979: Cuticular-Taxonomie der Gräser eines westafrikanischen Savannengebietes unter dem Aspekt der Futterpräferenzanalyse wildlebender Großsäuger. – *Trop. u. Subtrop. Pflanzenwelt* 30, 113.

Benedict F.A. 1957: Hair structure as a generic character in bats. – *Univ. Calif. Pub. Zool.* 59, 285–548

Benton T. 2017: *Solitary Bees*. Naturalist's Handbooks 33, Pelagic Publishing, 202.

Bernays E.A. & Chapman R.F. 1970: Food Selection by *Chorthippus parallelus* (Zetterstedt) (Orthoptera: Acrididae) in the Field. – *J. Anim. Ecol.* (Oxford) 39, 383–394

Berthold P. 1980: 4.7. Ernährungsbiologie. - In: Berthold P., Bezzel E. and ThielkeG. (eds.): Prakt. Vogelkund, p. 60-62. - Kilda Verlag.

Berthold P., Bezzel E. and Thielcke G. 1980: Praktische Vogelkunde. Ein Leitfaden für Feldornithologen. - Kilda-Verlag, Greven/Westfalen, 159.

Betz O., Koerner L., Dettner, K. (2018). The Biology of Steninae. In: Betz O., Irmler U., Klimaszewski J. (eds.). *Biology of Rove Beetles (Staphylinidae)*. Springer, Cham. p 229-283.

Bicudo, J. E. P. W., Buttemer, W. A., Chappell, M. A., Pearson, J. T., & Bech, C. (2010). Adaptations: Obtaining and Processing Food. In J. E. P. W. Bicudo, W. A. Buttemer, M. A. Chappell, J. T. Pearson, & C. Bech (Eds.), *Ecological and Environmental Physiology of Birds* (0). Oxford University Press. https://doi.org/10.1093/acprof:oso/9780199228447.003.0004

Birnie-Gauvin, K., Peiman, K. S., Raubenheimer, D., & Cooke, S. J. (2017). Nutritional physiology and ecology of wildlife in a changing world *Conservation Physiology*, *5*(1), cox030. https://doi.org/10.1093/conphys/cox030

Blösch M. 2012: *Digger wasps* – Illustrated catalog of the native species. NBB Scout, 200.

Bogenrieder A. 1993: Methods for the recording of vegetation. Chap. 2.1, 18-29. – In: Mühlenberg M. 1993: *Field Ecology* (3rd ed.). UTB at Quelle & Meyer Publishing

Boreham P.F.L. & Ohiagu C.E. 1978: The use of serology in evaluating invertebrate prey-predator relationships: a review. – *Bull. Entomol. Res.* 68, 171–194

Breton L.M. & Addicott J.F. 1992: Density-dependent mutualism in an aphid-ant interaction. – *Ecology* 73, 2175-2180

Brian M.V. 1952: The structure of a dense natural ant population. – *Journal of Animal Ecology* 21, 12–24

Brian M.V. 1955: Food collection by a Scottish ant community. – *Journal of Animal Ecology* 24, 336-351

Brian M.V. 1958: Interaction between ant populations. – *Proc. 10. int. Congress of Entomology, Montreal 1956*, 2, 781–784

Brian A.D. 1957: Differences in the flowers visited by four species of bumblebees and their causes. – *J. Anim. Ecol.* 26, 71–98

Brian M.V., Hibble J. & Stradling D. 1965: Ant pattern and density in a Southern English heath.v*Journal of Animal Ecology* 34, 544–555

Brian M. V., Hibble J. & Kelly A.F. 1966: The dispersion of ant species in a Southern English heath. – *Journal of Animal Ecology* 35, 281–290

Bristow C.M. 1984: Differential benefits from ant attendance to two species of Homoptera on New York ironweed. – *Journal of Animal Ecology* 53(3), 715-726

Brown R., Ferguson J., Lawrence M. & Leeds D. 1987: *Tracks and signs of the Birds of Britain and Europe*. – Christopher Helm, London, 232 p. (Deutsche Ausgabe 2003 (3rd ed.) AULA-Verlag, Wiesbaden, 336 p.)

Brown R., Ferguson J., Lawrene M. & Lees D. 2003 (3.ed.): *Feathers, Tracks & Signs of European Birds*. A Field Guide. Aula-Verlag Wiebelsheim, 336 p.

Brun, L., Schneider, J., Carrió, E. M., Dongre, P., Taberlet, P., de Waal, van, & Fumagalli, L. (2022). Focal vs. fecal: Seasonal variation in the diet of wild vervet monkeys from observational and DNA metabarcoding data. *Ecology and Evolution*, *12*(10), e9358. https://doi.org/10.1002/ece3.9358

Brunner H. & Coman B.J. 1974: *The Identification of Mammalian Hair*. – Inkata Press, Melbourne, 176 p.

Búrquez A. & Corbet S.A. 1991: Do flowers resorb nectar? – *Functional Ecology* 5, 369–379

Burrows R. 1968: *Wild Fox*. – David & Charles, 20.

Bzuder K.W. & Gypta A.PJ. 1972: Biology of the pavement ant, *Tetramorium caespitum* (Hymenoptera, Formicidae). – *Ann. Entomol. Soc. Am.* 65 (2), 358–367

Cammaerts-Tricot M.C. 1974: Piste et pheromone attractive chez la fourmi *Myrmica rubra*. – *J. comp. Physiol.* 88, 373–382

Carpenter S.R., , Cole J.J., , James R. Hodgson J.R., Kitchell J.F., Pace M.L., Bade D., Cottingham K.L., Essington T.E., Houser J.N. & Schindler D.E. 2001: Trophic Cascades, Nutrients, and Lake Productivity: Whole-Lake Experiments. – *Ecological Monographs* 71(2), 163–186

Cavia, R., Andrade, A., Zamero, M. E., Fernández, M. S., Muschetto, E., Cueto, G. R., & Suárez, O. V. (2008). Hair structure of small rodents from central Argentina: A tool for species identification, *72*(1), 35–43. https://doi.org/10.1515/MAMM.2008.009

Cherrill A. 1989: The diet of the wart biter, *Decticus verrucivorus* (L.). *Bulletin of the British Ecological Society* 20 (2), 115–118

Chiverton P.A. 1984: Pitfall-trap catches of the carabid beetle *Pterostichus melanarius*, in relation to gut contents and prey densities, in insecticide treated and untreated spring barley. – *Entomologia Experimentalis et Applicata* 36(1), 23-30

Ciucci P., Boitani L., Pelliccioni E.R. Rocco, M. & Guy I. 1996: A comparison of scat analysis methods to determine the diet of the wolf *Canis lupus* . – *Wildlife Biology* 2, 37-48

Clark M. 1983: *Mammal Watching*. – Hamlyn Paperbacks edition, Feltham, Middlesex, England, 175 p.

Clausen C.P. 1940: *Entomophagous insects*. – McGraw-Hill, London, 688 p.

Corbet C.B. 1969: *The identification of British Mammals*. – Brit. Mus. (Nat. Hist.)., 46 p.

Corbet S. 2003: Nectar sugar content: estimating standing crop and secretion rate in the field. *Apidologie* 34, 1-10. https://doi.org/10.1051/apido:2002049

Craig C.L. 1988: Insect perception of spider orb webs in three light habitats. – *Functional Ecology* 2, 277–282

Cui J., Zhang Y., Guo J., Wu N. & Zhou Y. 2023: Conflicting selection pressures on seed size and germination caused by carnivorous seed dispersers. – *Integrative Zoology* 18 (5), 799-816

Cummings S. 2024: How ants thwarted lions on the African savanna. Tiny insects have upended a Kenyan ecosystem. – *Science* 25 Jan 2024, doi: https://doi.org/10.1126/science.zrb84bz

Danforth B.N., Minckley R.L. & Neff J.L. 2019: *The Solitary Bees. Biology, Evolution, Conservation.* – Princeton University Press, 488 p.

Das, S., & Das, A. (2023). Ants are more than just curious bystanders to some flowers-they act as significant pollinators. *Frontiers in Insect Science*, *3*, 1145761. https://doi.org/10.3389/finsc.2023.1145761

Davies R.W. 1969: The production of antisera for detecting specific triclad antigens in the gut contents of predators. – *Oikos* 20, 248–260

Day M.G. 1966: Identification of hair and feather remains in the gut and faeces of stoats and weasels. – *J. Zool.* 148, 207–217

Deedrick D.W. & Koch S.L. 2004: Microscopy of Hair Part II: A practical guide and manual for animal hairs, Forensic Science Communications [Online]. Volume 6(3), https://www.researchgate.net/publication/318753108_Microscopy_of_Hair_Part_II_A_Practical_Guide_and_Manual_for_Animal_Hairs

Dempster J.P. 1960: A quantitative study of the predators on the eggs and larvae of the Broom beetle, *Phytodecta olivacea* Förster, using the precipitin test. – *J. Anim. Ecol.* 29, 149–167

Den Boer P.J. 1980: Exclusion or Coexistence and the Taxonomie or Ecological Relationship between Species. – *Netherlands Journal of Zoology* 30(2), 278–306

Denison D.F. & Hodkinson I.D. 1983: Structure of the predatory beetle community in a woodland soil ecosystem. I. Prey selection. – *Pedobiologia* 25, 109-115

Dickman C.R. 1986: A method of censusing small mammals in urban habitats. – *Journal of Zoology (London)* 210A, 631–636

Dickman C.R. 1988: Body Size, Prey Size, and Community Structure in Insectivorous Mammals. – *Ecology* 69 (3), 569–580

Dobrzanska J. 1958: Partition of foraging grounds and modes of conveying of Information among ants. – *Acta Biologiae Experimentalis* XVIII, 55–67

Drewitt E. 2024: *Bird Pellets. A complete photographic guide*. Pelagic Publishing, 258.

Dreyer J.H. 1966: A study of hair morphology in the family Bovidae. – *Onderstepoort J. Vet. Res.*, 33, 379–472

Düll R. & Kutzelnigg H. 1992: *Botanisch-ökologisches Exkursionstaschenbuch.* – Verl. Quelle & Meyer, Heidelberg, Wiesbaden, 546 p.

Dünn D.B., Sharma G.K. & Champbell C.C. 1965: Stomatal patterns of Dicotyledons and Monocotyledons. – *Amer. Midi. Nat.* 74, 185–195

Dyer L.A. & Letourneau D.K. 1999: Trophic cascades in a complex terrestrial community. – *PNAS Biological Sciences* 96 (9) 5072-5076. https://doi.org/10.1073/pnas.96.9.5072

Dyer L.A. & Letourneau D.K. 2013: Can Climate Change Trigger Massive Diversity Cascades in Terrestrial Ecosystems? *Diversity* 5, 479-504; doi:https://doi.org/10.3390/d5030479

Dziurdzik, B. 1973: Key to the identification of hairs of mammals from Poland. – *Acta Zool. Crac.* 18 (4), 73–91

Eberhard W.G. 1986: Effects of orb web geometry on prey interception and retention. In: Shear, W. A. (ed.): *Spiders: Webs, Behaviour, and Evolution*, 70100. – Stanford University Press, Stanford, California, 492 p.

Eisfeld D. 1984: Nahrungsbedarf des Rehwildes. – In: *Rehwild: Biologie, Hege*, 36–40. – Bericht einer Arbeitstagung, Bayer. Staatsministerium f. Ernährung, Landwirtsch. u. Forsten

Eisner T., Alsop R. & Ettershank G. 1964: Adhesiveness of spider silk. – *Science* 146, 1058–1061

Elbroch, M., Marks, E., & Boretos, C. D. (2001). *Bird Tracks & Sign : A Guide to North American Species* (First Edition.). Mechanicsburg, PA: Stackpole Books.

Elkinton J. S. & Liebhold A.M. 1990: Population dynamics of gypsy moth in North America. – *Annual Review of Entomology* 35, 571–596.

Elmes G.W. 1971: An Experimental Study on the Distribution of Heathland Ants. – *J. Anim. Ecol.* 40, 495–499

Else G.R. 1997: *Philanthus triangulum* (Fabr., 1775). - Bess Wasps & Ants Recording Society 2020 UK, https://bwars.com/wasp/crabronidae/philanthinae/philanthus-triangulum

Erfurt J. 2003: Identification of mammalian skulls in nutrition remnants and pellets. *Methods of field ecological mammal research* 2: 471-535, Halle/Saale, in German.

Evans H.E. 1966: *The comparative ethology and evolution of the Sand Wasps.* – Harvard University Press, Campbridge, Mass., 526 p.

Evans H.E. & West-Eberhard M.J. 1970: *The Wasps.* – University of Michigan Press, 265 p.

Ford H.A., Forde N. & Harrington S. 1982: Non- destructive methods to determine the diets of birds. – *Corella* 6, 6-10

Fraigneau C. 2021: *Feathers. An Identification Guide to the Feathers of Western European Birds.* – Lam Pp., 400 p.

Free J. & Butler C.G. 1959: *Bumblebees.* – The New Naturalist. Collins, London, 208 p.

Fründ H.-Ch. 1983: Untersuchungen zur Koexistenz verschiedener Chilopodenarten im Waldboden. Dissertation, Universität Würzburg, 164 p.

Funke W. & Weidemann G. 1971: O. Food and Energy Turnover of Phytophagous and Predator Arthropods.

Methods Used to Study Energy Flow. – *Ecol. Studies. Anal, and Synthesis* 2, 101–109

Gandar M.V. 1982: The Dynamics and Trophic Ecology of Grasshoppers (Acridoidea) in a South African Savanna. – *Oecologia* 54, 370–378

Ghazoul J. & Sheil D. 2017: *Tropical Rain Forest Ecology, Diversity, and Conservation.* – Oxford University Press, 516 p.

Gilbert F.S. 1981: Foraging ecology of hoverflies: morphology of the mouthparts in relation to feeding on nectar and pollen in some common urban species. – *Ecological Entomolgy* 6, 245–262

Gilbert F.S., Haines N. & Dickson K. 1991: Empty flowers. – *Functional Ecology* 5, 29–39

Gionfriddo J.P., Best L.B. & Giesler B.J. 1995: A Saline-Flushing Technique for Determining the Diet of Seed-Eating Birds. – *The Auk* 112 (3), 780-782

Glück E. 1985: Seed preference and energy intake of Goldfinches *Carduelis carduelis* in the breeding season. – *Ibis* 127, 1–9

Gokcezade J.F., Gereben-Krenn B.-A. & Neumayer J. 2018: *Field identification key for the bumblebees of Germany, Austria and Switzerland.* (2nd ed.). – Quelle & Meyer Publishing Wiebelsheim, 55 p.

Goulson D. & Sparrow K.R. 2008: Evidence for competition between honeybees and bumblebees; effects on bumblebee worker size. – *Journal of Insect Conservation* 13(2), 177-181

Gratton C. and Forbes A.E. 2006: Changes in 13 C stable isotopes in multiple tissues of insect predators fed isotopically distinct prey *Oecologia* 147, 615-624

Grimm-Seyfarth A., Zarzycka A. & Klenke R. 2017: Advantages of scat detection dogs: A case study from related mustelid species with identical diet. – 11 th International Conference on Behaviour, Physiology, and Genetics of Wildlife, *Wildlife Detection Dog Workshop*, Berlin, 10-13

Groom M.J., Meffe G.K. & Carroll C.R. 2006: *Principles of conservation biology* (3rd ed.). – Sinauer Associates, Inc. Sunderland, Massachusetts U.S.A., 793 p.

Hager, S. B., & Cosentino, B. J. (2006). An Identification Key to Rodent Prey in Owl Pellets from the Northwestern & Southeastern United States: Incisor Size To Distinguish Among Genera. *The American Biology Teacher*, *68*(9), 135-144. https://doi.org/10.1662/0002-7685(2006)68[135:AIKTRP]2.0.CO;2

Hangartner W. and Bernstein S. 1964: Über die Geruchsspur von *Lasius fuliginosus* zwischen Nest und Futterquelle. – *Experientia* 20, 392–393

Hansen W. & Oelke H. 1978: Bestimmungsbuch für Rupfungen und Mauserfedern. Teil 1 (6) u. (7). – *Beitr. Naturkunde Niedersachsen* 31, 53–121

Hansson, L. (2001). Diet Selection in Wild Animals. In J. B. Owen, J. L. Treasure, & D. A. Collier (Eds.), *Animal Models — Disorders of Eating Behaviour and Body Composition* (pp. 69–82). Dordrecht: Springer Netherlands. https://doi.org/10.1007/978-94-015-9662-6_4

Harder L.D. 1985: Morphology as a Predictor of Flower Choice by Bumblebees. – *Ecology* 66 (1), 198-210

Hayashida M. 2003: Seed dispersal of Japanese stone pine by the Eurasian Nutcracker. – *Ornithol. Sci* . 2, 33-40

Heiniger P.H. 1991: Adaptation strategies of the Snowfinch *Montifringilla nivalis* to the extreme environmental conditions of the high mountains. – *The Ornithological Observer* 88:193-207.

Heinrich B. 1976: The Foraging Specializations of Individual Bumblebees. – *Ecological Monographs* 46 (2), 105-128

Heinrich B. 1979: „Majoring" and „Minoring" by Foraging Bumblebees, *Bombus vagans:* An Experimental Analysis. – *Ecology* 60 (2), 245–255

Heleno R.H., Devoto, M. & Pocockc M. 2012: Connectance of species interaction networks and conservation value: Is it any good to be well connected? – *Ecological Indicators* 14, 7- 10. DOI: https://doi.org/10.1016/j.ecolind.2011.06.032

Hess C.A. 1997: Stomach Flushing: Sampling the Diet of Red-Cockaded Woodpeckers. – *The Wilson Bulletin* 109 (3), 535-539

Hirsch B.T., Kays R. & Jansen P.A. 2012: A telemetric thread tag for tracking seed dispersal by scatter-hoarding rodents. – *Plant Ecology* 213, 933-943

Höcherl, N., and J. Tautz. 2015. Nesting behavior of the paper wasp *Polistes dominula* in Central Europe – a flexible system for expanding into new areas. Ecosphere 6(12):262. https://doi.org/10.1890/ES15-00254.1

Hölldobler B. 1979: Territoriality in Ants. – *Proc. Amer. Philosoph. Soc.* 123, 211–218

Hölldobler B. & Wilson E.O. 1990: *The Ants* . – Springer-Verlag, 732 p.

Hoppe-Dominik B. 1984 : Etude du spectre des proies de la panthère, *Panthera pardus,* dans le Parc National de Tai en Côte d' Ivoire. – *Mammalia* 48 (4), 477–487

Howe H.F. & Westley L.C. 1988: *Ecological Relationships of Plants and Animals.* – Oxford University Press, New York, Oxford, 273 p.

Husson A.M. 1962: Flet determinera van Zoogdieren in Braakballen van Uilen (The determination of mammals in the pellets of owls). – *Zool. Bij. Uit. Rij. Mus. Nat. Hist. Leid.* 5, 1–63

Inouye D.W. 1978: Resource partitioning in bumblebees: experimental studies of foraging behavior. – *Ecology* 59 (4), 672-678

Jacobs J. 1985: Konkurrenz und Einnischung – Hat Konkurrenz um Ressourcen eine evolutionsbiologische Bedeutung für die Artenmannigfaltigkeit der Tiere. *Z. Zool Syst. Evolut.-forsch.* 23, 243–258

Jacobs W. & Renner M. 1988: *Biologie und Ökologie der Insekten* (2nd ed.). Gustav Fischer Stuttgart, New York, 690 p.

Jones C.E. & Little R.J. (eds.) 1983: *Handbook of Experimental Pollination Biology.* Van Nostrand Rheinhold, New York, 558 p.

Judd TM, Teal PE, Hernandez EJ, Choudhury T, Hunt JH. Quantitative differences in nourishment affect caste-related physiology and development in the paper wasp *Polistes metricus*. PLoS One 2015 Feb 23;10(2): e0116199. doi: https://doi.org/10.1371/journal.pone.0116199. PMID: 25706417; PMCID: PMC4338145.

Kamara D.N., Palmer T.M., Riginos C. & 2024: A New Ant Species on the Savannah Has Taken the Lions' Treecover. *Technology Networks Applied Sciences* Jan 26, 2024

Kamaru D.N., Palmer T.M., Riginos C., Ford A.T., Belnap J., Chira R.M., Githaiga J.M., Gituku B.C., Hays B.R., Kavwele C.M., Kibungei A.K., Lamb C.T., Maiyo N.J., Milligan P.D., Mutisya S., Ng'weno C.C., Ogutu M., Pietrek A.G., Wildt B.T. & Goheen J.R. 2024: Disruption of an ant-plant mutualism shapes interactions between lions and their primary prey. – *Science* 383, Issue 6681, 433-438. https://doi.org/10.1126/science.adg1464

Kaufmann T. 1965: Biological Studies on some Bavarian Acridoidea (Orthoptera), with special reference of their feeding habits. *Ann. Ent. Soc. Am.* 58, 791–801

Keiluhu H.J., Pangau-Adam M.Z., Maury H.K. and Waltert M. 2019: Effects of anthropogenic disturbance on a Victora crowned pigeon *Goura victoria* population in northern Papua, Indonesia, *Journal of Asia-Pacific Biodiversity* 12(4), 493-497

Klare U., Kamler J.F. & Macdonald D.W. 2011: A comparison and critique of different scat-analysis methods for determining carnivore diet. – *Mammal Review* 41(4), 294-312

Kochanek H.M. 1990: Ernährung des Turmfalken *(Falco tinnunculus):* Ergebnisse von Nestinhaltsanalysen und automatischer Registrierung. – *J. Orn.* 131 (3), 291–304

Koehler H. 1977: Nahrungsspektrum und Nahrungskonnex von *Pterostichus oblongopunctatus* (F.) und *Pterostichus metallicus* (F.) (Coleoptera, Carabidae). – *Verh. Ges. f. Ökol., Göttingen* 1976, 103–111

Koerner L., Gorb, S. N., Betz, O. (2012). Functional morphology and adhesive performance of the stick-capture apparatus of the rove beetles Stenus spp. (Coleoptera, Staphylinidae). *Zoology* 115(2): 117-127. https://doi.org/10.1016/j.zool.2011.09.006.

Koerner, L., Garamszegi, L. Z., Heethoff, M., Betz, O. (2017). Divergent morphologies of adhesive predatory mouthparts of Stenus species (Coleoptera: Staphylinidae) explain differences in adhesive performance and resulting prey-capture success. *Zoological Journal of the Linnean Society,* 181(3): 500-518. https://doi.org/10.1093/zoolinnean/zlx006

Kratochwil A. 1984: Pflanzengesellschaften und Blütenbesucher-Gemeinschaften: biozönologische Untersuchungen in einem nicht mehr bewirtschafteten Halbtrockenrasen (*Mesobrometum*) im Kaiserstuhl (Südwestdeutschland). – *Phytocoenologia* 11 (4), 455–669

Kratochwil A. 1985: Beobachtungen zur Blütenbesucher-Gemeinschaft (Hymenoptera, Lepidoptera, Diptera, Coleoptera) eines aufgelassenen Halbtrockenrasens im Kaiserstuhl (Südbaden) – ein Beitrag zur Bedeutung brachliegender Wiesen für den Naturschutz. – *Mitt. dtsch. Ges. allg. angew. Ent.* 4, 453–456

Kratochwil A. 1988: Zur Bestäubungsstrategie von *Pulsatilla vulgaris* MILL. – *Flora* 181, 261–324

Kratochwil A. & Kohl A. 1988: Pollensammel-Präferenzen bei Hummeln - ein Vergleich mit der Honigbiene. – *Mitt. bad. Naturkunde u. Naturschutz,* N. F. 14, 697–715

Kristin A. 1989: Diet of syntopic species Chiffchaff (*Phylloscopus collybita*) and dunnock (*Prunella modularis*). *Folia Zoologica -Praha* - 38(4), 349-362

Kruuk H. 1989: *The Social Badger.* Ecology and Behaviour of a Group-living Carnivore *(Meles meles).* – Oxford University Press, Oxford, New York, Tokyo, 155 p.

Kugler H. 1943: Hummeln als Blütenbesucher. – *Ergebn. d. Biologie* , 143–323

Kugler H. 1970: *Blütenökologie.* 2. Aufl. – G. Fischer-Verl. Stuttgart, 345 p.

Kunze E. 1971a: Mikroskopie der Säugetierhaare I. – *Mikrokosmos* 60 (6), 182–184

Kunze E. 1971b: Mikroskopie der Säugetierhaare II. – *Mikrokosmos* 60 (11), 346–348

Kunze E. 1972: Mikroskopische Untersuchungen an Säugetierhaaren III. – *Mikrokosmos* 61 (4), 123–127

Laitat E. 1983: Die mikrographische Losungsbestimmung zur Bestimmung der Nahrung von Rotwild und Rehwild. – *Z. Jagdwiss.* 29, 218–235

Levings S.C. & Traniello J.F.A. 1981: Territoriality, Nest Dispersion, and Community Structure in Ants. – *Psyche* 88, 265–319

Li Z., Lu J., Shi X., Duo L. Smith J.L.D. & Wang T. 2024: Competitive interactions and coexistence of sympatric flagship carnivores in Asia. – *Integrative Zoology* 19 (2), 183-199. https://doi.org/10.1111/1749-4877.12724

Liere H., Kim T.N., Werling B.P., Meehan T.D., Landis D.A. & Gratton C. 2015: Trophic cascades in agricultural landscapes: indirect effects of landscape composition on crop yield. – *Ecological Applications* 25 (3), 652-661

Link H. 1986: Untersuchungen am Habicht *(Accipiter gentilis).* – Habitatwahl, Ethologie, Populationsökologie. – *Deutscher Falkenorden (DFO) – Schriftenreihe Heft 2*, Nürnberg, 95 p.

Lockie J.D. 1959: The estimation of the food of foxes. – *J. Wildl. Manage.* 23 (2), 224–227

Loreau M. 1983: La regime alimentaire de *Abaxater* Vill. (Coleoptera, Carabidae). – *Acta Oecologica/Oecol. Gener.* 4, 253–263

Loreau M. 1986: Niche differentiation and community organization in forest Carabid beetles. – In: den Boer et al. (eds.): *Carabid Beetles*, 465–487. – G. Fischer, Stuttgart, New York, 551 p.

Loverty T.M. & Plowright R.C. 1988: Flower handling by bumblebees: a comparison of specialists and generalists. – *Animal Behaviour* 36 (3), 733-740

Lu C-H 2006: Roles of animals in seed dispersal of *Pinus* : A review. – *Chinese Journal of Ecology* 25(5), 557-562

Lundberg P. 1987: Breeding seasons of North Scandinavian Starlings (*Stumus vulgaris*): constrained by food or time? – *Journal of Animal Ecology* 56, 847–855

Mac Kenna M.A. & Thompson J.D. 1988: A Technique for Sampling and Measuring Small Amounts of Floral Nectar. – *Ecology* 69 (4), 1306–1307

Mader, E., Spivak, M., & Evans, E. (2010). *Managing Alternative Pollinators: A Handbook for Beekeepers, Growers and Conservationists*. Ithaca, NY and College Park, MD: Sustainable Agriculture Research and Education (SARE) and Natural Resource, Agriculture, and Engineering Service (NRAES). https://www.sare.org/resources/managing-alternative-pollinators/. Accessed 9 July 2024

Major R.E. 1990: Stomach Flushing of an Insectivorous Bird - an Assessment of Differential Digestibility of Prey and the Risk to Birds. – *Australian Wildlife Research* 17(6) 647-657

Malyshev S.I. 1968: *Genesis of the Hymenoptera and the phases of their evolution*. – Methuen & Co. Ltd., London, 319 p.

Mandelli G. 1961: Research on the food habits of foxes (*Vulpes vulpes* L.) in Gran Paradiso National Park: Preliminary investigations on the macroscopic characteristics of the hairs of their principal prey. – *Clin. Vet.* 83 (8), 235–243

Manning A. 1956: Some aspects of the foraging behaviour of bumblebees. – *Behaviour* 9, 164–201

Marone, L., Cueto, V. R., Lopez de Casenave, J., Zarco, A., & Camín, S. R. (2022). Plausible causes of seed preferences and diet composition in seed-eating passerines. *Journal of Avian Biology*, *2022*(3), e02875. https://doi.org/10.1111/jav.02875

Martin, T.E. 1987: Food as a limit on breeding birds: A life-history perspective. – *Ann. Rev. Ecol. Syst.* 18, 453–487

März R. 1987: *Gewöll- und Rupfungskunde*. – 3. neu von K. Banz bearbeitete Aufl. – Akademie-Verl. Berlin, 398 p.

Maschwitz U. & Mühlenberg M. 1975: On the hunting strategy of some oriental *Leptogenys* species (Formicidae: Ponerinae). – *Oecologia (Berl.)* 20, 65-83

May R.M. 1986: The search for Patterns in the Balance of Nature: Advances and Retreats. *Ecology* 67 (5), 1115–1126

McConkey K.R., Nathalang A., Brockelman W.Y., Saralamba C., Santon J., Matmoon U., Somnuk R. & Srinoppawan K. 2018: Different megafauna vary in their seed dispersal effectiveness of the megafaunal fruit *Platymitra macrocarpa* (Annonaceae). – *PLoS ONE* 13(7): e0198960, 18 p.

McKenney E.A. 2023: *Applied Ecology*. Chap. 10: Food Webs. – Raleigh, North Carolina State University. DOI:https://doi.org/10.52750/147069

Mee, L., & Barribeau, S. M. (2023). Influence of social lifestyles on host-microbe symbioses in the bees. *Ecology and Evolution*, *13*(11), e10679. https://doi.org/10.1002/ece3.10679

Meyer, W., Pohlmeyer, K., Schnapper, A., & Hülmann, G. (2001). Subgroup differentiation in the Cervidae by hair cuticle analysis. *Zeitschrift für Jagdwissenschaft*, *47*(4), 253–258. https://doi.org/10.1007/BF02240525

Meyer W., Schnapper A. & Hülmann G. 2002: The hair cuticle of mammals and its relationship to functions of the hair coat. – *Journal of Zoology* 256 (4), 489-494; Online version: 28 February 2006 https://doi.org/10.1017/S0952836902000535

Michener R.H. and Lajtha K. (eds.) 2007: *Stable isotopes in ecology and environmental science.* (2nd ed.) Blachwell, Oxford, 594 p.

Miller M.C., Chappell W.A., Gable W. & Bridges J.R. 1979: Evaluation of immunodiffusion and immunoelectrophoretic test using a broad spectrum anti-adult southern pine beetle serum. – *Ann. Entomol. Soc. Am.* 72, 99–104

Montgomery W.I. 1981: A removal experiment with sympatric populations of *Apodemus sylvaticus* (L.) and *A. flavicollis* (Melchior) (Rodentia: Muridae). – *Oecologia* 51, 123–132

Morrison M.L., Ralph J., Verner J. & Jehl Jr J.R. (eds.) 1990: *Avian Foraging: Theory, Methodology, and Applications*. – Studies in Avian Biology 13, 1-512. Allen Press, Inc., Lawrence, Kansas

Morse D.H. 1977a: Resource partitioning in bumblebees: the role of behavioral factors. – *Science* 197 (4304), 678-680

Morse D.H. 1977b: Estimating Proboscis Length from Wing Length in Bumblebees (*Bombus* spp.). – *Ann. Entomol. Soc. America* 70, 311–315

Morse D.H. 1978: Size-related foraging differences of bumble-bee workers. – *Ecological Entomology* 3, 189–192

Murie A. 1936: Following fox trails. – *Univ. Michigan, Mise, Publ.* 32, 1–45

Nagy, K. (2001). Food requirements of wild animals: predictive equations for free-living mammals, reptiles, and birds. https://escholarship.org/uc/item/18s7d943. Accessed 16 August 2024

Nagy, K. A., Girard, I. A., & Brown, T. K. (1999). Energetics of free-ranging mammals, reptiles, and birds. *Annual Review of Nutrition*, *19*(Volume 19, 1999), 247–277. https://doi.org/10.1146/annurev.nutr.19.1.247

Nentwig W. 1982a: Epigeic Spiders, Their Potential Prey and Competitors: Relationship Between Size and Frequency. *Oecologia* (Berl.) 55, 130–136

Nentwig W. 1982b: Nicht jedes Insekt geht der Spinne als Beute ins Netz. Der Vorteil hochentwickelter Augen und Reflexe. – *Mitteilungen der DFG, Forschung* (4), 1982, 6–8

Newton I. 1980: The role of food in limiting bird numbers. – *Ardea* 68, 11–30

Nield A.P., Enright N.J. & Ladd P.G. 2015: Study of seed dispersal by emu (*Dromaius novaehollandiae*) in the Jarrah (*Eucalyptus marginata*) forests of south-western Australia through satellite telemetry. *Emu 115*, 29–34

Nilsson S.G. 1987: Limitation and regulation of population density in the nuthatch *Sitta europaea* (Aves)

breeding in natural cavities. – *Journ. Anim. Ecology* 56, 921–937

Nyffeler M. 1982: Field studies on the ecological role of the spiders as insect predators in agroecosystems (abandoned grassland, meadows, and cereal fields). – Diss., ETH Nr. 7097, Swiss Federal Institute of Technology, Zürich, 174 p.

Nyffeler M. & Benz G. 1979: Zur ökologischen Bedeutung der Spinnen der Vegetationsschicht von Getreide- und Rapsfeldern bei Zürich (Schweiz). – *Z. angew. Ent.* 87, 348–376

Nyffeler M. & Benz G. 1981a: Einige Beobachtungen über die Flucht von adulten Lepidopteren aus den Netzen orbiteler Spinnen. – *Anz. Schädlingskde., Pflanzenschutz, Umweltschutz* 54, 113–114

Nyffeler M. & Benz, G. 1981b: Freilanduntersuchungen zur Nahrungsökologie der Spinnen: Beobachtungen aus der Region Zürich. – *Anz. Schädlingskde., Pflanzenschutz, Umweltschutz* 54, 33–39

Ohiagu C.E. & Boreham P.F.L. 1978: A simple field test for evaluating insect prey-predator relationships. – *Ent. exp. & appl. (Amsterdam)* 23, 40–47

Olberg G. 1959: *Das Verhalten der solitären Wespen Mitteleuropas.* – VEB Deutscher Verlag der Wissenschaften, Berlin, 401 p.

Olive C. 1980: Foraging specializations in orb-weaving spiders. – *Ecology* 61, 1133–1144

Olson L.-H. 2013: Tracks and Signs of the Animals and Birds of Britain and Europe. Princeton University Press, 273 p.

Orłowski, G., Wuczyński, A., Karg, J., & Grzesiak, W. (2017). The significance of seed food in chick development re-evaluated by tracking day-to-day dietary variation in the nestlings of a granivorous passerine. *Ibis, 159*(1), 124–138. https://doi.org/10.1111/ibi.12410

Ortmann S., Bradley B.J., Stolter C. & Ganzhorn J.G. 2006: Estimating the quality and composition of wild animal diets – acritical survey of methods, 395-418. – In: Hohmann G., Robbins M.M. & Boesch C. (eds): *Feeding Ecology in Apes and Other Primates.* Ecological, Physical and Behavioral Aspects. – Cambridge University Press, 540 p.

Owaga M.L. 1977: Comparison of analysis of stomach contents and faecal samples from Zebra. – *E. Afr. Wildl. J.* 15, 217–222

Oyer E.R. 1946: Identification of mammals from studies of hair structure. *Trans. Kansas Acad. Sci* 49 (2), 155–160

Pangau-Adam M. and Muehlenberg M. 2014a: Palm species in the diet of the Northern Cassowary (*Casuarius unappendiculatus*) in Jayapura, Papua, Indonesia, *Palms* 58 (1): 19-26.

Pangau-Adam M. and Muehlenberg M. 2014b: Dispersal of Terminalia seeds by the Northern cassowary in the lowland forest of Papua, Indonesia. – *Asian Journal of Conservation Biology* Dec 2014 Vol. 3 No. 2.

Pekkarinen A. 1976: Adaptation of the Mouthparts in Bumblebees. – *Zool. Scr.* 5, 189

Peterson B.J. and Fry B. 1987: Stable isotopes in ecosystem studies *Annual Review of Ecology and Systematics* 18, 293-320

Peterson R., Mountfort G. & Hollom P.A.D. 1985: *Die Vögel Europas.* - 14. Aufl. – Parey-Verl. Hamburg; Berlin, 535 p. – also 2001: *A Field Guide to the Birds of Britain and Europe.* - Houghton Mifflin, 261 p.

Pleasants J.M. & Zimmermann M. 1979: Patchiness in the Dispersion of Nectar Resources. – *Oecologia* (Berl.) 41 (3), 283–288

Pollen and Spores Image Database of the Georg-August-University Göttingen, n.d. Dept. Palynology and Climate Dynamics, Collections and Databases, with over 2300 specimens from all over the world.

Ponsard S. and Arditi R. 2000: What can stable isotopes 15 N and 13 C tell about the food web of soil macro-invertebrates? *Ecology* 81, 852-864

Potapov, A. M., Rozanova, O. L., Semenina, E. E., Leonov, V. D., Belyakova, O. I., Bogatyreva, V. Yu., et al. (2021). Size compartmentalization of energy channeling in terrestrial belowground food webs. *Ecology, 102*(8), e03421. https://doi.org/10.1002/ecy.3421

Potapov, A. M., Drescher, J., Darras, K., Wenzel, A., Janotta, N., Nazarreta, R., et al. (2024). Rainforest transformation reallocates energy from green to brown food webs. *Nature, 627*(8002), 116–122. https://doi.org/10.1038/s41586-024-07083-y

Proctor M. & Yeo P. 1973: *The Pollination of Flowers.* Collins, London, 418 p.

Prys-Jones O.E. & Corbet S.A. 1987: *Bumblebees.* – Naturalist's Handbooks 6. – Beimont press, Northampton, 86 p.

Pyke G.H. 1978: Optimal Foraging in Bumblebees and Coevolution with their Plants. – *Oecologia* 36, 281–293

Pyke G.H. 1980: Optimal Foraging in Bumblebees: Calculation on Net Rate of Energy Intake and Optimal Patch Choice. – *Theoretical Population Biology* 17, 232–246

Rader, R., Cunningham, S. A., Howlett, B. G., & Inouye, D. W. (2020). Non-Bee Insects as Visitors and Pollinators of Crops: Biology, Ecology, and Management. *Annual Review of Entomology, 65*, 391–407. https://doi.org/10.1146/annurev-ento-011019-025055

Radl G. 1987: Jahreszeitliche Änderungen in der Nahrungswahl dreier verschieden großer Antilopenarten im Comoé-Nationalpark (Elfenbeinküste) – Eine Analyse über Kotproben. – Diplomarbeit, Universität Würzburg, 85 p.

Raine, N. E., & Rundlöf, M. (2024). Pesticide Exposure and Effects on Non-Apis Bees. *Annual Review of Entomology, 69*, 551–576. https://doi.org/10.1146/annurev-ento-040323-020625

Rands, S. A., Whitney, H. M., & Hempel de Ibarra, N. (2023). Multimodal floral recognition by bumblebees. *Current Opinion in Insect Science, 59*, 101086. https://doi.org/10.1016/j.cois.2023.101086

Real L.A. (1981): Uncertainty and pollination-plant interactions: The foraging behaviour of bees and wasps on artificial flowers. – *Ecology* 62, 20–26

Real L. and Caraco Th. 1986: Risk and Foraging in Stochastic Environments. - *Ann. Rev. Ecol. Syst.* 17, 371–390

Reynolds J.C. & Aebischer N.J. 1991: Comparison and quantification of carnivore diet by faecal analysis: a critique, with recommendations, based on a study of the Fox *Vulpes vulpes*. – *Mammal Review* 21(3), 97-122

Riechert S.E. & Harp J.M. 1987: Nutritional Ecology of Spiders. – In: Slansky, F. & Rodriguez, J. G. (eds.): *Nutritional Ecology of Insects, Mites, Spiders and Related Invertebrates*, Chap. 20, 645–672. – John Wiley & Sons, New York, 1016 p.

Riechert S.E. & Luczak J. 1982: Spider foraging: behavioral responses to prey. – In: Witt, P. N. & Rovner, J. S. (eds.): *Spider Communication: Mechanisms and Ecological Significance* 353–385. – Princeton Univ. Press, Princeton, N. J., 440 p.

Robbins C.T. 1983: *Wildlife feeding and nutrition*. – Academic Press, New York, 356 p.

Ryder M.L. 1973: *Hair*. – Edward Arnold, 56 p.

Rypstra A. 1982: Building a better insect trap: an experimental investigation of prey capture in a variety of spider webs. – *Oecologia* 52, 31–36

Schaefer M. 1978 a: Some Experiments on the Regulation of Population Density in the Spider *Florinia bucculenta* (Araneida: Linyphiidae). – *Symp. Zool. Soc. Lond.* 42, 203–210

Schaefer M. 2018: Brohmer: *Fauna von Deutschland.* (25th ed.) – Quelle & Meyer Verlag Wiebelheim, 766 p.

Scherber C., Eisenhauer N., Weisser W., Schmid B., Voigt W., Fischer M., Schulze E.-D., Roscher C., Weigelt A., Allan E., Bessler H., Bonkowski M., Buchmann N., Buscot C., Clement L.W., Ebeling A., Engels C., Halle S., Kertscher I., Klein A.M., Koller R., Koenig S., Kowalski E., Kummer V., Kuu A., Lange M., Lauterbach D., Middelhoff C., Migunova V.D., Milcu A., Mueller R., Partsch S., Petermann J.S., Renker C., Rottstock T., Sabais A., Scheu S., Schumacher J., Temperton V.M. & Tscharntke T. 2010: Bottom-up effects of plant diversity on multitrophic interactions in a biodiversity experiment. – *Nature* 468, 553–556. https://doi.org/10.1038/nature09492

Scherney F. 1959: *Unsere Laufkäfer*. Kosmos-Verlag, 79.

Schmid-Hempel P. 1987: Efficient Nectar-Collecting by Honeybees. I. Economic Models. - J. Anim. Ecol. 56(1), 209–218

Scheu S. and Falca M. 2000: The soil food web of two beech foersts (Fagus sylvatica) of contrasting humus type: stable isotope analysis of a macro- and mesofauna-dominated community *Oecologia* 123, 285-296

Schoener T.W. 1983: Field Experiments on Interspecific Competition. – *The American Naturalist* 122 (2), 240–285

Schremmer F. 1941: Sinnesphysiologie und Blumenbesuch des Falters von *Plusia gamma* L. – *Zool. Jahrb.* 74 (5/6), 361–522

Schremmer F. 1949: *Die Wiese als Lebensgemeinschaft*. – Sammlung "Biologie". – Verl. Hollinek, Wien, 107 p.

Schremmer F. 1953: Blütenbiologische Beobachtungen an Labiaten (Nektar-und Pollendiebstahl). – *Österr. Botan. Zeitschr.* 100 (1/2), 8–24

Schremmer F. 1962a: Morphologische Anpassungen von Tieren – insbesondere Insekten – an die Gewinnung von Blumennahrung. – *Verh. Dtsch. Zool. Ges. Saarbrücken u. Leipzig*, 1961, 375–401

Schremmer F. 1962b: *Wasps and Hornets*. – The New Brehm Library 298, Kosmos Publishing Stuttgart, 104 p.

Schremmer F. 1973: *Blumen und Insekten*. – In: Grzimeks Tierleben. Ergänzungsband: Unsere Umwelt als Lebensraum. – Kindler Verlag, 99–132

Scotcher J.S.B. 1979: A review of faecal analysis techniques for determining the diet of wild grazing herbivores. – *Proc. Grassld. Soc. Sth. Afr.* 14, 131–136

Seifert L.I. & Scheu S. 2012: Linking aquatic and terrestrial food webs – Odonata in boreal systems. *Freshwater Biology*, doi: https://doi.org/10.1111/j.1365-2427.2012.02807.x, Blackwell Publ. Ltd., 9 p.

Sequeira D.M. 1980: Comparison of the Diet of the Red Fox (*Vutpes vulpes* L., 1758) in Gelderland (Holland), Denmark and Finnish Lapland. – In: Biogeographica 18: *The Red Fox*, ed. by E. Zimen. – Dr. W. Junk Publishers, The Hague, 35–51

Shannon, D. M., Richardson, N., Lahondère, C., & Peach, D. (2024). Mosquito Floral Visitation and Pollination. *Current Opinion in Insect Science*, 101230. https://doi.org/10.1016/j.cois.2024.101230

Simpson, S. J., & Raubenheimer, D. (2012). Applied Nutrition. In S. J. Simpson & D. Raubenheimer (Eds.), *The Nature of Nutrition: A Unifying Framework from Animal Adaptation to Human Obesity* (0). Princeton University Press. https://doi.org/10.23943/princeton/9780691145655.003.0009

Sinclair A.R E. 1975: The resource limitation of trophic levels in tropical grassland ecosystems. – *Journal of Animal Ecology* 44, 497–520

Siuda A., Zurowski W. & Siuda H. 1969: The food of the roe deer. – *Acta Theriol.* 14(18), 247-262

Skinner G.J. & Whittaker J.B. 1981: An experimental investigation of inter-relationships between the woodant (*Formica rufa*) and some tree-canopy herbivores. – *Journal of Animal Ecology* 50(1), 313-326

Southwood T.R.E. & Henderson P.A. 2000: *Ecological Methods* (3 rd ed.). – Blackwell Science, 575 p.

Sowig P. 1991: Die Erfassung und Analyse von Blütenbesucher-Gemeinschaften dargestellt am Beispiel der Hummeln (Hymenoptera: Apidae: Bombinae). – *Beihefte zu den Verhandlungen der Gesellschaft für Ökologie* 2, 129–144

Stains H.J. 1958: Field guide to guard hairs of Middle Western furbearers. – *J. Wild. Manage.* 22, 95–97

Steiner G. 1963: *Das Zoologische Laboratorium*. – E. Schweizerbart'sche Verlagsbuchhandlung, Stuttgart, 557 p.

Stewart D.R.M. 1965: The epidermal characters of grasses, with special reference to East African plain species. – *Bot. Jb.* 84 (1), 63–116

Stewart D.R.M. 1967: Analysis of plant epidermis in feaces: A technique for studying the food preferences of grazing hervibores. – *J. Appl. Ecol.* 4 (1), 83–111

Stoner KE, Riba-Hernandez P, Vulinec K. & Lambert JE. 2007: The role mammals in creating and modifying seedshadows in tropical forests and some possible consequences of their elimination. – *Biotropica* 39, 316-327

Stoves J.L. 1944: Morphology of mammalian hair. – *Nature Lond.* 153, 285 p.

Stowe M. 1986: Prey specialization in the Araneidae. – In: Shear W.A. (ed.): *Spiders, Webs, Behavior and Evolution*, 101–131. – Stanford University Press, Stanford, California, 492 p.

Strickler K. 1979: Specialization and Foraging Efficiency of Solitary Bees. – *Ecology* 60 (5), 998–1009

Svensson L. 1970: *Identification guide to European passerines*. – K. L. Bechman's Trykeri, 152 p.

Szlep R. & Jakobi T. 1967: The mechanism of recruitment to mass foraging in colonies of *Monomorium venustum* Smith, M. subopacum ssp. *phoenicum* Em., *Tapinana israelis* For. and *T. simothi v. phoenicum* Em. – *Insects Sociaux* 14 (1), 25–40

Taberlet P., Bonin A., Zinger L. & Coissac E. 2018: *Environmental DNA: For Biodiversity Research and Monitoring*. – Oxford University Press, 272 p.

Teerink, B. J. (1991). *Hair of West European Mammals: Atlas and Identification Key*. Cambridge England ; New York: Cambridge University Press.

Teräs J. 1976: Flower visits of bumblebees, *Bombus* latr. (Hymenoptera, Apidae) during one summer. – *Ann Zool. Fenn.* 13, 200–232

The Feather Atlas. (2023) Flight Feathers of North American Birds. – U.S. Fish & Wildlife Service. Department of the Interior | USA.gov.

Theuerkauf J., Waitkuwait WE., Guiro Y., Ellenberg H. & Porembski S. 2000: Diet of forest elephants and their role in seed dispersal in the Bossematié Forest Reserve, Ivory Coast. – *Mammalia* 64 (2), 447-459

Timmermann U. & Becker N. 2003: The impact of mosquito control on the diet of floodplain-dwelling bird species. – *Carolinea*, 61: 145-165, Karlsruhe.

Tiralla N. 2013: Feeding Ecology of the Mongolian Wolf (*Canis lupus*) in the Khentii Mountains, Mongolia. – Master thesis at the Georg-August-University Göttingen, 74 p.

Toldt K. 1935: *Aufbau und natürliche Färbung des Haarkleides der Wildsäugetiere*. – Deutsche Gesellschaft für Kleintier und Pelztierzucht GmbH. & Company, 291.

Tsuji, J., & Coe, L. (2014). Effects of Foliage Color on the Landing Response of Pieris rapae (Lepidoptera: Pieridae). Environmental Entomology, 43(4), 989–994. https://doi.org/10.1603/EN14084

Tsuji, J., Logan, T., & Russo, A. (2018). A Hierarchy of Cues Directs the Foraging of Pieris rapae (Lepidoptera: Pieridae) Larvae. Environmental Entomology, 47(6), 1485–1492. https://doi.org/10.1093/ee/nvy124

U.S. Fish & Wildlife Service, Forensics Laboratory. (2023, April 17). The Feather Atlas - Feather Identification and Scans - U.S. Fish and Wildlife Service Forensics Laboratory. https://www.fws.gov/lab/featheratlas/. Accessed 7 July 2024

Ueckermann E. 1971: *Die Fütterung des Schalenwildes*. – Parey, Hamburg, 89.

Van Leeuwen C.H.A., Villar N., Mendoza Sagrera I., Green A.J., Bakker E.S., Soons M.B., Galetti M., Jansen P.A., Nolet B.A. & Santamaria L. 2022: A seed dispersal effectiveness framework across the mutualism-antagonism continuum. – *Oikos* , e09254, 13 p.

Van Toor *M.L., O'Mara M.T., Abedi-Lartey M., Wikelski M., Fahr J., Dina K.N. and Dechmann D.K.N. 2019:* Linking colony size with quantitative estimates of ecosystem services of African fruit bats. – *Current Biology* 29 (7), R237-R238 https://doi.org/10.1016/j.cub.2019.02.033

Vanbergen A.J., Woodcock B.A., Heard M.S. & Chapman D.S. 2017: Network size, structure and mutualism dependence affect the propensity for plant–pollinator extinction cascades. – *Functional Ecology* 31, 1285–1293

Von Hagen E. 1994: *Bumblebees. Identification, Settlement, Breeding, Protection*. – Pub. Naturbuch Publishing, 320 p.

Wallis D.J. 1964: The foraging behaviour of the ant *Formica fusca*. – *Behaviour* 23, 149–175

WaloffN.&BlackithR.E.1962:Thegrowthanddistribution of the mounds of *Lasius flavus* (Fabr.) (Hymenoptera, Formicidae) in Silwood Park, Bershire. – *J. Anim. Ecol.* 31 (3), 421–437

Wang J., Wang G.G., Li R., Sun Y. Yu L., Zhu J. & Yan Q. 2022: Cascading effect of source limitation on the granivore-mediated seed dispersal of Korean pine (Pinus koraiensis) in secondary forest ecosystems. – *Ecological Processes* 11 (1), 1-13

Wanniarachchi, S., Swan, M., Nevil, P., & York, A. (2022). Using eDNA metabarcoding to understand the effect of fire on the diet of small mammals in a woodland ecosystem. *Ecology and Evolution*, *12*(11), e9457. https://doi.org/10.1002/ece3.9457

Warne, R. W., Baer, S. G., & Boyles, J. G. (2019). Community Physiological Ecology. *Trends in Ecology & Evolution*, *34*(6), 510–518. https://doi.org/10.1016/j.tree.2019.02.002

Wasmund N. 2011: Analysis of the causes of decline of the Red Kite (*Milvus milvus*). The importance of food shortage during the breeding season. – Annual report for the dissertation, Georg-August-University Göttingen, 11 p.

Wasmund N. 2013: The Red Kite (*Milvus milvus*) in Lower Eichsfeld. Breeding population, food ecology and causes of endangerment. – Dissertation at the University of Göttingen, 268 p.

Weidemann G. 1971: D. Food and Energy Turnover of Predatory Arthropods of the Soil Surface. – *Ecol. St. Anat. & Synth*. 2, 111–118

Weinreich E. 1968: Über den Klebfangapparat der Imagines von *Stenus* LATR. (Coleopt., Staphylinidae) mit einem Beitrag zur Kenntnis der Jugendstadien dieser Gattung. - *Zeitschrift für Morphologie und Ökologie der Tiere* 62, 162-210

Wiens J. A. 1989: *The ecology of bird communities*. Vol. 1: Foundation and patterns. – Cambridge University Press, Cambridge, New York, Melbome, 539 p.

Wignall V.R., Harry I.C., Davies N.L., Kenny S.D., McMinn J.K. & Ratnieks F.L.W. 2020: Seasonal variation in exploitative competition between honeybees and bumblebees. – *Oecologia* 192(2), 1-11

Wilder, S. M., Norris, M., Lee, R. W., Raubenheimer, D., & Simpson, S. J. (2013). Arthropod food webs become increasingly lipid-limited at higher trophic levels. *Ecology Letters*, *16*(7), 895–902. https://doi.org/10.1111/ele.12116

Williams, P. H., Thorp, R. W., Richardson, L. L., & Colla, S. R. (2014). *Bumble Bees of North America: An Identification Guide*. Princeton: Princeton University Press.

Williamson C.S. 1938: Aids to the identification of mole and shrew hairs with general comments on hair structure and hair determination. – *J. Wildl. Manage*. 2 (4), 239–250

Wilson E.O. 1971: *The Insect Societies*. – Belknap Press of Harvard University Press, Cambridge, Mass., 548 p.

Wilson K.-J. 2004: *Flight of the Huia. Ecology and conservation of New Zealand's frogs, reptiles, birds and mammals*. – Canterbury University Press, 411 p.

Wilson, J. S., & Carril, O. M. (2015). *The Bees in Your Backyard: A Guide to North America's Bees*. Princeton: Princeton University Press.

Wirtz, P., Szabados, M., Pethig, H., & Plant, J. (1988). An Extreme Case of Interspecific Territoriality: Male Anthidium manicatum (Hymenoptera, Megachilidae) Wound and Kill Intruders. *Ethology*, *78*(2), 159–167. https://doi.org/10.1111/j.1439-0310.1988.tb00227.x

Wise D.H. 1981: Inter-and Intraspecific Effects of Density Manipulations upon Females of Two Orb-Weaving Spiders (Araneae: Araneidae). – *Oecologia* (Ber.) 48, 252–256

Witt H. 1976: Untersuchungen zur Nahrungswahl von Füchsen *(Vulpes vulpes* Linne 1758) in Schleswig-Holstein. – *Zool Anzeiger* 197, 377–400

Witt H. 1980: The Diet of the Red Fox. Questions about method. – In: Biogeographica Vol. 18, *The Red Fox*, ed. by E. Zimen. – Dr. W. Junk bv Publishers, The Hague, 65–69

Witt R. 1998: *Wespen: beobachten, bestimmen*. – Naturbuch Verlag, Augsburg. 360.

Yeo P.F. & Corbet S.A. 1983: *Solitary Wasps*. – Naturalist's Handbooks 3, Cambridge University Press, Cambridge, 63.

8 Estimating Predation Pressure

Abstract

In a feeding relationship, the higher trophic level animals can be described as predators. There are two approaches to analyzing the effect of predators on lower-level organisms: exclusion of predators or direct counting of the loss of prey items by predators. In New Zealand, reducing the threat to endemic species by excluding invasive predator species is practiced on various scales. Experiments can be carried out with eggs laid as prey in artificial nests. The number of bites on prey reflects the strength of predation pressure. In the forests, browsing pressure is of economic importance. Measurement methods are addressed. However, large herbivores' browsing effect on vegetation can also be utilized for nature conservation. We also discuss the extent to which top carnivores control the density of prey populations.

Suppose we know how much and what kind of prey carnivorous animals have consumed as food. In that case, we can calculate their effect on the prey population's density, provided the prey population's production is known. Such results find practical application primarily in biological pest control (Caltagirone 1981; Franz and Krieg 1982; Fisher et al. 1999; Heimpel and Mills 2017; Abdel-Raheem 2022; Sedlag 2022). Also, due to the general decline of many species classified as "huntable," such as the partridge (see Sect. 1.9), quantitative knowledge of the prey spectrum of their potential predators is of great interest (e.g., Tapper 1976; Potts 1986). Global-scale estimates of the biomass of captured prey have been made by Nyffeler and Birkhofer (2017) for spiders and Nyffeler et al. (2018) for insectivorous birds.

8.1 Manipulation of Predator Presence

A direct way to estimate the influence of predators on prey populations quantitatively is by comparing the densities of prey with and without the presence of their predators or by experimentally altering the densities of prey (for vertebrate examples with the removal of the predator, see Taylor 1984, Parker 1984, and perspectives in Lennox et al. 2018 and Salo et al. 2010). The control experiments for the density comparison resemble the tests for determining competition (see Sect. 7.7, ants). Such experiments should ideally be carried out over several generations so that, for example, the effects of the predator on the mortality of certain age classes can be assessed in their impact on the development of the total population (e.g., Taylor 1984; FitzGibbon and Fanshawe 1989; Forrester and Wittmer 2013; Forsyth et al. 2019). Even if the numerical effect of the predator on the prey population may seem negligible, the predator can, for example,

M. Mühlenberg, T. Waßmer, *Field Ecology*, https://doi.org/10.1007/978-3-031-87547-2_8

significantly influence the age structure or the condition of the prey population (Gossow 1976; Wakeley 1982).

Holmes et al. (1979) used net coverings in North American deciduous forests to exclude insectivorous birds from foraging in undergrowth areas. The nets covered 6 m x 6 m x 2 m of undergrowth each. Ten areas separated by nets were compared with ten control areas. Each week, 400 leaves per test area were visually inspected for arthropods (Arachnida, Coleoptera, Homoptera, Hemiptera, and Lepidoptera larvae).

One can also protect trunk parts of individual trees in an orchard from birds that particularly search trunks for insect larvae and pupae (e.g., tree creepers and tits, Fig. 2.57), block off some trunks, and then, after a winter season, compare the trunks with and without bird protection by targeted searching of larvae and pupae. With such experiments, Solomon et al. (1976) determined a predation rate of about 90%.

In grassland, Joern (1986) kept birds as predators of grasshoppers away from foraging on specific plots by using gazebo coverings. The coverings consisted of 10 m × 10 m × 1 m plastic nets with a mesh size of about 2.5 cm held up by posts. Such nets are available in garden stores to cover fruit trees or shrubs against birds. Control areas of the same size were marked with posts. After the construction of the coverings and the selection of randomly distributed control areas in the study area, the areas were fenced from the outside with a ca. 1 m high plastic wall. In the summer, after 5–6 weeks, grasshoppers (Acrididae) were collected systematically with a net.

Clarke and Grant (1968) studied the role of spiders as predators within the leaf litter through control experiments. Four test areas were compared in terms of their prey density using Chilopods as spider prey and Diplopods as control, as they are not eaten by spiders, and Collembola as the most important prey for both the spiders and the Chilopods. Area A was not manipulated and served as an untreated control. Area B was sieved, all spiders were removed, and then redeployed and blocked against new intruders. Area C was treated and enclosed like B, but animals were not removed. Finally, area D was only sieved but not enclosed or freed from predators. D thus served as a control for a possible disturbance effect by manipulating sieving, and C as a possible disturbance by fencing (blocking by a plastic wall). Riechert and Bishop (1990) increased spiders in a garden by adding mulch and flowers to study the effect on so-called pest insects. The experiments were completed by systematically collecting and removing the spiders in equally manipulated areas.

Nyffeler and Birkhofer (2017) consider the role of spiders as natural enemies of insects to be highly significant worldwide. They estimate the annual prey killed by spiders worldwide to be 400–800 million tons (fresh weight), with insects and Collembola accounting for over 90%. Spiders in forests and grasslands account for over 95% of the annual prey kill of the global spider community.

8.2 Predator Reduction in New Zealand

In New Zealand, predation pressure from introduced predators on many native fauna species is particularly noticeable (cf. King 2019). As a result, many native species have become extinct in New Zealand (Wilson 2004). The main predators that have spread are, e.g., rats, weasels, ferrets, cats, and possums (Fig. 8.1).

Herbivorous species such as introduced rabbits and deer are also causing significant damage to the native flora and fauna. The introduced deer (Fig. 8.2) lead to a change in the native vegetation through their browsing, which has further negative consequences for the fauna.

The conservation efforts in New Zealand are trying to eliminate the invasive predators and other pest species as part of a large-scale experiment, and there are evident successes. Where predators were removed from smaller offshore islands, the endangered native bird species quickly recovered. Critically endangered species have been successfully released on predator-free islands (e.g., Kiwis, Coulbourne and Robertson 1997). In creating predator-free zones, the aim is

Fig. 8.1 (**a**) Brushtail possum (*Trichosurus vulpecula*). This species is native to Australia and feeds opportunistically on plants, invertebrates, and occasionally bird eggs. Possums threaten rare plants in forests and kiwi populations (Cowan 2001). (**b**) Stray cats, ferrets, and hedgehogs are the main predators. Showcase of the black stilt breeding station 2005 (see Fig. 8.6). Cats and ferrets were initially released to control rabbits

Fig. 8.2 Introduced deer mainly damage natural vegetation through browsing. While deer are to be reduced by hunting, other herbivores such as possums and rabbits are controlled by repellents, some of which are dispersed from aircraft (Department of Conservation 2009). New Zealand

to eradicate the introduced predators through mass trapping and poisoning (Fig. 8.3). The use of poison has been shown to be mostly beneficial in nature conservation (Innes and Barker 1999; Eason 2002). Methods are constantly being developed to administer poison as selectively as possible (e.g., Weser 2009, Fig. 8.4).

In 2016, New Zealand announced a national program, "Predator-free 2050," that has been achieved not only on numerous offshore islands but also in fenced sanctuaries (Johnson-Groh 2020) (Fig. 8.5). On Lundy Island, several seabird species (e.g., shearwaters, puffins) have been able to reestablish their colonies (McClure 2023). Native fauna has been rehabilitated on selected islands worldwide by hunting goats and trapping or poisoning invasive species (Greenfield 2022).

In rare cases, species close to extinction are kept in captivity for reproduction. The juveniles are then released back into the wild (see Sect. 13.10). One example is the black stilt (*Himantopus novaezelandiae*), which is endemic to New Zealand. Even captive breeding requires the most extraordinary precautions to prevent invasive predators from harming the birds (Figs. 8.6 and 8.7).

8.3 Estimating the Predation Pressure on Bird Nests

While habitat loss and direct persecution were seen as the most critical indicators for the decline of populations for a long time in nature conservation, several studies show that predation pressure among animals also plays an important role (e.g., Bellebaum 2002; Britton 2023; Devers et al. 2007; Evans 2004; Williams et al. 2024). However, it is not easy to determine the mortality

Fig. 8.3 (**a–c**) Three types of traps and lures with poison bait to suppress or eliminate exotic mammal populations. New Zealand, South Island

of vertebrates causally. However, the clarification of breeding losses in birds is experimentally testable.

This approach consists of the controlled exposure of potential prey. **Egg losses in bird nests** by natural enemies can be determined, for example, if artificial nests with chicken, pigeon, quail, or artificial songbird eggs are placed in hedges, forest edges, or on the ground in fields and checked weekly (cf. Gates and Gysel 1978;

Fig. 8.4 Keas (*Nestor notabilis*) are curious and potentially omnivorous. For this reason, the bait color most rejected by the keas is tested here in captivity. (Photo credit: Carolin Weser)

Fig. 8.5 Predator-proofed fence to protect a wildlife sanctuary (Maungatautari). (Photo credit: K-J. Wilson, NZ)

Fig. 8.6 The natural habitat of the black stilt is the braided rivers and wetlands of the Mackenzie Basin, South Island, N.Z., with Mount Cook in the background. The main predators of the black stilt population in the Mackenzie Basin are stray cats, ferrets, and hedgehogs; see Fig. 8.1b

Wilcove 1985; Andren and Angelstam 1988; Small and Hunter 1988; Møller 1989; Major and Kendal 1996; Barkow et al. 2001). These experiments are suitable for testing the level of predator pressure, for example, on different nest types at varying distances from the forest edge. In this way, the habitat value (see Sect. 10.15) of hedge structures can be compared (Barkow et al. 2001). To obtain reliable data, at least 200 artificial nests should be distributed in different hedge types. The artificial nests can each be exposed to a quail egg for a week during the breeding season.

Case Study: Predation on Artificial Bird Nests

One can place homemade eggs ("dummy eggs") or, for example, quail eggs purchased at markets (Fig. 8.8a, b) in an artificial nest and then recognize predation activity from egg losses or tracks on the eggs (Bayne et al. 1997; see Figs. 8.9 and 8.10).

Figures 8.9 compare natural nests with artificial nests.

Quail eggs can usually not be opened by small rodents, while the traces can be identified on the artificial Plasticine (play-dough) eggs. Møller (1989) was able to identify corvid predators (magpie, carrion crow, and jay) by their different beak marks (cf. Figure 8.11, on Plasticine eggs, laid out in artificial nests, Weber 2007).

Major (1991) suggests using automatic photography to detect nest predators at artificially deployed nests. A simple trigger mechanism by the nest predator at the nest itself triggers the remotely installed camera. Another method he recommends is attaching a double-sided adhesive tape to the nest to collect hairs or feathers from the nest predator. The artificial Plasticine or clay eggs can be attached to the nest floor with fine wire.

The time of nest loss can be determined by recording temperature sensors in the nest cavity. If the temperature recorded by a connected data logger (see Sect. 3.1) corre-

Fig. 8.7 (**a**) Facility for captive breeding of black stilts. A fence, which is electrified, serves as protection against predatory mammal species. The black stilt recovery program was started in 1981 when there were only 23 birds left. Artificially hatched chicks are released back into the wild. (**b**) The black stilt (*Himantopus novaezelandiae*) is endemic to New Zealand and critically endangered as a species. The population is estimated at only 156 wild adults (August 2023), Department of Conservation, N.Z. (Reed et al. 1993a, b). (**c**) Sign for visiting the black stilt station from the Department of Conservation. Behind it is the electrified protective fence. (**d**) Despite the protective fence of the black stilt breeding station, traps with bait are still set

sponds to the ambient temperature at that time of the day, the nest has been abandoned. Otherwise, the data logger should record more even temperatures closer to the animal's body temperature (Bellebaum 2001). The quantification of nest losses from controlled natural nests can be done according to Mayfield (1961, 1975). The author described breeding success as the probability that a nest is undisturbed the next day. For the standard deviation of this estimated "daily survival rate," Johnson (1979) provides a formula (cf. also Beintema and Müskens 1987). However, frequent visits to the nests can lead to increased loss due to predators, as some ground predators follow the observer's path (Major 1990). The behavior of the birds can also be influenced by frequent nest visits, as Fetterolf (1983) observed in the decreasing reproductive success of gulls.

Problems of reduced breeding success gain importance in nature conservation when, for example, habitats altered by

Fig. 8.8 (**a**) Quail eggs bought at a market. (**b**) Artificial egg replicas made of Plasticine

Fig. 8.9 (**a**) Nest of the Azure-winged Magpie (*Cyanopica cyanus*) with natural eggs. (Photo credit: Gantulga B., Northern Mongolia 2013). (**b**) Artificial nest, exposed in a birch-larch mixed forest for predation experiments. (Photo credit: Gantulga B., Northern Mongolia). (**c**) Natural nest of the black-faced bunting (*Emberiza spodocephala*), Northern Mongolia. (**d**) Natural nest of the thick-billed warbler (*Arundinax aedon*), Northern Mongolia. (**e**) Artificial nest for experiments in the bush habitat of black-faced bunting and thick-billed warbler

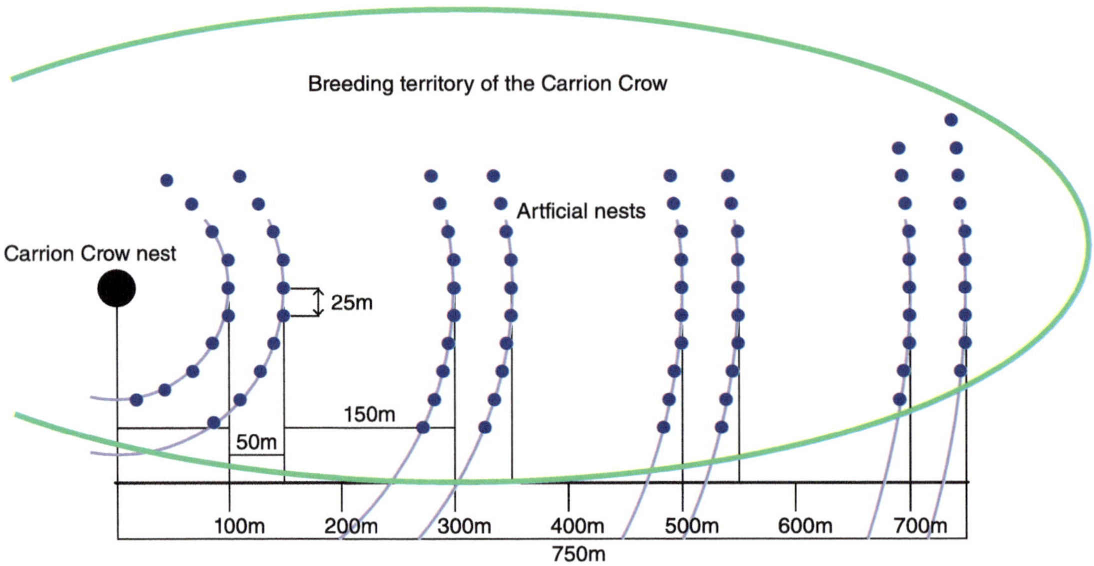

Fig. 8.10 Installation of artificial nests with Plasticine eggs for studying the nest-robbing behavior of a carrion crow. The nests were each equipped with two quail eggs and a Plasticine egg to identify the crow through the beak traces. The different distances of the nest rows served to test the hypothesis that the predation increases with proximity to the predator's location. (From Gantulga (2016))

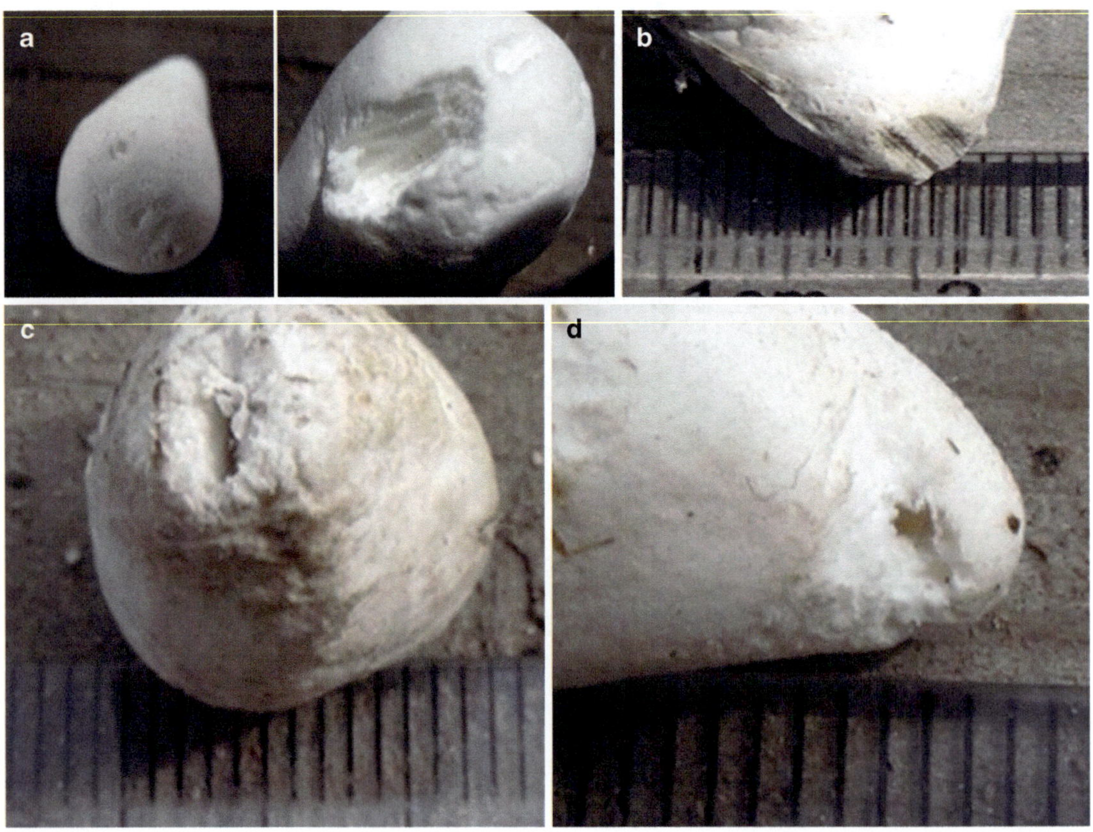

Fig. 8.11 (**a**) "Tweezer impressions" (left) and hacked hollow (right) by the tit *Parus major*. (**b**) Parallel gnawing teeth by Rodentia. (**c**) Bite of large incisors by chipmunks. (**d**) Bite with sharp canine teeth into the tip by Mustelidae

humans are to be compared with original habitats. Fragmentation of habitats creates more edge habitats (see Sect. 11.1). Therefore, it makes sense to compare the differences in predator pressure on bird nests between forest edges and the interior of a forest (e.g., Zegers et al. 2000). In addition, the predator pressure between ground nests and nests in the bush or higher strata can be qualitatively estimated (Pangau-Adam et al. 2006). In the case of an unknown spectrum of predators, it makes sense to install camera traps (WildCam; many models of wildlife available, current prices (2025) range between $/€ 50 and 150) to monitor the artificial nests. The camera traps are attached for the surveillance of nests in the treetops so that the nest-bearing tree trunk is in the focus of a camera trap. For example, raccoons in German forests increasingly contribute to breeding loss alongside natural predators such as martens (Gottschalk et al. 2019). They even attack bird broods in nesting boxes (Schwab et al. 2018).

Ground nests are also endangered by foxes in Europe's native fauna (Voigt and Meinecke 2009). Thus, the mortality of partridges due to foxes is high (Gottschalk and Beeke 2014). Controlling or reducing the predators demonstrated an increase in partridge density and their breeding success (Tapper et al. 1996). Bellebaum and Boschert (2003) provide some information to detect predators at nests of meadow breeders. Ramme et al. (2009) suggest installing thermologgers in the nest to capture the timing of possible predation. Analog Devices, Inc. (ADI) produces several types of coin-sized loggers called iButtons (https://www.ibuttonlink.com/collections/ibuttons).

The measurement accuracy can be varied between 0.0625 °C and 0.5 °C and the measurement interval between 1 and 30 minutes; see also Wassmer and Refinetti (2016). An overview of predation on ground-breeding bird species is provided by Langgemach and Bellebaum (2005).

In the case of **insects**, predation could be estimated if, e.g., cocoons of sawflies or pupae of butterflies are laid out. Their disappearance, their damage by small mammals, or their infestation by entomophagous insects are checked later (Buckner 1959, 1969; Frank 1967; East 1974; Cicero et al. 2024; Lövei and Ferrante 2017; Šipoš et al. 2013; Wang et al. 2022). The predators can be assigned to higher taxonomic units (e.g., family groups) based on traces left on the collected cocoons or directly determined by hatching the cocoons in gauze cages.

8.4 Frequency of Bite Injuries and Wounds in Small Animals

Indirectly, the frequency of bite injuries and wounds in prey can be relatively inferred from different predator influences. On butterflies, Carpenter (1937) and Shapiro (1974) inferred the frequency of bird attacks from wing features. A more comprehensive study of predation risk and wing damage to butterflies is provided by Molleman et al. (2020). Using 34 butterfly species in a tropical African forest, they showed that higher activity and flight speed most influenced the likelihood of predator attacks. On centipedes (Chilopoda, Lithobiidae), Fründ (1983) evaluated the frequency of scars as an indicator of predator pressure. 34% of the nearly 3000 Lithobiids showed apparent scars, indicating hostile encounters. In lizards, the abundance of tail autotomy cannot be readily inferred from the abundance of predators because effective predators kill their prey (Schoener 1979). Arnold (1984) discusses that tail shedding in lizards varies between species and individuals and is not simply an evolutionary consequence of selection pressure by predators. Further information regarding the relationship between tail breakage and predation in smaller reptiles can be found in

Feder and Lauder (1986), Greene (1988), and Arnold (1988).

8.5 Estimating Browsing Pressure at Woody Plant Sites

Changes in forest management and the alteration of the herb layer in forests due to eutrophication from emissions have led to a change in the population dynamics of roe deer (*Capreolus capreolus*) (Ellenberg 1986, 1988). Because the density of roe deer in our forests remains practically undeterminable (see also Sect. 5.2 on deer counts), it is more sensible to evaluate the use and overuse of both forestry-important and other vegetation (Gossow 1976). High economic losses are incurred in Central Europe at the expense of game animals (e.g., Meister et al. 1984), primarily due to browsing and bark-stripping damage to trees and shrub species (e.g., Figs. 8.12 and 8.13). Even in boreal forests, an excessive density of large herbivores, such as moose, changes forest development (Kolstad et al. 2018). In Central Europe, it is mainly roe deer that cause damage to forest growth. Branches, buds, and barks serve multiple functions for wildlife. They are essential for supplying crude protein, tannins, minerals, and dietary fiber and meeting water and energy needs (in winter) (Schauer 1973; Bubenik 1959, 1984; Eisfeld 1974; Ellenberg 1979). The need for this "tough forage" is about 60% of the total forage for roe deer annually and can increase to up to 80% in winter (Gossow 1976). The small rumen and the need for easily digestible food necessitate several foraging periods, alternating with rumination phases during the day.

For the quantitative recording of browsing damage in forestry, only regeneration areas and undergrowth areas can be selected. The area must have an average of one plant per 10 m^2 of at least 20 cm in height. The upper third of the young plants should still be reachable by game for browsing. Enough young plants must be present (at least 1000 pieces/ha), whose existing main shoot is unprotected (neither fence nor individual protection). For a sufficient assessment of the carrying capacity of the forest area of a hunting community (several thousand ha), at least 30 sample stocks must be recorded. Characteristics of a selected sample area include information such as the size of the regeneration field, the average age of the regeneration, estimated tree species proportions in stages, existing tree species in the fenced area closest to the regeneration area to be assessed (furthest distance 300 m), type and average distance to the nearest disturbance factor (e.g., hiking trail, barbecue area, road), and location in isolated forest areas under 5 ha or inside a larger forest complex (shortest distance to the forest-field border more than 500 m) or on the edge of a forest complex.

Fig. 8.12 Browsing by deer at the forest edge. Northern Limestone Alps, Bavaria

On a straight line through the regeneration area, the first ten individual plants (at least 20 cm high, upper third still reachable by wildlife) are assessed at five evenly distributed sample points (first point on the area boundary). The tree or shrub species and height of each individual plant are recorded. In addition, it is noted whether browsing damage occurred in the upper third and whether the main shoot itself was browsed (leads to losses in height growth; game leaves a rough, fibrous browsing site). For plants under 20 cm in height, the browsing percentage is estimated at each species' recording points. At each recording point, the average height of the growth-inhibiting competition vegetation (in tenths of a meter) is also recorded.

In the assessment of game browsing, the proportions of the regeneration of tree species are

Fig. 8.13 (**a**) Deer browsing on a pine, Northern Mongolia. (**b**) Mouse damage on a young *Pinus sibirica*

compared with the initial stocks. The growth relations of the tree species to each other are also shifted by game browsing. By comparing the existing tree species' proportions up to and greater than 20 cm in height, it becomes clear which tree species cannot grow into the layer above 20 cm due to the game's influence. The assessment of forest damage must be set concerning the browsing percentages to the absolute number of plants per unit area (determined using the Point-Centered-Quarter method; see Sect. 4.2.1) since only a certain number of individuals per ha is necessary for a tree species in terms of silviculture. The comparison of an unfenced regeneration with the nearest regeneration behind a fence shows the game's influence most clearly.

Since game in the forest prefers young, germinating trees for food, it can significantly influence the natural regeneration or rejuvenation of the forest. Woody species and herbaceous plants have been classified into stages of different "popularity" as roe deer forage (Ueckermann 1971; Esser 1958; Klötzli 1965; Siuda et al. 1969; Jauch 1987). Ellenberg et al. (1992) note that plants known as roe deer forage have high nitrogen indicator values, while less popular species have low nitrogen indicator values or have special protection against game browsing. As a result of the foraging pressure, the coverage of plant species wanted by deer is significantly reduced. From the occurrence of plant species in individual study areas (presence P) and their coverage (D), the difference can be calculated for the popularity groups (stages 2 + 3 + 4) (P–D): it reflects the foraging pressure (Ellenberg 1988). For such comparisons, the herbaceous plants should also be evaluated. A list of 100 forage plants with information on popularity and nitrogen indicator values can be found in Ellenberg (1988).

At a forest edge (or a hedge), game browsing per m can be quantified, especially during the

Fig. 8.14 (**a**) Elephants browsing on shrubs in the Miombo savannah, Zimbabwe. (**b**) Vegetation looks like browsing damage but can also be part of natural ecosystem dynamics if you calculate in long time frames. (**c**) Elephants have grazed shrubs for kilometers up to about 2 m in height, and only a few Miombo trees remain. Zimbabwe. (**d**) A baobab tree has been severely damaged by elephants that have eaten away the bark and can no longer regenerate. Zimbabwe

winter months. Hare browsing is characterized by a smooth, sharply defined, and diagonally running browsing spot, unlike roe deer browsing (see above). The browsing spots are counted per meter of the forest edge. The diameter of the twig end at the browsing spot can be measured; if necessary, these twig ends are brought to the lab for measurement. By comparing the undamaged twig tips of the respective branch, one can try to estimate the bud loss and the loss of twig substance (fresh weight) caused by browsing (Zwölfer et al. 1981). In Upper Franconia, Zwölfer et al. (1984) found that 160–690 sloe buds (*Prunus spinosa*) with a twig and bud fresh weight of 5–30 g/m had been eaten per meter of hedge edge. The peeling of bark is additionally recorded in the protocol taken.

Herbivore/vegetation interactions drive landscape change at all spatial scales. This is particularly noticeable in elephants due to the large area they influence (Fig. 8.14). An overpopulation of elephants in too-small national parks causes such damage. See also Kupika et al. (2014). However, the browsing pressure is tolerable if we assume natural restoration cycles of 100–200 years for such savannah ecosystems. Therefore, short-term interpretations of browsing can also be incorrect

Fig. 8.15 Upper panel: Example of a forest in Poland with artificial maintenance of forest openness through domesticated large herbivores. Lower panel: Heck cattle (*Bos primigenius* f. *taurus*) used in the forest for maintenance and diversification. Project near Göttingen, Germany

and would not require targeted management of large herbivores.

8.6 Artificial Wood-Pasture Systems (Silvopasture)

When large herbivores live in the forest, they change the original vegetation and reduce the regeneration potential of many woody species (Bunnel and Eastman 1976; Thomas et al. 1979; MacMahon 1981; Gill 2014; Vera 2000). Moderate grazing pressure increases the biodiversity of insects, other small animals, and birds (Hendriks 1986; Harding and Rose 1986). If economic goals regarding wood harvest do not determine forestry, browsing by large herbivores can be desirable. Therefore, there are projects with artificial grazing of forest parts by certain cattle breeds, ponies, pigs, goats, and sheep to thin out closed forests and thereby increase biodiversity (Cosijn and Hendriks 1983; Tree 2019) Fig. 8.15). Older projects, where farm animals were released for landscape maintenance, can be found, e.g., in southern England (McKay 2011; Tree 2019).

One line of argument is that the megaherbivores that became extinct in the Quaternary in temperate America, Europe, and China should be replaced by artificial grazing with domestic breeds, thereby achieving a mosaic of vegetation structures with patchy promotion of herbaceous and shrub layers, which, for example, would benefit pollinators (Berzaghi et al. 2023; Gill 2014; Teng et al. 2023; Tree 2019; Vera 2000). The individual species groups must be compared regarding species number and frequency between grazed and ungrazed forests; however, if one were to let a non-artificially grazed forest also partially age, a natural mosaic of succession stages would be created as well (Remmert 1991; Peterken 1996), which supports forest-specific species. Conservation should be guided by which species are most endangered and can survive in the area (see Sect. 10.9, Mühlenberg et al. 1996). Such a question also arises in the management concept of open spaces near forests or shrubs. Without management, for example, an open

Fig. 8.16 (**a**) Without management, the open spaces in Central Europe (above and below) would become overgrown due to succession, first with shrubs and then trees. (**b**, **c**) Stonechat (*Saxicola torquata*), in its natural open habitat

space on the edge of the Hainich National Park in Thuringia, Germany, would become overgrown by trees, and thus the large population of the Stonechat (*Saxicola torquata*) depending on these open landscapes would decrease significantly according to Grüneberg et al. (2016) (Fig. 8.16).

8.7 About the Top-Down Control of Population Densities

The mere presence and absence of predators can influence the frequency of other species of **vertebrates.** However, clear observations of this regulation are relatively rare (Skogland 1991). Soulé et al. (1988) and Wilcove (1985) observed a substantial increase in so-called medium predators in the absence of large predators ("*mesopredator release*"), which increasingly damage bird populations (see also Douglas et al. 2023). A similar system has been studied in mammals (Palomares et al. 1995).

The effect of predators is not limited to changes in prey density. Slagsvold (1984) examined the reaction of prey to losses due to predators: he manipulated the egg count of the first clutches of great tit broods and then artificially removed the 15-day-old chicks from the nest. Birds usually respond to nest losses with subsequent or replacement broods, but in different ways depending on their investment in the first brood. Multiyear experiments are needed to determine whether consumers regulate the next lower trophic level species, which involves the general ecological question of whether there is a

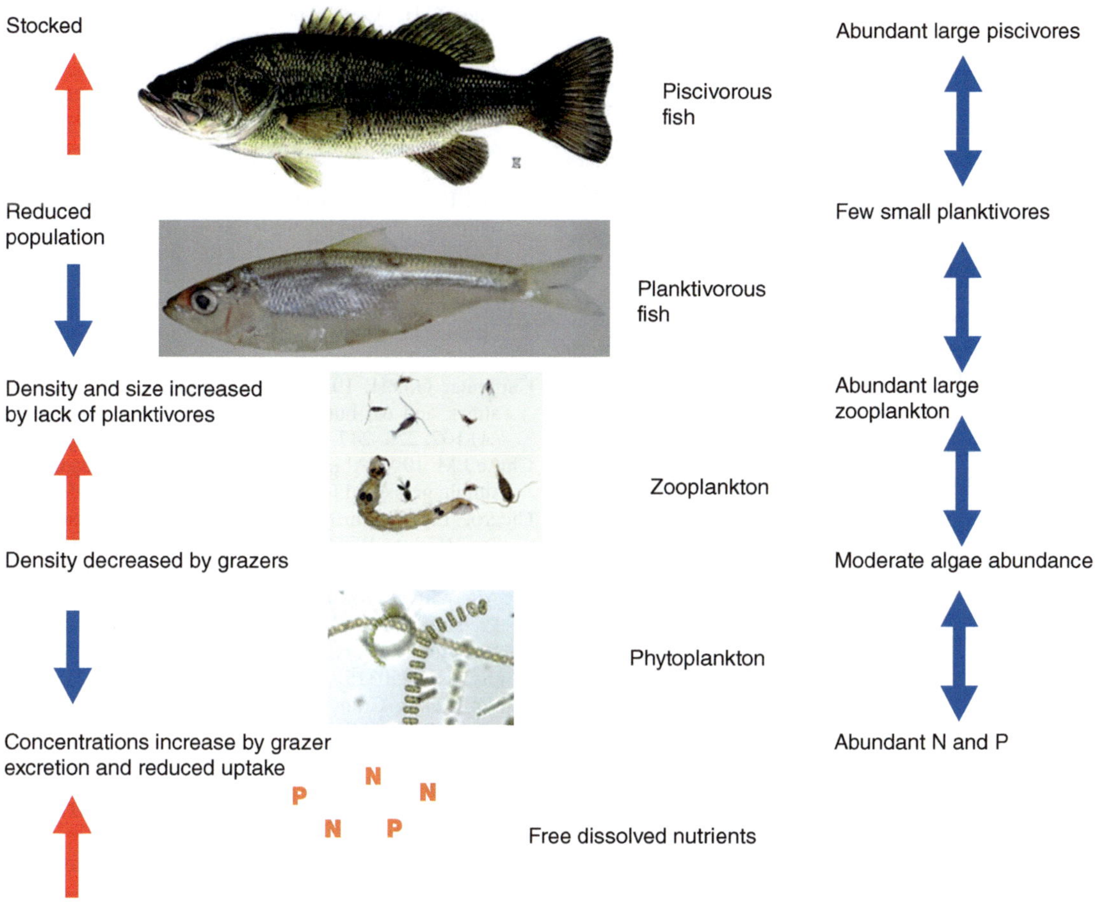

Fig. 8.17 Trophic cascade with top-down control can be used to bioremediate algal blooms caused by eutrophication. The top-down control starts with (re)stocking the top predator, a fish-eating large fish, which then controls planktivorous fish, lowering their abundance and size, which trickles down to having more and larger zooplankton, which is then able to control blooms of phytoplankton

top-down or bottom-up control in a food web (Martin 2002; Begon et al. 2006; Zhang and Adams 2011; Scherber et al. 2010). The principle of such experiments is to exclude first-degree consumers (herbivores) or second-degree consumers (carnivores) from the food web. Field cages often achieve this with and without the exclusion of carnivores. Such studies have shown both cascade effects (impact of carnivores on primary producers) and dependencies on the nutrient content of the primary producers. Exemplary works have been provided by Fretwell (1987), Joern (1992), Chase (1996), Atlegrim (1989), Marquis and Whelan (1994), and Letourneau and Dyer (1998). Top-down or bottom-up controls play a crucial role in interpreting cyclic population developments of herbivores and carnivores in northern countries. There are also numerous works on this, e.g., Jedrzejewski and Jedrzejewska (1996), Sheftel (1989, 2010), Oksanen (1990), Hanski et al. (1991), Cynthia and Anthony (2003), Sinclair et al. (2006), Scherber et al. (2010). Above all, Krebs et al. (2001) have shown that investigating the relationships of population dynamics must be planned for at least ten years. The Kluane project also clearly shows which questions large-scale experimental areas (manipulated 1-km^2 blocks) are advantageous or necessary (Krebs et al. 2001). Top-down control of populations is also common in freshwater habitats and is often used to manage unsustainable algae populations due to eutrophication (biomanipulation) (Fig. 8.17) (Guo et al. 2015; Mao et al. 2020; Xie and Liu 2001; see also Sect. 14.3).

References

Abdel-Raheem M. 2022: *Biological Control of pests and its impact on rural development.*—Lap Lambert Academic Publishing, 76 p.

Andren H. & Angelstam P. 1988: Elevated Predation Rates as an Edge Effect in Habitat Islands: Experimental Evidence.—*Ecology* 69 (2), 544–547

Arnold E.N. 1984: Evolutionary aspects of tail shedding in lizards and their relatives.—*Journal of Natural History* 18, 127-169

Arnold, E. N. 1988. Caudal anatomy as a defense. Chapter 3, 235-274.—In: C. Gans and R.B. Huey (eds.) *Biology of the Reptilia*, Volume 16, Ecology B. Defense and Life History.—Branta Books Box 3457, Ann.Arbor. MI 48106, Defense p. 1-330.

Atlegrim O. 1989: Exclusion of birds from bilberry stands: impact on insect larval density and damage to the bilberry.—*Oecologia* 79, 136-139

Barkow A., Bairlein F. & Mühlenberg M. 2001: Prädationsraten an Vogelnestern in Hecken.—In: Gottschalk E., Barkow A. & Settele J. (eds.): *Naturschutz und Verhalten*. UFZ-Bericht 2/2001: 111-115, Leipzig

Bayne E.M., Hobson K.A. & Fargey P. 1997: Predation on artificial nests in relation to forest type: contrasting the use of quail and plasticine eggs.—*Ecography* 20: 233-239

Begon M., Townsend C.R. & Harper J.L. 2006: *Ecology. From Individuals to Ecosystems*.—Blackwell Publishing, 738 p.

Beintema A.J. & Müskens G.J.D.M. 1987: Nesting success of birds breeding in Dutch agricultural grasslands.—*Journal of Applied Ecology* 24**,** 743–758

Bellebaum J. 2001: Prädation auf Wiesenbrüter in Brandenburg: Untersuchungsmethoden und erste Ergebnisse. - In: Gottschalk E., Barkow A. and Settele J. (eds.): *Naturschutz und Verhalten*. UFZ-Bericht 2/2001: 117-122, Leipzig

Bellebaum J. 2002: Prädation als Gefährdung bodenbrütender Vögel in Deutschland—eine Übersicht.—*Ber. Vogelschutz* 39, 95-117

Bellebaum J. & Boschert M. 2003: Bestimmung von Prädatoren an Nestern von Wiesenvögeln.—*Vogelwelt* 124, 83-91

Berzaghi, F., Bretagnolle, F., Durand-Bessart, C., & Blake, S. (2023). Megaherbivores modify forest structure and increase carbon stocks through multiple pathways. *Proceedings of the National Academy of Sciences*, *120*(5), e2201832120. https://doi.org/10.1073/pnas.2201832120

Britton, J. R. (2023). Contemporary perspectives on the ecological impacts of invasive freshwater fishes. *Journal of Fish Biology*, *103*(4), 752–764. https://doi.org/10.1111/jfb.15240

Bubenik A.B. 1959: *Grundlagen der Wildernährung*.—Deutsch. Bauernverlag, Berlin, 300 p.

Bubenik A.B. 1984: *Ernährung, Verhalten und Umwelt des Schalenwildes*.—BLV- Verlagsgesellschaft, München, Wien, Zürich, 272 p.

Buckner C.H. 1959: The assessment of larch sawfly cocoon predation by small mammals. –. *Can Ent*. 91, 275–282

Buckner C.H. 1969: The common shrew *(Sorex araneus)* as a predator of the winter moth *(Operophtera brumata)* near Oxford, England.—*Canad. Entom*. 101, 370–375

Bunnel F. & Eastman D. 1976: Effects of forest management practices on wildlife in the forests of British Columbia.—*16. IUFRO World Congr./Oslo* , 631-689

Caltagirone L.E. 1981: Landmark examples in classical biological control.—*Annual Review of Entomology* 26, 213–232

Carpenter G.D.H. 1937: Further evidence that birds do attack and eat butterflies.—*Proc. Zool. Soc. London (A)* 107, 223–247

Chase J.M. 1996: Abiotic controls of trophic cascades in a simple grassland food chain.—*Oikos* 77, 495-506

Cicero, L., Chavarín-Gómez, L. E., Pérez-Ascencio, D., Barreto-Barriga, O., Guevara, R., Desneux, N., & Ramírez-Romero, R. (2024). Influence of Alternative Prey on the Functional Response of a Predator in Two Contexts: With and without Intraguild Predation. *Insects*, *15*(5), 315. https://doi.org/10.3390/insects15050315

Clarke R.D. & Grant P.R. 1968: An Experimental Study of the Role of Spiders as Predators in a Forest Litter Community. Part I.—*Ecology* 49, 1152–1154

Cosijn R. & Hendriks J.L.J. (1983): Increasing the natural value of forest, also for mammals.—*Skin and hair, 2* (3), 147–153.

Coulbourne R.M. & Robertson H.A. 1997: Successful translocations of little spotted kiwi (*Apteryx owenii*) between offshore islands of New Zealand.—*Notornis* 44, 253-258

Cowan P.E. 2001: Advances in New Zealand mammalogy 1900-2000: Brushtail possum.—*Journal of the Royal Society of New Zealand* 31, 15-29

Cynthia J.Z. & Anthony R.G. 2003 (eds.): *Mammal Community Dynamics*. Management and Conservation in the Coniferous Forests of Western North America.—Cambridge University Press, 709 p.

Department of Conservation (2009). *Deer repellent on 1080*. Retrieved July 24, 2025, from https://www.doc.govt.nz/our-work/national-predator-control-programme/

Devers, P. K., Stauffer, D. F., Norman, G. W., Steffen, D. E., Whitaker, D. M., Sole, J. D., et al. (2007). Ruffed Grouse Population Ecology in the Appalachian Region. *Wildlife Monographs*, *168*(1), 1–36. https://doi.org/10.2193/0084-0173.168

Douglas D.J.T., Tománková I., Gullett P., Dodd S.G., Brown D., Clift M., Russell N., Warnock N., Smart J. & Sanders S. 2023: Varying response of breeding waders to experimental manipulation of their habitat and predators.—*Journal for Nature Conservation* 72, https://doi.org/10.1016/j.jnc.2023.126353

Eason C. 2002: Sodium monofouroacetate (1080) risk assessment and risk communication.—*Toxicology* 181-182, 523-530

East R. 1974: Predation on the soil-dwelling stages of the winter-moth at Wytham Woods, Berkshire.—*J. Anim. Ecol.* 43, 611–626

Eisfeld D. 1974: Der Proteinbedarf des Rehes *(Capreolus capreolus* L.) zur Erhaltung.—*Z. f. Jagdwiss.* 20 (1), 43–48

Ellenberg H. 1979: Frühjahrs-und Sommernahrung bestimmen die Dynamik von Rehpopulationen.—*Verh. Ges. f. Ökol., Münster* 1978, 439–452

Ellenberg H. 1986: Immissionen—Produktivität der Krautschicht—Populationsdynamik des Rehwildes: Ein Versuch zum Verständnis ökologischer Zusammenhänge.—*Natur und Landschaft* 61 (9), 335–340

Ellenberg H. 1988: Eutrophierung—Veränderung der Waldvegetation—Folgen für den Rehwild-Verbiss und dessen Rückwirkungen auf die Vegetation.—*Schweiz. Ztschr. f. Forstwesen* 139 (4), 261–282

Ellenberg H., Weber H.E., Düll R., Wirth V., Werner W. & Paulissen D. 1992: *Zeigerwerte von Pflanzen in Mitteleuropa*.—Erich Goltze, Göttingen, 262 p.

Esser W. 1958: Beitrag zur Untersuchung der Äsung des Rehwildes.—*Jagdwissen* 4, 1–41

Evans, K. L. (2004). The potential for interactions between predation and habitat change to cause population declines of farmland birds. *Ibis*, *146*(1), 1–13. https://doi.org/10.1111/j.1474-919X.2004.00231.x

Feder M.E. & Lauder G.V. (eds). 1986: *Predator-Prey Relationships*. Perspectives and approaches from the study of lower vertebrates.—Univ. Chicago Press, 208 p.

Fetterolf P.M. 1983: Effects of investigator activity on ring-billed gull behavior and reproductive performance.—*Wilson Bulletin* 95, 23–41

Fisher T., Bellows T., Caltagirone L., Dahlsten D., Huffaker C. & Gordh G. 1999: *Handbook of Biological Control*. Principles and Applications of Biological Control.—Academic Press, 1046 p.

FitzGibbon C.D. & Fanshawe J. 1989: The condition and age of Thomson's gazelles killed by cheetahs and wild dogs.—*Journal of Zoology* 218, 99-107

Forrester, T. D., & Wittmer, H. U. (2013). A review of the population dynamics of mule deer and black-tailed deer *Odocoileus hemionus* in North America. *Mammal Review*, *43*(4), 292–308. https://doi.org/10.1111/mam.12002

Forsyth, D. M., Latham, A. D. M., Davis, N. E., Caley, P., Letnic, M., Moloney, P. D., et al. (2019). Interactions between dingoes and introduced wild ungulates: concepts, evidence and knowledge gaps. *Australian Mammalogy*, *41*(1), 12–26. https://doi.org/10.1071/AM17042

Frank J.H. 1967: The effect of pupal predators on a population of winter moth, *Operophtera brumata* (L.) (Hydriomenidae).—*J. Anim. Ecol.* 36, 611–621

Franz J.M. & Krieg A. 1982: *Biologische Schädlingsbekämpfung* (3. Aufl.)—Pareys Studientexte 12. Verl. P. Parey, Hamburg, Berlin, 222 p.

Fretwell S.D. 1987: Food chain dynamics: the central theory of ecology?—*Oikos* 50, 291-301

Fründ H-.-Ch. 1983: Untersuchungen zur Koexistenz verschiedener Chilopodenarten im Waldboden. - Dissertation, Universität Würzburg, 164p.

Gantulga B. 2016: *Cooperative breeding and anti-predator strategies of the Azure-winged Magpie in Northern Mongolia*.—Dissertation Georg-August-University Göttingen, 127 p.

Gates J.E. & Gysel L.W. 1978: Avian Nest Dispersion and Fledging Success in Field-Forest Ecotones.—*Ecology* 59 (5), 871–883

Gill, J. L. (2014). Ecological impacts of the late Quaternary megaherbivore extinctions. *New Phytologist*, *201*(4), 1163–1169. https://doi.org/10.1111/nph.12576

Gossow H. 1976: *Wildökologie*.—BLV-Verlagsgesellschaft, München, 316 p.

Gottschalk E. & Beeke W. 2014: How can the drastic decline in the Grey Partridge (*Perdix perdix*) be stopped? Lessons from ten years of the Grey Partridge Conservation Project in the district of Göttingen.—*Reports on Bird Protection* 51, 95-116 (in German)

Gottschalk E., Bayoh R., Kamrad M. & Wasmund N. (2019): Mortality of young Red Kites *Milvus milvus* in the nest—Extent and causes.—*Vogelwelt* 139, 155-160

Greene H.W. 1988: Antipredator mechanisms in reptiles. Chapter 1, 1-152.—In: Gans C. & Huey R.B. (eds.) *Biology of the Reptilia*, Volume 16, Ecology B. Defense and Life History.—Branta Books Box 3457, Ann. Arbor. MI 48106, Defense p. 1-330

Greenfield P. 2022: Driving out invasive species on islands has high success rate and big benefits—study.—*theguardian.org* 10 Aug 2022.

Grüneberg C., Bauer H.-G., Haupt H., Hüppop O., Ryslavy T. & Südbeck P. 2016: Red List of breeding birds in Germany. 5. Edition, November 30, 2015.—*Reports on Bird Protection* 52: 19–67.

Guo, L., Wang, Q., Xie, P., Tao, M., Zhang, J., Niu, Y., & Ma, Z. (2015). A non-classical biomanipulation experiment in Gonghu Bay of Lake Taihu: control of icrocystis blooms using silver and bighead carp. *Aquaculture Research*, *46*(9), 2211–2224. https://doi.org/10.1111/are.12375

Hanski I., Hansson L. & Henttonen H. 1991: Specialist predators, generalist predators, and the microtine rodent cycle.—*Journal of Animal Ecology* 364, 232-235

Harding P. & Rose F. 1986: Pasture woodlands in lowland Britain. A review of their importance for wildlife conservation.—*Inst. Terr. Ecology/Huntingdon*, 89 p.

Heimpel G.E. & Mills N.J. 2017: *Biological Control. Ecology and Applications*.—Cambridge University Press, 386 p.

Hendriks J. 1986: Nature conservation in Dutch forest.—*Nat. Forest Service/Utrecht* , MS: 7p.

Holmes R.T., Bonney R.E. Jr. & Pacala S.W. 1979: Guild Structure of the Hubbard Brook Bird Community: A Multivariate Approach.—*Ecology* 60 (3), 512–520

Innes J. & Barker G. 1999: Ecological consequences of toxin use for mammalian pest control in New Zealand:

An overview.—*New Zealand Journal of Ecology* 23, 111-127

Jauch E. 1987: *Der Einfluss des Rehwildes auf die Waldvegetation in verschiedenen Forstrevieren Baden-Württembergs.* – Dissertation, Universität Hohenheim, Inst, f. Landeskultur und Pflanzenökologie, 187 p.

Jedrzejewski W. & Jedrzejewska B. 1996: Rodent cycles in relation to biomass and productivity of ground vegetation and predation in the Palearctic.—*Acta Theriologica* 41 (1), 1-34

Joern A. 1986: Experimental study of avian predation on coexisting grasshopper populations (Orthoptera: Acrididae) in a sandhills grassland.—*Oikos* 46 (2), 243–249

Joern A. 1992: Variable impact of avian predation on grasshopper assemblies in sandhills grassland.—*Oikos* 64, 458-463

Johnson D.H. 1979: Estimating nesting success: the Mayfield method and an alternative.—*The Auk* 96, 651–661

Johnson-Groh M. 2020: The Quest to Purge New Zealand of Invasive Predators.—*undark.org* 05.25.2020

King C.M. 2019: *Invasive Predators in New Zealand. Disaster on Four Small Paws.*—Palgrave Macmillan, 364 p.

Klötzli F. 1965: Qualität und Quantität der Reh Äsung in Wald- und Grünlandgesellschaften des Schweizer Mittellandes.—*Veröff. Geobot. Inst. Stiftg. Rübel, ETH Zürich*, 38, 128 p.

Kolstad A., Austrheim G., Solberg E.J., De Vriendt L. & James S. 2018: Pervasive moose browsing in boreal forests alters successional trajectories by severely suppressing keystone species.—*Ecosphere* 9(10): e02458. https://doi.org/10.1002/ecs2.2458.

Krebs C.J., Boutin S. & Boonstra R. 2001 (eds.): *Ecosystem Dynamics of the Boreal Forest. The Kluane Project.*—Oxford University Press, 511 p.

Kupika O.L., Shakkie S., Gandiwa E. & Gumbi A. 2014: Impact of African elephants on baobab (*Adansonia digitata* L.) population structure in northern Gonarezhou National Park, Zimbabwe.—*Tropical Ecology* 55(2), 159-166

Langgemach T. & Bellebaum J. 2005: Prädation und der Schutz bodenbrütender Vogelarten in Deutschland.—*Vogelwelt* 126, 259-298

Lennox, R. J., Gallagher, A. J., Ritchie, E. G., & Cooke, S. J. (2018). Evaluating the efficacy of predator removal in a conflict-prone world. *Biological Conservation*, *224*, 277–289. https://doi.org/10.1016/j.biocon.2018.05.003

Letourneau D.K. & Dyer L.A. 1998: Experimental test in a lowland tropical forest shows top-down effects through four trophic levels.—*Ecology* 79, 1678-1687

Lövei, G. L., & Ferrante, M. (2017). A review of the sentinel prey method as a way of quantifying invertebrate predation under field conditions. *Insect Science*, *24*(4), 528–542. https://doi.org/10.1111/1744-7917.12405

MacMahon J. 1981: Successional processes: comparison among biomes with special reference to probable roles of and influence on animals, 277-304.—In: West DC, Shugart H & Botkin D (eds.): *Forest succession. Concepts and application.* Springer Advanced Texts, Springer Verlag, 517 p.

Major R.E. 1990: The effect of human observers on the intensity of nest predation. - *Ibis* 132 (4), 608-612

Major R.E. 1991: Identification of nest predators by photography, dummy eggs, and adhesive tape.—*The Auk* 108 (1), 190–195

Major R.E. and Kendal C.E. 1996: The contribution of artificial nest experiments to understanding avian reproductive success: a review of methods and conclusions. - *Ibis* 138, 298-307

Mao, Z., Gu, X., Cao, Y., Zhang, M., Zeng, Q., Chen, H., et al. (2020). The Role of Top-Down and Bottom-Up Control for Phytoplankton in a Subtropical Shallow Eutrophic Lake: Evidence Based on Long-Term Monitoring and Modeling. *Ecosystems*, *23*(7), 1449–1463. https://doi.org/10.1007/s10021-020-00480-0

Marquis R.J. & Whelan C.J. 1994: Insectivorous birds increase growth of white oak through consumption of leaf-chewing insects.—*Ecology* 75, 2007-2014

Martin K. 2002: *Ökologie der Biozönosen.*—Springer, 325 p.

Mayfield H. 1961: Nesting success calculated from exposure.—*Wilson Bulletin* 73, 255–261

Mayfield H.F. 1975: Suggestions for calculating nest success.—*Wilson Bulletin* 87, 456–466

McClure T. 2023: She kills to be kind: the mastermind ecologist eliminating invasive predators.—https://www.wmil.co.nz/

McKay I. (ed.) 2011: *A New Forest Reader: A Companion Guide to the New Forest, its History and Landscape*, Hatchet Green Publishing

Meister G., Schutze Ch. & Sperber G. 1984: *Die Lage des Waldes.* Ein Altas der Bundesrepublik. Daten, Analysen, Konsequenzen.—GEO im Verlag Grüner & Jahr, Hamburg, 352 p.

Molleman F., Javoiš J., Davis R.B., Whitaker M.R.L., Tammaru T., Prinzing A., Õunap E., Wahlberg N., Kodandaramaiah U., Aduse-Poku K., Kaasik A. & Carey J.R. 2020: Quantifying the effects of species traits on predation risk in nature: a comparative study of butterfly wing damage.—*Journal of Animal Ecology* 89 (3), 716-729

Møller A.P. 1989: Nest site selection across field-woodland ecotones: the effect of nest predation.—*Oikos* 56 (2), 240–246

Mühlenberg M., Henle K., Settele J., Poschlod P. , Seitz A. & Kaule G. 1996: Studying species survival in fragmented landscapes: The approach of the F.I.F.B. - In: Settele J., Margules C., Poschlod P. & Henle K.: *Species Survival in fragmented landscapes.* - 152-160, The GeoJournal Library Vol. 35, Kluwer Academic Publishers, Dordrecht. 381 p.

Nyffeler M. & Birkhofer K. 2017: An estimated 400–800 million tons of prey are annually killed by the global spider community.—*Sci. Nat.* 104, 30. https://doi.org/10.1007/s00114-017-1440-1

Nyffeler M., Şekercioğlu Ç.H. & Whelan C.J. 2018: Insectivorous birds consume an estimated 400-500 million tons of prey annually.—*Naturwissenschaften* 105(7-8): 47. doi: https://doi.org/10.1007/s00114-018-1571-z.

Oksanen T. 1990: *Predator-prey dynamics in small mammals along gradients of primary productivity.*—Doctoral Dissertation at Umeå University , Sweden. 184 p.

Palomares F., Gaona P., Ferreras P. & Delibes M. 1995: Positive effects on game species of top predators by controlling smaller predator populations: An example with Lynx, Mongooses, and Rabbits.—*Conserv. Biol.* 9 (2), 295-305

Pangau-Adam M.Z., Waltert M., Mühlenberg M. 2006: Nest predation risk on ground and shrub nests in forest margin areas of Sulawesi, Indonesia.—*Biodiversity and Conservation* 15: 4143-4158

Parker H. 1984: Effect of corvid removal on reproduction of Willow Ptarmican and Black Grouse.—*J. Wild. Manage.* 48 (4), 1197–1205

Peterken G.F. 1996: *Natural Woodland. Ecology and Conservation in Northern Temperate Regions.*—Cambridge University Press, 522 p.

Potts G.R. 1986: *The Partridge. Pesticides, Predation and Conservation.*—Collins, London, 274 p.

Ramme S., Düttmann H. & Ehrnsberger R. 2009: *Telemetrie an Kiebitz- und Uferschnepfenküken in der Stollhammer Wisch* (Landkreis Wesermarsch) 2009. 94 p.

Reed C.E.M., Murray D.P. & Butler D.J. 1993a: *Black stilt recovery plan (Himantopus novaezealandiae).*—Department of Conservation, Wellington

Reed C.E.M., Nilsson R.J. & Murray D.P. 1993b: Cross-fostering New Zealand's black stilt.—*Journal of Wildlife Management* 57, 608-611

Remmert, H. (Ed.). (1991). *The Mosaic-Cycle Concept of Ecosystems* (Vol. 85). Berlin, Heidelberg: Springer. https://doi.org/10.1007/978-3-642-75650-4

Riechert S.E. & Bishop L. 1990: Prey control by an assemblage of generalist-predators: spiders in garden test systems.—*Ecology* 71 (4), 1441–1450

Salo, P., Banks, P. B., Dickman, C. R., & Korpimaki, E. (2010). Predator manipulation experiments: impacts on populations of terrestrial vertebrate prey. *Ecological Monographs, 80*(4), 531–546. https://doi.org/10.1890/09-1260.1

Schauer Th. 1973: Wieviel Äsung braucht das Wild?—*Pirsch-DDJ* 25, 349–354

Scherber C., Eisenhauer N., Weisser W., Schmid B., Voigt W., Fischer M., Schulze E.-D., Roscher C., Weigelt A., Allan E., Bessler H., Bonkowski M., Buchmann N., Buscot C., Clement L.W., Ebeling A., Engels C., Halle S., Kertscher I., Klein A.M., Koller R., Koenig S., Kowalski E., Kummer V., Kuu A., Lange M., Lauterbach D., Middelhoff C., Migunova V.D., Milcu A., Mueller R., Partsch S., Petermann J.S., Renker C., Rottstock T., Sabais A., Scheu S., Schumacher J., Temperton V.M. & Tscharntke T. 2010: Bottom-up effects of plant diversity on multitrophic interactions in a biodiversity experiment. – *Nature 468*, 553–556. https://doi.org/10.1038/nature09492

Schoener T.W. 1979: Inferring the Properties of Predation and Other Injury-Producing Agents from Injury Frequencies.—*Ecology* 60 (6), 1110-1115

Schwab T., Fischer S. & Arndt E. 2018: Der Waschbär *Procyon lotor* als Prädator des Trauerschnäppers *Ficedula hypoleuca* in einem Nistkastenrevier in Sachsen-Anhalt.—*Vogelwelt* 138: 177-184

Sedlag U. 2022: *Biological Pest Control.*—De Gruyter Oldenbourg, Scientific Pocketbooks. 204 p.

Shapiro A.M. 1974: Beak-mark frequency as an index of seasonal predation intensity on common butterflies.—*Am. Nat.* 98, 50–58

Sheftel B.I. 1989: Long-term and seasonal dynamics of shrews in Central Siberia.—*Ann. Zool. Fennici.* 26(4), 357-369

Sheftel B.I. 2010: Role of different mechanisms in type determination of population dynamics for small mammals from boreal forestry zone, 130-143.—In: Spellerberg I., Slowik J., Mühlenberg M. & Dgebuadze Y.Y.: *Biological Diversity and Nature Conservation: Theory and Practice for teaching.* K.M.K. Scientific Press Ltd., Moscow, 266 p.

Sinclair A.R.E., Fryxell J.M. & Caughley G. 2006: *Wildlife Ecology, Conservation, and Management.*—Blackwell Publishing, 469 p.

Šipoš, J., Drozdová, M., & Drozd, P. (2013). Assessment of trends in predation pressure on insects across temperate forest microhabitats. *Agricultural and Forest Entomology, 15*(3), 255–261. https://doi.org/10.1111/afe.12012

Siuda A., Zurowski H. & Siuda H. 1969: The food of the Roe deer.—*Acta Theriologica* 14, 247–262

Skogland T. 1991: What are the effects of predators on large ungulate populations?—*Oikos* 61, 401-411

Slagsvold T. 1984: Clutch size Variation of birds in relation to nest predation: on the cost of reproduction.—*Journal of Animal Ecology* 53, 945–953

Small M.F. & Hunter M.L. 1988: Forest fragmentation and avian nest predation in forested landscapes.—*Oecologia* 76, 62–64

Solomon M.E., Glen D.M., Kendall D.A. & Milsom N.F. 1976: Predation of overwintering larvae of codling moth (*Cydia pomonella* L.) by birds.—*J. Appl. Ecol.* 13, 341-353

Soulé M.E., Bolger D.T., Alberts A.C., Wright J., Sorice M. & Hill, S. 1988: Reconstructed dynamics of rapid extinctions of chaparral-requiring birds in urban habitat islands.—*Conservation Biology* 2, 75–92

Tapper S.C. 1976: The diet of weasels, *Mustela nivalis* and stoats, *Mustela erminea* during early summer, in relation to predation on gamebirds.—*Journal of Zoology (London)* 179, 219–224

Tapper S.C., Potts G.R. & Brockless M.H. 1996: The effect of an experimental reduction in predation pressure on the breeding success and population density of Grey Partridges *Perdix perdix.*—*J. of Appl. Ecol.* 33, 965-978

Taylor R.J. 1984: *Predation, Population and Community Biology*.—Chapman and Hall, New York, London, 166 p.

Teng, S. N., Svenning, J.-C., & Xu, C. (2023). Large mammals and trees in eastern monsoonal China: anthropogenic losses since the Late Pleistocene and restoration prospects in the Anthropocene. *Biological Reviews*, *98*(5), 1607–1632. https://doi.org/10.1111/brv.12968

Thomas J.R., Miller R., Anderson R. & Carter B. 1979: Plant communities and successional stages.—In: Thomas: Wildlife habitats in managed forests.—*U.S.D.A. Forest Serv., Agric. Handbook* 553: 22-39

Tree, I. (2019). *Wilding: The Return of Nature to a British Farm* (Main Market edition.). Picador.

Ueckermann E. 1971: *Die Fütterung des Schalenwildes*.—Parey, Hamburg, 89 p.

Vera, F. W. M. (2000). *Grazing Ecology and Forest History* (1st edition.). Wallingford, Oxon ; New York, NY: CABI.

Voigt U. & Meinecke B. 2009: *Literaturstudie zur Prädation bei den Niederwildarten Feldhase, Rebhuhn und Fasan*.—Institut für Wildtierforschung. Hannover, 48 p.

Wakeley J.S. (ed.) 1982: *Wildlife Population Ecology*.—The Pennsylvania State University Press, University Park and London, 385 p.

Wang, Z., Tang, P., Shi, M., Huang, J., & Chen, X. (2022). Flowering plants and entomophagous arthropods in the agricultural landscape: a practise-oriented summary of a complex relationship. *Frontiers of Agricultural Science and Engineering*, *9*(1), 63–74. https://doi.org/10.15302/J-FASE-2021427

Wassmer, T., & Refinetti, R. (2016). Daily activity and nest occupation patterns of fox squirrels (*Sciurus niger*) throughout the year. PLOS One, 11(3), e0151249. https://doi.org/10.1371/journal.pone.0151249

Weber M. 2007: Untersuchungen zur Nestprädation an boden- und buschbrütenden Vögeln in Khonin Nuga, West Khentey, Mongolei.—Diplomarbeit an der Georg-August-University Göttingen, 75 p.

Weser C. 2009: Colour preferences and the effect of deer repellent on feeding behaviour of kea (*Nestor notabilis*): implications for deterring birds from carrot baits.—Master thesis of Georg-August-Universität Göttingen, Germany, and Lincoln University, New Zealand, 91 p.

Wilcove D.S. 1985: Nest predation in forest tracts and the decline of migratory songbirds.—*Ecology* 66, 1211–1214

Williams, S., Hebblewhite, M., Martin, H., Meyer, C., Whittington, J., Killeen, J., et al. (2024). Predation risk drives long-term shifts in migratory behaviour and demography in a large herbivore population. *Journal of Animal Ecology*, *93*(1), 21–35. https://doi.org/10.1111/1365-2656.14022

Wilson K.-J. 2004: *Flight of the Huia. Ecology and conservation of New Zealand's frogs, reptiles, birds and mammals*.—Canterbury University Press, 411 p.

Xie, P., & Liu, J. (2001). Practical Success of Biomanipulation using Filter-Feeding Fish to Control Cyanobacteria Blooms: A Synthesis of Decades of Research and Application in a Subtropical Hypereutrophic Lake. *The Scientific World Journal*, *1*(2), 276487. https://doi.org/10.1100/tsw.2001.67

Zegers D.A., May S. & Goodrich L.J. 2000: Identification of Nest Predators at Farm/Forest Edge and Forest Interior Sites.—*Journal of Field Ornithology* 71(2), 207-216

Zhang, Y., & Adams, J. (2011). Top-down control of herbivores varies with ecosystem types. *Journal of Ecology*, *99*(2), 370–372. https://doi.org/10.1111/j.1365-2745.2010.01770.x

Zwölfer H., Bauer G. & Heusinger G. 1981: *Ökologische Funktionsanalyse von Feldhecken*—Tierökologische Untersuchungen über Struktur und Funktion biozönotischer Komplexe.—Lehrstuhl Tierökologie, Universität Bayreuth, 422 p.

Zwölfer H., Bauer G., Heusinger G. & Stechmann D. 1984: *Die tierökologische Bedeutung und Bewertung von Hecken*.—Akademie für Naturschutz u. Landespflege, Beiheft 3 (2), 155 p.

9 Habitat Selection

Abstract

The fundamental question of why we encounter certain animals in a given place is related to their habitat selection. To study habitat selection, we need to define variables we can measure in the landscape. Conversely, species lists can be suitable for describing the quality of habitats. Different approaches are discussed to analyze the factors that determine the habitat choice of species or individuals, such as correlations, key factor search, and experimental changes. The procedure for using animal focal sampling to analyze habitat use is illustrated with examples. The factors determining habitat selection for a species can also change with the year's seasons and can also depend on other species. Agroforestry systems are managed to achieve the highest possible biodiversity in land-use systems. The extent to which animals accept agroforestry systems is discussed. If the habitat choice of an animal species is known, its presence can also be used as an indicator for a specific habitat type. We provide examples of bioindication for birds and beetles.

9.1 Which Variables Are Used to Assess the Habitat Selection of Animal Species?

The measure of the selective use of a resource or habitat requires comparing the frequency of use of a particular resource or habitat and its availability in the environment (Fryxell et al. 2014). The chosen habitat variables depend on the investigated species and the survey scale. Habitat variables used in the analyses of breeding-site selection in large birds, e.g., vultures, can be (Gavashelishvili and McGrady 2006, see also Fig. 9.1):

- Distance from the nest site to certain specific landscape features.
- Elevation above ground.
- Slope (steepness [°] and aspect [compass direction of exposure]).
- Percentage of forest cover or rather percentage of open areas within a fixed radius.
- Number of humans living within a certain radius (e.g., a 20-km radius).
- Mean distance to roads within a certain radius.

M. Mühlenberg, T. Waßmer, *Field Ecology*, https://doi.org/10.1007/978-3-031-87547-2_9

Fig. 9.1 Nest (eyrie) of a Steppe eagle (*Aquila nipalensis*) with nearly fledged young birds in Hustai National Park, Central Mongolia. Here, habitat variables for nest site selection can be determined on a large scale

The habitat selection of ground-dwelling arthropods can be assessed by the description of the following environment characteristics around a pitfall trap (e.g., Sroka and Finch 2006):

- Light intensity of the sub-plot (measured with a light meter).
- Downed dead wood (three size classes for wood: <2, 2–7, and >7 cm in diameter).
- Total of dead wood within a defined radius.
- Height (cm) and vegetation cover (%) in the tree-, shrub-, and field-layer.
- Litter depth (cm) and cover (%).
- Number of holes within 1 m radius (holes of mice or voles).
- Distance to the forest edge.

9.2 How Do you Compare Two Species Lists?

Qualitative species lists without information on (relative) frequencies are only used to show the presence or absence of certain species. If, for example, you want to compare a more species-rich community with a less species-rich one for a habitat assessment, the comparison is only reliable with an utterly equal sample size. The assessment of the similarity of two species communities and their computational description is discussed in Sect. 15.11. Quantitative survey methods for estimating the frequency of individual species are described in Chaps. 1, 2, and 5.

When comparing different habitats, population sizes and species compositions are critical, and the number of species also plays a role. Diversity measures consider the number of species and the frequency of individual species. A more detailed treatment of diversity measures is provided in Sect. 15.11.

9.3 Habitat Preference

Mere species lists for a particular habitat cannot be used to infer that the species in question preferentially reside, breed, or have chosen the location for foraging. For example, so-called forest species can sometimes be found in meadow habitats (Mühlenberg and Werres 1983). Through the multiyear standardized recording of spiders and ground beetles in five similar meadows and five similar beech forest habitats in the Steigerwald, Bavaria, we were able to identify the individual species clearly as "meadow" or "forest species," even though forest species were repeatedly encountered in the meadow. One possibility is that these encounters represent dispersal stages that spread across a meadow. A more differentiated categorization addresses habitat preference according to the frequency (e.g., over 75% in forests: silvicolous (silvicole) or on meadows: pratinicolous (pratinicole), or in general terms, stenopotent vs. eurypotent). Habitat preferences were described for some animal groups; summaries for arthropods are, for example, Thiele (1977), Tretzel (1952, 1954), Heydemann (1955), Huhta (1965), Braun (1969), Braun and Rabeler (1969), Toft (1976), Luczak (1979), Blumenthal (1981), Kühn (1982), Schweizerischer Bund für Naturschutz (1987), Koch (1989) and Wassmer (1995).

A species is, with high probability, a member of the species community of a habitat if it reproduces there. Eggs or nonflying juvenile stages are thus a clear indication of the affiliation of this species to the examined habitat. The origin of dragonflies hunting in the meadow or at the forest edge can, therefore, be determined by the collection of their exuviae (cf. Gerken 1984). The same applies to Plecoptera from certain river sections (Fig. 9.2).

Fig. 9.2 (**a**) Exuvia of a Plecoptera (*Isoperla* sp.) outside of the water is considered evidence of the hatching site. (**b**) Exuvia of an Odonata (*Ophiogomphus* sp.), shortly before hatching, Mongolia

Mühlenberg and Slowik (1997) described three methods to investigate the relationships between habitat and animals:

A. Correlating the occurrence and frequency of species with habitat factors chosen by the researcher (example below).
B. Studying the population biology of a species and searching for the key factors limiting the spread and fitness of a species in the habitat.
C. Experimental alteration of certain biotic and abiotic factors and comparisons of the reaction of the species of interest, including corresponding controls (see the example in Sect. 10.21).

Method A is very common and has experienced a new upswing through the development of computer programs for multivariate statistics. The correlation method assumes that the habitat quality is exceptionally high where the animals are particularly frequent.

Suppose one compares different habitats for their different population densities and simultaneously records the characteristic features of these habitats. In that case, key factors that determine the presence of the species can be identified with correlation analyses. Such a quantitative description of the relationship between population density and habitat factors forms the basis for "Habitat-Suitability-Index-models"(see below Sect. 9.5, and, e.g., Pearsall et al. 1986; Schamberger and O'Neil 1986; Roloff and Kernohan 1999; Schröder and Reineking 2004; Badry et al. 2024; Brooks 1997; Lõhmus et al. 2020; Seccombe and Salguero-Gómez 2024). However, many animals use different habitats for their life requirements and would be classified as generalists. The assignment to a particular habitat type would characterize specialists (cf. Wolff 2003).

A thorough literature evaluation is necessary to measure the "correct"selection of the habitat parameters. Problems with the correlation method remain, such as the exact density determinations (see Chaps. 1 and 2) and the possibly too large fluctuations of individual factors in microhabitats. A high population density does not always indicate natural conditions. Studies in the primary forests in Central Europe showed a low bird density (Tomiałojć et al. 1984; Wesołowski et al. 1987, 2002; Wesołowski 2007). With the correlation method, results are always obtained since the occurrence and frequency naturally correlate with some habitat factors. However, the biological meaningfulness of this correlation is not guaranteed: Birds may, for example, be more frequent in a high hedge than in a differently structured one. However, they could be more frequent in the first hedge only because it connects to a forest. The result will be biased if only hedge structures and not their spatial relationship to forests are included in the statistical analysis. Method A is vulnerable to misinterpretation if a decisive, unknown factor is not considered.

9.4 On the Relationship Between Habitat Quality and Animal Distribution

Although one can statistically show that certain species are significantly more frequent in one habitat, this does not explain why these differences occur. Ceresa et al. (2012) show how statistical methods can determine the importance of various habitat variables to Red-backed Shrike occurrence through stepwise regression. The authors used generalized linear models (Chi^2) and ANOVA (see Sect. 15.10) to evaluate the different factors of habitat types. A more comprehensive analysis of habitat selection and breeding success of the Red-backed Shrike was also conducted by Bloche et al. (2023) in Denmark. They compared the distribution of breeding pair points in the landscape classified by habitat with random points using the Chi^2-test.

Examples of different frequencies in tropical butterflies between different land use systems can be found in Fermon et al. (2003, 2005), Bobo et al. (2006), and for birds in Waltert et al. (2004, 2005). Even if the differences in species frequencies are not justified beyond the correlation with specific habitat characteristics, such results are valuable for assessing the conservation value and possibly also for recommendations for nature conservation management. For the latter, it is also helpful to inform oneself about the geographical distribution and previous studies about the habitat selectivity of target species. Manuals are usually too broadly designed, but there is often helpful literature, e.g., Larsen (1999) on Afrotropical butterflies.

Case Study
Comparison of Two Habitats with Flight Interception Traps at Two Different Height Levels

Six collection points (sample station) were randomly selected in one habitat, and two traps were hung vertically, one above the other at different heights (Fig. 9.3). Different types of flight interception traps were presented in Sect. 2.11.

We want to estimate the effort required for such a collection design.

The sampling effort is calculated using the following scheme: Two vertical traps per sampling station, six sampling stations per habitat, two habitats = $2 \times 6 \times 2 = 24$ traps.

One day is planned to check all traps (= 12 traps) in one habitat. Two days of field work are necessary for a total sample from both habitats and strata. In 40 rainless days, you have 20 repetitions of a total sample, resulting in 960 insect point counts (6 traps per stratum $\times$ 2 strata $\times$ 2 habitats $\times$ 40 days = 960). Extra days must be planned for processing the catches, for example, one day after a total sample of 2 days. With three days for a total sample and 20 repetitions, you then need 60 days of continuous work. Breaks must also be planned so that three months should be scheduled for the study.

The danger of misinterpretation according to method A (see above) can be largely avoided by using method B because, in this approach, one tries to understand and interpret the environment from the perspective (or from the senses) of the

Fig. 9.3 Study design of 6 × 2 flight interception traps at two heights in two different habitats. The red symbol for the trap stands for the lower trap, blue for the upper trap

respective species ("individual analysis"). The importance of a precise analysis of the habitat requirements of a species is shown by the example of the Ortolan Bunting (*Emberiza hortulana*) (Fig. 9.4d). In Franconia, Germany, the Ortolan prefers orchards and forest edges as habitats in plateau-like landscapes (Lang et al. 1990). While it occurs on small-parceled fields in Franconia, it also uses bushy, dry grassland areas in Switzerland (Menz et al. 2009). In Greater Poland, the species prefers to live on the edge of large fields, often measuring over 100 ha, and in old tree rows ("*shelter belts*"). In Northern Sweden, on the other hand, the bird is found on wooded islands in wet lowlands.

A comparison between the different areas within the species'geographical area shows that the species depends not on forests or fruit trees but on suitable singing perches within each habitat. Since birds usually utilize more than one habitat type, their habitat requirements vary according to the regional ratio of different biotope types (Collins 1983). As density increases, individuals move into suboptimal areas (Wiens et al. 1987, see also Sect. 10.13). In these animals, one may observe other preferred habitat features than in individuals living in optimal habitats. Even among insects, we know of regionally different habitat preferences. While some ground beetle species in Upper Bavaria at the foothills of the Alps are pronounced forest animals, the same species live on open alpine pastures at higher elevations (Förster 1986). Similar shifts are also known among beetle species when comparing Southern and Northern German locations. For an insect, a certain sum of heat encountered in the south in dense and high vegetation can guarantee development. In contrast, the same heat sum is only reached on sparsely vegetated south-facing slopes in the north of its distribution. However, it can also be that geographically separated populations differ genetically to such an extent that their habitat preference expresses diverging evolution.

The assessment of habitat preference for a species also becomes problematic when a species is forced back into certain regions on a large scale. An example is the Alpine Chough (Red-billed Chough, *Pyrrhocorax pyrrhocorax*), which in Central Europe almost only occurs on alpine meadows as relic populations due to having access to cattle dung. In contrast, in Great Britain, it occurs in remnant populations on the coast, where it lives in vegetation kept short by sheep or in Coastal grassland (*maritime turf*) where it can find invertebrates (Bullock et al. 1983; Bignall and Curtis 1988). It is even more different in Mongolia, where this species occurs widely, as livestock dung is abundant in the steppe (Fig. 9.4a).

Fig. 9.4 (**a**) Red-billed Chough *Pyrrhocorax pyrrhocorax* common in the Mongolian steppe but rare and endangered in Europe. (**b**) *Dendrocopos leucotos*, Northern Mongolia. (**c**) The hoopoe (*Upupa epops*) has a crooked beak similar to the Chough and uses it to poke around in the dung for insects. (**d**) The Ortolan Bunting (Emberiza hortulana) likes to use elevated singing perches. Photo courtesy of Sergey Neklyudov, CC BY 4.0 (https://creativecommons.org/licenses/by/4.0/)

The White-backed Woodpecker (*Dendrocopos leucotos*) is also essentially pushed back in Europe (Virkkala et al. 1993; Wesołowski 1995) but is common in Northern Mongolia (Fig. 9.4b). For this species, in addition to a sufficiently large area (at least 1 km^2 per breeding pair), a supply of softwood for the construction of the nesting hole is necessary, as this Woodpecker has a weaker beak than, for example, the Great Spotted Woodpecker (Meyer et al. 1993). The availability of a critical factor is often crucial for an animal's occurrence. For the White-backed Woodpecker, this is the availability of suitable cavity trees and softer deadwood; for the Alpine Chough it is food supply. The hoopoe (*Upupa epops*) also relies on dung for food, as it can safely find insects in it even in rainy weather (Fig. 9.4c).

Case Study on Ortolan Territories

The misleading results that the correlation method can lead to when rigidly applied can be illustrated using a thought experiment on the Ortolan: The Ortolan occupies territories with an approximate diameter of 200 m (Bandorf and Laubender 1982; Lang et al. 1990) in the middle of which there is a high singing perch with adjacent farmland. With further progressing land consolidation, the situation in Franconia could arise where the only satisfactory singing perches can be found at the edges of forests. If one now records the equipment of the habitats, one concludes that forest is one of the necessary requisites. Conservation measures for the species would only start here and would not capture the diverse possibilities in the cultural landscape.

In Fig. 9.5, the possible distribution of territories in orchards and at the edges of forest islands are shown schematically. With a very commonly used method of grid evaluation, one could show that the Ortolan is a forest bird with a four-field table (unit square with or without forest share, Ortolan territory present or missing). This is certainly not the case.

A percentage breakdown of different landscape structures (e.g., 20% grassland, 50% arable land, 10% fallow, 15% forest, and 5% water) is also unsuitable for describing habitat requirements, as the corresponding area proportions can be distributed and used differently. Here, the ongoing observation of individuals and recording of their locations, as carried out in a habitat use analysis, is more helpful. The detailed statements about nest site selection and dependent breeding success that can be obtained through individual analysis are described by Lang et al. (1992) also using the example of the Ortolan.

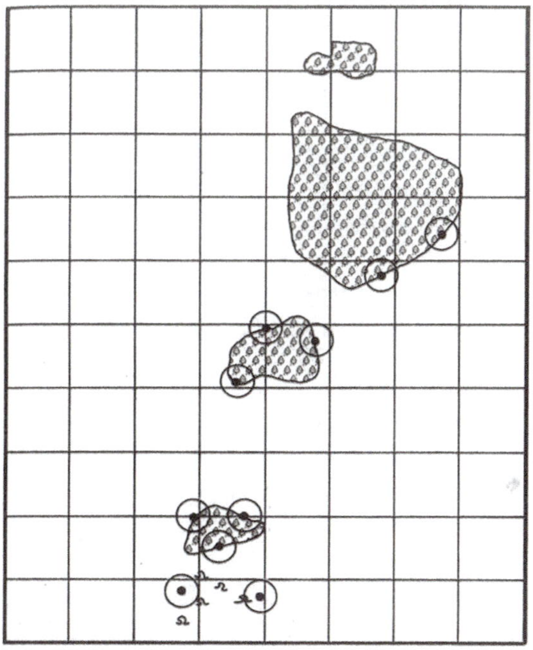

Fig. 9.5 Grid map with Ortolan territories. The patterned areas represent forest, while Ω denotes individual fruit trees

Case Study

Focal Animal Sampling Through Observations on the Great Grey Shrike

A similar situation is also emerging for the Great Grey Shrike (*Lanius excubitor*) (Fig. 9.6), which has now almost wholly vacated the typical agricultural landscape in which it used to be a characteristic inhabitant (Flade and Schwarz 1999; Hagemeijer and Blair 1997; Günther et al. 2005). For a suitable habitat, perches for foraging are crucial (Rothhaupt and Vogel 1996).

In the individual analysis, one seeks out the individual animals and directly analyzes their locations. These data are then compared using an *Analysis of Variance (ANOVA)* (see, e.g., Flandorfer 2022, and Sect. 15.10) with the characteristics of randomly selected locations within the study area as a control. As an example, for the "focal animal sampling" technique (see Altmann 1974; Dawkins (2007) for analyses of the application of this method), observations on the Great Grey Shrike (*Lanius excubitor*) are provided. In our case, the equipment consists of binoculars, a spotting scope, and a dictaphone. It is essential to standardize the observations as much as possible to keep subjective influences as low as possible. However, a certain error rate is hardly avoidable. Nevertheless,

Fig. 9.6 The Great Grey Shrike, *Lanius excubitor*, needs scattered high bushes or trees as perches. It preys preferably on large insects and small mammals. Gobi, Mongolia

Eberhardt and Thomas (1991) point out that direct observations in ecological research (not limited to the visual aspect), combined with experiments, are very important.

As a first step towards standardization, the duration of each observation series is determined, e.g., 20 or 30 minutes. The timing starts at the first visual contact with the animal to be observed. The time until then also provides valuable clues and should be recorded. Now, specific parameters are noted for each observation. Create a checklist to record what you want to be registered for each observation. In this context, an individual observation is the sequence from the beginning to the end of a behavior. The checklist could, for example, record the following data:

- Start of observation.
- Duration of observation.
- Behavior.
- Location (exact description or assignment on an aerial photo).
- Reason for behavior change.

In the observations of the Shrike, a behavior change is often associated with a perch and prey flight change. One could also record the type (on the ground or in the air) and the distance of the prey flight.

Each location and the location changes should be recorded in an aerial photo or on Google Earth map (see Sect. 4.5). The observation series should be systematically distributed over the hours of a day and the year's seasons to prevent observing only certain phases, e.g., the midday rest. By entering the area to be studied from different directions, it is ensured that each part of the habitat receives the same attention.

Follow the animal continuously from the first contact and terminate the observation series after the predetermined time. A significant problem with observations is the overrepresentation of noticeable behaviors. One way to minimize (not eliminate!) this error is to include follow-up observations in the evaluations. These are observations where the start of the respective behavior is clearly defined. The initial observations are, therefore, not considered. However, this significantly increases the amount of data needed.

Further evaluation of the material is done with the computer. Various locations where the animal was observed in varying frequencies are typically obtained. These points, mostly perches in the case of the Shrike, can now be assigned to various characteristics such as tree species, height, distance to the nearest road or forest, surrounding vegetation, vegetation density, distance to the nest, and exposure. With the help of the PCQ method (Sect. 4.2.1), it is possible to determine a value for the density of trees for each point. Many parameters can be conveniently taken from aerial photos or aerial photo maps including drone – or Google Earth Pro acquired photography or maps (see Sect. 4.3). On the computer, it is possible to calculate the specific use for each point or each "community" of points by comparing them to each other, the entire dataset, or random points.

To find out, for example, what role disturbances play in the habitat use of the Shrike, one could group the observation points according to their distance from a source of disturbance, e.g., a road or a parking lot. The supply of perches in the respective distance category is determined using the aerial photo. Suppose the road does not influence the choice of perch. In that case, the distribution of observations under otherwise equal conditions should correspond to the offer of perches in the respective distance categories. Whether any significant differences may occur can be tested, for example, with the Chi^2 goodness of fit test (see Sect. 15.9).

Significant in this example is the restriction "under otherwise equal conditions." The perches in the different distance groups may also have a different height composition. If, for example, larger trees were felled close to the road, the under-proportional use of this region can also be explained by the Shrike's preference for taller trees. It is, therefore, essential to ensure that only the factor currently under consideration varies. In many cases, however, this is not possible. Here, multivariate methods can help (see Sect. 15.10).

Determining the x- and y-coordinates of the observation points also allows the determination of movement patterns or the representation of the temporal variation of the action space size (see Kenward 1987; White and Garrott 1990). These evaluations are facilitated by special computer programs (cf. Kenward 2001), which were initially intended for telemetry data (see Sect. 4.6) but fit our requirements very well. Further evaluation possibilities could include, for example, the specific use of certain types of perches and their distances from roads or other landscape features, certain plant sociological associations (Sect. 4.1), height preferences, and time budgets. The method can be expanded with the help of various technical tools. For example, Leugger (1991) installed artificial perches equipped with scales in Red-backed Shrike (*Lanius collurio*) territories, which allowed investigations about the birds' fitness throughout the breeding cycle. In addition, the combination with telemetry (see Sect. 4.6) opens up even more possibilities.

Case Study
Focal Animal Sampling on Fox Squirrels

To verify that changes in collar temperature in our study on the activity patterns of fox squirrels (*Sciurus niger*) (Wassmer and Refinetti 2016) were indeed related to changes in behavior and location, I (Thomas Wassmer) conducted short-term visual observations of individual collared squirrels and correlated the temperature records with the behavioral records. As I followed a single individual squirrel as long I could, I conducted a focal animal sampling (Altmann 1974) in which I recorded all occurrences of the previously defined (inter)actions of the individual (ethogram) during the sample period and their timing (begin, end, and duration). As my study question related to the activity patterns of fox squirrels, the behaviors I distinguished related closely to the continuum between activity and rest. I did not attempt to capture all specific types of behavior a fox squirrel might exhibit. Table 9.1 provides a list of behaviors and modifiers that I recorded. My behavioral observations were conducted using the iPhone application Animal Behaviour Pro 1.7.1 (School of Anthropology and Conservation, University of Kent, Canterbury, UK, Fig. 9.7).

In addition, I also recorded the modifier arboreal vs. terrestrial. While some behaviors could only occur in one modifier stage and were automatically applied with the behavior to the recording (e.g., basking, chilling, in-nest), others were defined as optional and had to be specified during the observation.

The focal sampling begins when the observer confirms the entries with the OK button. After confirming any new mutually exclusive behavior, the previous behavior is concluded, and the new behavior begins. This leads to a table of the beginning and end times of behaviors—and the duration of each behavior. An example is shown in Table 9.2.

For the original purpose of this focal animal sampling, the behavior log was then aligned with the collar temperature recordings (Fig. 9.8).

Table 9.1 Ethogram of relevant behaviors of fox squirrels. Mutually exclusive tells the program whether a second behavior can be selected simultaneously. Events are instantaneous, while states have considerable durations. Target refers to the receiver of a behavior

		Description	Mutually exclusive?	State or event?	Target
Behavior	"Basking"	Lies flat in the sun	YES	State	None
Behavior	"Chilling"	Lies flat in the shade	YES	State	None
Behavior	"Feeding"	Ingests food	YES	State	None
Behavior	"Food Caching"	Digs or uncovers nuts	YES	State	None
Behavior	"Grooming"	Cleans itself	YES	State	Self
Behavior	"In Nest"	Disappeared into drey	YES	State	None
Behavior	"Locomotion"	Walks, runs, climbs or jumps	YES	State	None
Behavior	"Mobbing"	Vocally and behaviorally harassing other species	YES	State	Other
Behavior	"Not visible"	Lost from sight	YES	State	None
Behavior	"Other"	Any behavior not covered	YES	State	None
Behavior	"Resting"	Not moving other than basking and chilling	YES	State	None
Behavior	"Social"	Interacting with conspecific	YES	State	Other
Behavior	"Vigilance"	Eyes and head following another animal	NO	State	None
Modifier	"Arboreal"	The squirrel is on a tree	"Mobbing", "Chilling", "Locomotion", "Other", "Resting", "Grooming", "Social", "Feeding", "In Nest", "Not visible", "Vigliance", "Basking"		
Modifier	"Terrestrial"	The squirrel is on the ground	"Feeding", "Other", "Grooming", "Social", "Food Caching", "Mobbing", "Vigliance", "Locomotion"		
Subject		Eartag number or collar color			

It can be seen that changes in the collar temperature Tc are primarily associated with moving and grooming, whereas stable Tc is associated with resting and vigilance. One of the squirrels (Panel A) remained outside the nest for the one-hour observation period, and Tc never exceeded 35 °C. The other squirrel (Panel B) entered the nest after approximately 40 min. A steep elevation in Tc can be seen afterward, reaching past 38 °C after about 15 minutes, which is expected as the squirrel nest insulates the animal from the environmental temperature, and by curling up for resting, the collar temperature reaches values around the average core body temperature of the squirrel.

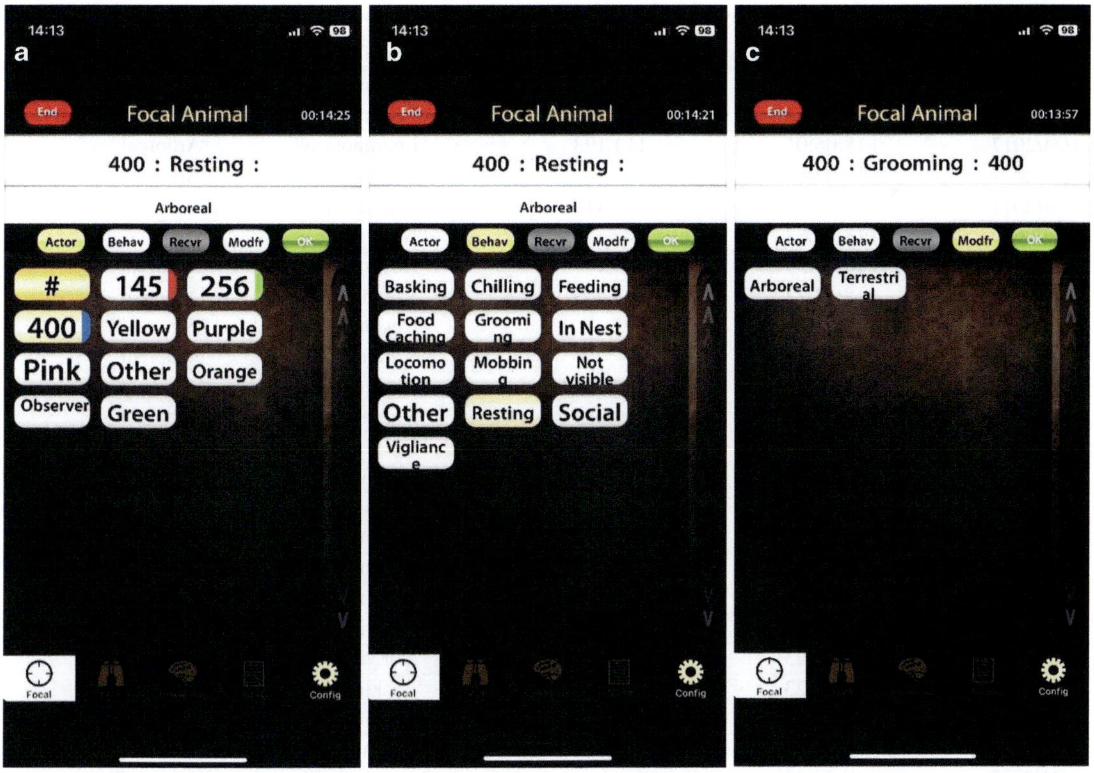

Fig. 9.7 Focal sampling of the behaviors of an individual fox squirrel using the iPhone app Animal Behaviour. (**a**) Selection of the focal animal (actor). (**b**) Selection of the behavior—here, Resting. (**c**) Selection of the modifier = here of the behavior Grooming that can occur when on trees (arboreal) or when on the ground (terrestrial)

9.5 Habitat Suitability Indices

Up to this point, the investigation of habitat use is a purely descriptive method. The causal analysis must be done through targeted experiments. In the Great Grey Shrike example above, one could try to block paths for visitor traffic and then record the changes in habitat use. Furthermore, it could be checked whether parts of the habitat not previously used by the Shrike are included in the action space if additional (artificial or natural) perches are provided.

Joern (1982) and Larson and Bock (1986) provide further examples for field crickets and birds, respectively. Habitat characteristics are measured only in a small radius around the observed bird individual (“*bird-centered vegetation sampling*”) and where the bird was first seen. For each bird species, at least 30 localizations are needed. The “Habitat Suitability Indices (HSI)” (Pearsall et al. 1986) are based on the critical components of the habitat relevant to the respective species. For example, we have found three habitat parameters to be particularly important for a species through individual analysis: Degree of coverage with deciduous shrubs, degree of humidity, and number of a specific type of flower per area. Then, the species is studied in various habitats to determine by regression how its density is related to changes in these parameters or where the optima lie. Each parameter is examined for its entire expression spectrum regarding its suitability for the species in question.

On this basis, three transformation curves can be created. The three habitat parameters (abscissa) are plotted individually against the respective population densities determined (ordinate). The density axis is converted into an index axis

Table 9.2 Example of a focal sampling using the iPhone app Animal Behaviour Pro

Date_ymd	Time_Absolute_hms	Duration_s	Behavior	Modifier_1
4/24/2017	18:04:59	61.972	"Locomotion"	"Terrestrial"
4/24/2017	18:06:01	0	"Locomotion"	"Terrestrial"
4/24/2017	18:06:01	114.193	"Locomotion"	"Arboreal"
4/24/2017	18:07:55	0	"Locomotion"	"Arboreal"
4/24/2017	18:07:55	218.418	"Not visible"	
4/24/2017	18:11:33	0	"Not visible"	
4/24/2017	18:11:33	333.135	"Resting"	"Arboreal"
4/24/2017	18:13:22	224.609	"Vigliance"	"Arboreal"
4/24/2017	18:17:06	0	"Resting"	"Arboreal"
4/24/2017	18:17:06	0	"Vigliance"	"Arboreal"
4/24/2017	18:17:06	1034.738	"Basking"	"Arboreal"
4/24/2017	18:32:42	99.408	"Vigliance"	"Arboreal"
4/24/2017	18:34:21	0	"Basking"	"Arboreal"
4/24/2017	18:34:21	0	"Vigliance"	"Arboreal"
4/24/2017	18:34:21	359.717	"Basking"	"Arboreal"
4/24/2017	18:40:21	0	"Basking"	"Arboreal"
4/24/2017	18:40:21	224.332	"Basking"	
4/24/2017	18:44:05	0	"Basking"	
4/24/2017	18:44:05	197.5	"Not visible"	
4/24/2017	18:47:23	0	"Not visible"	
4/24/2017	18:47:23	168.367	"Basking"	"Arboreal"
4/24/2017	18:50:11	0	"Basking"	"Arboreal"
4/24/2017	18:50:11	32.852	Locomotion"	"Arboreal"
4/24/2017	18:50:44	0	Locomotion"	"Arboreal"
4/24/2017	18:50:44	61.587	"Feeding"	"Arboreal"
4/24/2017	18:51:18	27.369	"Vigliance"	"Arboreal"

("HSI"), where the habitats with the highest densities receive the index value 1. The HSI can, therefore, assume values between 0 and 1. Suppose one wants to determine the HSI of a species for a specific area. In that case, the expression of the three relevant habitat parameters must be determined, and then, the indices for the individual components are read off the transformation curves. The HSI is then obtained by multiplying the individual indices because each of the three habitat parameters is equally important for the species. In many HSI models, however, the key components must be summarized differently: limiting (if a component exceeds a particular value, another becomes worthless), cumulative, or compensatory (a low value for one component can be compensated with a high value of another). HSIs can be understood as a proportional measure of carrying capacity but must be considered a hypothesis about the species-habitat relationship according to their formation through correlations (Morrison et al. 2006). If, for example, it can be documented with the help of the HSI that the supply of suitable habitats is greater than the current extent of the stands, then specific causes of the species' decline can be investigated. HSI values are more valuable for comparing habitats and less suitable for predicting the viability of species (Roloff and Kernohan 1999).

By studying the lifestyle of a species, it is also possible to identify the decisive factors for the reproduction of this species in the studied habitat, e.g., growth form and location of the larval food plant of a butterfly (see Fig. 10.13 on *Melitaea*, also see Sect. 10.11). In addition to method A, where we only compare occurrence and frequency, method B should be used to add fitness characteristics (e.g., reproduction) to assess different habitats. For the biology and survival of a species, this is of crucial importance (see Sect.

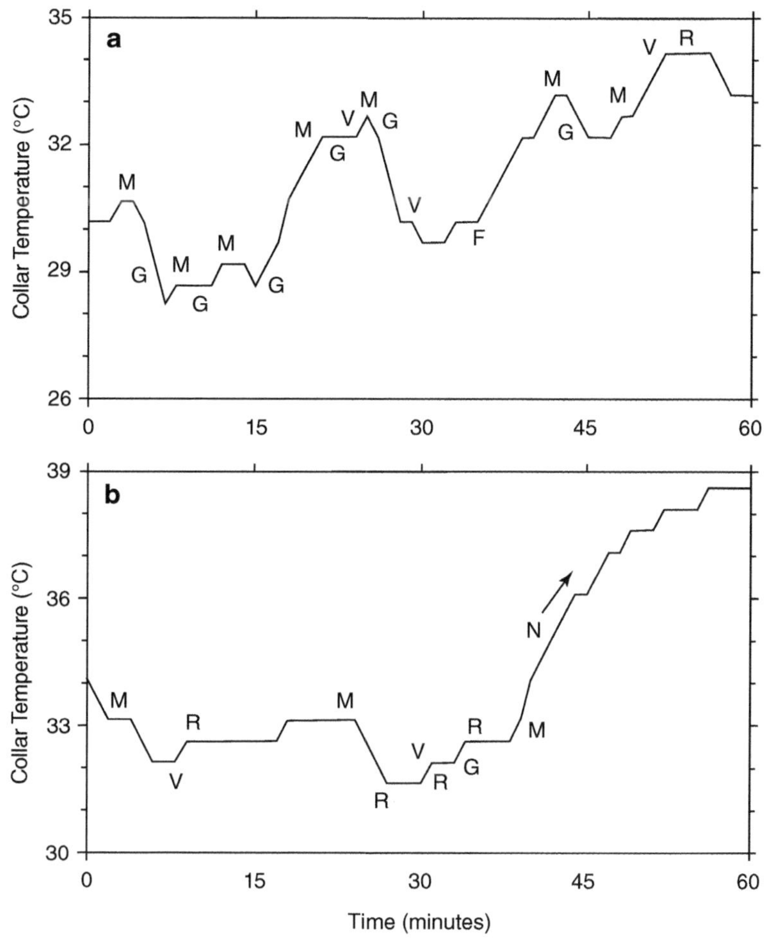

Fig. 9.8 Relationship between collar temperature and animal behavior in one-hour segments of the records of a female (**a**) and a male (**b**) fox squirrel. The ambient temperature was 15 °C in both cases. *F* feeding, *M* moving, *N* inside the nest, *G* grooming, *R* resting, *V* vigilance

10.13 on "good" and "bad" habitats). With method A, this aspect is practically never considered.

9.6 Which Characteristics Determine the Habitat Quality for an Animal Species?

Although the Paradise Flycatcher *(Tchitrea rufiventer)* prefers more or less intact timber forests (Hochwald), many nests are also found in degraded forests and even in plantations (Figs. 9.9 and 9.10). Here, nest checks showed that breeding success in the suboptimal habitats was higher than in the timber forest, which had more nest predators (Waltert 2000). This example also shows that a species classification based on habitat choice can be unreliable. Urban et al. (1997) state, for example, the following about the Paradise Flycatcher: "... *inhabits primary and secondary lowland and montane forests, also forest-grassland mosaic, oil-palm groves in forest clearings, gallery forest and thickets in savanna, cocoa plantations, and large gardens; may also enter mangroves* (e.g., *N Angola*)...". The best scientific results about the habitat choice of individual species can only be obtained through comparative studies of different investigation areas in a region.

A well-analyzed example of which features determine habitat quality from the perspective of a species was provided by Warren et al. (1986) on the Wood White (*Leptidea sinapis*), a widespread butterfly species. The authors found that the

Fig. 9.9 (**a**) Preserved timber forest area in Foret Bossematié, where *Tchitrea rufiventer* predominantly breed (Waltert 2000). Here, the species is three times more abundant than in degraded forests. Ivory Coast . (**b**). Degraded forest area in the Foret de Bossematié, Cote d'Ivoire. The flycatcher also occurs and propagates here

annual fluctuations in this butterfly mainly depend on the weather; wet and cold weather during the flight period and the early larval stages lead to fewer adults in the following year. The number of eggs laid at suitable locations and the survival of the young larval stages were also important factors causing annual fluctuations. The authors also found a connection between the shading of forest paths and the long-term trend in population development through the number of eggs laid. The incidence of light on the forest paths determined their quality as suitable egg-laying sites. The degree of direct shading on forest paths depended on the paths' width and direction, as well as the height of the adjacent trees. The abundance of adults was highest when the direct shading was 20–50%. Less than 20% and more than 50% direct shading made paths less suitable. For example, the number of Wood Whites increased sharply along railway lines when the bushes were removed. The researchers found no correlation between the degree of shading and the frequency of the food plant (*Lathyrus pratensis*) nor between the number of eggs and the frequency of this plant. The largest food plant biomasses could be correlated with the lowest egg numbers and vice versa. The artificial displacement of the larvae showed that they could survive equally well in the different degrees of shading. Female butterflies preferred to lay their eggs on tall food plants that stood out from the surrounding vegetation. The food plants could be made significantly more attractive for egg-laying by experimentally cutting away the surrounding vegetation. Ultimately, egg-laying is not influenced by the shading as such but by the growth form of the food plant under different shading conditions. The structure of the food plant in relation to the surrounding vegetation has emerged in this example as the critical factor for the suitabil-

Fig. 9.10 (**a**) Paradise Flycatcher (*Tchitrea rufiventer*) caught for ringing, Ivory Coast. (**b**) The nest of the Paradise Flycatcher hangs freely in the undergrowth. (**c**) Habitat offer and use of the Paradise Flycatcher in the Bossematié forest

ity of the habitat, i.e., its quality for the survival of a butterfly species. Against this background, it also becomes clear that the form of forest management influences the degree of shading and thus also determines how many individuals of these butterflies can live in it.

Ravenscroft (1994) showed that the larval food plant's growth form and nutrient content determine the distribution of the Chequered Skipper (*Carterocephalus palaemon*) in Scotland. Vogel and Johannesen (1996) demonstrated similar conditions for the Spotted Fritillary (*Melitaea didyma*), where the growth form of the food plant is also the critical factor for its population biology (see Sect. 10.11 key factors for a butterfly). The stinging nettle (*Urtica dioica*) is the larval food plant for several butterfly species (Small Tortoiseshell (*Aglais urticae*), Map Butterfly (*Araschnia levana*), and Peacock Butterfly (*Inachis io*)). However, each species chose different abiotic conditions and thus different locations of this widespread plant (Martin 2002).

Gottschalk (1996) was able to show experimentally that the development of a long-horned grasshopper (Ensifera), the Western Bush-cricket (*Platycleis albopunctata*), is controlled by the microclimate at the egg-laying site, in this case by the high temperature (see below Sect. 10.13). However, fewer seeds are available on southern slopes with loose vegetation as protein-rich food parts. Hence, the habitat choice of this grasshopper represents a compromise between the preference for warm locations on the one hand and areas rich in seed supply on the other.

The most conclusive method to assess the importance of a habitat factor for a specific species is experimental manipulation (method C). Experiments with the leaf litter (see Sect. 10.21) are still relatively simple. Key factors such as "perch" for a bird or "cavity supply" for nesting can also be easily manipulated. The territories of the Woodlark (*Lullula arborea*) could be extended in the border area to the Skylark, which prefers short vegetation, through artificial perches (Sommerkamp 1994; Vogel 1998). Often, it is not easy to establish an experimental control in the field. As soon as one works side by side or even at further apart locations, unforeseeable new factors are added. In nature, no two locations are the same.

9.7 Different Land Use Systems and Agroforestry

There are hardly any intact primary forests left in the tropics (e.g., Ghazoul and Sheil 2010; Corlett 2019). Land use for agricultural purposes is one of the main reasons for the loss of tropical rainforests and, thus, the loss of biodiversity. By combining forestry with crops (agroforestry in a strict sense) or pasture farming (silvopastoral systems), agroforestry practices are believed to be able to preserve some of the biodiversity while producing food. Agroforest areas are managed to maintain as much biodiversity as possible (Schroth et al. 2004; Santos et al. 2019; Boinot et al. 2022; Elevitch et al. 2018; Udawatta et al. 2019; Rahman et al. 2023; Santos et al. 2022; Tscharntke et al. 2015; Wurz et al. 2022; Yang et al. 2024). To what extent is this successful must be checked in each case (cf. Green et al. 2005; Cervantes-Lopez and Morante-Filho 2024; Torralba et al. 2016). The term agroforestry is also used when hedgerows, for example, are managed in agricultural landscapes (Boinot et al. 2022). This is particularly the case in Europe. There is no doubt that hedges or individual trees increase biodiversity in the agricultural landscape (e.g., Mühlenberg and Slowik 1997; López-Sánchez et al. 2016; Alignier et al. 2020; Boinot et al. 2020; see also Sect. 10.15). However, scientific proof of increased biodiversity in a cultivated landscape managed with agroforestry is often difficult. For comparison, a similarly managed cropland must be investigated, located in the same landscape context, and there must be a sufficient distance between agroforestry and cropland samples.

In the tropics, one must include tropical forests as naturally as possible (primary forests) and areas managed with annual crops (crop fields) and fallow land. The recording of vertebrates in the comparison of habitats is described with an example in Sect. 1.14. Species numbers alone are not meaningful for nature conservation (Harris and Silva-Lopez 1992; Meijaard et al. 2005). To evaluate the individual species, one needs a classification of the threat status, geographical distribution (endemic or widespread), and known classification as a forest-dependent species (or forest specialist species) or an agricultural generalist species. Also, a classification of the feeding habits of the found species provides valuable information, e.g., in specialized understory bird species in large frugivores. For a simple representation of the differences, an estimated accumulation curve of selected taxa is useful, where the collection effort (number of survey points, trap locations, or collection time) is plotted on the x-axis, and the determined species number is plotted on the *y*-axis (e.g., Thiollay 1995).

Multivariate statistical methods are needed to analyze the sensitivity of individual species or species groups concerning forest change through logging and fragmentation. Since most variables are rather categorical than numerical, a categorical principal components analysis is recommended (e.g., Data Theory Scaling System Group (DTSS) 1999). As dependent variables for the occurrence of a species relative to the independent variable "influence of logging" can be categorized as "intolerant" (density of species decreases by >20% in the first year after logging), "neutral" (no effects on species density after logging), or "tolerant" (logging increases density by >20% within the first year after logging) (Meijaard et al. 2005). More information is needed than just a mere increase or decrease in the population size of a species to derive recommendations for better forestry practices from the studies. Then, the ecology of each species must also be related to habitat use and forest change.

9.8 Are Species Useful as Bioindicators for Valuable Habitats?

Since the complete recording of species communities or animal groups will always be a laborious task, in practical nature conservation, one often strives to use only the presence of individual animal species as indicators for entire species communities or certain qualities of the habitat. (e.g., Schulze et al. 2005). However, a prerequisite is a very detailed knowledge of the requirements of the respective animal species and their habitat selectivity. If such knowledge is available, the size of the animal's habitat, requirements, and detectability determine which animal species are better suited as habitat indicators than others. The yellow wagtails (*Motacilla flava* group) are an example of the preferred breeding habitats studied in great detail in relationship of the plant assemblage (Artemyeva 2024). The author concluded with much information from different regions that *Motacilla* species are a good indicator for steppe-like grasslands, but the four species differ somewhat in preference (Fig. 9.11).

Almost every expert of a certain animal group claims that "his/her" animals are particularly well-suited as indicators. In addition, in almost every systematic animal group, there is a spectrum of species that ranges from representatives with very wide tolerances (euryoecious/eurypotent) to specialists with very narrow requirements (stenoeceous/stenopotent), be it orders like harvestmen and Passeriformes or families like ground beetles or classes like amphibians (e.g., Rainio and Niemelä 2003; Brodeur and Candioti 2017; Aguillón-Gutiérrez 2019). In aquatic habitats, it is easier to use organisms to indicate a certain water quality, see Sect. 6.2.6.5.

Compared to plants (e.g., indicator values, Ellenberg 1974; Ellenberg and Leuschner 2010; Tichý et al. 2023), the presence or absence of an animal species may depend on a larger amount of complex factors to capture so that not too far-reaching conclusions can be drawn from the observations. Indicator species are, therefore, of limited use and limited suitability for protecting animal species (Plachter 1980). Changes in their densities usually do not allow for compelling conclusions about the density changes of other species.

Grouse (Tetraoninae) are often considered bioindicators for particular forests (Fig. 9.12). They require large, contiguous habitats, but the species prefer different food plants throughout the seasons. Therefore, the composition of tree

Fig. 9.11 (**a**) *Motacilla flava* in the grassland of Mongolia. (**b**) *Motacilla citreola* favors more damp meadows (floodplain habitats) for breeding (Artemyeva 2024). Tajikistan

Fig. 9.12 (**a**) The Hazel Grouse (*Tetrastes (Bonasa) bonasia*) is typical for forests with earlier stages of succession but also lives in old forests when smaller clearings are present due to natural disturbances. Northern Mongolia. (**b**) The Capercaillie (*Tetrao parvirostris*) lives more in forests that host older stages of succession but likes to breed at the edge of the forests. See Fig. **c** and **d** below left (blue arrows to the nest) and below right. Northern Mongolia. (**c**) Forest edge with capercaillie nest on the ground in the foreground. Northern Mongolia. (**d**) Breeding capercaillie female on the nest

species in the forests is also crucial for their occurrence (Thüringer Landesanstalt für Umwelt Jena 1995; Bergmann et al. 1996; Müller et al. 2009). Only a natural landscape with many stages of succession is sufficient for the capercaillie as a habitat, as they require very different habitat types during their life cycle (Schroth 1994) (Fig. 9.12c, d).

Animals have another value in bioindication for environmental monitoring, i.e., for assessing anthropogenic, negative impacts on landscape and community (Schubert 1985; Ellenberg et al. 1985; Holt and Miller 2010; Wurster 2015; a global view is provided by Padayachee et al. 2022). When development trends of organisms and biocenoses are observed, this control can serve as a basis for promptly introducing appropriate protection measures.

Sedentary vertebrate species with large home ranges are suitable. Compared to plants, they integrate pollutant loads not only over time but also over significant spatial dimensions. Ellenberg et al. (1985) suggest Tawny Owl, Goshawk, Carrion Crow, and Magpie as suitable biomonitors for Central Europe and have successfully documented emission loads of heavy metals on Goshawk feathers and loads of polychlorinated biphenyls (PCBs) on Magpie eggs. Goshawk feathers (see also Ellenberg et al. 1986) and Magpie eggs can be collected in a standardized manner. Thanks to the nationwide distribution of these birds, the evaluation can also be carried out on a large scale. For “remote sensing” of distant breeding areas, certain geese, swans, cranes, and waders would be suitable at their late summer resting places or wintering areas (Ellenberg et al. 1985).

Even at the exact location, there are often significant differences in contamination with pollutants between individual species. This is not primarily due to the species-specific physiology of the animals but to the individually different diets (May and Ellenberg 1985). For standardized comparisons, therefore, a more precise knowledge of the diet of the selected species is necessary. Species with the same diet spectra at different locations should be preferred for biomonitors.

The eggs and nestling carcasses of tree swallows in the North American Great Lakes were successfully analysed for the biomonitoring of the degree of contamination with PCBs and heavy metals in water bodies (Custer et al. 2012; 2021). Ankley et al. (2023) present a broad-based study on the detection methodology and the risk effects on various organisms (birds, fish, invertebrates) due to contamination with various toxic

Case Study

Dung Beetles as Bioindicators for Pasture Systems

Bioindicators are species, communities, or biological processes whose function, population, or status can be used to assess the quality of a habitat, ecosystem, or environment (Holt and Miller 2010; Wikipedia 2024) (Fig. 9.13).

After finishing my diploma thesis on dung beetle ecology (Wassmer 1991; Wassmer and Sowig 1994) and working for another year on one of the most diverse dung beetle communities in Europe (Wassmer 1994, 1995), I published an article on the suitability of dung beetles as bioindicators for the conservational integrity and value of pasture ecosystems (Waßmer 1995). I came to this suggestion after realizing the tight juncture between dung beetles and the pasture ecosystem due to the beetles' dependency on herbivore dung for food, habitat, and parental care. It seemed logical that a pasture system that features various habitats (forest, grassland, different dung types, and topography, Fig. 9.14), is located in a climate suitable for inhabitation for most of the year and is set at the crossroads of floral, faunal, and historical realms, will have higher species diversity than areas without these features. I furthermore proposed that the biodiversity of many other plants and animals should follow the same trend as that of dung beetles.

At about the same time, Gonzalo Halffter and Mario E. Favila proposed dung beetles as bioindicators for tropical forests and anthropogenously derived formations (Favila and Halffter 1997; Gonzalo Halffter and Favila 1993). Since then, the vast majority of studies about dung beetles as bioindicators come from the Meso- and South American tropics and subtropics (Arellano et al. 2008; Audino et al. 2014; Cajaiba et al. 2018; Davis et al. 2004; Escobar et al. 2008; Filgueiras et al. 2015; Gonzalez-Vainer et al. 2012; Halffter and Arellano 2002; Hayes et al. 2009; Liberal et al. 2011; McGeoch et al. 2002; Rodrigues et al. 2013; Ros et al. 2012; Viegas et al. 2014).

Only a few studies reported on the usability of dung beetles as bioindicators from other areas, including subtropical South Africa (Davis et al. 2004; Jacobs et al. 2010; McGeoch et al. 2002; Stanbrook et al. 2021) and Indonesia (Davis et al. 2001)—some of which take a comparative view on similarities between the Neotropic and Afrotropic regions (Davis et al. 2004). Much rarer are studies from warm temperate climates such as in Australia (Gollan et al. 2011, 2013) and are still lacking from cold-temperate climates besides my 1995 study (Waßmer 1995). Nichols and Gardner (2011), and Spector (2006) provide com-

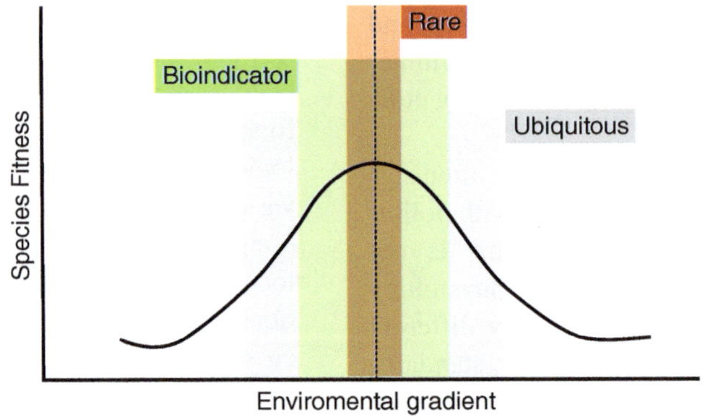

Fig. 9.13 Three types of species classified as ubiquitous, rare, and a (potential) bioindicator based on their environmental tolerance. Reproduced with permission from Tryjanowski and Morelli (2017), based on Holt and Miller (2010). The dashed line in the center indicates where the biological entity's fitness or abundance is optimal. In contrast, the orange, green, and gray rectangles around the dashed line and the corresponding parts of the bell curve show the optimum range for the different classifications. Good bioindicators have a moderate tolerance range compared to rare and ubiquitous species. This tolerance variability indicates sensitivity to environmental change while still being able to endure moderate changes

Fig. 9.14 High biodiversity pasture system utilizing mixed livestock grazing (cows and sheep) in open grassland with forest patches. Picture credit: with permission from Greg Brann Consulting, Gregbrann.com, Adolphus Kentucky

parative reviews on the usefulness of dung beetles as bioindicators.

The species composition and species richness of a dung beetle community also depends on the grazing density (Allan et al. 2015; Laliberté et al. 2010; Noriega et al. 2023), the grazing animal species (Wassmer 1995, 2014; Whipple and Hoback 2012) and the history of land use of a pasture area (Barragan et al. 2011; Filgueiras et al. 2015; Wassmer et al. 1994; Waβmer 1995) and reflect factors that influence the biodiversity of an ecosystem as a whole and thus indicate its conservation priority.

The preservation of extensively managed pasture areas, with their significantly higher biodiversity compared to modern intensive pastures, is becoming increasingly important for conservation and landscape management (Anderle et al. 2023; Guariento et al. 2023; Šálek et al. 2022; Shipley et al. 2024).

substances. This meta-analysis also refers to the North American Great Lakes.

References

Aguillón-Gutiérrez D.R. 2019: *Amphibians as biological indicators.*—In: Jenkins O.P. (ed.): Advances in Animal Science and Zoology, Vol. 14, 61–82. Nova science publishers, New York. https://www.academia.edu/41461673/Amphibians_as_biological_indicators

Alignier A., Uroy L., Aviron S. 2020: The role of hedgerows in supporting biodiversity and other ecosystem services in intensively managed agricultural landscapes.—In: Bàrberi P., Moonen A-C. (eds.): *Reconciling agricultural production with biodiver-*

sity conservation. Cambridge, UK: Burleigh Dodds Science Publishing; 2020. p. 177–204.

Allan, E., Manning, P., Alt, F., Binkenstein, J., Blaser, S., Blüthgen, N., et al. (2015). Land use intensification alters ecosystem multifunctionality via loss of biodiversity and changes to functional composition. *Ecology Letters*, *18*(8), 834–843. https://doi.org/10.1111/ele.12469

Altmann, J. (1974). Observational Study of Behavior: Sampling Methods. *Behaviour*, *49*(3–4), 227–266. https://doi.org/10.1163/156853974X00534

Anderle, M., Brambilla, M., Hilpold, A., Matabishi, J. G., Paniccia, C., Rocchini, D., et al. (2023). Habitat heterogeneity promotes bird diversity in agricultural landscapes: Insights from remote sensing data. *Basic and Applied Ecology*, *70*, 38–49. https://doi.org/10.1016/j.baae.2023.04.006

Ankley, G. T., Corsi, S. R., Custer, C. M., Ekman, D. R., Hummel, S. L., Kimbrough, K. L., et al. (2023). Assessing Contaminants of Emerging Concern in the Great Lakes Ecosystem: A Decade of Method Development and Practical Application. *Environmental Toxicology and Chemistry*, 42(12), 2506–2518. https://doi.org/10.1002/etc.5740

Arellano, L., Leon-Cortes, J. L., & Halffter, G. (2008). Response of dung beetle assemblages to landscape structure in remnant natural and modified habitats in southern Mexico. *Insect Conservation and Diversity*, *1*(4), 253–262. https://doi.org/10.1111/j.1752-4598.2008.00033.x

Artemyeva E.A. 2024: Yellow Wagtails (Passeriformes, Motacillidae) are indicators of steppe and meadow landscapes.—*Rivista Italiana Di Ornitologia - Research in Ornithology* 94 (1), 7–30. https://doi.org/10.4081/rio.2024.712

Audino, L. D., Louzada, J., & Comita, L. (2014). Dung beetles as indicators of tropical forest restoration success: Is it possible to recover species and functional diversity? *Biological Conservation*, *169*, 248–257. https://doi.org/10.1016/j.biocon.2013.11.023

Badry, N. A., MacMillan, G. A., Stern, E. R., Landry-Cuerrier, M., Hickey, G. M., & Humphries, M. M. (2024). Boundary Spanning Methodological Approaches for Collaborative Moose Governance in Eeyou Istchee. *Environmental Management*, *74*(1), 132–147. https://doi.org/10.1007/s00267-023-01918-6

Bandorf H. & Laubender H. (Hrsg.) 1982: *Die Vogelwelt zwischen Steigerwald und Rhön*. Band 1 und 2.—Schriftenreihe des Landesbundes für Vogelschutz in Bayern, Münnerstadt und Schweinfurt, 1071 p.

Barragan, F., Moreno, C. E., Escobar, F., Halffter, G., & Navarrete, D. (2011). Negative Impacts of Human Land Use on Dung Beetle Functional Diversity. *Plos One*, *6*(3). https://doi.org/10.1371/journal.pone.0017976

Bergmann H.-H., Klaus S., Müller F., Scherzinger W., Swenson J.E. & Wiesner J. 1996: *Die Haselhühner*.—Die Neue Brehm-Bücherei Bd. 77, Westarp Wissenschaften, Magdeburg, 278 p.

Bignall E. & Curtis D.J. (eds.) 1988: *Choughs and land-use in Europe*.—Scottish Chough Study Group. Tarbert.

Bloche D.A.F., Thorup K., Olsen K., Ekberg P., Larsen P.E., Strange K-E. & Tottrup A.P. 2023: Breeding biology of Red-backed Shrikes (*Lanius collurio*): distribution, performance and post-fledging survival in Denmark.—*Ornis Fennica* 100, 1–15

Blumenthal C.L. 1981: Einheimische *Carabus*-Arten als Bioindikatoren.— *Jber. naturwiss Ver. Wuppertal* 34, 70–77

Bobo K.S., Waltert M., Fermon H., Njokagbor J. & Mühlenberg M. 2006: From forest to farmland: butterfly diversity and habitat associations along a gradient of forest conversion in Southwestern Cameroon.—*Journal of Insect Conservation* 10, 29–42

Boinot S., Mézière D., Poulmarc'h J., Saintilan A., Lauri P-E., Sarthou J-P. 2020: Promoting generalist predators of crop pests in alley cropping agroforestry fields: farming system matters *Ecol. Eng*. 158: 106041. https://doi.org/10.1016/j.ecoleng.2020.106041.

Boinot, S., Barkaoui, K., Mézière, D., Lauri, P.-E., Sarthou, J.-P., & Alignier, A. (2022). Research on agroforestry systems and biodiversity conservation: what can we conclude so far and what should we improve? *BMC Ecol Evol*, *22*(1), 24. https://doi.org/10.1186/s12862-022-01977-z

Braun R. 1969: Zur Autökologie und Phänologie der Spinnen (Araneida) des Naturschutzgebietes, "Mainzer Sand". .— *Mz. naturwiss Arch*. 8, 193–288

Braun R. & Rabeler W. 1969: Zur Autökologie und Phänologie der Spinnenfauna des nordwestdeutschen Altmoränen-Gebiets –. *Abh Senckenberg. naturforsch. Ges*. 522, 1–89

Brodeur J.C. & Candioti J.V. 2017: Impacts of Agriculture and Pesticides on Amphibian Terrestrial Life Stages: Potential Biomonitor/ Bioindicator Species for the Pampa Region of Argentina.—In: Larramendy M.L., Waters M.D. & Marrs T. (eds.): Ecotoxicology and Genotoxicology: Non-traditional Aquatic Models. Chap. 8, 163–194.—Issues in Toxicology No. 33. Royal Society of Chemistry, 300 p. https://doi.org/10.1039/9781788010573-00163

Brooks, R. P. (1997). Improving habitat suitability index models. *Wildlife Society Bulletin*, *25*(1), 163–167.

Bullock I.D., Drewett R.R. & Mickelburgh S.P. 1983: The Chough in Britain and Ireland.—*Brit. Birds* 76, 377–401

Cajaiba, R. L., Périco, E., da Silva, W. B., Leote, P., & Santos, M. (2018). Are Small Dung Beetles (Aphodiinae) useful for monitoring neotropical forests' ecological status? Lessons from a preliminary case study in the Brazilian Amazon. *Forest Ecology and Management*, *429*, 115–123. https://doi.org/10.1016/j.foreco.2018.07.005

Ceresa F., Bogliani G., Pedrini P. & Brambilla M. 2012: The importance of key marginal habitat features for birds in farmland: An assessment of habitat preferences of Red-backed Shrikes *Lanius collurio* in the Italian Alps .—*Bird Study* 59(3), 327–334

Cervantes-Lopez, M. D J., & Morante-Filho, J. C. (2024). A global meta-analysis on patterns of amphibian and reptile diversity in agroforestry systems. *Global Ecology And Conservation*, *51*, e02914. https://doi.org/10.1016/j.gecco.2024.e02914

Collins S. 1983: Geographic variation in habitat structure of the Black-throated Green Warbler (*Dendroica virens*). *Auk* 100, 382–389

Corlett R.T. 2019: *The Ecology of Tropical East Asia* (3rd ed.).—Oxford University Press, 320 p.

Custer, T. W., Custer, C. M., Thogmartin, W. E., Dummer, P. M., Rossmann, R., Kenow, K. P., & Meyer, M. W. (2012). Mercury and other element exposure in tree swallows nesting at low pH and neutral pH lakes in northern Wisconsin USA. *Environmental Pollution*, 163, 68–76. https://doi.org/10.1016/j.envpol.2011.12.017

Custer, C. M., Custer, T. W., & Dummer, P. M. (2021). Using tree swallows to assess reductions in PCB exposure as a result of dredging at Great Lakes Restoration Initiative (GLRI) sites in the Upper Midwest, USA. *Ecotoxicology*, 30(6), 1116–1125. https://doi.org/10.1007/s10646-021-02420-7

Data Theory Scaling System Group (DTSS): 1999. CATPCA Version 1.0. *Faculty of Social and Behavioral Sciences. Leiden University*, The Netherlands, Leiden, the Netherlands.

Davis, A. J., Holloway, J. D., Huijbregts, H., Krikken, J., Kirk-Spriggs, A. H., & Sutton, S. L. (2001). Dung beetles as indicators of change in the forests of northern Borneo. *Journal of Applied Ecology*, *38*(3), 593–616. https://doi.org/10.1046/j.1365-2664.2001.00619.x

Davis, A., Scholtz, C. H., Dooley, P. W., Bham, N., & Kryger, U. (2004). Scarabaeine dung beetles as indicators of biodiversity, habitat transformation and pest control chemicals in agro-ecosystems : review article. *South African Journal of Science*, *100*(9), 415–424. https://doi.org/10.10520/EJC96297

Dawkins M.S. 2007: *Observing Animal Behaviour. Design and Analysis of Quantitative Data.*— Oxford University Press, 158

Eberhardt L.L. & Thomas J.M. 1991: Designing environmental field studies.—*Ecological Monographs* 61 (1), 53–73

Elevitch, C. R., Mazaroli, D. N., & Ragone, D. (2018). Agroforestry Standards for Regenerative Agriculture. *Sustainability*, *10*(9), 3337. https://doi.org/10.3390/su10093337

Ellenberg, H. (1974). Zeigerwerte der Gefässpflanzen Mitteleuropas. Göttingen: Goltze

Ellenberg, H., & Leuschner, C. (2010). Vegetation Mitteleuropas mit den Alpen: In ökologischer, dynamischer und historischer Sicht. Stuttgart: UTB, Stuttgart

Ellenberg H., Dietrich J., Gast F., Hahn E. & Mayr R. 1985: Vögel als Biomonitoren für die Schadstoffbelastung von Landschaftsausschnitten—ein Überblick.— *Z. f Jagdwissenschaft* 31, 22–33

Ellenberg H., Dietrich J., Stoeppler M. & Nürnberg H.W. 1986: Habicht-Mauserfedern als hochintegrierende Biomonitoren für die Schadstoffbelastung von Landschaftsausschnitten.—*Allg Forst–Zeitschrift* 1/2, 23–25

Escobar, F., Halffter, G., Solis, A., Halffter, V., & Navarrete, D. (2008). Temporal shifts in dung beetle community structure within a protected area of tropical wet forest: a 35-year study and its implications for long-term conservation. *Journal of Applied Ecology*, *45*(6), 1584–1592. https://doi.org/10.1111/j.1365-2664.2008.01551.x

Favila, M. E., & Halffter, G. (1997). The use of indicator groups for measuring biodiversity as related to community structure and function. *Acta Zoológica Mexicana (nueva serie)*, *72*, 1–25. https://doi.org/10.21829/azm.1997.72721734

Fermon H., Waltert M. & Mühlenberg M. 2003: Movement and vertical stratification of fruit-feeding butterflies in a managed West African rainforest.—*Journal of Insect Conservation* 7, 7–19

Fermon H., Waltert M., Vane-Wright R.I. & Mühlenberg M. 2005: Forest use and vertical stratification in fruit-feeding butterflies of Sulawesi, Indonesia: impacts for conservation.—*Biodiversity and Conservation* 14, 333–350

Filgueiras, B. K. C., Tabarelli, M., Leal, I. R., Vaz-de-Mello, F. Z., & Iannuzzi, L. (2015). Dung beetle persistence in human-modified landscapes: Combining indicator species with anthropogenic land use and fragmentation-related effects. *Ecological Indicators*, *55*, 65–73. https://doi.org/10.1016/j.ecolind.2015.02.032

Flade M. & Schwarz J. 1999: Current status and new results from German Common Birds Census (DDA Monitoring Program).—In: Flade M. & Helbig A.J. (eds.): *Die Vogelwelt—Beiträge zur Vogelkunde.*— Wiebelheim (Aula-Verlag), 47–52

Flandorfer P. (2022, November 11): ANOVA - Conducting and interpreting variance analysis.—*Scribbr*. Retrieved on January 19, 2023, from https://www.scribbr.de/statistik/anova-varianzanalyse/

Förster D. 1986: Zur Habitatbindung von Laufkäfern (Coleoptera: Carabidae) in verschiedenen Höhenlagen und Nutzungstypen.—Dissertation University of Würzburg

Fryxell J.M., Sinclair A.R.E. & Caughley G. 2014: *Wildlife Ecology, Conservation, and Management* (3rd ed.)—Wiley-Blackwell, 528 p.

Gavashelishvili A. & McGrady M.J. 2006: Breeding site selection by bearded vulture (*Gypaetus barbatus*) and Eurasian griffon (*Gyps fulvus*) in the Caucasus.—*Animal Conservation* 9 (2), 159–170

Gerken B. 1984: Die Sammlung von Libellen-Exuvien. Hinweise zur Methodik der Sammlung und zum Schlüpfort von Libellen. *Libellula* 3 (3/4), 59–72

Ghazoul J. & Sheil D. 2010: *Tropical Rain Forest Ecology, Diversity, and Conservation.*— Oxford University Press, 516 p.

Gollan, J. R., Reid, C. A. M., Barnes, P. B., & Wilkie, L. (2011). The ratio of exotic-to-native dung beetles can indicate habitat quality in riparian restoration. *Insect*

Conservation and Diversity, *4*(2), 123–131. https://doi.org/10.1111/j.1752-4598.2010.00115.x

Gollan, J. R., de Bruyn, L. L., Reid, N., & Wilkie, L. (2013). Monitoring the ecosystem service provided by dung beetles offers benefits over commonly used biodiversity metrics and a traditional trapping method. *Journal for Nature Conservation*, *21*(3), 183–188. https://doi.org/10.1016/j.jnc.2012.12.004

Gonzalez-Vainer, P., Morelli, E., & Defeo, O. (2012). Differences in Coprophilous Beetle Communities Structure in Sierra de Minas (Uruguay): a Mosaic Landscape. *Neotropical Entomology*, *41*(5), 366–374. https://doi.org/10.1007/s13744-012-0062-8

Gottschalk E. 1996: Population Vulnerability of the Gray Bush Cricket, *Platycleis albopunctata* (Goeze 1778) (Ensifera: Tettigoniidae).—In: Settele J., Margules C., Poschlod P. & Henle K. (eds.): *Species survival in fragmented landscapes* . 324–328, Kluwer Verlag, Dordrecht, Boston, London, 381 p.

Green R.E., Cornell S.J., Scharlemann J.P.W. & Balmford A. 2005: Farming and the fate of wild nature.—*Science* 307, 550–555

Guariento, E., Rüdisser, J., Fiedler, K., Paniccia, C., Stifter, S., Tappeiner, U., et al. (2023). From diverse to simple: butterfly communities erode from extensive grasslands to intensively used farmland and urban areas. *Biodiversity and Conservation*, *32*(3), 867–882. https://doi.org/10.1007/s10531-022-02498-3

Günther A., Nigmann U., Achtziger R. & Gruttke H. 2005: Analyse der Gefährdungsursachen planungsrelevanter Tiergruppen in Deutschland.—*Naturschutz und Biologische Vielfalt* 21, Bundesamt für Naturschutz Bonn-Bad Godesberg, 605 p.

Hagemeijer W. & Blair M.J. 1997: *The EBCC Atlas of European Breeding Birds*—Their Distribution and Abundance.—London, Poyser Natural History, 903 p.

Halffter, G., & Arellano, L. (2002). Response of dung beetle diversity to human-induced changes in a tropical landscape. *Biotropica*, *34*(1), 144–154. https://doi.org/10.1111/j.1744-7429.2002.tb00250.x

Halffter, Gonzalo, & Favila, M. E. (1993). The Scarabaeinae (Insecta: Coleoptera) an animal group for analysing, inventorying and monitoring biodiversity in tropical rainforest and modified landscapes. *Biology International*, *27*, 15–21.

Harris L.D. & Silva-Lopez G. 1992: Forest fragmentation and the conservation of biological diversity, 197–237 —In P. L. Fiedler, and S. K. Jain, editors. *Conservation Biology*. Chapman and Hall, London, UK.

Hayes, L., Mann, D. J., Monastyrskii, A. L., & Lewis, O. T. (2009). Rapid assessments of tropical dung beetle and butterfly assemblages: contrasting trends along a forest disturbance gradient. *Insect Conservation and Diversity*, *2*(3), 194–203.

Heydemann B. 1955: Carabiden der Kulturfelder als ökologische Indikatoren. — *Ber. ü. d Wandervers. Dtsch. Entomologen* 7, 172–185

Holt, E. A., & Miller, S. W. (2010). Bioindicators: Using Organisms to Measure Environmental Impacts | Learn Science at Scitable. *Nature Education Knowledge*, *3*(10), 8.

Huhta V. 1965: Ecology of spiders in the soil and litter of Finnish forest. —*Ann Zool. Fennici* 2, 260–308

Jacobs, C. T., Scholtz, C. H., Escobar, F., & Davis, A. L. V. (2010). How might intensification of farming influence dung beetle diversity (Coleoptera: Scarabaeidae) in Maputo Special Reserve (Mozambique)? *Journal of Insect Conservation*, *14*(4), 389–399. https://doi.org/10.1007/s10841-010-9270-x

Joern A. 1982: Vegetation structure and microhabitat selection in grasshoppers (Orthoptera, Acrididae).—*The Southwestern Naturalist* 27 (2), 197–209

Kenward R. 1987: *Wildlife Radio Tagging, Equipment, Field Techniques and Data Analysis*.—Academic Press, Inc. (London) Ltd. (Publ.), Cambridge University Press, Cambridge, 222 p.

Kenward R.E. 2001: *A Manual for Wildlife Radio Tagging*,—San Diego, CA: Academic Press, 350 p.

Koch K. 1989: *Die Käfer Mitteleuropas. Ökologie*. Band 1.—Verl. Goecke & Evers, Krefeld, 440 p.

Kühn I. 1982: *Faunistisch-Ökologische Untersuchungen an epigäischen Spinnen (Araneae)* unter besonderer Berücksichtigung ihrer bioindikatorischen Bedeutung.—Fachbeitrag im Rahmen der Modellstudie „Zoologischer Artenschutz in Bayern“, Fabrikschleichach, 151 p.

Laliberté, E., Wells, J. A., DeClerck, F., Metcalfe, D. J., Catterall, C. P., Queiroz, C., et al. (2010). Land-use intensification reduces functional redundancy and response diversity in plant communities. *Ecology Letters*, *13*(1), 76–86. https://doi.org/10.1111/j.1461-0248.2009.01403.x

Lang M., Bandorf H., Dornberger W., Klein H. & Mattern U. 1990: Verbreitung, Bestandsentwicklung und Ökologie des Ortolans (*Emberiza hortulana*) in Franken.— *Ökologie der Vögel* 12, 97–126

Lang M., Bandorf H., Dornberger W., Klein H. & Mattern U. 1992: Der Ortolan (*Emberiza hortulana*) in Franken: Möglichkeiten eines strukturellen Biotopschutzes.—*Ornithologen-Kalender 1992*, AULA-Verlag Wiesbaden, 178–187

Larsen T.B. 1999: *Butterflies of West Africa*.—Origins, Natural History, Diversity and Conservation.—Larsen CD-ROM. Torben.larsen@hotmail.com. Manila.

Larson D.L. & Bock C.E. 1986: Determining avian habitat preference by bird-centered Vegetation sampling.—In: Verner J., Morrison M.L. & Ralph C.J. (eds.): *Wildlife 2000. Modeling Habitat Relationships of Terrestrial Vertebrates*, 37–43.—University of Wisconsin Press, Wisconsin u. London, 470 p.

Leugger U. 1991: Einfluss der Habitatqualität auf den Verlauf des Brutgeschäftes und den damit verbundenen Aufwand der Eltern beim Neuntöter.—Vortrag auf der 124. *Jahresversammlung der Deutschen Ornithologen-Gesellschaft* in Interlaken/Schweiz

Liberal, C. N., de Farias, A. M. I., Meiado, M. V., Filgueiras, B. K. C., & Iannuzzi, L. (2011). How habitat change and rainfall affect dung beetle diversity in

Caatinga, a Brazilian semi-arid ecosystem. *Journal of Insect Science*, *11*, 1–11.

Lõhmus, A., Kont, R., Runnel, K., Vaikre, M., & Remm, L. (2020). Habitat Models of Focal Species Can Link Ecology and Decision-Making in Sustainable Forest Management. *Forests*, *11*(7), 721. https://doi.org/10.3390/f11070721

López-Sánchez A., San Miguel A., Dirzo R., Roig S. 2016: Scattered trees and livestock grazing as keystones organisms for sustainable use and conservation of Mediterranean dehesas.—*J. Nat. Conserv.* 33, 58–67. https://doi.org/10.1016/j.jnc.2016.07.003.

Luczak J. 1979: Spiders in agrocoenoses. .— *Pol. ecol Stud.* 5, 152–200

Martin K. 2002: *Ökologie der Biozönosen.*— Springer, 325 p.

May R. & Ellenberg H. 1985: Ein Freilandexperiment zur Ökologie der Schadstoff-Kontamination von Vögeln und Folgerungen für die Verwendung von Organismen als Biomonitoren.— *Ökol. Vögel (Ecol. Birds)* 7, 97–112

McGeoch, M. A., Van Rensburg, B. J., & Botes, A. (2002). The verification and application of bioindicators: a case study of dung beetles in a savanna ecosystem. *Journal of Applied Ecology*, *39*(4), 661–672. https://doi.org/10.1046/j.1365-2664.2002.00743.x

Meijaard E., Sheil D., Nasi R., Augeri D., Rosenbaum B., Iskandar D., Setyawati T., Lammertink M., Rachmatika I., Wong A., Soehartono T., Stanley S. & O'Brien T. 2005: *Life after logging* . Reconciling wildlife conservation and production forestry in Indonesian Borneo.—CIFOR and UNESCO, Jakarta, 370 p

Menz M.H.M., Mosimann-Kampe P. and Arlettaz R. 2009: Foraging Habitat Selection in the Last Ortolan Bunting Emberiza hortulana Population in Switzerland: Final Lessons before Extinction. - *Ardea* 97, 323–333. https://doi.org/10.5253/078.097.0308

Meyer W., Bartels T., Neurand K. & Lange N. 1993: Specific structural features of the woodpecker's beak, as revealed by direct magnifying microfocal radiography (Aves, Piciformes) .—*European Archives of Biology* 104, 149–155

Morrison M.L., Marcot B.G. & Mannan R.W. 2006: *Wildlife-Habitat Relationships: Concepts and Applications* (3rd ed.).— Island Press, 520 p.

Mühlenberg M. & Slowik J. 1997: *Kulturlandschaft als Lebensraum.*—UTB 1947 für Wissenschaft, Quelle & Meyer Verlag Wiesbaden, 312 p.

Mühlenberg M. & Werres W. 1983: Lebensraumverkleinerung und ihre Folgen für einzelne Tiergemeinschaften. Experimentelle Untersuchung auf einer Wiesenfläche.—*Natur und Landschaft* 58 (2), 43–50, Stuttgart

Müller D., Schröder B. & Müller J. 2009: Modelling habitat selection of the cryptic Hazel Grouse *Bonasia bonasia* in a montane forest.— *Journal Ornithology* 150, 717–732

Noriega, J. A., Hortal, J., deCastro-Arrazola, I., Alves-Martins, F., Ortega, J. C. G., Bini, L. M., et al. (2023). Dung removal increases under higher dung beetle functional diversity regardless of grazing intensification. *Nature Communications*, *14*(1), 8070. https://doi.org/10.1038/s41467-023-43760-8

Padayachee K., Amar A. and Reynolds C. 2022: A dangerous pesticide isn't being monitored in key bird of prey populations - we're shedding light on that gap.— *The Conversation*, Paris. Published: December 4, 2022

Pearsall S.H., Durham D. and Eagar D.C. 1986: Evaluation methods in the United States. p. 111–133—In: Usher M.B. (ed.), *Wildlife Conservation Evaluation.* Chapman and Hall Ltd, London, 396 p.

Plachter H. 1980: Grundsätze und Praxis des Tierartenschutzes in Bayern.— *Schriftenreihe Naturschutz und Landschaftspflege* 12, 7–15

Rahman, M. H. U, Ahrends, H. E., Raza, A., & Gaiser, T. (2023). Current approaches for modeling ecosystem services and biodiversity in agroforestry systems: Challenges and ways forward. *Frontiers in Forests and Global Change*, *5*. https://doi.org/10.3389/ffgc.2022.1032442

Rainio J. and Niemelä J. 2003: Ground beetles (Coleoptera: Carabidae) as bioindicators.—*Biodiversity and Conservation* 12, 487–506. https://doi.org/10.1023/A:1022412617568

Ravenscroft N.O.M. 1994: The ecology of the chequered skipper butterfly (*Caterocephalus palaemon*) in Scotland: 1. Microhabitat. .— *J. of Appl. Ecol.* 31 (4), 613–622

Rodrigues, M. M., Uchoa, M. A., & Ide, S. (2013). Dung beetles (Coleoptera: Scarabaeoidea) in three landscapes in Mato Grosso do Sul, Brazil, *Brazilian Journal of Biology*, *73*(1), 211–220.

Roloff G.J. & Kernohan B.J. 1999: Evaluating reliability of habitat suitability index models.—*Wildlife Society Bulletin* 27 (4), 973–985

Ros, M., Escobar, F., & Halffter, G. (2012). How dung beetles respond to a human-modified variegated landscape in Mexican cloud forest: a study of biodiversity integrating ecological and biogeographical perspectives. *Diversity and Distributions*, *18*(4), 377–389. https://doi.org/10.1111/j.1472-4642.2011.00834.x

Rothhaupt G. and Vogel B. 1996: Survival of Birds in Fragmented Landscapes, 230–236.—In: Settele J., Margules C.R., Poschlod P. & Henle K. (eds.), *Species Survival in Fragmented Landscapes.*— Kluwer Academic Publishers, 381 p.

Šálek, M., Bažant, M., Żmihorski, M., & Gamero, A. (2022). Evaluating conservation tools in intensively-used farmland: Higher bird and mammal diversity in seed-rich strips during winter. *Agriculture, Ecosystems & Environment*, *327*, 107844. https://doi.org/10.1016/j.agee.2021.107844

Santos P.Z.F., Crouzeilles R. & Sansevero J.B.B. 2019: Can agroforestry systems enhance biodiversity and ecosystem service provision in agricultural landscapes? A meta-analysis for the Brazilian Atlantic Forest.—*Forest Ecol Manag.* 433, 140–145. https://doi.org/10.1016/j.foreco.2018.10.064.

Santos, M., Cajaiba, R. L., Bastos, R., Gonzalez, D., Petrescu Bakış, A.-L., Ferreira, D., et al. (2022). Why Do Agroforestry Systems Enhance Biodiversity? Evidence From Habitat Amount Hypothesis Predictions. *Frontiers in Ecology and Evolution, 9*. https://doi.org/10.3389/fevo.2021.630151

Schamberger M. & O'Neil J. 1986: Concepts and constraints of habitat model testing, 5–10.—In: Verner J., Morrison M. & Ralph C.J. (eds): *Wildlife 2000: Modeling habitat relationships of terrestrial vertebrates*.— University of Wisconsin Press, 470 p.

Schulze, E.-D., Wirth, C., Mollicone, D., & Ziegler, W. (2005). Succession after stand replacing disturbances by fire, wind throw, and insects in the dark Taiga of Central Siberia. Oecologia, 146(1), 77–88. https://doi.org/10.1007/s00442-005-0173-6

Schröder B. & Reineking B. 2004: Modellierung der Art-Habitat-Beziehung—ein Überblick über die Verfahren der Habitatmodellierung, 5–25.—In: Dormann C.F., Blaschke T., Lausch A. Schröder B. & Söndgerath D. (eds.): *Habitatmodelle—Methodik, Anwendung, Nutzen*.—Tagungsband zum Workshop UFZ Leipzig, UFZ Berichte 9/2004, 178 p.

Schroth K.-E. 1994: Zum Lebensraum des Auerhuhns (*Tetrao urogallus*) im Nordschwarzwald.—*Mitt. Forstl. Veruchs- u. Forschungsanst. Baden-Württ./ Freiburg* 178, 133 p.

Schroth G., da Fonseca G.A.B., Harvey C.A., Gascon C., Vasconcelos H.L. & Izac A-M. (eds.) 2004: *Agroforestry and Biodiversity Conservation in Tropical Landscapes*.—Bibliovault OAI Repository, the University of Chicago Press, 575 p.

Schubert R. (Hrsg.) 1985: *Bioindikation in terrestrischen Ökosystemen*.—Fischer-Verl., Stuttgart, 327 p.

Schweizerischer Bund für Naturschutz (SBFN) 1987: *Tagfalter und ihre Lebensräume*. Arten—Gefährdung—Schutz.—Holliger Verlag, Basel, 516 p.

Seccombe, E., & Salguero-Gómez, R. (2024). An improved habitat suitability index for the great crested newt Triturus cristatus. *Herpetological Journal, 34*(2), 55–67.

Shipley, J. R., Frei, E. R., Bergamini, A., Boch, S., Schulz, T., Ginzler, C., et al. (2024). Agricultural practices and biodiversity: Conservation policies for semi-natural grasslands in Europe. *Current Biology, 34*(16), R753–R761. https://doi.org/10.1016/j.cub.2024.06.062

Sommerkamp K. 1994: Untersuchungen zur Konkurrenz zwischen Heidelerche und Feldlerche um Habitate.—Unveröffentlichte Staatsexamensarbeit (Lehramt Gymnasien), Universität Würzburg, 82 p.

Sroka K. & Finch O-D. 2006: Ground beetle diversity in ancient woodland remnants in north-western Germany (Coleoptera, Carabidae).—*J. Insect Conservation* 10, 335–350

Stanbrook, R., Norrey, J., Kisingo, A. W., & Jones, M. (2021). Dung Beetle Diversity and Community Composition Along a Land Use Gradient in a Savannah Ecosystem of North Western Tanzania. *Tropical Conservation Science, 14*, 19400829211008756. https://doi.org/10.1177/19400829211008756

Thiele H.-U. 1977: *Carabid Beetles in Their Environments*. A study on habitat selection by adaptations in physiology and behaviour.—Springer-Verl. Berl., Heidelberg., New York, 369 p.

Thiollay J.M. 1995: The role of traditional agroforests in the conservation of rain forest bird diversity in Sumatra.—*Conservation Biology* 9, 335–353

Thüringer Landesanstalt für Umwelt Jena (ed.) 1995: *Ökologie und Schutz der Rauhfußhühner*.—Naturschutzreport 10, 295 p.

Tichý, L., Axmanová, I., Dengler, J., Guarino, R., Jansen, F., Midolo, G., et al. (2023). Ellenberg-type indicator values for European vascular plant species. Journal of Vegetation Science, 34(1), e13168. https://doi.org/10.1111/jvs.13168

Toft S. 1976: Life-histories of spiders in a Danish beech wood.— *Natura Jutlandica* 19, 5–40

Tomiałojć L., Wesołowski T. & Walankiewicz W. 1984: Breeding bird community of a primaeval temperate forest (Białowieża National Park, Poland).— *Acta Ornithologica* 20 (3), 242–310

Torralba, M., Fagerholm, N., Burgess, P. J., Moreno, G., & Plieninger, T. (2016). Do European agroforestry systems enhance biodiversity and ecosystem services? A meta-analysis. *Agriculture Ecosystems & Environment, 230*, 150–161. https://doi.org/10.1016/j.agee.2016.06.002

Tretzel E. 1952: Zur Ökologie der Spinnen (Araneae). Autökologie der Arten im Raum von Erlangen.—. *Sitzungsber. Phys.-Med Sozietät Erlangen* 75, 36–131

Tretzel E. 1954: Reife-und Fortpflanzungszeit bei Spinnen.— *Z Morph. Ökol. der Tiere* 42, 634–691

Tryjanowski, P., & Morelli, F. (2017). Suitable Methods for Monitoring HNV Farmland Using Bird Species. In F. Morelli & P. Tryjanowski (Eds.), *Birds as Useful Indicators of High Nature Value Farmlands: Using Species Distribution Models as a Tool for Monitoring the Health of Agro-ecosystems* (pp. 53–68). Cham: Springer International Publishing. https://doi.org/10.1007/978-3-319-50284-7_4

Tscharntke, T., Milder, J. C., Schroth, G., Clough, Y., DeClerck, F., Waldron, A., et al. (2015). Conserving Biodiversity Through Certification of Tropical Agroforestry Crops at Local and Landscape Scales. *Conservation Letters, 8*(1), 14–23. https://doi.org/10.1111/conl.12110

P. Udawatta, Rankoth, R.L., & Jose, S. (2019). Agroforestry and Biodiversity. *Sustainability, 11*(10), 2879. https://doi.org/10.3390/su11102879

Urban E.K., Fry C.H. & Keith S. (1997): *The birds of Africa*, Vol. V.—Academic Press, San Diego, London, 669 p.

Viegas, G., Stenert, C., Schulz, U. H., & Maltchik, L. (2014). Dung beetle communities as biological indicators of riparian forest widths in southern Brazil. *Ecological Indicators, 36*, 703–710. https://doi.org/10.1016/j.ecolind.2013.09.036

Virkkala R., Alanko T., Laine T. & Tiainen J. 1993: Population Contraction of the White-backed Woodpecker *Dendrocopos leucotos* in Finland as a Consequence of Habitat Alteration. *Biol Conserv*. 66, 47–53

Vogel B. 1998: Habitatqualität oder Landschaftsdynamik - Was bestimmt das Überleben der Heidelerche (*Lullula arborea*)?—Dissertation Universität Würzburg, 136 p.

Vogel K. & Johannesen J. 1996: Research on population viability of *Melitaea didyma* (Lepidoptera, Nymphalidae).—In: Settele J., Margules C., Poschlod P. & Henle K. (eds.): *Species survival in fragmented landscapes*. 262–267, Kluwer Verlag, Dordrecht, Boston, London, 381 p.

Waltert M. 2000. Forest management and the distribution of understorey birds in the Bossematié Forest, eastern Ivory Coast .—*Ostrich* 71 (1 & 2): 295–299

Waltert M., Mardiastuti A. &. Mühlenberg M. 2004: Effects of land use on bird species richness in Sulawesi, Indonesia.—*Conservation Biology* 18, 1339–1346

Waltert M, Bobo Ks, Sainge Nm, Fermon H, & Mühlenberg M (2005): From Forest to Farmland: Habitat Effects on Afrotropical Forest Bird Diversity.—*Ecological Applications* 15(4), 1351–1366

Warren M.S., Pollard E. & Bibby T.J. 1986: Annual and long-term changes in a population of the Wood White Butterfly *Leptidea sinapis* .— *J. of Animal Ecology* 55, 707–719

Wassmer, T. (1991, April). *Einfluß einiger abiotischer und biotischer Faktoren, insbesondere des Wassergehaltes und der Faecesgröße, auf die Artengemeinschaft coprophager Käfer der Familien Scarabaeidae und Hydrophilidae in Schafsexkrementen einer alten Weidelandschaft mit sommerlicher Beweidungspause*. (Ph. D. Thesis). Albert-Ludwigs Universität, Freiburg.

Wassmer, T. (1994). Seasonality of Coprophagous Beetles in the Kaiserstuhl Area near Freiburg (Sw Germany) Including the Winter Months. *Acta Oecologica-International Journal of Ecology*, *15*(5), 607–631.

Wassmer, T. (1995). Selection of the spatial habitat of Coprophagous beetles in the Kaiserstuhl area near Freiburg (SW-Germany). *Acta Oecologica-International Journal of Ecology*, *16*(4), 461–478.

Wassmer, T. (2014). Seasonal Occurrence (Phenology) of Coprophilous Beetles (Coleoptera: Scarabaeidae and Hydrophilidae) from Cattle and Sheep Farms in Southeastern Michigan, USA. *The Coleopterists Bulletin*, *68*(3), 603–618. https://doi.org/10.1649/072.068.0327

Wassmer, T., & Refinetti, R. (2016). Daily Activity and Nest Occupation Patterns of Fox Squirrels (Sciurus Niger) throughout the Year. *PLOS One*, *11*(3), e0151249. https://doi.org/10.1371/journal.pone.0151249

Wassmer, T., & Sowig, P. (1994). Die coprophagen Käfer der Schafsweide: "Flachsland" am Schönberg bei Freiburg. *Veröffentlichungen für Naturschutz und Landschaftspflege in Baden-Württemberg*, *68/69*, 355–376.

Wassmer, T., Himmelsbach, W., & Himmelsbach, R. (1994). Dungbewohnende Blatthornkäfer (Scarabaeoidea) und Wasserkäfer (Hydrophilidae) aus dem Hessental bei Schelingen im Kaiserstuhl. *Mitteilungen des Badischen Landesvereins für Naturkunde und Naturschutz*, *16*, 75–83.

Waβmer, T. (1995). Mistkäfer (Scarabaeidae und Hydrophilidae) als Bioindikatoren für die naturschützerische Bewertung von Weidebiotopen. *Zeitschrift für Ökologie und Naturschutz*, *4*, 135–142.

Wesołowski T. 1995: Value of Białowieża Forest for the Conservation of White-backed Woodpecker *Dendrocopos leucotos* in Poland.— *Biol Conserv*. 71, 69–75

Wesołowski T. 2007: Primeval conditions—what can we learn from them? —*Ibis* 149 (2), 64–77

Wesołowski T., Tomiałojć L. & Stawarczyk T. 1987: Why low numbers of *Parus major* in Bialowieza Forest—removal experiments.— *Acta Ornithologica* 23 (2), 303–316

Wesołowski T., Tomiałojć L., Mitrus C, Rowiński P. & Czeszczewik D. 2002: The Breeding Bird Community of a Primaeval Temperate Forest (Białowieża National Park, Poland) at the End of the 20th Century.— *Acta Ornithologica*, 37(1), 27–45

Whipple, S. D., & Hoback, W. W. (2012). A Comparison of Dung Beetle (Coleoptera: Scarabaeidae) Attraction to Native and Exotic Mammal Dung. *Environmental Entomology*, *41*(2), 238–244. https://doi.org/10.1603/En11285

White G.C. & Garrott R.A. 1990: *Analysis of Wildlife Radio-Tracking Data*.— Academic Press Inc., San Diego, California, 383. p

Wiens J.A., Rotenberry J.T. & Van Horne B. 1987: Habitat occupancy patterns of North American shrub-steppe birds: the effects of spatial scale.—*Oikos* 48, 132–147

Wikipedia. (2024 , April 19). Bioindicator. In *Wikipedia*. https://en.wikipedia.org/w/index.php?title=Bioindicator&oldid=1219673783. Accessed 2 May 2024

Wolff J.O. 2003: An evolutionary and behavioral perspective on dispersal and colonization of mammals in fragmented landscapes, 614–630.—In: Zabel C.J. & Anthony R.G. (eds.): *Mammal Community Dynamics*. Management and Conservation in the Coniferous Forests of Western North America.— Cambridge University Press, 709 p.

Wurster C. 2015: *DDT Wars*: Rescuing Our National Bird, Preventing Cancer, and Creating the Environmental Defense Fund.— Oxford University Press, 265 p.

Wurz, A., Tscharntke, T., Martin, D. A., Osen, K., Rakotomalala, A. A. N. A., Raveloaritiana, E., et al. (2022). Win-win opportunities combining high yields with high multi-taxa biodiversity in tropical agroforestry. *Nature Communications*, *13*(1), 4127. https://doi.org/10.1038/s41467-022-30866-8

Yang, Y., Xiong, K., & Xiao, J. (2024). A review of agroforestry biodiversity-driven provision of ecosystem services and implications for karst desertification control. *Ecosystem Services*, *67*, 101634. https://doi.org/10.1016/j.ecoser.2024.101634

Habitat Evaluation for Conservation

10

Abstract

Habitats need to be assessed for each type of landscape planning. Our focus is on habitat characteristics that are important for animals. Assessment criteria can relate to species communities (e.g., number of species, biodiversity, trophic levels) or to the fitness of individual species. Problems with assessment according to the frequency or occurrence of animals are discussed. With long-term observations, we can break down changes in habitat characteristics (e.g., successions), and long-term monitoring can also be used to recognize the effects of climate change. The concept of target species is discussed in detail, focusing on selected species, followed by examples of how to analyze a population's future development and viability. Animals can experience good conditions for reproduction in one type of habitat and poor conditions in another (source-sink habitats).

Individual landscape structures are treated separately, such as hedges in the cultivated landscape, forest edges, and strips of land that are either excluded from cultivation or created in a planned manner. The manipulation of habitat components is also illustrated using mown meadows and leaf litter in forests.

10.1 Assessment for Landscape Planning

The increasing consumption of natural areas and the intensive agricultural use by humans have endangered many animal and plant species or have already caused them to become extinct (e.g., Hochkirch et al. 2023). In order to protect the threatened species, it would be desirable to record the populations as thoroughly and as comprehensively as possible. When planned construction measures and other anthropogenic interventions are carried out, these population assessments can be used to prioritize protection areas or suggest alternatives based on the knowledge of the species' ecology, including replacement properties or other compensatory measures. Protected areas harbor more biodiversity than unprotected areas (Cooke et al. 2023). The most important seems to be that the population densities of threatened animal species in protected areas are higher (Gray et al. 2016). In addition, maintenance plans must also be created for established protected areas. Today, such assessments are dictated mainly by planning dates and time pressure. From an ecologist's point of view, however, it cannot be accepted that interventions take place at arbitrarily determined times (e.g., any month of the year) but are instead assessed on an ecological

M. Mühlenberg, T. Waßmer, *Field Ecology*, https://doi.org/10.1007/978-3-031-87547-2_10

basis developed into future measures. A broad overview of species and habitat protection, compensation areas, current evaluation criteria, and practices of nature conservation and landscape planning in Germany are given by Kaule (1991) and Usher and Erz (1994) and by (Dyke and Lamb 2020; Groves and Game 2015; Lopez et al. 2024) for North America.

10.2 Problems with Species Lists

An inventory precedes a habitat evaluation. Usually, species lists are created for individual plant and animal groups, and then some habitat requirements of the listed species are indicated from the literature. The number of recorded "Red List" species (Blab et al. 1984, internationally (Baillie and Zedan 2004) since then updated online, latest (International Union for Conservation of Nature and Natural Resources 2024)) is highlighted separately. This approach is only suitable when the habitats under consideration become rare and are inhabited by specialized organisms (Bauer 1989).

The problem begins with the fact that only a tiny part of all organism groups is considered in the inventory. Furthermore, one cannot conclude one animal group's diversity from another's. Another critical point is the uncertainty of whether a higher number of species alone is sufficient for a positive evaluation. Shouldn't the species be evaluated individually according to their characteristics, population size, and degree of threat? One more problem often discussed is the predatory influence of top carnivores on protected species (e.g., Goshawk on Black Grouse, Biosphere Reserve Rhön 2007; Prüter and Wübbenhorst 2004; Siebert et al. 2020); Tost et al. 2022; Przewalski horses and wolves, Van Duyne et al. 2009). While in large natural areas, predators can hardly limit their prey but instead change the age class composition and the average condition (see Sect. 8.7); they often become a strongly limiting or even extinguishing factor for prey in isolated and small locations.

Again and again, conflicts of objectives arise from the fact that the preference of one animal species or group can disadvantage another. For example, are the warbler species and the Red-backed Shrike more critical in a bushy, dry area than some thermophilic ant and grasshopper species of the open spaces? The conflict of objectives, for which species management measures should be carried out at the expense of other species, should be decided according to criteria that are initially independent of the biology of the individual species. We want to suggest seven criteria for such decisions:

1. Size of the area, compared to the area requirement of the relevant animal population.
2. Surroundings of the area.
3. Foreseeable disturbances of the area (traffic, tourism, entry of pollutants).
4. Stability in the future, i.e., constancy of the habitat in the foreseeable future development.
5. Uniqueness of the habitat on a larger geographical scale, e.g., on a European or North American scale.
6. Existing habitat diversity (moisture gradients, inclusion of north and south slopes).
7. Presence of different succession stages, which depends very much on criterion 1 (Remmert 1984, 1985, 1988).

10.3 Habitat Quality Evaluation Criteria

The habitat quality must be measured regarding the future viability of the population; this means that not only is the occurrence of a species important, but so is the density, offspring production, and survival of the individuals (Van Horne 1983).

A habitat is, for example, highly valued when it hosts a resting or breeding population, the size of which exceeds a certain proportion of the total population of a larger area (Balmford et al. 2002). For birds, we usually have the necessary information and can, for example, set 1% of the biogeographic population (at least 100 individuals) as a criterion, as has been successfully and biologically meaningfully practiced, especially for waterfowl in the designation of significant areas (Erwin 2002, Ramsar Sites https://www.ramsar.

org (Convention on Wetlands 2021)). E. Glück (oral comm.) determined the habitat quality for great and blue tits (*Parus major* and *P. caeruleus*) by hanging up nest boxes in the area and monitoring the success of raising the young. The resource offer is measured by the weight gain of the fed young birds (benefit) and the flight performance of the adult birds (cost). Richner (1991) showed how the growth of the nestlings in carrion crows differs between rural and urban habitats, determining the breeding success of the next generation. Suppose the breeding success—as in previous practice—is indicated by the number of fledged nestlings of the local breeders, and only this parameter is used to assess the quality of the habitat. In that case, the difference does not come out sufficiently: The local offspring is not primarily ensured by the number of fledged nestlings but by their quality. In the specific case of the carrion crow, the animal's size decides the possibility of establishing breeding territories. Newton and Marquiss (1991) showed the value of the individual experience in sparrowhawks by removing the territories' owners. Even those sparrowhawk individuals, which had taken over the known good, artificially "freed" territories, were far behind in breeding success compared to the older and more successfully hunting animals.

Decisions at the species level, therefore, require more extensive information than is usually available from the inventories described above. Important biological data on the degree of threat to individual species are also important (see "Population Vulnerability Analysis (PVA)," according to Gilpin and Soule 1986). The corresponding questions are:

1. How large is the population of the species considered?
2. What population fluctuations are known or are to be expected? In arthropods, populations often fluctuate by a factor 100 and more (cf. Remmert 1984; Van Dijk and den Boer 1992; Den Boer 1977; Den Boer 1986; Müller 1978; Cook 1985; Kendeigh 1979), in snails by a factor 5 (Williamson et al. 1977; Dempster and Pollard 1981), in small birds around a factor 3 (Berthold et al. 1986; Berndt and Henss 1967), and in small mammals by a factor 10 (e.g. Eurola et al. 1984).
3. Does the (sub)population living in the examined habitat have a connection to other populations, and is it part of a metapopulation (Shaffer 1985, Gilpin 1987)?
4. How high is the degree of isolation of the population?
5. How significant is the colonization ability of this species?
6. How large are the home ranges of the individuals? In mammals, you could also differentiate between sexes and age groups, e.g., there are home range differences in roe deer by a factor of 4 (Ellenberg 1978).
7. How strong is the species' habitat dependency, and what are the critical resources for bottlenecks occurring throughout the year (for resource sampling see e.g. Gregoire and Valentine 2008)?
8. How large is the degree of polymorphism and, depending on this, probably the adaptability? (This question becomes vital in vertebrate populations with less than 50 animals, Franklin 1980, Loeschke 1988, Frankham et al. 2004).

When a habitat with its wildlife is destroyed by changed land use (e.g., road construction), the question of compensating for such an intervention arises (see also Sect. 13.2). Other species cannot compensate for a lost species. If the lost species are not supported elsewhere, legally required compensation can be sought to replace certain functions. If, for example, a forest is assigned the function of a filtering effect, the same tree crown volume would have to be offered again by new plantings in the event of deforestation. Compensation for a hectare-sized area might take about 100 years until a regrowing forest redevelops. A grazing effect or a predatory influence can be quantified for certain animals that need compensation. It is not enough to simply provide an apparently suitable habitat of a matching size as the colonization ability of the animals is usually not known, and successful colonization remains uncertain (Mühlenberg 1982); see also Sect. 13.1.

Size, distance from roads, and the frequency of agricultural or forestry activities are also crucial for assessing habitat quality (Landesamt für Umweltschutz 1996, Selva et al. 2011, Kapustka 2004). A comparison between a small-scale structured cultural landscape in Northern Bavaria and a large-scale cultivated landscape in Poland showed that the species richness of birds is higher in Poland. In addition, more species that are categorized as endangered in Central Europe live there. In contrast, insect biomass and diversity hardly differed between the two cultivated landscapes (Mühlenberg and Slowik 1997).

10.4 Problems in Estimating Habitat Quality

If no accurate habitat mapping is available, habitat quality is estimated using quadrats and is based on small-scale topographic maps (e.g., 1:5000) according to previously established habitat classes in percentage shares (see Fig. 4.9 and Sect. 4.3). Such an evaluation of land use also serves to compare home range sizes (see Sect. 4.7) with habitat quality (see Sect. 10.2). If one wants to know which parts of a habitat are preferred, i.e., are most used, and if one cannot record the habitat type or the microhabitat directly, then the quadrats are sorted according to the frequency of their visits, and the corresponding habitat parts (percentage shares) are determined. A result could be that 60% of the visits are in quadrats with the highest proportion of grassland. The habitat classes are thus weighted according to their use by the animals. Deviations of the encounter frequency of the animals in certain parts of the habitat from their percentage total share of the home ranges (distribution of land use forms in the residential area) are tested with the Chi-square test (Sect. 15.9). They can be interpreted as the animal's specific habitat preferences. For grassland areas, it is essential to distinguish on-site between fertilized meadows that mature one to three times after the appropriate mowing and continuous short-grass biotopes due to grazing. For the occurrence of animals, these habitat differences are of essential importance (see Sect. 10.19).

However, there are no universally valid criteria for "good" habitats. Dry grassland provides a habitat for certain grasshopper species when it contains open, vegetation-poor sites, but for some bird species only when it is rich in bushes. A general and frequently used criterion for a "good" habitat is a high number of plant and animal species as a measure of biodiversity. Apart from the fact that the number of species is always a measure that strongly depends on the effort of recording (see Sects. 15.6 and 15.11), there is no possibility of setting threshold values or limits for a decision about habitat use. Should a grassland area, therefore, be released for a housing development if it contains only 60 species of ground beetles instead of 80? Are we satisfied that we could only identify 50 species in a survey? The change in the number of species moves along a continuum, and any drawing of boundaries remains arbitrary (see Fig. 10.1).

There is no biological reason to rate a habitat with great diversity higher than one with fewer species. Why should, for example, a wetland with high bird diversity be "better" than a forest with fewer bird species? Peatlands, for example, have dramatically lost area in Northern Europe and are worth protecting, even though they are species-poor compared to other biotopes (Succow 2001; Humpenöder et al. 2020; Muller 2018; Murdiyarso et al. 2019; Nordbeck and Hogl 2023). However, some "peatland-specific" animal species are hardly found elsewhere (Reißmann et al. 2005; Kratochwil and Schwabe 2001; Alekseev et al. 2024; Hammerich et al. 2022). Lowland and highland peatlands are protected under the German Federal Nature Conservation Act and included as biotope types in the *European Council Directive on the Conservation of Natural Habitats and of Wild Fauna and Flora*, with highland peatlands classified as "priority habitats" (Riecken et al. 1994). The German Federal Climate Protection Act (2021) also enshrined Peatland protection.

A diversity measure can, at best, describe species communities (see Sect. 15.11), not evaluate them. This makes sense if you want to represent change in a habitat quantitatively. If fewer species are in a habitat type than before or compared

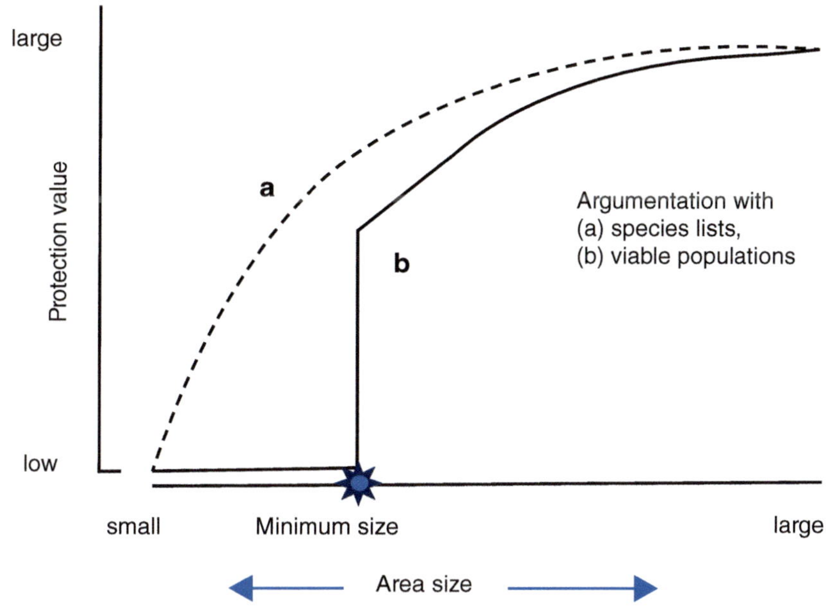

Fig. 10.1 Quantitative estimation of the value of protected areas or outstanding habitats according to their size. While in the past, species lists (**a**) were used to establish a continuous increase in quality, the target species concept (**b**) (see Sect. 10.8), defines clear boundaries, below which specific populations cannot survive. From Mühlenberg and Hovestadt (1992), modified after Usher (1986)

to another environment, then you can specifically look for ecological explanations.

A habitat is rated higher if it increases an endangered species's survival probability. This, however, can only be verified for some populations corresponding to the "target species," as explained in Sect. 10.8. When orienting towards a target species, you will probably also find other endangered species for which the conservation measures are beneficial.

10.5 What Animal Groups Should Be Recorded?

In many countries, the monitoring of animals must be approved by the respective higher nature conservation authority.

Due to the easy processability and accessibility of ecological data in the literature, the following animal groups have come to the fore in nature conservation practice and landscape planning in Europe (see, e.g., Usher 1986, Usher and Erz 1994): large mammals, small mammals (bats, shrews, dormice), birds in general, reptiles only by some species (e.g., the wall lizard, smooth snake, adder), amphibians especially concerning breeding water bodies, ground beetles (Carabidae), bumblebees (*Bombus* spp.), butterflies, grasshoppers (Caelifera and Ensifera), dragonflies, spiders (Araneae, predominantly web-building species) and mollusks (shell snails). Solitary bees (Apoidea), digger wasps (Sphecidae), hoverflies (Syrphidae), flower and soft bugs (Anthocoridae, Miridae), harvestmen (Opiliones), and other beetle groups (longhorn beetles, weevils, and lamellicorn beetles) should also receive increased attention.

The adults of the various animal groups have their main activities at different times of the year. Suppose you want to collect the most meaningful animal data as possible for Central Europe within a forced minimal solution (cf. Rümer and Mühlenberg 1988). In that case, we suggest the schedule with a survey time of at least 6 weeks per animal group indicated in Table 10.1. If you also need to restrict the selection of animal groups, choose animal groups in addition to songbirds according to the following criteria: representatives of phytophages and carnivores, species with minimal mobility and species with very high mobility (Table 10.1).

The inventory recording has nothing to do with the evaluation; it merely describes the "current state" and is thus a prerequisite for the evaluation.

Table 10.1 Compilation of observation times and survey methods for various animal groups for a minimal survey in Central Europe

Animal group	p	C	Mobility			Months												Survey method
			–	±	+	1	2	3	4	5	6	7	8	9	10	11	12	
Mammals	x	x		x	x	–												Trace analysis in the snow
						–	–											Nest box control for dormice
Birds of prey nests		x			x		–	–										Inspections of forests and shrubs before leafing
Songbirds (breeding birds)	x	x			x					–	–							Mapping of territory indicating behavior between 6–8 h
Butterflies	x				x					–	–	–	–	–*				Visual observation on transects and supplementary net catches.
Bumblebees	x			x							–	–	–					Net catches, observations at flowers
Ground beetles (Carabidae)		x		x					–	–	–		–	–				Pitfall traps
Field grasshoppers (Caelifera)	x		x										–	–				Sweep netting on transects, in sunshine between 13–17 h
Bird nests in shrubs								–								–	–	Systematic search
Forest arthropods (species inventory)	x	x		x			–	–	–	–	–	–	–	–	–	–		Photoeclectors
Moths	x				x						–	–		–				Light trapping

P = Phytophage, C = Carnivore, Mobility: — = low, ± = medium, + = high
*Especially for Lycaenids, make sure to capture them also in August!

The evaluation should not be done simply by assigning habitat requirements to each listed species according to the literature. Instead, it should ultimately be based on assessing the survival probability of selected animal populations (see Sect. 10.8). An extensive inventory is required as a first step in selecting these animal populations.

In an ecosystemic approach to the question, not only are species inventories needed, but also the acquisition of quantitative area- and time-related data from a whole spectrum of animal groups. For forest ecosystems, Grimm et al. (1975) propose the combined use of three types of traps as a minimum program for the recording of arthropods: ground eclectors, tree photoeclectors, and pitfall traps (see Chap. 2). Over the course of a year, up to 10,000 arthropods were caught from 1 m^2 ground surface of a beech forest with the ground eclector. Of the 94 species of the beetle family Staphylinidae, 83% were only recorded with the soil eclectors. Species with a more significant hatching density are sufficiently well-recorded with a catch area of 5 m^2.

Five sampling years are not enough for the description of the "ecosystem-typical density"(Grimm et al. 1975) of arthropod species (question: how often is a species "normally" found in a particular ecosystem?). Due to fluctuations, standard recordings for more populations for up to 10 years are recommended!

10.6 Long-Term Observations in Nature Conservation

Since targeted conservation measures are associated with costs, the success of the measures must be monitored (see Sect. 10.8). Establishing protected areas is also associated with a great effort; the official steps alone usually take several years. An effectiveness check is therefore appropriate (cf. Haarmann 1985). In addition to the economic justification of long-term observation as success control, the future stability of selected target species forces us to monitor the population development continuously (see Sect. 10.8). Population density and size changes can also result from slow changes in overall conditions. Climate and soil changes and the life activities of dominant organisms can lead to an aperiodic transformation of an ecosystem (succession). Humans cause transformations of communities, even without directly changing individual habitats, by producing excessive eutrophication and pollution (Ellenberg 1986a, 1987, Kratochwil 1989). Such disturbances, which have a regional and widespread effect through immissions, must be quantitatively tracked in their effect through biomonitoring. They can render local individual measures in nature conservation pointless (cf. Spellerberg 2005).

Trend analyses of population developments are the basis for including species in the Red Lists (Blab et al. 1984). Population developments are only recognized through long-term observations as a trend if they can be distinguished from annual fluctuations. Here, another problem arises: A locally limited permanent observation does not consider the spatial shifts of the animal population in successive years (see, e.g., Taylor and Taylor 1977; Wiens 1989). Therefore, the monitoring of a species should take place simultaneously at many locations within a larger region or country (Pollard 1991a; Baillie 1991; Conner et al. 2001; Jeffress et al. 2013; Kamp et al. 2021; Brlík et al. 2021; Bowler et al. 2021). Similar or opposing trends at different locations provide essential clues about climate influences or regional changes. Exemplary monitoring programs from Great Britain are presented by Goldsmith (1991) and Clarke (1986), Slocombe 1992; Jensen and Bourgeron 2001; Rapport et al. 2002 providing overviews of the various ecological monitoring methods.

Since standardized observations of all possible animal groups cannot be carried out, the question arises to which species or relevant research parameters we can reduce our long-term observations. For vegetation, Sukopp et al. (1986) and Pfadenhauer et al. (1986) have developed models and practical suggestions for long-term observation. If faunistic data are to be included, we must first clarify the goals of the long-term observation (Mühlenberg 1990).

By recording dominant animal groups at multiple trophic levels, energy flows in the ecosystem can be measured. However, monitoring widespread species requires many samples and much evaluation time, as all animal individuals must usually be assigned to the species level. Nevertheless, energy flow does not have to depend on a specific species. The ecological role of an animal species can certainly be taken over by other animal species when its numbers decrease, but predictions are not always possible. The takeover of the function of an animal species by others stabilizes the course of ecological processes in the ecosystem. Therefore, a causal relationship between a high species diversity and the system's stability has been seen repeatedly (summary by Pimm 1984; see also Byrnes et al. 2014). However, high species numbers alone do not guarantee the return of all system variables to their initial values after the ecosystem has been disturbed. In this sense, stable ecosystems require little care, but the focus is on dominant, usually not threatened, animal species.

There are many years of data available monitoring butterflies (Pollard 1977, 1991a, b) and birds (Baillie 1991). Butterflies are monitored using the transect method (Sect. 2.3), whereas the counting method for birds depends on the observed area and habitat (Sects. 1.7 and 1.8). Where nationwide monitoring programs exist, you get data on the general trend of frequency change and can also compare individual sites in their development with each other. Changes in habitat preference (e.g., nesting sites) of small local populations can be signs of adaptation to the cultural landscape, as Tast (1968) was able to show for the linnet (*Linaria cannabina*). Synchronous data from different locations allow the investigation of possible reasons for population decreases or increases, such as local habitat changes, weather factors, climate change, or shifts in spatial distribution due to immigration or emigration (Stafford 1971; Pollard 1991b; see also Sect. 2.28).

If long-term observation of various, extensive animal groups is too time-consuming for practical reasons, we need to look for representative species for the considered ecosystem. While in vegetation science, the description of plant communities via indicator and character species is well developed (Wilmanns 1989), comparable ordering principles in zoology are more complicated (cf. Kratochwil 1987, 1991, Blab and Riecken 1989; Chase et al. 2000; Funke 1993; Peters et al. 2016).

As an example of the selection of character species, which are supposed to represent communities of specific habitats in Central Europe, we refer to the work of Kratochwil (1989). He assigns certain butterfly species as typical for the following plant communities (Table 10.2).

In this context, a character species (in the sense of an indicator species) is a species characteristic of specific plant communities and whose presence indicates intact conditions of these habitats. From their presence, it is inferred that many other habitat-typical species are present and form a characteristic species community (see also Erhardt 1985).

From an animal ecological perspective, the problem remains as to how clearly communities can be separated from each other. For a biologically meaningful demarcation, the following prerequisites would have to be met (Collier et al. 1974): Coincidence of the occurrence or disappearance of species groups along an environmental continuum,

(a) due to co-adaptation and agreement of tolerance limits to environmental factors,
(b) due to dominant species influencing their environment, discontinuities arise.

We can contrast two possibilities in the distribution of communities (A and B in Fig. 10.2):

According to pattern A, we can distinguish communities from each other in the species distribution, but animal species permanently inhabit several such communities (example a). Suppose the species are distributed randomly or noncoincidentally along an environmental gradient (pattern B). In that case, the selection of an animal species does not describe a definable community, but rather, the area of interest along the environmental gradient is delimited (examples b and c). The site area is not the starting point of the investigation, but the animal spe-

Table 10.2 Examples of butterflies as character species of plant communities (according to Kratochwil 1989; Kratochwil and Schwabe 2001)

Plant Community	Lepidoptera	Degree of Threat (Red List BRD 1997)
Forests		
Thermophilic mixed oak forests	*Euphydryas maturna, Lopinga achine*	2
Quercion pubescenti-petraeae	*Coenonympha hero*	2
Hornbeam-dominated forest paths and shrubs	*Apatura* species,	3
	Limenitis species	2 and 3
Carpinion	*Pararge aegeria*	Least concern
Meadows, grassland		
Mesobrometum	*Melitaea phoebe*	2
	Lysandra bellargus	
Agricultural grassland		
Molinio-Arrhenatheretea:		
Nutrient-rich wet meadows	*Lycaena dispar*	2
Calthion pipe grass meadows	*Minois dryas*	2
Molinion (partly fallow)	*Maculinea nausithous,*	3
	Maculinea teleius	3

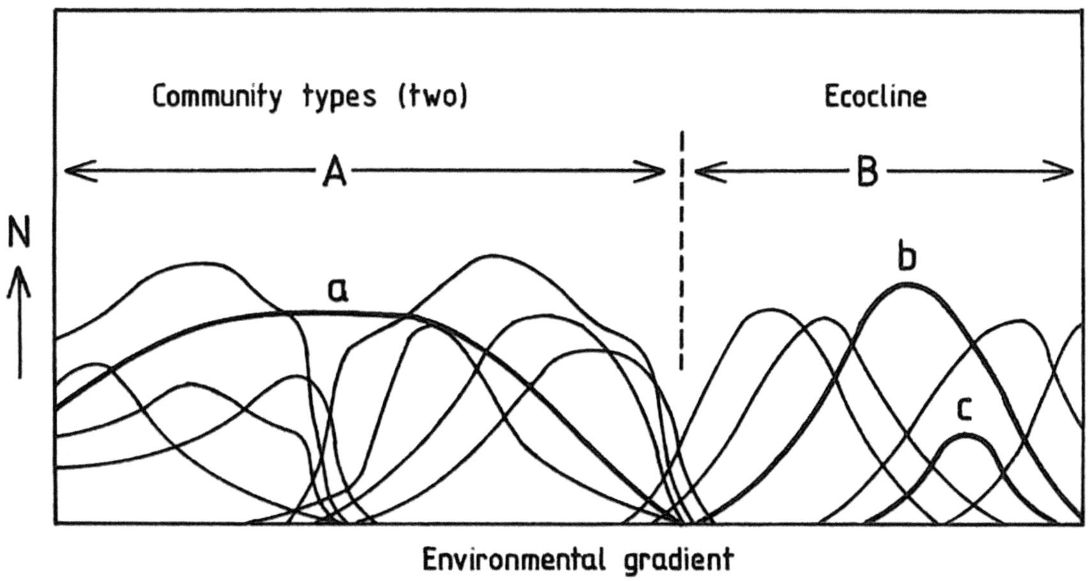

Fig. 10.2 Possible distribution of different species along an environmental gradient. In area (**a**), two communities can be separated from each other for most species. In (**b**), each species has its own spatial distribution (Modified after Collier et al. 1974)

cies itself, selected as a "character population." With the choice of a character population, an observation space is thus delimited, with species b having a larger area than with species c according to Fig. 10.2. We speak of a target species when the population is studied explicitly for the purpose of its long-term future stability (see Sect. 10.8).

The scale of measures (municipal level, federal level) influences the selection and monitoring of target species. Monitoring of target species of habitat protection not only includes the recording of the population development (determination of population size and density) but also a routine control of the genetic variability of the populations (determination of the degree of polymorphism) and the demographic characteristics (changes in the sex ratio and the birth or death rate, proportion of reproducing individuals in the total population). In addition, the control of the exchange rate between animal populations must be added, as it can be used to estimate the possibility of maintaining genetic diversity and supporting locally declining populations. Such extensive monitoring is only possible to do for a few species. However, safeguarding the target species also protects particular habitats with further species (see below Sect. 10.8).

Ecologically significant species for monitoring and management should be selected according to Soulé and Kohm (1989) from the following three categories of species:

1. Keystone species, whose disappearance leads to the extinction of many other species.
2. Mobile link species, which form critical functional components between several food chains or animal-plant associations.
3. Indicator species of particular interest for environmental monitoring or management measures.

Examples of (1) could be individual woodpecker species in the Central European fauna and for (2) the red wood ant (*Formica rufa*) or a beaver. Long-term observation in nature conservation can also serve general research objectives. Essential insights about the importance of breeding successes for preserving local populations can only be gained from long-term monitoring of known individuals. Siefke (1985), for example, was able to document the 15-year "family history" of a ringed plover female (*Charadrius hiaticula*) through a long-term ringing program. Almost all of the questions addressed in this book would only be satisfactorily answered through long-term studies, i.e., meaningfully in periods of about 10 years. The critical results for species protection that can emerge from long-term observations were demonstrated in the 24-year study of the California checkerspot butterfly (Ehrlich and Murphy 1987). We need further case studies on the dispersal behavior of animal species, which are necessary to replace locally extinct populations (see Sect. 11.4). Less studied are species that live in an environmental gradient, e.g., of moisture, depending on the weather in the respective year (dry, hot summer, or wet spring), and optimize their primary location through spatial relocation. Reproduction and survival in these different microhabitat areas determine the population fluctuations (Jeffress et al. 2013). Topographical heterogeneity can also enable species such as the pikas to withstand climate change longer (see Fig. 10.3).

Similarly, the change in microclimate due to excessive nutrient inputs has so far been given too little attention, although the decline of many animal species can be explained by this (see Ellenberg 1986b, Beaumelle et al. 2023; Brain et al. 2015; Forbes et al. 2016; Schaap et al. 2023). The intensification of agriculture and land use changes also lead to changes in the population density of common species, which can only be recognized through many years of comparison. Shifts in territories and decreases in breeding success can be indicative, as Schläpfer (1988) and Jenny (1990) have demonstrated for the skylark.

Another research approach lies in the long-term observation of microsuccessions. Animals often play a crucial role in the spatial distribution and preservation of plant diversity: Often, the place where a single plant grows is determined by animals. We know this relationship from all seed-dispersing animals. As examples, we mention tree squirrels (Gurnell 1987), jays (Bossema 1979), and some ant species (Culver and Beattie 1978, Beattie 1983, 1985) as seed dispersers. The nutcracker, as another example, is mentioned above in Sect. 7.10. Much less known are the relationships between the local destruction of the plant cover by animals and newly created colonization opportunities for plant species. Small sites are distributed in a mosaic pattern within a community, constantly undergoing micro successions (see Remmert 1985). A detailed example is the work of Goldberg and Gross (1988) in a North

Fig. 10.3 (**a**) Pika *Ochotona hyperborea* lives in a stable colony (over 20 years) on a northern mountain slope with rocks, moss cushions, and many hiding places. The heterogeneity of microhabitats allows pikas to be quite resistant to climate change. (**b**) Pikas of the same species in a black phase in another colony nearby. Both Northern Mongolia.

American grassland. Moles were responsible for 52%, and ants for 19% of the dm^2-sized open spaces in the grassland, where new grass species could settle. Field mice can play a similar key role in semi-dry grassland (see Zielinski 1982). In the Mongolian steppe, rodents also expose the soil on a small scale, enabling new plant species to settle (Fig. 10.4).

The influence of the steppe mice is often interpreted negatively: the hypothesis is that a substantial proliferation of the mice is responsible for the original tall-grass prairie becoming a short-grass steppe and subsequently narrowing the food base for grazing animals. However, another hypothesis can also be put forward: very high grazing pressure by domestic animals on the tall-grass prairie leads to the short-grass steppe, and then mice are able to reproduce more strongly. This is again an example that only long-term observations and comparisons of different regions can lead to scientifically sound statements.

Beech seedlings usually dry up in the forest because they are kept from the mineral soil by the thick layer of fallen leaves. Only the activity of game, which scrapes away the leaves in places or churns up the soil, allows the young seedlings to reach the mineral soil and thus also the connection to the water supply. Herbaceous plants also have the opportunity to spread over such open spaces in the forest. If forest insects cause some trees to die off by defoliation, other plant species could also grow at the location due to the changed light conditions. Animal species that spread the

Fig. 10.4 (**a**) (left panel): Brandt's vole (*Microtus brandti*) in the steppe of Mongolia creates small patches of ground free of vegetation. (**b**) (right panel): Only through small, randomly occurring open spaces is the colonization of a natural grassland by pioneer trees possible

seeds of dominant plants also play a crucial role in the system. The idea that new colonization opportunities are constantly arising at individual locations through random processes is crucial for maintaining a high diversity of the relevant community (Pickett and White 1985).

Succession in the vegetation and the associated animal community occurs naturally or is caused by disturbance. Disturbances are an integral part of ecosystems. First, it is important to recognize whether ecosystem changes result from natural, i.e., system-inherent processes (*intrinsic*) or anthropogenic disturbances (*extrinsic*). Experimental disturbances of different degrees and extents would be a method to obtain sufficient data about the recovery mode of the studied ecosystems (Calahan 1984). From the long-term observation of different "recovery patterns," we learn complex properties of ecosystems that also allow us to make specific predictions about the effect of anthropogenic disturbances (Jentsch et al. 2002, Wohlgemuth et al. 2022). However, interactions between different disturbances make predictions more difficult (Burton et al. 2020).

Experimental disturbances should be oriented toward possible natural disturbances (see Pickett and White 1985). In a meadow, for example, 25 cm × 25 cm areas can be manipulated; in the forest, square meter areas can be treated by soil exposure, disruption, compaction, or shading. Small meadow sections could also be flooded. Removal (relocation) of animals or plants would be further disturbance effects for an analysis of the recovery of the systems.

Another type of experimental disturbance can be carried out through maintenance measures in biotopes within certain succession stages. Open dry locations often become overgrown without maintenance. Removal of the shrubs and exposure of the soil would often be the prerequisite for some endangered species, for which the corresponding protected area was designated, to return or recover (see also Sect. 8.6, Fig. 8.16, example Stonechat *Saxicola torquata*). Only in large areas are the natural living conditions to be found in all succession stages, and therefore, a high species diversity is secured in the long term, even without maintenance measures. Even Yellowstone National Park in North America, with about 9000 km^2 area, is considered too small for a natural self-regulation of the large carnivores and herbivores and for a sufficient natural disturbance regime (Brussard 1991).

10.7 What Measurable Short-Term Characteristics of a Species Community Are Essential?

While it is usually possible to set up species-specific protection programs for larger animal species (mammals, birds) based on the knowledge of their individual biology and to carry them out when the species are reliably identified, for small animals (e.g., insects and spiders) this is only possible in a few individual cases. In those cases, only the protection of habitats or individual species will be safeguarded by the corresponding measures (e.g., Kudrna 1986). For small animals (arthropods), the endangerment of individual species is usually only recognized through the overall assessment of changing species communities. Key figures can characterize species communities. The value of a key figure depends on the accuracy and the transferability of the measurement. A compilation and critical assessment can be found in Table 10.3 (from Mühlenberg 1983). For the most important and measurable key figures to evaluate habitats for conservation issues in a short time, we suggest the number of species, evenness, and dominance structure with listing the five most common species. Calculations of species richness are presented in Sect. 15.11.

10.8 What Ecological Properties or Interactions Make Species Valuable for Habitat Assessment?

In conservation practice, the "Red Lists" of endangered animal species (Blab et al. 1984; Federal Agency for Nature Conservation 1998; International Union for Conservation of Nature and Natural Resources 2024) have become essential tools for discussion. The classification of animals as "critically endangered" or "endangered" should theoretically be based on an assessment of population development over the last 20 years. However, in many countries, including Germany, studies or knowledge about long-term population dynamics are only available for a few individual species and almost exclusively for vertebrate species. For other animal groups, natural rarities, therefore, have often been classified as endangerment. For arthropods, the problem is that a decline from 10,000 to 5000 individuals is much harder to detect than a decline from 100 individuals to 50 animals in a mammal or bird species.

Moreover, it is easier to assess an animal group consisting of long-lived adults (mostly vertebrates) than a group of short-lived adults with a very high birth rate and a correspondingly high death rate (Kudrna 1986). The author, therefore, suggests using a biogeographical assessment rather than the species' frequency as a criterion for its evaluation. For butterflies, he has developed a chorology index composed of the area size, composition, and affinity. The highest rating is given to a species that is restricted to a minimal area or is only known from a single location. However, the area size may, by definition, occupy, e.g., about 1% of the area of Europe.

A corresponding example was provided by the study of Fermon et al. 2000. While the number of species and individuals of Nymphalinae butterflies does not differ between managed forests and control forests (unmanaged), there were significant differences at the species level (distribution, habitat preference) (Fig. 10.5).

In terms of the spatial distribution of species, the highest level of a chorology index is reached when the distribution exists only in the form of isolated, widely separated colonies or only in a single "large colony." In Europe, the highest area affinity level is held by endemic species that have not been found outside of Europe so far. Habitats with a species community for which a high chorology index is calculated as the sum of the species values or a high average value can also be classified as regionally, supra-regionally, or nationally significant in comparison. A similar principle is also applied in the definition of so-called Endemic Bird Areas (EBAs)(Brooks et al. 2006).

When species with large home ranges are encountered (see also Sect. 4.7), the habitat will be highly rated, as a certain quality must be widely available. Generally, species at the top of the food pyramid have the most extensive home ranges (Tucker et al. 2014). The presence of a characteristic top predator is often used as a criterion for the area requirement of a largely self-regulating community (Soulé 1987). In practice, however, one does not often start from a top predator population but from the area requirements of a single breeding pair. Nevertheless, this can vary widely depending on the habitat quality (e.g.,

Table 10.3 Overview of essential parameters of species communities and their detectability, suitability, and application

Parameter	For Calculation recommended index	Sufficiently accurate detection	Dependence on the sample size	Comparability		Reaction to habitat change and environmental stress	Suitability for evaluation procedures of habitats		
				between different locations	between different animal groups		Advantages	Disadvantages	Overall judgment
Number of species	S (sum of species)	Possible for comparisons	Yes: Duration, spatial extent	Well suited	Hardly feasible	Sensitive	As rel. Value well captured	Only limited comparability due to sample size and collection method	Best value in direct comparison
Abundance	Individuals per area; also "activity density" (depending on mobility)	Difficult, only relative	No, but temporal Fluctuations	Only simultaneously	Not possible	Sensitive	Threat only recognizable through changes in abundance	Only meaningful in the long term	Most crucial value; development trends only recognizable after 5 years
Diversity	H's according to SHANNON and SIMPSON (see Sect. 15.11)	Possible	Low	Well suited	Not very suitable	Less sensitive	Well captured	Information content as sole value unclear	Good sample value for detecting stresses, disturbances
Evenness	E = max H's/In S $0 \le E \le 1$	Possible	Largely independent	Well suited	Suitable to a limited extent	Less sensitive	Well captured	Only differences of over 10% are usable	Meaningful addition to the number of species
Dominance structure	Histogram with species names	Possible	Little, if only the structure of the dominant species is asked	Well suited	Not very suitable	Sensitive	Bioindicator statement	Knowledge of species, at least of the dominants necessary	Important characterization of a habitat
Similarity index	Kw (product of RENKONEN dominant identity and JACCARD's number) ($0 \le Kw \le 100$) (Wainstein 1967; Huhta 1979)	Possible with some effort	Yes: Same sample size necessary for comparison	Well suited	Not possible	Describes quantitatively the degree of change	Quantitative measure of the change in fauna	Not easy to capture, species knowledge necessary	Useful for comparing faunas

Fig. 10.5 (**a**) Bossematié secondary forest, SE Ivory Coast. (**b**) *Euphedra* sp., Nymphalidae, Ivory Coast. (**c**) Mabi forest, unmanaged, SE Ivory Coast. A similar species is shown on Wikipedia: https://en.wikipedia.org/wiki/Euphaedra_xypete#/media/File:Euphaedra_(Xypetana)_xypete_MHNT.jpg

food supply), which, e.g., for the sparrowhawk varies by a factor of 2 (Glutz v. Blotzheim et al. 1971). In this species, the average distance between the nests in closed forests varies by a factor of 4 (Newton 1986). The home range of the buzzard varies by a factor of 2–3 and in the American red-tailed hawk and the golden eagle, approximately by a factor of 4 (Newton 1979). According to Brüll (1977), the golden eagle claims a home range of 8000–14,000 ha for the rearing of young animals. Home range sizes often differ between summer and winter stays, e.g., in the sparrowhawk, more than a hundred-fold (Newton 1986). More precise information on the variation of the density of birds of prey depending on the food supply can be found in Newton (1979). Once the top predator has a chance of survival, it is concluded that the species lower in the food pyramid also have their requisites in sufficient quantity. However, this only applies to a limited extent. Smaller species may have special requirements for their habitat or need to have escape options during extreme weather conditions that are not needed for the territory of the top predator (e.g., overwintering places, sunny places for development, see Sect. 9.6 (Ehrlich and Murphy 1987; Weiss et al. 1988).

If one speaks about securing the functioning of natural ecosystems (e.g., in the German Federal Nature Conservation Act § 1), species that essentially determine the energy flow in the ecosystem have a crucial position. According to this view, this would usually be the most common species, e.g., the dominant species in the group of decomposers, which cannot easily be replaced by other saprophagous species.

As a rule, ecosystems undisturbed by humans are valued the highest. Anthropogenic impairment disrupts the interrelationships between species, with host-parasite or prey-predator complexes usually failing first. The presence of specialized entomophagous or brood parasitic insects can, therefore, be considered to assess how intact a habitat still is. Among the wild bees (Apoidea), these are the brood-parasitic cuckoo bees (genera *Sphecodes, Nomada, Epeoloides, Coelioxys, Stelis*) and parasitic bumblebees (genus Psithyrus) and brood parasitic wasps (e.g., Leucospis gigas) (see Fig. 10.6). The presence of cuckoo birds as brood parasites also increases the rating in conservation assessment.

Also, among the spider wasps (Pompilidae) and the digger wasps (Sphecidae), there are brood parasitic species (*Nysson* spp., *Pompilus* spp., *Anoplius* sp.). The thick-headed flies (Conopidae), the bee flies (Bombyliidae, Figs. 10.6 and 10.7, and the Ichneumonoidea are parasitoids. Natural enemies of xylobiont insects are Ichneumonidae (cf. *Rhyssa* sp.), which reach the larvae in the wood with the long ovipositor (Fig. 10.8). The presence of ichneumon, chalcid, and cuckoo wasps must be valued equally high, especially the hyperparasites among them. However, in practice, they are challenging to evaluate due to the problems of species determination.

Fig. 10.6 (**a**) Cuckoo bumblebee *Psithyrus* sp., brood parasite on bumblebees *Bombus* sp. (**b**) Cuckoo bee *Nomada* sp., brood and kleptoparasite in the bees *Andrena* sp., Germany. (**c**) Bee-flies (Bombyliidae, *Bombylius* sp.), larvae are entomophagous in the bees *Andrena* sp. (**d**) Leucospis gigas (Chalcididae) ovipositing on the nest of Chalicodoma sp., Megachilidae)

Fig. 10.7 (**a**) A bee-fly (Anthracinae, Bombyliidae) hovers in front of the entrances of old beetle boreholes in the wood, which are used as breeding tubes by solitary bees (see Fig. 10.6e right), West Papua. (**b**) The bee fly throws eggs into the bee's breeding tube with the tip of its abdomen. The eggs are dusted with sand for throwing (Mühlenberg 1971). The hatching bee-fly larvae are entomophagous on the bee larvae. (**c**) A solitary leafcutter bee (Megachilidae) carries pollen into the breeding tube for its larvae, West Papua

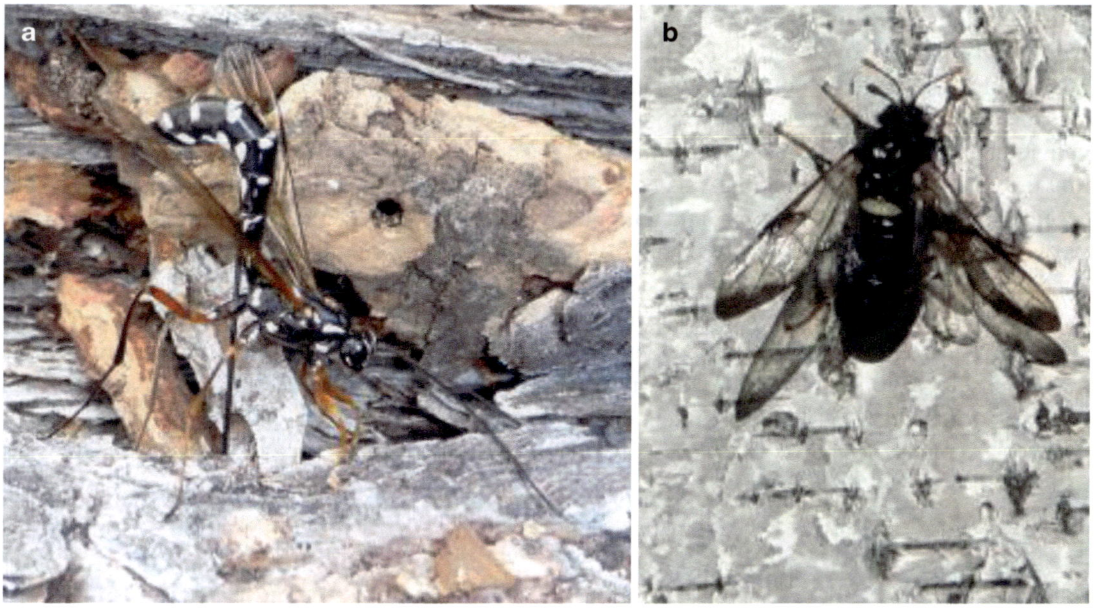

Fig. 10.8 (**a**) The Ichneumonid wasp (cf. *Rhyssa* sp.) lays its eggs in the brood tunnels of a wood wasp. It drills into the wood trunk with its long ovipositor to reach the prey larva. Northern Mongolia. (**b**) Wood wasps (Siricidae) as hosts are rarely seen in managed forests. The larvae are also wood feeders. Northern Mongolia

Fig. 10.9 Female oil beetle (*Meloe proscarabaeus*, family Meloidae blister beetles), Germany

Beetles of the family Meloidae (blister beetles, Fig. 10.9) lay many eggs in the soil, which develop parasitically. The tiny larvae climb onto flowers and cling to solitary bees carrying them into their nests. In the nest of solitary bees (e.g. *Anthophora, Andrena*), they eat the host egg and later the pollen nectar mash (Jacobs and Renner 1988).

Species with differentiated biotope relationships (Blab 1986) or dependence on a combination of factors should also be highly valued. These species usually require several habitats or different requisites for their development: our native amphibians claim a spawning ground, a summer, and a winter quarter for the annual habitat. The butterfly Large White Admiral (*Limenitis populi*) lives as a caterpillar on trembling and black poplars of forest edges, sucks as a butterfly on carrion and excrements, e.g., on unpaved forest paths, and meets for mating at dominant treetops (Blab 1986, Fig. 10.10).

The Large Blue (*Maculinea nausithous*, Lycaenidae) develops in the first larval stages in the flower heads of the Great Burnet (*Sanguisorba officinalis*), then leaves the food plant and migrates as a larva into the nests of the red ant *Myrmica* sp., where it completes its further larval development as a kleptoparasite of the ant brood (cf. Maschwitz and Fiedler 1988; Settele and Geissler 1988, similar to the butterfly *Maculinea teleus*, Woyciechowski et al. 2006, Fig. 10.11). Although the great burnet and knotted ant are both widespread when considered individually, anthropogenic changes often disrupt the coexistence of these otherwise common species. It is usually the spatial and temporal combination of several requisites, disturbed by altered economic practices that threaten the diversity of species. The lycaenid butterfly *Maculinea arion* shows a similar dependence on the presence of *Myrmica* ant species. Their abundance also depends on the management of the meadow (Thomas 1991a, b).

Fig. 10.10 *Limenitis populi* (Nymphalidae), northern Mongolia

10.9 Target Species

High costs for protective measures require success control. The quality of the protective measures can only be quantitatively and clearly evaluated through monitoring a selected population. This is known as habitat protection through safeguarding a target species (also focal or priority species) (Mühlenberg et al. 1991; Mühlenberg and Slowik 1997). In a protection strategy to do "something for all species" in a limited area, examining to what extent this goal has been achieved or should be continuously further improved remains an open discussion. In contrast, the future prognosis for a selected target species can serve as a quantitative evaluation standard (*population viability analysis*, Beissinger and McCullough 2002; Morris & Doak 2002; Chaudhary and Oli 2020; Fieberg and Ellner 2001; Manlik et al. 2018). The criteria for selecting target species and their application for landscape planning are described in (Mühlenberg 1993, Mühlenberg et al. 1991, 1996; Brambilla et al. 2011; Gibson Vega et al. 2024). The monitoring also serves as a basis for potential improvements in the area (for maintenance measures, see Sect. 14.2) if this increases the survival chances of the selected target species (Ruggiero et al. 1994). The orientation toward a target species has the advantage that, at least for one endangered species, the evaluation or mea-

Fig. 10.11 (**a**) Scarce large blue butterfly (Lycaenidae, *Maculinea teleius*); IUCN Red List: near threatened species; additionally, as a larva predatory in ant broods. (**b**) Larva of *Maculinea teleius* (in the centre) in an ant nest of *Myrmica*. The larva preys upon the white larvae of ants but is not attacked by ants due to its desired secretion

surement is carried out with the perspective of its future survival using quantitative statements and remaining scientifically verifiable (see Sect. 14.1).

For a risk analysis, the following information must be collected and evaluated:

1. Population size.
2. Area requirement.
3. Population biology.
4. Habitat quality.
5. Assessment of future development.

The assessment of the future development of a population depends on the probability of the population's survival.

As an example, questions on population and habitat quality are listed below (see also Marcot 1986; Mühlenberg 1993):

1. Current distribution and abundance of the target species in question: How many breeding pairs/individuals does the population in the observed area comprise in the current breeding period? How large is the area being studied? Is the entire population or only a specific part being monitored?
2. Population development over the last decades: Where have occurrences disappeared and—as far as known—for what reasons? Have occurrences in the neighborhood disappeared? How far away are the studied species' nearest current (and former) occurrences? What are the suspected causes for the observed population trends? Where have new or re-colonizations taken place?
3. Deterministic threats such as habitat destruction or direct exploitation.
4. Primary habitat and substrate for reproduction and rearing: What "props" (requisites) are essential for a suitable habitat?
5. Type, quantity, quality, and arrangement of microhabitats that promote survival and reproduction. How does the winter habitat differ from the breeding habitat?
6. Habitat qualities used for foraging.
7. Habitat dynamics: Are habitats constantly lost, for example, as part of successions, or when and where do new ones emerge?

10.10 Quick Forecast of Survival of a Population (Threat-Level)

So far, detailed threat-level analyses exist only for a few species. The example of the Northern spotted owl from California (*Strix occidentalis caurina*) was groundbreaking (Lande 1988, Noon and Blakesley 2006). The studies by Murphy et al. (1990) on a butterfly species are exemplary for insects. Threat level analyses require many years of fieldwork to estimate the parameters in their variability. Walters (1991) showed for the North American woodpecker *Picoides borealis* (a relative of the European three-toed woodpecker *Picoides tridactylus*) that the search for critical factors limiting the population can be far more important for developing a protection strategy than the calculation of a population model that is based on weakly estimable demographic parameters even after thorough research, see also Conner et al. (2001). Research on population biology and the existing requisites in the area should be initiated parallel to practical conservation measures (see also Brussard 1991). As a first step, we suggest a so-called quick forecast (Hovestadt et al. 1991; Mühlenberg et al. 1991). With this method, the future chances of a population are initially estimated without a significant field program so that an immediate action catalog can be created within a short time. As part of a quick forecast, all quickly recordable rough data regarding occurrence, distribution, and habitat quality are assessed in the light of all available knowledge about the species in question. Literature reviews and rough data collection in the field, which do not last longer than half a year, form the basis for a quick forecast. Local information and a well-prepared written survey for relevant clubs, experts, and amateurs can significantly expand the knowledge necessary to record a population nationwide and to inform about neighboring populations (see also Sect. 2.28). The questions for the survey about a specific species are selected from the information collected from a risk analysis set up above.

The planning ends with a calculation of the anticipated costs. Expenditures can reach millions of Dollars/Euros to secure a population against as many risks as possible. The financial expenditures for the protective measures include the areas of "Habitat Design" (purchase of areas, structural changes in agriculture and forestry), maintenance measures (maintenance of certain successional stages), and accompanying research (risk analyses, monitoring). The expenditures are, in turn, dependent on a suitable legal toolkit for effective nature conservation.

It is easier to justify high costs if they are spent simultaneously on broad goals, i.e., for many species. However, their use can only be estimated based on the selected target species (see above). Other endangered species can also benefit from the protective measures for target species, and their survival can be improved but not optimized simultaneously. If you optimize, for example, the habitat for the Ortolan bunting in Europe, the living conditions for other species with the requisite fruit trees (Lesser Grey Shrike, Red-backed Shrike, Wryneck, Garden Dormouse, some butterfly species) and field margins (various beetle and spider species) will also improve. We, therefore, practice habitat protection through the future viability of target species.

10.11 The Steps of a Population Viability Analysis

It is best to proceed with the below steps when carrying out a risk analysis:

1. Study of literature.
2. Selection of target species.

This information lets you initiate the first actions to protect the selected species. Steps (1) and (2) take about a month.

3. You can now formulate hypotheses about essential dependencies of the target species on environmental factors, which we assume to be "key factors" of population dynamics.
4. Now, you examine the reaction (response) of the target species to these identified key factors.

With the knowledge gained, you can now support the target species through specific measures. Steps (3) and (4) could be dealt with in a master's thesis, i.e., in one season.

5. Continue investigating the key factors and the population dynamics in the following years.
6. Comparison between habitats that differ in size, isolation, and quality.
7. Study of reproductive success in optimal and marginal habitats.
8. Assessment of the target population's future survival chances (viability) through modeling.

Steps (5) to (8) could be the basis for a Ph.D. thesis, which would take at least 3 years. According to the objective (how high is the probability of the population's survival?), the corresponding area requirement of the population can then also be determined, considering the habitat quality, probable population dynamics, and dispersal ability (Hovestadt et al. 1991).

Fig. 10.12 The Woodlark (*Lullula arborea*) often forages in patches devoid of plants. Croatia

Case Study: Analysis of Key Factors for a Songbird

Using the example of the **Woodlark** (*Lullula arborea*), we want to demonstrate how one could proceed in the search and analysis of key factors (steps (1)–(6)). In Germany, the Woodlark (see Fig. 10.12) is considered endangered to critically endangered (Bundesamt für Naturschutz 1998) and is strictly protected. In Europe, the species is on the "pre-warning list" (protection category V ("*Lower Concern*"), BirdLife International 2021). The population trend is decreasing.

The species profile (Artensteckbrief 2014–2023) describes the breeding habitat as follows: breeding habitats are located in semi-open landscapes, especially on warm, dry sandy soils and in sunny slope locations. Significant are thinned forest stands (especially pines) with low herb and shrub layers, singing perches, and vegetation-free areas for foraging, preferably at forest edges. It avoids closed forests and treeless agricultural landscapes. To locate them, consult the literature on local distribution. For many regions, there are local fauna lists for birds (e.g., Zang and Heckenroth 2001). The Woodlark is relatively common in Germany, for example, in the Dübener Heide Nature Park (Eastern Germany, Saxony-Anhalt and Saxony) and the Lüneburg Heath (Northern Germany). Here, you can find pine afforestation, clear-cuts, dry open land biotopes, and military training areas, all satisfying the requirements of woodlark breeding habitats. The literature indicates that individual shrubs or low pines as singing perches are essential, but also open, predominantly sandy soil areas for foraging. One can use sound decoys to help locate woodlarks (see above Sect. 1.9, and Parker 1991). Once woodlarks have been found, map the territories (see Fig. 1.9 in Sect. 1.7 Inventory Recording). Then, spatial parameters are quantitatively estimated within the territories and within the radius of random points in the same landscape space. The territories reach a size of 2–5 ha (Vogel 1998a, b). According to the literature, the following parameters seem promising: perches, height, and density of individual trees (e.g., in pine plantations), percentage of open soil areas, and proximity to closed forest and agricultural areas. The methodology of mapping vegetation formations and landscape elements via satellite images, aerial photos, and topographic maps is described in Chap. 4. The areal proportions of the respective formations are then measured on the created map with a planimeter[1] or with tools included in digital maps and GIS systems. In the case of clear-cutting and pine reforestation, the age of the individual trees is important as it relates to their height. For example, if one plots different-aged pine reforestation (size at least 3 ha, age stages on the x-axis, number on the y-axis) with the occupation by woodlarks in a simple bar chart, it becomes clear that the woodlark prefers younger plantations (under 5 years). Pine forest is also significantly more occupied by woodlarks than mixed and deciduous forest plantations. Large-scale mapping of open land structures, patchy vegetation, and dense vegetation also clearly show how much open land and sparse vegetation are important for the occupation of woodlarks (for landscape mapping, see Sect. 4.5). Once such key factors for the territory formation of the woodlark have been identified (see above steps 3 and 4), the study is continued in the following years (step 5). For example, one could demonstrate a shift of the territories with the aging of the pine reforestation or the overgrowth of open spaces. One could also offer artificial perches in different habitats to study the occurrence of

[1]Mechanical planimeters can be purchased from around 150 euros, digital planimeters cost more than double (Internet research 2023)

Fig. 10.13 (**a**) The Spotted Fritillary *Melitaea didyma* on a Knapweed during nectar intake. (**b**) Habitat of the checkered butterfly near Hammelburg, Northern Bavaria. (**c**) *Stachys germanica* on calcareous oligotrophic grassland. The butterfly uses this plant for egg-laying and caterpillar food and occurs in the habitat in small-scale succession stages with very loose shrubbery

the woodlark. For artifical perches, it is recommended to use wooden posts of at least 10 cm in diameter and up to 2.5 m in height. The posts can then be set up in a 25 m grid from the forest edge into the open landscape parts. In step (6), one compares the habitats in their occupation with further factors, such as the spatial extent of the vegetation formations in the region, the next occurrence of another (sub)population of the woodlark, and comparisons to territory occupations of a related bird species such as the skylark (*Alauda arvensis*) or the tree pipit (*Anthus trivialis*). To estimate the reproductive success (step (7), one must find and check nests, which is tricky. The nests of the woodlark are hidden on the ground and can best be found by observing feeding adult birds.

Case Study: Analysis of Key Factors for a Butterfly

In another example, we present the search and analysis of critical factors for an insect species. We are moving at a different scale than with bird studies. As a target species for which some prior knowledge exists, we choose a checkered butterfly, the spotted fritillary *Melitaea didyma* (Vogel 1998a, b, Fig. 10.13). This species is spread over Southern Europe and North Africa to Central Asia. In the north, it reaches the border of Belgium and the Baltic States. In Germany, the butterfly is extinct in some federal states, while in others it is endangered to severely endangered (see Red Lists of the individual federal states). To find the species, we studied first the regional distribution maps, which exist for individual federal states, e.g., for Bavaria in the species protection mapping (Bavarian State Office for Environmental Protection), for Baden-Württemberg, see Ebert (1993), or Kudrna (1993). Regarding habitat requirements and host plants (caterpillar food plants), we collected information from numerous summarizing works about butterflies, like Higgins and Riley (1978), Dennis (1992), Bink (1992), Pullin (1995), Settele et al. (2015).

We were then interested in already known factors influencing the population, i.e., what causes fluctuations of the population size. In butterflies, regulation is expected from the lower trophic level (Dempster 1983; Stiling 1988), with not only the occurrence of the host plant but also its growth form and microhabitat environment (Thomas 1991a, b). The butterfly female receives information about the plant's vitality for oviposition and the microclimate of the location (Chew and Robins 1984). Another factor for habitat quality, or habitat capacity, is the nectar offered to butterfly females, which also affects the fecundity of the females, i.e., the possible number of eggs laid.

For estimating the probability of survival in a study area, we compiled the following questionnaire according to Vogel (1998a, b) (see also Sect. 10.8 Risk analysis above):

1. How large is the population under consideration in the study area?
2. How much does the population size fluctuate?
3. Are the individual colonies isolated?
4. What are the habitat quality requirements of *Melitaea didyma*?
5. Which factors of habitat quality affect the population dynamics?
6. How much suitable habitat is available in the study area?
7. How will the viability of *M. didyma* be assessed if the status quo of the habitat offer is maintained?
8. What could management for the study area look like to maintain *Melitaea didyma* long-term?

With the knowledge gained from the literature (see above), investigating the key factors can include sunshine duration. Information on sunshine duration in the study areas can be obtained from the nearest weather stations (Sect. 3.1, Data logger). It can be related to the estimated population size between the years: longer sunshine duration leads to faster development of the caterpillars, a higher survival rate, and a more considerable weight of the caterpillars in the spring (Vogel 1998a, b). Earlier hatching of the butterflies allows them a longer time span for egg-laying. In the field, caterpillars can be counted on the food plants in the area. The duration of sunshine during the flight time of the butterflies is also related to the population size (Pollard and Yates 1993; Pollard et al. 1995). To estimate the population size of butterflies, see Sect. 2.3.

Host plants are counted along transect lines. At the host plants, the ground cover in a circle of 50 cm is estimated, and the coverage degree is expressed as one of four classes (0–25%, 25–50%, 50–75%, and 75–100%). The number of laid eggs is counted on individual plants, and the height of the plants is also measured. Individual plants were also brought into the field for an experiment. We planted some of them in low ground cover and an equal part in dense ground cover while increasing the ground cover artificially around the plant. The plants were placed in individual tubes. The influence of temperature on the development time is best estimated in the laboratory, where fresh egg clutches (observations in the field) are allowed to develop at different temperatures (e.g., in three classes, constant <15 °C, at approx. 20 °C, and 25 °C).

For the factor of nectar supply, flower usage in the field is observed (see also Sect. 7.9). Binoculars were used to avoid disturbing the butterflies. Quantification is facilitated if transects (e.g., 5 m wide) are marked for observation. The number of nectar plants was counted along the transects to estimate the flower offer. Post-diapause larvae are collected from the field and allowed to hatch in cages. Freshly mated females are kept separately in three approaches: feed solutions with water, sugar solution, and sugar solution plus amino acids. The egg count can be determined by the number of laid eggs and additional abdomen dissection; some females were dissected immediately after hatching to determine the existing egg count (Vogel 1998a, b). In this way, an estimate is made of how the nectar supply contributes to the fecundity of the females.

10.12 Relationship between Species Densities and Habitat Quality

If you find substantial differences in the frequency of a particular species when comparing two habitats, this might indicate differences in the quality of the habitat (Fig. 10.14).

A comparison of the abundance of birds between a commercial forest and a mature forest (forest with old succession stages) shows the carrying capacity of suitable habitats compared to managed ones (e.g., Struhsaker 1997). The population density of typical forest bird species can be up to 20 times higher (M. Waltert, oral communication on investigations in the Ivory Coast 1999, Waltert 2000). For example, the densities of a turaco, a duiker, and a primate were significantly lower in the overused Bossematié forest in the southeast of the Ivory Coast than in an intact African rainforest (see Table 10.4).

Another relationship between abundance and habitat quality results from territories in forest parcels disturbed by management being abandoned more often than in undisturbed forest areas. This can be demonstrated by the capture and recapture of territorial bird species (Winker et al. 1995). In addition, weight data at the time of capture provide information about the condition of the individuals (Waltert and Mühlenberg 2001). Despite more frequent captures, the managed habitat for the Greenbul (*Andropadus latirostris*) is lower rated (Fig. 10.15).

10.13 Source and Sink Habitats

The occurrence of a species in a habitat does not necessarily indicate a "good" habitat for the species (see also Hoopes and Harrison 1998). Even if a species has reproductive success in various habitats, these can be greater in one habitat than another. If the annual reproduction outweighs the average mortality, we speak of "source habitats." If the reproductive performance does not compensate for mortality, we have "sink habitats"

Fig. 10.14 (**a**) The Red Colobus (*Colobus badius*) occurs in degraded forests, Ivory Coast. (**b**) Foraging Red Colobus. (**c**) Closed rainforest, where the Red Colobus occurs ten times more frequently than in (**a**), Tai National Park, Ivory Coast

Table 10.4 Density estimates using the distance sampling method to compare between an overused tropical rainforest in West Africa (Bossematié Forest, Ivory Coast) and less used, more intact rainforest areas in the vicinity of Bossematié or neighboring countries of Africa. The density estimates were obtained through community-based monitoring (see above, Sect. 1.15). The Bossematié forest will be described in more detail in Sect. 14.1.1 (rehabilitation of forests)

Turaco *Tauraco macrorhynchus*	Density in Bossematié, overexploited forest (ind./km^2)	Density in Mabi forest, largely intact forest near Bossematié	Estimated maximum capacity, e.g., virgin forest
	1995: 4.0 1996: 3.7 1997: 4.9	1997: 11.4	28 (Gabon, Brosset and Erard 1986)
Duiker *Cephalophus monticola*	1995: 0.8 1996: 1.6 1997: 1.6	1996: 4.8 in Songan forest, close to Bossematié	11.8 in Nigeria, Adeyemi et al. 2018, 15,5 in Nigeria (Jimoh et al. 2013), 24.5 (Ghana), 8,3 Cameroon (Viquerat et al. 2012),
Mona monkey *Cercopithecus mona*	1995: 0.4 1996: 0.4 1997: 1.0	1997: 1.4 (Songan, compare Waltert et al. 2002)	14 (Cameroon, Korup NP, Waltert verbally, unpublished)

Fig. 10.15 Relationship between abundance and catch rate using the example of the Yellow-whiskered Greenbul (*Andropadus latirostris*). (**a**) Upper left panel: Secondary forest (overgrown farms), SW Cameroon, (**b**) Upper right panel: The Yellow-whiskered Greenbul (*Andropadus latirostris*), caught alive in the mist-net for banding and weight determination, (**c**) Lower panel: Near primary forest in SW Cameroon

(Ens et al. 1992). In "source habitats," an excess of individuals is produced, exemplarily studied in the Oystercatcher *Haematopus ostralegus* (Ens et al. 1992). The dichotomy between sources and sinks is even noticeable in their corresponding microhabitats (Fig. 10.16). Gottschalk (1996) was able to investigate this using controlled egg development of the Western Bush-cricket (*Platycleis albopunctata)* in the laboratory. Whether a habitat is optimal for a species can only be assessed if the reproductive performance is also examined.

A drastic example of a sink habitat is the large-scale agricultural landscape in Poland dur-

Fig. 10.16 (**a**) Slope areas with patchy vegetation are the optimal habitats for the Western bush cricket (*Platycleis albopunctata*). Cultural landscape in Lower Franconia (Northern Bavaria), Germany. (**b**) The endangered *Platycleis albopunctata.* (**c**) The eggs develop faster in patchy vegetation, thus determining the overall population development of *Platycleis albopunctata.* These differences are already evident in the direction of exposure (slope position). Only the south-facing wall sides are suitable microhabitats for oviposition and grasshopper development (Gottschalk 1996)

Fig. 10.17 Agricultural landscape in Central Poland, 1983. Due to uniform mowing, many corncrakes land in the green areas to breed but lose all their clutches. Thus, these are sink habitats

Fig. 10.18 Meadows and pastures in the Narew lowlands in Northeastern Poland. The small-scale farming at different times allows corncrakes to breed successfully. Thus, these habitats are source habitats

ing the month of May for the corncrake *Crex crex* (Fig. 10.19), which attracts the animals for breeding (Fig. 10.17). However, simultaneous mowing at the end of May destroys the broods (Schäffer 1984, 1999; Schäffer and Weisser 1996). Source habitats are the natural meadows in the swampy terrain of the Narew lowlands (National Park in north-eastern Poland, Fig. 10.18).

Perlut et al. (2008) show similar issues of agricultural practices for North American grassland songbirds. However, here, it is not the centralized growing and harvesting schedule in the old socialist planned economy as in the example of Poland above, but the mowing of grassland for hay production that, in most cases, happens before the conclusion of the breeding cycle of the bobolink (*Dolichonyx oryzivorus*), Fig. 10.20 and other North American grassland songbirds.

Fig. 10.19 Corncrake (*Crex crex*) on its nest in cover within the grassland. The corncrake is one of Europe's most endangered species. Photo credit N. Schäffer

10.14 Terrain and Vegetation Parameters and Animal Ecological Evaluation Criteria

This chapter outlines how habitats can be evaluated from an animal ecological perspective and what should be considered when creating maintenance plans. Different species and habitats require different evaluation criteria and management measures. Before an evaluation based on the animal's ecology, the criteria by which its populations should be assessed must be defined.

The basic knowledge for an animal ecological evaluation comes from comparing these criteria

Fig. 10.20 (**a**) Left panel: Mature male, Bobolink perching near Lancaster, PA, USA. Picture credit: Andrew Weitzel on Wikimedia, CC BY-SA 2.0. (**b**) Bobolink habitat near Adrian, Michigan, USA. The top panel shows the landscape of the tall grass meadow (approx. three ft. high) before mowing, surrounded and separated by tree rows, fences, and woods. (**c**) The lower picture shows the same landscape three days later after mowing

with ecological parameters determined by the terrain and the vegetation. Examining these parameters' influence on the values, which depend on terrain and vegetation, is essential to describe an animal's ecological situation. Suppose a parameter change significantly increases or decreases the numerical value of an evaluation criterion; in that case, this parameter is suitable for measuring animal ecological evaluations in practice.

Forests are structurally rich ecosystems and their vegetation can be stratified at different scales. Many studies have shown that structurally richer forests also harbor more birds (MacArthur and MacArthur 1961; Recher 1969; Roth 1976; Mühlenberg 1980; Ćosović et al. 2020). A general correlation exists between more diverse resource offers and higher species diversity for many other animal groups (e.g., Currie 1991). With modern technology, the forest structure can also be measured from the air using laser beams (Müller et al. 2014; Carrasco et al. 2019). Such a system is called LIDAR, or light detection and ranging. This system can also be used with drones. You may find these two YouTube videos helpful (Neon Science[2]); also, see Sect. 4.5.

Significant correlations between bird fauna characteristics and vegetation structure parameters are often determined by the class formation, i.e., the degree of fineness of the grid in the vegetation description. Structure parameters with a coarse grid will always correlate with a change in the animal community. Comprehensive or "macro-habitat" features are associated with the bird species community as a whole (Anderson 1981). It must be considered whether the habitat features should later be assigned to individual species to derive biologically meaningful statements from a correlation between vegetation structure and animal species c determined via maps and aerial photographs ommunity. If so, microhabitat features are essential (see Sect. 4.2). The vegetation structure will be divided as finely as possible on this "micro-level." Horizontal and vertical structural components should be analyzed separately. Meaningful microhabitat features at the level of individuals and single species are often recognized when the vegetation structures in the occupied territories of birds are recorded and, in subsequent years, when territories are spatially shifted and are described again. A shift in territories can be the response to changing habitats, competition, or predator-pressure conditions.

The method of "bird-centered vegetation sampling" (Sect. 9.4, Habitat suitability and use, example Great Grey Shrike) is better suited to study actual habitat preferences.

Tree cavities and cavity-nesting bird species are an example of a direct connection between a vegetation parameter and an associated animal group. Quantitatively, tree cavities in a forest stand are counted by directly scanning tree trunks. In a selected hectare area, two points are systematically chosen, at which the trees >5 cm DBH (*diameter at breast height*) are determined, and their properties are recorded, e.g., DBH, dead or alive, with or without broken branches (see Table 4.3 in Sect. 4.2.4). Each recorded tree is then searched for cavities from three sides, for deciduous trees ideally in autumn after leaf fall (Bai et al. 2003). The data from each collection point can then be expressed per area unit (e.g., 1 ha).

To answer whether the cavity supply is sufficient for and used by secondary cavity breeders, the cavity trees must be checked for breeding occurrences in the following spring. In natural forests of the boreal and temperate zones, less than 10% of the cavity supply is used (Bai et al. 2003; Carlson et al. 1998; Wesolowski 1989). Woodpeckers must be recorded in early spring (March–April) by mapping, drumming, and calling birds. Breeding evidence is provided by observation of begging young birds or feeding adult birds in May/June. Woodpeckers can be more reliably located with trained detection dogs (H. Hondong, Conservation Detection Canines, see Sect.1.4).

Such a recording of the forest structure and availability of trees, which are essential for nature conservation, provides an objective assessment of forest areas. Excellent habitats are the so-called "wildlife trees" that are relevant for several species and do not longer occur in conventional commercial forests (see Fig. 10.21). They are added to the list of suitable habitats without further investigations.

[2]Neonscience.org in North America. NEON is a continental-scale ecological observation facility, sponsored by the National Science Foundation. How Does LiDAR Remote Sensing Work? https://www.youtube.com/watch?v=EYbhNSUnIdU Drone LIDAR explained: https://www.youtube.com/watch?v=xS4kOWcoDPY.

Fig. 10.21 (**a**) This old Siberian larch *Larix sibirica* with partly broken branches represents a typical "wildlife tree." (**b**) Wildlife trees with tree cavities are the essential habitat resource for the Siberian Flying Squirrel (*Pteromys volans*)

10.15 Evaluation of Hedges

Hedges are a cultural landscape component that can be used for nature conservation. Their value for biodiversity and endangered individual species depends on the design and maintenance of the hedges. Figures 10.22 and 10.23 depict examples of hedges and woody strips in cultural landscapes, representing very different values for nature conservation objectives.

Hedges are phytosociologically autonomous mantle-edge complexes on forest edges (Wilmanns 1989). As edge biotopes, they often have an increased species diversity (García de León et al. 2021; Kratschmer et al. 2024; Montgomery et al. 2020).

Zwölfer et al. (1981) have shown how to analyze individual evaluation criteria of hedge characteristics regarding the occurrence and reproduction of selected animal groups. He summarised this in a matrix (Table 10.5). Each crossing point in this matrix identifies a separate study. Furthermore, Zwölfer et al. (1981, 1984) provided an analysis of the function of hedges and an evaluation scheme. They have examined the role of hedge parameters for (a) phytophagous insects, (b) entomophagous insects and spiders, (c) breeding birds, (d) refuges and food of wildlife, and (e) phytophagous-entomophagous complexes. Graham et al. (2018) summarized results on the function of hedges in a meta-analysis through a systematic literature review and narrative synthesis.

From the results of such comprehensive investigations, some criteria can then be selected as needed since priorities for protection and maintenance measures are usually determined depending on the part of the landscape to be assessed (e.g., breeding birds, amphibian protection, insect pollinators). Planners ideally want easily measurable parameters that can be used directly in an evaluation scheme based on the selected investigation results. For the hedge evaluation, Zwölfer et al. (1981) suggested an index composed of the

Fig. 10.22 (**a**) Landscape section in New Zealand, South Island with an artificially fragmented cultural landscape through exotic hedges in the upper half. In the lower half, a wilderness-like landscape with a meandering river system and largely undisturbed growth of shrubs. (**b**) Hedges with exotic plant species do not support the native fauna; New Zealand. (**c**) Agricultural landscape in Central Germany (Southern Lower Saxony) with interspersed native hedges and strips of wood

product of a shrub species factor, the age class factor of the hedge installation, and a surface density factor of the hedges in the area. If the quality of habitat is expressed in numbers, these numerical values must under no circumstances be offset against other measures, e.g., the area of the habitat (Kaule 1991). An area of quality five, e.g., is never equivalent to half the size of an area of quality 10. The disadvantages of such an evaluation are always the blanket estimates of several criteria, where in individual cases, no reliable statement can be made about the protective effect for a specific, threatened animal population. Including as many characteristics as possible (with or without evaluation values) will not scientifically achieve a higher level of objectivity.

Fig. 10.23 (**a**) Cultural landscape in New Zealand, Banks Peninsula, where an attempt is being made to create a network of forests and hedges by promoting natural shrubs. (**b**) Natural forest formation in New Zealand, South Island. (**c**) Hedges in Central Germany (Southern Lower Saxony). On the left side of the road, it is in high-quality condition for birds living in hedges, and on the right side, too much is trimmed towards the field side. (**d**) A hedge trimmed on two sides offers only a few corvids (e.g., magpies) a nesting opportunity in the agricultural landscape. Central Germany

10.16 Forest Edges and Edge Effect

The vegetation at forest edges is often structured characteristically (Wilmanns 1989): The forest is shielded from open spaces by a shrub strip, the mantle. In front of the shrub mantle is a strip dominated by perennials called the fringe (border). Natural forest edge structures reflect site characteristics, which differ between south and west exposure with warm and dry and north and east exposure with cooler and fresh-moist conditions. Faunistically, both pure stands can be represented in the transition zone (e.g., between forest and meadow). Edges are inhabited by many plant and animal species, and the population densities of animals are often higher than in the adjacent habitats. This phenomenon is called the **"edge effect"** (Smith et al. 2018; Harper et al. 2005). For the influence of forest edges on species in the forest interior, see also Sect. 11.1, fragmentation. The increased diversity in edge areas can have various causes: (1) New combination of biotic and abiotic factors resulting from the steep abiotic gradient and the special vegetation conditions (mantle, edge) forming new resources. They facilitate the evasion of animals in case of weather-related disturbances and thus "stabilize" the populations (cf. Schaefer 1974b). (2) In the edge area, animals from both adjacent living areas find a habitat. (3) The bor-

Table 10.5 Compilation of terrain and vegetation parameters (rows) and animal ecological evaluation criteria (columns). - Ideally, data from the field would have to be collected for each table cell in the matrix

Magnitude of an abiotic gradient (humidity) in the area	Degree of mixing of different succession stages	Successional stage	Trophic state (low nitrogen - high in nitrogen)	Surroundings and anthropogenic influence	Size of area to be evaluated	Degree of isolation of the habitat	Main wood species and numbers	Presence of special small structures[a]	Structural richness of vegetation	Plant diversity	Terrain and Vegetation-Parameter / Animal Ecological Evaluation Criteria
											Number of species and other key figures for a selected animal group
											Proportion of predatory species in the animal community
											Density of breeding birds
											Particular plant food supply[b]
											Particular animal food supply[c]
											Refuges and cover opportunities for vertebrates
											Presence of so-called "Red List" species[d]
											Presence of a top carnivore
											Number and completeness of food chains
											Distance to the nearest subpopulation for less mobile species

[a] Rock piles, soil exposures, caves, ditches, springs, vegetation-free soil spots, tree stumps
[b] e.g., buds, shrubs—berries, perennials—seeds
[c] e.g., high mosquito density, high density of invertebrates in mud or silt
[d] Political argument

der zone offers "visitors" a temporally limited resource (berry food, hibernation). Leipold (1985) demonstrated the edge effect on spiders and ground beetles in the Northern Steigerwald (Lower Franconia) using pitfall traps. In just the mantle and edge, she found 45 spider species (27% of all caught spider species). The increased diversity on the edge might be caused by a better food supply and increased structure and microclimate factors. Experimental removal of the mantle and partial artificial roofing (with tarpaulins, for example, Fig. 10.24) can help separate the effects of structure and microclimate factors (D. Leipold, unpublished). After a strip has been cleared, it can be observed from which areas the recolonization takes place (cf. Schaefer 1974a). Although the species in the mantle and especially in the fringes can also occur independently of the forest edge, in our cultural landscape, often only forest edges offer these species opportunities to live. This also applies to specialized plants (Wilmanns 1989). Various designs of a forest edge are shown in Fig. 10.25.

Preserving the forest edge causes little conflicts of interest with other forms of use regarding area requirements. An intact forest edge offers many ecological advantages for biodiversity: higher species diversity, constant habitat for independent species, a refuge for beneficial organisms of the adjacent areas (e.g., forest edge breeders) or overwintering guests, higher food supply through greater plant diversity, continuous transition of abiotic conditions between forest and open land (temperature, wind). Regarding the use of forest edges by endangered species in Germany, some structural features were more important than others (Hondong et al. 1993). These include southeast to southwest exposure, high age of the transition area, wide shrub layer, and blackberry pre-mantles; nutrient-poor, dry perennial edge, interlocking with elements of the lean pastures and dwarf shrub heaths; sporadic occurrence of open soil spots, vertical earth cracks as well as short-grass or sparse vegetation; participation of old, site-native deciduous trees in the formation, occurrence of pioneer tree species like aspen and willow in the mantle.

Fig. 10.24 South-facing forest edge in Steigerwald, Lower Franconia. The forest edge is still "intact" in the left half of the picture, showing mantle and fringe. In contrast, to the right (in continuation of the intact forest edge), the mantle was artificially removed and partially covered with tarpaulins for experimental purposes. Southern Germany, 1985

10.17 Faunal Shifts in an Environmental Gradient

We want to investigate which animal species specifically inhabit the forest edge area and whether one can speak of a separate community. The problem is whether zoocoenoses can be distinguished from each other. On the one hand, the species within a biocoenosis are, by definition, linked to each other through interactions. On the other hand, the composition of the fauna often changes continuously from one area to another. Discontinuities can then occur along a gradient when dominant species, which overlap to some extent, influence the abiotic environmental gradients so that other species depend on their distribution (see above Fig. 10.2, Possible distribution of different species along an environmental gradient).

Interactions are especially expected between the members of different trophic levels, less so in a systematically defined group (taxocoenosis), which belongs to the same trophic level (Müller 1978).

An environmental gradient can be documented more easily by the sequence of plant communities than for the animal world. For a large part of the animal species, however, the question remains whether the animal species are strictly bound to

Fig. 10.25 (**a**) Forest edge without mantle, shrub layer, and perennial fringe in Lithuania. (**b**) Forest edge with mantle and fringe in Lithuania. (**c**) Forest edge at Hainich National Park, Thuringia, Central Germany, with pronounced forest mantle

plant associations. The relationships between animal communities and plant communities were analyzed by Kratochwil (1987) using the example of flower visitor communities of a vegetation complex (Gehu 1979; Schmithüsen 1959; Schwabe 1989; Tüxen 1973). While flower-visiting insect species of a particular faunistic area (e.g., Eurosiberian, sub-Mediterranean, subcontinental) prefer those plant species that belong to the same distribution type (geo-element), many species change the plant community seasonally or even diurnally depending on resource availability or activity phase (foraging, rest) (Kratochwil 1987; Maier and Waldbauer 1979). The zoocoenosis does not necessarily change with the plant community. We therefore ask the following questions:

- Which environmental gradient determines the natural zoning?
- How do the species combinations change according to the environmental gradient?
- Are species groups that can be defined along an environmental gradient associated with equal adaptations and similarities of tolerance limits?
- Do discontinuities occur in a habitat gradient caused by the dominance of individual spe-

cies, or are the dominant species so euryoecious/eurypotent that they are still dominant in the neighboring areas represented?

- Does the distribution of resident species (rare to scattered) along the gradients correspond to the distribution of the dominant species?
- Can a qualitative correlation be established in the different zones between the amount of food supply (plant parts, insect prey) and the density of the consumer population (phytophages, predators)?
- How strongly are individual species associated?
- Do vertebrates seek out the forest edge for perching, cover, or foraging?
- In this respect, how do forest edges with and without a shrub layer differ?

The easiest way to determine the distance between the edge of the shrubbery to the "forest interior" is to determine the plants in the herb layer. The following measurements are taken to describe the environmental gradient:

The gradient can be most easily described with evaporimeters (Sect. 3.5.2) at approximately 2 m intervals on a transect from the open area into the forest. Recordings with thermohygrographs require using several devices simultaneously (e.g., five devices distributed on an open area, perennial fringe, shrub mantle, the beginning of the high forest, and the forest interior). The expensive use of many thermohygrographs can be partially circumvented by spot measurements of temperature with electrical temperature sensors (thermistors, Chap. 3) and the simultaneous use of several PICHE-evaporimeters (Sect. 3.5.2) along the transect through the forest edge. The temperature measurements must be carried out quickly, one after the other at the same ground height, and repeated, totaling at least four approximately simultaneous measurements per zone in a 1 m radius. With the latest developments in data loggers, these measurements can be carried out more efficiently using data loggers (see Sect. 3.1). For suitable devices, see e.g., https://meteorologyshop.eu/en/products/verdunstungswaage-nach-dr-klausing or https://www.licor.com/products/water-node. However, evaporimeters are quite expensive (starting at EUR/$1500 per unit). For many applications, a combined temperature and humidity data logger, such as the Onset HOBO MX series (around $200 per unit), might suffice (https://www.onsetcomp.com/products/data-loggers/mx2301a).

The distribution of species along a zonation is graphically represented with a histogram. The abscissa is divided into classes and used for the environmental gradients or the different zones, the ordinate for the respective percentage share of the total number of recorded individuals of the examined species (Fig. 10.26).

The test of whether two species occur randomly or not randomly together is carried out by an association analysis using the 2×2 chi-square test (see, e.g., https://www.socscistatistics.com/tests/chisquare/, Sect. 15.9). Calculations of the association coefficients for some species according to Cole, Southwood and Halbach are presented e.g., by Ratcliff 1982) (see Sect. 15.11, faunal similarity).

A simple way to separate groups based on their members' frequency of joint occurrence and thus to define zonation biocoenoses is to calculate the *recurrent groups* (Fager 1957, see Sect. 15.11).

More complex calculations using computer programs for the delimitation of species communities have been described by McCoy et al. (1986) and by Ludwig and Cornelius (1987). The required basic information is the presence or absence of a particular species at a collection point for McCoy et al. (1986). At the same time, Ludwig and Cornelius (1987) also consider the quantitative distribution of a species along the gradient-oriented collection points in their calculation procedure.

10.18 Strips Along Field Margins and Stream Banks

Like the forest edge, the edges of cultivated agricultural fields can be taken out of use creating biodiverse habitats. Many countries, including Germany and the US, financially support field edge strips (field borders and filter strips) about 5 m wide to compensate for the possible economic yield loss for the farmer ("agri-environmental

Fig. 10.26 Zonation from forest to meadow plants along a gradient of increased light, temperature, and mowing intensity. Each transect point is subdivided into four 0.25 m × 0.25 m quadrats in which each species' presence is noted in a score between 0 and 4 (occurrence in zero, one, two, three or all four quadrats). Resorting the species list on the left allows us to identify and group specialist species for the forest (transect points 1–5), edge species (transect points 7 and 8), and meadow species (transect points 9 and 10). In addition, we can identify generalists that are found throughout the zonation. Picture credit Kimberly Chicosky, Terrestrial Ecology student at Siena Heights University in 2019

Fig. 10.27 (a) Field edge strip, which was exempted from fertilization and pesticide treatment. (b) Field margins between the narrow plots in an agricultural landscape in southern Poland, Pieniny Mountains. The field margins, although created by property rights between the field strips, are in this form beneficial to many small animals

Fig. 10.28 A fallow field was established to enrich the landscape with insects (sign reads: bee-friendly farmer). Such areas support ecological networking. Southern Lower Saxony, Germany near Göttingen

schemes"). In field edge strips (Fig. 10.27), wild herbs and many small animals develop, which can be both protected and economically advantageous as so-called beneficial organisms (e.g., aphid eaters) for the surrounding area (Denys et al. 1997; Thies et al. 1997, 2000a, b; Tscharntke et al. 2002; Schmidt et al. 2003).

Fig. 10.29 Prairie strips for conservation. With permission from Iowa State University STRIPS, nrem.iastate.edu/research/STRIPS/

In England, creating new "*grass banks*" as narrow strips right through the grain fields is recommended. The sowing should be done with tussock-forming grasses. This promotes many beneficial arthropods that can overwinter in these linear habitats (Thomas and Wratten 1988; Boatman et al. 1990). Wider strips can establish an "ecological networking" in intensive agricultural sites (see Sect. 11.4, Corridors) (Birnkammer 1993) (Fig. 10.28).

Iowa State University creates science-based trials of row crops integrated with prairie strips Fig. 10.29.

In Germany, road verges and watercourse edge strips (riparian strips) generally receive special attention from nature conservation. For example, Radtke (2014) provides transparent information about their significance and legal basis in Germany. The strips are not agricultural land and, therefore, must also be considered by the Federal Nature Conservation Act. In riparian strips, the development of shrubs is desirable (Fig. 10.30), also see Sect. 10.2. The ministry compensates yield loss via EU funds if edge strips are excluded from possible use. The path widths are determined via maps and aerial photographs of the corre-

Fig. 10.30 Example of a riparian strip removed from use. On the right of the trees runs the stream. The woody vegetation is desirable. Pesticides from arable farming (left of the picture) are largely prevented from entering the stream. Northern Bavaria

sponding landscape section. The legal basis for riparian edge strips is the proper maintenance and improvement of the ecological functions of above-ground waters. Strips at the roadside and the water's edge are also affected by fertilizer- and plant protection laws. Protected riparian strips filter pesticides and fertilizers from the surrounding agriculture, thus reducing the potential input into adjacent water bodies (Aguiar Jr et al. 2015).

Case Study: What Are the Consequences of Reducing the Use of Pesticides in an Agroecosystem?

The use of pesticides in agroecosystems is considered the main driver of biodiversity loss (van Lexmond et al. 2015; Hayes and Hansen 2017; Nath et al. 2023; Sanchez-Bayo and Wyckhuys 2019). Pesticides harm not only flower pollinators, other invertebrates, and birds (e.g., Chiron et al. 2014, Tassin de Montaigu and Goulson 2020; Tscharntke et al. 2012; Chagnon et al. 2014). It is, therefore, vital to examine the effects of pesticides on wildlife and how biodiversity loss can be counteracted (Dudley and Alexander 2017). Sattler et al. (2021) have investigated whether reducing pesticide use reduces crop yields in rice fields. Rice fields are significant food resources in Southeast Asia. Attempts are being made to increase crop yields through the intensive use of pesticides and fertilizers (Normile 2013). In their approach, Sattler et al. (2021) compared insect diversity and crop yield between three rice-growing areas in Cambodia: (I) fields that received no pesticide as a control, (II) fields not treated with pesticides but with non-rice crops planted in the surrounding (Fig. 10.31) and (III) fields that were treated with pesticides according to conventional cultivation methods.

Arthropods were collected on the three different experimental plots by standardized sampling with a sweep net (catch per unit effort, see Sect. 2.3). The arthropods were identified by family and assigned to guilds. A distinction was made between predators, parasitoids, herbivores, detritivores, and pollinators. The yields in the trial plots without pesticides were no less than in the control plots and the fields with non-rice crops planted in the surrounding area. A cost-benefit analysis revealed the high value of the control plots and those surrounded by other crops. In the latter, the proportion of beneficial insects (parasitoids) was higher than in the pesticide-treated fields. The results indicate that sustainable management can be achieved with a mixture of rice paddy cultivation and other crops in the landscape.

10.19 Created Flower Strips

Even more compelling than excluding an edge strip from the management is the sowing of flower strips, which benefit birds of the field (such as in Central Europe, the threatened Partridge, finch species, and smaller mammals) (Köppl et al. 2014) as well as many insects (Haenke et al. 2009) (Figs. 10.32 and 10.33). To

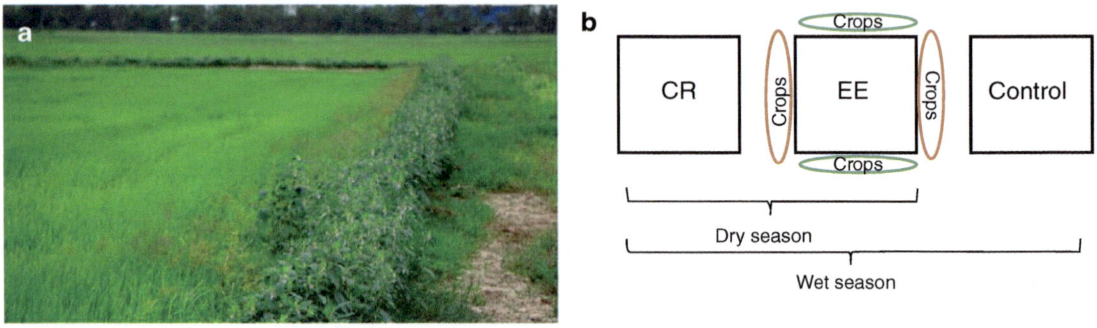

Fig. 10.31 (**a**) Example of an ecologically engineered rice field (EE). (**b**) Study design during the dry and wet season. Conventionally farmed rice fields (CR) were treated with pesticides, EE fields were not treated with pesticides and cultivated with additional plants on the rice bunds (earthen mound between rice fields to keep water level in the fields), and fields with no pesticide application and no bund plants served as control fields. (from Sattler et al. 2021, Photo credit Cornelia Sattler)

Fig. 10.32 (**a**) Flowering strip established for the protection of partridges. Photo credit E. Gottschalk. (**b**) Established field edge strip, approximately 5 m wide, seeded with many wildflowers and grasses, providing habitat for pollinators and are also used by birds of the open countryside

Fig. 10.33 (**a**) Flowering strip, established with a seed mixture, summer aspect. (**b**) Wildflowers in a flowering strip at the edge of the field also promote many insects and counteract the general "insect die-off" in the agricultural landscape. Seeds for the wildflower herb mixture are available, among other places, at saatgut-shop.de (information also via info@wildacker.de). For the US, consider https://www.prairiemoon.com

evaluate the biodiversity obtained, we need to compare groups of animals using standardized samples. Hoverflies can be caught for comparison on transects with 100 net strokes (see Sect. 2.3) and determined. In addition, you can specifically net hoverflies in an area for 30 min. For comparison with a field edge or field edge strip, you must then sample numerous pairs of similar areas (flower strips and each field edge). Suction samples of insects in created flower strips (see Sect. 2.8) yielded four times the number of species and 12 times the number of individuals as on field edges (Gottschalk and Beeke 2014). Bird densities can be determined in pairs per area (flower strip and section of a wheat field of the same size) by each 30-min observation unit. Especially for the protection of the partridge (*Perdix perdix*), three variants of flower strips are suggested: (a) annual, (b) structurally rich, and (c) perennial flower strips (Gottschalk and Beeke 2014). Corresponding seed mixtures were developed as part of the Partridge Protection Project and can be purchased from the company "Camena Samen" in D-31867 Lauenau; the sowing quantity is about 7 kg per ha. Requirements for management and specifications for the species composition of the individual seed mixtures are given in a guideline "Agricultural Environmental Measures AUM 2014" in Lower Saxony (Gottschalk and Beeke 2011). A list of the most effective flowering strip plants is provided by Meindl et al. (2018). The benefit of the flowering strips for the partridge chicks was also demonstrated in the field experiment by Gottschalk and Beeke (2014). The authors let hand-reared partridge chicks run free in grain fields and flowering strips for 1 h each and weighed the chicks before and after the run: it was found that only in the flowering strips did the chicks gain weight from the food.

10.20 Estimating the Influence of Mowing on the Fauna

Measurements of vegetation structure are addressed in Sect. 4.2.6. Density measurements of the grass layer could be supplemented by counting all contact points of grasses and herbs (recorded separately) with a horizontal rod at 5 cm, 20 cm, and 40 cm height, which is pushed undisturbed into the vegetation (Fig. 4.8 in Sect. 4.2). Such a sample requires 15–20 min. Since many meadow plant species tend to form tussocks, many measurements are necessary due to the high sample variance. The advantages of such a supplementary habitat description are the possibilities of quantitative evaluation.

Since mowing is a very radical intervention in the animal community of a meadow, it is obvious to leave strips of meadow unmowed as protection for juvenile stages or as a refuge for adult animals.

Mowing changes the plants' structure and quality concerning their nitrogen content (e.g., Quetier et al. 2007). Protein determination on the plant material would be necessary, which need to be done in the laboratory.

We want to investigate to what extent mowing of a meadows influences the presence of animal species mainly through structural changes:

- Does the epigaeic arthropod fauna differ between mowed and unmowed parts of the meadow regarding species and individual numbers?
- Do the arthropods retreat into the unmowed strips during mowing?
- Do the population developments show temporal parallels to the normal mowing rhythm?
- Are the old grass strips preferred places for some of the orb-weaving spiders?
- Are old grass strips needed for the development of young animals?
- Are old grass strips preferred for overwintering?
- Which habitats do the old grass strip dwellers prefer outside the meadow?
- Do the grasshoppers prefer to eat in mowed or unmowed meadows?

Experimental habitat manipulation of meadows can be achieved by partially mowing small areas or leaving out strips of meadow (2–5 m wide) when mowing the rest of the area. Animals are sampled using barber traps (pitfall traps) and quadrate samples. Spiders and grasshoppers are especially suitable for the evaluation as both animal groups are highly dependent on the vegetation structure.

The ground beetle fauna of the grassland is less vertically stratified. While mowing can promote phytophagous groups, more species and individuals of predatory and saprophagous groups are found in unmowed parts (shown for the beetle fauna by Morris and Rispin 1987). Suppose the pitfall traps remain in place over the vegetation period (at least from early May to mid-September). In that case, the catch results can be evaluated phenologically, and the population development can be compared with the regular, i.e., long-maintained mowing rhythm in this region. Orb-weaving spiders are counted using the quadrate method (see Sect. 2.3). Cocoons and young animals in the unmown parts are considered an essential function of the old grass strips if they only occur in those strips when the meadows are usually mowed. In autumn, cocoons and other developmental stages in the old grass strips indicate overwintering in these parts of the meadow, provided these stages are not found or simultaneously found significantly less in the same sample within the mowed parts of the meadow. About twice as many species of web-building spiders were found in the old grass strips, but also about ten times as many cocoons and "nurseries" (Ritschel-Kandel 1986). The evidence that temperatures in old grass strips (*long grass*) are milder on cold nights in winter was provided by Bossenbroek et al. (1977a, b) for England. In the border zone of grazed grassland with a dense turf and litter layer, Maelfait et al. (1988) found densities of up to 250 spider individuals and up to 900 carabids per m^2 during the winter. During the summer, the densities on the pastures reached 150–200 spiders and 30–40 ground beetle individuals per m^2. In addition, Sotherton (1985) pointed out the importance of field edges as overwintering sites for predatory beetles. He determined the density of ground and rove beetles (Carabidae and Staphylinidae) in a hedge bank during the winter months at 1239 beetles per m^2, while in comparison, he found about 200 beetles/m^2 on the forest floor, and 50–80 predatory beetles per m^2 in an established grassland, while only25–40 animals per m^2 were found in a one-year-old meadow (Sotherton 1984).

In many respects, mowed meadows resemble grazed, unmowed areas of grassland. Morris (1969) highlighted the higher species diversity in ungrazed grassland for a broader spectrum of arthropod groups and discussed the management of grassland reserves in England in Morris (1971). According to long-term studies by pitfall traps on the spider fauna, the grazing of alpine grassland had only marginal effects in terms of species diversity, but it changed the dominance ratios of some species (Höfer et al. 2010).

Fauna comparisons between differently managed meadows were also conducted on a broad basis by Southwood and van Emden (1967), Duffey (1978), and Morris and Rispin (1987). While old grass strips positively affect species' diversity, reproduction, and overwintering, this does not necessarily apply to flower visitors. Wells (1971) noted a decrease in flower production in uncut grassland. Kricke et al. (2014) demonstrated the relationship between mowing concepts and butterfly fauna, Hiller and Betz (2014) for the grasshopper fauna, and Ade et al. (2012) for the beetle fauna. Just as plant species react differently to mowing, insects that live in or on the plants are also affected differently. Völkl et al. (1993) analyzed the effect of mowing on the population development of insects in flower heads of thistles. The authors found that individual species reacted quite different.

The different value of unmown vs. mowed parts of meadows is also made clear by the visit of foraging young birds (Bezzel 1980). Carnivorous ground birds benefit from the accessibility of earthworms, soil arthropods, and snails in the mowed area, herbivorous ground and tree birds from the seed supply in the fall, especially of tall perennials in the unmowed area. Individual bird visits were recorded in time logs (minute samples) for comparison. These records can then be summed up per month as the number of individuals per 10 min. and presented as histograms and differentiated for different purposes, e.g., for foraging (see Bezzel 1980 and Sect. 9.4).

10.21 Habitat Manipulation with Leaf Litter

We may ask the following questions:

- Can the dependence of individual animal species on vegetation structure parameters be tested experimentally?
- How does the litter layer thickness influence the spider and centipede fauna?
- Are the spiders on the forest floor more dependent on the vegetation structure or more on the prey supply?

As an experiment, litter areas of 1 m^2 can be demarcated with plastic walls (rigid PVC 30–35 cm wide) buried 5–10 cm into the ground and protruding approximately 25 cm above the ground. Before burying the PVC ring, the leaf litter is collected by hand in a ring 11 cm wide around the outside of this arena (random sample of m^2, 25 cm × 25 cm). The litter layer in this arena can then be increased by adding collected, untreated litter or by litter devoid of macrofauna by heat extraction. After waiting as long as possible (many months up to 2 years), the macrofauna of the arena is examined by hand-picking or heat extraction (see Sect. 2.20).

According to Uetz (1979), five different litter depths can be produced experimentally in sample areas of 10 m × 10 m in early spring (March): (1) areas swept empty, (2) removal of half the leaf litter, (3) control with typical litter depth, (4) addition of leaf litter from (2) resulting in double the litter depth and (5) addition of leaf litter from (1) resulting in double the litter depth. After these habitat manipulations, the sample areas require at least 3 weeks to recover. Pitfall traps, emptied once a week, are then used to monitor the development of the species communities of the "wandering spiders" guild: Gnaphosidae, Clubionidae, Lycosidae, Thomisidae, some Agelenidae, and Hahniidae. In addition, many small spiders from the Linyphiidae family are also caught. The collembolans, homopterans, small dipterans, and coleopterans that are also caught in the ground traps provide a relative measure of the prey spectrum (cf. Nentwig 1982), which can be supplemented by square samples at the same location.

Whether the litter layer thickness or the prey supply plays a more significant role can be tested using partial correlation (Sect. 15.10). Artificial structural elements can also be introduced into the habitat, e.g., cubes with an edge length of approximately 30 cm with horizontal and vertical ledges or macramé jute threads (cf. Robinson 1981). The colonization of these structures by spiders can be compared with the colonization of non-manipulated (hollow notch) samples in the same habitat. The cubes placed in the field are collected after about 14 days and shaken out or blown out over a white plastic tub. The spiders are then preserved in alcohol for species identification or are counted only according to their size and are immediately released. If additional colonization data are required, the cubes are checked at shorter intervals, e.g., at 1, 2, 4, 8, and 16-day intervals.

Schaefer (1978) found through experimental density changes in a web spider (Linyphiidae, *Florinia bucculenta*) that habitat structures, not the food supply, are density-regulating factors. Scheu & Schaefer (1998) experimentally demonstrated bottum-up regulation of soil fauna by manipulating food resources.

If you cover certain test plots with coarse-meshed netting for several years, you can compare the litter year by year and study the layer structure. In central European forests, beech leaves take 5–6 years to decompose macroscopically into the humus layer. Primary decomposers are fungi (especially in an acidic environment) and soil animals such as earthworms, slugs, and woodlice (especially in the neutral to slightly acidic environment of calcareous beech forests) (Beck 1984). Most secondary decomposers are fungivores, which increase the overall decomposition rate by grazing on the fungal flora. This can be determined by placing fresh fallen leaves in net bags with different mesh sizes (Beck 1984): Fine-meshed bags allow fungi and bacteria to enter, bags of medium mesh size allow collembolans and horned mites to enter, and animals of the macrofauna only gain access through coarse-meshed bags. Herlitzius and Herlitzius (1977) covered plastic rings 5 cm high with gauze at mesh sizes of 50 μm, 1120 μm, and 6000 μm, respectively. The experimental containers were checked every 2 months. After 2 years, the weight loss of the fallen leaves in the medium and coarse mesh sizes was around 70%, and in the fine mesh bags, only around 30–35%. The actual remineralization was mainly carried out by bacteria, which in turn were controlled by micro-fauna such as protozoa and nematode species (Beck 1984).

Adding caterpillar droppings also accelerated leaf litter decomposition (Herlitzius and Herlitzius 1977). Leaves freshly picked in September were air-dried, weighed, and distributed to litter containers. The authors obtained phytophage feces from collected lepidopteran caterpillars fed with the corresponding leaf species in the laboratory. Fecal particles can also be collected from the canopy in the field using fabric tarpaulins. Phytophage feces were added in 33% of the amount of leaf litter; the amount for the test preparations was calculated according to a dry weight of 290 g of fallen leaves per m^2 in the beech forest.

References

Ade J., Wolf-Schwenninger K. & Betz O. 2012: Auswirkungen der Wiesenmahd auf verschiedene Käferarten ausgewählter Grünflächen im Stadtgebiet Tübingens. – *Jh. Ges. Naturkde. Württemberg* 168, 199–215, Stuttgart

Adeyemi A.A., Edet D.I. & Okoronkwo A.E. 2018: Status and Distribution of Duiker Species in Okomu National Park, Nigeria. – *Journal of Forestry Research and Management*. 15(2), 136–148; ISSN 0189-8418 www.jfrm.org.ng

Aguiar Jr. T.R., Bortolozo F.R., Hansel F.A., Rasera K. & Ferreira M.T. 2015: Riparian buffer zones as pesticide filters of no-till crops. – *Environ. Sci. Pollut. Res.* https://doi.org/10.1007/s11356-015-4281-5

Alekseev, V., Napreenko, M., & Napreenko-Dorokhova, T. (2024). Ecological Groups of Coleoptera (Insecta) as Indicators of Habitat Transformation on Drained and Rewetted Peatlands: A Baseline Study from a Carbon Supersite, Kaliningrad, Russia. *Insects*, *15*(5), 356. https://doi.org/10.3390/insects15050356

Anderson S.H. 1981: Correlating habitat variables and birds. - Studies in Avian Biology 6, 538–542

Artensteckbrief: www.Artensteckbrief.de aus Internetportal www.MultiBaseCS.de. Copyright © 2014–2023, 34u GmbH in Kooperation mit dem Sächsischen Landesamt für Umwelt, Landwirtschaft und Geologie

Bai M.-L., Wichmann F. & Mühlenberg M. 2003: The abundance of tree holes and their utilization by hole-nesting birds in a primeval boreal forest of Mongolia. – *Acta Ornithologica* 38, 95–102.

Baillie S.R. 1991: 7. Monitoring terrestrial breeding bird populations. – In: Goldsmith F.B. (ed.): *Monitoring for Conservation and Ecology*, 112–132. – Chapman and Flall, London, New York, Tokyo, Melbourne, Madras, 275 p.

Baillie, J. E. M., & Zedan, H. (2004). *A 2004 IUCN Red list of threatened species: A Global Species Assessment*. (C. Hilton-Taylor & S. N. Stuart, Eds.). Gland, Switzerland: World Conservation Union.

Balmford A., Bruner A., Turner R.K. & 16 authors 2002: Economic Reasons for Conserving Wild Nature. – *Science* 297, Issue 5583, 950–953. https://doi.org/10.1126/science.1073947

Bauer G. 1989: Grenzen des „Rote Liste Instruments" und Möglichkeiten einer alternativen Bewertung von Biotopen. – *Schriftenreihe für Landschaftspflege und Naturschutz* 29, 95–106

Beattie A.J. 1983: Distribution of ant-dispersal plants. – *Sonderbände des Naturwissenschaftl. Vereins Hamburg*. 7, 249–270.

Beattie A.J. 1985: *The evolutionary ecology of ant-plant mutualisms*. – University Press, Cambridge, 181 p.

Beaumelle, L., Tison, L., Eisenhauer, N., Hines, J., Malladi, S., Pelosi, C., et al. (2023). Pesticide effects on soil fauna communities—A meta-analysis. *Journal of Applied Ecology*, *60*(7), 1239–1253. https://doi.org/10.1111/1365-2664.14437

Beck L. 1984: Bodentiere im Laub des Buchenwaldes. Entscheidende Rolle im Naturhaushalt. – *Forschung-Mitteilungen der DFG* 2/84, 15–18

Beissinger S.R. & McCullough D.R. (eds.) 2002: *Population viability analysis*. – University of Chicago Press, Chicago I, 593 p.

Berndt R. & Henss M. 1967: Die Kohlmeise, *Parus major*, als Invasionsvogel. – *Die Vogelwarte* 24, 17–37

Berthold P., Fliege G., Querner U. & Winkler H. 1986: Die Bestandsentwicklung von Kleinvögeln in Mitteleuropa: Analyse von Fangzahlen. – *J. Orn.* 127, 397–437

Bezzel E. 1980: Beobachtung zur Nutzung von Kleinstrukturen durch Vögel. – *Berichte der Bayerischen Akademie f. Naturschutz und Landschaftspflege* (ANL) 4, 119–125, https://www.zobodat.at/publikation_volumes.php?id=45925

Bink F.A. 1992: *Ecologische Atlas van de Dagvlinders van Noordwest-Europa*. – Schuyt & Co Uitgevers en Importeurs bv, Haarlem, 512 p.

Biosphärenreservat Rhön 2007: Das Birkhuhn in der Rhön. – Broschüre https://www.biosphaerenreservat-rhoen.de

BirdLife International 2021: *European Red List of Birds*. Luxembourg: Publications Office of the European Union, 71 p.

Birnkammer H. 1993: Gezielte Streifeneinsaat und ökologische Vernetzung auf intensiv genutzten Ackerbaustandorten – eine Alternative zur konventionellen Flächenstilllegung, 175–179. – In: Kohler A. & Böcker R., *Die Zukunft der Kulturlandschaft*. – 25. Hohenheimer Umwelttagung. Verlag Josef Margraf, Weikersheim.

Blab J. 1986: *Grundlagen des Biotopschutzes für Tiere*. – Kilda-Verlag, Greven, 257 p.

Blab J. & Riecken U. 1989: Konzept und Probleme einer Biotopgliederung als Grundlage für ein Verzeichnis der gefährdeten Tier-Lebensstätten in der Bundesrepublik Deutschland. – *Schriftenreihe für Landschaftspflege und Naturschutz* 29, 78–94

Blab J., Nowak E., Trautmann W. & Sukopp H. (eds.) 1984: *Rote Liste der gefährdeten Tiere und Pflanzen in der Bundesrepublik Deutschland*. – Kilda-Verlag, Greven, 270 p.

Boatman N.D., Dover J.W., Wilson P.J., Thomas M.B. & Cowgill S.E. 1990: Modification of farming practice at field margins to encourage wildlife, 299–311. – In: Buckley G.P. (ed.): *Biological habitat reconstruction*. Belhaven Press, London, 363 p.

Bond M.L., Lee D.E., Bradley C.M. & Hanson C.T. 2009: Influence of Pre-Fire Tree Mortality on Fire Severity in Conifer Forests of the San Bernardino Mountains, California. – *The Open Forest Science Journal* 2, 41–47

Bossema I. 1979: Jays and Oaks: An eco-ethological study of a symbiosis. – *Behaviour* 70, 1–117

Bossenbroek Ph., Kessler A., Liem A.S.N. & Vlijm L. 1977a: 1. The significance of plant growth forms as „shelter“ for terrestrial animals. – *Journal of Zoology* 182, 1–6

Bossenbroek Ph., Kessler A., Liem A.S.N. & Vlijm L. 1977b: An experimental analysis of the significance of tuft structures as a shelter for invertebrate fauna with respect to wind-velocity and temperature. – *J. Zool.* 182, 7–16

Bowler D., Richter R.L., Eskildsen D., Kamp J., Moshøj C.M., Reif J., Strebel N., Trautmann S. & Voříšek P. 2021: Geographic variation in the population trends of common breeding birds across central Europe. – *Basic and Applied Ecology* 56, 72–84

Brain, R. A., Teed, R. S., Bang, J., Thorbek, P., Perine, J., Peranginangin, N., et al. (2015). Risk assessment considerations with regard to the potential impacts of pesticides on endangered species. *Integrated Environmental Assessment and Management, 11*(1), 102–117. https://doi.org/10.1002/ieam.1572

Brambilla, M., Gustin, M., & Celada, C. (2011). Defining favourable reference values for bird populations in Italy: setting long-term conservation targets for priority species. *Bird Conservation International, 21*(1), 107–118. https://doi.org/10.1017/S0959270910000407

Brlík V, Šilarová E, Škorpilová J, Alonso H, Anton M, Aunins A, Benkö Z, Biver G, Busch M, Chodkiewicz T, Chylarecki P. 2021: Long-term and large-scale multispecies dataset tracking population changes of common European breeding birds. – *Scientific data*, 8(1), 21 p.

Brooks T.M., Mittermeier R.A., da Fonseca G.A.B., Gerlach J., Hoffmann M., Lamoreux J.F., Mittermeier C.G., Pilgrim J.D. & Rodrigues A.S.L. 2006: Global Biodiversity Conservation Priorities. – *Science* 313, 58–61. https://doi.org/10.1126/science.1127609

Brosset A. & Erard C. 1986: Les oiseaux des régions forestières du Nord-Est du Gabon. Vol. 1. Ecologie et comportement des espèces. – *Rev. Ecol. (Terre Vie)*, Suppl. 3, 1–297

Brüll H. 1977: *Das Leben europäischer Greifvögel*. 3. Aufl. – G. Fischer Verl., Stuttgart, New York, 315 p.

Brussard P.F. 1991: The role of ecology in biological Conservation. – *Ecological Applications* 1 (1), 6–12

Buddle C. M., Spence J.R. & Langor D.W. 2000: Succession of boreal forest spider assemblages following wildfire and harvest. – *Ecography* 23, 424–436

Bull, E.L. 2002: The value of coarse woody debris to vertebrates in the Pacific Northwest. – General technical report PSW-GTR-181, 883–892. U.S. Department of Agriculture Forest Service, Albany, California

Bund-Länder-Ziel Vereinbarung zum Klimaschutz durch Moorbodenschutz. BRD 2021, 13 p.

Bundesamt für Naturschutz 1998: *Rote Liste gefährdeter Tiere Deutschlands*. – Schriftenreihe für Landschaftspflege und Naturschutz, Heft 55, 434 p.

Burton P.J., Jentsch A. & Walker L.R. 2020: The Ecology of Disturbance Interactions. – *BioScience* 70 (10), 854–870

Byrnes J.E.K., Gamfeldt L., Isbell F., Lefcheck J.S., Griffin J.N., Hector A., Cardinal B.J., Hooper D.U., Dee L.E. & Duffy J.E. 2014: Investigating the relationship between biodiversity and ecosystem multifunctionality: challenges and solutions. – *Methods Ecol. Evol.* **5**, 111–124. https://doi.org/10.1111/2041-210X.12143

Calahan J.T. 1984: Long-Term Ecological Research. – *Biol. Science* 34, 363–367

Carlson A., Sandström U. & Olsson O. 1998: Availability and use of natural tree holes by hole-nesting birds in a Swedish deciduous forest. – *Ardea* 86, 109–119

Carrasco L., Giam X., Papes M. & Sheldon K.S. 2019: Metrics of Lidar-Derived 3D Vegetation Structure Reveal Contrasting Effects of Horizontal and Vertical Forest Heterogeneity on Bird Species Richness. – *Remote Sens*. 2019, 11(7), 743; https://doi.org/10.3390/rs11070743

Chagnon M., Kreutzweiser D., Mitchell E.A., Morrissey C.A., Noome D.A. & Van der Sluijs J.P. 2014: Risks of large-scale use of systemic insecticides to ecosystem functioning and services. – Environ. Sci. Pollut. Res. 22, 119–134. https://doi.org/10.1007/s11356-014-3277-x

Chase, M. K., Kristan III, W. B., Lynam, A. J., Price, M. V., & Rotenberry, J. T. (2000). Single Species as Indicators of Species Richness and Composition in California Coastal Sage Scrub Birds and Small Mammals. *Conservation Biology, 14*(2), 474–487. https://doi.org/10.1046/j.1523-1739.2000.98312.x

Chaudhary, V., & Oli, M. K. (2020). A critical appraisal of population viability analysis. *Conservation Biology, 34*(1), 26–40. https://doi.org/10.1111/cobi.13414

Chew F.S. & Robins R.K. 1984: Egg-laying in butterflies. – In: Van-Wright R.I. & Ackery P.R.: *The biology of butterflies*. – Academic Press, London, 65–78

Chiang J., McEwan R.W., Yaussy D.A. & Brown K.J. 2009: The effects of prescribed fire and silvicultural thinning on the aboveground carbon stocks and net primary production of overstory trees in an oak-hickory ecosystem in southern Ohio. – *Forest Ecology and Management* 255, 1584–1594

Chiron F., Chargé R., Julliard R., Jiguet F. & Muratet A. 2014: Pesticide doses, landscape structure and their relative effects on farmland birds. – *Agriculture, Ecosystems and Environment* 185, 153–160. https://doi.org/10.1016/j.agee.2013.12.013.

Clarke R. (ed.) 1986: *The Handbook of Ecological Monitoring*. – Clarendon Press, Oxford, 298 p.

Collier B.D., Cox G.W., Johnson A.W. & Miller P.C. 1974: *Dynamic Ecology*. – Prentice/Hall International, Inc., London, 563 p.

Conner R.N., Rudolph D.C. & Walters J.R. 2001: *The Red-cockaded Woodpecker*. Surviving in a Fire-maintained Ecosystem. – University of Texas Press, Austin, 363 p.

Convention on Wetlands. 2021: Global guidelines for peatland rewetting and restoration. Ramsar Technical Report No. 11. Gland, Switzerland: Secretariat of the Convention on Wetlands.

Cook L.M. (ed.) 1985: *Case studies in population biology*. – Manchester University Press, Manchester/USA, 218 p.

Cooke, R., Mancini, F., Boyd, R. J., Evans, K. L., Shaw, A., Webb, T. J., & Isaac, N. J. B. (2023). Protected areas support more species than unprotected areas in Great Britain, but lose them equally rapidly. *Biological Conservation, 278*, 109884. https://doi.org/10.1016/j.biocon.2022.109884

Ćosović, M., Bugalho, M. N., Thom, D., & Borges, J. G. (2020). Stand Structural Characteristics Are the Most Practical Biodiversity Indicators for Forest Management Planning in Europe. *Forests, 11*(3), 343. https://doi.org/10.3390/f11030343

Culver D.C. & Beattie A.J. 1978: Myrmecochory in *Viola*. – *J. Ecol.* 66, 53–72

Currie D.J. 1991: Energy and large-scale patterns of animal and plant species richness. – *American Naturalist* 137, 27–49

Dempster J.P. 1983: The natural control of populations of butterflies and moths. – *Biological Reviews* 58, 461–481

Dempster J.P. & Pollard E. 1981: Fluctuations in Resource Availability and Insect Populations. – *Oecologia* 50, 412–416

Den Boer P.J. 1977: *Dispersal Power and Survival. Carabids in a cultivated countryside*. – Miscell. Papers 14. –Veenman & Zonen, Wageningen, 190 p.

Den Boer P.J. 1986: What Can Carabid Beetles Tell us about Dynamics of Populations? – In: Den Boer et al.: *Carabid Beetles*, 315–330. Gustav Fischer, Stuttgart, New York, 551 p.

Dennis R.L.H. (ed) 1992: *The Ecology of Butterflies in Britain*. – Oxford Science Publications, 368 p.

Denys C., Thies C., Fischer R. & Tscharntke T. 1997: Die ökologische Bewertung von Ackerrandstreifen im integrierten Landbau. – *Mitteilungen der NNA* (Alfred Töpfer Akademie für Naturschutz) 3/97: 2–11

Drapeau P.A., Nappi A., Giroux J.-F., Leduc A. & Savard J.-P. L. 2002: Distribution patterns of birds associated with snags in natural and managed eastern boreal forests. – In: Laudenslayer, W.F. et al. (eds.): *Proceedings of the symposium on the ecology and management of dead wood in western forests*. USDA Forest Service General technical report PSW-GTR-181, U.S. Department of Agriculture Forest Service, Albany, California, 193–205

Dudley N. & Alexander S. 2017: Agriculture and biodiversity: a review. – *Biodiversity* 18 (2–3), 45–49. https://doi.org/10.1080/14888386.2017.1351892

Duffey E. 1978: Ecological Strategies in Spiders Including some Characteristics of Species in Pioneer and Mature Habitats. – *Symp. Zool. Soc. London* 42, 109–123

Dyke, F. V., & Lamb, R. L. (2020). *Conservation Biology: Foundations, Concepts, Applications* (3rd ed. 2020 edition.). Cham, Switzerland: Springer.

Ebert G. 1993: *Die Schmetterlinge Baden-Württembergs*. Band 1: Tagfalter I. – Eugen Ulmer, Stuttgart, 552 p.

Ehrlich P.R. & Murphy D.D. 1987: Conservation Lessons from Long-Term Studies of Checkerspot Butterflies. – *Conservation Biology* 1 (2), 122–131

Ellenberg H. 1978: *Zur Populationsökologie des Rehes* (*Capreolus capreolus* L, Cervidae) in Mitteleuropa. – Spixiana (München) 2, 211 p.

Ellenberg H. 1986a: Immissionen – Produktivität der Krautschicht – Populationsdynamik des Rehwildes: Ein Versuch zum Verständnis ökologischer Zusammenhänge. – *Natur und Landschaft* 61 (9), 335–340

Ellenberg H. 1986b: Sind Neuntöter durch Pestizide gefährdet? – *Laufener Seminarbeiträge* 5/86, Akademie f. Natursch. u. Landschaftspflege, 26–37

Ellenberg H. 1987: Floristic changes due to eutrophication. – *Ammonia and Acidification* 6, 301–308

Ens B.J., Kersten M., Brenninkmeijer A. & Hulscher J.B. 1992: Territory quality, parental effort and reproductive success of Oystercatchers (*Haematopus ostralegus*). – *Journ. of Anim. Ecol.* 61, 703–715

Erhardt A. 1985: Diurnal Lepidoptera: Sensitive Indicators of Cultivated and Abandoned Grassland. – *Journal of Applied Ecology*, 22, 849–861. https://doi.org/10.2307/2403234

Erwin R.M. 2002: Integrated management of waterbirds: beyond the conventional. – *Waterbirds* 25 (suppl. 2), 5–12

Eurola S., Kyllonen H. & Laine K. 1984: Plant Production and Its Relation to Climatic Conditions and Small Rodent Density in Kilpisjarvi Region (69°05' N,

20°40' E), Finnish Lapland. – In: Merrit J.F. (ed.): *Winter Ecology of Small Mammals*. – Carnegie Museum of Natural History Spec. Publ. (Pittsburgh) 10, 121–30

Fager E.W. 1957: Determination and Analysis of Recurrent Groups. – *Ecology* 38 (4), 586–95. JSTOR, https://doi.org/10.2307/1943124. Accessed 22 Jan. 2023.

Federal Agency for Nature Conservation 1998: *Red List of endangered animals in Germany*. – Series of publications for landscape maintenance and nature conservation, Issue 55, 434 p.

Federal-State Target Agreement for Climate Protection through Peatland Protection. FRG 2021, 13 p. See also https://www.bundesumweltministerium.de/en/ 2022: Federal Action Plan on Nature-based Solutions for Climate and Biodiversity – key issues paper –, 8p.

Fermon H., Waltert M., Larsen T.B., Dall'Asta U. & Mühlenberg M. 2000: Effects of forest management on diversity and abundance of fruit-feeding nymphalid butterflies in south-eastern Côte d'Ivoire. – *Journal of Insect Conservation* 4, 173–189

Fieberg, J., & Ellner, S. P. (2001). Stochastic matrix models for conservation and management: a comparative review of methods. *Ecology Letters*, *4*(3), 244–266. https://doi.org/10.1046/j.1461-0248.2001.00202.x

Fischborn M. 2011: Influence of Forest Structure on Fire Severity after a Forest Fire in North Mongolia. – Master thesis at the Georg-August University Göttingen, 92 p.

Forbes, V. E., Galic, N., Schmolke, A., Vavra, J., Pastorok, R., & Thorbek, P. (2016). Assessing the risks of pesticides to threatened and endangered species using population modeling: A critical review and recommendations for future work. *Environmental Toxicology and Chemistry*, *35*(8), 1904–1913. https://doi.org/10.1002/etc.3440

Foster D. R. & Orwing D. A. 2006: Preemptive and salvage harvesting of New England forests: When doing nothing is a viable alternative. – *Conservation Biology* 20 (4), 959–970

Frankham R., Ballou J.D. & Briscoe D.A. 2004: *A Primer of Conservation Genetics*. – Cambridge University Press, 220 p.

Franklin I.A. 1980: Evolutionary change in small populations. – In: Soulé, M. E. & Wilcox, B. A. (eds.): *Conservation Biology: an evolutionary ecological perspective*, 135–150. – Sinauer Associates, Sunderland, MA., 395 p.

Funke, W. (1993). Animals as indicators of alterations of the environment - a survey. *PHYTOCOENOLOGIA*, *23*, 363–375.

García de León, D., Rey Benayas, J. M., & Andivia, E. (2021). Contributions of Hedgerows to People: A Global Meta-Analysis. *Frontiers in Conservation Science*, *2*. https://doi.org/10.3389/fcosc.2021.789612

Gehu, J.-M. (1979). Pour une approche nouvelle des paysages végétaux: la symphytosociologie. *Bulletin de la Société Botanique de France. Lettres Botaniques*, *126*(2), 213–223. https://doi.org/10.1080/01811797.1979.10824391

Gibson Vega, A., Ridley, A. R., Burbidge, A. H., Hall, M. L., & Cowen, S. J. (2024). Population viability analysis informs Western Grasswren translocation: Multi-population sourcing of 112 founder individuals needed to meet success criteria. *Ornithological Applications*, *126*(1), duad057. https://doi.org/10.1093/ornithapp/duad057

Gilpin M.E. 1987: Spatial structure and population vulnerability. – In: Soulé, M. E. (ed.): *Viable populations for Conservation*, 125–139. – Cambridge University Press, Cambridge, 189 p.

Gilpin M. E. & Soulé M.E. 1986: Minimum Viable Populations: the processes of species extinctions. – In: Soulé M.E. (ed.): *Conservation Biology: The Science of Scarcity and Diversity*, 13–34. – Sinauer Associates, Mass, 584 p.

Glutz v. Blotzheim U.N., Bauer K.M. & Bezzel E. 1971: *Handbuch der Vögel Mitteleuropas*. Bd. 4. – Akad. Verlagsgesellschaft. Wiesbaden, 943 p.

Goldberg D.E. & Gross K.L. 1988: Disturbance Regimes of Midsuccessional Old Fields. – *Ecology* 69 (6), 1677–1688

Goldsmith F.B. (ed.) 1991: *Monitoring for Conservation and Ecology*. – Chapman and Hall, London, New York, Tokyo, Melbourne, Madras, 275 p.

Gottschalk E. 1996: Population vulnerability of the gray bush cricket, *Platycleis albopunctata* (Goetze 1778) (Ensifera: Tettigoniidae), 324–328. – In: Settele J., Margules C., Poschlod P. & Henle K. (eds.), *Species survival in a fragmented landscape: from the local to the landscape level*. – Kluwer Academic Publishers, Dordrecht, Boston, London, 381 p.

Gottschalk E. & Beeke W. 2011: Ein kurzer Leitfaden für ein Rebhuhnschutzprojekt nach unseren Erfahrungen im Landkreis Göttingen. – Universität Göttingen, 15 p., https://www.rebhuhnschutzprojekt.de/files/Leitfaden-Rebhuhnschutz-vor-Ihrer-Haustuer-2021.pdf, auch https://northsearegion.eu/partridge

Gottschalk E. & Beeke W. 2014: How can the drastic decline in the Grey Partridge (Perdix perdix) be stopped? Lessons from ten years of the Grey Partridge Conservation Project in the district of Göttingen. – *Berichte zum Vogelschutz* 51, 95–116 (in German)

Graham L., Gaulton R., Gerard F., Staley J.T. 2018: The influence of hedgerow structural condition on wildlife habitat provision in farmed landscapes. – *Biol. Conserv*. 220, 122–31

Gray C.L., Hill S.L.L., Newbold T., Hudson L.N., Börger L., Contu S., Hoskins A.J., Ferrier S., Purvis A. & Scharlemann J.P.W. 2016: Local biodiversity is higher inside than outside terrestrial protected areas worldwide. – *Nature Communications* 7, 12306. https://doi.org/10.1038/ncomms12306

Greenberg C.H., Neary D.G. & Harris L.D. 1995: A comparison of bird communities in burned and salvaged-logged clearcut, and forested Florida Sand Pine Scrub. – *Wilson Bulletin* 107, 40–54

Gregoire T.G. & Valentine H.T. 2008: *Sampling Strategies for Natural Resources and the Environment*. –

Chapman & Hall/CRC, Boca Raton and London, 474 p.

Grimm R., Funke W. & Schauermann J. 1975: Minimalprogramm zur Ökosystemanalyse: Untersuchungen an Tierpopulationen in Wald-Ökosystemen. – *Verh. Ges. f. Ökol., Erlangen* 1974, 77–87

Grosenbaugh L.R. 1952: Plotless timber estimates new, fast, easy. – J. Forestry 50, 32–37

Groves, C. R., & Game, E. T. (2015). *Conservation Planning: Informed Decisions for a Healthier Planet* (1st edition.). W. H. Freeman.

Gurnell J. 1987: *The Natural History of Squirrels.* – Christopher Helm, London, 201 p.

Haarmann K. 1985: Zustand und Effizienz der Naturschutzgebiete in der Bundesrepublik Deutschland am Beispiel von fünfzehn „Vogelfreistätten". – *Vogelwelt* 106, 216–224

Haenke S., Scheid B., Schaefer M., Tscharntke T., Thies C. (2009): Increasing syprhid fly diversity and density in sown flower strips within simple vs. complex landscapes. – *Journal of Applied Ecology* 46, 1106–1114.

Hammerich, J., Dammann, C., Schulz, C., Tanneberger, F., Zeitz, J., & Luthardt, V. (2022). Assessing mire-specific biodiversity with an indicator based approach. *Mires And Peat, 28.* https://doi.org/10.19189/MaP.2021.SJ.StA.2205

Harper K.A., MacDonald S.E., Burton P.J., Chen J., Brosofske K.D., Saunders S.C., Euskirchen E.S., Roberts D., Jaiteh M.S. & Esseen P.-A. 2005: Edge influence on forest structure and composition in fragmented landscapes. – *Conservation Biology* 19, 768–782

Hayes, T. B., & Hansen, M. (2017). From silent spring to silent night: Agrochemicals and the anthropocene. *ELEMENTA-SCIENCE OF THE ANTHROPOCENE, 5*, 57. https://doi.org/10.1525/elementa.246

Herlitzius R. & Herlitzius H. 1977: Streuabbau in Laubwäldern. Untersuchungen in Kalk-und Sauerhumusbuchenwäldern. – *Oecologia (Berl.)* 30, 147–171

Higgins L.G. & Riley N.D. 1978: *Die Tagfalter Europas und Nordwestafrikas.* Ein Taschenbuch für Biologen und Naturfreunde. – Parey, Hamburg, Berlin, 377 p.

Hiller D. & Betz O. 2014: Auswirkungen verschiedener Mahdkonzepte auf die Heuschreckenfauna städtischer Grünflächen. – *Naturschutz und Landschaftsplanung* 46 (8), 241–246

Hochkirch, A., Bilz, M., Ferreira, C. C., Danielczak, A., Allen, D., Nieto, A., et al. (2023). A multi-taxon analysis of European Red Lists reveals major threats to biodiversity. *PLOS ONE, 18*(11), e0293083. https://doi.org/10.1371/journal.pone.0293083

Höfer H., Blick T., Muster C. & Paulsch D. 2010: Artenvielfalt und Diversität der Spinnen (Araneae) auf einem beweideten Allgäuer Grasberg (Alpe Einödsberg) und unbeweideten Vergleichsstandorten im Naturschutzgebiet Allgäuer Hochalpen. – *Andrias* 18, 53–82

Hondong H., Langner S. & Cock T. 1993: *Untersuchungen zum Naturschutz an Waldrändern.* – Bristol-Stiftung, Zürich, Ruth und Herbert Uhl-Forschungssstelle für Natur- und Umweltschutz, FL Schaan, 196 p.

Hoopes M.F. & Harrison S. 1998: Metapopulation, source-sink and disturbance dynamics. p. 135–151 – In: Sutherland W.J. (ed.): *Conservation Science and Action.* Blackwell Science Ltd, 363 p.

Hovestadt T., Roeser J. & Mühlenberg M. 1991: *Flächenbedarf von Tierpopulationen.* – Forschungszentrum Jülich GmbH, Berichte aus der ökologischen Forschung, Band 1, 277 p.

Hughes M., Hopwood P., Dolan, M. & Dolan B. 2021: Applications of thermal imaging for bird surveys: examples from the field. – *Ringing & Migration*, 36:2, 78–81, https://doi.org/10.1080/03078698.2022.2123026

Huhta V. 1979: Evaluation of Different Similarity Indices as Measures of Succession in Arthropod Communities of the Forest Floor after Clear-Cutting. – *Oecologia* (Berl.) 41, 11–23

Humpenöder, F., Karstens, K., Lotze-Campen, H., Leifeld, J., Menichetti, L., Barthelmes, A., & Popp, A. (2020). Peatland protection and restoration are key for climate change mitigation. *Environmental Research Letters, 15*(10), 104093. https://doi.org/10.1088/1748-9326/abae2a

International Union for Conservation of Nature and Natural Resources. (2024). The IUCN Red List of Threatened Species. *IUCN Red List of Threatened Species.* https://www.iucnredlist.org/en. Accessed 17 July 2024

Jacobs W. & Renner M. 1988: Biologie und Ökologie der Insekten. 2nd ed. – Gustav Fischer Verlag, 690 p.

Jeffress, M.R., Rodhouse T.J., Ray C., Wolff S. & Epps C.W. 2013: The idiosyncrasies of place: geographic variation in the climate–distribution relationships of the American pika. - *Ecological Applications* 23(4), 864–878

Jenny M. 1990: Territorialität und Brutbiologie der Feldlerche *Alauda arvensis* in einer intensiv genutzten Agrarlandschaft. – *J. Orn.* 131 (3), 241–265

Jensen, M. E., & Bourgeron, P. S. (Eds.). (2001). *A Guidebook for Integrated Ecological Assessments.* New York, NY: Springer. https://doi.org/10.1007/978-1-4419-8620-7

Jentsch A., Beierkuhnlein C. & White P.S. 2002: Scale, the Dynamic Stability of Forest Ecosystems, and the Persistence of Biodiversity. – *Silva Fennica* 36(1) discussion papers, 393–400

Jimoh S.O., Ikyaagba E.T., Alarape A.A., Adeyemi A.A. and Waltert M. 2013: Local depletion of two larger Duikers in the Oban Hills Region, Nigeria. - *African Journal of Ecology* 51(2), 228–234

Jonášová M. & Prach K. 2008: The influence of bark beetles' outbreak vs. salvage logging on ground layer vegetation in Central European Mountain spruce forests. – *Biol. Cons.* 141, 1525–1535

Kamp J, Frank C, Trautmann S, Busch M, Dröschmeister R, Flade M, Gerlach B, Karthäuser J, Kunz F, Mitschke A & Schwarz J. 2021: Population trends of common breeding birds in Germany 1990–2018. – *Journal of Ornithology* 162, 1–15

Kapustka, L. (Ed.). (2004). *Landscape Ecology and Wildlife Habitat Evaluation: Critical Information for Ecological Risk Assessment, Land-Use Management Activities, and Biodiversi.* West Conshohocken, Pa: ASTM International.

Kaule G. 1991: *Arten- und Biotopschutz.* 2nd ed. – Verlag Eugen Ulmer, Stuttgart, 519 p.

Kendeigh S.Ch. 1979: *Invertebrate Populations of the Deciduous Forest*: Fluctuations and Relations to Weather. – Illinois Biological Monographs 50, 107 p.

Köppl A., Schönfeld F. & Wagner C. 2014: Der Einfluss von Blühflächen auf den Niederwildbestand in einer intensiv genutzten Agrarlandschaft in Südostbayern. – *Schriftenreihe der Bayerischen Landesanstalt für Landwirtschaft* 1/2014, 127–128

Kramer H. & Akça A. 1995: Leitfaden zur Waldmesslehre. 3rd ed., Sauerländer Verlag, Frankfurt, 266 p.

Kratochwil A. 1987: Zoologische Untersuchungen auf pflanzensoziologischem Raster. – Methoden, Probleme und Beispiele biozönologischer Forschung. – *Tuexenia* 7, 13–51

Kratochwil A. 1989: Biozönotische Umschichtungen im Grünland durch Düngung. – *NNA-Berichte* 2 (1), 46–58

Kratochwil A. 1991: Die Stellung der Biozönologie in der Biologie, ihre Teildisziplinen und ihre methodischen Ansätze. – *Beihefte zu den Verhandlungen der Gesellschaft für Ökologie* 2, 9–44

Kratochwil A. & Schwabe A. 2001: Ökologie der Lebensgemeinschaften. Biozönologie. – UTB 8199 Eugen Ulmer, Stuttgart, 756 p.

Kratschmer, S., Hauer, J., Zaller, J. G., Duerr, A., & Weninger, T. (2024). Hedgerow structural diversity is key to promoting biodiversity and ecosystem services: A systematic review of Central European studies. *Basic and Applied Ecology*, *78*, 28–38. https://doi.org/10.1016/j.baae.2024.04.010

Kricke C., Bamann T. & Betz O. 2014: Einfluss städtischer Mahdkonzepte auf die Artenvielfalt der Tagfalter. – *Naturschutz und Landschaftsplanung* 46 (2), 052–058

Kudrna O. 1986: *Grundlagen zu einem Artenschutzprogramm für die Tagschmetterlingsfauna* in Bayern und Analyse der Schutzproblematik in der Bundesrepublik Deutschland. – Nachr. ent. Ver. Apollo, Frankfurt, 6, 1–90

Kudrna O. 1993: Verbreitungsatlas der Tagfalter (Rhopalocera) der Rhön. – *Oedippus* 6, 1–138

Lande R. 1988: Demographic models of the Northern Spotted Owl (*Strix occidentalis caurina*). – *Oecologia* 75, 601–607

Landesamt für Umweltschutz Mecklenburg-Vorpommern 1996: Die Bedeutung unzerschnittener, störungsarmer Landschaftsräume für Wirbeltierarten mit großen Raumansprüchen – ein Forschungsprojekt. – Schr.-R. d. Landesamtes für Umwelt und Natur Mecklenburg-Vorpommern 1, 82 p.

Leipold D. 1985: Untersuchungen an Spinnen und Laufkäfern zum Randeffekt im Steigerwald, 87–89. – In: Mühlenberg M. & Linsenmair K.E.: Die Ökologische Station der Universität Würzburg in Fabrikschleichach, Haßfurt, 43 p.

Lindenmayer D., Burton P.J. & Franklin J.F. 2008: *Salvage Logging and Its Ecological Consequences.* – Island Press, UBC Press Canada. 277 p.

Loeschke V. 1988: Populationsgenetik und Artenschutz. – *Naturw. Rundschau* 41 (8), 310–314

Lopez, R. R., Beaver, J., Parker, I. D., & Morrison, M. L. (2024). *Applied Wildlife Habitat Management, Second Edition* (2nd edition.). College Station: Texas A&M University Press.

Ludwig J.A. & Cornelius J.M. 1987: Locating Discontinuities Along Ecological Gradients. – *Ecology* 68 (2), 448–450

MacArthur R.H. & MacArthur J.W. 1961: On Bird Species Diversity. – *Ecology* 42, 594–598

Maelfait J.-P., Desender K. & De Keer R. 1988: The Arthropod Community of the Edge of an Intensively Grazed Pasture. – In: Schreiber K.-F. (ed.): *Connectivity in Landscape Ecology.* Proceedings of the 2nd International Seminar of the „International Association for Landscape Ecology“. – Münstersche Geographische Arbeiten 29, 115–117

Magurran A. E. 1988: Ecological Diversity and Its Measurement. – Croom Helm, Sydney, London, 179 p.

Maier Ch.T. & Waldbauer G.P. 1979: Diurnal Activity Patterns of Hoverflies Flies (Diptera: Syrphidae) in an Illinois Sand Area. – *Ann. Entomol. Soc. of America* 72 (2), 237–245

Manlik, O., Lacy, R. C., & Sherwin, W. B. (2018). Applicability and limitations of sensitivity analyses for wildlife management. *Journal of Applied Ecology*, *55*(3), 1430–1440. https://doi.org/10.1111/1365-2664.13044

Marcot B.G. 1986: Concepts of risk analysis as applied to viable population assessment and planning. – In: Wilcox B.A., Brussard P.F. & Marcot B.G. (eds.): *The management of viable populations*: theory, applications, and case studies. Center for Conservation Biology, Stanford, C.A., 101 p.

Maschwitz U. & Fiedler K. 1988: Koexistenz, Symbiose, Parasitismus: Erfolgsstrategien der Bläulinge. – *Spektrum der Wissenschaft* 5, 56–66

McCoy E.D., Bell S.S. & Walthers K. 1986: Identifying Biotic Boundaries Along Environmental Gradients. – *Ecology* 67 (3), 749–759

Meindl P., Schmidt R. & Drapela T. 2018: Die 55 wichtigsten Blühstreifenpflanzen. – FIBL Forschungsinstitut für biologischen Landbau und LFI Ländliches Fortbildungsinstitut Österreich, Wien 2018. (https://www.fibl.org/en/shop-en)

Merganičová K., Merganič J., Svoboda M., Bače R. & Šebeň V. 2012: Deadwood in Forest

Ecosystems. – p. 82–108 in: Blanco J.A. (ed.): *Forest Ecosystems – More than just trees*. InTech, Available from: http://www.intechopen.com/books/forest-ecosystems-more-than-just-trees/deadwood

Montgomery, I., Caruso, T., & Reid, N. (2020). Hedgerows as Ecosystems: Service Delivery, Management, and Restoration. *Annual Review of Ecology, Evolution, and Systematics, 51*(Volume 51, 2020), 81–102. https://doi.org/10.1146/annurev-ecolsys-012120-100346

Morissette J.L., Cobb T.P., Brigham R.M. & James P.C. 2002: The response of boreal forest songbird communities to fire and post-fire harvesting. – *Canadian Journal of Forest Research* 32, 2169–2183

Morris M.G. 1969: Differences between the Invertebrate Faunas of grazed and ungrazed Chalk Grassland. III: The Heteropterous Fauna. – *Journal of applied Ecology* 6, 475–487

Morris M.G. 1971: Differences between the invertebrate faunas of grazed and ungrazed chalk grassland. IV. Abundance and diversity of Homoptera-Auchenorhyncha. - *Journal of applied ecology* 1971, 37–52

Morris W.F. & Doak D.F. 2002: *Quantitative Conservation Biology: Theory and Practice of Population Viability Analysis*. – Sinauer Associates, Sunderland, MA. 480 p.

Morris M.G. & Rispin W.E. 1987: Abundance and Diversity of the Coleopterous Fauna of a Calcareous Grassland under Different Cutting Régimes. – *J. Appl. Ecol.* 24, 451–465

Mühlenberg M. 1971: Die Abwandlung des Eilegeapparates der Bombyliidae (Diptera). Eine funktionsmorphologische Studie, - *Z. Morph. Tiere* (Springer) 70, 1–72

Mühlenberg M. 1980: Kann man mit Hilfe von Vegetationsstruktur-Messungen Veränderung von Tiergemeinschaften bestimmen? – *Berichte d. Internat. Symposium d. intern. Vereinigung f. Vegetationskunde,* Rinteln 1979, 309–327, J. Cramer, Vaduz

Mühlenberg M. 1982: Artenverlust - trotz ökologischer Planung? – *Natur und Landschaft* 57 (9), 295–296

Mühlenberg M. 1983: Bewertung von Kenngrößen zur Beschreibung von Artengemeinschaften. – Fachbeitrag im Rahmen der Modellstudie „Zool. Artenschutz in Bayern", Fabrikschleichach, 30 p.

Mühlenberg M. 1990: Langzeitbeobachtung für Naturschutz — Faunistische Erhebungs- und Bewertungsverfahren. – *Berichte der Bayerischen Akademie für Naturschutz und Landschaftspflege* (ANL) 14, 79–100

Mühlenberg M. 1993: *Freilandökologie*. 3. Aufl. - UTB für Wissenschaften 595, Quelle & Meyer Verlag Wiesbaden, 512 p.

Mühlenberg M. & Hovestadt T. 1992: Das Zielartenkonzept. – *NNA Berichte*, Schneverdingen, 5(1), 36–41

Mühlenberg M. & Slowik J. 1997: *Kulturlandschaft als Lebensraum*. – UTB für Wissenschaften 1947, Quelle & Meyer Verlag Wiesbaden, 312 p.

Mühlenberg M., Hovestadt T. & Röser J. 1991: Are there Minimal Areas for Animal Populations? 227–264. – In: Seitz A. & Loeschke V.: *Species Conservation: A Population-Biological Approach*. – Birkhäuser, Basel, 281 p.

Mühlenberg M., Henle K., Settele J., Poschlod P., Seitz A. & Kaule G. 1996: Studying Species Survival in Fragmented Landscapes: the Approach of the FIFB. – In: Settele J., Margules C., Poschlod P. & Henle K. (eds.): *Species survival in fragmented landscapes*. 152–160, Kluwer Verlag, Dordrecht, Boston, London, 381 p.

Mühlenberg M., Hondong H., Dulamsuren D. & Gadow K. 2004: Large-scale biodiversity research in the southern Taiga, Northern Mongolia. – In: Creating a Legacy for Sustainable Science-based Forest Management: Lessons learned from field experiments (Szaro, R.C., Peterson, C.E., v. Gadow, K., Kräuchi, N., eds.) *Forest, Snow and Landscape Research* 78 (1/2), 93–118

Müller H.J. 1978: Strukturanalyse der Zikadenfauna (Homoptera Auchenorrhyncha) einer Rasenkatena Thüringens (Leutratal bei Jena). – *Zool. Jb. Syst.* 105, 258–334

Muller, F. (2018). Strategies for peatland conservation in France - a review of progress. *MIRES AND PEAT, 21*, 06. https://doi.org/10.19189/MaP.2016.OMB.218

Müller J., Bae S., Röder J., Chao A. & Didham R.K. 2014: Airborne LiDAR reveals context dependence in the effects of canopy architecture on arthropod diversity. – *For. Ecol. Manag.* 312, 129–137

Murdiyarso, D., Lilleskov, E., & Kolka, R. (2019). Tropical peatlands under siege: the need for evidence-based policies and strategies. *Mitigation and Adaptation Strategies for Global Change, 24*(4), 493–505. https://doi.org/10.1007/s11027-019-9844-1

Murphy D.D., Freas K.E. & Weiss S.B. 1990: An environment-metapopulation approach to population viability analysis for a threatened invertebrate. – *Conservation Biology* 4 (1), 41–51

Nappi A., Drapeau P. & Savard J.-P.L. 2004: Salvage logging after wildfire in the boreal forest: Is it becoming a hot issue for wildlife? – *The Forestry Chronicle* 80 (1), https://doi.org/10.5558/tfc80067-1

Nath, R., Singh, H., & Mukherjee, S. (2023). Insect pollinators decline: an emerging concern of Anthropocene epoch. *Journal of Apicultural Research, 62*(1), 23–38. https://doi.org/10.1080/00218839.2022.2088931

Nentwig W. 1982: Epigeic Spiders, Their Potential Prey and Competitors: Relationship Between Size and Frequency. – *Oecologia (Berl.)* 55, 130–136

Newton I. 1979: *Population Ecology of Raptors*. – T. & A.D. Poyser Ltd, Berkhamsted, Hertfordshire, England, 399 p.

Newton I. 1986: *The Sparrowhawk*. – Poyser, Calton, 396 p.

Newton I. & Marquiss M. 1991: Removal experiments and the limitation of breeding density in Sparrowhawks. – *Journal of Animal Ecology* 60, 535–544

Noon B.R. & Blakesley J.A. 2006: Conservation of the Northern Spotted Owl under the Northwest Forest Plan. – *Conservation Biology* 20, 288–296

Nordbeck, R., & Hogl, K. (2023). National peatland strategies in Europe: current status, key themes, and challenges. *Regional Environmental Change*, *24*(1), 5. https://doi.org/10.1007/s10113-023-02166-4

Normile D. 2013: Vietnam Tums Back a 'Tsunami of Pesticides'. – *Science* 341, 737–738

Parker T.A. 1991: On the use of tape recorders in avifaunal surveys. – *The Auk* 108, 443–444

Pausas J.G., Ouadah N., Ferran A., Gimeno T. & Vallejo R. 2003: Fire severity and seedling establishment in *Pinus halepensis* woodlands, eastern Iberian Peninsula. – *Plant Ecology* 169, 205–213

Perlut, N. G., Strong, A. M., Donovan, T. M., & Buckley, N. J. (2008). Grassland Songbird Survival and Recruitment in Agricultural Landscapes: Implications for Source–Sink Demography. *Ecology*, *89*(7), 1941–1952. https://doi.org/10.1890/07-0900.1

Peters, V. E., Campbell, K. U., Dienno, G., García, M., Leak, E., Loyke, C., et al. (2016). Ants and plants as indicators of biodiversity, ecosystem services, and conservation value in constructed grasslands. *Biodiversity and Conservation*, *25*(8), 1481–1501. https://doi.org/10.1007/s10531-016-1120-z

Pfadenhauer J., Poschlod P. & Buchwald R. 1986: Überlegungen zu einem Konzept geobotanischer Dauerbeobachtungsflächen für Bayern. Teil I. Methodik der Anlage und Aufnahme. – *Ber. ANL (Laufen)* 10, 41–60

Pickett S.T.A. & White P.S. 1985: *The ecology of natural disturbance and patch dynamics*. – Academic Press, Orlando, Florida, USA, 472 p.

Pielou E. C. 1975: Ecological Diversity. – John Wiley & Sons, New York, London, Sydney, Toronto, 165 p.

Pimm S.L. 1984: The complexity and stability of ecosystems. – *Nature* 307, 321–326

Pollard E. 1977: A method for assessing changes in the abundance of butterflies. – *Biol. Conserv.* 12 (2), 115–134

Pollard E. 1991a: 6. Monitoring butterfly numbers. – In: Goldsmith F.B. (ed.): *Monitoring for Conservation and Ecology*, 87–111.– Chapman and Hall, London, New York, Tokyo, Melbourne, Madras, 275 p.

Pollard E. 1991b: Synchrony of population fluctuations: the dominant influence of widespread factors on local butterfly populations. – *Oikos* 60 (1), 7–10

Pollard E. & Yates T.J. 1993: *Monitoring butterflies for ecology and conservation*. – Chapman & Hall, London, 274 p.

Pollard E., Moss D. & Yates T.J. 1995: Population trends of common British butterflies at monitored sites. – *Journal of Applied Ecology* 32, 9–16

Pooter L., Craven D., Jakova C.C., van der Sande M., et al. 2021: Multi-dimensional tropical forest recovery. – *Science*, https://doi.org/10.1126/science.abh3629

Prüter J. & Wübbenhorst J. 2004: Zur Situation des Birkhuhns (*Tetrao tetrix*) im Naturschutzgebiet Lüneburger Heide. – *Jb. Naturw. Verein Fstm. Lbg.* 43, 73–82

Pullin A.S. (ed.) 1995: *Ecology and Conservation of Butterflies*. – Springer Dordrecht, 363 p. (Softcover published 2012)

Quetier F., Thebault A. & Lavorel, S. 2007: Plant traits in a state and transition framework as markers of ecosystem response to land-use change. – *Ecological Monographs* 77, 33–52

Radtke M. 2014: Wegraine und Gewässerrandstreifen. Bedeutung und rechtliche Grundlagen. – *BUND LV Niedersachsen* e.V., Hannover, 35 p.

Ramsar 1971: https://www.ramsar.org

Rapport, D. J., Lasley, B. L., Rolston, D. E., Nielsen, N. O., Qualset, C. O., & Damania, A. B. (Eds.). (2002). *Managing for Healthy Ecosystems*. Boca Raton: CRC Press. https://doi.org/10.1201/9781420032130

Ratcliff R.D. 1982: A correction of Cole's C7 and Hurlbert's C8 coefficients of interspecific association. – *Ecology* 63, 1605–1606

Recher H. 1969: Bird species diversity and habitat diversity in North America and Australia. – *Amer. Natur.* 103, 75–80

Reißmann R., Gebert J. & Schmidt J. 2005: Laufkäfer (Coleoptera: Carabidae)., 224–260. In: Günther A., Nigmann U., Achtziger R. & Gruttke H.: *Analyse der Gefährdungsursachen planungsrelevanter Tiergruppen in Deutschland*. – Naturschutz und Biologische Vielfalt 21, Bundesamt für Naturschutz, Bonn - Bad Godesberg, 605 p.

Remmert H. 1984: *Ökologie*. Ein Lehrbuch. – Springer-Verlag, Berlin, Heidelberg, New York, 334 p.

Remmert H. 1985: Was geschieht im Klimax-Stadium? – *Naturwissenschaften* 72, 505–512

Remmert H. 1988: Gleichgewicht durch Katastrophen. – *Aus Forschung und Medizin*, Schering, 3 (1), 7–17

Richner H. 1991: Identifikation der ethologisch und ökologisch bedeutungsvollen Faktoren im Arten-und Habitatschutz. – *Der Ornithologische Beobachter* 88 (3), 243–252

Riecken U., Ries U. & Ssymank A. 1994: *Rote Liste der gefährdeten Biotoptypen der Bundesrepublik Deutschland*. – Schriftenreihe für Landschaftspflege und Naturschutz 41, Bonn, Kilda-Verlag, 184 p.

Ritschel-Kandel G. 1986: Hilfsprogramm für Spinnen und Insekten. Ungedüngte Altgrasstreifen. – *Abh. Naturwiss. Ver. Würzburg*, 25 (Jhrg. 1984), 1–28

Robinson J.V. 1981: The Effect of Architectural Variation in Habitat on a Spider Community: An Experimental Field Study. – *Ecology* 62 (1), 73–80

Roth R.R. 1976: Spatial Heterogeneity and Bird Species Diversity. – *Ecology* 57 (4), 773–782

Ruggiero L.F., Hayward G.D. & Squires J.R. 1994: Viability Analysis in Biological Evaluations: Concepts of Population Viability Analysis, Biological

Population, and Ecological Scale. – *Conservation Biology* 8 (2), 364–372

Rümer H. & Mühlenberg M. 1988: Kritische Überprüfung von „Minimalprogrammen" zur zoologischen Bestandserfassung. – *Schriftenreihe Bayer. Landesamt für Umweltschutz*, München, 83, 151–157

Samuelsson J., Gustafsson L. & Ingelög T. 1994: *Dying and dead trees*. A review of their importance for biodiversity. – Swedish Threatened Species Unit, Uppsala, Sweden, 109 p.

Sanchez-Bayo, F., & Wyckhuys, K. A. G. (2019). Worldwide decline of the entomofauna: A review of its drivers. *Biological Conservation, 232*, 8–27. https://doi.org/10.1016/j.biocon.2019.01.020

Sattler C., Schrader J., Flor R.J., Keo M., Chhun S., Choun S., Hadi B.A.R. & Settele J. 2021: Reducing Pesticides and Increasing Crop Diversification Offer Ecological and Economic Benefits for Farmers—A Case Study in Cambodian Rice Fields. – *Insects* 12 (3). https://doi.org/10.3390/insects12030267

Schaap, I., Buedenbender, L., Johann, S., Hollert, H., & Dogruer, G. (2023). Impact of chemical pollution on threatened marine mammals: A systematic review. *Journal of Hazardous Materials, 459*, 132203. https://doi.org/10.1016/j.jhazmat.2023.132203

Schaefer M. 1974a: Experimentelle Untersuchungen zur Bedeutung der interspezifischen Konkurrenz bei 3 Wolfspinnenarten (Araneida, Lycosidae) einer Salzwiese. – *Zool. Jb. Syst.* 110, 213–235

Schaefer M. 1974b: Auswirkung natürlicher und experimenteller Störungen in Grenzzonen von Ökosystemen, untersucht am Beispiel der epigäischen Arthropodenfauna. – *Pedobiol.* 14, 51–60

Schaefer M. 1978: Some Experiments on the Regulation of Population Density in the Spider *Florinia bucculenta* (Araneida: Linyphiidae). – *Symp. Zool. Soc. Lond.* 42, 203–210

Schäffer N. 1984: Habitatwahl und Partnerschaftssystem von Tüpfelrallen (*Porzana porzana*) und Wachtelkönig (*Crex crex*). Diss. University of Würzburg.

Schäffer N. 1999: Habitatwahl und Partnerschaftssystem von Tüpfelralle *Porzana porzana* und Wachtelkönig *Crex crex*. - *Ökol. Vögel (Ecol. Birds) 21, 1–267*

Schäffer N. and Weisser W.W. 1996: Modell für den Schutz des Wachtelkönigs *Crex crex*. - *J. Ornithol.* 137, 53–75. https://doi.org/10.1007/BF01651499

Schläpfer A. 1988: Populationsökologie der Feldlerche *Alauda arvensis* in der intensiv genutzten Agrarlandschaft. – *Der Ornithologische Beobachter* 85 (4), 309–371

Scheu S. and Schaefer M. 1998: Bottom-up control of the soil macrofauna community in a beechwood on limestone: Manipulation of food resourceS. - *Ecology 79*, (5), 1573–158

Schmidt M.H., Lauer A., Purtauf T., Thies C., Schäfer M. & Tscharntke T. 2003: Relative importance of predators and parasitoids for cereal aphid control. – *Proceedings of the Royal Society of London B* 270: 1905–1909.

Schmiegelow F.K.A., Stepnisky D.P., Stambaugh C.A. & Koivula M. 2006: Reconciling salvage logging of boreal forests with a natural-disturbance management model. – *Conservation Biology* 20 (4), 971–983

Schmithüsen, J. (1959). *Allgemeine Vegetationsgeographie* (Reprint 2019 edition.). De Gruyter.

Schulze E.-D., Wirth C., Mollicone D. & Ziegler W. 2005: Succession after stand replacing disturbances by fire, wind throw, and insects in the dark Taiga of Central Siberia. – *Oecologia* 146, 77–88

Schwabe, A. (1989). Vegetation complexes of flowing-water habitats and their importance for the differentiation of landscape units. *Landscape Ecology, 2*(4), 237–253. https://doi.org/10.1007/BF00125094

Selva, N., Kreft, S., Kati, V., Schluck M., Jonsson B-G., Mihok B., Okarma H. and Ibisch P.L. 2011: Roadless and Low-Traffic Areas as Conservation Targets in Europe. - *Environmental Management* 48, 865–877 (2011). https://doi.org/10.1007/s00267-011-9751-z

Settele J. & Geissler S. 1988: Schutz des vom Aussterben bedrohten Blauschwarzen Moorbläulings durch Brachenerhalt, Grabenpflege und Biotopverbund im Filderraum. – *Natur und Landschaft* 63 (11), 467–470

Settele J., Steiner R., Reinhardt R., Feldmann R. & Hermann G. 2015: *Schmetterlinge*: Die Tagfalter Deutschlands. –Eugen Ulmer Verlag, 256 p.

Shaffer M. L. 1985: The metapopulation and species conservation: the special case of the northern spotted owl. – In: Gutierrez & Carey (eds.): *Ecology and Management of the Spotted Owl in the Pacific Northwest*, 86–99. – Portland. – OR: U.S.D.A. Forest Service, 119 p. : https://doi.org/10.2737/PNW-GTR-185

Shewring M.P. & Vafidis J.O. 2021: Using UAV-mounted thermal cameras to detect the presence of nesting nightjar in upland clear-fell: A case study in South Wales, UK. – *Ecological Solutions and Evidence* 2 (1), 1–10

Siebert U., Voigt U., Keuling O. & Tost D. 2020-24: *Occurrence and space usage of black grouse predators* in the Lüneburg Heath. – Research project of the Stiftung Tierärztliche Hochschule Hannover. www.tiho-hannover.de

Siefke A. 1985: Hiddensee 8013 8862 – der 15jährige Lebenslauf eines Sandregenpfeifer (*Charadrius hiaticula*)-Weibchens. – *Ber. Vogelwarte Hiddensee* 6, 59–64

Sklodowski J. 2022: Abandonment of salvage logging in windthrown stands is key to protection of late-succession carabid assemblages. 10.21203/rs.3.rs-1966688/v1

Slocombe, D. S. (1992). Environmental monitoring for protected areas: Review and prospect. *Environmental Monitoring and Assessment, 21*(1), 49–78. https://doi.org/10.1007/BF00400057

Smith I.A., Hutyra L.R., Reinmann A.B. et al. 2018: Piecing together the fragments: Elucidating edge effects on forest carbon dynamics. – *Frontiers in Ecology and the Environment* 16, 213–221

Sotherton N.W. 1984: The distribution and abundance of predatory arthropods overwintering on farmland. – *Ann. appl. Biol.* 105, 423–429.

Sotherton N.W. 1985: The distribution and abundance of predatory Coleoptera overwintering in field boundaries. – *Annals of Applied Biology* 106, 17–21

Soulé M.E. 1987: Where do we go from here?- In: Soulé M.E. (ed.): Viable populations for conservation. - Cambridge University Press, Cambridge, 189.

Soulé M.E. & Kohm K.A. (eds.) 1989: *Research Priorities for Conservation Biology*. – Island Press (Critical Issues Series), Washington, D.C., 97 p.

Southwood T.R.E. & van Emden H.F. 1967: A Comparison of the Fauna of Cut and Uncut Grassland. – *Z. angew. Ent.* 60, 188–198

Spellerberg I.F. 2005: *Monitoring Ecological Change* (2nd ed.). – Cambridge University Press, 391 p.

Staupendahl K. 1997: Ein neues Stichprobenverfahren zur Erfassung und Beschreibung von Naturverjüngung. – In: Jahrestagung der Sektion Forstliche Biometrie und Informatik des Deutschen Verbandes Forstlicher Forschungsanstalten in Freiburg i.Br., 24.-26.9.1997, 32–49

Stiling P. 1988: Density-dependence processes and key factors in insect populations. – *Journal of Animal Ecology* 57, 581–593

Struhsaker T.T. 1997: *Ecology of an African rain forest*: Logging in Kibale and the conflict between conservation and exploitation. – University Press of Florida, 15 N.W. 15th St., Gainesville, Florida 32611-2079, USA. 456 p.

Succow M. 2001: Wiederversumpfung von Niederungslandschaften, 167–178 p. – In: Succow M., Jeschke L. & Knapp H.D.: *Die Krise als Chance – Naturschutz in neuer Dimension.* Findling, Neuenhagen, 256 p.

Sukopp H., Seidel K. & Böcker R. 1986: Bausteine zu einem Monitoring für den Naturschutz. – *Ber. ANL* 10, 27–39

Tassin de Montaigu C. & Goulson D. 2020: Identifying agricultural pesticides that may pose a risk for birds. – *PeerJ* 8:e9526 https://doi.org/10.7717/peerj.9526

Tast J. 1968: Changes in the distribution habitat requirements and nest-sites of the Linnet *Carduelis cannabina* (L.) in Finland. – *Ann. Zool. Fenn.* 5, 159–178

Taylor L.R. & Taylor R.A.J. 1977: Aggregation, migration and population mechanics. – *Nature* 265, 415–421

Thies C., Denys C., Ulber B. & Tscharntke T. 1997: Effects of field margins and fallows on parasitism of rape pollen beetles. – *Boletin de la Asociacion Espaniola de Entomologia* 21: 90.

Thies C., Denys C. & Tscharntke T. (2000a). Randstreifen als Lebensraum: Vegetation, Artenreichtum und Schädlings-Nützlings-Interaktionen in Raps und Getreide. – In: Steinmann H.H. & Gerowitt B. (eds): *Ackerbau in der Kulturlandschaft - Funktionen und Leistungen*, pp. 135–156. Mecke, Duderstadt.

Thies C., Denys C. & Tscharntke T. (2000b). Die Förderung der biologischen Schädlingskontrolle durch Ackerrandstreifen und Ackerbrachen. – In: Nentwig W. (ed): *Streifenförmige ökologische Ausgleichsflächen in der Kulturlandschaft*. pp. 219–227. Verlag Agrarökologie, Bern, Hannover.

Thomas C.D. 1991a: Spatial and temporal variability in a butterfly population. – *Oecologia* 87, 577–580

Thomas J.A. 1991b: Rare species conservation: case studies of European butterflies. – p. 144–197 In: Spellerberg I.F., Goldsmith F.B. & Morris M.G. (eds.): *The Scientific Management of Temperate Communities for Conservation*. Oxford, Blackwell Scientific Publications, 566 p.

Thomas M.B. & Wratten S.D. 1988: Manipulating the arable crop environment to enhance the activity of predatory insects. – *Aspects of Applied Biology* 17, 57–66

Tost, D., Ludwig, T., Strauss, E., Jung, K., & Siebert, U. (2022). Habitat selection of black grouse in an isolated population in northern Germany—the importance of mixing dry and wet habitats. PeerJ, 10, e14161. https://doi.org/10.7717/peerj.14161

Tscharntke T., Steffan-Dewenter I., Kruess A. & Thies C. 2002: The contribution of small habitat fragments to the conservation of insect communities of grassland-cropland landscape mosaics. – *Ecological Applications* 12: 354–363

Tscharntke T., Clough Y., Wanger T.C., Jackson L., Motzke I., Perfecto I., Vandermeer J. & Whitbread A. 2012: Global food security, biodiversity conservation and the future of agricultural intensification. – *Biological Conservation* 151, 53–59

Tucker, M. A., Ord, T. J., & Rogers, T. L. (2014). Evolutionary predictors of mammalian home range size: body mass, diet and the environment. *Global Ecology and Biogeography*, *23*(10), 1105–1114. https://doi.org/10.1111/geb.12194

Tüxen, R. (1973). Vorschlag zur Aufnahme von Gesellschaftskomplexen in potentiell natürlichen Vegetationsgebieten. *Acta botanica Hungarica*, *19*(1–4), 379–384.

Uetz G.W. 1979: The influence of Variation in litter habitats on spider communities. – *Oecologia* 40, 29–42

Usher M.B. (ed.) 1986: *Wildlife conservation evaluation*. – Chapman & Hall, London, 394 p.

Usher M.B. & Erz W. (eds.) 1994: Erfassen und Bewerten im Naturschutz. – UTB für Wissenschaft, Quelle & Meyer Heidelberg, 340 p.

van Duyne C., Ras E., de Vos A.E.W., de Boer W.F., Henkens R.J.H.G. & Usukhjargal D. 2009: Wolf Predation among Reintroduced Przewalski Horses in Hustai National Park, Mongolia. – *The Journal of Wildlife Management* 73 (6), 836–843. http://www.jstor.org/stable/20616729, Accessed 22 Jan. 2023.

van Horne B. 1983: Density as a misleading indicator of habitat quality. – *J. Wildl. Manage.* 47 (4), 893–901

van Lexmond M.B., Bonmatin J.M., Goulson D. & Noome D. 2015: Worldwide integrated assessment on systemic pesticides. – Environ. Sci. Pollut. Res. 22, 1–4 https://doi.org/10.1007/s11356-014-3220-1

Viquerat S.M.A., Bobo K.S., Müller M., Kiffner C. and Waltert M. 2012: Estimating forest duiker (Cephalophinae) density in Korup National Park: a case study on the performance of three line transect methods. - *South African Journal of Wildlife Research* 42(1), 1–10

Vogel B. 1998a: Habitatqualität oder Landschaftsdynamik: Was bestimmt das Überleben der Heidelerche (*Lullula arborea*)? – Dissertation Universität Würzburg, Cuvillier Göttingen, 136 p.

Vogel K. 1998b: Sonne, Ziest und Flockenblumen: Was braucht eine überlebensfähige Population des Roten Scheckenfalters (*Melitaea didyma*)? – Dissertation an der Universität Würzburg, Cuvillier Göttingen, 134 p.

Völkl W., Zwölfer H., Romstöck-Völkl M. & Schmelzer C. 1993: Habitat Management in Calcareous Grasslands: Effects on the Insect Community Developing in Flower Heads of *Cynarea*. – *Journal of Applied Ecology* 30 (2), 307–315

Wainstein B.A. 1967: Some methods of evaluation of similarity of biocoenoses. – *Zool. Z.* 46, 981–986. (russ.)

Waldrop T.A. & Brose P.H. 1999: A comparison of fire intensity levels for stand replacement of Table Mountain pine (*Pinus pungens* Lamb.). – *Forest Ecology and Management* 113, 155–166

Walters J.R. 1991: Application of ecological principles to the management of endangered species: the case of the Red-cockaded Woodpecker. – *Annual Review of Ecology and Systematics* 22, 505–523

Waltert M. 1999, oral communication, unpublished

Waltert M. 2000: Forest management and the distribution of understorey birds in the Bossematié Forest, eastern Ivory Coast. – *Ostrich* 71 (1 & 2): 295–299.

Waltert M. & Mühlenberg M. 2001: Zur Beziehung zwischen Abundanz und Habitatqualität: Fangraten territorialer Gelbbartbülbüls *Andropadus latirostris*. – In: Gottschalk E., Barkow A., Mühlenberg M. & Settele J.: *Naturschutz und Verhalten*. UFZ-Bericht 2/2001, 103–110, Leipzig

Waltert M., Yaokokore-Beibro K.H., Mühlenberg M. & Waitkuwait W.E. 1999: Preliminary checklist of the birds of the Bossematié area, SE Ivory Coast. – *Malimbus* 21 (2), 93–109

Waltert M., Lien K.F. and Mühlenberg M. 2002: Further declines of threatened primates in the Korup Project Area, south-west Cameroon. - *Oryx* 36(3), 257–265

Weiss S.B., Murphy D.D. & White, R.R. 1988: Sun, Slope, and Butterflies: Topography Determinants of Habitat Quality for *Euphydryas editha*. – *Ecology* 69(5), 1486–1496

Wells T.C.E. 1971: A comparison of the effects of sheep grazing and mechanical cutting on the structure and botanical composition of chalk grassland. – In: Duffey, E. & Watt, A. S. (eds): *The Scientific Management of Animal and Plant Communities for Conservation*. 11th Symposium British Ecological Society: 497–515.

Wesolowski T. 1989: Nest-sites of hole-nesters in a primeval temperate forest (Bialowieza National Park, Poland). – *Acta Ornithol.* 25, 321–351

Wiens J.A. 1989: *The ecology of bird communities*. Vol. 1: Foundation and patterns. – Cambridge University Press, Cambridge, New York, Melbome, 539 p.

Williamson P., Cameron R.A.D. & Carter M.A. 1977: Population dynamics of the landsnail *Cepaea nemoralis* L.: A six-year study. – *J. Anim. Ecol.* 46, 181–194

Wilmanns O. 1989: *Ökologische Pflanzensoziologie*. 4. Aufl. – Quelle & Meyer, Heidelberg, Wiesbaden, 378 p.

Winker K., Rappole J.H. & Ramos M.A. 1995: The use of movement data as an assay of habitat quality. – *Oecologia* 101, 211–216

Wohlgemuth T., Jentsch A. & Seidl R. (eds.) 2022: *Disturbance Ecology*. – Springer International Publishing, XI, Landscape Series 32, 439 p.

Woyciechowski M., Slowik J. & Mühlenberg M. 2006: Hosts of the Butterfly *Maculinea teleius* among *Myrmica* Ants in Northern Mongolia (Lepidoptera: Lycaenidae; Hymenoptera: Formicidae). – *Sociobiology* 48: 493–502

Zang H. & Heckenroth H. 2001: *Die Vögel Niedersachsens, Lerchen bis Braunellen*. – Naturschutz Landschaftspflege Niedersachsen B, Heft 2.8, 260 p.

Zielinski Y. 1982: The effect of Brandt Vole (*Microtus brandti* Radde, 1861) colonies upon Vegetation in the *Caragana* steppe of central Eastern Mongolia. – *Pol. ecol. Stud.* 8, 41–56

Zwölfer H., Bauer G. & Heusinger G. 1981: *Ökologische Funktionsanalyse von Feldhecken* – Tierökologische Untersuchungen über Struktur und Funktion biozönotischer Komplexe. – Lehrstuhl Tierökologie, Universität Bayreuth, 422 p.

Zwölfer H., Bauer G., Heusinger G. & Stechmann D. 1984: *Die tierökologische Bedeutung und Bewertung von Hecken*. – Akademie für Naturschutz u. Landespflege, Beiheft 3 (2), 155 p.

11 Landscape Ecology

Abstract

Habitat destruction and fragmentation are the most frequently cited reasons for bringing species to the brink of extinction, raising the crucial question of how much space animal populations need to survive. The first approach to this question is to study species spectra and population development on islands of different sizes. Forest islands and patches of particular locations on the mainland are comparable to natural islands in the ocean. They can be analyzed in terms of size and distance to similar patches and correlated to the number of plant and animal taxa. Experiments with the artificial reduction of habitat islands and its consequences are also discussed. Studies on individual plant species and their populations as habitat islands are summarized for both woody and herbaceous plants. In experiments, the isolation factor can be manipulated, as isolation is crucial in a network of biotopes. Corridors connect habitats in a fragmented landscape. Studies on small-scale and transnational landscape corridors are presented. The colonization of habitat patches can also be studied experimentally, e.g., in artificial water containers.

11.1 Area Requirement of Animal Populations

Urban sprawl, road construction, and agricultural measures continuously lead to the fragmentation of originally contiguous habitats (Harris 1984; Burgess and Sharpe 1981; Mühlenberg 1985; Fahrig 2003; Mullu 2016; Cardinale et al. 2020). In the conflict over the use of our landscape, the area requirements of animal populations have become a central question (Pe'er et al. 2014). It can theoretically be determined by the size of the home ranges of reproductive units (pair or social group) and the minimum size of a viable population (minimum viable population, MVP) (Hovestadt 1990; Mühlenberg et al. 1991). While the sizes of the home ranges are influenced by habitat quality (see Sect. 10.7), external and internal stochastic changes must be considered for the MVP, which may lead to the local extinction of small and isolated populations. The risk hedging against the local extinction of patchily distributed populations (interaction groups, den Boer 1977) consists mainly of increasing the number of suitable areas as habitats and reducing the distance between the areas (see Sect. 10.17 Fallow Areas).

M. Mühlenberg, T. Waßmer, *Field Ecology*, https://doi.org/10.1007/978-3-031-87547-2_11

Decreasing area, and thus usually also decreasing resource supply, and increasing isolation escalate the probability of species extinction but also lead to a decrease in individual density for many species (Mühlenberg et al. 1977; Mühlenberg and Werres 1983; Dreyer and Kirsch 1987). The probability of species extinction increases with decreasing population size and increasing coefficient of population size variation (Thomas 1990). Pimm et al. (1988) provided concrete examples of bird species from the English coastal islands. Species represented by 1–2 pairs went extinct on average after 1.6 years, whereas with 6–12 pairs, the average survival time extended to 7.5 years. However, when comparing large and small species with the same average population size, the risk of extinction increased for the large species when they reached high individual numbers.

Fragmentation of habitats leads to more edge zones and apparently increases diversity (see Sect. 10.15 Forest Edges). However, species that live in the core area of habitats, such as forest interior birds, are disadvantaged and have a lower reproductive success, either due to degradation of habitat quality or increased predator pressure (Harris 1988; Temple and Cary 1988; Lampila et al. 2005, see also Van Dyke 2008; Cardinale et al. 2020). Also, the relationships between plants and their pollinators worsen in fragmented landscapes, and the plants produce fewer seeds (Jennersten 1988, Tscharntke et al. 2012).

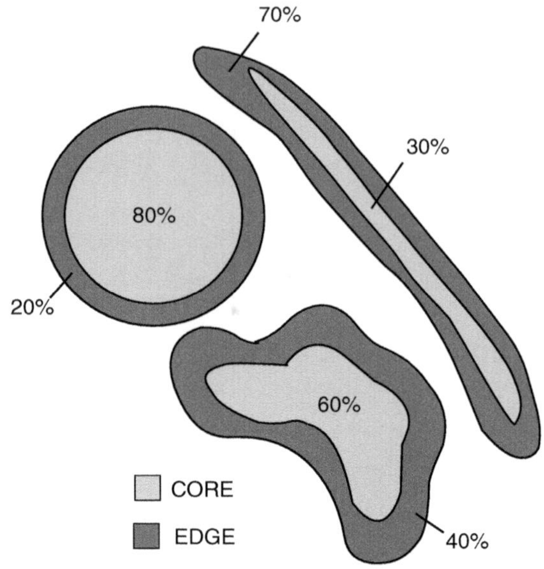

Fig. 11.1 Patch shape and edge. The edge-to-interior ratio of a habitat patch is affected by patch shape. A more convoluted, irregular, or linear patch will have a higher proportion of edge, thus increasing the number of edge species and decreasing the number of interior species. Picture credit: (Kennedy et al. 2003). Copyright© 2003, Environmental Law Institute®, Washington, DC. Reprinted with permission from ELI®

11.2 Comparison of Forest Islands and Dry Locations as Habitat Islands

We can investigate the faunal dependence on area size by comparing the fauna of large habitat islands to similar, small habitat islands. As a rule of thumb, it can be assumed that in one-tenth of the area, half of the species are present under undisturbed conditions and approximate equilibrium (see MacArthur and Wilson 1967). It is not easy to find habitat islands for comparison that differ only in size and not significantly in their resource offer (see Hortal et al. 2009; Sfenthourakis and Triantis 2009). With a smaller area, the edge zones increase and also contain species not present in the core of the habitat island (Mader 1983, see also edge effect in Sect. 10.15). Certain species are missing in narrow, elongated habitat islands despite offering a sufficiently large total area since the necessary core areas are absent (Temple 1986) (Fig. 11.1).

If habitat islands of different sizes are available for study, guilds (Sect. 12.1) can be compared regarding species and individual numbers. **Forest islands** of different sizes (see Fig. 11.2) can be easily examined for their bird fauna (for methods, see Sect. 1.8). Point counts are suitable. For a comparison of species numbers, equal-sized samples must be observed thoroughly. The identified species can be compared in groups according to their habitat binding (species of the forest interior, clearings, edges, and open spaces). For birds, sufficient literature is available, but more difficult are reliable and transferable literature references about the habitat binding of

Fig. 11.2 (**a**) Forest islands in the cultural landscape of the Steigerwald, Northern Bavaria, Germany. (**b**) Forest islands in New Zealand, created by the surrounding land use. (**c**) An artificial habitat island created through deforestation within the rainforest, Papua. (**d**) Artificial forest island, planted as a "plantation forest" in New Zealand

arthropods. The butterfly diversity is, for example, much more dependent on specific habitat requirements than on habitat size (Murphy and Wilcox 1986). Through dynamic occupancy models (on occupancy models, see above Sect. 1.6), Robles and Ciudad (2012) demonstrated for the Middle-Spotted Woodpecker (*Dendrocopos medius*) in Northwestern Spain that the habitat quality measured, e.g., by the number and thickness of the oaks, is more critical for stable settlement than the habitat size.

Schrader et al. (2019) demonstrated that in addition to habitat size, habitat quality played an essential role in the number of species using woody plants on isolated islands of different sizes. However, the size of the samples was also decisive for the comparison. Results from samples of woody plants at a local scale are more dependent on habitat quality and random processes than when comparing samples of the entire island area. The work occurred in a natural West Papua archipelago (see Fig. 11.3a). In addition, Schrader et al. (2020) were able to show that the number of species on small islands below a specific size was no longer determined by the size of the island ("small-island effect," Niering 1963; Triantis and Sfenthourakis 2012). Individual species groups, such as angiosperms, also differed in this relationship. It is, therefore, likely that different animal groups do not have to show the same relationships between diversity, habitat island size, and isolation.

Natural forest islands (e.g., Fig. 11.3a, b) are best suited to verify the predictions of island biogeography (MacArthur and Wilson 1967; Simberloff and Abele 1976).

The extinction of species in habitat islands can be slow and depends on the ecology of the individual species groups. Only long-term monitoring can document the effects of fragmentation. A

Fig. 11.3 (**a**) Naturally forested islands of various sizes in the bay of the island Gam, West Papua. (Photo credit: Fabian Mühlberger). (**b**) Natural forest islands in the West African tree savannah of the Comoé National Park, Ivory Coast. In the background runs the gallery forest of the Comoé River

large-scale field experiment in the tropical Amazon rainforest has demonstrated the different effects on various animal groups. It was shown that edge effects play an essential role (Bierregaard et al. 1992; Lovejoy et al. 1986; Laurance 2004a; Laurance et al. 2011; Luther et al. 2020).

With the loss of individual species due to reduced habitat area, the remaining species can also change their behavior in using their resources. Figure 11.4 shows, for example, how the Red Colobus uses all strata in an island forest in the absence of other *Colobus* species, thus expanding its niche. Niche width and species richness are negatively correlated (e.g., Granot and Belmaker 2020; Carscadden et al. 2020).

Even in large landscapes, different habitat types can be scattered (Fig. 11.5). Smaller habitat islands are often dry sites (Figs. 11.6 and 11.7a). They can also be found in the cultural landscape, such as quarries, sinkholes, rocks, lean grasslands, rock heaths, and dunes, which often have sparse vegetation. Here, solitary wasps (e.g., digger wasps), ants, and grasshoppers can be compared in their species community. Many grasshopper species are extremely resident and show a philopatry to their habitat (Köhler 1996). Even small sites (e.g., up to 30 m × 30 m in size) can have their own grasshopper fauna. A typical species for dry grassland habitats is, for example, *Metrioptera bicolor* (Tettigoniidae, Fig. 11.7b, c). With mark-release-recapture methods (Sect. 5.3), the movement distance of marked animals can be documented (e.g., Samietz et al. 1996).

A persistence in small habitat spots was also found in a metapopulation of the grasshopper *Bryoderma* (Fig. 11.7c), which is also bound to open and dry habitat patches (Stelter et al. 1997).

11.3 Individual Plants as Habitat Islands

When comparing two habitat islands, problems can arise with area size and different resource supplies. This problem does not arise when comparing individual plants or small stands con-

Fig. 11.4 (**a**) Average locations of different primate species in a closed forest. Tai National Park, Ivory Coast. (**b**) Average locations of different primate species in a forest island. Pirang forest, The Gambia. Pictures credit (**a**) and (**b**) Galat-Luong (1988)

cerning their inhabitants. One can compare individual standing trees with a tree population of the same species or differently sized herb plant populations. With the variables island size (number of plants), distance (distance of the isolated standing plants to the next, larger population of the same species), and age (time of isolation or establishment of the plant subpopulation), the following island ecological questions can be addressed:

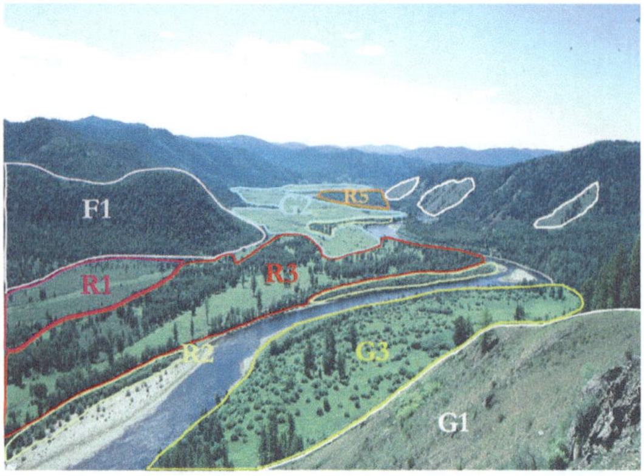

Fig. 11.5 Mosaic-like vegetation distribution in Northern Mongolia's forest-steppe: (Separating different habitats marked with abbreviations illustrating their patchy distribution. shows a profile of the vegetation units. G1 = mountain dry steppe, G2 = herb-rich terrace meadow, G3 = terrace meadow with *Padus asiatica* and *Salix* shrub, R3 = loose *Larix sibirica—Betula platyphylla*—floodplain forest, F1 = *Larix sibirica—Betula platyphylla* forest on north slope, R2 = Salix shrub on the riverbank, R1 = dense *Betula fusca* shrub with *Salix* sp

- Which phytophagous species live on the individual plants in the large plant population?
- How does leaf consumption or the number of leaf mines differ between individual plants and the large population?
- Did specialists (monophagous species) or generalists (polyphagous species) first colonized the individual plants?
- Do generalists predominate on individual plants?
- What distance is covered in what time by the phytophagous species, i.e., how significant is their colonization ability?
- Are small plant populations colonized more efficiently than individual plants (at ± the same distance to the large population)?
- Are plants between the large population and the individual plants used as stepping stones for colonization?
- Do predators and parasites of the phytophagous species appear with a time delay?

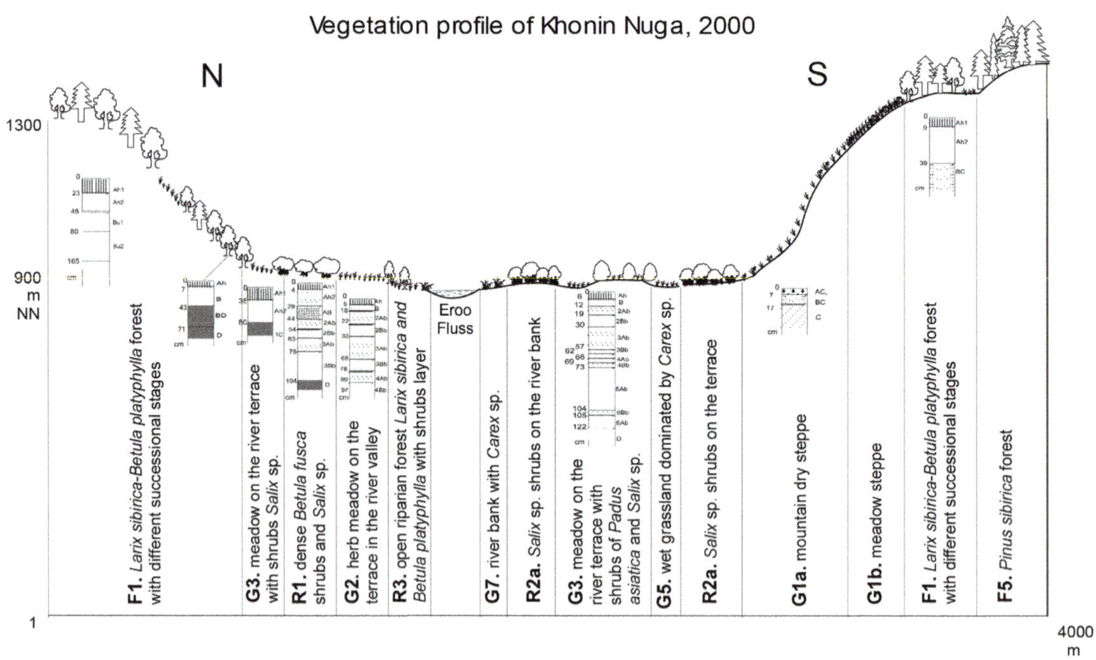

Fig. 11.6 Example of a vegetation profile and corresponding site classification of the Khonin Nuga area, Khentey mountains (Dulamsuren et al. 2005 The figure shows a profile of the vegetation units. G1 = mountain dry steppe, G2 = herb-rich terrace meadow, G3 = terrace meadow with *Padus asiatica* and *Salix* shrub, R3 = loose *Larix sibirica*—Betula platyphylla—floodplain forest, F1 = Larix sibirica—Betula platyphylla forest on north slope, R2 = Salix shrub on the riverbank, R1 = dense Betula fusca shrub with Salix sp)

Fig. 11.7 (**a**) Mountain dry steppe, a patchily distributed habitat (G1 in Figs. 11.5 and 11.6) in the mountain-forest-steppe of Northern Mongolia. Suitable habitat patches for the grasshopper are scattered like islands and separated by habitats with denser vegetation. (**b**) Marked male of *Metrioptera bicolor* (Tettigoniidae). Northern Mongolia. Photo credit by J. Munkhbat. (**c**) *Bryodema tuberculosa*, another grasshopper, which is bound to dry locations

- Do the phytophagous species go extinct more easily on individual plants than in large populations, and what are the mortality factors?

Individual Woody Plants and Woody Plant Populations

If a specific woody plant is chosen as a habitat island, then leaf consumption or the number of leaf mines from phytophagous species can be compared. On the trees or shrubs isolated by different distances, 100–1000 leaves were randomly collected and examined in the laboratory for leaf mines or other feeding traces (Faeth 1984). The samples must be the same size to compare individual plants and plant populations. If the original leaf shape is used as a basis, the amount of consumed leaf mass is estimated in percentage classes. In the lab, a surface-measuring device can help refine the estimate. Schrader et al. (2021) developed a simple method to estimate leaf area. The relative consumption portion, i.e., the visible loss of leaf surface, is expressed in % of the potential leaf area: Consumption area/[Consumption area + Remaining area] (Zwölfer et al. 1981). This measurement can be done with computer programs like WinFOLIA (Regent Instruments Inc., Canada, https://regent.qc.ca/assets/winfolia_about.html). Since the consumption areas grow with the leaves, correction calculations must be made when stating the actually consumed leaf portions (Lange in Zwölfer et al. 1981).

Both, leaf growth and primary consumption occur mainly during the period from April to June. However, many woody plants in Central Europe have a second leaf sprout, the "midsummer sprout," between June and July. If all plant species are examined for consumption losses in July or August, the consumption of the first leaf generation can be lost in the annual calculation. Different phytophagous species leave different consumption traces, such as mines, edges, or hole consumption. With the knockdown method (Sect. 2.6), a significant part of the phytophagous guild and some of their predators can be directly detected, and the species community can be compared between individual plants and plant populations. Another control of leaf-eating phytophages consists of the collection of twig samples and their careful examination in the lab. The sample twigs are selected according to random coordinates (horizontal distance and height). Collect at least ten twig samples from shrubs, each about 40 cm long. The reference size for the individual density (number of insect larvae, for example) is the number of leaves or buds on the twig sample (twig samples in winter).

As described in the following chapter (see below), the parasite fauna of leaf miners can also be determined by covering them in mesh bags. Usually, phytophagous species on woody plants have a larger parasite fauna than on herbaceous plants, and trees in old succession stages (forest) have a larger fauna than trees in earlier succession stages like in hedges (Askew 1980).

11.3.1 Case Study: Manipulation with Herbaceous Plants as Habitat Islands

The choice of herbaceous plants as island habitats for phytophagous insects and their predators allows experimental manipulation of the variables distance (distance between individual plants and plant population) and age (time of isolation or replanting) through adding or moving plants. Suitable are common, easily cultivable plants with a large spectrum of permanent residents, such as nettles (*Urtica* sp. Fig. 11.8a), thistles (*Cirsium* sp. Fig. 11.8b, *Carduus* sp. Fig. 11.8c), and cornflowers (*Centaurea* sp.). The examination of thistle heads was already presented in Sect. 2.25 (Case study). The insects living on thistles (*Cirsium* sp.) are conveniently compiled with an identification key of European species by Redfern (2011). The resource also provides all the necessary information on the biology of the insects living on the thistles. For their study, it is worth distinguishing different phytophagous guilds: (1) the flower visitors, (2) the inhabitants of the thistle heads, (3) the stem borers and leaf miners, and (4) the insects on the surface of the plant.

- Which phytophagous guild colonizes the isolated thistles first?

To study guilds (2) and (3), the corresponding plant parts must be cut off in the field and carefully examined in the lab. The plant parts can be stored in plastic bags in the refrigerator for several days. For particular questions and, in some cases, for breeding, the found larvae or pupae are needed for accurate species identification. Larvae still eating plant material must be fed accordingly in containers, whereby old plant material should be removed from the breeding containers. The different stages of development are kept in containers on moistened sand. In the field, it is worth enclosing plant parts cage-like with a small sleeve of gauze so that the emerging insects are caught (Redfern 1968, 2011, Askew 1980).

There is also a vivid compilation of the insects on nettles (*Urtica* sp.) (Davis 1983). Nettles have the advantage over thistles in that they are more common and easier to transplant for experiments. Potted nettles can be easily moved to create an isolated habitat island as a nettle patch. It is easiest to transplant whole plants or rhizome parts. From August to early spring, you can find fresh young rhizomes on or just below the soil surface, where an old rootstock of nettles grows. Since transferred plant material can already contain insect stages, it is recommended that you shake the plants, free the rhizomes from all soil material, and dip them in a bucket with a garden insecticide (Davis 1983). Davis (1975) used 57 ml Malathion plus 57 ml Dimethoat in 14 L water. Another agent would be, for example, Fenitrothion. Clayey or peat soil is most suitable for rooting; phosphate fertilizer can be added. The plants must initially be kept relatively moist. Nettles can also be easily grown from soaked seeds. However, flowers and rhizomes are not obtained in the first season, and nettles grown from rhizomes are usually more robust. When creating artificial habitat islands, you can create isolated small populations (up to 3×3 m^2) with and without corridors. The corridors are single plants or small groups, placed like stepping stones between the small and large nettle populations. Following this methodology, Davis (1975)

Fig. 11.8 (**a**) A small habitat island of a nettle stand (*Urtica* sp.) in August. Northern Germany. (**b**) The bull thistle *Cirsium vulgare*, Northern Germany. Photo credit Eva Mühlenberg-Horn. (**c**) The welted thistle *Carduus crispus*, Northern Germany. Photo credit Eva Mühlenberg-Horn

found 18 insect species more or less host-specific on nettles and 21 polyphagous or predatory species in England. In the first year, 28 species colonized the nettle populations up to 75 m away, and within 3 years, 39 species. If the populations are several kilometers isolated from other nettle occurrences, one or more species of the usual nettle inhabitants are often missing; 9–16 species of Heteroptera and Coleoptera were found.

The species are most effectively collected with the D-Vac suction device (see Sect. 2.8), with the collection supplemented by direct searching for caterpillars, galls, and leaf mines. The study of the species' dispersal and flight activity, for example, by setting up sticky traps (see Sect. 2.13), must be phenologically controlled.

When comparing populations, standardized equal-sized samples are a prerequisite. It is also necessary to ensure that the compared populations have approximately the same site conditions for the abovementioned questions concerning island ecology. For example, the insect fauna on nettles differs significantly between sunny and shady locations (Davis 1983).

The relationship between the number of plants and the number of phytophagous arthropod species has also been described for the cosmopolitan bracken fern (*Pteridium aquilinum*) (Lawton 1976).

The relationship between the number of species and the corresponding habitat size (number of plants, area size of the habitat island) is investigated using the area-species curve. The curve is described by a formula according to MacArthur and Wilson (1967) (see Sect. 15.11), in which the slope of the line in a double-logarithmic representation is expressed by the "z" value. "z" varies only within relatively narrow limits, the most common values are between 0.2 and 0.3 (e.g., Triantis et al. 2012). The higher the value, the more isolated the habitat islands are and the less colonization ability the animal groups have. Even if only two or a few differently sized habitat islands can be examined, the equation should be calculated, and the z-value should be compared with literature data for discussion but not yet considered as a scientific result.

A comparison of the number of species does not consider their ecological characterization as stenotopic or eurytopic. Often, with a decreasing area, stenotopic species are replaced by eurytopic species that can better tolerate the boundary conditions of the habitat. In addition to comparing species numbers, the densities of individuals should always be taken into consideration as they can decrease with decreasing area (Mühlenberg and Werres 1983). In some cases, the loss of species is compensated for by an increased density of the remaining species (MacArthur et al. 1972; George 1987). A broader discussion about changes in the density of island populations can be found in Faeth (1984) and Blondel (2000).

The parameter isolation can be investigated by planting or moving individual plants at variable distances to the nearest plant population. The size effect can be studied by excluding differently sized partitions from a managed meadow:

- How do partitions of different sizes (e.g., 2 m^2, 8 m^2, 32 m^2) that are no longer mowed or grazed develop in respect to the plant and animal population?

Quinn and Robinson (1987) found that while there were more plant species in larger quadrats than in smaller ones, the total number of several small quadrates exceeded the number of species of an equally large, undivided large quadrat. In nature conservation, there was a lifely debate based on the theory of island biogeography about whether many small habitats harbor more species than a few large ones (single large or several small, SLOSS debate, Diamond 1975). Patterson (1987) discussed this issue for isolated faunas as different conservation strategies can be drawn from both scenarios (see also Wilcox and Murphy 1985; Quinn and Harrison 1988; Fahrig and Storch 2020; Fahrig et al. 2022).

One must proceed experimentally to demonstrate an area effect for a species community. On a meadow, this can be achieved practically without much effort (Mühlenberg and Werres 1983, Fig. 11.9). For example, one captures the epigaeic soil fauna of spiders and ground beetles (see Sect. 2.16) for a year on the entire study area and after reduction by plowing to, for example, a fifth of the area in the following year. A reduced number of ground beetle and spider species after the second year in the remaining areas can indicate a species loss due to area reduction. There is also immigration of new species in the second year. These are then ecologically evaluated (known as capable of flight and with high dispersal ability or without these properties). The average size of the species can also be compared to their respective abundance. A comparison meadow without treatment in the 2 years must also be sampled in the same way to control for natural fluctuations between the years.

Fig. 11.9 Experimental reduction of the meadow to three sections by plowing up the original large meadow area. Northern Bavaria. (Mühlenberg and Werres 1983)

In freshwater ecology, the colonization of artificial habitat islands can be studied using experimental ponds, see Sect. 11.6.

11.4 Networks of Biotopes

The reduction of original habitats and the division of the remaining residual area into increasingly isolated pieces is intended to be compensated by increasing networking of biotopes (Noss 1987; Mader 1985). In smaller areas, the probability of extinction of populations and the species turnover increases (Mühlenberg and Werres 1983). However, this does not necessarily apply to plants (Schrader et al. 2023). Locally extinct subpopulations must be rebuilt through re-establishment. Measures to facilitate the recolonization of habitat islands include creating additional similar habitat islands in the nearby area (increase of colonization sources), the extensification of land use (the environment of the habitat islands becomes less hostile), and the networking of these biotopes (spread of habitat-bound individuals along these corridors) (see Harris and Scheck 1991; Noss and Daly 2006; Hilty et al. 2006). Landscape corridors are also attempted to be achieved through agroforestry management in the tropics (Laurance 2004b).

This raises the question of whether corridors can be used for the dispersal of specific animal species. A research program on the dispersal of animals can be subdivided into several parts:

One approach is to compare the genetic similarity between isolated populations and populations connected by a landscape corridor. The genetic similarity can be checked after determining the number of alleles for isozymes and analyzing their respective frequencies with laboratory methods such as electrophoresis (Selander et al. 1971; Slatkin 1985; Seitz 1988).

Case Study: Function of Corridors

A second research approach involves propagation studies in the field through direct monitoring of the migration of individuals. This approach is developed in four stages:

1. First, evidence must be provided that species from two habitat islands, biotope A and B, can also be found in the landscape patches connecting them, the corridor C. Selected arthropod groups can be captured with pitfall traps, net catches, or Malaise traps, and birds can be observed at their locations. However, detecting species in C will only show that these species can also occur in C and use the corridor as a habitat.
2. For further analysis, it needs to be checked whether the movement patterns of individual animals are influenced by corridor C. This evidence may show that the animals move more in the longitudinal direction of the corridor and do not leave it. A corridor can support the spread of the animals in a channeling way. Direct observation of animals is possible with birds, butterflies, and dragonflies (e.g., Saunders and Rebeira 1991; Wood and Samways 1991). Genetic methods can also be used with environmental DNA to analyze animal dispersal on a larger spatial scale (Barnes and Turner 2016). Another method is the recapture of marked animals. For a more effective recapture, trapping trenches are used in the ground (Sect. 2.17). Monitoring of the movement pattern is carried out with the same trap arrangement in the extended area ("normal area"). Strips, 1–2 m wide, can be blocked off for epigaeic animals using cross-placed trapping trenches. Similar trapping trenches can also run parallel to the sideline of the strip to sample animals leaving the corridor. These trapping trenches should be placed 30–50 cm and 2 m from the corridor. In the evaluation of the catches, evidence of the habitat binding of the examined species is obtained at the same time. It makes sense to compare two closely related species in a meadow corridor, a eurytopic and a stenotopic species. Such species pairs exist in ground beetles (e.g., genus *Poecilus*) and grasshoppers (e.g., genus *Chorthippus*).
3. Only the third stage shows if animals migrating through the corridor are in the dispersal phase (*dispersal*). By comparing traps with no catches and the control of the re-advancement of individuals of the observed species, indications of the current spread of a species can be obtained. Again, the corridor must be compared with the normal area. A temporary increase in the recolonization of the controlled site can be a sign of the actual dispersal processes of the corresponding population. It is also possible

to estimate the dispersal pressure (or the "re-advancement") of migrating species by catching them at the corridor entrance (best with completely blocking access to the extended area, e.g., by a partially buried 30 cm high plastic wall in connection with ground traps; see Sect. 2.17). Again, controls must be carried out in the regular area and at its borderline to the other habitat.

4. Finally, for successful dispersal, evidence of mating or reproduction of the migrating animals in the new area must be provided. Despite the emigration of sexually mature animals, gene flow is not yet guaranteed (example in Ehrlich and Murphy 1987).

In Fig. 11.10, trap arrangements are shown that answer some of the questions raised, especially for epigaeic animals, and provide the necessary checks. When one creates corridors through fragmented habitats, their biological significance can be experimentally demonstrated (Nicholls and Margules 1991; Margules 1996). Additionally, the recolonization in habitat islands, where individual species have locally disappeared, can be compared by creating many experimental plots with and without corridor connection to the original habitat.

Only through an extensive series of individual markings and recaptures can the dispersal behavior of the selected species be conclusively clarified. Such knowledge would be a prerequisite for networking grassland biotopes, assessing the ecological role of meadow paths and field margins (Way and Greig-Smith 1986), and their usefulness when decommissioning or extensifying arable land (Mühlenberg 1988).

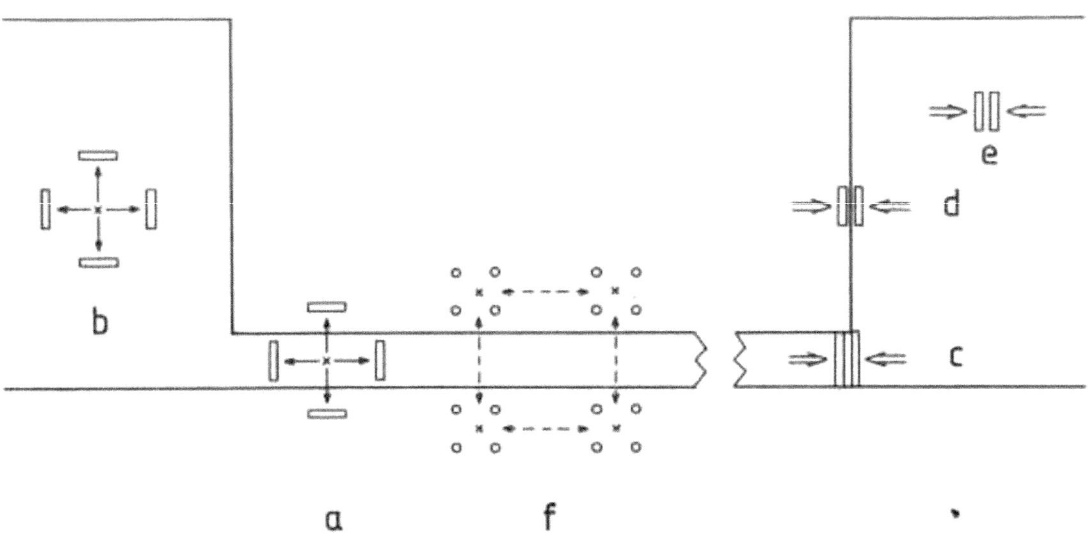

Fig. 11.10 Diagram of a corridor connecting two similar, large-scale habitat islands, with the placement of possible trap arrangements for studying the corridor function for nonflying animals. Traps include ordinary pitfall traps (circles) or catch trenches (rectangles, see Sect. 2.16). With arrangement (**a**) to investigate the question of whether the animals are influenced in their movement direction by the corridor structure, (**b**) serves as a control in the extended comparison area. With (**c**), one analyzes the immigration rate into the corridor, and (**d**) and (**e**) represent the controls. With arrangement (**f**), the problem can be addressed whether the corridors also represent mobility barriers for certain species; the trap sets are each equally far apart. (from Mühlenberg 1993)

Corridors can have another positive effect: They usually offer a steep gradient of abiotic factors through their border strips and, thus, the possibility for animal populations to better evade weather-related "disturbances." This effect can be verified by observing changes in the distribution pattern of arthropods, e.g., with the dispersion index (Sect. 15.11). Schaefer (1974) has demonstrated this buffering and stabilizing effect in border zones between salt marshes and dry grasslands. Weiss et al. (1988) pointed out the importance of such topographic diversity for the population development of butterflies (see Sect. 10.17). The creation of fallow lands or field edge strips (see above) can also support dispersing insects and small vertebrates.

The political framework conditions for land strips and areas that can serve as corridors are summarized in Kettunen et al. (2007). However, as the function of corridors can only be interpreted on a species-specific basis, focal species should be selected to plan corridors (Noss 2007). The main issue here is the spatial scale.

Wildlife-crossing structures built to mitigate the barrier effect of roads can also be considered corridors (see Sect. 13.2).

Because of the connectivity of forest fragments in Ontario, the Southern flying squirrel can expand its range further north due to climate change (Bowman et al. 2005; Deslauriers 2021). While patch size does not play a significant role for the flying squirrel, it was necessary for immigration and inter-patch colonization that none of the forest patches exceeded 300 m of isolation from another (Deslauriers 2021). Dispersal and immigration are crucial for the survival of flying squirrels (Fig. 11.11) (Brommer et al. 2017). Intact, contiguous boreal forests will best keep up with rapidly changing climate conditions because of high connectivity and abundant propagule pro-

Fig. 11.11 (**a**) The Siberian flying squirrel *Pteromys volans* is the Eurasian representative of the North American flying squirrel *Glaucomys volans*. (**b**) The North American Southern flying squirrel *Glaucomys volans*. Picture credit: Cephas on Wikimedia. In Southside, Etowah County, Alabama. CC BY-SA 4.0

duction. The slowed rate of change [by moderating local climate and providing alternate habitats] is expected to aid in adapting boreal wildlife to climatic shifts. It is more critical for the viability of many forest species to maintain intact forests so that a range shift due to climate change remains possible.

11.5 Landscape Corridors

Landscape corridors are usually on a different scale, and are considered for larger wildlife and longer distances (e.g., Hilty et al. 2006). Figure 11.12 shows a natural landscape where river valleys with riparian forest strips connect different parts.

The Selous-Niassa Wildlife Corridor is a cross-country corridor in East Africa (Fig. 11.13, Baldus et al. 2003; Baldus and Hahn 2009). It connects the Selous Game Reserve in Southern Tanzania (Baldus 2009) with the Niassa Game Reserve in Northern Mozambique over a length of about 180 km and an area of about 10,000 km^2. The total area of the transboundary natural ecosystem covers an area of 154,000 km^2 (Bloesch and Hahn 2007). The habitats consist of Miombo woodland, wetlands, open woodland, and riparian forests and benefit the African elephant (*Loxodonta africana*), which is now on the Red List (https://www.iucnredlist.org/species/181008073/223031019), and many other wildlife such as wild dogs, leopards, lions, buffalos, crocodiles, eland, impala, greater kudu, Lichtenstein's hartbeest, Roosevelt sable antelope, hippopotamus, Niassa wildbeest and others. The corridor area also serves resting and breeding birds.

A corridor that is supposed to connect continents is the *Jaguar Corridor Initiative* in Central

Fig. 11.12 The Tuul River Valley connects far-apart landscape patches with gallery forests and river meadows. Smaller stream-accompanying woods (blue line) connect the valley with the north-exposed slope forests (purple line). The photo depicts a natural landscape with a habitat mosaic that is not fragmented by people. Khentii, Northern Mongolia

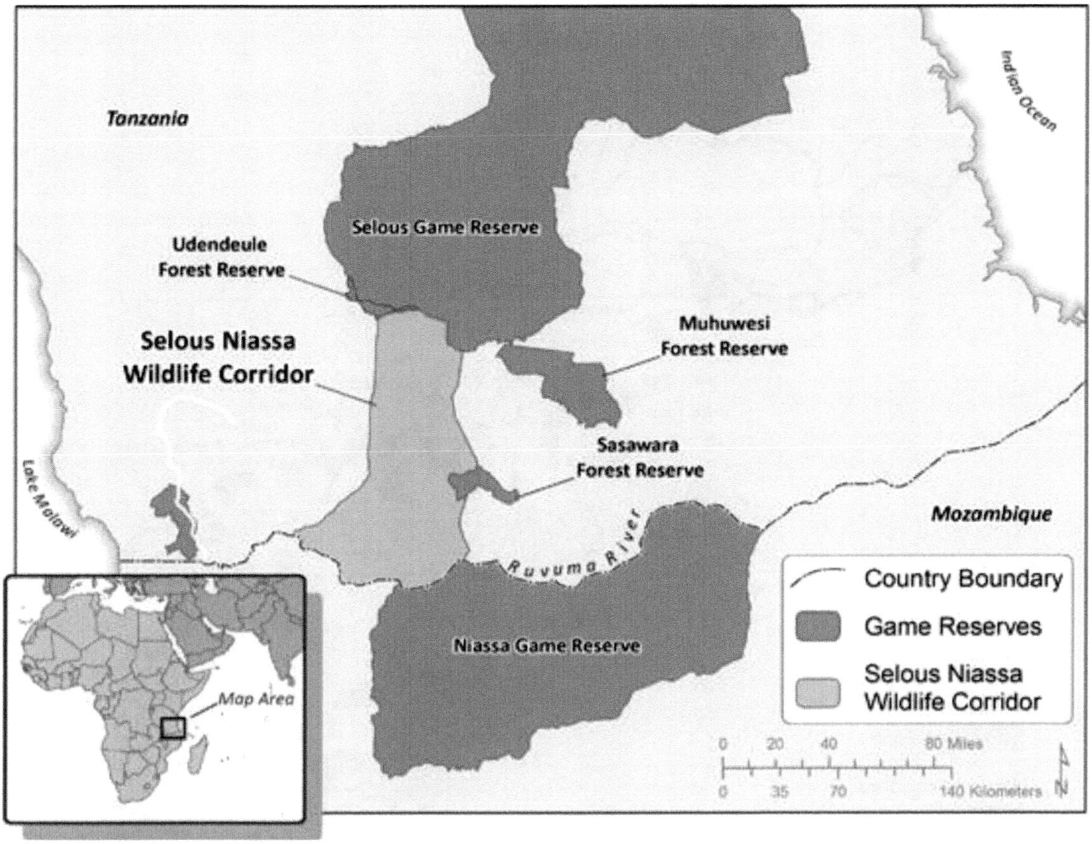

Fig. 11.13 The Selous-Niassa Wildlife Corridor (SNWC) in East Africa (Picard 2015. Reproduced with kind permission from Springer Nature)

and South America (Rabinowitz and Zeller 2010) (see Fig. 11.14). "*While good jaguar habitat, where the cats can live and breed, has decreased by 50 percent since the 1900s, the habitat a jaguar can use to travel through has decreased only by 16 percent. Most of it is intact and contiguous. We were writing these places off because they're not habitat where a permanent jaguar population can live. Now they're turning out to be crucial.*" (Rabinowitz 1986, see also Monsalve 2009). Landscape corridors also help species relocate or migrate to new habitats due to climate change (e.g., Bencharif 2010; Beier 2012, Center for large Landscape Conservation 2024).

Habitat restoration can also improve landscape connectivity (see also Sect. 14.1) and recreate functional landscape corridors (Hodgson et al. 2011). One more recent example is the studies on the Asian elephant in China (Li et al. 2024).

11.6 Experiments with Artificial Small Bodies of Water

Experiments with artificial small bodies of water can also be set up with simple water containers placed in the landscape. The containers need not be larger than 1–2 m in diameter. Experimental small bodies of water can also be set up as *plastic wading pools* with, for example, 80 L to 300 L size. You can also dig out shallow depressions (30–60 cm) in suitable sediment. You must then place sediment samples into these mesocosms so that existing organisms in the sediment can hatch. Only then can you be sure what new faunal elements have migrated into the water tank through colonization (see Frisch and Green 2007). Questions are: Is the initial colonization random or always the same? Does it depend on the environment (see Laan and Verboom 1990; Lutz 2010)? Do predatory species prevent the estab-

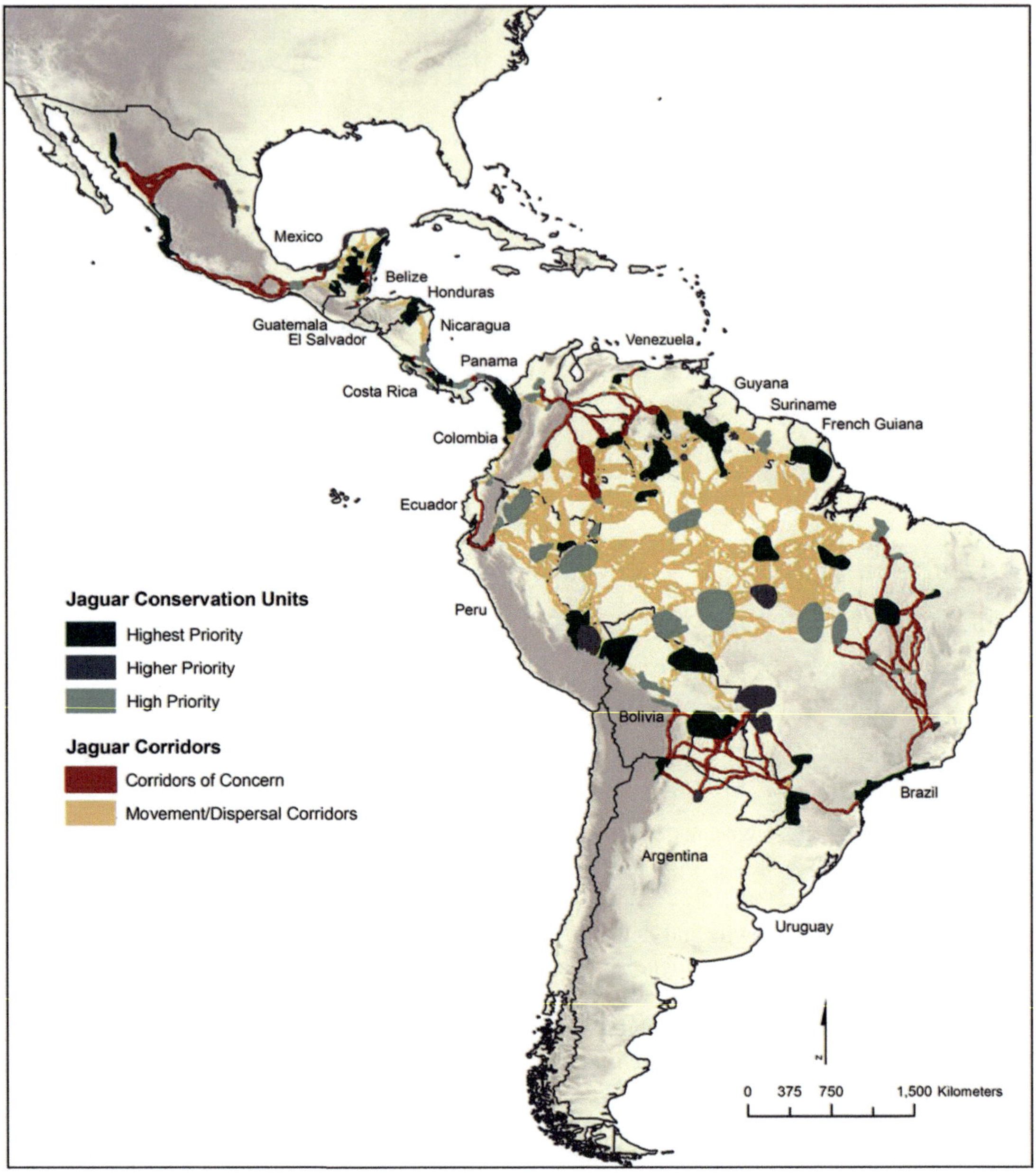

Fig. 11.14 Jaguar corridors in Central and South America: Jaguar Conservation Units and their level of prioritization from the latest range-wide priority-setting exercise (Zeller 2007) are displayed along with the least-cost jaguar corridors and corridors of concern from our jaguar connectivity model. Source: Rabinowitz and Zeller (2010). Reproduced with kind permission from Springer Nature

lishment of subpopulations of phytophagous species or particle feeders?

Often, artificial smaller bodies of water are created to compensate for lost wetland biotopes. Through the same setup, colonization can be studied, but the compensatory measures are usually about amphibians (Baker and Halliday 1999; Pechmann et al. 2001; The created ponds are about 16 m in diameter, leading to an area of 200 m^2, with a depth of up to 1 m

(Pechmann et al. 2001). The colonization of aquatic insects can also be studied in artificial small bodies of water (Cononnelly 2010).

It can be interesting to compare the colonization of small water bodies between neighboring regions.

- Can the organisms fly, such as dragonflies, beetles, and some other water bugs?
- Are they carried as developmental stages by birds, and are they thus widely spread as suggested by the species community of similar ponds over a large region (Poloczek et al. 2012)?

It need to be kept in mind that the initialsettlers can influence the colonization of subsequent species (Pintar and Resetarits Jr 2020).

Another recent example is the complex of four man-made experimental watersheds, each consisting of an interconnected pond, stream, and wetland constituting the Notre Dame Linked Experimental Ecosystem Facility. Scientists use the facility to conduct experiments in a field-like setting but in a more controlled environment than one can find in nature. More information is available on their website https://environmental-change.nd.edu/resources/nd-leef/, including several short videos such as Biofilms and the Fate of Environmental DNA (eDNA): https://www.youtube.com/watch?v=7o05GFIKe8I and through their published work, e.g., (Shogren et al. 2017, 2020).

References

Askew R.R. 1980: The diversity of insect communities in leafmines and plantgalls. – *Journal of Animal Ecology* 49, 817–829

Baker J.M.R. & Halliday T.R. 1999: Amphibian colonization of new ponds in an agricultural landscape. – *Herpetological Journal* 9, 55–63

Baldus R.D. (ed.) 2009: *Wild Heart of Africa. The Selous Game Reserve in Tanzania.* – Rowland Ward Publications, Johannesburg. 268 p.

Baldus R.D. & Hahn R. 2009: *The Selous – Niassa Wildlife Corridor in Tanzania:* Biodiversity Conservation from the Grassroots. Practical Experiences and Lessons from Integrating Local Communities into Transboundary Natural Resources Management.—Joint publication of FAO and CIC. Budapest. 48 p.

Baldus R.D., Hahn R., Mpanduji D.G. and Siege L. 2003: The Selous-Niassa Wildlife Corridor. - In: Baldus R.D. and Siege L. (eds.): *Tanzania Wildlife Discussion Paper* No. 34, 19 p.

Barnes M.A. & Turner C.R. 2016: The ecology of environmental DNA and implications for conservation genetics. – *Conservation Genetics* 17, 1–17

Bateman J. 1982: *Animal Traps and Trapping*. (2nd ed.) – Biddles Limited for David & Charles (Publishers) Limited, Newton Abbot, London, 286 p.

Beier P. 2012: Conceptualizing and Designing Corridors for Climate Change. – *Ecological Restoration* 30, (4), 312–19. http://www.jstor.org/stable/43441504.

Bencharif S-T. 2010: *Climate Change and connectivity: Are corridors the solution?* – ResearchGate, 40 p.

Bierregaard R.O., Lovejoy T.E., Kapos V., dos Santos A.A. & Hutchings R.W. 1992: The biological dynamics of tropical rainforest fragments. – *BioScience* 42, 859–866

Birks J.D.S. & Linn I.J. 1982: Studies of Home Range of the Feral Mink, *Mustela vison*. – *Symp. zool. Soc. Lond.* 49, 231–257

Bloesch U. & Hahn R. 2007: Biodiversity Values, Threats and Conservation Strategies of the Selous – Niassa Wildlife Corridor, Tanzania. – *Proceedings of the 6th TAWIRI Scientific Conference,* 3–6 December, Arusha, Tanzania

Blondel J. 2000: Evolution and ecology of birds on islands: Trends and prospects. – *Vie Milieu* 50, 205–220.

Jeff Bowman, Gillian L Holloway, Jay R Malcolm, Kevin R Middel, and Paul J Wilson. 2005: Northern range boundary dynamics of southern flying squirrels: evidence of an energetic bottleneck. – *Canadian Journal of Zoology* 83(11), 1486–1494. https://doi.org/10.1139/z05-144

Brommer J.E., Wistbacka R. & Selonen V. 2017: Immigration ensures population survival in the Siberian flying squirrel. – *Ecology and Evolution* 7 (6), 1858–1868

Bub H. 1977: Vogelfang und Vogelberingung, Teil. II. – A. Ziemsen-Verlag, Wittenberg Lutherstadt, 182 p., Teil I (1978), 169 p.

Burgess R.L. & Sharpe D.M. (eds.) 1981: *Forest Island Dynamics in Man-Dominated Landscapes.* – Ecological Studies 41. –Springer-Verlag, New York, Heidelberg, Berlin, 310 p.

Cardinale B.J., Primack R.B. & Murdoch J.D. 2020: *Conservation Biology*. – Oxford University Press, Sinauer Associates. 584 p.

Carscadden K.A., Emery N.C., Arnillas C.A., Cadotte M.W., Afkhami M.E., Gravel D., Livingstone S.W. & Wiens J.J. 2020: Niche Breadth: Causes and Consequences for Ecology, Evolution, and Conservation. – *The Quarterly Review of Biology* 95 (3), 179–287

Center for Large Landscape Conservation 2024: Ecological Corridors - A Vital Nature-based Solution to Climate Change -. Center for Large Landscape Conservation, Bozeman, MT (USA). https://large-landscapes.org › wp-content › uploads

Connelly R. 2010: Colonization by Aquatic Insect Larvae in Experimental Ponds: Reproductive Decisions Based on Biotic and Abiotic Factors. – Master Thesis at Coastal Carolina University, *Electronic Theses and Dissertations*. 66. https://digitalcommons.coastal.edu/etd/66

Davis B.N.K. 1975: The colonization of isolated patches of nettles (*Urtica dioica* L.) by insects. – *Journal of Applied Ecology* 12, 1–14

Davis B.N.K. 1983: *Insects on nettles*. – Cambridge University Press, Cambridge, 64 p.

Den Boer P.J. 1977: *Dispersal Power and Survival. Carabids in a cultivated countryside*. –Miscell. Papers 14., Veenman & Zonen, Wageningen, 190 p.

Deslauriers S. 2021: Effects of Forest Characteristics and Landscape Configuration on Flying Squirrel Occurrence and Abundance in Rouge National Urban Park. – Master thesis at University of Toronto, 40 p.

Diamond J.M. 1975: The island dilemma: Lessons of modern biogeographic studies for the design of natural reserves. – *Biological Conservation* 7 (2), 129–146

Dreyer W. & Kirsch N. 1987: Der Insektenkomplex des Mädesüß *(Filipendula ulmaria* L.). – *Zool. Anz.* 218 (1/2), 49–64

Dulamsuren Ch. Hauck, M. & Mühlenberg, M. 2005: Vegetation at the taiga forest-steppe borderline in the western Khentej Mountains, northern Mongolia. – *Annales Botanici Fennici* 42, 411–426

Ehrlich P.R. & Murphy D.D. 1987: Conservation Lessons from Long-Term Studies of Checkerspot Butterflies. – *Conservation Biology* 1 (2), 122–131

Ellenberg H. & Dreifke R. 1988: Ursachen für Verteilung und Häufigkeit, einschließlich des Fehlens, von Elstern auf zwei Probeflächen in Schleswig-Holstein. 2. Jahresbericht zu einem Forschungsauftrag im Rahmen des BMFT-Projekts (Mühlenberg, Kaule, Ellenberg) über Minimalareale für Tierpopulationen aus Naturschutzsicht, Manuskript, 35 p.

Faeth S.H. 1984: Density Compensation in Vertebrates and Invertebrates: A Review and an Experiment, 491–509. – In: Strong D.R. jr., Simberloff D., Abele L.G. & Thistle A.B. (eds.): *Ecological Communities: Conceptual Issues and Evidence*. – Princeton University Press, Princeton, New Jersey, 613 p.

Fahrig L. 2003: Effects of Habitat Fragmentation on Biodiversity. – *Annu. Rev. Ecol. Evol. Syst.* 34, 487–515. https://doi.org/10.1146/annurev.ecolsys.34.011802.132419

Fahrig L. & Storch D. 2020: Why do several small patches hold more species than few large patches? – *Global Ecology and Biogeography* 29, 615–628. https://doi.org/10.1111/geb.13059

Fahrig L., Watling J.I., Arnillas C.A., Arroyo-Rodríguez V., Jörger-Hickfang T., Müller J., Pereira H.M., Riva F., Rösch V., Seibold S., Tscharntke T. & May F. 2022: Resolving the SLOSS dilemma for biodiversity conservation: a research agenda. – *Biological Reviews* 97(1), 99–114

Frisch D. & Green A.J. 2007: Copepods come in first: rapid colonization of new temporary ponds. – *Fundamental and Applied Limnology* 168 (4), 289–297

Galat-Luong A. 1988: Monkeys in the Pirang forest. – In: *Pirang, Ecological Investigations in a Forest Island in The Gambia*. – Ed. by Stiftung Walderhaltung in Afrika & Bundesforschungsanstalt für Forst- und Holzwirtschaft, Hamburg, 187–208

George T.L. 1987: Greater land bird densities in islands vs. mainland: relation to nest predation level. – *Ecology* 68, 1393–1400.

Granot I. & Belmaker J. 2020: Niche breadth and species richness: correlation strength, scale and mechanisms. – *Global Ecology and Biogeography* 29, 159–170

Harris L.D. 1984: *The Fragmented Forest*. Island Biogeography Theory and the Preservation of Biotic Diversity. – The University of Chicago Press, Chicago, London, 211 p.

Harris L.D. 1988: Edge effects and conservation of biotic diversity. – *Conservation Biology* 2, 330–332

Harris L.D. & Scheck J. 1991: From implications to applications: the dispersal corridor principle applied to the conservation of biological diversity. – In: Saunders D.A. & Hobbs R.J.: *Nature Conservation 2: The role of corridors*, 189–220. – Surrey Beatty & Sons Pty Limited, Chipping Norton, Australia, 442 p.

Heller N.E. & Zavaleta E.S. 2009: Biodiversity management in the face of climate change: A review of 22 years of recommendations. – *Biological Conservation* 124(1), 14–32. https://doi.org/10.1016/j.biocon.2008.10.006.

Hilty J.A., Lidicker Jr. W.Z. & Merenlender A.M. 2006: *Corridor Ecology. The Science and Practice of Linking Landscapes for Biodiversity Conservation*. – Island Press Washington, London, 323 p.

Hodgson J.A., Thomas C.D., Cinderby S., Cambridge H., Evans P. & Hill J.K. 2011: Habitat re-creation strategies for promoting adaptation of species to climate change. – *Conservation Letters* 4. 289–297

Hofman M.P.G., Hayward M.W., Heim, M., et al. 2019: Right on track? Performance of satellite telemetry in terrestrial wildlife research. – *PLoS One* 14, e0216223.

Holmes A. 2022: Researchers put radio backpacks on bumblebees in Tasmania in insect-tracking trial. – https://www.abc.net.au/news/2022-10-10/bumblebees-radio-transmitters-research-insect-tracking/101515280

Holohil (2020). Holohil Transmitters List. Available at: https://www.holohil.com/transmitters/#tableall

Hortal J., Triantis K.A., Meiri S., Thebault E. & Sfenthourakis S. 2009: Island species richness increases with habitat diversity. – *Am. Nat.* 174 (6), E205–17. https://doi.org/10.1086/645085.

Hovestadt T. 1990: Possibilities and criteria for the determination of "minimal areas" of animal populations and ecosystem stocks. – *Series of Publications for Landscape Conservation and Nature Protection* 32, 175–186, Bonn-Bad Godesberg

Jennersten O. 1988: Pollination in *Dianthus deltoides* (Caryophyllaceae): Effects of Habitat Fragmentation on Visitation and Seed Set. – *Conservation Biology* 2 (4), 359–366

Kennedy, C. M., Wilkinson, J., & Balch, J. (2003). *Conservation Thresholds for Land Use Planners* (No. 003101) (p. 55). Washington DC: The Environmental Law Institute. https://www.eli.org/research-report/conservation-thresholds-land-use-planners. Accessed 23 August 2024

Kenward R.E. 1982: Techniques for Monitoring the Behaviour of Grey Squirrels by Radio. – *Symp. zool. Soc. Lond.* 49, 175–196

Kenward R. 1987: *Wildlife Radio Tagging, Equipment, Field Techniques and Data Analysis.* – Academic Press, Inc. (London) Ltd. (Publ.), Cambridge University Press, Cambridge, 222 p.

Kenward R. 2001: *A manual for wildlife radio tagging*, 2nd ed. – New York: Academic Press. 350 p.

Kettunen M., Terry A., Tucker G. & Jones A. 2007: *Guidance on the maintenance of landscape features of major importance for wild flora and fauna* - Guidance on the implementation of Article 3 of the Birds Directive (79/409/EEC) and Article 10 of the Habitats Directive (92/43/EEC). Institute for European Environmental Policy (IEEP), Brussels, 114 pp. & Annexes.

Köhler G. 1996: The ecological background of population vulnerability in Central European grasshoppers and bush crickets: a brief review. p. 290-298. – In: Settele J., Margules C.R., Poschlod P. & Henle K. (eds.): *Species Survival in Fragmented Landscapes.* Kluwer Academic Publishers, 381 p.

Kruuk H. 1978: Spatial Organisation and territorial behaviour of the European badger *Meles meles.* – *Journal of Zoology*, London, 184, 1–19

Laan R. & Verboom B. 1990: Effects of pool size and isolation on amphibian communities. – *Biological Conservation* 54, 251–262

Lampila P., Monkkonen M. & Desrochers A. 2005: Demographic responses by birds to forest fragmentation. – *Conservation Biology* 19, 1537-1546

Laurance S.G. 2004a: Responses of understory rain forest birds to road edges in central Amazonia. – *Ecol. Appl.* 14(5), 1344–1357. https://doi.org/10.1890/03-5194

Laurance S.G.W. 2004b: Landscape Connectivity and Biological Corridors, 50-64. – In: Schroth G., da Fonseca G.A.B., Harvey C.A., Gascon C., Vasconcelos H.L. & Izac AM. N. (eds.): *Agroforestry and Biodiversity Conservation in Tropical Landscapes.* Island Press, 537 p.

Laurance W.F., Camargo J.L.C., Luizão R.C.C., Laurance S.G., Pimm S.L., Bruna E.M., Stouffer P.C., Williamson G. Bruce, Benítez-Malvido J., Vasconcelos H.L., Van Houtan K.S., Zartman C.E., Boyle S.A., Didham R.K., Andrade A. & Lovejoy T.E. 2011: The fate of Amazonian forest fragments: a 32-year investigation. – *Biol. Conserv.* 144 (1), 56–67, https://doi.org/10.1016/j.biocon.2010.09.021

Lawton J.H. 1976: The structure of the arthropod community on bracken. – *Bot. J. Linn. Soc.* 73, 187–216

Li W., Liu P., Yang N., Chen S., Guo X., Wang B. & Zhang L. 2024: Improving landscape connectivity through habitat restoration: application for Asian elephant conservation in Xishuangbanna Prefecture, China. – *Integrative Zoology* 19 (2), 319–335. https://doi.org/10.1111/1749-4877.12713

Lovejoy, T.E., Bierregaard R., Rylands A., Malcolm J.R., Quintela C.E., Harper L.H, Harper Jr, K.S., Powell G.V.N., Schubart H.O.R. & Hays M.B.. 1986: *Edge and other effects of isolation on Amazon forest fragments.* – In: Soulé M.E. (ed.): Conservation Biology: The Science of Scarcety and Diversity, 257–285. Sinauer Associates, Sunderland, M.A.

Luther D.A., Cooper W.J., Wolfe J.D., Bierregaard Jr R.O., Gonzales A. & Lovejoy T.E. 2020: Tropical forest fragmentation and isolation: Is community decay a random process? – *Global Ecology and Conservation* 23, September 2020, https://doi.org/10.1016/j.gecco.2020.e01168.

Lutz D. 2010: The great pond experiment. A seven-year experiment shows that pond communities bear a lasting imprint of random events in their past. – Newsroom in The Source, Washington University in St. Louis. https://source.washu.edu/2010/05/the-great-pond-experiment/

MacArthur R.H. & Wilson E.O. 1967: *The Theory of Island Biogeography.* – Princeton University Press, Princeton, N.J., 203 p.

MacArthur R.H., Diamond J.R. & Karr J.R. 1972: Density compensation in island faunas. – *Ecology* 53, 330–342

MacDonald D.W., Ball F.G. & Hough N.G. 1980: The evaluation of home range size and configuration using radio tracking data. – In: Amlaner C.J & MacDonald D.W. (eds.) 2013: *A handbook on biotelemetry and radio tracking.* Proceedings of an International Conference on Telemetry and Radio Tracking in Biology and Medicine, Oxford, 20–22 March 1979, 405–424. Pergamon Press, Oxford, 824 p.

Mader H.-J. 1983: Warum haben kleine Inselbiotope hohe Artenzahlen? – *Natur und Landschaft* 58 (10), 367–370

Mader H.-J. 1985: Welche Bedeutung hat die Vernetzung für den Artenschutz? – *Schr.-R. Deutscher Rat für Landespflege* 46, 631–634

Margules C.R. 1996: Experimental Fragmentation. p. 128-137. – In: Settele J., Margules C.R., Poschlod P. & Henle K. (eds.): *Species Survival in Fragmented Landscapes.* Kluwer Academic Publishers, 381 p.

Miles M.A., de Souza A.A. & Povoa M.M. 1981: Mammal tracking and nest location in Brazilian forest with an improved spool-and-line device. – *Journal of Zoology*, London, 195, 331–347

Millspaugh J. & Marzluff J. 2001: *Radio Tracking and Animal Populations.* – New York, Harcourt Publishers. 400 p.

Monsalve J.L.P. 2009: Imaginaries, Hunts and Trade of Jaguar in the Basins of the Caquetá-Japurá and Putumayo-Içá (Amazonia), During the 20th

Century. – Thesis Universidad Nacional de Colombia, Bogota, 145 p.
Mühlenberg M. 1985: Verkleinerung der Lebensräume von Pflanzen und Tieren durch Zerschneidung der Kulturlandschaft. – *Forschung zur Raumentwicklung (Bonn)* 14, 93–104
Mühlenberg M. 1988: Konzeptentwicklung und Möglichkeiten praktischer Umsetzung von Biotopverbundsystemen. – In: Biotopvernetzung in der Kulturlandschaft II, *Schriftenreihe Angewandter Naturschutz*, Lieh, 5, 14–31
Mühlenberg M. 1993: *Freilandökologie*. - UTB für Wissenschaft 595, 3rd ed.. Quelle & Meyer Verlag Heidelberg, Wiesbaden, 512 p.
Mühlenberg M. & Werres W. 1983: Lebensraumverkleinerung und ihre Folgen für einzelne Tiergemeinschaften. Experimentelle Untersuchung auf einer Wiesenfläche. – *Natur und Landschaft* 58 (2), 43–50
Mühlenberg M., Leipold D., Mader H.-J. & Steinhauer B. 1977: Island Ecology of Arthropods. I. Diversity, Niches and Resources on Some Seychelles Islands. – *Oecologia* (Berlin) 29, 117–134
Mühlenberg M., Hovestadt T. & Röser J. 1991: Are there minimal areas for animal populations? – In: Seitz A. & Loeschke V. (eds.): *Species Conservation: A Population-Biological Approach*, 227–264. – Birkhäuser Verlag, Basel, 281 p.
Mullu D. 2016: A Review on the Effect of Habitat Fragmentation on Ecosystem. – *Journal of Natural Sciences Research* 6 (15), 1–15
Murphy D.D. & Wilcox B.A. 1986: Butterfly Diversity in Natural Habitat Fragments: A Test of the Validity of Vertebrate-Based Management. – In: Verner, J., Morrison, M. L. & Ralph, C. J. (eds.): *Wildlife 2000. Modeling Relationships of Terrestrial Vertebrates*, 287–292. – University of Wisconsin Press, Wisconsin u. London, 470 p.
Naef-Daenzer B. 1993: A New Transmitter for Small Animals and Enhanced Methods of Home-Range Analysis. – *Journal of Wildlife Management* 57(4), 680-689
Nicholls A.O. & Margules C.R. 1991: The design of studies to demonstrate the biological importance of corridors. – In: Saunders, D. A. & Hobbs, R. J.: *Nature Conservation 2: The role of corridors*, 49–61. Surrey Beatty & Sons Pty Limited, Chippping Norton, Australia, 442 p.
Niering W.A. 1963: Terrestrial Ecology of Kapingamarangi Atoll, Caroline Islands. – *Ecological Monographs* 33 (2), 131–160. https://doi.org/10.2307/1948559
Noss R.F. 1987: Corridors in Real Landscapes: A Reply of Simberloff and Cox. – *Conservation Biology* 1 (2), 159–164
Noss R.F. 2007: Focal Species for Determining Connectivity Requirements in Conservation Planning. p. 263–279. – In: Lindenmayer D.B. & Hobbs R.J. (eds.): *Managing and Designing Landscapes for Conservation: Moving from Perspectives to Principles*. – Blackwell Publishing, 587 p.
Noss R.F. & Daly K.M. 2006: Incorporating connectivity into broad-scale conservation planning. Pages 587–619: – In Crooks K.R. & Sanjayan M. (eds.) *Connectivity Conservation*. Conservation Biology Vol. 14, Cambridge University Press, 732 p.
Patterson B.D. 1987: The Principle of Nested Subsets and Its Implications for Biological Conservation. – *Conservation Biology* 1 (4), 323–334
Pe'er G., Tsianou M.A., Franz K.W., Matsinos Y.G., Mazaris A.D., Storch D., Kopsova L., Verboom J., Baguette M., Stevens V.M. & Henle K. 2014: Toward better application of minimum area requirements in conservation planning. – *Biological Conservation* 170, 92–102
Pechmann J.H.K., Estes R.A., Scott D.E. & Gibbons J.W. 2001: Amphibian colonization and use of ponds created for trial mitigation of wetland loss. – *Wetlands* 21 (1), 93–111
Picard C. (2015). Conserving Tanzania's Wildlife: What is the Policy Problem? - In: Clark S., Hohl A., Picard C., Thomas E. (eds.): *Large-Scale Conservation in the Common Interest*. - Springer Series on Environmental Management. Springer, Cham. https://doi.org/10.1007/978-3-319-07419-1_8
Pimm S.L., Jones H.L. & Diamond J. 1988: On the risk of extinction. – *The American Naturalist* 132 (6), 757–785
Pintar M.R. & Resetarits Jr. W.J. 2020: Aquatic beetles influence colonization of disparate taxa in small lentic systems. – *Ecol. Evol.* 10 (21), 12170–12182
Poloczek A., Mühlenberg M. & Stürmer I.W. 2012: Uniformity in a Diversity of Landscapes—Branchiopod Communities in Eastern and Central Mongolia. – *Research on biological resources of Mongolia (Halle/Saale)* 12, 275–286
Quinn J.F. & Harrison S.P. 1988: Effects of habitat fragmentation and isolation on species richness: evidence from biogeographic patterns. – *Oecologia* (Berl.) 75, 132–140
Quinn J.F. & Robinson G.R. 1987: The effects of experimental subdivision on flowering plant diversity in a California annual grassland. – *Journal of Ecology* 75, 837–856
Rabinowitz A. 1986: Jaguar: One Man's Struggle to Establish the World's First Jaguar Preserve. New York: Arbor House, 368 p.
Rabinowitz A. & Zeller K.A. 2010: A range-wide model of landscape connectivity and conservation for the jaguar, *Panthera onca*. – *Biological Conservation* 143, 939–945
Randolph S.E. 1977: Changing spatial relationships in a population of *Apodemus sylvaticus* with onset of breeding. – *Journal of Animal Ecology* 46, 653–676
Redfern M. 1968: The natural history of spear thistle-heads. – *Field Studies, London*, 2 (5), 669–717

Redfern M. 2011: *Insects and Thistles*. – Naturalists' Handbook, 4, Vol. 4., Richmond Publishing Co Ltd, 70 p.

Reeve N.J. 1982: The home range of the Hedgehog as revealed by a radio tracking study. – *Symp. zool. Soc. Lond.* 49, 207–230

Robles H. & Ciudad C. 2012: Influence of Habitat Quality, Population Size, Patch Size, and Connectivity on Patch-Occupancy Dynamics of the Middle-Spotted Woodpecker. – *Conservation Biology* 26, 284–293

Samietz J., Berger U. & Köhler G. 1996: A population vulnerability analysis of the stripe-winged grasshopper, *Stenobothrus lineatus* (Caeifera: Acrididae). p. 299-311. – In: Settele J., Margules C.R., Poschlod P. & Henle K. (eds.): *Species Survival in Fragmented Landscapes*. Kluwer Academic Publishers, 381 p.

Saunders D.A. & Rebeira C.P. 1991: Values of corridors to avian populations in a fragmented landscape. – In: Saunders D.A. & Hobbs R.J.: *Nature Conservation 2: The role of corridors*, 221–240. Surrey Beatty & Sons Pty Limited, Chipping Norton, Australia, 442 p.

Schaefer M. 1974: Auswirkung natürlicher und experimenteller Störungen in Grenzzonen von Ökosystemen, untersucht am Beispiel der epigäischen Arthropodenfauna. – *Pedobiol.* 14, 51–60

Schrader J., Moeljono S., Keppel G. & Kreft H. 2019: Plants on small islands revisited: the effects of spatial scale and habitat quality on the species–area relationship. – *Ecography* 42, 1405–1414

Schrader J., König C., Triantis K.A., Trigas P., Kreft H. & Weigelt P. 2020: Species–area relationships on small islands differ among plant growth forms. – *Global Ecology and Biogeography* 29, 814–829

Schrader J., Shi P., Royer D.L., Peppe D.J., Gallagher R.V., Li Y., Wang R. & Wright I.J. 2021: Leaf size estimation based on leaf length, width and shape. – *Annals of Botany*, 128, 395–406. https://doi.org/10.1093/aob/mcab078

Schrader J., Wright I.J., Kref, H., Weigelt P., Andrew S.C., Abbott I. & Westoby M. 2023: ETIB-T: An Equilibrium Theory of Island Biogeography for plant traits. – *Journal of Biogeography* 50(1), 223–234. https://doi.org/10.1111/jbi.14526

Seitz A. 1988: Microgeographic Variation of Genetic Polymorphism in *Argyresthia mendica* (Lep.: Argyresthiidae). – In: de Jong G. (ed.): *Population Genetics and Evolution*, 202–208. – Springer-Verlag, Berlin, Heidelberg, 288 p.

Selander R.K., Smith M.H., Yang S.Y., Johnson W.E. & Gentry J.B. 1971: Biochemical polymorphism and systematics in the genus *Peromyscus*. I. Variation in the old-field mouse.—In: *Studies in Genetics* 6, 49–90. (Univ. Tex. Publ. 7013.)

Sfenthourakis S. & Triantis K.A. 2009: Habitat diversity, ecological requirements of species and the Small Island Effect. – *Diversity and Distributions* 15 (1), 131–140. https://doi.org/10.1111/j.1472-4642.2008.00526.x

Shogren, A. J., Tank, J. L., Andruszkiewicz, E., Olds, B., Mahon, A. R., Jerde, C. L., & Bolster, D. (2017). Controls on eDNA movement in streams: Transport, Retention, and Resuspension. *Scientific Reports*, *7*(1), 5065. https://doi.org/10.1038/s41598-017-05223-1

Shogren, A. J., Tank, J. L., Hanrahan, B. R., & Bolster, D. (2020). Controls on fine particle retention in experimental streams. *Freshwater Science*, *39*(1), 28–38. https://doi.org/10.1086/707396

Simberloff D.S. & Abele L.G. 1976: Island biogeography theory and conservation practice. – *Science* 191, 285–286

Slatkin M. 1985: Gene flow in natural populations. – *Ann. Rev. Ecol. Syst.* 16, 393–430

Stelter C., Reich M., Grimm V. & Wissel C. 1997: Modelling persistence in dynamic landscapes: lessons from a metapopulation of the grasshopper *Bryodema tuberculosa*. – *Journal of Animal Ecology* 66, 508–518

Stubbe C., Ahrens M., Stubbe M. & Goretzki J. 1995: *Live Capture of Wild Animals*. Capture techniques, methods, experiences. – German Agricultural Publishing House Berlin, 208 p.

Temple S.A. 1986: Predicting Impacts of Habitat Fragmentation on Forest Birds: A Comparison of Two Models. – In: Verner J., Morrison M.L. & Ralph C. J. (eds.): *Wildlife 2000, Modeling Habitat Relationships of Terrestrial Vertebrates*, 301–304. – University of Wisconsin Press, Wisconsin u. London, 470 p.

Temple S.A. & Cary J.R. 1988: Modeling Dynamics of Habitat-Interior Bird Populations in Fragmented Landscapes. – *Conservation Biology* 2 (4), 340–347

Thomas C.D. 1990: What do real population dynamics tell us about minimum viable population sizes? – *Conservation Biology* 4 (3), 324–327

Tomkiewicz S.M., Fuller M.R., Kie J.G., & Bates K.K. 2010: Global positioning system and associated technologies in animal behaviour and ecological research. – *Philosophical Transactions of the Royal Society of London B: Biological Sciences* 365, 2163–2176.

Triantis K.A. & Sfenthourakis S. 2012: Island biogeography is not a single-variable discipline: The small island effect debate. – *Diversity and Distributions* 18, 92–96. https://doi.org/10.1111/j.1472-4642.2011.00812.x.

Triantis K.A., Guilhaumon F. & Whittaker R.J. 2012: The island species-area relationship: biology and statistics. – *Journal of Biogeography* 39 (2), 215–231

Tscharntke T., Tylianakis J., Rand T., Didham R., Fahrig L., & Batary P., Bengtsson J., Clough Y., Crist T., Dormann C., Ewers R., Fründ J., Holt R., Holzschuh A., Klein A., Kleijn D., Kremen C., Landis D., Laurance W. & Westphal C. 2012: Landscape moderation of biodiversity patterns and processes-eight hypotheses. – *Biological reviews* 87(3), 661–685. https://doi.org/10.1111/j.1469-185X.2011.00216.x

Van Dyke F. 2008: *Conservation Biology. Foundations, Concepts, Applications*. 2nd ed.—Springer, 477 p.

Way J.M. & Greig-Smith P.W. 1986: *Field Margins*. – BCPC Monograph Nr. 35. BCPC Publications, Thomton Heath, 128 p.

Weiss S.B., Murphy D.D. & White R.R. 1988: Sun, Slope, and Butterflies: Topography Determinants of Habitat

Quality for *Euphydryas editha*. – *Ecology* 69(5), 1486–1496

White, G.C. & Garrott R.A. 1990: *Analysis of Wildlife Radio-Tracking Data*. – Academic Press Inc., San Diego, California, 383 p.

Wilcox B.A. & Murphy D.D. 1985: Conservation strategy: the effect of fragmentation on extinction. – *Amer. Nat*. 125, 879–887

Wildlife Drones Minimercial.mp4

wildlifedrones.net, The main address: CEAT, Gould Building, Daley Road, Acton, ACT 2601, Australia;

Wood P.A. & Samways M.J. 1991: Landscape element pattern and continuity of butterfly paths in an ecologically landscaped botanic garden, Natal, South Africa. – *Biological Conservation* 58, 149–166

Zeller K.A. 2007: *Jaguars in the New Millenium Data Set Update: The State of the Jaguar in 2006*. - Wildlife Conservation Society, Bronx, New York, 1–77

Zwölfer H., Bauer G. & Heusinger G. 1981: *Ökologische Funktionsanalyse von Feldhecken* – Tierökologische Untersuchungen über Struktur und Funktion biozönotischer Komplexe. – Lehrstuhl Tierökologie, Universität Bayreuth, 422 p.

Niche Differentiation Between Related Species

12

Abstract

Niches describe how a species utilizes existing resources. Niches are divided into different dimensions. Guilds are used to group species that use resources similarly. Examples of how birds search for food in vegetation are discussed. A distinction is made between micro-locations and foraging behavior. A classification of foraging behavior is presented. Closely related and ecologically similar pairs of species offer themselves as objects of observation and comparison. Species pairs of sparrows and primates are selected as examples. The niche differentiation of damselflies at water bodies is discussed.

12.1 Niches and Guilds

The chapters on feeding ecology ecology (Chap. 7) and habitat selection (Chap. 9) also addressed the topic of competition. The underlying question is how various but seemingly similar species coexist. This question can be examined by comparing the "niches" of the coexisting species. The niche of a species describes its location in its environment and its use of the available resources, which corresponds to a multidimensional space (detailed explanations can be found, for example, in Begon et al. 2006, Carscadden et al. 2020). A niche can be quantified by dividing it into different, individually accessible dimensions. The entire multidimensional space cannot be captured. We want to focus on a spatial dimension (location) and a dimension that describes food intake (food resource). In the spatial dimension, we must structurally divide the species' habitat into different units. To capture the food resource, we must quantitatively describe the supply and use (temporally and quantitatively). If the resource is limited, ecologically similar species can only coexist if they differ in how they use resources. This is referred to as "niche differentiation."

We want to approach this topic with bird observations in forests. Species that exploit the same resources in a similar way are commonly grouped into "guilds." In guilds, it is usually only one niche dimension that differentiates the species' resource use. This makes investigations easier.

Bird species can be grouped and evaluated regarding their food ecology. According to Wartmann and Furrer (1978), a guild division separates carnivorous and herbivorous ground birds, trunk climbers, carnivorous and herbivorous tree birds, perch hunters for vertebrates and insects, and flight hunters. Another classification distinguishes between insectivores, nectarivores, frugivores, omnivores, carnivores, and granivores (Shafie et al. 2023). In aquatic habitats, you can add a piscivorous and planktivorous food guild. Gonzáles-Salazar et al. (2014) arrived at an even finer classification of the food guilds by considering the main food items, the substrate,

M. Mühlenberg, T. Waßmer, *Field Ecology*, https://doi.org/10.1007/978-3-031-87547-2_12

the technique of food acquisition, and the birds' activity patterns.

Apparent differences in their vegetation structure requirements will be shown, for example, between the guilds of trunk climbers (forest and garden creepers, nuthatches, and woodpecker species) and the carnivorous tree birds ("foliage gleaners," like the tit species, goldcrests, and leaf warbler species). Both carnivorous and herbivorous food choices can be further divided into, e.g., arthropods/mollusks or vertebrates, and dry seeds or fleshy fruits, respectively. Food search techniques can be differentiated differentiated as presented in Sect. 12.2.

Open nesters (canopy, low shrub layer, or ground nesters), cavity nesters, and resident and migratory birds are possible further classification criteria for the bird fauna to investigate specific questions and enabling quantitative habitat description. The application and usefulness of the guild concept for capturing species communities were discussed by Simberloff and Dayan (1991), Andersson and McPhearson (2018), Blaum et al. (2011), Branco et al. (2023), Gonçalves-Souza et al. (2023), and Voigt et al. (2007).

A causal analysis of the occurrence of bird guilds based on specific vegetation structures can be initiated when individual birds can be observed foraging at small structures on different tree species , e.g., bark structure, branches, seed and bud supply, tall or short herbaceous vegetation covers, and bare soil areas. Birches and pines provide birds food with their seeds, cherry trees with their buds, and alders with their fruits from previous years (see Fig. 12.1). A popular food supply at certain times of the year is the finely structured branchwork of birches in which blue tits and chiffchaffs search for aphids and other small arthropods. Tall herbaceous vegetation is preferred by seed-eating finches in September and October (Bezzel 1980). The observed microstructures offer only some of the requisites of bird individuals, and the seasonal aspect of their availability must be particularly considered (Bezzel 1980).

In North American forests, a significant correlation between the vegetation structure and the occurrence of birds cannot always be expected (e.g., Brush and Stiles 1986). However, the authors found a significant dependence of insectivorous bird densities on arthropod biomass, determined by branch-knocking sampling (Sect. 2.6). They also observed that this foraging guild reacted to changing arthropod biomass by changing their habitats and seeking out areas with a better food supply. The arthropod density in forests depends clearly on the dominance of a few plant species and the phenology of the trees to which birds adjust their activity schedule accordingly (Van Balen 1973, see also Naef-Daenzer and Keller 1999). We gain more knowledge for species protection when we compare individuals of selected bird species in terms of their fitness in differently structured habitats.

12.2 Ecological Segregation of Songbirds in Foraging for Food

From observations of the vegetation structure and the bird fauna, we can infer that a richer resource offer is correlated with a more diverse animal species community. This correlation was already quantified by MacArthur (1972) (see also Sect. 11.2). We want to know which mechanisms enable the coexistence of ecologically similar species. The mechanisms leading to the ecological isolation of species can consist of spatial or temporal segregation or different uses of available resources. For example, we may want to study different resource usage in the food choices of insectivorous songbirds. Such differences can occur in the same area and at the same time but vary in the following: (a) morphological differentiation (e.g., different beak lengths), lengths: a famous example are the Darwin's finches on Galapagos, which differ mainly in the shape of their beaks (Grant and Grant 2024, Fig. 12.2), (b) different behavior (e.g., food search by slowly scanning a place or by quick and frequent change of location) and (c) food search in different microhabitats. We try to address the following questions through quantified field observations:

- Which insectivorous songbird species co-occur simultaneously in the study area?
- Based on what criteria regarding their food acquisition can the birds be assigned to different ecological groups?

Fig. 12.1 (**a**) In winter, Eurasian Siskins (*Carduelis spinus*) feed on Black Alder (*Alnus glutinosa*) seeds from the fruits of the previous year (**b**) Male Eurasian Siskin on the fruits of the Black Alder, Lower Saxony, Germany

Fig. 12.2 (**a**) Green Warbler Finch (*Certhidea olivacea*) from Galapagos Islands. Photo credit Eva Mühlenberg-Horn. (**b**) Woodpecker Finch (*Cactospiza pallida*) from Galapagos Islands. Photo credit Eva Mühlenberg-Horn

- Where do individual species seek food within their habitat?
- How long do the observed individuals stay at certain places while foraging?
- How often do they change location while foraging?
- What distance does an individual typically cover when it moves from one place to another while foraging?
- Does the distance depend on the current success of prey acquisition?
- What different behaviors do closely related species show while seeking their food?
- Do the differences in spatial separation outweigh those of selecting different prey?

The methods to develop an inventory of the songbird fauna were described above (Sect. 1.8). For our current research questions we will be focusing our observations on a few closely related species pairs consulting the literature for their ecological characterization, e.g., in central Europe, the great and blue tit (*Parus major* and *P. caeruleus*), coal tit and blue tit (*Parus ater* and *P. caeruleus* (cf. Partridge 1976, Alatalo and Moreno 1987, Tjernshaugen 2021), and the willow warbler and wood warbler (*Phylloscopus trochilus* and *P. sibilatrix*), and garden warbler and blackcap (*Sylvia borin* and *S. atricapilla*) (e.g. Gibb 1954, Newton 1972, Perrins 1979, Moore 1983).

Additionally, morphological differences can be studied in advance using relevant bird handbooks. On their significance for ecological differentiation, consult, for example, Leisler (1980a, b), Leisler and Thaler (1982), and Leisler et al. (1987).

The study area is divided into different zones. The individual strata of the habitat serve as the basis for this division. A finely graded division is desirable. Trees are differentiated by species and divided into various vertical and horizontal zones, considering the growth form (e.g., leaf-free trunk zone, old leaf zone, zone of young shoots) (examples in MacArthur 1958, Alatalo and Alatalo 1979, Herrera 1978).

The behavior of the birds during foraging should be quantitatively recorded by dividing it into different behavioral patterns (see Dawkins (2007) on the quantification of behavioral patterns). For example, one could choose foraging patterns such as (a) short flights to catch passing insects (flycatching) or (b) catching insects moving on a substrate (perch hunters, or sallying), (c) hovering in the air in front of plant parts (hovering), (d) collecting live insects on the top and bottom of leaves (foliage gleaning), (e) searching for prey by digging, pecking or flushing it out from hiding places (unearthing, insistent-pecking). A distinction is also made between long-lasting foraging at one place or quick location changes during foraging with small or large distances. The above categories can be grouped into two strategies of prey acquisition: active searching (c, d, e) or waiting in ambush (sit-and-wait strategy; a, b).

It is best to take voice notes during observations using a recorder or dictaphone. You can also use a voice recorder on a smartphone or specialized apps to record animal behaviors, such as Animal Behaviour Pro (https://research.kent.ac.uk/lprg/software, only for iPhones) or Behayve (https://www.behayve.com, available for iPhone and Android phones). If no such device is available, tables with corresponding rows and columns can be prepared in advance so that only lines or numbers need to be entered during the observations. The case study on the sampling of squirrel behavior in Sect. 9.4 provides an example of focal animal sampling using a smartphone. The observation times of all behaviors relating to the acquisition of food are summed up for later comparisons. For statistical evaluations, the time units serve as samples. Suppose food search as a behavioral element is to be considered in connection with other activities. In that case, the observations of individual animals must be broken down into time units, e.g., 2- or 5-minute sections, depending on how long on average can an individual be continuously observed. The denser the vegetation is, the shorter the time unit. However, comparing data between different observers becomes more difficult with shorter time units. To capture the distribution of primary activities, such as resting, eating, grooming, and wandering, longer time units (5 or 10 min) are more suitable. For the time samples, it is best to move along a transect. Longer observations from one point often lead to an overemphasis on particularly noticeable behaviors (systematic error). In evaluating the morphological adaptation to food search, it must be considered that the adaptation ensures reproductive success in the summer months and survival in the winter months with often changed feeding habits and varying time spent in different habitats (Berthold 1976, Laursen 1978, Martin 1987). Davies (1977) has also demonstrated a behavior change in the food search of the Spotted Flycatcher (*Muscicapa striata*), depending on the amount of food supply. In addition, the food supply of flying insects is weather-dependent (cf. Lewis and Taylor 1964).

An evaluation of the observations can be calculated as niche values (width and overlap), as described by Colwell and Futuyma (1971), Pielou (1972), Hanski and Koskela (1977), summarized in Southwood and Henderson (2000 and Krebs (1998, 2014) (see Sect. 15.11, Niche calculation). Spatial subdivisions of a habitat (different strata, tree areas, see above) and classified behavior patterns (searching leaves, rummaging in the bark or on the ground, waiting for passing insects, see above) constitute the niche dimensions habitat and behavior. Chapter 7, Foraging Ecology, informs about the possibilities for identifying food objects.

Another interesting question is how migratory birds niche into the tropical and subtropical bird communities in winter and coexist with the resident avifauna. Does competition with resident species from the same guild occur? While evidence of competition is difficult to obtain (see Sect. 7.7), the calculation of niche overlap between Palaearctic or Nearctic migrants and similar resident species allows conclusions about differences in foraging flexibility (Salewski et al. 2003). To measure niches, behavioral elements of the selected species are quantified. For this purpose, individual observations of foraging behavior are summed up. As suggested above, food acquisition is divided into classes (duration of sallying, hovering, foliage gleaning, etc.). Another example of niche separation between four closely related syntopic herons has been studied in the tropical wetlands in India (Abbasi and Khan 2023).

12.3 Studies of Species Pairs Concerning Niche Differentiation

The two closely related sparrow species *Passer domesticus* (House Sparrow) and *Passer montanus* (Tree Sparrow) are holarcticly distributed and occasionally occur together. Therefore, this species pair can be studied for ecological differences in different regions. While there is little difference in food intake, the Tree Sparrow (both sexes similarly colored) usually lives in permanent monogamy, while the House Sparrow (sexes differently colored) lives in polygamy. Both species are cavity nesters and accept nest boxes. They can be found in settlements and their edges to agricultural landscapes, parks, and loose forests (Fig. 12.3).

In a selected study area where both species occur, the following questions could be asked:

1. Which species nests more frequently near houses?
2. Where do the two species forage for food?

Fig. 12.3 Tree Sparrows (*Passer montanus*) near a small settlement in the north of Mongolia

3. Does one of the two species live more frequently in groups?
4. Are there places where both species look for food?
5. Are there signs of competition between the two species?
6. If only the House Sparrow occurs, which landscape parts (habitats) are missing?

1. If possible, we observe the places where the sparrows nest during the breeding and feeding season. A systematic approach would be to hang, for example, four suitable nest boxes on a house wall in early spring (from mid-March on) and then at distances of 20 m, 40 m, and 60 m from any house wall. Nest boxes should be at least 5 m apart from each other. Nest boxes with a flight hole width of 30 × 45 mm (oval) are particularly suitable for sparrows. Sparrows usually dominate the nest boxes before tits can take them over. Suppose sparrows occupy many nest boxes. In that case, a statistical check of the hypothesis that Tree Sparrows nest further away from houses is easy to carry out (see statistical note Sect. 15.9).
2. The "focal animal sampling" method can be used for individuals and the "group scan sampling" is used for groups - both in fixed time units (Altmann (1974), Dawkins (2007) and see Sect. 9.4) to observe birds while they are

foraging for food enabling quantitative analyzes. The location references to the settlement edge, microhabitats, and plant parts or soil surface are also prepared as categories in a recording table (or as categories in one of the above mentioned apps). If in a time unit (e.g., 30 min, 20 min, or 10 min) the focal individuals or the focal group gets out of sight, this time must be registered as "time out" until the individuals are seen again. Only the total times of direct observations are compared. For statistics, only the average number of feeding behaviors in a particular category matters.
3. In the observation protocol (see (2)), it is also recorded whether the observed individual is searching for food alone or whether other food-seeking individuals of the same species surround it.
4. Procedure as in (2).
5. Competition arises when a resource used by two or more species is limited. Actual competition is difficult to verify, as one must conduct experiments with different food supplies or different densities of animal individuals (e.g., Wise 1981; Jacobs 1985; May 1986). A model study of this kind was conducted by Schaefer (1978) on spiders (see Sect. 7.7). More straightforward than indicating competition would be to observe a possible dominance of one species over the other, e.g., house sparrow dominates field sparrow (hypothesis). For all observations where both species are seen simultaneously, one notes the frequency of neutral behavior, species A displaces B, or B displaces A (Martin and Bateson 2007). However, observations of aggression at the food source alone are also insufficient to indicate competition. Often, one species is displaced by another in using a resource or even spatially, without the second seeming to have a visible advantage over the absence of the first species. We addressed similar issues in Sect. 7.9 on insect floral visitation. Field experiments can be easily performed to analyze niche differentiation, competition, and dominance with artificial food sources for ants. Such studies were presented in Sect. 7.7.
6. Classify the environment of the settlement edge into different habitat categories: buildings, paths or roads, vegetated road edges, cultivated farmland and fallow, field edge strips, gardens with and without tree populations, hedges, green spaces with tall grass or mowed. A quantification is described in Sect. 4.1.

12.4 Studies on Trophic Niche Overlap of Two Sympatric Primates

Another example of the investigation of niche differentiation can be primates in the tropics. The habitats and food resources of two related species often overlap, raising the question of whether sympatry is made possible by different food utilization, an abundant food supply, or whether cohabitation is enforced by avoiding predation pressure. In order to estimate the availability of food resources, the composition of the forest must be described in terms of tree species diversity. Data on the two selected primate species are obtained by focal sampling (focal animal sampling as described above and in Sect. 9.4). The focal sampling results lead to the quantified observation of individuals from different groups. Data on home range size, utilization of the vertical stratification of the forest layers, activity patterns, and food choice are obtained from the observation hours with individual animals. From this, niche width and overlap values are calculated using standard methods (see above and Sect. 15.11).

Hadi studied groups of two sympatric colobine species on Mentawai Island on the West coast of Sumatra (Hadi et al. 2012). He showed that the smaller species (*Presbytis potenziani*) occupies a more extensive home range, forages more in the upper canopy layer of the forest, and eats more fruits than the heavier species (*Simias concolor*). There was only a slight niche overlap in food utilization but a complete spatial and temporal overlap of the home ranges in the forest. The results are interesting as both species are endemic to the island, and their populations are endangered.

12.5 Niche Segregation in a Damselfly Community

Several species of damselflies (Zygoptera) often occur simultaneously and sympatrically at a small body of water with vegetation. The composition of the species community depends heavily on the conditions under which the larvae can develop in the waters. Differences occur between waters with and without predatory fish (e.g., McPeek and Brown 2000; Siepielski et al. 2010). Individual adults can easily be caught with an insect net, identified to species level and released. Once the species (up to about five species, e.g., Borisov 2006) have been identified, you can observe whether the individuals show interspecific and/or intraspecific interactions. For the biology and identification of Zygoptera, see, e.g., Corbet (1962, 1999), Dreyer (1986), Needham et al. (2000), Askew (2004), Dijkstra et al. (2020), and Boudot et al. (2021). The species are usually encountered during the egg-laying season, which, in temperate zones, occurs between May and July. Dragonflies and damselflies lay their eggs on aquatic plants. The males wait in places that are suitable for the females to lay their eggs and seize the arriving females for mating (Fig. 12.4).

The central questions for this study are:

- How do several species with the same physique and feeding behavior coexist in the same place?
- Do some males temporarily maintain small territories through intraspecific interactions?
- Is interspecific competition detectable through observation of interspecific aggression?
- Do males recognize the females of their species when they approach (see Miller and Miller 1981; Monetti et al. 2002; Fincke et al. 2007)?

Fig. 12.4 *Coenagrion puella* (Zygoptera): the males guard the egg-laying females

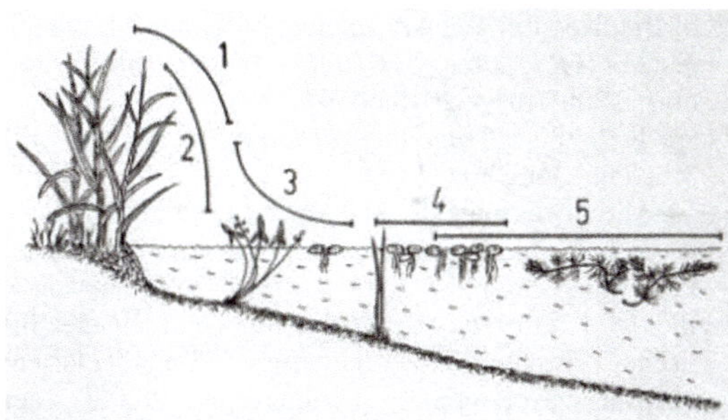

Fig. 12.5 Activity zones of male damselflies (Zygoptera) at a pond. (1) *Coenagrion pulchellum* and *C. puella*. (2) *Pyrrhosoma nymphula*. (3) *Platycnemis pennipes*. (4) *Cercion lindeni*. (5) *Erythromma najas* and *E. viridulum*. Figure credit Wolfgang Dreyer (1986)

- Do intraspecific interactions consist of capturing females and keeping them for mating again?

Wellenreuther and Sánchez-Guillén (2016) were able to show that in some species of damselflies, the ecological differences are quite negligible, but that the species can be recognized by their different genital structures and are thus reproductively isolated (see also Robertson and Paterson 1982; Robinson and Allgeyer 1996).

A sketch of the water's edge can be very helpful for the topographic representation of observations (Fig. 12.5). For an individual analysis of the zygopteran community, adults can be marked with a spot of color applied to their wings (see Sect. 5.3.1). A pair of binoculars that allow focusing up to 2 m away facilitates the observations.

References

Abbasi F. & Khan M.S. 2023: Spatial, temporal and trophic resource partitioning among the four egret species (Aves: Pelecaniformes: Ardeidae) in a tropical wetland ecosystem, India. - *Journal of Threatened Taxa* 15 (11), https://doi.org/10.11609/jott.8429.15.11.24201-24211

Alatalo R.V. & Alatalo R.H. 1979: Resource partitioning among a flycatcher guild in Finland. – *Oikos* (Copenhagen) 33, 46–54

Alatalo R.V. & Moreno J. 1987: Body Size, Interspecific Interactions, and Use of Foraging Sites in Tits (Paridae). – *Ecology* 68 (6), 1773–1777

Altmann, J. (1974). Observational Study of Behavior: Sampling Methods. Behaviour, 49(3–4), 227–266. https://doi.org/10.1163/156853974X00534

Andersson, E., & McPhearson, T. (2018). Making Sense of Biodiversity: The Affordances of Systems Ecology. *FRONTIERS IN PSYCHOLOGY*, *9*. https://doi.org/10.3389/fpsyg.2018.00594

Askew R.R. 2004: *Dragonflies of Europe*. Rev/e2. – Brill Academic Pub, 308 p.

Begon M., Townsend C.R. & Harper J.L. 2006: *Ecology. From Individuals to Ecosystems.* 4th ed. – Blackwell Publishing, 738 p.

Berthold P. 1976: Animalische und vegetabilische Ernährung omnivorer Singvogelarten: Nahrungsbevorzugung, Jahresperiodik der Nahrungswahl, physiologische und ökologische Bedeutung. – *J. Orn.* 117, 145–209

Bezzel E. 1980: Beobachtung zur Nutzung von Kleinstrukturen durch Vögel. – *Akademie f. Naturschutz und Landschaftspflege, Laufen,* Berichte 4, 5–11

Blaum, N., Mosner, E., Schwager, M., & Jeltsch, F. (2011). How functional is functional? Ecological groupings in terrestrial animal ecology: towards an animal functional type approach. *BIODIVERSITY AND CONSERVATION*, *20*(11), 2333–2345. https://doi.org/10.1007/s10531-011-9995-1

Borisov S.N. 2006: Ecological niches of species of the genus *Ischnura* (Odonata, Coenagrionidae) in oases of the Pamir-Alai Mountains. – *Entomological Review* 86, 623–631

Boudot J.-P., Doucet G. & Grand D. 2021: *Dragonflies and Damselflies of Britain and Western Europe:* A Photographic Guide. – Bloomsbury Publishing PLC, 160 p.

Branco, C., Fintelman-Oliveira, E., & Miranda, V. (2023). A review of functional approaches for the study of freshwater communities with a focus on zooplankton. *HYDROBIOLOGIA*, *850*(21), 4719–4744. https://doi.org/10.1007/s10750-023-05227-1

Brush T. & Stiles E.W. 1986: Using Food Abundance to Predict Habitat Use by Birds. – In: Verner J., Morrison M.L. & Ralph C.J. (eds.): *Wildlife 2000. Modeling Habitat Relationships of Terrestrial Vertebrates*, 57–63. University of Wisconsin Press, Wisconsin u. London, 470 p.

Carscadden K.A., Emery N.C., Arnillas C.A., Cadotte M.W., Afkhami M.E., Gravel D., Livingstone S.W. & Wiens J.J. 2020: Niche Breadth: Causes and Consequences for Ecology, Evolution, and Conservation. – *The Quarterly Review of Biology* 95 (3), 179–287

Colwell R. K. & Futuyma D. J. 1971: On the measurement of niche breadth and overlap. – *Ecology* 52, 567–576

Corbet P.S. 1962: *Biology of Dragonflies* (Aspects of Zoology). – Witherby, 264 p.

Corbet, P.S. 1999: *Dragonflies: behaviour and ecology of Odonata*. – Comstock Pub. Associates, 829 p.

Davies N.B. 1977: Prey selection and the search strategy of the Spotted Flycatcher *(Muscicapa striata):* a field study on optimal foraging. – *Anim. Behav.* 25, 1016–1033

Dawkins MS. 2007: *Observing Animal Behaviour. Design and Analysis of Quantitative Data*. – Oxford University Press, 158 p.

Dijkstra K-D.B., Schröter A. & Lewington R. 2020: *Field Guide to the Dragonflies of Britain and Europe* (2nd ed.). – Bloomsbury Publishing, London, 336 p.

Dreyer W. 1986: *Die Libellen*. – Gerstenberg Verlag, Hildesheim, 219 p.

Fincke O., Fargevieille A. & Schultz T. 2007: Lack of innate preference for morph and species identity in mate-searching *Enallagma* damselflies. – *Behavioral Ecology and Sociobiology* 61, 1121–1131

Gibb J. 1954: Feeding ecology of tits, with notes on Treecreeper and Goldcrest. – *Ibis* 96, 513–543

Gonçalves-Souza, T., Chaves, L., Boldorini, G., Ferreira, N., Gusmao, R., Perônico, P., et al. (2023). Bringing light onto the RaunkiÆran shortfall: A comprehensive review of traits used in functional animal ecology. *ECOLOGY AND EVOLUTION*, *13*(4). https://doi.org/10.1002/ece3.10016

Gonzáles-Salazar C., Martinez-Meyer E. and López-Santiago G. 2014: A hierarchical classification of trophic guilds for North American birds and mammals. - *Revista Mexicana de Biodiversidad* 85(3), 931–941

Grant P.R. and Grant B.R. 2024: 40 Years of Evolution: Darwin's Finches on Daphne Major Island . (New edition) - Princeton University Press, 464p

Hadi S., Ziegler T., Waltert M., Syamsuri F., Mühlenberg M. & Hodges J.-K. 2012: Habitat Use and Trophic Niche Overlap of Two Sympatric Colobines, *Presbytis potenziani* and *Simias concolor*, on Siberut Island, Indonesia. – *Intern. Journ. of Primatology* 33: 218–232

Hanski I. & Koskela H. 1977: Niche relations among dung-inhabiting beetles. – *Oecologia* 28, 203–231

Herrera C.M. 1978: Ecological correlates of residence and no residence in a Mediterranean passerine bird Community. – *J. Anim. Ecol*. 47, 871–890

Jacobs J. 1985: Competition and niche – Does competition for resources have an evolutionary biological significance for the species diversity of animals? – *Z. zool. Syst. Evolut.-forsch*. 23, 243–258

Krebs, C. J. (1998). Ecological Methodology (2nd edition.). Menlo Park, Calif: Benjamin Cummings.

Krebs, C. J. (2014). Ecological methodology (3rd ed.). Vanvouver: Charles Krebs. https://www.zoology.ubc.ca/~krebs/books.html

Laursen K. 1978: Interspecific relationships between some insectivorous passerine species, illustrated by their diet during spring migration. – *Ornis scand.* 9, 178–192

Leisler B. 1980a: Morphological aspects of ecological specialization in bird genera. – *Ökol. Vögel* 2, 199–220

Leisler B. 1980b: Ökomorphologische Freiland-und Laboratoriumsuntersuchungen. – *Acta XVII Congr. Intern. Orn.*, 202–208

Leisler B. & Thaler E. 1982: Differences in morphology and foraging behaviour in the goldcrest *Regulus regulus* and firecrest *R. ignicapillus.* – *Ann. Zool. Fennici* 19, 277–284

Leisler B., Ley H.W. & Winkler H. 1987: Die Rolle von Lokomotion und Morphologie in der Habitatwahl der Rohrsänger *(Acrocephalus).* – *J. Orn.* 128, 114–117

Lewis T. & Taylor L.R. 1964: Diurnal periodicity of flight by insects. – *Trans. R. ent. Soc. Lond.* 116, 393–476

MacArthur R.H. 1958: Population Ecology of Some Warblers of Northeastern Coniferous Forests. – *Ecology* 39, 599–619. http://campus.lakeforest.edu/menke/PDFs/Bio373/MacArthur_1958_Ecology.pdf

MacArthur R.H. 1972: *Geographical Ecology: Patterns in the Distribution of Species.* – Harper & Row, Publ., New York, 269 p.

Martin T.E. 1987: Food as a limit on breeding birds: A life-history perspective. – *Ann. Rev. Ecol. Syst.* 18, 453–487

Martin P. & Bateson P. 2007: *Measuring Behaviour. An introductory Guide.* – Cambridge University Press, 176 p.

May RM. 1986: The search for Patterns in the Balance of Nature: Advances and Retreats. – *Ecology* 67 (5), 1115–1126

McPeek M.A. & Brown J.M. 2000: Building a regional species pool:: Diversification of the Enallagma damselflies in eastern North America. – *Ecology* 81 (4), 904–920

Miller P.L. & Miller C.A. 1981: Field observations on copulatory behaviour in Zygoptera, with an examination of the structure and activity of male genitalia. – *Odonatologica* 10, 201–218

Monetti L., Sanchez-Guillen R.A. & Cordero-Rivera A. 2002: Hybridization between *Ischnura graellsii* (Vander Linder) and *I. elegans* (Rambur) (Odonata: Coenagrionidae): are they different species? – *Biological Journal of the Linnean Society* 76, 225–235

Moore A. 1983: *A field guide to the warblers of Britain and Europe.* – Oxford University Press, 145 p.

Naef-Daenzer B. & Keller L. 1999: The foraging performance of great and blue tits (*Parus major* and *P. caeruleus*) in relation to caterpillar development, and its consequences for nestling growth and fledging weight. – *Journal of Animal Ecology* 68, 708–718

Needham J.G., Westfall Jr. M.J. & May M.L. 2000: *Dragonflies of North America.* – Scientific Pub, 939 p.

Newton I. 1972: *Finches.* – Collins, London, Limpback edition 1985, 288 p.

Partridge L. 1976: Field and laboratory observations on the foraging and feeding techniques of Blue Tits *(Parus caeruleus)* and Coal Tits *(P. ater)* in relation to their habitats. – *Anim. Behav.* 24, 534–544

Perrins C.M. 1979: *British Tits.* – Collins, London, 304 p.

Pielou E. C. 1972: Niche width and niche overlap: a method for measuring them. – *Ecology* 53 (4), 687–692

Robertson H.M. & Paterson H. 1982: Mate recognition and mechanical isolation in *Enallagma* damselflies (Odonata: Coenagrionidae). – *Evolution* 36, 250

Robinson J.V. & Allgeyer R. 1996: Covariation in life-history traits, demographics and behaviour in ischnuran damselflies: The evolution of monandry. – *Biological Journal of the Linnean Society* 58(1), 85–98

Salewski V., Bairlein F. & Leisler B. 2003: Niche partitioning of two Palearctic passerine migrants with Afrotropical residents in their West African winter quarters. – *Behavioral Ecology* 14(4), 493–502

Schaefer M. 1978: Some Experiments on the Regulation of Population Density in the Spider *Florinia bucculenta* (Araneae: Linyphiidae). – *Symp. Zool. Soc. Lond.* 42, 203–210

Shafie, N.J., Anuar, H., David, G., Ahmad A. and Abdullah M.T. 2023:. Bird species composition, density and feeding guilds in contrasting lowland dipterocarp forests of Terengganu, Peninsular Malaysia. - *Trop Ecol* 64, 238–248 (2023). https://doi.org/10.1007/s42965-022-00267-501863.x

Siepielski A.M., Hung K-L., Bein E.E.B. and McPeek M.A. (2010). Experimental evidence for neutral community dynamics governing an insect assemblage. - Ecology 91(3), 847–857

Simberloff D. & Dayan T. 1991: The guild concept and the structure of ecological communities. – *Ann. Rev. Ecol. Syst.* 22, 115–143

Southwood T.R.E. & Henderson P.A. 2000: *Ecological Methods.* 3rd ed. – Blackwell Science Ltd, 575 p.

Tjernshaugen A. 2021: *Das verborgene Leben der Meisen*, 3rd ed. – Insel Verlag, Berlin. 235 p.

van Balen J.H. 1973: A comparative study of the breeding ecology of the Great Tit, *Parus major,* in different habitats. – *Ardea* 61, 1–93

Voigt, W., Perner, J., & Jones, T. (2007). Using functional groups to investigate community response to environmental changes: two grassland case studies. *GLOBAL CHANGE BIOLOGY, 13*(8), 1710–1721. https://doi.org/10.1111/j.1365-2486.2007.01398.x

Wartmann B. & Furrer R.K. 1978: Zur Struktur der Avifauna eines Alpentales entlang des Höhengradienten. II. Ökologische Gilden. – *Orn. Beob.* 75, 1–9

Wellenreuther M. & Sánchez-Guillén R.A. 2016: Nonadaptive radiation in damselflies. – *Evolutionary Applications* 9, 103–118

Wise DH. 1981: Inter- and Intraspecific Effects of Density Manipulations upon Females of Two Orb-Weaving Spiders (Araneae: Araneidae). – *Oecologia* (Ber.) 48, 252–256

Part III

Focus on Management and Planning

Managing Landscapes and Populations

13

Abstract

Anthropogenic changes in the landscape require an environmental impact assessment of these interventions' effects on the environment, fauna, and flora. Road construction projects are used as examples to show the consequences they can have for endangered species. Compensatory measures are shown. If studies on the future development of endangered populations cannot be made in advance, decision analysis methods are recommended, suggesting the probability of consequences. The adverse effects of tourism, wind farms, removal of deadwood from disturbed forests, and prescribed burning in forest management are discussed using examples.

The management of ecosystems also includes the harvesting of non-timber forest products, which should be sustainable. Examples of fruit collection and bushmeat hunting are provided. Trophy hunting is also part of this type of intervention in animal populations and is discussed with an example. Methods for restoring depleted wildlife populations are demonstrated based on captive breeding and reintroduction.

The management of forest ecosystems determines forest structure and biodiversity. In Central Europe, nature conservation has long favored within-stand habitat heterogeneity for high biodiversity. However, a comprehensive study of the biodiversity of various taxa (e.g., bats, birds, spiders, beetles, Hymenoptera, bugs) shows that biodiversity is higher in even-aged forests than in structured permanent forests (uneven-aged forests with high within-stand heterogeneity) (Schall et al. 2018). Biodiversity was also compared with a nearby national park (unmanaged forest). Diversity accumulation curves and various diversity measures were used for the biodiversity comparison (see also Sect. 15.11). Biodiversity was determined within several 100 x 100 m plots distributed over 400 km^2 areas in each forest type. Within each plot, arthropods were collected using two pitfall traps and four flight-interception traps at two different heights. Where possible, the taxa were separated into forest specialists and non-forest specialists. Another study also showed that forest management (timber harvesting, mixed forest stands) positively affects the long-term population trends of birds (Schulze et al. 2019; King and DeGraaf 2000). However, a distinction must be made between migratory and nonmigratory forest birds. The abundance of migratory bird species is decreasing for other reasons. The long-term trends of bird populations in Central Europe were obtained from the European Bird Census Council (2018).

M. Mühlenberg, T. Waßmer, *Field Ecology*, https://doi.org/10.1007/978-3-031-87547-2_13

13.1 Environmental Impact Assessment

In may countries, an environmental impact assessment (EIA) is required by law before changes in the landscape are planned through construction projects. In Germany, this is regulated by §8 Federal Nature Conservation Act, for the EU, the EIA Directive 2011/92/EU as amended by 2014/52/EU applies (European Commission 2022). In the USA, the legislation that considers environmental changes dates back to the National Environmental Policy Act of 1969 (NEPA), effective as law since 1970, leading to the environmental impact statement (EIS), which is part of the planned actions of the federal agency. However, existing state laws also determine whether an EIS must be applied before a measure. An EIS must be circulated between government agencies and public venues (NGOs, private citizens' associations) for commenting (Glasson and Therivel 2019; Sanford and Holtgrieve 2022; Struthers et al. 2023). The EU directive applies to European countries and is then legally formulated in each country. In each country, it is necessary for an EIA to determine separately which legal requirements (statutory requirements) are created. Questions to be answered are:

- Do international agreements also apply?
- Are protected areas and/or protected species included?
- How is the term "significant" officially interpreted?
- Is monitoring prescribed?

There is also an international organization for EIA, which can be found online at http://www.iaia.org. It can vary between countries whether an EIA is generally required before a measure or whether the EIA is only required when it is believed in advance that the project will lead to "significant" effects on the environment. In the UK, both options exist (Spellerberg 2005). Spellerberg also presents the New Zealand Resource Management Act of 1991 in detail with its fourth schedule: Assessment of Effects on the Environment that also considers effects on the neighborhood of the project area (see below the European Union Habitats Directive on protected areas, which also includes the surroundings of protected areas). "Significant" is legally challenging to formulate. For biologists, a significant impairment would always apply if, through a project, a species has a higher probability of extinction or is threatened with extinction (see Sect. 10.9).

When plans for a project are submitted, and private individuals or groups (e.g., NGOs) suspect that the project will have negative impacts, they can legally enforce a detailed EIS with an objection. An EIS leads to widespread and clearly defined procedures with public participation and possible legal challenges of the EIS. In an EIS, the consequences of interventions must be described, and alternatives must be provided. Although this is only one of several parts of an EIS, it is the essential part for field biologists (see Van Dyke 2008). An EIS is built with a clear description of the proposed action, a detailed description of the environmental change, an estimate of the (long term) effects on the affected environment, and possible alternatives if the planned measure is not carried out (Petts 1999; Mareddy 2017). For biologists, species lists, habitat surveys, extent of protected areas, economically important species, and indicator species are essential within the EIA. At the same time, she or he must present alternatives and mitigation measures, and make predictions based on existing knowledge. If a scientifically based forecast becomes too complex, the path of decision analysis can be taken (see Sect. 13.3 below).

From the various areas that make up an EIS, the field ecologist collects data showing which parts of the flora and fauna will be affected by the planned measure. A planned measure is almost always underpinned by a detailed geographical map, which forms the basis for estimating and discussing the extent (scope, scale) of the environmental effects. Based on the available maps, the mapping of biological objects (e.g., animal species) for the EIS can be planned (for mapping methods, see Sect. 4.4).

The crucial point is that in a lawsuit, i.e., an objection against the planned measures, the argument must adhere to the law's wording and be considered by a court through scientific substantiation. The court can then weigh whether "higher arguments" can justify ignoring a protection provision for a biological species or a habitat, which in most cases only happens when considering human safety or health. For a scientific argument, quantitative results or forecasts are necessary. Since these can almost only refer to individual species (see Sect. 10.9 target species concept), the field ecologist in the EIA will deal with selected species or habitats of legal status. For this purpose, the Red Lists (IUCN), collected in the appendices of the directives (e.g., in Europe, NATURA 2000), are relevant.

The goal is to avoid or compensate for interventions that significantly or sustainably affect the natural balance (Kaule 1991; Usher and Erz 1994). What is considered "significant" or "sustainable" must be adapted to the wording of the respective national laws. It is difficult to predict the effect of the planned measures on the "natural balance" (which also includes the functions of an ecosystem) before an intervention occurs. First, target objects for the examination must be selected. These species are usually "worthy of protection" according to the law, thus worth preserving. The risk analysis (PVA, population viability analysis, see in Sect. 10.11) then assesses a possible threat, which is the basis for estimating the consequences of a change quantitatively and on a scientific basis. If a lawsuit against the intended measures leads to a court case, then judges are bound to the existing law, e.g., a law for protecting a species and scientific forecasts.

We want to illustrate the procedure and argumentation in an EIA using examples of bypass roads and road extensions. More examples can be found in the database International Case Studies on Biodiversity and Impact Assessment (Secretariat of the Convention on Biological Diversity 2024).

Case Study: Effects of Bypass Roads
A bypass road for a town is to be built, which in the planning would run through landscape sections that are also in the interest of nature conservation. The first thing to do is identify sections in the terrain to be visited and map species protected by law along the planned bypass road. In Europe, one should deal with species listed in Annex II and IV of the EU Directive (NATURA 2000, Habitats Directive, Federal Agency for Nature Conservation 2021). Annex II lists animal and plant species of community interest that require particular protection areas for their conservation, while Annex IV lists strictly protected animal and plant species. These species are threatened in the EU or rare, i.e., their populations are small and at least indirectly or potentially threatened, or they are endemic, i.e., they live in limited geographical regions.

For example, if amphibian habitats are affected, amphibian species must be able to move unhindered between spawning and hibernation sites. Suppose a specific butterfly population (species according to Annex IV) lives in the planned road region and can only survive because of immigration/emigration of individuals to and from a neighboring population (gene flow). In that case, it must be ensured that this connection is not interrupted by the new road, which would usually be the case because roads cut populations and cannot be overrun or flown over by small animals (e.g., Mader 1979, 1984; Mader et al. 1990; Weidemann and Reich 1995). Mark-recapture methods or genetic fingerprinting (Gerlach and Musolf 2000) must verify the exchange between the subpopulations (see Sect. 5.3).

The example below shows how a planned bypass road would affect several conservation-relevant species and how mitigation measures would be difficult to achieve. The planned bypass road touches a nature reserve (included in NATURA 2000 according to the EU_FFH-Directive n.d., Directive 92/43/EEC on habitats, flora and fauna, and Directive 79/409/EEC on the protection of birds). It also impacts another adjacent nature reserve (Fig. 13.1a). The

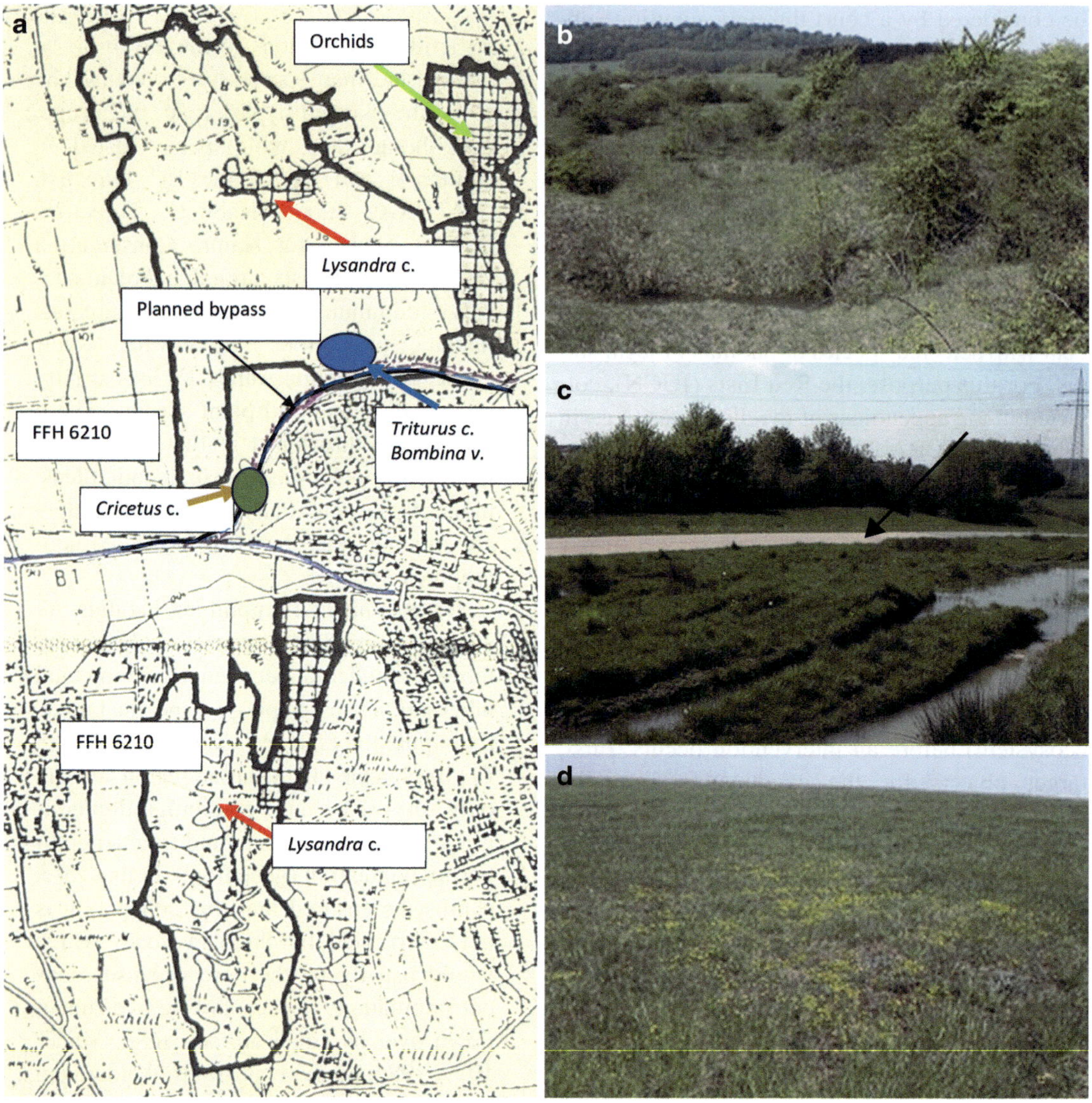

Fig. 13.1 (**a**) Cartographic representation of a planned bypass between two Habitats Directive protected areas (FFH 6210), marked occurrences of conservation-relevant animal and plant species, and photos of the characteristic habitats **b**–**d**. (**b**) Calcareous dry grassland, *Festuco Brometalia,* with Orchids (*Orchis mascula*), *Stachys germanica*, and foodplant *Hippocrepis* for the Lycaenid butterfly. (**c**) Warty newt habitat and Yellow-bellied toad occurrence. The arrow indicates the street proposed to be converted to a 50 m wide bypass. (**d**) Stepping stone for the Chalkhill Blue butterfly and habitat of the European hamster (*Cricetus cricetus*)

EU Directive also includes that changes outside a protected area are considered a compensable intervention if adverse effects on the actual protected area are expected.

For a nature reserve, there are species lists that justify the protected area's designation. From these lists, one must select species and small biotopes that could be affected by the measure. In the EIS, possible mitigation measures must then be presented and discussed. A brief outline of the possible argumentation based on the selected, affected species is presented here (see Fig. 13.1a–d):

Limestone grassland can theoretically be transplanted or newly created, but it takes over a decade to become biotope-typical. A renaturation is, therefore, not possible in the short term (Kiehl 2009). The crested newt (*Triturus cristatus*) is represented in Annexes II and IV of the EU Habitats Directive and is, therefore, strictly protected. It has only a minimal range of action. The planned route would destroy part of the aquatic habitat and the connection to the nearby winter quarters (Fig. 13.1c). Increased mortality is to be expected on the road. A possible underpass for the protection of amphibians does not guarantee the conservation objective of the crested newt population. The likewise threatened yellow-bellied toad (*Bombina variegata*) is also a species protected under Annex IV (Habitats Directive) and is therefore subject to the necessary protection measures. Compensation through the artificial creation of spawning habitats remains uncertain for both species, as they are known not to be easily redirected.

For the Chalkhill Blue (*Lysandra corigon*), the road construction would lead to losing a vital stepping stone biotope, which connects the two subpopulations. The two protected areas must be considered as a network between the habitat types, as a hitherto possible exchange of individuals maintains the respective subpopulations. Also, with further traffic, the open land habitat of this butterfly is additionally endangered by increased harmful and eutrophicating substances (e.g., Günther et al. 2005). The impairment of the exchange of butterfly populations would increase the probability of extinction. For butterflies, it was also shown that individuals from smaller populations are more prone to migrate, leading to an increased probability of allele exchange with another subpopulation compared to significant subpopulations and that these movements are hindered by landscape barriers (e.g., Roland et al. 2000). A more general overview of animal movements and resistances through landscape structures was provided by Zeller et al. (2012).

The European hamster (*Cricetus cricetus)* occurring in the project area is also a "strictly protected species of community interest" (Habitats Directive Annex IV). The project would also destroy the habitat of the local hamster population. It remains uncertain whether relocation to a suitable habitat would be successful in the event of the destruction of its habitat (Mammen and Mammen 2003). For a more detailed coverage of the hamster in the agricultural landscape, see the case study "Conservation of hamsters" below in Sect. 13.2.

The project area is also a "priority" area according to the EU directive because it contains orchids, which would be affected by the planned measures.

Now, we highlight how a large-scale road expansion could severely impair a section of a stream: The work requires cartographic maps of the surrounding area, biotope types, and the usage structure. In addition to determining that, in this case, a "priority" habitat (classified as 91EO) exists, it is again necessary to first characterize the endangered stream section, including the bank fauna, and determine whether species of the Habitats Directive are present. Annex I of the EU Directive lists the habitat types of community interest and, therefore, require particular protection areas for their conservation. The characteristic species that live in the respective priority habitat types are also listed in the European Natura 2000 (Federal Agency for Nature Conservation, Bundesamt für Naturschutz 2021). In our example, floodplain areas are affected, where characteristic species live, such as the protected shore beetles of the genus *Elaphrus* (Carabidae). The planned construction would destroy an essential connection between priority habitats. The evidence of a corridor function in the targeted stream section can be demonstrated by the occurrence of endangered species (in this case, the shore beetles). Floodplain habitats are subject to dynamics, and local extinction events can only be compensated for by recolonization from other local populations via corridors or suitable stepping-stone habitats (see Sect. 11.4). An example of a similar EIS from the US is reported by (Gunther and Winfrey 2005).

EISs should not only refer to one animal group but include as many different orders or animal classes as possible for better argumentation. Thus, bats (Chiroptera) should also be considered in the above case of possible fragmentation of a floodplain landscape. In Europe, almost all bats are protected. Several species patrol along streams for feeding. If these flights are disturbed, the populations of bats are also in danger. Bat hunting flights can be investigated by recording their ultrasonic echolocation sounds, see Sect. 1.12.

Fig. 13.2 Compensation measures for newts include artificial pond facilities (left) and hibernation sites (right)

Fig. 13.3 Wildcat (*Felis silvestris*) rubbing a lure rod. From Viviani et al. (2024). Reproduced by permission

13.2 Compensation Measures

What happens when a new road cuts through the habitat of a wild animal species? In this case, the aim is to show how compensation measures after an EIA can support the affected species.

In some cases, replacement biotopes can be created. For the newts mentioned above in Sect. 13.1, ponds and hibernation habitats must be constructed, as shown in Fig. 13.2.

For larger wild animals, it is not possible to create new habitats but rather to directly try to compensate for effects that impair animals in their habitat. One example is the wildcat (*Felis silvestris*)(Fig. 13.3), whose roaming areas are often affected by new roads. The wildcat is considered "critically endangered" in Germany (Red List category 2). One standard compensation measure are wildlife bridges (Fig. 13.4).

The fragmentation of the habitat for wild cat species by larger roads has already been extensively studied (e.g., Crooks 2002; Herrmann and Mathews 2007; Huck et al. 2010; Dobbert et al. 2021). Road kills on heavily trafficked roads can be reduced by erecting wildcat-specific fences and providing crossing points (Klar et al. 2009). Wildcats are very patchily distributed in Europe. It is, therefore, essential to connect the distributed habitats for genetic exchange (e.g., Klar et al. 2012; Hartmann et al. 2013).

Since constructing a wildlife bridge (Figs. 13.4 and 13.5a) is very expensive (e.g., >2 million Dollars/Euros), it must be proven whether wildcats are present in the area and traverse the region affected by the road construction. Wildcat roaming areas are detected in several steps: (1) Telemetric studies of movement patterns. For this technique, individual animals must be caught alive and tagged (see Sect. 4.6). (2) Installation of wildlife cameras as camera traps (see Sect. 1.2).

As photos cannot clearly distinguish between wild and feral domestic cats, camera traps alone are insufficient. (3) Placement of so-called lure rods in a larger grid. The lure rods consist of rough wooden rods smeared with valerian (*Valeriana officinalis*) extract, attracting the males, who roam around looking for partners (Fig. 13.3). When the cat rubs against the rods, hairs remain stuck, which are then collected and used for genetic analysis (Steyer et al. 2013), which can clearly distinguish between domestic and wild cats (see Oliveira et al. 2007; Hertwig et al. 2009). In addition, genetic analysis provides information about the individual animal and its affiliation with nearby or more distant subpopulations (Eckert et al. 2010).

Fig. 13.4 (**a**) Construction of a wildlife bridge to connect two forest areas, which are part of the wildcat's habitat. FFH Forest of Göttingen, Germany. (**b**) Wildlife bridge over the 4-lane Trans-Canada highway heading eastbound in Banff National Park. Picture credit: Coolcaesar on Wikimedia, CC BY-SA 4.0

Monitoring must be conducted after constructing a wildlife bridge to assess its success. For this purpose, camera traps can easily be installed on the bridge. Figure 13.5b shows an example of success for the overpass over Singapore's 6-lane Bukit Timah Expressway. The construction and subsequent monitoring of this wildlife bridge are described in detail by Chan et al. (2023).

Temporary artificial walls and buckets are erected along roads to minimize the risk of roadkills for migrating amphibians (Beebee 2013, Fig. 13.6). This method is regularly used to protect toads and frogs from the risky crossing of roads in spring. If no helpers are available to carry the amphibians caught in the buckets across the road, the Anurans can also be guided into tunnels by the fences to cross the road safely (e.g., Jochimsen et al. 2004, Schmidt and Zumbach 2008, Hill et al. 2018, Smith et al. 2018, Helldin and Petrovan 2019).

In many cases, there are large-scale barriers to animal migrations. One example should be highlighted: the division of a gazelle population by a railway line in Mongolia. The Mongolian gazelle (*Procapra gutturosa*) lives partly nomadic in large herds with over a thousand animals in the Mongolian steppe (Fig. 13.7) and

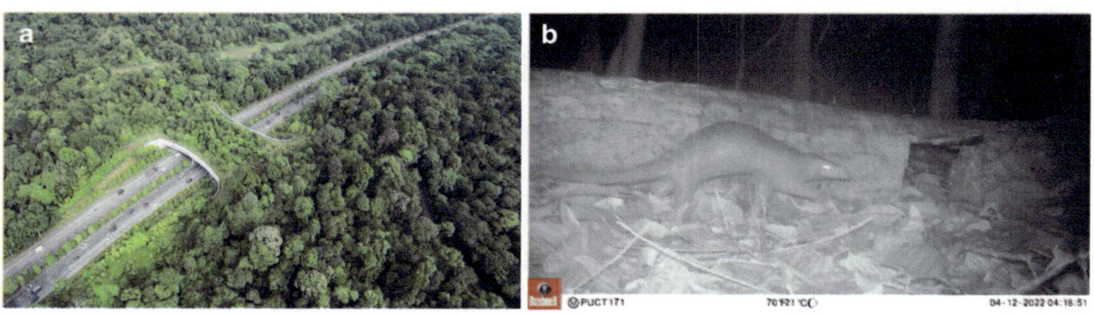

Fig. 13.5 (**a**) Wildlife bridge (Eco-Link) over Singapore's 6-lane Bukit Timah Expressway (BKE) in 2021—eight years after construction began. Picture credit: Tong Tran Son @charlietongphotographer. Reproduced by permission. (**b**) Camera trap capturing a pangolin crossing the overpass over Singapore's 6-lane BKE. Picture credit: National Parks Board, Singapore. Reproduced by permission

Fig. 13.6 (**a**) A plastic wall as a barrier for Anurans, which should not cross the road during their spring migration from the forest to their spawning waters. Central Germany. (**b**) A bucket buried as a pitfall serves as a collection container for the Anurans migrating along the fence. A stick in the bucket allows mice and other small animals that have fallen into the bucket to escape again

Fig. 13.7 Mongolian gazelle (*Procapra gutturosa*) in the Eastern Steppe of Mongolia

migrates seasonally to better grazing grounds and water sources (Bernard 2022; Dejid et al. 2022). Fences secured the railway track on both sides to prevent larger animals from reaching a railway line running through the country. These are generally insurmountable for the gazelles, so the species, originally widespread in Mongolia, was divided into two populations (Fig. 13.8a). The smaller population (west of the railway line) is now threatened with extinction. Some underpasses have been built along the railway line to reduce the separation of the populations (Fig. 13.9). Camera traps, direct observations, and/or tagged animals must show whether the gazelles pass through the underpasses.

Fig. 13.8 (**a**) The fenced railway line separates the two populations of the Mongolian gazelle. To the left is the Western population (yellow arrow); in the background, the Eastern population is on the other side of the railway (red arrows). (**b**) The gazelles do not cross the railway line (right in the background) and return

Case Study: Conservation of the European (Black-Bellied, Common) Hamster

The European hamster *Cricetus cricetus* L. is a Guinea pig-sized hamster with a sizeable zonal distribution occupying a belt between 43° and 58° N and 5° to 90° E (Fig. 13.10). Across most of this vast area, the species used to be widely distributed, tapering out into highly disjuncted insular areas in Western Central Europe. However, in 2020, the species was classified as Critically Endangered across its entire range in the IUCN Red List of Threatened Species (Banaszek et al. 2020a, b). Compared to the previous version of the Red List in 2016 (Kryštufek et al. 2016), the protection status of the European hamster was elevated by four levels from "Least Concern" to "Critically Endangered" in just 4 years.

Due to the unique distribution of the species covering the entire spectrum from dense zonal distribution to disjunct isolated habitat islands, and the corresponding implications for conservation biology and management, the European hamster attracted a lot of genetic studies during the last 15 years, e.g. (Banaszek et al. 2009a, b; Banaszek and Ziomek 2012; Feoktistova

Fig. 13.9 An underpass under the railway track provides a possible passage for larger animals. If accepted by the animals, the passage could help to decrease the isolation of the populations

Fig. 13.10 (**a**) Portrait of the European hamster in an agricultural field in Lubin, Poland. Note the expanded left pouch (white fur color) that hamsters use to transport food into their burrows. Picture credit: Agnieszka Szeląg on Wikimedia, CC BY-SA 3.0. (**b**) Distribution map of *Cricetus cricetus* specimens showing the extent of the species' distribution ranging from Central Europe to Western Siberia. Picture credit: Alexander Kürthy on Wikimedia. Data according to IUCN version 2020.2 compiled by IUCN SSC Small Mammal Specialist Group 2020 (Banaszek et al. 2020a, b)

et al. 2016, 2019; Feoktistova et al. 2023a, b; Kryštufek et al. 2016; La Haye et al. 2017; Melosik et al. 2017).

The International Hamster Workgroup (IHWG, http://institut-faunistik.de/Hamster-Workgroup/) was founded in 1994 to promote the exchange of research and conservation efforts to facilitate the species' survival. The group includes members of all Western and Eastern European countries in which the hamster occurs, and it had its 31th meeting in 2024 in Cologne, Germany.

The species is listed in the European Council Directive 92 / 43 / EEC of 21 May 1992 on the conservation of natural habitats and of wild fauna and flora, Annex IV: Animal and Plant Species of Community Interest in Need of Strict Protection (European Council 1992). The most likely reasons for the dramatic decline of the European hamster—at least in France are changes in agriculture, especially the expansion of maize (corn) monocultures, and climate change, especially increasing winter rainfall (Tissier et al. 2016). Solutions for the dramatic decline of the European hamster are mainly based on land management improvements (La Haye et al. 2017; Tissier et al. 2021). Meinig et al. (2020a, b) list the following conservation measures in Germany:

- Provide alfalfa and other perennial forage crops, and make sure that one part of the hamster habitat area always contains a perennial forage crop rotated at least every 5 years.
- Promote organic farming on the best agricultural soils.

- Establish a species conservation program if not yet developed.
- Adjust agricultural practices in the following ways:
  - Leave harvest strips distributed over the entire field amounting to approx. 2% of the field size and leave a strip as wide as the cultivation width for each ha of the field.
  - Leave stubbles at least 35 cm in height for at least 6 weeks after the crop harvest to protect hamsters from predators. Depending on the type of crop in the rotation, remove stubbles in a staggered fashion rather than in one incident.
  - Do not plow deeper than 25 cm.
  - Cultivate the highest possible proportion of cereals in the crop rotation.
  - Cultivate winter and summer cereals in a balanced ratio.
  - Use as little pesticide as possible.
  - Do not keep farmland as a black fallow (no ground cover).
  - Cultivate the lowest possible proportion of root crops (sugar beet, potatoes, etc.) and a low proportion of temporary cover crops such as maize.
  - If needed, restrict the application of sewage sludge, liquid manure, and dry poultry manure to the winter months.
  - Consistent enforcement of §13 of the Plant Protection Act to protect the European hamster from the use of rodenticides. In addition to the direct intoxication of European hamsters, indirect risks for hamsters develop if voles are excluded as prey and predators shift to hamsters.
- Adjustment of forestry practices.
  - Avoid afforestation of areas colonized by the European hamster.

As conservation measures like the above are often expensive, their acceptance depends on explaining why and how they are taken. The key to approval is not in providing scientific data but in engaging in communication that addresses emotions. Conviction and motivation are generally not based on directives and statistics but on fascinating personal experiences. Conservation professionals should not fear emotions but instead seek to arouse them (Cyriacks 2014).

After the European hamster was perceived as an agricultural pest for centuries, it is on the brink of extinction in many countries today. Only now do we realize that the species also provides essential and invaluable ecosystem services, and its conservation provides multiple synergies for preserving biodiversity and agroecological productivity at large.

13.3 Decision Analysis

An EIA is bound by time and money. For a scientifically justified forecast of the consequences of a planned intervention, one would have to either refer to relevant literature or start a scientific investigation. However, one cannot conduct experiments on the reactions of animal populations to changes in the landscape that are only planned. Problems often arise as the various interest groups have conflicting objectives. The anticipated results of alternative actions will always involve a degree of uncertainty. So, one must argue with probabilities (cf. Harwood 2000). A well-founded prediction about the consequences of a project can be achieved through the consultation of an expert panel. In the USA, a biological assessment of the consequences of specific management measures is formalized. It is submitted to the Bureau of Reclamation of the U.S. Department of the Interior, e.g., Reclamation, https://www.usbr.gov/recman/, passed in 2004. Here, the decision depends on compiling a detailed and very structured questionnaire.

Consider this example: A dam construction could threaten a water snake already classified as endangered (Mathews 1989). The Concho water snake has been listed as threatened under the Endangered Species Act (ESA) since 1986. Under the ESA, possible action plans for the rescue of threatened species are required to be added to the classification of an endangered species.

A structured questionnaire for the risk assessment of the probability of survival of the water snake in the event of a dam construction could look like this (answers as percentage values):

Scenario 1 No dam is built. What is the probability,

- that water pollution occurs
  - that the snake goes extinct when water pollution occurs
- that the riverbed dries up
  - that the snake goes extinct when the riverbed dries up
- that snakes of this species are killed by humans
  - that the snake goes extinct when humans kill snakes of this species
- that a disease affects the snake
  - that the snake goes extinct when it is affected by a disease
- that the habitat is degraded
  - that the snake goes extinct when its habitat is degraded

Scenario 2 The dam is built. What is the probability

- that pollution occurs
  - that the snake goes extinct in case of pollution.

Following the same pattern as in Scenario 1, the occurrence of drought, killing of snakes by humans, occurrence of diseases for the snakes, and habitat destruction are queried again.

Scenario 3 The dam is built. What measures will reduce or compensate for the probability of extinction? What are the probabilities that the following measures will happen, and to what probabilities of extinction will they lead to under successful management: Construction of artificial rapids (how long do they last?), reintroduction of snakes into side arms of the river (successful colonization?), construction of new habitats for the snakes (suitable habitat?).

It has been shown that the percentage values in the answers to the questionnaire differ only slightly between experts.

Another approach to decision analysis is the application of a cost-benefit analysis (CBA). An example is the question if the translocation of rhino individuals could strengthen, for example, a smaller, more endangered population. Such actions are very costly and there is a substantial risk involved that the translocation will not succeed (see Linklater et al. 2011; Pfannerstill et al. 2023) and whether the original population could be weakened as a result (e.g., Drechsler and Burgman 2004; Støen et al. 2009; Emslie et al. 2009). On the other hand, there is also a risk that both populations will go extinct. Again, a decision tree would help provide possible answers to the structure and more clarity and transparency. Similar problems arise, for example, when a small population of an endangered species lives in a forest at risk of fire. What is the probability that fire will wipe out the population? Is the risk lower if the population is divided into two smaller subpopulations, but is a fundamentally higher risk of extinction accepted with the smaller subpopulation? A decision tree of risk can facilitate the decision for the most effective management option (Van Dyke 2008). Dividing an endangered population into two subpopulations is an option for many highly endangered species to reduce the risk of extinction. Another example could be the endangered whooping crane (*Grus americanus)* (Maguire 1986 in Van Dyke 2008).

One could also calculate models with existing data to shorten an elaborate risk analysis. However, the problem is whether the results from other case studies are applicable to other issues.

13.4 Impact Through Tourism

Protected areas are also designated because they become economically interesting due to emerging tourism. In principle, there are evaluations for wildlife as a "recreation service" within ecosystem services, calculated in dollars per year (Costanza et al. 1997). However, tourism can also impair the original conservation goals (Forst et al. 2019; Strasdas 2002; Lei et al. 2023). Fei et al. (2023) reviewed the literature on conservation and tourism. The best way to verify the effect of tourists scientifically is by the changed living conditions of selected species. For example, the African chameleon in a protected area in Southern Greece shifted its breeding season due to disturbances from tourism. As a result, the hatched animals became smaller (Schoenemann 2009). In birds, the flight initiation distance is often measured as an indication of disturbances, see Fig. 13.11. Goodship et al. (2022) have compiled flight distances for many bird species caused by various disturbances from the literature and derived recommendations for protection and buffer zones. They also considered various habitats and seasonal activity cycles, such as the birds' breeding and migration times. Apex predators are particularly affected by roads that allow human access to natural habitats in protected areas (Carter et al. 2020; Quintana et al. 2022). The effect of roads on top predators is assumed to reach at least 5 km on each side of the road (Benítez-López et al. 2010). Of special interest are the species that are part of the protection goal of the area. Tourist disturbances can occur through displacements from parts of the habitat (e.g., Pienkowski 1992; Thiel et al. 2008). As an increasingly popular recreational activity, mountain biking negatively affects vegetation and wildlife (Kuwaczka et al. 2023).

More decisive for the affected animal population would be evidence of a decline in the reproduction rate, as McClung et al. (2004) were able to show for the Yellow-eyed penguin (*Megadyptes antipodes*, see also Fig. 13.12). A conservation strategy that combines tourism with the preservation of the penguin population consists of releasing part of the population for the observing tourists in exchange for keeping another part of the population in a restricted zone without access.

Fig. 13.11 (**a**) Sanderlings (*Callidris alba*) fly off due to disturbance by a visitor. The flight distance can easily be measured. (**b**) Black-tailed Godwit (*Limosa limosa*) flies over its breeding area. (**c**) Black-tailed Godwit has landed in the meadow and runs to its hidden ground nest. (**d**) In this pair of Oystercatchers (*Haematopus ostralegus*), the female is sitting on the nest and is incubating. West Denmark

Fig. 13.12 When disturbed by tourists, the adult of the Yellow-eyed Penguins in New Zealand delay their return and feed their young later in the day, causing the young to grow slower and not achieve sufficient weight (McClung et al. 2004).

Such a strategy has also been practized in some national parks. A minor part is released for tourism, while a major part is closed to all access. A spatial and temporal separation between tourists and sensitive species was, for example, recommended by Gundersen et al. (2019). National parks in India practice strict zoning management: Critical tiger habitats are excluded from access and should support a viable tiger population with prey (National Tiger Action Plan India 2011-2022, https://globaltigerforum.org/wp-content/uploads/2017/04/india.pdf, and https://www.corbettnationalpark.in/safari-zones-in-corbett.htm). Only about 20% of the core zone is accessible to tourists. Zoning of use in protected areas is common practice (e.g., Rotich 2012).

Studies on meadow breeders in England show that far less disturbances occur when well-marked paths guide visitor flows (Pearce-Higgins et al. 2007). The nests are first searched for with binoculars and a spotting scope from the edge of the study area by looking for breeding birds or birds returning to the nests. Found nests are marked at a distance of about 3 m with a bamboo stick, and the coordinates for the position of the clutch are determined by a GPS device (see Sect. 4.4). Nests for checking breeding successes can also be found with small drones, as Israel and Reinhard (2017) were able to show with the Lapwing (*Vanellus vanellus*). Tourism can also increase the predation rate (predator pressure) on waders, thereby reducing breeding success (Tjørve and Underhill 2008). A meta-analysis showed that public access can reduce breeding success, but the effects on each species must be considered separately (Showler et al. 2010).

If you want to determine the breeding success, for example, of lapwings, you must determine the number of fledged young birds. This number is then divided by the number of recorded breeding pairs, and the result is: "fledged young birds per breeding pair and year." Values above 1 indicate a stable population. Daily survival probability can be calculated according to Mayfield (1975). The breeding area should be checked at least once a week. If no chicks are visible, but the adult birds are intensely alerting, one assumes at least one young bird (Michael-Otto-Institut im NABU 2019). Weekly inspection does not seem to disturb the breeding birds in their breeding business (Fletcher et al. 2005).

13.5 Impact of Wind Farms

Wind turbines are an effective way to reduce fossil fuels usage to generate electrical energy. However, they affect wildlife directly through collisions and indirectly through noise pollution, habitat loss, and reduced survival or reproduction (Fig. 13.13) (e.g., Latham et al. 2017; Dähne et al. 2013; https://tinyurl.com/windfarmimpact.; Goodale 2018). The main species affected are birds and bats (e.g., Tesfahunegny et al. 2020; MIT Climate 2023; Kumara et al. 2022; Nilsson et al. 2023; Thaxter et al. 2017; De Lucas et al. 2008; Garvin et al. 2011; Tolvanen et al. 2023). Forest specialist bats avoid wind turbines in forests by keeping distances of over several hundred meters (Ellerbrok et al. 2022). The activity of bats was recorded via call sequences. The acoustic activity of narrow-space foraging bat species (*Myotis, Plecotus*) decreased by 77% with increasing wind speed and rotating wind turbines, but not with the increased wind speed without wind turbines (Ellerbrok et al. 2024). The US-based manufacturer of bioacoustics equipment, Wildlife Acoustics, offers virtual

Fig. 13.13 Impacts of wind farms. (**a**) Birds keeping their distance from wind farms (background) - a behavior corresponding to a loss of habitat that needs to be assessed, especially for birds of conservation concern (e.g., Rees 2012). (**b**) Many seabirds and waterfowl are large, long-lived species with low reproduction rates and poor maneuverability. Even a few collisions can have severe effects on populations

courses to record bats at wind farms (https://www.wildlifeacoustics.com/resources/training-courses/virtual-community-training), see also Sect. 1.12. The company cooperated with wind farm operators in the development of Acoustic-Activated Curtailment allowing wind farms to use acoustic monitoring to detect bats and signal the operators in near real-time to feather turbine blades and pause rotation. As bats are protected by European law, their impairment must be considered when installing and using wind turbines. The night-time warning lighting on wind turbines also has an impact on bats. Bats are attracted to red light, while they search more for food at warm-white LEDs (Voigt et al. 2018). Tests were conducted using ultrasonic recorders with 10 min light-on/light-off intervals and additional dark controls. The behavior of songbirds is also influenced in the vicinity of wind turbines (e.g., Zwart et al. 2016; Szymański et al. 2017)

An environmental impact assessment of wind farms requires measures to mitigate or compensate for adverse effects (e.g., Garthe and Hüppop 2004; Madders and Whitfield 2006). In addition to mitigating the technical constructions (Conkling et al. 2021; Tolvanen et al. 2023; Williams et al. 2024), there are also proposed solutions for creating new habitats for wildlife in the vicinity of wind turbines (Wilson and Elliott 2009).

13.6 Salvage Logging

When a forest has many dead trees following a natural disturbance or due to anthropogenic interventions, the question may arise whether to remove the dead trees from the forest or leave them as natural structures. Depending on whether deadwood is removed or not, we may further ask which forest type is to be rated higher regarding nature conservation? Standing deadwood (snags) provides many animals with nests, food, and roosting habitats (Samuelsson et al. 1994; Bull 2002; Drapeau et al. 2002). However, salvage logging is seen to be harmful in some reports, while in others positive for wildlife (e.g., Morissette et al. 2002; Greenberg et al. 1995; Nappi et al. 2004; Schmiegelow et al. 2006; Lindenmayer et al. 2008). Jonášová and Prach (2008) showed that salvage logging harms the species community in a mountain spruce forest and has a more substantial impact on the ground vegetation than forest damage caused by an outbreak of bark beetles (see Fig. 13.14b). Since every natural disturbance of a forest system depends on on-site factors, the context in the landscape, and the respective disturbance history, local studies should be carried out to investigate the regeneration of forests and changes in the species community (Buddle et al. 2000; Buddle 2001; Merganičová et al. 2012). The conditions

Fig. 13.14 (**a**) Windthrow in a beech forest, northern Bavaria, Germany. (**b**) Bark beetles (Scolytinae, *Ips typographus*) destroyed large areas of spruce trees in the Harz Mountains. In the National Park, many affected areas are left without management ("allowing natural disturbances"). Over decades, a mixed forest will emerge. The foxglove (*Digitalis purpurea*) in the foreground shows the development of an herbaceous meadow in the forest clearing. National Park Harz, Germany. (**c**) Left: Natural disturbances by bark beetles in the National Park Harz, Germany

for recruitment dynamics for the main tree species in the boreal forests of North America have been summarized by Greene et al. (1999).

In a study by Staupendahl (1997), all seedlings and saplings of tree species were counted within circular plots of 1.78 m radius (circle area 10 m^2) to measure the rejuvenation of the forest in different forest locations. All trees with a diameter < 10 cm were considered "regeneration." The sapling species closest to the center of the sample plot was considered representative. The sample plots were randomly distributed within a forest stratum. The density of tree species in a forest stratum was determined by the angle count method ("plotless technique," see Fig. 4.3 in Sect. 4.2.4).

Removing dead trunks is also recommended to keep damage from fire or windthrow as low as possible (e.g., Chiang et al. 2009). However, Foster and Orwing (2006) recommend the opposite. The boreal forests frequently experience fire or windthrow as natural events (Schulze et al. 2005). Nevertheless, the frequency of fires has significantly increased in recent decades due to anthropogenic activities. The question arises whether removing deadwood, which is valuable in nature conservation, reduces fire damage. To investigate this, at least two different forests are compared for fire damage to individual trees (Fischborn 2011). The number and condition of trees in sample circles of 10–13 m radius are counted along randomly selected transects using the previously described plotless method. Tree species, trunk diameter, height, and number in the sample area are determined. To describe the fire damage, burn marks (fire scorch) on the trunk, their height, crown damage in classes (0–25%, 25–50%, etc., see also Fig. 13.15c above), and the tree condition as alive or dead are recorded (Waldrop and Brose 1999; Pausas et al. 2003).

Some studies suggest that fire damage is not less common in lighter or thinned forests (Fischborn 2011; Bond et al. 2009). In the boreal Southern taiga, this is also because the grass layer that emerges in lighter forest stands spreads the fire faster, and the lighter forest stands are drier. Snags, downed logs, fallen branches, and stumps are important for rejuvenation (Fig. 13.16).

13.7 Prescribed Burning and Fire Effects in Forests

Fire is a natural occurrence in boreal forests. However, the frequency of fires has increased significantly due to human activities. Since fires can cause significant damage in many forest types, some management is being considered to reduce the adverse effects of forest fires. Since deadwood and remnants of previous fires are believed to exacerbate current fires, one management approach is to set smaller fires in a controlled manner to reduce flammable material such as coarse woody debris in the forest. Such management is called "prescribed burning". As fires also thin out the forests, additional benefits are seen for the wildlife in burned forests (Block et al. 2016; Fawcett et al. 2021; Fazzio 2022). However, whether this promotes carbon storage in the forests remains questionable (Campbell et al. 2012).

We may ask the following questions about this complex:

- Is the damage in the forest greater when wildfires occur less frequently?
- Does the strength of a fire depend on the type of forest?
- What wildlife benefits from forest fires?
- Are there species that depend on fire in their life cycle?
- In what way do forest fires harm animals or plants?

Do fires spread evenly or heterogeneously through the forest?

Case Study: Fire Effects in Boreal Forests

While many scientists in North America favor controlled fire management (prescribed fires, e.g. Neary et al. (1999); Fernandes (2018)), this practice cannot be transferred to the boreal forests of Siberia and Northern Mongolia. The lack of management and materials in many Asian countries often does not allow for proper setup of controlled fires. How can this complex problem be studied? We can first compare faunal elements between longtime unburned forest patches and freshly burned forest patches (cf. Sect. 2.16). Fischer (2010) found few significant differences in

Fig. 13.15 (**a**) *Picea-Abies* forest burned, Northern Mongolia. (**b**) View inside a burnt *Picea-Abies* stand, Northern Mongolia. (**c**) Light taiga forest, dominated by *Larix sibirica* and *Betula platyphylla* with fire scorches on tree stems, Northern Mongolia. (Photo credit M. Fischborn)

the bird fauna between a long unburned and a recently burned birch forest in Northern Mongolia (Fig. 13.17). However, samples from two forest sites must be spatially far apart. In these vast natural habitats, birds roam widely. The decisive question is which animal species are disadvantaged by the fires and which species profit from fires. For nature conservation, the focus shall be on endangered species.

Fig. 13.16 (**a**) Rejuvenation in the light taiga, Northern Mongolia. (**b**) More rejuvenation is seen between remaining (staying) dead wood. Northern Mongolia

Some xylobiont beetle species benefit from fires in boreal forests and can be classified as pyrophilous (Wikars 1992, 1997; Saint-Germain et al. 2008). They are attracted by fire and are the first to lay their eggs in burnt trees. Some species specialize in larch, birch, or pine (see, e.g., Fig. 13.18). These beetle species have become rare in Central Europe (Müller et al. 2013). In order to promote them, small forest plots should be burned where it is technically possible (Müller et al. 2010; Kouki et al. 2012). Pyrophilous beetles perceive fire from many kilometers away (e.g., Schmitz and Bleckmann 1998; Schütz et al. 1999; Evans 2005; Müller et al. 2008). Old burnt trees should, therefore, be left standing even after post-fire salvage logging (Kaila et al. 1997; Ehnström 2001; Müller et al. 2013).

Other studies could focus on the natural regeneration and recolonization of burned

Fig. 13.17 (**a**) A birch-larch forest burned by low-intensity fire. Khentey, Northern Mongolia. Photo E.-L. Fischer. (**b**) A birch-larch forest burned by high-intensity fire. Khentey, Northern Mongolia. Photo E.-L. Fischer

Fig. 13.18 The pyrophilous Black Fire Beetle (*Melanophila acuminata*, family Buprestidae) is initially attracted by guaiacol and its derivatives in smoke perceived by chemical receptors in the antennae even if several kilometers away. They also perceive fire through infrared sensory organs on the mid-coxae of their legs and lay their eggs in the bast of dead trees. Photo: AG Prof. Schmitz on Wikimedia, CC BY-SA 2.5

areas after a fire event. Forest regeneration measurements are discussed above in Sect. 13.6. In mature forests, regeneration can be recorded, e.g., by counting all trees with a diameter <10 cm in plots with a 12 m radius as regeneration. The regeneration trees could be divided into 6–9 cm classes according to their stem thickness. The plot center should be marked with a metal plug, and the GPS coordinates recorded. For all trees with a diameter >10 cm, document species, height, diameter, and polar coordinates. The average diameter and length of lying deadwood in the plot are also estimated.

The time period between fires plays a decisive role in animal recolonization. In studies on forest regeneration in Siberia, Schulze et al. (2005) found that natural intervals between forest fires can be up to 400 years. Fire intervals can be studied dendrologically on tree slices for at least 300 years (Hessl et al. 2011; Byambasuren 2011). The Khentey Mountains in Northern Mongolia represent the Southern foothills of the Siberian taiga. Altitude and soil are the main factors determining the local forest types, which can be roughly divided into "dark taiga" (Fig. 13.19) and sub-taiga forest ("light taiga," Fig. 13.16a) (Dulamsuren et al. 2005). The fire intervals are significantly longer in the dark taiga than in the light taiga. Even if one speaks of "stand replacement fires" in the dark taiga, the control of regeneration shows that the typical tree species, such as the Siberian fir (*Abies sibirica*), also rejuvenate most frequently in the dark taiga (Krause, oral presentation 2011, not published), corresponding well to stand self-replacement. In boreal forests in Alaska, self-replacement similarly only occurs when there are few low-intensity fires (Johnstone et al. 2008). Only with frequent fires do the regeneration stages pass through different forest formations (succession stages). The moisture and permafrost conditions in the soil are decisive for this (Johnstone and Chapin 2006).

Fischborn (2011) compared the damage to individual trees in the two forest types in Mongolia (see Sect. 13.6). The samples were randomly selected, as fires cause a highly heterogeneous microhabitat distribution (Fig. 13.19b). Fire damage was no less frequent in sparse forests than in dense forests containing more wood. Thus, prescribed burning would not be an adequate protection against severe fires. An old-growth forest has not been disturbed by fire or altered by management for a long time (>100 years). Such forests have a complex structure and present different stages of succession in small areas due to fallen trees (Fig. 13.19a), as expected by the mosaic cycle concept (Remmert 1991). This concept describes the simultaneous presence of various stages of development in a climax forest.

Fig. 13.19 (**a**) Recruitment in an undisturbed dark taiga - a *Pinus sibirica* forest with *Abies sibirica*. Khentey, Mongolia. (**b**) Heterogeneity in a burnt forest on the edge of the dark taiga. The photo shows burned and unburned trees side by side. Khentey, Mongolia.

Although the general attention on fire events and their effects is more focused on the boreal and temperate zones, there are more fires in the tropics (Cochrane 2009). Unlike forests, the savannah burns almost every year. However, the herbaceous biomass determines fire activity (Kahiu and Hanan 2018). As grass grows relatively quickly, native people have set fires in the savannahs to attract wild animals for hunting or graze their domestic animals for centuries (Bird et al. 2024; Bowman et al. 2011; Sheuyange et al. 2005; Takacs et al. 2021). The regrowing grass contains more protein than the old grass. Adaptations of the savannah vegetation to fire must be interpreted differently than in forests: Fire-resistant trees are primarily species that have been selected by fire. However, they would also reproduce or grow without fire (Fig. 13.20). Like plants, most animal species in the savannah can survive the fires (Fig. 13.21). They have their own specific adaptations. For example, reed frogs of the genus *Hyperolius* survive the savannah's dry season by attaching themselves to plants and changing their skin, so it primarily reflects the sun's rays. They perceive approaching fires acoustically and flee (Kobelt and Linsenmair 1992; Spieler 1997; Linsenmair 1998; Grafe et al. 2002). Early fires at the beginning of the dry season are not as destructive as late fires at the end of the dry season because the vegetation has not yet completely dried out. Targeted management should take this into account. In Australia, some tree species can only germinate after fire, such as the Austral grass-tree *Xanthorrhoea* sp. (https://www.bushheritage.org.au/species/grass-trees). As eucalyptus trees catch fire quickly and burn vigorously, most fires in Australia also damage the vegetation and biodiversity and are becoming more frequent and intensified by climate change (see the devastating Australian wildfires of 2019–2020 (Dickman 2021; Haque et al. 2023; Khan 2021), influence of climate change: Abram et al. (2021), Cai et al. (2022)).

Case Study: Nut Collection in the Dark Taiga

The Siberian stone pine (*Pinus sibirica*) is a dominant tree in parts of the dark taiga (Fig. 13.23). Its cones contain valuable nuts highly sought after by animals and humans (e.g., Craig 2018). People harvest the nuts in the fall, mainly in October. The nuts are used for personal consumption and sale (e.g., https://siberianpinenutoil.org). Several tons of nuts are harvested each year. Although it is often argued that the harvesting of secondary forest products is sustainable, several questions need to be asked about sustainability to substantiate or refute this statement (see also Newton and Freyfogle 2005; Miina et al. 2020; Unterschultz et al. 2003; Pereira et al. 2015; Holt 2020; Secretariat of the Convention on Biological Diversity 2024):

- Does the annual harvest affect subsequent harvests in later years?
- How many seeds do *Pinus sibirica* trees produce?
- Do the trees produce approximately the same amount of fruit each year?

Fig. 13.20 (**a**) Savannah fire in Comoé National Park, Ivory Coast. (**b**) Most trees in the savannah have thick bark to resist fire. The photo shows *Butyrospermum paradoxum*. Comoé NP, Ivory Coast

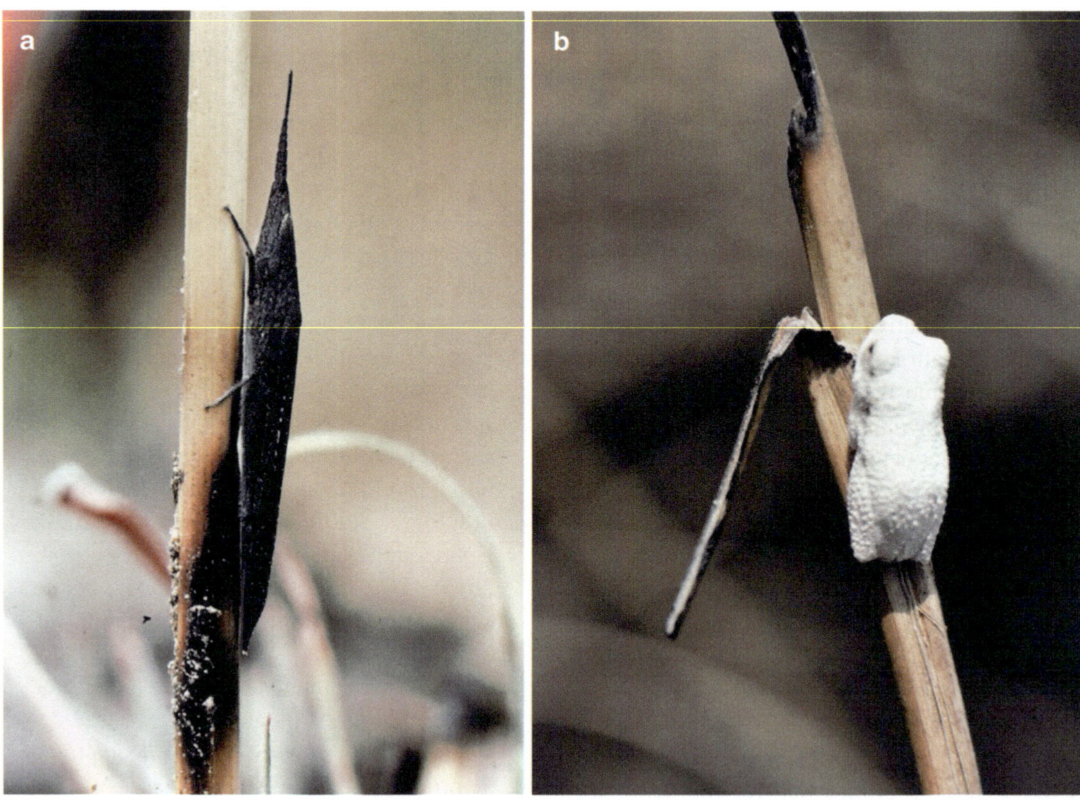

Fig. 13.21 (**a**) A grasshopper (*Acrida* sp.) is mimetically adapted to burnt grass by its shape, coloration, and posture. Comoé NP, Ivory Coast. (**b**) Reed frogs of the genus *Hyperolius* attach themselves to plants and change the color of their skin in the dry season. Comoé NP, Ivory Coast

- Does the annual harvest affect the animals in the forest?
- Does human harvesting compete with animal harvesting?
- Which animal species depend on the nuts in the *Pinus sibirica* forest?
- Does the type of harvesting method cause long-term damage to the harvested trees?
- Does harvesting in the forest harm the flora and fauna of the forest?
- How can the goal of sustainably harvesting the nuts be achieved?

Almost all dark taiga sites in the Mongolian wilderness can only be reached on foot or on horseback (Fig. 13.22). This is why during pine nuts harvest time, people stay in the forest for several weeks.

People harvest the cones by hitting the trees with a heavy wooden hammer (approximately 60 kg). Ripe cones fall to the ground and are collected. The cones are then put into a simple shredding machine, which separates the seeds and scales from the cones. The seeds are then sieved out and packed in bags for transportation (Fig. 13.22). About 10% of the total nut supply in Mongolia is exported (Figs. 13.23 and 13.24).

The harvest success is estimated after the number of bags of nuts collected is determined. However, it is challenging to determine cone production from the number of nuts. By visual, only about half of the cones of each tree can be counted. This results in a comparable number of cones knocked down from the tree. An average value for the seed quantity and different cone sizes must also be determined. In addition, it is unknown how many nuts have been eaten by birds while on the tree and how many fallen cones were used by other animals as food. Fruit traps can be set up under selected trees, determined as randomly as possible (see Fig. 7.16b). Like

Fig. 13.22 Travel to the dark taiga to inspect the condition of the *Pinus sibirica* forest site and supply the harvesters in the forest. (Author MM with employee and packhorse)

Fig. 13.23 *Pinus sibirica* old-growth forest, representing dark taiga. Khentey Mountains, North Mongolia

many other trees, the Siberian stone pine has mast years and years of nut shortages, i.e., it does not produce the same amount of fruit every year.

More than 100 vertebrate species in the taiga live off the pine nuts (e.g., Eletu 2022). These include species of Insectivora (shrews), Lagomorpha (pika and hare), Rodentia (squirrels, chipmunks, voles, lemmings), Carnivora (wolverine, bear, sable, stoat, weasel, lynx), Artiodactyla (wild boar, deer, moose), and many bird species (jays, woodpeckers, nutcrackers, tits, thrushes, finches, buntings) (Fig. 13.25, see also Fig. 7.19 with birds as seed eaters and distributors, Sect. 7.10). Most of them need the nuts as an energy reserve for the cold winters. Harvested trees hit with a mallet show pressure wounds on the bark through which fungi can penetrate. A comparison of harvested forest sites with those without harvesting shows more dying and dead trees in the harvested forest. The harvesting groups stay approximately 3–4 weeks in the forest, which leads to disturbances of both wildlife and the natural regeneration of the forests. Especially the brown bear (*Ursus arctos*) is disturbed by the presence of humans as he also feeds on pine nuts (Fig. 7.8c). The disturbance effects of humans on wildlife are also discussed in Sect. 13.4 (see above). Lee et al. (2024) detail behavioral changes in animals caused by human disturbance.

A suggestion to decrease the potential conflict between people and conservation is to strive for a sustainable harvest, for example, by granting a limited amount of family groups priority licenses, marking individual harvest trees, and periodically establishing a legally binding export ban (e.g., every 2 years), or change the harvest area from year to year. In order to strive for sustainability, economic and social aspects must be considered in addition to environmental aspects (see, e.g., UCLA: https://www.sustain.ucla.edu/what-is-sustainability/, Dai et al. 2024; Raman et al. 2023; Scoones et al. 2020; United Nations 2024). An overview of the sustainable use of non-timber forest product (NTFP) can be found in de Mello et al. (2020) and Pandey et al. (2016). Sustainable management in tropical forests is described, for example, by Ros-Tonen (2000).

13.8 Subsistence Hunting of Bushmeat

Animals are a resource humans harvest in addition to plants, but their usage should follow specific guidelines, especially in protected areas (e.g., Roth and Merz 1997). Thus, non-timber forest products (NTFPs) also include forest wildlife as a resource that can be utilized economically. The term "Bushmeat" indicates the hunting of wild animals for food and originates in Africa, where it includes pangolins, duikers, large rodents, and day-active primates (Fig. 13.26). Subsistence hunting is a long tradition in developing tropical countries. However, population growth and commercialization mean that bushmeat hunting is no longer sustainable and, in some cases, is leading to a dramatic decline in wildlife populations (Fa et al. 2003; Ripple et al. 2016). The use of a wild species is likely to be sustainable if (1) it does not reduce the future use potential of the target population or impair its long-term viability, (2) it is compatible with the maintenance of the long-term viability of supporting and dependent ecosystems, and (3) it does not reduce the future use potential or impair the long-term viability of other species.

There are two main perspectives when approaching this topic: the biological and the sociological vantage points (e.g., Davies and Brown 2007).

The field ecologist must analyze the extent to which wildlife populations are declining. Methods for estimating the population density of

Fig. 13.24 (**a**) Heavy wooden hammer for hitting the Siberian stone pine so that the cones fall down. The hammer is over a man's height and requires two people to hit it. (**b**) Wound on the Siberian stone pine caused by hammer blows. (**c**) Simple shredder hand machine for separating seeds and cone scales. (**d**) Sieving out the seeds after shredding the cones. (**e**) Sale of cleaned *Pinus* seeds on public markets

Fig. 13.25 (**a**) The Spotted Nutcracker (*Nucifraga caryocatactes*) regularly visits *Pinus sibirica* forests and eats and caches the pine nuts. (**b**) A Siberian musk deer (*Moschus moschiferus*) in the dark taiga also forage on the pine nuts. (**c**) The Siberian Chipmunk (*Tamias sibiricus*) collects and caches the pine nuts for overwintering. (**d**) A brown bear has dug for cached pine nuts. All pictures from Mongolia.

Fig. 13.26 (**a**) Open market for bushmeat in Cameroon. (**b**) Bullet casings in the rainforest of Cameroon indicate the activities of poachers.

larger mammals were described in Chap. 1. For example, Viquerat et al. (2012) compared three methods using line transects to estimate duiker abundance in the rainforest: diurnal, nocturnal, and indirect dung surveys. The coefficient of variation was highest for nocturnal surveys. Dung surveys yielded lower abundance estimates and correlation coefficients above 100%. Therefore, it was confirmed that diurnal transects give the most reliable estimates, even for shy duikers. Hunting forest wildlife can also have negative consequences for other species in the forest, e.g., on seed dispersal (Pangau-Adam et al. 2012; Nasi et al. 2011).

The sociologist must analyze to what extent the local population depends on the protein supply of bushmeat and, if possible, alternatives that can be developed to meet protein requirements. Through questionnaires, the sociologist also aims to clarify how important the socioeconomic component is for rural livelihoods (e.g., Obioha et al. 2012). The data were collected using participatory rural appraisal tools (for the methodology, see Milner-Gulland and Rowcliffe 2007).

Large carnivores in the rainforest can also be affected by the decline in prey through exploitative competition (Henschel et al. 2011). The diet of leopards was determined in this study by analyzing their scats (see Sect. 7.3, hair atlas of prey remains). The density of leopards and the abundance of bushmeat hunters were recorded by camera traps (photos per 100 trap days, leopard individuals/100 km^2). The prey choice was described as the dietary niche breadth (see Sect. 15.11, niche calculations), and the equal utilization of prey by leopards and hunters by the niche overlap. However, there are signs that large carnivores can switch to other prey species when one species declines (Ghoddousi et al. 2017). The authors determined the diet of the leopards by analyzing the scats and the prey offtake by poachers using seizure records. Prey density was determined using line transect sampling, camera trapping, point counts, and dung counts (see Chap. 1). In addition, they conducted interviews with local poachers. Ziegler et al. (2016) identified the areas in Central Africa with the greatest hunting pressure by studying the bushmeat offtake and the number of animals hunted. A map of these high-risk areas for wildlife can be useful for large-scale conservation planning.

13.9 Trophy Hunting and Conservation

Another related issue is the controversial question of sustainability in trophy or recreational hunting. Recreational hunting in rich wildlife areas usually has a extremely high price tag. If the animal populations are well known and hunting licenses are only sold according to prescribed quotas, then the income can benefit the local population and be used for conservation purposes. In order to make this work, years of effort must be put into building the appropriate structures in the social sector. For example, a buffalo hunted under an expensive licence, is much more valuable to the local community than a poached buffalo. There are examples of good practice, mainly from East and South Africa (e.g., the CAMPFIRE (*The Communal Areas Management Programme for Indigenous Resources*) Project in Zimbabwe, Safari Club International 2015). Campfire aims to utilize wildlife as an essential source of revenue for local communities. The communities will invest in environmental conservation if they can exploit these resources sustainably for their benefit (Murindagomo 1990). However, problems with local institutional and economic forces can also lead to confrontation with CAMPFIRE (e.g., Alexander and McGregor 2002; Tichaawa and Mhlanga 2015; Shereni and Saarinen 2021). Hunting large carnivores is particularly tricky. Most top predators are globally on the decline. This can affect the entire ecosystem negatively (e.g., Groom et al. 2006).

The work of the field ecologist consists of counting the animal populations, verifying their age structure, and thus creating a basis for a quota system. However, hunting areas should be separated from areas used by tourists for sightseeing (photography). Photography tourism is another type of revenue.

The trophy hunting of lions (*Panthera leo*) is an example of how population development can be maintained sustainably if the population's age structure is strictly observed despite the removal of individual animals. Whitman et al. (2004) simulated population trends in lions from Northern Tanzania when only males above 6 years of age are authorized for hunting. The prerequisite for this is the accurate determination of the sex and age of the animals permitted for hunting. There are clear instructions for recognizing the age of lions (Whitman and Packer 2006). The coloration of the nose is a relatively simple and reliable method for age recognition (Fig. 13.27). Males with a nose that is at least 60% pigmented are about 6 years old. However, as the age of six years cannot always be determined with certainty, Loveridge et al. (2023b) suggest only permitting the hunting of lions aged 8 years and older

However, if quotas are fixed and there is no trophy monitoring or age restriction, trophy hunting can lead to a sharp decline in the population (Lindsey et al. 2013). With poor management of trophy hunting and additional threats to lion populations caused by humans (poaching, retaliatory killing), most populations are endangered and can, therefore, not be hunted sustainably (Loveridge et al. 2016a, b, 2023a). In addition, ethical considerations are a partially connected and partially separate aspect widely discussed (e.g. Ghasemi 2021; Ghasemi et al. 2023; Kansky and Kidd 2024; Latombe et al. 2022).

Saker falcons (*Falco cherrug*) are highly valued as hunting aids (falconry) in some Arab countries. In 2010, up to 12,000 US dollars were paid for individual animals. Although saker falcons are protected in Mongolia, some are still sold (Stretesky et al. 2018). A reliable population estimate and continuous monitoring of the population development are required to set a quota for the release of falcons. Saker falcons also hunt in the dry steppe (Fig. 13.28a) but hardly find breeding sites. A key factor for the population is the availability of nesting sites. The carrying capacity of the steppe for saker falcons is not reached in terms of the rich prey supply (mainly rodents). Nest boxes have, therefore, been installed in the open steppe to increase the number of breeding pairs (Fig. 13.28b) (Dixon et al. 2008). This measure is not only a benefit for the saker falcon population but also makes it possible to increase the quota for the sale of falcon individuals (Dixon et al. 2011).

13.10 Reintroduction to Reestablish Wild Populations

Reintroduction is the intentional process of reinstating an organism into an ecosystem previously occupied by that species. If wild populations are severely weakened, additional individuals can be added by translocation from other safe populations as a conservation measure. Releasing captive-bred individuals can also reinforce or establish a new population in the wild. It can be seen as a form of *ex situ* conservation (e.g., Mühlenberg and Spellerberg 2010; Cardinale et al. 2020). Such support measures have been carried out for both invertebrates and vertebrates of various classes (e.g., Kleiman 1989; Griffiths and Pavajeau 2008; Sutherland et al. 2015; Department for Environment Food and Rural Affairs 2021; Amphibian and Reptile Conservation 2023). Reintroduction programs of this kind are subject to strict rules, e.g.,

- The species must have previously lived in the habitat where animals are released (Maunders and Byers 2005; IUCN/SSC 2013).
- The factors that led to the local extinction of the species must be known and eliminated or compensated.
- Individuals released from captivity must be prepared for potential predators through behavioral trials (e.g., Buchholz 2007; Reidler 2016; Greggor et al. 2019, 2021).
- The habitat conditions of the area for reintroduction must be suitable for the managed species.
- The development of the released individuals must subsequently be monitored. Post-release activities include monitoring and investigating mortality causes and mortality rates. Close collaboration with land managers should guarantee the program's long-term financial and political support (Maunders and Byers 2005).

Fig. 13.27 (**a**) Male lion in Ruaha National Park, Tanzania. (**b**) With a nasal pigmentation of ~40%, the age of this lion can be estimated at 3–4 years. (**c**) Lion in Virunga National Park. In this male, the nasal pigmentation is about 50%. The age can be estimated to be between 5 and 6 years, République Démocratique du Congo. (**d**) This male from Leipzig Zoo has a nose pigmentation of over 60% and is thus over 6 years old

Fig. 13.28 (**a**) A pair of saker falcons (*Falco cherrug*) has preyed in the dry steppe. Southern Mongolia. (**b**) Nest box set up in the open steppe as a breeding opportunity for saker falcons, Southern Mongolia

In reintroductions, the tasks for field ecologists are diverse: risk analysis (what were the causes of the decline—what risks still exist?, see Sect. 10.11), habitat assessment (see Chap. 10), monitoring (see Chaps. 1 and 2), and tracking the movements of released animals (see Sect. 4.6).

There are many successfull reintroduction programs, but also problematic cases and failures. The reintroduction of the Black-footed ferret (*Mustela nigripes*) became a textbook example for recovery programs (Miller et al. 1996; Biggins and Godbey 2003; Biggins et al. 2006). The

Californian condor (*Gymnogyps californianus*) is another example of a complex but successful reintroduction (National Park Service 2017; Kelly et al. 2015). Reintroduction programs for Old World vultures were also successful in Europe (e.g., Cerri et al. (2024). Wynhoff (1998) provided an example of successful reintroduction trials with an endangered butterfly, and Sutherland et al. (2015) analyzed many examples of reintroductions of different animal groups to determine whether they were successful or ineffective. For example, the rescue attempts of critically endangered rhinos through translocation are striking (Amslie et al. 2009; Knight 2021; Beckmann and Soorae 2022, see also Chap. 14, Decision Analyses, Støen et al. 2009). On the other side, many years were necessary to reintroduce the Przewalski's horse in Mongolia before the animals formed independent social groups, and foals were successfully born in the wild (Kaczensky et al. 2011) (Fig. 13.29).

Fig. 13.29 An established social unit of the Przewalski horse (*Equus przewalskii*) with wild-born foals of released individuals, National Park Hustai, Mongolia

The flightless bird Kagu has become restricted to remote forest areas in New Caledonia, with only about 650 birds remaining (Perrins 2003; BirdLife International 2024). The protection in only one National Park cannot guarantee the survival of the bird. The captive breeding of the Kagu is an exemplary project because it is following most IUCN guidelines: (a) breeding close to the natural habitat in the capital of the country of the species' origin, (b) keeping partnerships with other breeding stations in the world as a safeguard, (c) reintroducing captive-reared individuals into the wild, and (d) controlling roaming exotic carnivores as they are currently the main threat to the wild population. The Vogelpark Walsrode in Lower Saxony, Germany, has been part of the breeding program since 2001 (Fig. 13.30).

Fig. 13.30 The flightless bird Kagu (*Rhynochetos jubatus*) in a captive breeding program, Vogelpark Walsrode, Germany

The introduced Brown Tree snake (*Boiga irregularis*) (Fig. 13.31) is responsible for the extinction or extirpation of nine of 12 species of native forest birds on Guam and other islands of the Commonwealth of the Northern Mariana Islands (CNMI) in the Southeast Pacific. The snake was most likely introduced by Fighter jets coming from air bases in the original distribution of the snake. As there were no tree snakes on the Mariana Islands before, the birds of the archipelago did not have any antipredator behaviors and fell easy prey to the snakes.

Especially the endemic White-Eyes (genus *Zosterops*) are of primary conservation concern (Fig. 13.32), leading to the foundation of the Marianas Avifauna Conservation (MAC) project, consisting of biologists with the CNMI Division of Fish and Wildlife (DFW) and U.S. Fish and Wildlife Service (USFWS) and the Association of Zoos and Aquariums (AZA). The MAC project focused on *conservation introductions*, an attempt to establish a species, for conservation purposes, outside its recorded distribution but within an appropriate habitat and ecogeographical area (IUCN Species Survival Commission

Fig. 13.31 Brown tree snake (*Boiga irregularis*). Picture credit: Pavel Kirillov on Wikimedia, CC BY-SA 2.0

Fig. 13.32 Bridled White-eye (*Zosterops conspicillatus*) looking for bugs on a tree of brightly colored flowers. Peter on Wikimedia, CC BY-SA 2.0

(SSC) and IUCN Reintroduction Specialist Group 1998), as well as *translocations*, deliberate and mediated movements of wild individuals or populations from one area to another (IUCN Commission on Environmental Policy, Law and Administration (CEPLA) et al. 1987; IUCN Species Survival Commission (SSC) and IUCN Reintroduction Specialist Group 1998), e.g., (Radley and Amidon 2008). However, several breeding programs at zoos in the US are on their way to breed the white-eyed species and reintroduce them to one of the islands (Smith and Wassmer 2016).

References

Abram, N. J., Henley, B. J., Sen Gupta, A., Lippmann, T. J. R., Clarke, H., Dowdy, A. J., et al. (2021). Connections of climate change and variability to large and extreme forest fires in southeast Australia. *Communications Earth & Environment, 2*(1), 1–17. https://doi.org/10.1038/s43247-020-00065-8

Alexander J. & Mcgregor J. 2002: Wildlife and politics: CAMPFIRE in Zimbabwe.—*Development and Change* 31(3), 605—627. DOI:https://doi.org/10.1111/1467-7660.00169

Amphibian and Reptile Conservation UK 2023: *Bringing back species: Reintroductions, translocations and captive breeding*.—https://www.arc-trust.org/reintroductions-and-captive-breeding.

Amslie R., Rajan A. & Kock R. 2009: *Guidelines for the in situ re-introduction and translocation of African and Asian rhinoceros*.—IUCN Publication, Monographic Series no.039, 115 p. https://www.iucn.org › resources

Banaszek, A., & Ziomek, J. (2012). Genetic variation and effective population size in an isolated population of the common hamster, Cricetus cricetus . *Folia Zoologica, 61*(1), 34–43. https://doi.org/10.25225/fozo.v61.i1.a6.2012

Banaszek, A., Jadwiszczak, K. A., Ratkiewicz, M., & Ziomek, J. (2009a). Low genetic diversity and significant structuring of the common hamster populations Cricetus cricetus in Poland revealed by the mtDNA control region sequence variation. *Acta Theriologica, 54*(4), 289–295. https://doi.org/10.4098/j.at.0001-7051.021.2009

Banaszek, A., Ziomek, J., & Jadwiszczak, K. (2009b). Morphometric Differences Between the Phylogeographic Lineages of the Common Hamster Cricetus Cricetus in Poland. *Zoologica Poloniae, 54–55*. https://doi.org/10.2478/v10049-010-0002-2

Banaszek, A., Bogomolov, P. L., Feoktistova, N., La Haye, M. J. J., Monecke, S., Reiners, T. E., et al. (2020a). IUCN Red List of Threatened Species: Cricetus cricetus. *IUCN Red List of Threatened Species*. https://www.iucnredlist.org/en. Accessed 18 February 2024

Banaszek A, Bogomolov P., Feoktistova N., La Haye M., Monecke S., Reiners T.E., Rusin M., Surov A., Weinhold U. & Ziomek J. 2020b: *Cricetus cricetus*. IUCN Red List Threat Species. 2020; 235:Annex IV. doi: https://doi.org/10.2305/IUCN.UK.2020-2.RLTS.T5529A111875852.en.

Beckmann K.M. & Soorae P.S. 2022: Conservation translocations from the 'Global Reintroduction Perspectives' series: Disease and other biological problems.—*Ecological Solutions and Evidence* 2022;3:e12163, 7p. British Ecological Society. https://doi.org/10.1002/2688-8319.12163

Beebee T.J.C. 2013: Effects of road mortality and mitigation measures on amphibian populations. *Conservation Biology* 27(4), 657–668. doi: https://doi.org/10.1111/cobi.12063.

Benítez-López A., Alkemade R., Verweij P. A. 2010: The impacts of roads and other infrastructure on mammal and bird populations: A meta-analysis. *Biol Conserv.* 143, 1307–1316

Bernard M. 2022: Always on the move: *How the Mongolian Gazelle challenges nature conservation.*—Press release from Goethe University Frankfurt. https://www.goethe-university-frankfurt.de/118649670.pdf

Biggins D.E. & Godbey J.L. 2003: Challenges to reestablishment of free-ranging populations of black-footed ferrets.—*CR Biol.* 326, 104–111

Biggins D.E., Miller B.J., Clark T.W. & Reading R.P. 2006: Restoration of an Endangered Species. The Black-footed Ferret. p. 581–585—In: Groom M.J., Meffe G.K. & Carroll C.R. 2006: *Principles of Conservation Biology*, 3rd ed., Sinauer Associates, Inc., 793 p.

BirdLife International (2024) Species factsheet: *Rhynochetos jubatus*. Downloaded from https://datazone.birdlife.org/species/factsheet/kagu-rhynochetos-jubatus on 19/05/2024.

Block W.M., Conner L.M., Brewer P.A., Ford P., Haufler J., Litt A., Masters R.E., Mitchell L.R. & Park J. 2016: Effects of Prescribed Fire on Wildlife and Wildlife Habitat in Selected Ecosystems of North America.—*The Wildlife Society Technical Review* 16-01. The Wildlife Society, Bethesda, Maryland, USA. 69 p.

Bond M.L., Lee D.E., Bradley C.M. & Hanson C.T. 2009: Influence of Pre-Fire Tree Mortality on Fire Severity in Conifer Forests of the San Bernardino Mountains, California.—*The Open Forest Science Journal* 2, 41–47

Buchholz R. 2007: Behavioural biology: an effective and relevant conservation tool.—*Trends Ecol. Evol.* 22(8), 401–407. doi: https://doi.org/10.1016/j.tree.2007.06.002.

Buddle C. M., Spence J.R. & Langor D.W. 2000: Succession of boreal forest spider assemblages following wildfire and harvest.—*Ecography* 23, 424–436

Buddle, C.M. 2001: Spiders (Araneae) associated with downed woody material in deciduous forest in central Alberta, Canada. – iAgricultural and Forest Entomology /i 3, 241-251

Bird, M. I., Brand, M., Comley, R., Fu, X., Hadeen, X., Jacobs, Z., et al. (2024). Late Pleistocene emergence of an anthropogenic fire regime in Australia's tropical savannahs. Nature Geoscience, 17(3), 233–240. https://doi.org/10.1038/s41561-024-01388-3

Bowman, D. M. J. S., Balch, J., Artaxo, P., Bond, W. J., Cochrane, M. A., D'Antonio, C. M., et al. (2011). The human dimension of fire regimes on Earth. Journal of Biogeography, 38(12), 2223–2236. https://doi.org/10.1111/j.1365-2699.2011.02595.x

Bull, E.L. 2002: The value of coarse woody debris to vertebrates in the Pacific Northwest. – General technical report PSW-GTR-181, 883-892. U.S. Department of Agriculture Forest Service, Albany, California

Bundesamt für Naturschutz (ed.) 2021: Das europäische Schutzgebietssystem Natura 2000. BfN-Handbuch zur Umsetzung der Fauna-Flora-Habitat Richtlinie und der Vogelschutzgebiete. Bd. 2.1 Symank A. et al., Heft 172, 800 p. https://www.eea.europa.eu/themes/biodiversity/natura-2000

Byambasuren O. 2011: Fire and stand dynamics in different forest types of the West Khentey Mountains, Mongolia.—Dissertation at the Georg-August-Universität Göttingen, 116 p.

Cai, D., Abram, N. J., Sharples, J. J., & Perkins-Kirkpatrick, S. E. (2022). Increasing intensity and frequency of cold fronts contributed to Australia's 2019–2020 Black Summer fire disaster. *Environmental Research Letters*, *17*(9), 094044. https://doi.org/10.1088/1748-9326/ac8e88

Campbell J.I., Harmon M.E. & Mitchell S.R. 2012: Can fuel-reduction treatments really increase forest carbon storage in the western US by reducing future fire emissions?—*Frontiers in Ecology and the Environment* 10 (2), 83–90

Cardinale B.J., Primack R.B. & Murdoch J.D. 2020: *Conservation Biology* .—Sinauer Associates, Oxford University Press. 584 p.

Carter N, Killion A, Easter T, Brandt J & Ford A. 2020: Road development in Asia: Assessing the range-wide risks to tigers.—*Science Advances* 2020 Apr 29;6(18):eaaz9619. doi: 10.1126/sciadv.aaz9619. PMID: 32494684; PMCID: PMC7190336.

Cerri J., De Rosa, D., Fozzi I., Terraube J., Tavares J., Lörcher F., Secci D., Muzzeddu M., Seguin J.-F. & Berlinguer F. 2024: Movements of translocated Griffon Vultures (*Gyps fulvus*) from Sardinia to Corsica call for the transboundary management of an expanding vulture population.—*Rivista Italiana Di Ornitologia*, *94*(1). https://doi.org/10.4081/rio.2024.714

Chan L., Ng D and Lim L.J. 2023: *Handbook on Habitat Restoration. General Principles and Case Studies in Singapore.* - National Parks Board, 340 p. https://www.nparks.gov.sg/-/media/handbook-on-habitat-restoration/handbook-on-habitat-restoration-(full-book).pdf

Chiang J., McEwan R.W., Yaussy D.A. & Brown K.J. 2009: The effects of prescribed fire and silvicultural thinning on the aboveground carbon stocks and net primary production of overstory trees in an oak-hickory ecosystem in southern Ohio.—*Forest Ecology and Management* 255, 1584–1594

Cochrane M.A. 2009: Fire in the tropics.—In: *Tropical Fire Ecology*. Springer Praxis Books. Springer, Berlin, Heidelberg. https://doi.org/10.1007/978-3-540-77381-8_1

Conkling, T. J., Loss, S. R., Diffendorfer, J. E., Duerr, A. E., & Katzner, T. E. (2021). Limitations, lack of standardization, and recommended best practices in studies of renewable energy effects on birds and bats. *Conservation Biology*, *35*(1), 64–76. https://doi.org/10.1111/cobi.13457

Costanza R., d'Arge R., de Groot R., Farber S., Grasso M., Hannon B., Limburg K., Naeem S., O'Neill R.V.,

Paruelo J., Raskin R.G., Sutton P. & van den Belt M. 1997: The value of the world's ecosystem services and natural capital.—*Nature* 387, 253–260

Craig E. 2018: Mongolia's obsession with pine nuts.—https://www.bbc.com/travel/article/20180114-mongolias-obsession-with-pine-nuts

Crooks K.R. 2002: Relative Sensitivities of Mammalian Carnivores to Habitat Fragmentation.—*Conservation Biology* 16(2), 488–502

Cyriacks, P. (2014). Baustopper oder Schutzobjekt? Natur und Landschaft, 89(8), 364–369. https://doi.org/10.17433/8.2014.50153291.364-369

Dähne M., Gilles A., Lucke K., Peschko V., Adler S., Krügel K., Sundermeyer J. & Siebert U. 2013: Effects of pile-driving on harbour porpoises (*Phocoena phocoena*) at the first offshore wind farm in Germany.—*Environ. Res. Lett.* 8, 025002, 16p. doi:https://doi.org/10.1088/1748-9326/8/2/025002

Dai, J., Ahmed, Z., Alvarado, R., & Ahmad, M. (2024). Assessing the nexus between human capital, green energy, and load capacity factor: Policymaking for achieving sustainable development goals. *Gondwana Research*, *129*, 452–464. https://doi.org/10.1016/j.gr.2023.04.009

Davies G. & Brown D. (eds.) 2007. *Bushmeat and Livelihoods: Wildlife Management and Poverty Reduction*.—Conservation Science and Practice Series Vol. 2, Blackwell Publishing, Malden, USA, 274 p.

De Lucas M., Janss G.F.E., Whitfield D.P. & Ferrer M. 2008: Collision fatality of raptors in wind farms does not depend on raptor abundance.—*Journal of Applied Ecology* 45, 1695-1703. https://doi.org/10.1111/j.1365-2664.2008.01549.x

de Mello N.G.R., Gulinck H., Van den Broeck P. & Parra C. 2020: Social-ecological sustainability of non-timber forest products: A review and theoretical considerations for future research.—*Forest Policy and Economics* 112, 2020, 102109, ISSN 1389-9341, https://doi.org/10.1016/j.forpol.2020.102109

Dejid N., Olson K., Stratmann T.S.M. & Mueller T. 2022: An Extraordinary 18,000-km-Long Journey of a Gazelle through the Steppes of Mongolia.—*Ecology* e3660. https://doi.org/10.1002/ecy.3660

Department for Environment Food & Rural Affairs 2021: *Reintroductions and other conservation translocations*: code and guidance for England. Version 1.1, 77 p. https://www.gov.uk/government/publications/reintroductions-and-conservation-translocations-in-england-code-guidance-and-forms

Dickman, C. R. (2021). Ecological consequences of Australia's "Black Summer" bushfires: Managing for recovery. *Integrated Environmental Assessment and Management*, *17*(6), 1162–1167. https://doi.org/10.1002/ieam.4496

Dixon A., Batbayar N., Etheridge M., Purev-Ochir G. & Gombobaatar S. 2008: Development of the artificial nest project in Mongolia.—*Falco* 32, 7–10. https://www.researchgate.net/publication/270217609

Dixon A., Batbayar N., Purev-Ochir G. & Fox N. 2011: Developing a sustainable harvest of Saker Falcons (*Falco cherrug*) for falconry in Mongolia. Pages 363–372—In: Watson R.T., Cade T.J., Fuller M., Hunt G. & Potapov E. (eds.): *Gyrfalcons and Ptarmigan in a Changing World*, Volume II. The Peregrine Fund, Boise, Idaho, USA. https://doi.org/10.4080/gpcw.2011.0315

Dobbert S., Pape R. & Löffler J. 2021: Continent-wide Modelling of Potential Habitats and Connectivity for the Eurasian Lynx (*Lynx lynx*) in Europe.—*Erdkunde* 75 (2), 69–85

Drapeau P.A., Nappi A., Giroux J.-F., Leduc A. & Savard J.-P. L. 2002: Distribution patterns of birds associated with snags in natural and managed eastern boreal forests.—In: Laudenslayer, W.F. et al. (eds.): Proceedings of the symposium on the ecology and management of dead wood in western forests. USDA Forest Service General technical report PSW-GTR-181, U.S. Department of Agriculture Forest Service, Albany, California, 193–205

Drechsler M. & Burgman M.A. 2004: Combining population viability analysis with decision analysis.—*Biodiversity and Conservation* 13, 115–139

Dulamsuren Ch., Hauck M. & Mühlenberg, M. 2005: Vegetation at the taiga forest-steppe borderline in the western Khentej Mountains, northern Mongolia.—*Annales Botanici Fennici* 42: 411–426

Eckert I., Suchentrunk F., Markov G. & Hartl G.B. 2010: Genetic diversity and integrity of German wildcat (*Felis silvestris*) populations as revealed by microsatellites, allozymes, and mitochondrial DNA sequences.—*Mamm. Biol.* 75(2):160–174. doi:https://doi.org/10.1016/j.mambio.2009.07.005

Ehnström B. 2001: Leaving dead wood for insects in boreal forests—suggestions for the future. *Scand J. For. Res*. 16, 91–98

Eletu A. 2022: What Animals Eat Pine Cones?— https://www.animalquarters.com/animals-that-eat-pine-cones/. October 31st, 2022

Ellerbrok J.S., Delius A., Peter F., Farwig N. & Voigt C.C. 2022: Activity of forest specialist bats decreases towards wind turbines at forest sites.—*Journal of Applied Ecology* 2022;00:1–10. DOI:https://doi.org/10.1111/1365-2664.14249

Ellerbrok J.S., Farwig N., Peter F. & Voigt C.C. 2024: Forest bat activity declines with increasing wind speed in proximity of operating wind turbines.—*Global Ecology and Conservation* 49, https://doi.org/10.1016/j.gecco.2023.e02782

Emslie R.H., Amin R. & Kock R. (eds.) 2009: *Guidelines for the in-situ Re-introduction and Translocation of African and Asian Rhinoceros*.—Gland, Switzerland: IUCN. vi+115 p.

Energy Management, Office of Renewable Energy Programs n.d. OCS Study BOEM 2017-014.

EU_FFH-Directive n.d.: *Council Directive 92/43/EEC of 21 May 1992 on the conservation of natural habitats and of wild fauna and flora*

European Bird Census Council 2018: Pan-European Common Bird Monitoring Scheme. https://pecbms.info/ Accessed 10 Sept 2018

European Commission, Directorate-General for Environment, Environmental impact assessment of projects: rulings of the Court of Justice of the European union, Publications Office of the European Union, 2022, https://data.europa.eu/doi/10.2779/21

European Council. Directive 92 / 43 / EEC of 21 May 1992 on the conservation of natural habitats and of wild fauna and flora. , Pub. L. No. L 206 / 7 (1992a). https://eur-lex.europa.eu/legal-content/EN/TXT/?uri=celex%3A31992L0043

European Council 1992b: Council Directive 92/43/EEC of 21 May 1992 on the conservation of natural habitats and of wild fauna and flora.—Official Journal of the European Communities, No. I. 206/7

Evans W.G. 2005: Infrared radiation sensors of *Melanophila acuminata* (Coleoptera: Buprestidae): A thermopneumatic model.—*Annals of the Entomological Society of America* 98, 738–746

Fa J.E., Currie D. & Meeuwig J.J. 2003: Bushmeat and food security in The Congo Basin: Linkages between wildlife and people's future.—*Environmental Conservation* 30, 71–78

Fawcett J., Moorman C. & Sharpe T. 2021: *Using Fire to Improve Wildlife Habitat.*—NC State Extension Publications, AG-630, April 7, 2021; https://content.ces.ncsu.edu/using-fire-to-improve-wildlife-habitat#section_heading_7688

Fazzio C. 2022: *Prescribed Burns and Wildlife Concerns.*—April 28, 2022 https://www.nativeresourcepreservation.com/blog/prescribed-burns-and-wildlife-concerns

Federal Agency for Nature Conservation (ed.) 2021: *The European Protected Area System Natura 2000.* BfN Handbook for the implementation of the Fauna-Flora-Habitat Directive and the Bird Protection Areas. Vol. 2.1 Symank A. et al., Issue 172, 800 p.

Fei G., Xiong K., Fei G., Zhang H. & Zhang S. 2023: The conservation and tourism development of World Natural Heritage sites: The current situation and future prospects of research.—*J. f. Nature Conservation* 72, https://doi.org/10.1016/j.jnc.2023.126347

Feoktistova, N. Yu., Meschersky, I. G., Surov, A. V., Bogomolov, P. L., Tovpinetz, N. N., & Poplavskaya, N. S. (2016). Genetic structure of urban population of the common hamster (Cricetus cricetus). *Russ J. Genet.*, *52*(2), 194–203. https://doi.org/10.1134/S1022795416020046

Feoktistova, N. Yu., Meschersky, I. G., Bogomolov, P. L., Meschersky, S. I., Poplavskaya, N. S., Chunkov, M. M., et al. (2019). Genetic Structure of Urban and Suburban Populations of Common Hamster (Cricetus cricetus) in Ciscaucasia. *Russian Journal of Genetics*, *55*(3), 337–348. https://doi.org/10.1134/S1022795419020054

Feoktistova, Natalia Yu., Meschersky, I. G., Shenbrot, G. I., Puzachenko, A. Yu., Meschersky, S. I., Bogomolov, P. L., & Surov, A. V. (2023a). Phylogeography of the common hamster (Cricetus cricetus): paleoclimatic reconstructions of Late Pleistocene colonization. *Integrative Zoology*, *18*(3), 581–599. https://doi.org/10.1111/1749-4877.12687

Feoktistova N.Yu., Meschersky I.G., Katsman E.A., Meschersky S.I., Gureeva A.V., Kupriyanov V.P., Bogomolov P.L., Kropotkina M.V., Davydov I.D., Leonov A. G., Surov A.V. 2023b: Genetic differentiation within two large phylogroups of the common hamster (*Cricetus cricetus*) (Cricetidae, Rodentia), the boundary between them, and the population abundance *Povolzhskiy Journal of Ecology*, 2023, no. 4, 481–497 (in Russian). https://doi.org/10.35885/1684-7318-2023-4-481-497

Fernandes, P. M. (2018). Scientific support to prescribed underburning in southern Europe: What do we know? Science of The Total Environment, 630, 340–348. https://doi.org/10.1016/j.scitotenv.2018.02.214

Fischborn M. 2011: Influence of Forest Structure on Fire Severity after a Forest Fire in North Mongolia.—Master thesis at the Georg-August University Göttingen, 92 p.

Fischer E.-L. 2010: The influence of fire severity on bird species composition and density in a natural boreal forest of West Khentey, Mongolia. – Diplom Thesis at Georg-August-Universität Göttingen, 88 p.

Fletcher K., Warren P. & Baines D. 2005: Impact of nest visits by human observers on hatching success in Lapwings *Vanellus vanellus*: a field experiment.—*Bird Study* 52, 221–223

Forst R., Porzelt M. & Scherfose V. (eds.) 2019: *Konflikte durch Erholungsnutzung in Großschutzgebieten und deren Entschärfung durch innovatives Besuchermanagement.*—Bundesamt für Naturschutz, BfN-Skripten 520, Bonn, 183 p.

Frost G.H. & Bond I. 2006. *CAMPFIRE and the payment for environmental services.*—International Institute for Environment and Development, London. 26 p. https://www.cifor-icraf.org/publications/pdf_files/15503IIED.pdf at www.iied.org or from sustainablemarkets@iied.org.

Garthe S. & Hüppop O. 2004: Scaling possible adverse effects of marine wind farms on seabirds: developing and applying a vulnerability index.—*Journal of Applied Ecology* 41 (4), 724–734. https://doi.org/10.1111/j.0021-8901.2004.00918.x

Garvin J.C., Jennelle C.S., Drake D. & Steven M. Grodsky S.M. 2011: Response of raptors to a windfarm.—*Journal of Applied Ecology* 48, 199–209. https://doi.org/10.1111/j.1365-2664.2010.01912.x

Gerlach, G., & Musolf, K. (2000). Fragmentation of Landscape as a Cause for Genetic Subdivision in Bank Voles. *Conservation Biology*, *14*(4), 1066–1074. https://doi.org/10.1046/j.1523-1739.2000.98519.x

Ghasemi, B. (2021). Trophy hunting and conservation: Do the major ethical theories converge in opposition to trophy hunting? *People and Nature*, *3*(1), 77–87. https://doi.org/10.1002/pan3.10160

Ghasemi, B., Kyle, G., Sell, J., & Varner, G. (2023). Conservation scholars' perspectives on the morality of trophy hunting for the sake of conservation. *People and Nature*, *5*(6), 2061–2073. https://doi.org/10.1002/pan3.10539

Ghoddousi A., Soofi M., Hamidi A. Kh., Lumetsberger T., Egli L., Ashayeri S., Khorozyan I., Kiabi B.H. & Waltert M. 2017: When pork is not on the menu: Assessing trophic competition between large carnivores and poachers.—*Biological Conservation* 209, 223–229. https://doi.org/10.1016/j.biocon.2017.02.032

Glasson, J., & Therivel, R. (2019). *Introduction To Environmental Impact Assessment* (5th edition.). London: Routledge.

Goodale M. 2018: Cumulative adverse effects of offshore wind energy development on wildlife.—Doctoral Dissertations. 1343. https://doi.org/10.7275/12512053, https://scholarworks.umass.edu/dissertations_2/1343

Goodship N.M. & Furness R.W. (MacArthur Green) 2022: Disturbance Distances Review: An updated literature review of disturbance distances of selected bird species.—*NatureScot Research Report* 1283.

Grafe T.U., Döbler S. & Linsenmair K.E. 2002: Frogs flee from the sound of fire.— *Proc. R. Soc. Lond. B.* 269:999–1003

Greenberg C.H., Neary D.G. & Harris L.D. 1995: A comparison of bird communities in burned and salvaged-logged clearcut, and forested Florida Sand Pine Scrub.—*Wilson Bulletin* 107, 40–54

Greene D.F., Zasada J.C., Sirois L., Kneeshaw D., Morin H., Charron I. & Simard M.J. 1999: A review of the regeneration dynamics of North American boreal forest tree species.—*Canadian Journal of Forest Research* 29(6), 824–839.

Greggor A.L., Price C.J. & Shier D.M. 2019: Examining the efficacy of anti-predator training for increasing survival in conservation translocations: a systematic review protocol.—*Environ. Evid.* 8 (Suppl 1), 11. https://doi.org/10.1186/s13750-019-0154-6

Greggor A.L., Masuda B.M., Sabol A.C. & Swaisgood R.R. 2021: Designing anti-predator training to maximize learning and efficacy assessments.—*bioRxiv* 2021.11.30.470590; doi: https://doi.org/10.1101/2021.11.30.470590

Griffiths R.A. & Pavajeau L. 2008: Captive Breeding, Reintroduction, and the Conservation of Amphibians.—*Conservation Biology* 22(4), 852–861. https://doi.org/10.1111/j.1523-1739.2008.00967.x

Groom M.J., Meffe G.K. & Carroll C.R. 2006: *Principles of conservation biology* (3rd ed.).—Sinauer Associates, Inc. Sunderland, Massachusetts U.S.A., 793 p.

Gundersen V., Vistad O.I., Panzacchi M., Strand O. & van Moorter B. 2019: Large-scale segregation of tourists and wild reindeer in three Norwegian national parks: Management implications.—*Tourism Management* 75, 22–33

Gunther, P. M., & Winfrey, J. (2005). Biodiversity issues for road reconstruction adjacent to a river with a federally listed threatened fish species. *Impact Assessment and Project Appraisal*, *23*(1), 17–27. https://doi.org/10.3152/147154605781765788

Günther A., Nigmann U., Achtziger R. & Gruttke H. (eds.) 2005: *Analysis of the causes of endangerment of species relevant to planning.*—Conservation and Biological Diversity, Volume 21, Federal Agency for Nature Conservation, Bad Godesberg, 605 p.

Haque, K. M. S., Uddin, M., Ampah, J. D., Haque, Md. K., Hossen, Md. S., Rokonuzzaman, Md., et al. (2023). Wildfires in Australia: a bibliometric analysis and a glimpse on 'Black Summer' (2019/2020) disaster. *Environmental Science and Pollution Research*, *30*(29), 73061–73086. https://doi.org/10.1007/s11356-023-27423-1

Hartmann S.A., Steyer K., Kraus R.H.S., Segelbacher G. & Nowak C. 2013: Potential barriers to gene flow in the endangered European wildcat (*Felis silvestris*).—*Conservation Genetics* 14, 413–426

Harwood J. 2000: Risk assessment and decision analysis in conservation. - *Biological Conservation* 95, 219–226

Henschel P., Hunter L.T.B., Coad L., Abernethy K.A. & Mühlenberg M. 2011: Leopard prey choice in The Congo Basin rainforest suggests exploitative competition with human bushmeat hunters.—*Journal of Zoology* 285 (1), 11–20

Herrmann M. and Mathews A. 2007: Wirkung von Barrieren auf Säuger und Reptilien.—Deutscher Jagdverband: Verbände-Vorhaben *„Überwindung von Barrieren“*, gefördert vom Bundesamt für Naturschutz, 64 p. https://www.jagdverband.de/sites/default/files/herrmann_endberichtdok20_0.pdf

Hertwig S.T., Schweizer M., Stepanow S., Jungnickel A., Bohle U.R. & Fischer M.S. 2009: Regionally high rates of hybridization and introgression in German wildcat populations (*Felis silvestris* , Carnivora, Felidae).—J. Zool. Syst. Evol. Res. 47(3), 283–297. doi:https://doi.org/10.1111/j.1439-0469.2009.00536.x

Hessl A., Ariya U., Brown P., Byambasuren O., Green T., Jacoby G., Sutherland E., Baatarbileg N., Maxwell S., Pederson N., De Grandpré L., Saladyga T. & Tardif J. 2011: Reconstructing fire history in central Mongolia from tree-rings.—*International Journal of Wildland Fire* 21. https://doi.org/10.1071/WF10108.

Hill I.D.C., Rossi C.A., Petrovan S.O., Hartup M., Clark F. & Downie J.R. 2018: Mitigating the effects of a road on amphibian migrations: a Scottish case study of road tunnels.—*The Glasgow Naturalist* 27(Supp),25–37. The Amphibians and Reptiles of Scotland.

Holt G. 2020: Non-Timber Forest Products.—*The Boreal Forest Newsletter* Spring 2020, 1–5, University of Alaska Fairbanks

Huck M., Jedrzejewski W., Borowik T., Milosz-Cielma M., Schmidt K., Jedrzejewska B., Nowak S. & Myslajek R.W. 2010: Habitat suitability, corridors and dispersal barriers for large carnivores in Poland.—*Acta Theriologica* 55, 177–192

Israel M. & Reinhard A. 2017: Detecting nests of lapwing birds with the aid of a small unmanned aerial vehicle with thermal camera.—In: IEEE *International Conference on Unmanned Aircraft Systems*, (ICUAS),

1199–1207. IEEE. Miami, USA. doi: https://doi.org/10.1109/ICUAS.2017.7991393.

IUCN Commission on Environmental Policy, Law and Administration (CEPLA), IUCN Species Survival Commission (SSC), & IUCN Commission on Ecology. (1987). *The IUCN position statement on translocation of living organisms : introductions, re-introductions and re-stocking*. Gland, Switzerland: IUCN. https://portals.iucn.org/library/node/6507. Accessed 24 July 2024

IUCN Species Survival Commission (SSC), & IUCN Reintroduction Specialist Group. (1998). *IUCN guidelines for re-introductions*. Gland, Switzerland: IUCN. https://portals.iucn.org/library/node/7405. Accessed 24 July 2024

IUCN/SSC 2013:. *Guidelines for Reintroductions and Other Conservation Translocations. Version 1.0.*—Gland, Switzerland: IUCN Species Survival Commission, viiii + 57 pp.

Jochimsen D.M., Peterson C.R., Andrews K.M. & Whitfield Gibbons J. 2004: A literature review of the effects of roads on amphibians and reptiles and the measures used to minimize those effects.—*Idaho Fish and Game Department and USDA Forest Service Report.* 2004. https://idfg.idaho.gov/old-web/docs/wildlife/collisionAmphibRep.pdf

Johnstone J.F. & Chapin F.S. 2006: Effects of soil burn severity on postfire tree recruitment in boreal forests.—*Ecosystems* 9, 14–31

Johnstone J.F., Hollingsworth T.N. & Chapin F.S. 2008: *A key for predicting postfire successional trajectories in Black Spruce Stands of Interior Alaska.*—Gen. Tech. Rep. PNW-GTR-767, Portland, OR, Department of Agriculture, Forest Service, Pacific Northwest Research Station, 37 p.

Jonášová M. & Prach K. 2008: The influence of bark beetles' outbreak vs. salvage logging on ground layer vegetation in Central European mountain spruce forests. *Biol Cons.* 141, 1525–1535

Kaczensky P, Ganbataar O, Altansukh N, Enkhsaikhan N, Stauffer C, & Walzer C. 2011: The Danger of Having All Your Eggs in One Basket—Winter Crash of the Re-Introduced Przewalski's Horses in the Mongolian Gobi.—*PLoS ONE* 6(12): e28057, 8. doi:https://doi.org/10.1371/journal.pone.0028057

Kahiu M.N. & Hanan N.P. 2018: Fire in sub-Saharan Africa: The fuel, cure and connectivity hypothesis.—*Global Ecology and Biogeography* 2018;00:1–12. https://doi.org/10.1111/geb.12753

Kaila L., Martikainen P. & Punttila P. 1997: Dead trees left in clear cuts benefit saproxylic Coleoptera adapted to natural disturbances in boreal forest. *Biodivers Conserv.* 6, 1–18

Kansky, R., & Kidd, M. (2024). Putting yourself in an animal's shoes - empathy and intangible benefits drive tolerance towards wildlife in Namibian communal conservancies. *BIOLOGICAL CONSERVATION, 293*, 110588. https://doi.org/10.1016/j.biocon.2024.110588

Kaule G. 1991: *Arten- und Biotopschutz.* 2. Aufl..—UTB für Wissenschaft, Verlag Eugen Ulmer, 519 p.

Kelly, T. R., Rideout, B. A., Grantham, J., Brandt, J., Burnett, L. J., Sorenson, K. J., et al. (2015). Two decades of cumulative impacts to survivorship of endangered California condors in California. *Biological Conservation, 191*, 391–399. https://doi.org/10.1016/j.biocon.2015.07.012

Khan, S. J. (2021). Ecological consequences of Australian "Black Summer" (2019–20) fires: A synthesis of Australian Commonwealth Government report findings. *Integrated Environmental Assessment and Management, 17*(6), 1136–1140. https://doi.org/10.1002/ieam.4469

Kiehl K. 2009: Renaturierung von Kalkmagerrasen.—In: Zerbe, S., Wiegleb, G. (eds): *Renaturierung von Ökosystemen in Mitteleuropa.*—Spektrum Akademischer Verlag. https://doi.org/10.1007/978-3-8274-2161-6_10

King D.I. and DeGraaf R.M. 2000: Bird species diversity and nesting success in mature, clearcut and shelterwood forest in northern New Hampshire, USA.—*Forest Ecol. Manag.* 129, 227–235

Klar N., Herrmann M. and Kramer-Schadt S. 2009: Effects and Mitigation of Road Impacts on Individual Movement Behavior of Wildcats.—*Journal of Wildlife Management* 73(5), 631–638. https://doi.org/10.2193/2007-574

Klar N., Herrmann M., Henning-Hahn M., Pott-Dörfer B., Hofer H. & Kramer-Schadt S. 2012: Between ecological theory and planning practice: (Re-) Connecting forest patches for the wildcat in Lower Saxony, Germany.—*Landscape and Urban Planning* 105 (4), 376–384

Kleiman D.G. 1989: Reintroduction of Captive Mammals for Conservation.—*BioScience* 39 (3), 152–161. https://doi.org/10.2307/1311025

Knight M. 2021: Moving African rhinos: what it takes to translocate an endangered species.—*The Conservation*, published: December 14, 2021 3.30pm CET.

Kobelt R.F. & Linsenmair K.E. 1992: Adaptations of the reed frog *Hyperolius viridiflavus* (Amphibia, Anura, Hyperoliidae) to its arid environment: VI. The iridophores in the skin of *Hyperolius viridiflavus taeniatus* as radiation reflectors.—*Journal of Comparative Physiology* B, 162, 314–326

Kouki J., Hyvärinen E., Lappalainen H., Martikainen P. & Similä M. 2012: Landscape context affects the success of habitat restoration: large-scale colonization patterns of saproxylic and fire associated species in boreal forests.—*Divers. Distrib.* 18, 348–355

Kryštufek, B., Pozdnyakov, A. A., Ivajnšič, D., & Janžekovič, F. (2016). Low phenotypic variation in eastern common hamsters Cricetus cricetus.— *Folia Zool., 65*(2), 148–156

Kumara H.N., Babu S., Rao G.B., Mahato S., Bhattacharya M., Ranga Rao N.V., Tamiliniyan D., Parengal H., Deepak D., Balakrishnan A. & Bilaskar M. 2022: Responses of birds and mammals to long-established wind farms in India.—*Scientific Reports* 12, 1339. https://doi.org/10.1038/s41598-022-05159-1

Kuwaczka L.F., Mitterwallner V., Audorff V. & Steinbauer M.J. 2023: Ecological impacts of (electrically assisted) mountain biking.—*Global Ecology and Conservation* 44, https://doi.org/10.1016/j.gecco.2023.e02475

La Haye, M. J. J., Reiners, T. E., Raedts, R., Verbist, V., & Koelewijn, H. P. (2017). Genetic monitoring to evaluate reintroduction attempts of a highly endangered rodent. *Conservation Genetics*, *18*(4), 877–892. https://doi.org/10.1007/s10592-017-0940-z

Latham P., Fiore W., Bauman M. & Weaver J. 2017: *Effects Matrix for Evaluating Potential Impacts of Offshore Wind Energy Development on U.S. Atlantic Coastal Habitats.*—Final Report to the U.S. Department of the Interior, Bureau of Ocean

Latombe, G., Lenzner, B., Schertler, A., Dullinger, S., Glaser, M., Jarić, I., et al. (2022). What is valued in conservation? A framework to compare ethical perspectives. *NeoBiota*, *72*, 45–80. https://doi.org/10.3897/neobiota.72.79070

Lee, S.X.T., Amir, Z., Moore, J.H., Gaynor K.M. and Luskin M.S. 2024: Effects of human disturbances on wildlife behaviour and consequences for predator-prey overlap in Southeast Asia. - *Natures Communications* 15, 1521 (2024). https://doi.org/10.1038/s41467-024-45905-9

Lei B., Zheng Z., Cui J., Zhao J., Newman C. & Zhou Y. 2023: Ecotourist trail-use affects the taxonomic, functional and phylogenetic diversity of mammals in a protected area: lessons for conservation management . - *Integrative Zoology* 18 (4), 647–660

Lindenmayer D., Burton P.J. & Franklin J.F. 2008: *Salvage Logging and Its Ecological Consequences.*—Island Press, UBC Press Canada. 277 p.

Lindsey P.A,. Balme G.A., Funston P., Henschel P., Hunter L., Madzikanda H., Midlane N. & Nyirenda V. 2013: The Trophy Hunting of African Lions: Scale, Current Management Practices and Factors Undermining Sustainability.—*PloS ONE* 8(9): e73808. https://doi.org/10.1371/journal.pone.0073808

Linklater W.L., Adcock K., du Preez P., Swaisgood R.R., Law P.R., Knight M.H., Gedir J.V. & Kerley G.I.H. 2011: Guidelines for large herbivore translocation simplified: black rhinoceros case study.—*Journal of Applied Ecology* 48(2), 493–502. http://www.jstor.org/stable/20869965

Linsenmair K.E. 1998: Risk-spreading and risk reducing tactics of West African anurans in an unpredictably changing and stressful environment. – pp. 221–242 in: Newberry D.M., Prins H.H.T. & Brown N.D. (eds.): *Dynamics of Tropical Communities.*—London (Blackwell Science)

Loveridge A.J., Valeix M., Chapron G., Davidson Z., Mtare G. & Macdonald D.W. 2016a: Conservation of large predator populations: Demographic and spatial responses of African lions to the intensity of trophy hunting.—*Biological Conservation* 204, Part B, 247–254. https://doi.org/10.1016/j.biocon.2016.10.024

Loveridge A.J., Valeix M., Elliot N.B. & Macdonald D.W. 2016b: The landscape of anthropogenic mortality: how African lions respond to spatial variation in risk.—*Journal of Applied Ecology* 54 (3), 815–825

Loveridge A.J., Wijers M., Mandisodza-Chikerema R., Macdonald D.W. & Chapron G. 2023a: Anthropogenic edge effects and aging errors by hunters can affect the sustainability of lion trophy hunting.—*Sci. Rep.* 13, 95. https://doi.org/10.1038/s41598-022-25020-9

Loveridge A.J. et al. 2023b: Raising the minimum age of trophy hunted lions could make hunting more sustainable, new study finds. - https://www.ox.ac.uk/news/2023-01-13-raising-minimum-age-trophy-hunted-lions-could-make-hunting-more-sustainable-new. News & Events, University of Oxford, 13 Jan 2023.

Madders M & Whitfield D.P. 2006: Upland raptors and the assessment of wind farm impacts.—*Ibis* 148 (s1), 43–56

Mader H.-J. 1979: Die Isolationswirkung von Verkehrsstraßen auf Tierpopulationen untersucht am Beispiel von Arthropoden und Kleinsäugern der Waldbiozoenosen.—*Schriftenreihe für Landschaftspflege und Naturschutz* 19, 1–119

Mader H.-J. 1984: Animal habitat isolation by roads and agricultural fields.—*Biological Conservation*, 29, 81–96

Mader H.-J., Schell C. and Kornacker P. 1990: Linear barriers to arthropod movements in the landscape.—*Biological Conservation* 54, 209–222

Maguire L.A. 1986: Using decision analysis to manage endangered species populations.—*Journal of Environmental Management* 22, 345–360

Mammen K. & Mammen U. 2003: Möglichkeiten und Grenzen der Umsiedlung von Feldhamstern (*Cricetus cricetus*),—In: Stubbe M. & Stubbe A. (eds.): *Methoden feldökologischer Säugetierforschung* 2, 461–470. Halle/Saale. http://www.oekotop-halle.de/pubs/mammen_mammen_umsiedlung_feldhamster_2003.pdf

Mareddy A.R. 2017: *Environmental Impact Assessment. Theory and Practice.*—Butterworth-Heinemann, 632 p.

Mathews A.E. 1989: Conflict, controversy, and compromise: The Concho water snake (*Nerodia harteri paucimaculata*) versus the Stacy Dam and Reservoir.—*Environmental Management* 13, 297–307

Maunders M. & Byers O. 2005: The IUCN Technical Guidelines on the Management of Ex Situ Populations for Conservation: reflecting major changes in the application of ex situ conservation.—*Oryx* 39, 95–98

Mayfield H. 1975: Suggestions for calculating nest success.—*Wilson Bull.* 87, 456–466

McClung M.R., Seddon P.J., Massaro M. & Setiawan A.N. 2004: Nature-based tourism impacts on yellow-eyed penguins *Megadyptes antipodes*: does unregulated visitor access affect fledging weight and juvenile survival?—*Biological Conservation* 119, 279–285

Meinig, H., Weinhold, U., Dalbeck, L., Zimmermann, M., Kayser, A., Köhler, U., et al. (2020a). Cricetus cricetus - Feldhamster. *Cricetus cricetus - Feldhamster*.

Meinig H.U., Boye P., Dähne M., Hutterer R. & Lang J. 2020b: Rote Liste der Tiere, Pflanzen und Pilze Deutschlands. Säugetiere.—In: *Rote Liste und Gesamtartenliste der Säugetiere (Mammalia) Deutschlands* (p.1–73). -Publisher: Bundesamt für Naturschutz (BfN) (Hrsg.).

Melosik, I., Ziomek, J., Winnicka, K., Reiners, T. E., Banaszek, A., Mammen, K., et al. (2017). The genetic characterization of an isolated remnant population of an endangered rodent (Cricetus cricetus L.) using comparative data: implications for conservation. *Conservation Genetics*, *18*(4), 759–775. https://doi.org/10.1007/s10592-017-0925-y

Merganičová K., Merganič J., Svoboda M., Bače R. & Šebeň V. 2012: Deadwood in Forest Ecosystems.—p. 82–108 in: Blanco J.A. (ed.): *Forest Ecosystems—More than just trees*. InTech, Available from: http://www.intechopen.com/books/forest-ecosystems-more-than-just-trees/deadwood. www.intechopen.com

Michael-Otto-Institut im NABU. (2019, November 20). Lapwing Conservation.org | English information. *Lapwing Conservation.org*. https://lapwingconservation.org/english-information/. Accessed 18 August 2024

Miina J., Kurttila M., Calama R., de-Miguel S. & Pukkala T. 2020: Modelling Non-timber Forest Products for Forest Management Planning in Europe.— *Curr. Forestry Rep.* 6, 309–322. https://doi.org/10.1007/s40725-020-00130-7

Miller B., Reading R. & Forrest S. 1996: *Prairie night: Black-footed Ferrets and Recovery of Endangered Species.*—Smithsonian Institution Press, Washington, D.C., 254 p.

Milner-Gulland E.J. & Rowcliffe M. 2007: *Conservation and Sustainable Use. A Handbook of Techniques.*—Oxford University Press, 310 p.

MIT Climate: Do wind turbines kill birds?—Climate Portal, https://climate.mit.edu/ask-mit/do-wind-turbines-kill-birds. Updated December 12, 2023

Morissette J.L., Cobb T.P., Brigham R.M. & James P.C. 2002: The response of boreal forest songbird communities to fire and post-fire harvesting.—*Canadian Journal of Forest Research* 32, 2169–2183

Mühlenberg M. & Spellerberg I. 2010: Nature Conservation as a science., p. 74–94 In: Spellerberg I., Slowik J., Mühlenberg M., & Dgebuadze,Y.Y. (eds): *Biological Diversity and Nature Conservation: Theory and Practice for Teaching*. Moscow. 266 p.

Müller M., Olek M., Giersig M. & Schmitz H. 2008: Micromechanical properties of consecutive layers in specialized insect cuticle: the gula of *Pachnoda marginata* (Coleoptera, Scarabaeidae) and the infrared sensilla of *Melanophila acuminata* (Coleoptera, Buprestidae).—*J. exp. biol.* 211 (2008): 2576–2583

Müller J., Reed N., Bussler H. & Brandl R. 2010: Learning from a "benign neglect strategy" in a national park: response of saproxylic beetles to dead wood accumulation.—*Biol. Conserv.* 143, 2559–2569

Müller J., Jarzabek-Müller A. & Bussler H. 2013: Some of the rarest European saproxylic beetles are common in the wilderness of Northern Mongolia.—*Insect Conservation* 17, 989–1001

Murindagomo F. 1990: Wildlife management in Zimbabwe: The CAMPFIRE programme.—In: *Living with wildlife*, Washington, D.C., World Bank, 1990. https://www.fao.org/4/U5200E/u5200e06.htm

Nappi A., Drapeau P. & Savard J.-P.L. 2004: Salvage logging after wildfire in the boreal forest: Is it becoming a hot issue for wildlife?—*The Forestry Chronicle* 80 (1), https://doi.org/10.5558/tfc80067-1

Nasi R., Taber A. and Van Vliet N. 2011: Empty forests, empty stomachs? Bushmeat and livelihoods in The Congo and Amazon Basins.—*International Forestry Review*, 13 (3), 355–368

National Park Service 2017: *California Condor Reintroduction & Recovery*. –https://www.nps.gov/articles/california-condor-recovery.htm

Neary, D. G., Klopatek, C. C., DeBano, L. F., & Ffolliott, P. F. (1999). Fire effects on belowground sustainability: A review and synthesis. Forest Ecology and Management, 122(1–2), 51–71. https://doi.org/10.1016/S0378-1127(99)00032-8

Newton J.L. & Freyfogle E.T. 2005: Sustainability: a Dissent—*Conservation Biology* 19(1), 23–32. https://doi.org/10.1111/j.1523-1739.2005.538_1.x

Nilsson A.L.K., Molværsmyr S., Breistøl A. & Systad G.H.R. 2023: Estimating mortality of small passerine birds colliding with wind turbines.—*Scientific Reports* 13, 21365 (2023). https://doi.org/10.1038/s41598-023-46909-z

Obioha E.E., Isiugo P.N., Jimoh S.O., Ikyaagba E., Ngoufo R., Bobo K.S. & Waltert M. 2012: Bushmeat harvesting and human subsistence nexus in the Oban Hill Communities of Nigeria.—*Journal of Human Ecology* 38, 49–64

Oliveira R., Godinho R., Randi E., Ferrand N. & Alves P.C. 2007: Molecular analysis of hybridisation between wild and domestic cats (*Felis silvestris*) in Portugal: implications for conservation.—*Cons. Genet.* 9(1), 1–11

Pandey A.K., Tripathi Y.C. & Kumar A. 2016: Non Timber Forest Products (NTFPs) for Sustained Livelihood: Challenges and Strategies.—*Research Journal of Forestry* ISSN 1819-3439. DOI: https://doi.org/10.3923/rjf.2016.

Pangau-Adam M., Noske R. & Muehlenberg M. 2012: Wildmeat or Bushmeat? Subsistence Hunting and Commercial Harvesting in Papua (West New Guinea), Indonesia. – *Hum. Ecol.* https://doi.org/10.1007/s10745-012-9492-5; published online Springer.

Pausas J.G., Ouadah N., Ferran A., Gimeno T. & Vallejo R. 2003: Fire severity and seedling establishment in *Pinus halepensis* woodlands, eastern Iberian Peninsula.—*Plant Ecology* 169, 205–213

Pearce-Higgins J.W., Finney S.K., Yalden D.W. & Langston R.H.W. 2007: Testing the effects of recreational disturbance on two upland breeding waders.—*Ibis* 149(s1), 45–55

Pereira S., Prieto A., Calama R. & Diaz-Balteiro L. 2015: Optimal management in *Pinus pinea* L. stands combining silvicultural schedules for timber and cone production.—*Silva Fennica* 49 no. 3 article id 1226. https://doi.org/10.14214/sf.1226

Perrins C. (ed.) 2003: *The New Encyclopedia of Birds*.—Oxford University Press

Petts J. (ed.) 1999: *Handbook of Environmental Impact Assessment*. Vol. I Principles, Process and Methods, Vol. II Practice: Impact and Limitations.—Blackwell Science Ltd., Vol. I, 496 p., Vol. II 449 p.

Pfannerstill V., Balkenhohl N., Bennitt E., Maboga O.S. & Scheumann M. 2023: Assessing the potential of conspecific playbacks as a post-translocation management tool for white rhinoceros.—*Conservation Science and Practice*, 14 p., e12996. https://doi.org/10.1111/csp2.12996

Pienkowski M.W. 1992: The impact of tourism on coastal breeding waders in western and southern Europe: an overview.—*WaderStudyGroupBull.* 68: 92–96

Quintana I, Cifuentes EF, Dunnink JA, Ariza M, Martínez-Medina D, Fantacini FM, Shrestha BR & Richard FJ. 2022: Severe conservation risks of roads on apex predators.—*Scientific Reports* 2022 Feb 21;12(1):2902. doi: 10.1038/s41598-022-05294-9. PMID: 35190573; PMCID: PMC8861172.

Radley, P., & Amidon, F. (2008). *Conservation Introduction of the Bridled White-eye (Zosterops conspicillatus saypani) from the Island of Saipan to the Island of Sarigan, Commonwealth of the Northern Mariana Islands.* Northern Mariana Islands: Division of Fish and Wildlife, Commonwealth of the Northern Mariana Islands. https://doi.org/10.13140/RG.2.1.1358.4484

Raman, R., Nair, V. K., Shivdas, A., Bhukya, R., Viswanathan, P. K., Subramaniam, N., & Nedungadi, P. (2023). Mapping sustainability reporting research with the UN's sustainable development goal. *Heliyon*, *9*(8). https://doi.org/10.1016/j.heliyon.2023.e18510

Rees E.C. 2012: Impacts of wind farms on swans and geese: A review.—*Wildfowl* 62, 37–72

Reidler K. 2016: California Condors, Better Births, Incredible Intervention.—*Organikos*, https://organikos.net/2016/02/26/california-condors-better-births-incredible-intervention/ 2016/02/26

Remmert, H. (Ed.). (1991). *The Mosaic-Cycle Concept of Ecosystems* (Vol. 85). Berlin, Heidelberg: Springer. https://doi.org/10.1007/978-3-642-75650-4

Ripple W.J., Abernethy K., Betts M.G., Chapron G., Dirzo R., Galetti M.., Levi T., Lindsey P.A., Macdonald D.W., Machovina B., Newsome T.M., Peres C.A., Wallach A.D., Wolf C. & Young H. 2016: Bushmeat hunting and extinction risk to the world's mammals.—*R. Soc. Open Sci.* 2016 Oct 19;3(10):160498. https://doi.org/10.1098/rsos.160498. PMID: 27853564; PMCID: PMC5098989.

Roland J., Keyghobadi N. & Fownes S. 2000: Alpine *Parnassius* Butterfly Dispersal: Effects of Landscape and Population Size.—*Ecology* 81 (6), 1642–1653

Ros-Tonen M.A.F. 2000: The role of non-timber forest products in sustainable tropical forest management.—*Holz als Roh- und Werkstoff* 58, 196–201, Springer-Verlag

Roth H.H. & Merz G. (eds.) 1997: *Wildlife Resources. A Global Account of Economic Use*.—Springer-Verlag, 403 p.

Rotich D. 2012: Concept of Zoning Management in Protected Areas.—*Journal of Environment and Earth Science* 2, (10), 173–183

Safari Club International Sept. 21, 2015: Hunting is Conservation: The CAMPFIRE Program in Zimbabwe. https://safariclub.org/hunting-is-conservation-the-campfire-program-in-zimbabwe/

Saint-Germain M., Drapeau P. & Buddle C.M. 2008: Persistence of pyrophilous insects in fire-driven boreal forests: population dynamics in burned and unburned habitats.—*Divers. Distrib.* 14, 713–720

Sheuyange, A., Oba, G., & Weladji, R. B. (2005). Effects of anthropogenic fire history on savanna vegetation in northeastern Namibia. Journal of Environmental Management, 75(3), 189–198. https://doi.org/10.1016/j.jenvman.2004.11.004

Samuelsson J., Gustafsson L. and Ingelög T. 1994: *Dying and dead trees*. A review of their importance for biodiversity.—Swedish Threatened Species Unit, Uppsala, Sweden, 109 p.

Sanford, R. M., & Holtgrieve, D. G. (2022). *Environmental Impact Assessment in the United States* (1st edition.). Routledge.

Schall P., Gossner M.M., Heinrichs S., Fischer M., Boch S., Prati D., Jung K., Baumgartner V., Blaser S., Böhm S., Buscot F., Daniel R., Goldmann K., Kaiser K., Kahl T., Lange M., Müller J., Overmann J., Renner S.C., Schulze E-D., Sikorski J., Tschapka M., Türke M., Weisser W.W., Wernheuer B., Wubet T. & Ammer C. 2018: The impact of even-aged and uneven-aged forest management on regional biodiversity of multiple taxa in European beech forests.—*J. Appl. Ecol.* 55, 267–278. https://doi.org/10.1111/1365-2664.12950

Schmidt B.R. & Zumbach S. 2008: Amphibian road mortality and how to prevent it: a review. – p. 157–167 in: Mitchell J.C., Jung Brown R.E., Bartolomew B. (eds.): *Urban Herpetology.* St. Louis: Society for the Study of Amphibians and Reptiles.

Schmiegelow F.K.A., Stepnisky D.P., Stambaugh C.A. & Koivula M. 2006: Reconciling salvage logging of boreal forests with a natural-disturbance management model.—*Conservation Biology* 20 (4), 971–983

Schmitz H. & Bleckmann H. 1998: The photomechanic infrared receptor for the detection of forest fires in the beetle *Melanophila acuminata* (Coleoptera: Buprestidae).—*J. Comp. Physiol. A* 182, 647–657

Schoenemann S. 2009: Nest site distribution and reproductive timing of the African chameleon (*Chamaeleo africanus*) in a high intensity tourism area in the southwestern Peloponnese, Greece. – Diploma thesis at the University Goettingen.

Schulze E.-D., Wirth C., Mollicone D. & Ziegler W. 2005: Succession after stand replacing disturbances by fire, wind throw, and insects in the dark Taiga of Central Siberia.—*Oecologia* 146, 77–88

Schulze, E.D., Craven, D., Durso, A.M., Reif J., Guderle M., Kroiher F., Hennig P., Weiserbs A., Schall P., Ammer C. & Eisenhauer N. 2019: Positive association between forest management, environmental change, and forest bird abundance .— *Forest Ecosystems* 6, 3. https://doi.org/10.1186/s40663-019-0160-8

Schütz S., Weissbecker B., Hummel H.E., Apel K-H., Schmitz H. & Bleckmann H. 1999: Insect antenna as a smoke detector.—*Nature* 398, 298–299. doi: https://doi.org/10.1038/18585

Scoones, I., Stirling, A., Abrol, D., Atela, J., Charli-Joseph, L., Eakin, H., et al. (2020). Transformations to sustainability: combining structural, systemic and enabling approaches. *Current Opinion in Environmental Sustainability*, *42*, 65–75. https://doi.org/10.1016/j.cosust.2019.12.004

Secretariat of the Convention on Biological Diversity. (2024). Case Studies - Impact Assessment. *Case Studies on Biodiversity and Impact Assessment*. Data Base, Secretariat of the Convention on Biological Diversity (SCBD). https://www.cbd.int/programmes/cross-cutting/impact/search.aspx. Accessed 23 July 2024

Shereni N.C. & Saarinen J. 2021: Community perceptions on the benefits and challenges of community-based natural resources management in Zimbabwe.—*Development Southern Africa* (Taylor & Francis Journals) vol. 38(6), 879–895, November.

Showler D.A., Stewart G.B., Sutherland W.J. & Pullin A.S. 2010: What is the impact of public access on the breeding success of ground-nesting and cliff-nesting birds?—CEE review 05-010 (SR16). *Collaboration for Environmental Evidence*: www.environmentalevidence.org/SR16.html

Smith, O., & Wassmer, T. (2016). An ethogram of commonly observed behaviors of the endangered Bridled White-eye (Zosterops conspicillatus) in a Zoo Setting. *The Wilson Journal of Ornithology*, *128*(3), 647–653. https://doi.org/10.1676/1559-4491-128.3.647

Smith R.K., Meredith H. & Sutherland W.J. 2018: Amphibian conservation, pp. 9–65.—In: Sutherland W.J., Dicks L.V., Ockendon N., Petrovan S.O., Smith R.K., (eds.): *What Works in Conservation.* - Cambridge: Open Book Publishers, 604 p.

Spellerberg I.F. 2005: *Monitoring Ecological Change* (2nd ed.).—Cambridge University Press, 391 p.

Spieler M. 1997: Anpassungen ausgewählter Froschlurche an Trockenstress und Räuberdruck in einer westafrikanischen Savanne.—*Salamandra* 33(2), 133–152

Staupendahl K. 1997: Ein neues Stichprobenverfahren zur Erfassung und Beschreibung von Naturverjüngung.—In: Jahrestagung der Sektion Forstliche Biometrie und Informatik des Deutschen Verbandes Forstlicher Forschungsanstalten in Freiburg i.Br., 24.-26.9.1997, 32–49

Steyer K., Simon O., Kraus R.H.S., Haase P. & Nowak C. 2013: Hair trapping with valerian-treated lure sticks as a tool for genetic wildcat monitoring in low-density habitats.—*Eur. J . Wildl. Res*. 59(1),39–46. https://doi.org/10.1007/s10344-012-0644-0

Støen O.-G., Pitlagano M.L. & Moe S.R. 2009: Same-site multiple releases of translocated white rhinoceroses *Ceratotherium simum* may increase the risk of unwanted dispersal.—*Oryx* 43 (4), 580–585, DOI: https://doi.org/10.1017/S0030605309990202

Strasdas W. 2002: *The ecotourism training manual for protected area managers*.—German Foundation for International Development (DSE) / Centre for Food, Rural Development and the Environment (ZEL). Zschortau, 58 p.

Stretesky, P. B., McKie, R. E., Lynch, M. J., Long, M. A., & Barrett, K. L. (2018). Where have all the falcons gone? Saker falcon (falco cherrug) exports in a global economy. *Global Ecology and Conservation*, *13*, e00372. https://doi.org/10.1016/j.gecco.2017.e00372

Struthers, C. L., Murenbeeld, K. J., & Williamson, M. A. (2023). Environmental impact assessments not the main barrier to timely forest management in the United States. *Nature Sustainability*, *6*(12), 1542–1546. https://doi.org/10.1038/s41893-023-01218-1

Sutherland W.J., Dicks L.V., Ockendon N. & Smith R.K. (eds.) 2015: *What Works in Conservation.*—Cambridge UK: Open Book Publishers, 374 p. https://doi.org/10.11647/OBP.0060

Szymański P., Deoniziak K., Łosak K. & Osiejuk T.S. 2017: The song of Skylarks *Alauda arvensis* indicates the deterioration of an acoustic environment resulting from wind farm start-up.—*Ibis* 159 (4), 769–777. https://doi.org/10.1111/ibi.12514

Tesfahunegny W., Datiko D., Wale M., Hailay G.E. & Hunduma T. 2020: Impact of wind energy development on birds and bats: the case of Adama wind farm, Central Ethiopia .— *The Journal of Basic and Applied Zoology* 81, Article Number 41. https://doi.org/10.1186/s41936-020-00171-1

Thaxter C.B., Buchanan G.M., Carr J., Butchart S.H.M., Newbold T., Green R.E., Tobias J.A., Foden W.B., O'Brien S. & Pearce-Higgins J.W. 2017: Bird and bat species' global vulnerability to collision mortality at wind farms revealed through a trait-based assessment.—*Proceedings of the Royal Society B* 284, 20170829. https://doi.org/10.1098/rspb.2017.0829

Thiel D., Jenni-Eiermann S. & Jenni L. 2008: Effects of recreation activities on flushing behaviour, habitat use and stress physiology of Western Capercaillie.—*Ornithol. Beob*. 105 (1), 85–96 (in German)

Tichaawa T.M. & Mhlanga O. 2015: Community perceptions of a community-based tourism project: a case study of the CAMPFIRE programme in Zimbabwe.—, *African Journal for Physical, Health Education, Recreation and Dance* 21, 55–67. https://api.semanticscholar.org/CorpusID:79122257

Tiger Conservation Plan 2014: Corbett Tiger Reserve, Core Zone. 316 p. + 37 p. annexes. https://forest.uk.gov.in/uploads/news/1701171601.pdf

Tissier, Mathilde L., Handrich, Y., Robin, J.-P., Weitten, M., Pevet, P., Kourkgy, C., & Habold, C. (2016). How maize monoculture and increasing winter rainfall have brought the hibernating European hamster to the verge of extinction. *Scientific Reports*, *6*(1), 25531. https://doi.org/10.1038/srep25531

Takacs, S., Schulte to Bühne, H., & Pettorelli, N. (2021). What shapes fire size and spread in African savannahs? Remote Sensing in Ecology and Conservation, 7(4), 610–620. https://doi.org/10.1002/rse2.212

Tissier, Mathilde Louise, Kletty, F., Robin, J.-P., & Habold, C. (2021). Sustainable Agriculture: Nutritional Benefits of Wheat–Soybean and Maize–Sunflower Associations for Hibernation and Reproduction of Endangered Common Hamsters. *Sustainability*, *13*(24), 13521. https://doi.org/10.3390/su132413521

Tjørve K.M.C. & Underhill L.G. 2008: Influence of Disturbance and Predation on Breeding Success of the African Black Oystercatcher, *Haematopus moquini*, on Robben Island, South Africa.—*Waterbirds* (The International Journal of Waterbird Biology) 31 (1), 83–96

Tolvanen, A., Routavaara, H., Jokikokko, M., & Rana, P. (2023). How far are birds, bats, and terrestrial mammals displaced from onshore wind power development? - A systematic review. *Biological Conservation*, *288*, 110382. https://doi.org/10.1016/j.biocon.2023.110382

United Nations. (2024). THE 17 GOALS | Sustainable Development. *THE 17 GOALS*. https://sdgs.un.org/goals. Accessed 24 July 2024

Unterschultz J., Boxall P., Murray G. & Boxall P. 2003: Non-timber forest products from the Canadian boreal forest: An exploration of aboriginal opportunities.—*Journal of Forest Economics*. 9. 75–96. https://doi.org/10.1078/1104-6899-00027.

Usher M.B. & Erz W. 1994 (eds.): *Erfassen und Bewerten im Naturschutz*.—UTB, Quelle & Meyer, Heidelberg, Wiesbaden, 340 p.

Van Dyke F. 2008: *Conservation Biology. Foundations, Concepts, Applications*. (2nd ed.)—Springer Science +Business Media B.V., 477 p.

Viquerat S.M.A., Bobo K.S., Müller M., Kiffner C. & Waltert M. 2012: Estimating forest duiker (Cephalophinae) density: a case study on the performance of three line transect methods.—*South African Journal of Wildlife Research* 42, 1–10. https://doi.org/10.3957/056.042.0110

Viviani, L., Nonnis Marzano, F., Mattucci, F. et al. 2024: Non-invasive monitoring of the European wildcat in a recently colonized area in the Northern Apennines (Italy). *Eur J Wildl Res* 70, 22. https://doi.org/10.1007/s10344-024-01777-4

Voigt C.C., Rehnig K., Lindecke O. & Pētersons G. 2018: Migratory bats are attracted by red light but not by warm-white light: Implications for the protection of nocturnal migrants.—*Ecol. Evol.* 2018; 00:1–9. https://doi.org/10.1002/ece3.4400

Waldrop T.A. & Brose P.H. 1999: A comparison of fire intensity levels for stand replacement of Table Mountain pine (*Pinus pungens* Lamb.).—*Forest Ecology and Management* 113, 155–166

Weidemann G. & Reich M. 1995: Zur Wirkung von Straßen auf die Tierwelt der Kalkmagerrasen unter besonderer Berücksichtigung der Rotflügeligen Schnarrschrecke (*Psophus stridulus*) und des Schachbretts (*Melanargia galathea*).—*Beiheft Veröffentlichung Naturschutz und Landschaftspflege Baden-Württemberg* 83, 407–424

Whitman K.L. & Packer C. 2006: *A Hunter's Guide to Aging Lions in Eastern and Southern Africa*.—Conservation Force, USA. Library of Congress Cataloging-in-Publication Data, 47 p. https://cbs.umn.edu/sites/cbs.umn.edu/files/migrated-files/downloads/Lion_Aging_Guide-1.pdf

Whitman K.L., Starfield A.M., Quadling H.S. & Packer C. 2004: Sustainable trophy hunting of African lions.—*Letters to nature*, Nature Publishing Group, 1–4. doi:https://doi.org/10.1038/nature02395 | www.nature.com/nature

Wikars L.-O. 1992: Forest fires and insects.—*Ent Tidskr* 113, 1–11

Wikars L.-O. 1997: Brandinsekter i Orsa Finnmark: biologi, utbredning och artbevarande .—. *Ent Tidskr*. 118, 155–169

Williams, K. A., Gulka, J., Cook, A. S. C. P., Diehl, R. H., Farnsworth, A., Goyert, H., et al. (2024). A framework for studying the effects of offshore wind energy development on birds and bats in the Eastern United States. *Frontiers in Marine Science*, *11*. https://doi.org/10.3389/fmars.2024.1274052

Wilson J.C. & Elliott M. 2009: The Habitat-creation Potential of Offshore Wind Farms.—*Wind Energy* 12, 203–212. DOI: 10.1002/we

Wynhoff I. 1998: Lessons from the reintroduction of *Maculinea teleius* and *M. nausithous* in The Netherlands.—*Journal of Insect Conservation* 2, 47–57

Zeller K.A., McGarigal K. & Whiteley A.R. 2012: Estimating landscape resistance to movement: a review.—*Landscape Ecology* 27, 777–797

Ziegler S., Fa J.E., Wohlfart C., Streit B., Jacob S., & Wegmann M. 2016: Mapping Bushmeat Hunting Pressure in Central Africa.—*Biotropica*, *48*(3), 405–412. https://www.jstor.org/stable/48574988

Zwart M.C., Dunn J.C., McGowan P.J.K. & Whittingham M.J. 2016: Wind farm noise suppresses territorial defense behavior in a songbird.—*Behav. Ecol.*, 27 (1), 101–108. https://doi.org/10.1093/beheco/arv128

14 Restoration Ecology

Abstract

The restoration of ecosystems can be divided into terrestrial and aquatic habitat restoration. Different approaches are required to restore tropical, boreal, and temperate forests. We also include the rewilding of home gardens and backyards. In each case, examples with corresponding programs are provided. For aquatic habitats, river and lake restorations are explained separately. Procedures for restoring river systems include the management of banks, alignment, and pollution removal. For lakes, we discuss how to assess and improve the habitat quality of lakeshores and how poor water quality can be improved using aeration and nanobubbles. Wetlands are located between terrestrial and aquatic biotopes. We showcase coastal areas of lakes, fens, and bogs and how these ecosystems can be restored. Specific examples of their destruction and measures for their restoration process are discussed. Raised bogs are exceptional biotopes threatened by peat extraction and drainage. A simple measure using backwatering is presented.

The term "restoration ecology" refers to "*the establishment of the plant and animal community at a disturbed site that existed there before the disturbance*" (Allaby 2004). Books and overviews on this topic include Harker et al. (1999), Falk et al. (2006), Clewell and Aronson (2007), Greipsson (2011), Florentine et al. (2024), Holl (2020), and Palmer et al.(2016). In the USA, the law requires restoration (Berger 1991). In addition to the reactive component, the Endangered Species Act (ESA) of 1973 also contains a proactive component. The international Bonn Convention on Migratory Species of Wild Animals (Convention on the Conservation of Migratory Species of Wild Animals, CMS, http://www.cms.int/) also calls for measures to improve the habitats visited by the animals and the status of the endangered species (Bonn Convention 1979/1983). All states that have signed the convention are listed in the CMS. Recovery plans for animal species are always species-specific. They can include habitat restoration, captive breeding, and reintroduction or translocation for critically endangered species. Examples are provided in general textbooks of conservation biology (e.g., Cardinale et al. 2020; Groom et al. 2006; Van Dyke 2008; Hambler 2004, see also Sect. 13.10). We will only highlight a few examples that relate to ecosystem restoration.

14.1 Restoration of Forests

Planting trees is generally seen as beneficial to the environment, e.g., measures against climate change, including carbon sequestration, protection against erosion, protection against desertification and sandstorms, improvement of water

M. Mühlenberg, T. Waßmer, *Field Ecology*, https://doi.org/10.1007/978-3-031-87547-2_14

quality, and biodiversity conservation. However, care must be taken to restore only where there was previously forest or a forest has been degraded. Grantham et al. (2020) provide a global overview of the state of the forests and develop a forest integrity index. If the restoration goes wrong, it could change the ecosystem with negative consequences for fauna and flora. Parr et al. (2024) found through a comparison of ecosystem mapping (Dinerstein et al. 2017) and the localization of tree planting projects that tree planting is also widespread in non-forested ecosystems. Ly (2024) comments that such restoration initiatives can destroy natural grasslands and savannahs, which is particularly relevant in Africa.

Another problem is that the internationally accepted definition of forest includes at least 10% canopy cover (Chazdon et al. 2016). Forests have been drastically reduced worldwide, which is especially concerning as forests harbor the highest terrestrial biodiversity and provide many ecosystem services (e.g., Cardinale et al. 2020). Tree diversity is essential in this context (Ampoorter et al. 2020). Therefore, many reasons exist to restore degraded forest ecosystems (e.g., Braat and de Groot 2012; IUCN 2020). Forests can be restored by either passive or active restoration (Greipsson 2011). Depending on the climate zones, management must adapt to boreal, temperate, or tropical forests. If the forest destruction is not too substantial, passive restoration is recommended. The focus is on natural forest regrowth, which can be supported by restoration efforts if necessary.

Fig. 14.1 Abandoned coffee plantation, Ivory Coast. About a third of the African continent is severely damaged by land degradation

14.1.1 Tropical Forests

In order to restore tropical forests there must still be forest sites with original seed sources in the neighborhood. As most tree species in the tropics are spread by animals (see Sect. 7.10), even single trees in an abandoned plantation can attract birds, e.g., as perches, and promote seed dispersal (see Fig. 14.1) (e.g., Poorter et al. 2021). In this way, scattered trees and forest islands contribute substantially to reforestation (Greipsson 2011).

Heymann et al. (2019) showed that the small tamarin primates in Peru effectively carry seeds from primary forests into secondary forests. Through observations of feeding and defecation in *Leontocebus nigrifrons* and *Saguinus mystax* and genetic analyses, the authors, the authors provided evidence that the seeds germinate and form young plants that contribute to the natural regeneration of secondary forests.

Disturbances in the tropical rainforest also have a negative impact on seed dispersal by animals and, thus, on natural regeneration. Pangau-Adam et al. (2021) showed that logging reduces the number and diversity of fruit trees in Northern Papua. They surveyed five habitat types with a gradient of increasing disturbance. In each habitat type, they walked 12 transects (each 1.5 km long) at peak fruiting time. Seventy-one species of fruits were counted, of which 89% were larger fruits consumed by large-gaped frugivores such as Cassowaries and Hornbills (a hornbill is seen in Fig. 5.4). Together with crowned pigeons (Fig. 7.22d), they are the only seed dispersers for these fruits in Papua and depend on this fruit supply. Brancalion et al. (2019) and Brancalion (2023) give a broader overview of forest restoration in the tropics.

Harrison and Swinfield (2015) conducted experiments in restoring degraded tropical forests in Sumatra. Their main aim was to answer the following questions: Can succession, i.e., natural regeneration, be accelerated through selective removal of pioneer trees, enrichment planting, or direct seeding? Monitoring was conducted on the ground in randomly distributed plots and via canopy analysis using unmanned aerial vehicles (drones). Standard surveys of tree species composition were conducted in plots of different sizes: 1 ha plots for forest composition, and 20 × 20 m vegetation plots to study growth, survival of individual tree individuals, and total basal area (see also Sect. 4.2 for methods to describe forest structure). Mist-netting for understorey birds, camera traps for small mammals, malaise traps for arthropods, and metagenomic (barcoding) techniques were used to assess faunal biodiversity. Monitoring was carried out before the experiment, 1 year after treatment, and at 3-year intervals afterward.

In Southeast Asia, extended ranges of rainforests have been converted into oil palm plantations (Koh et al. 2011). A massive loss of biodiversity accompanies this. Forest islands have been successfully created to restore oil palm plantations and increase biodiversity (Zemp et al. 2023). Projects are also beginning to convert oil palm plantations into secondary rainforests through reforestation (Leibnitz 2019; Rahmani et al. 2022; Zemp et al. 2019). The establishment of corridors to neighboring protected areas can accelerate this process. One of the main objectives is to increase the habitat of the endangered orangutan in Borneo (Axelsson et al. 2024).

Rinaudo (2008), see also Reij et al. (2009a, b) developed a technique for farmers to revegetate degraded forest stands in dry Africa: "Farmer Managed Natural Regeneration (FMNR)." There is also an educational film and several YouTube videos about this (https://www.youtube.com/@WorldVisionDE, in German, https://www.youtube.com/@fmnrhub4194, e.g., *Tony Rinaudo: "The Niger I came To"* (https://www.youtube.com/watch?v=afjVaehQRxg). This method of revegetating depleted and deforested soils does not involve planting trees. It utilizes the remaining underground roots of uprooted trees to sprout again. Deserted soils in the Sahel can also be made fertile again with the help of soil organisms. The main actors here are termites, which are attracted to holes dug into the hardened soil and filled with dung or compost and whose activities turn over and aerate the soil (Tilch 2015; Müller 2015; Natter 2018). This method is based on the work of Yacouba Sawadogo, an illiterate farmer from Burkina Faso, who combined the traditional agricultural practices of the Sahel region called Zaï (Cheik and Jouquet 2020; Kaboré and Reij 2004; Reij et al. 2009a, b; Roose et al. 1999) with his observations of termites (Kaiser 2014; Kaiser et al. 2015, 2017; Padonou et al. 2020) (Fig. 14.2). The 2010 feature documentary, "The Man Who Stopped the Desert" (https://www.imdb.com/title/tt1694580/), portrays Yacouba Sawadogo and his work. Zaï methods were recently also used in reforestation projects in the Sahel (Bayen et al. 2020).

Case Study: Rehabilitation of an Overexploited Tropical Forest

Rehabilitation is a form of restoration. The aim is not to restore the original structure and function of the degraded ecosystem but to restore the essential structures and functions so that typical forest inhabitants can rebuild secure populations. These targets can be supported by placing parts of the forest system under complete protection, i.e., excluding them from human utilization. An example for rehabilitation is the recovery of the Francolin *Francolinus lathami* (Phasianidae) in the overexploited Bossematié Forest in the SE Ivory Coast monitored by density estimation (Table 14.1).

Long-term monitoring of selected forest species is essential for this process. A monitoring program involving local villagers was presented in Sect. 1.15; see also Mühlenberg et al. (1995, 1997, and 1999). Here, a trained team monitored around 50 bird and mammal species (hornbills, turacos, duikers, and monkeys) monthly on 16 transects, each 2 km long (Fig. 14.3). Using the distance sampling method (see Sect. 5.1), the population sizes in the forest and the probability of survival of the controlled species can be estimated. The varying density of animal species in different forest segments is also the basis for recommending protection zones within the forest. About 20% of the area was proposed to be designated as a special management zone and excluded from any utilization.

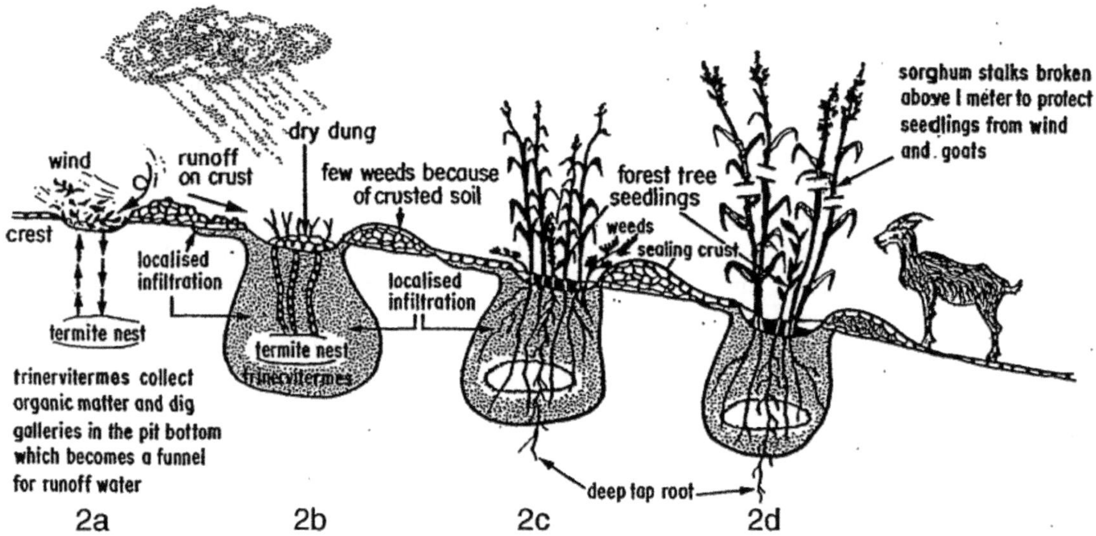

Fig. 14.2 Zaï practice for soil and green cover rehabilitation. 2a: Basins 20–40 cm wide and 10–15 cm deep are dug into the dry season to collect runoff water. 2b: Adding 1–2 handfuls of dry dung attracts termites, building galleries at the pit's bottom. Water runoff can penetrate deep, forming moisture pockets that are protected from evaporation. 2c: Sowing of 12 sorghum seeds per pit, which break through the crust and root deep. Weeds and tree seedlings from seeds in dung develop between sorghum stalks. 2d: Sorghum stalks are broken 1 m above the soil to hide tree seedlings from goats and protect them from the wind. Picture credit (Roose et al. 1999). Reproduced by permission

Table 14.1 Impact of forest management on the population of *Francolinus lathami* (Forest Francolin, Phasianidae) in the Bossematié Forest, SE Ivory Coast, with a total area of 222 km^2. The density of the francolin was estimated using distance sampling and found to be 2.1 ind/km^2 in untreated parts of the forest and 0.5 ind/km^2 in utilized parts of the forest

		Area (km^2)	Density (Ind/km^2)	Estimated population size
Option 0	No management	222	2.1	**466**
Option 1	Nature reserve (20%)	44	2.1	92
	40% treated area	89	0.5	45
	40% unmanaged area	89	2.1	187
	Sum in total forest	222		**324**
Option 2	Nature reserve (20%)	44	2.1	92
	80% treated area	178	0.5	89
	Sum in total forest	222		**181**

14.1.2 Boreal Forests

The boreal forest is the second largest terrestrial biome on Earth (Fig. 14.4). It represents about 25% of the Earth's forested area and is disproportionally affected by climate change (Girona et al. 2023).

As the growth of trees in the boreal zone is particularly slow, planting efforts require a correspondingly longer period of management. The annual growth (volume increment) in boreal forests is around 1 m^3 per ha and year (Crisp et al. 2004; Stokland 2021) compared to a growth rate between 4 m^3 and 9 m^3 in central European countries (Tomter et al. 2016). In Mongolia, where annual precipitation is very low, it is particularly important to distinguish between affor-

Fig. 14.3 (**a**) Satellite image of the Forêt Classée de la Bossematié with location of 16 permanent transects (white lines). Pale patches constitute abandoned cocoa/banana plantations now covered by tree plantations. With 16 transects in the forest area of 216 sq. km, a surface of about 6% is surveyed. SE Côte d'Ivoire. (**b**) Overexploited forest area, where *Chromolaena* sp. bushes grow in the resulting open spaces. Bossematié, SE Côte d'Ivoire. (**c**) Part of the forest still intact. Bossematié, SE Côte d'Ivoire

estation and planting in existing forests (reforestation). Afforestation at sites where no forest has ever grown is generally unsuccessful (e.g., Fig. 14.5, Mühlenberg et al. 2006). Planting in existing forests, even if they have been disturbed by fire or logging, is only successful if there are still enough trees to provide shade. Natural regrowth is most likely to be successful, see also Sect. 13.6, Fig. 13.16).

14.1.3 Temperate Forests

In Central Europe, forestry guidelines recommend keeping the forest canopy closed as far as possible even when timber is harvested ("high or timber forest management"). In sustainable forestry programs, maintaining continuous forest cover will allow faster regeneration. However, Central European forests have been altered by foreign plantations throughout history, so there is hardly any natural forest left (Ellenberg and Leuschner 2010). One example are the Harz Mountains in central Germany. After the area was declared a National Park, efforts were made to convert the forest maintained for timber production into a local natural forest and to restore it in this way (Fig. 14.6).

With the use of drones (see Sect. 4.5), damaged forest areas can also be restored more quickly by scattering tree seeds (SKYSEED GmbH, Rollbergstraße 28a, 12053 Berlin, Germany, www.skyseed.eco/en, https://vimeo.com/999228709).

Deadwood, also known as coarse woody debris (CWD) in forests, is a critical component that promotes biodiversity (Grove and Meggs 2003), see also Sect. 4.2.3, Deadwood Recording, and Sect. 13.6, Salvage Logging). It is also recommended to enrich CWD in the forest through management and restore forests depleted of CWD due to management or disturbance events (Sullivan et al. 2012; Manning et al. 2013). Positive effects have been demonstrated for small mammals and reptiles.

Fig. 14.4 Part of the extensive boreal forest in central Siberia in the middle Yennisei region (1994). Even though the forest in this area is spared from human influence, it naturally shows a heterogeneous landscape

14.1.4 Rewilding of Home Gardens and Backyards (Homegrown National Park)

The US-American ecologist, and conservationist entomologist Douglas Tallamy advocates for residential backyards, home gardens, and public and institutional landscaping that bridges the fragmented parks and preserves by providing habitat for native species. He calls our present gardens and yards "ecological deserts" as they are often lawn monocultures and dominated by non-native plants providing poor habitats that only enable very limited biodiversity (Fig. 14.7).

He specifically suggests that we do the following: (some of the measures are shown in Figs. 14.8 and 14.9)

1. shrink our lawns, which we do not really need or use, and replace at least half of them with native plants that sustain more animal life. The conversion also reduces water, pesticide, and fertilizer use.
2. Remove introduced invasive plants that crowd out the native flora, such as bush honeysuckle, oriental bittersweet, multiflora rose, and kudzu.
3. Allocate some of your property as no-mow zones to allow insects to complete their life cycle. Put mulch, a native ground cover, or stacks of cut brush around the base of your trees to accommodate insects. This will benefit birds and insects, such as moths and butterflies.
4. Reduce lights during nighttime as they attract many moths and other insects and disturb their behavior. If you need nightlights, use yellow LED lights (attract fewer insects) activated by a motion sensor.
5. Plant native keystone species that contribute more to the food web than others, such as native oaks, cherries, cottonwoods, willows, and birches.
6. Plant native pollinator gardens and build insect hotels to support pollinators. Pollinator magnets are goldenrods, asters, sunflowers, and evening primrose.

Fig. 14.5 (**a**) A 13-year-old *Pinus sylvestris* afforestation in West Khentey, North Mongolia. (**b**) Planting on the edge of a previously burnt forest is only partially successful. Northern Mongolia. (**c**) Natural regeneration of deciduous trees can shelter under-planted conifers, accelerate natural succession, and be harvested after ~50 years. Northern Mongolia (**d**) Unsuccessful planting experiment as a "green belt" in the Gobi desert, Mongolia

7. Do not use pesticides and harsh chemicals. Most weeds can be dug up or controlled by mowing or with vinegar. Mosquitoes can be controlled by packets containing *Bacillus thuringiensis* and will be naturally reduced by the increasing biodiversity in your backyard and garden. If you have a pond, consider stocking it with mosquito-eating native fish, such as shiners and keep the water surface in motion with a fountain or waterfall.

More information about these concepts can be found on the Homegrown National Park website (Adler 2024; Homegrown National Park, Inc. 2024). This concept is well founded on scientific evidence, e.g., (Burghardt et al. 2010; Kramer et al. 2019; Narango et al. 2017, 2020; Narango et al. 2018; Richard et al. 2019; Tallamy 2017; Tallamy and Shriver 2021).

14.2 River Restoration

River restoration includes various modifications of river channels and their riparian zones and floodplains as well as alterations of the input of water, sediment, and solutes into riv-

Fig. 14.6 Conservation management and reforestation in the Harz Mountains, Germany. (**a**) Spruce forests still dominate large parts of the area, although the predominant natural vegetation at those locations would be deciduous forests. Spruce trees were planted to meet the timber needs of the historic mining industry. (**b**) Spruce trees are deliberately felled and removed to convert the forest into a mixed-beech forest. (**c**). Increasing dominance of beech at lower altitudes indicates the development to the potential natural vegetation. (**d**) Rejuvenation with native beech saplings inside the planted spruce forest

ers. These adjustments aim to improve a river's hydrology, geomorphology, and ecology within a compromised watershed by replacing lost, impaired, or compromised parts of a river's natural systems. Techniques include limited modifications such as bank stabilization, riparian fencing, manipulating plants and animals, and entire ecosystem processes (Wohl et al. 2005, 2015).

14.2.1 Physical River Restoration

Fig. 14.10 shows the twenty most common river restoration methods in the Upper Midwest of the USA. Bernhardt et al. (2005) are synthesizing the river restoration efforts in the U.S..

A sand trap or, more generally, a sediment trap is one option for removing excess sediment accumulated in the stream channel. This technique reduces or prevents the embeddedness of river bottoms (Fig. 6.7) such as gravel that are essential for spawning fish. Most sediment traps are, in principle, deep excavations dug into the middle or to the side of a river channel. As water passes over them, its velocity decreases, and suspended sediment settles down. If needed, the trapped sediment can be removed from the trap after a certain time.

Rip-rap (rip-rap, shot rock, rock armor, or rubble) are cobbles, boulders, gravels, graded stones, or crushed rocks that stabilize shoreline or riverbed structures against water, wave, or ice erosion (Fig. 14.11a).

Fig. 14.7 A conventional backyard before the transformation. Arrows and annotations indicate the problems with such spaces for biodiversity

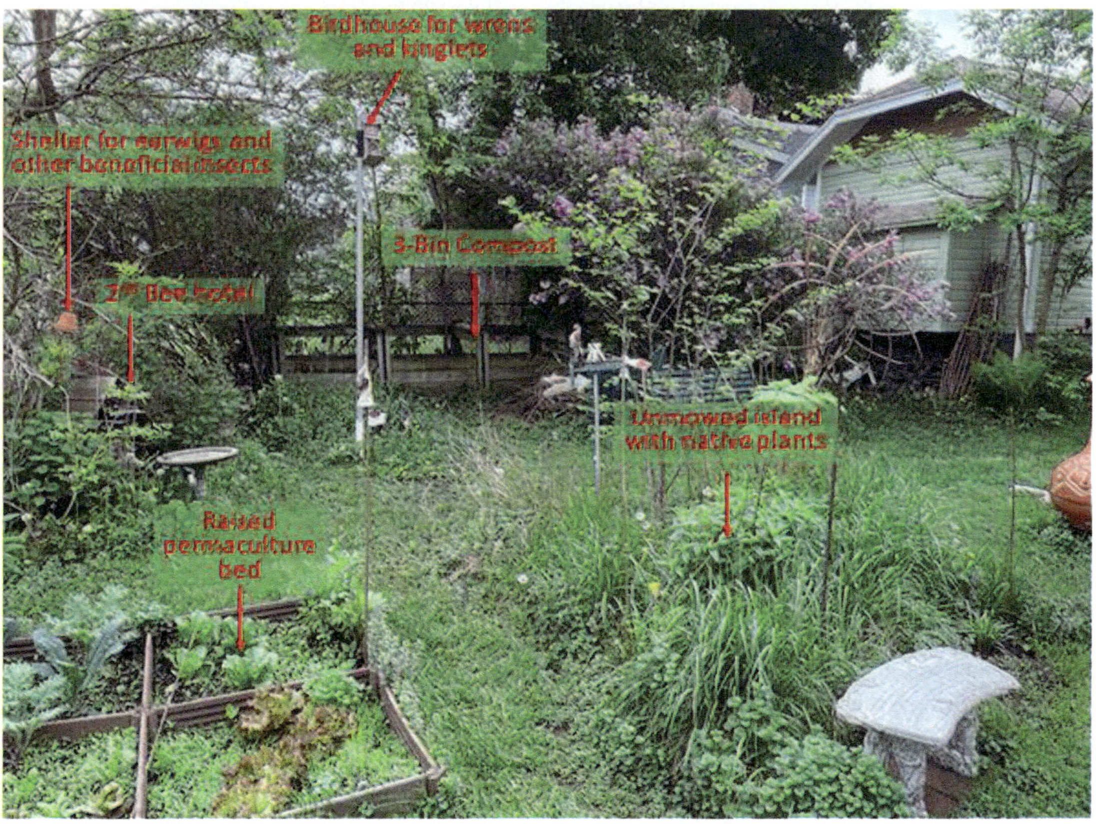

Fig. 14.8 Part of the same backyard after converting parts of the "ecological desert" into a more biodiverse backyard

Fig. 14.9 A small piece of prairie established on the previous lawn

Large woody debris (LWD) includes fallen trees, logs, and branches at least 4 inches wide and 6 feet long (Fig. 14.11b). In river restorations, LWD is placed into streams to provide cover and habitat complexity for fish and other aquatic organisms. LWD increases the diversity of macroinvertebrates in streams (O'Connor 1991; Zilli and Fernández 2023).

Deflectors are structures added to the banks that narrow the channel locally, causing variations in currents and patterns of erosion and deposition (Fig. 14.12). This creates different habitats for wildlife and fish.

A Lunker (or as it is an acronym, LUNKER, which stands for Little Underwater Neighborhood Keepers Encompassing Rheotactic Salmonids) is a crib-like structure placed in streams to provide bank stability and cover for fish (Fig. 14.13). According to Jim Schmiedeskamp, Lunker structures constructed like the ones shown in Fig. 14.14 may no longer be the best practice—at least in Wisconsin due to heavy rain events in recent years significantly damaging traditional lunker structures. In a multiyear, five-phase Weister Creek project in Wisconsin's Driftless Area, the project moved from the depicted traditional Lunker structure design to one with more natural rocks and logs that could withstand extreme rain events.

Tree revetments are made by anchoring entire trees and LWD along a stream bank (Fig. 14.14). They are inexpensive and effective in minimizing and reversing bank erosion. The tree branches slow down the current along the eroding bank, decreasing erosion and depositing sediment along the bank. Seeds of river trees, such as cottonwood and sycamore, can sprout and grow in the deposited sediment, spreading their roots throughout the revetment and stream bank. By the time the trees used in the revetment decay, the roots of the living trees stabilize the banks and provide excellent fish and wildlife cover.

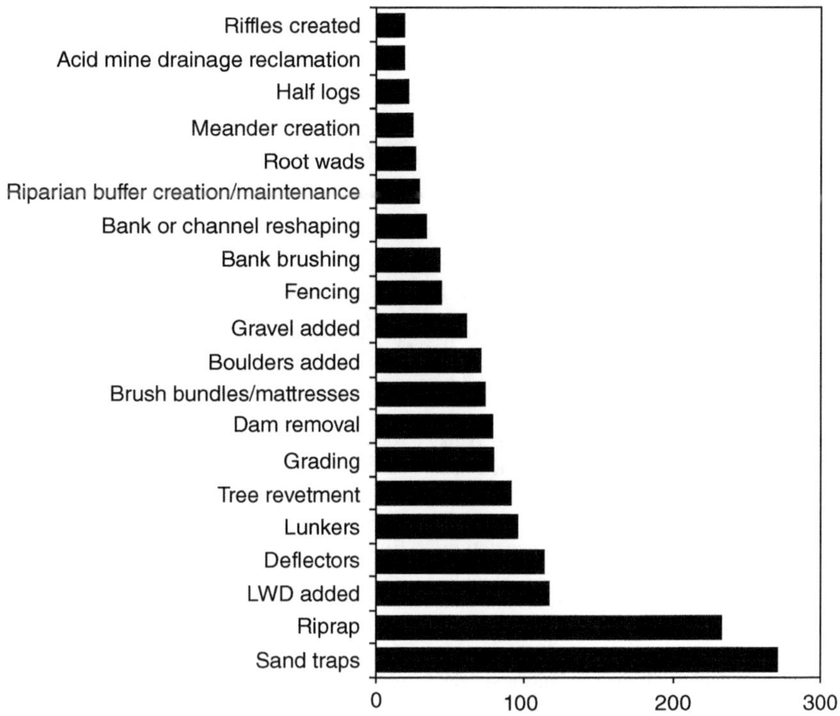

Fig. 14.10 The 20 most common project activities listed in the National River Restoration Scientific Synthesis (NRRSS) summary database for the Upper Midwest region. The number of projects reported for each activity is shown on the horizontal axis. Using sand traps and riprap stand out as the dominant techniques in stream restoration in the Upper Midwest region. Source: Alexander and Allan (2006). Reproduced by permission

A brush mattress is a layer of interlaced live branches (mattress) placed on a bank face, often with a bundle of live brushwood (fascines) and rock at the base (Fig. 14.15). Live branches originate from woody plants that sprout roots from stems, such as willows (*Salix spec.*), some shrub dogwood (*Cornus spec.*), and alder species (*Alnus spec.*). The mattress and brush bundles are attached by wire, robe, or stakes. Brush bundles and mattresses armor eroding riverbanks as they lower flow velocities along the bank and accumulate sediment. Especially with alive sprouting plants, brush mattress develops a network of interlocking roots and plant stems.

14.2.2 Meander Creation

Meanders are broad, sinuous (winding) bends in a riverbed. Meandering follows a landscape's topography/grading and pedology/geology, leading to a decreased reach slope. The highly variable spatial and temporal depth patterns, flow rates, bed materials, and land covers lead to diverse hydrologic conditions and increased biodiversity. Meander restoration reconstructs meandering riverbeds that humans have straightened to facilitate boating, mills, agriculture, and the construction of roadways, bridges, and settlements. However, the channeling of rivers often led to erosion due to excessive slopes and a loss in biodiversity and other ecosystem services of rivers.

In many cases, ecological functions can be restored by restoring a river's previous alignment or providing a new meandering riverbed (Fig. 14.16). However, new erosion along the outer (concave) eroding (cut) bank is likely. Site-specific problems such as flooding are also possible, depending upon local flow and drainage patterns and soils encountered by the new channel (Bertalan et al. 2018).

Fig. 14.11 Physical river restoration: (**a**) Stream bank stabilization. Streams naturally meander or cut back and forth. Eroding stream banks constitute a significant source of sediment in streams. Stream bank erosion is often accelerated due to a lack of deep roots from vegetation, which holds the soil together, and by livestock grazing, which can directly cause erosion due to cattle walking on the banks. Stream bank stabilization traditionally consists of grading the slope from 2 feet horizontal to 1 foot vertical and then armoring the bank with rip-rap (various-sized rocks). This process improves water quality by reducing stream bank erosion and sedimentation into streams and also improves habitat for terrestrial and aquatic life. Picture credit: Farm Production and Conservation Business Center and Natural Resources Conservation Service (NRCS) Iowa, https://allamakeeswcd.org. (**b**) River restoration specialists install large woody debris to stabilize a riverbank. Picture credit: Salix River & Wetland Services Limited, https://www.salixrw.com. Both pictures reproduced by permission

Fig. 14.12 Use of deflector to increase flow variations in rivers. (**a**) Using a gravel berm. Picture credit: Fishtrack Limited, https://fishtrack.co.uk. (**b**) Using logs. Picture credit: Yorkshire Dales Rivers Trust, https://www.ydrt.org.uk. Both pictures are reproduced by permission

The success of river restoration strategies might depend on intentions and perception (Jähnig et al. 2011; Wright 2021) and need to be based on standards (Palmer et al. 2005; Pander and Geist 2013). While some studies report improvements (Atristain et al. 2024; Bernhardt et al. 2007), some are more critical of the outcomes (Brettschneider et al. 2023; Sinclair et al. 2023). There is more consensus that more adaptive management studies are needed to evaluate the effectiveness of restoration efforts and improve and fine-tune current methods (Burrow and Lance 2022; Jenkinson et al. 2006; Polvi et al. 2020). A new approach is the biomic river restoration, which argues that if river restoration is to reverse long-standing declines in river func-

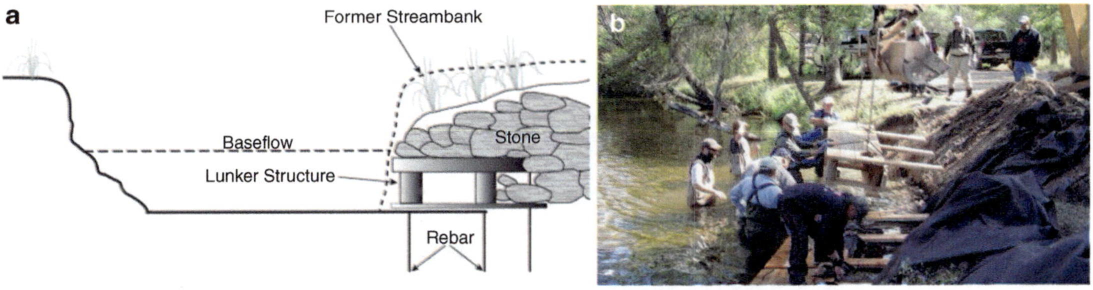

Fig. 14.13 A Lunker provides both bank stabilization and improved fish habitat. (**a**) Principle placement at the toe of a riverbank. (**b**) Premade wooden Lunker construction was placed in the river by a crew of volunteers from the Mason Griffith Trouts Unlimited chapter in Grayling, Michigan, under the supervision of the Michigan Department of Natural Resources. Picture and illustration credits: (**a**) Courtesy of Jim Schmiedeskamp. (**b**) Brian Kozminski, True North Trout, https://truenorthtrout.com. Both pictures are reproduced by permission

Fig. 14.14 Tree revetments. (**a**) Conceptual drawing of tree revetments to protect stream banks from erosion. (**b**) On the left, the situation before the revetment when the streambank of the Vermillion River in Linear Park in Hastings, Minnesota eroded multiple feet yearly. (**c**) After the revetment, spruce and cedar brush now line the bank to rebuild the riverbank naturally. Picture credits: (**a**) United States Department of Agriculture (USDA) 1999. (**b**) Friends of the Mississippi River (https://fmr.org). Both pictures are reproduced by permission

tions, it is necessary to recognize the influence of biology on river forms and processes and re-envisage what it means to restore a river. This entails shifting the focus of river restoration from designing and constructing stable channels that mimic natural forms to reconnecting streams within balanced and healthy biomes and levering biology's power to influence river processes (Johnson et al. 2020).

14.2.3 Cleanup of Polluted Rivers

As described above, sediment traps can remove suspended particulate pollutants such as phosphorus, hydrophobic chemicals, and microplastic with the trapped sediment (Bellucci et al. 2016; García-Haba et al. 2023; Wu et al. 2023). Please refer to Ahmed and Marhaba (2017) and Kondor et al. (2024) for information about riverbank filtration. In the following, we will focus on phytoremediation (Fletcher et al. 2020; Ielpi et al. 2022; Rofkar et al. 2014; Wang et al. 2011).

Aquatic phytoremediation is a treatment method for polluted waters using the ability of macrophytes along and within waterways (Fig. 14.17) to absorb and decompose inorganic (metals and fertilizers), organic (hydrocarbons, pesticides, and algal toxins), and biological (pathogens and harmful algae) (Fletcher et al. 2020). It is explicitly conducted by macrophytes (macroscopic freshwater angiosperms and pteri-

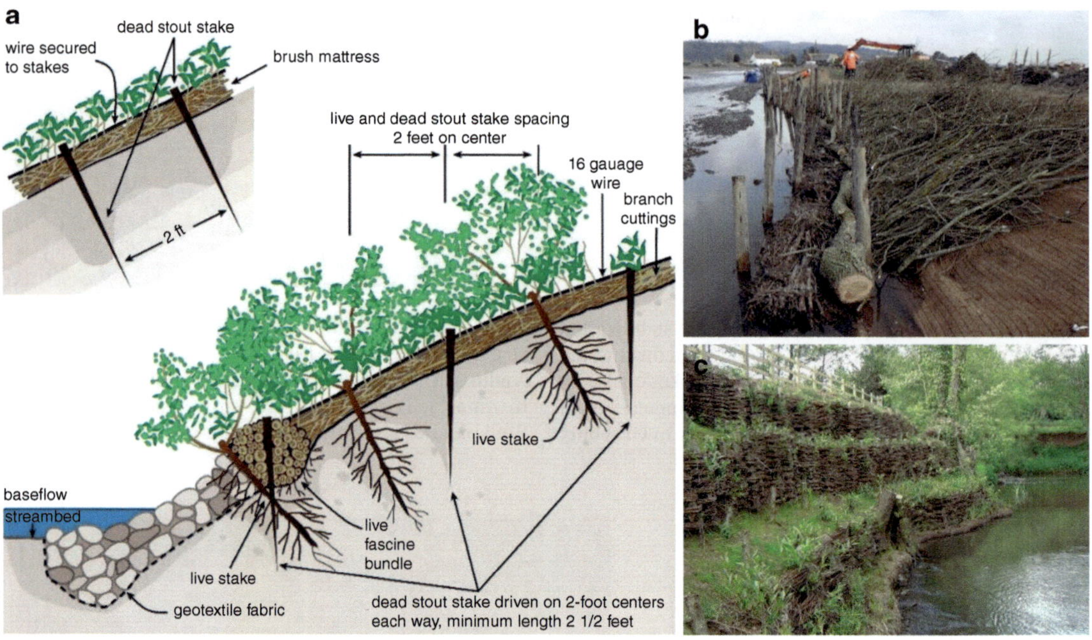

Fig. 14.15 Brush mattresses and brush bundles. (**a**) Details of brush mattress technique with stone toe protection. Picture credit: Federal Interagency Stream Restoration Working Group (FISRWG 1998). (**b**) Installing a live brush mattress revetment with whole tree toe. (**c**) Willow spiling installed. Picture credit for B and C: Salix River & Wetland Services Ltd., https://www.salixrw.com. All pictures reproduced by permission

dophytes). Macrophytes are classified into three growth forms: emergent (Fig. 14.18, far right), floating (Fig. 14.18, center), and submerged (Fig. 14.18, far left).

Emergent macrophytes occupy the margins of water bodies and are rooted in the substrate but have significant shoot growth above the water level. This group includes *Typha* (cattail, reedmace), *Scirpus* (bullrush), *Sagittaria* (arrowhead or duck potato), *Pontederia* (pickerelweed), and *Phragmites* (common reed). Floating macrophytes occupy the water surface and include rooted *Nymphaea* and *Nuphar* (water lilies) or free-floating taxa, including genera such as *Lemna* and *Spirodela* (duckweeds) and *Azolla* (water fern). Submerged macrophytes grow primarily below the water surface. They are mainly rooted to the substrate, such as the genera *Potamogeton* (pondweed), *Myriophyllum* (water milfoil), and *Elodea* (waterweed), although some can be unrooted, e.g., *Ceratophyllum* (hornwort). Due to their different growth forms, a combination of the three groups facilitates the removal of pollutants from both the water column and the sediment (Newete and Byrne 2016). We intentionally excluded non-native species for North America due to their problematic use in restorations, where they can become invasive. In other locations, different native species can perform the same services. In addition to macrophytes, microbes also offer great potential for stream restorations (Hilderbrand et al. 2023).

Rivers can also be improved in terms of water quality by changing the riparian vegetation. An example from the Harz National Park in Central Germany is shown in Fig. 14.19. The unnaturally planted spruce forest led to acidification and shading of the stream. The acidification could be significantly reduced by converting the neighboring spruce forest to a natural mixed deciduous forest (see also Sect. 14.1.3) (T. Späth, 2012, oral presentation).

Exercises:

1. Watch the following two videos:
   (a) River Restoration on the Hampshire Avon https://youtu.be/H5MW9aB2Av0 (6:10).

a

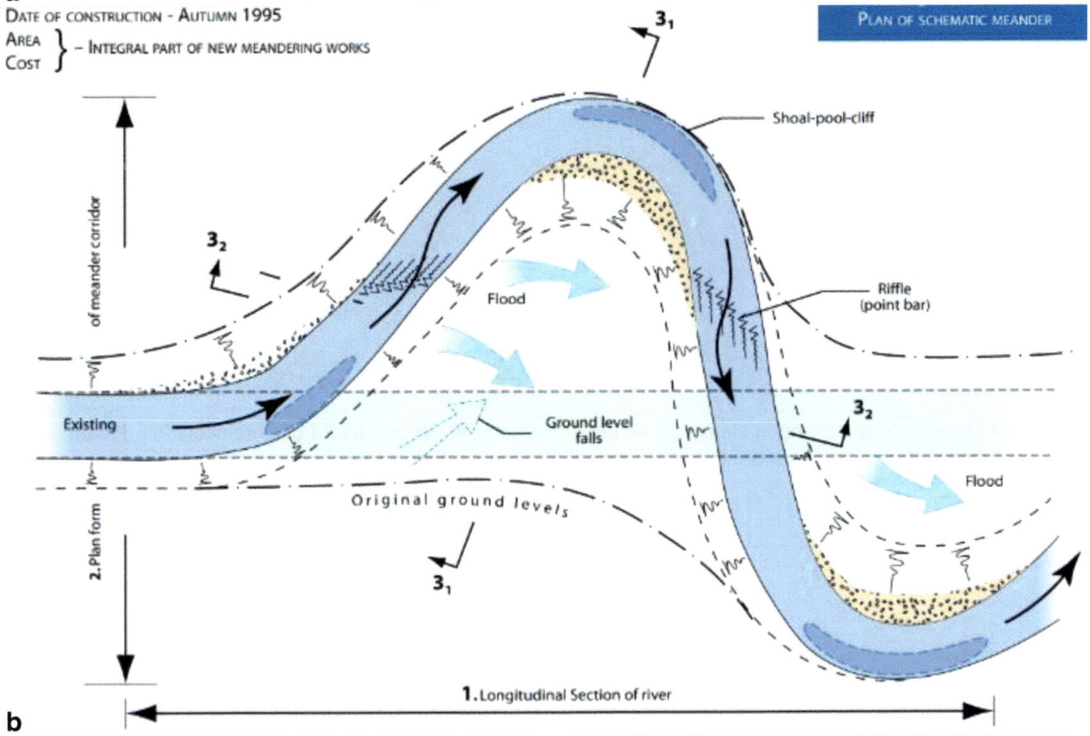

b

Fig. 14.16 Meander restauration. (**a**) Profiling of land between meanders. An idealized meander is where the outer bank is eroding and the inner bank is accreting, thereby generating a slow migration of the meander down the river valley. The creation of the new meanders reflected this process by reprofiling land within the meandering corridor. If this had not been done and works had been limited to simply excavating a sinuous channel, then erosive forces would have been unnaturally high due to the confinement of the river until the river had itself adjusted to the more balanced form depicted. Picture credit: Manual of River Restoration Techniques, 6.2 Cole Skerne, (The River Restoration Centre 2020) (**b**) Blackwell—Spring Brook Wetland & Creek Restoration meandered the waterway (in the foreground) — replacing a straight ditch created decades ago (the diagonal line in the background). Picture credit: Forest Preserve District of DuPage County, https://www.dupageforest.org. Both pictures are reproduced by permission

Fig. 14.17 Examples of vegetation along and within meandering rivers capable of phytoremediation. Picture credit: Fishtrack Limited, https://fishtrack.co.uk. Both pictures are reproduced by permission

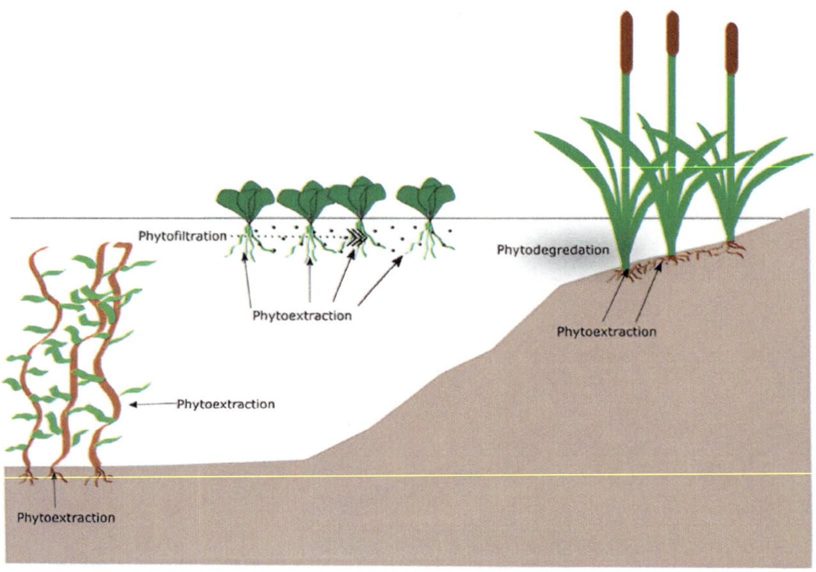

Fig. 14.18 Phytoremediation mechanisms used to degrade/remove waterborne pollutants by growth form. Picture credit: (Fletcher et al. 2020). Reproduced by permission

(b) Kinnickinnic River Restoration—Great Lakes Now https://youtu.be/bndTXeK-G0Po approx. 8 min.

2. Explain the benefits of river restoration. Describe four river restoration techniques shown in the videos in your own words.

Search for three good examples of before/after pictures for river restoration: Google keywords such as: "river, restoration, before, after."

14.3 Restoration of Lakes

The ecological restoration of lakes can be subdivided into the recreation of natural shorelines and the remediation of the water quality in eutrophicated and polluted lakes.

Fig. 14.19 (**a**) Stream in the spruce forest, Harz National Park, 1980. (**b**) the same stream after conversion of the spruce plantation into a mixed deciduous forest. 1994

14.3.1 Lakeshore Restoration

Loss of habitat through lakeshore or shoreline development is considered to be the most significant problem degrading the health of lakes in the United States (Kaufmann et al. 2014; U.S. Environmental Protection Agency 2022) and a major cause of declining biodiversity across North America (Rahel 2002). Unfortunately, the conversion of natural shorelines to degraded lakeshores optimized for human recreation continued between the National Lakes Assessment 2012 and 2017. The percentage of lakes in fair condition decreased, and those with poor conditions (high levels of lakeshore disturbance) increased from 22% to 29%. These changes were statistically significant (U.S. Environmental Protection Agency 2022). In the following, we will focus on the Midwest of the USA as we are most familiar with the developments in this region. However, many countries worldwide can expect the same developments and outcomes (Boromisza et al. 2014; Lorenz et al. 2015; Schmieder 2004).

Since the 1940s, rural landscapes in the Midwest of the United States have experienced spectacular changes due to residential development known as "Urban Sprawl" (Radeloff et al. 2005). This residential development led to fragmentation and the destruction of wildlife habitats (Theobald et al. 1997), and threatening biodiversity (Czech et al. 2000; Wilcove et al. 1998). The construction of primary or secondary homes in and around natural areas booms because they provide recreation and scenery (Schnaiberg et al. 2002). Freshwater ecosystems have especially attracted development for a long time (Naiman 1996; Riera et al. 2001). In many Midwestern states, residential development has increased by over 200% along lakeshores in recent decades (Gonzalez-Abraham et al. 2007; Radeloff et al. 2005). Many studies comparing low- and high-development lakes documented declines in flora and fauna on the more developed lakeshores (e.g., Elias and Meyer (2003); Lindsay et al. (2002); Woodford and Meyer (2003) along low- and high-residential development lakeshores. This development goes hand-in-hand with a loss of ecosystem services such as erosion and runoff control (Elias and Meyer 2003) and overall biodiversity loss (Braunss et al. 2011; Pejchar et al. 2015) (Fig. 14.20).

14.3.2 Assessment Methods

A healthy lakeshore consists of several natural elements (soil, rocks, plants, decaying trees, and other woody debris) that form a functional unit and an ecosystem. Human shoreline alterations may damage or remove some of these components, and the ecosystem may thus lose essential biological and

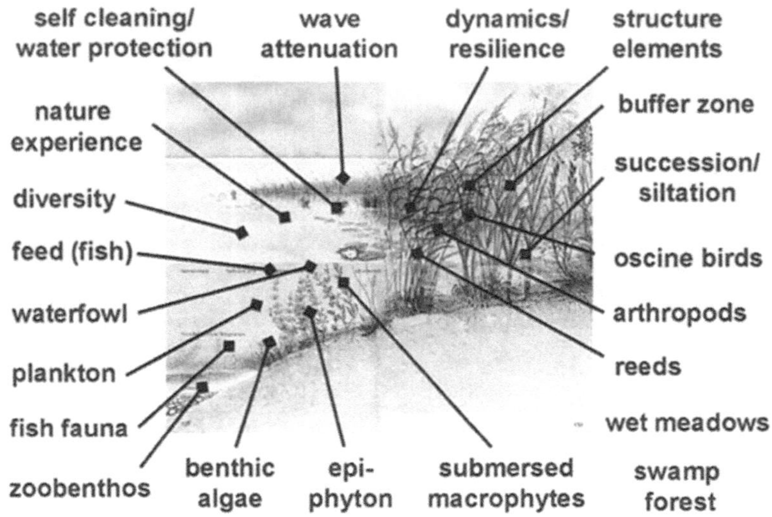

Fig. 14.20 Ecological functions and plant and animal communities of lakeshores. Picture credit: (Schmieder 2004). Reproduced by permission

physical connections between the upland and the water. From the variety of assessment tools, we are especially fond of the "Score Your Shore" protocol developed by the Minnesota Department of Natural Resources (Fig. 14.21) (Perleberg et al. 2012), which provides an objective and systematic method to assess the type, quantity, and quality of the existing shore and adjacent water habitat. It allows the assessment of remaining habitat at developed sites, raises awareness of what makes a high-quality functioning shoreline buffer, and provides a system to recognize landowners with functioning shoreline buffers.

14.3.3 Restoration Methods

Due to the scarcity of scientific literature about this topic, we base this part on the excellent materials shared on the Michigan Natural Shoreline Partnership website (Michigan Natural Shoreline Partnership 2024), the ARCGIS storymap bioengineered shoreline protection on Inland Lakes (Michigan Department of Environment, Great Lakes, and Energy 2024), the 2007 guidebook for shoreline property owners "Understanding, Living With, & Controlling Shoreline Erosion" (Tip of the Mitt Watershed Council 2007), an older review on the ecology of freshwater shore zones (Strayer and Findlay 2010), and Chap. 16 of the 1996 Engineering Field Handbook on Streambank and Shoreline Protection (Natural Resources Conservation Service, U.S. Department of Agriculture 1996). Preventing erosion is the most straightforward and least expensive approach to maintaining a healthy shoreline. Erosion can be minimized by building houses no closer than 30 m from the lake, keeping or growing as many native trees, shrubs, forbs, and grasses as possible, and limiting turf grass on your property, especially at the lake edge (Fig. 14.22). Furthermore, impervious surfaces should be avoided, especially in the buffer zone, and the shore should be shared with native aquatic plants when designing boat access and swimming areas. However, there are many situations where shorelines are still eroding and need stabilization. Seawalls and other "hard armoring" methods rarely stop erosion as they can erode from behind, underneath, or above by runoff, wave action, scouring, ice shearing, and repeated thawing-refreezing cycles. In addition, they are not creating habitats for biodiversity and preventing the ecosystem services natural lakeshores provide.

In contrast, "soft-armoring or "bioengineering" can solve shoreline erosion without compromising the lake ecosystem. Some techniques are similar to the methods to reduce riverbank erosion described above. A fundamental design element of every technique is the reintroduction of

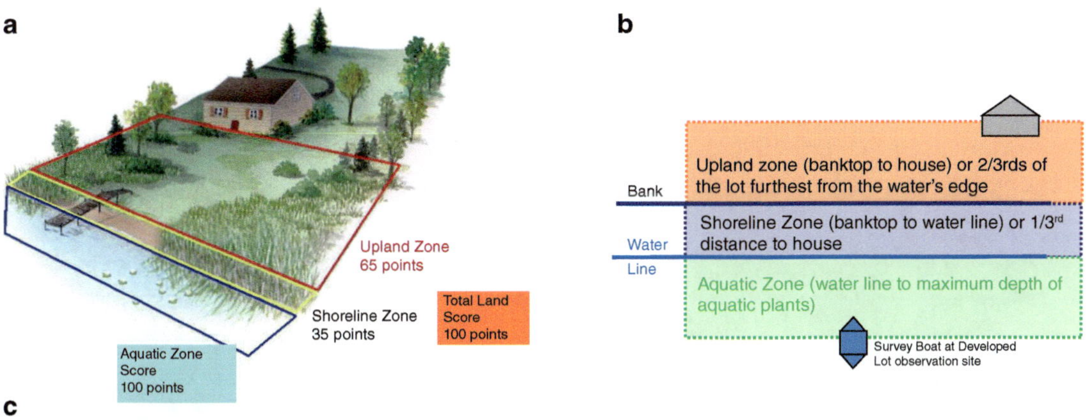

c

1. Upland Zone (65 pts max)			% of lot
A.	Upland tress		75-100
			50-74
			25-49
			1-24
			0
B.	Upland shrubs		75-100
			50-74
			25-49
			1-24
			0
C.	Upland ground cover	Naturally vegetated Ground	75-100
			50-74
			25-49
			1-24
			0
		Mowed, Bare or Impervious Surface	75-100
			50-74
			25-49
			1-24
C			0

2. Shoreline Zone (35 pts max)			% of lot
A.	Shoreline Trees/Shrubs		75-100
			50-74
			25-49
			1-24
			0
B.	Shoreline Ground Cover	Naturally Vegeated Ground	75-100
			50-74
			25-49
			1-24
			0
		Mowed, Bare or Impervious Surface	75-100
			50-74
			25-49
			1-24
			0

3. Aquatic Zone (100 pts max)			
A.	Emergent and Floating-leaf Plants	Percent of Shoreline frontage	75-100
			50-74
			25-49
			1-24
			0
B.	Submerged Plants	abundant plant growth	
		scattered plants	
		absent	
		can't see	
C.	Openings in plant beds	no unnatural openings	
		human-made channels/openings	
D.	Overhead woody habitat	shoreline vegetation hangs over water	
		no overhanging vegetation	
E.	Downed woody habitat	dead tree limbs and branches	
		no woody habitat present	

Fig. 14.21 The "Score Your Shore" assessment tool developed by the Minnesota Department of Natural Resources. (**a**) Scoreable habitat features in each lakeshore zone. (**b**) Surveyors navigate by boat to the midpoint of the lot within 50 feet (15 meters) of shore to view the site. (**c**) Scorecards for upland, shoreland, and aquatic zones. Reprinted with Permission. Minnesota Department of Natural Resources. (Perleberg et al. 2012). Reproduced by permission

deep-rooted native plants to mimic naturally stable shorelines.

Other design elements include coir (coconut fiber) logs and erosion control blankets (Fig. 14.23). Coir logs are rolled coconut fiber blankets held together with coconut threads. They can be created in various lengths, thicknesses, and densities. Coir logs are most commonly used at the toe of the shoreline but can also stabilize other parts of the shore. Their primary purpose is to provide a protective shield for reintroduced native plants to create a robust root system that can withstand the erosive forces at the shoreline. Within a few years, the coconut fiber will biodegrade and leave behind a well-established root system that will naturally stabilize the shoreline. One limitation of coir logs is that they cannot withstand excessive wave energy and ice push without extra protection, such as appropriately sized and placed rock. Erosion control blankets are manufactured from jute, coconut husk, or wood-fiber blankets and are used on sloped shorelines to hold the soil in place until native plantings become established. To serve their purpose, they should be biodegradable. The fiber blankets are placed flat over an area and fixed in place by stakes. Native plants are planted through natural gaps in the fabrics or through slits applied to denser fabrics. Erosion control blankets should not be used as the only technique on steep slopes but should be accompa-

Fig. 14.22 Two examples of native deep-rooting emergent shore plants that provide a soft erosion barrier. (**a**) Beaver Island, MI, along the Eastern coast. (**b**) Unknown lakeshore location in Michigan. Picture credits: (**a**) Elizabeth Riggs, Huron River Watershed Council. (**b**) Jane Herbert, Michigan Natural Shoreline Partnership. Both pictures are reproduced by permission

Fig. 14.23 Coir (coconut fiber) logs and erosion control blankets. (**a**) Coir (coconut fiber) logs can be planted with tubelings to create a stable vegetated streambank. (**b**) Erosion control blankets. Picture credits: (**a**) Coleman Moore Company (https://www.colemanmoorecompany.com), (**b**) (Michigan Natural Shoreline Partnership 2024). Both pictures are reproduced by permission

nied by other techniques, such as coir logs, rocks, or rip-rap, to ensure the slope is adequately stabilized.

14.3.4 Cleanup of Polluted Lakes

Like in rivers, many pollutants in lakes can be cleaned up by phytoremediation using the same or similar species of macrophytes described above for rivers. Shallow lakes that are most prone to becoming eutrophic can be triggered to self-clean by successfully reestablishing and maintaining submerged macrophytes since these plants can shift the entire lake ecosystem from turbid to clear water states (Liu et al. 2020). Another technique to improve the water quality of shallow lakes that became eutrophic can be the

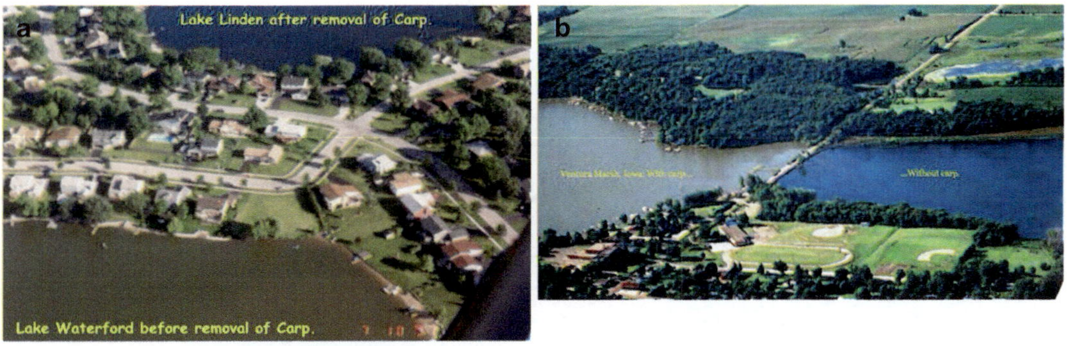

Fig. 14.24 Improvement of the water quality by removing carps. (**a**) Improvement of the water quality in one of two neighboring shallow eutrophic lakes following the removal of carp in one of the lakes (Lake Linden). (**b**) Improvement of the water quality in one of two portions of Clear Lake, a shallow eutrophic lake in Iowa, following carp removal in the portion to the right of the road. Figure credits: (**a**) Village of Lindenhurst, Illinois, (**b**) CLEAR Project, City of Clear Lake, Iowa. Both reproduced by permission

removal of certain fish species, mainly the common carp (*Cyprinus carpio*), which was documented to lead to an improvement of the water quality and sediment phosphorus (Fig. 14.24) (Huser et al. 2022).

Another technique to improve the water quality of eutrophic lakes is aeration (Kibuye et al. 2021; Koszelnik 2015; Preece et al. 2019). Adequate aeration consists of circulating oxygen throughout a body of water to avoid low dissolved oxygen levels. Aeration can also reduce or even prevent the sedimentation of large amounts of undecomposed organic matter, typically addressed as muck, and the accompanying issue of foul odor due to anaerobic decomposition, as well as improve water quality and fish habitat. In addition, aeration and the associated movement of the water body and surface by air bubbles can also help to reduce mosquitos.

Artificial circulation and hypolimnetic aeration are especially effective techniques to improve water quality and avoid problems such as algal blooms, and fish kills (Fig. 14.25). Artificial circulation infuses diffused air into lower water strata and pumps water from one lake stratum to another or creates turbulence on the surface using axial-flow pumps. Unlike this, hypolimnetic aeration mainly affects the lower strata of a lake (Figs. 14.25 and 14.26). Both techniques lower the concentrations of reduced compounds, thus improving water quality in lakes. Hypolimnetic aeration is more beneficial in preserving the natural lake stratification but may lead to low dissolved oxygen concentrations in the metalimnion. One downside of all aeration methods is that they may cause supersaturation of nitrogen, which may lead to gas bubble disease in fish.

A relatively new approach to restoring water quality in eutrophic lakes is using oxygen nanobubbles to remediate low oxygen conditions at the lake sediment/water interface (Pan et al. 2019; Waters et al. 2022). Oxygen nanobubbles are oxygen-filled bubbles with a diameter of less than 1000 nm that can form spontaneously at the interface of solid surfaces and aqueous solutions such as the lake sediment and the water column (Lyu et al. 2019). These bubbles have higher density than water and stay on the bottom of lakes, persist for extended periods, and facilitate greater oxygen solubility than ordinary oxygen bubbles. Therefore, they can remediate hypolimnion and sediment hypoxia/anoxia and the uptake of nutrients from the lake sediments to the water column under hypoxic conditions without the costs and limitations of aeration techniques described above. Nanobubbles can be generated at the lake sediment water body interface by geoengineering the sediment. The slow and continuous release of oxygen then removes algal blooms, restores the submersed vegetation and improves the overall water quality (Figs. 14.27 and 14.28).

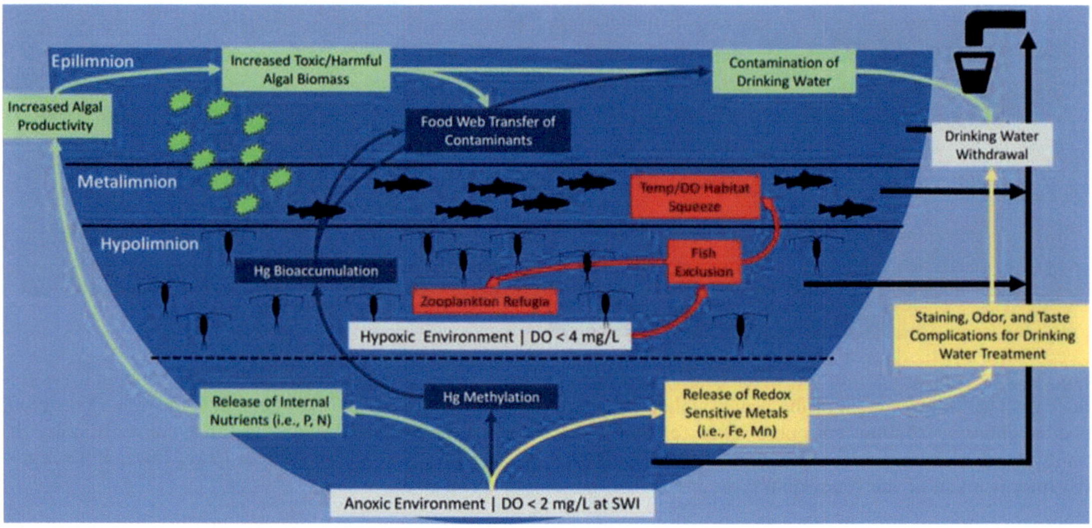

Fig. 14.25 Water quality issues in a eutrophic lake with low oxygen levels. Picture credit: A review of the biological and chemical effects of hypolimnetic oxygenation. Preece, E. P., Moore, B. C., Skinner, M. M., Child, A., & Dent, S. *Lake and Reservoir Management*, copyright © 2019 North American Lake Management Society, reprinted by permission of Taylor & Francis Ltd., https://www.tandfonline.com on behalf of North American Lake Management Society. (Preece et al. 2019)

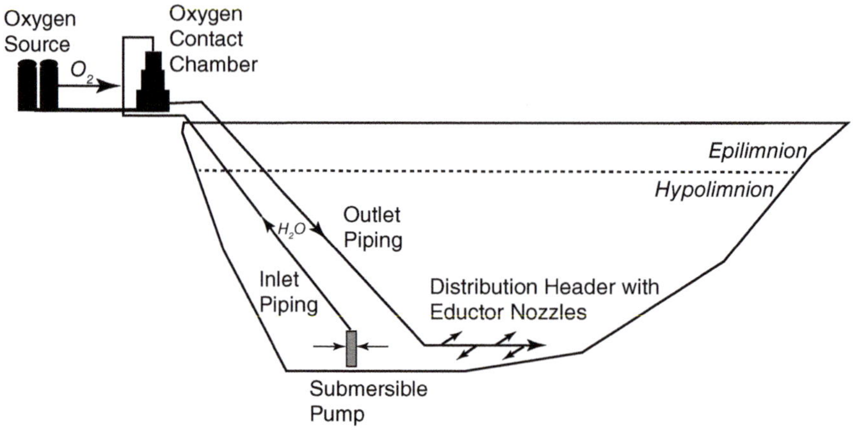

Fig. 14.26 Hypolimnetic aeration. Hypolimnetic water is drawn by a submersible pump into the inlet piping and transported into the oxygen contact chamber on the surface. Oxygen gas is injected into the water, resulting in supersaturated dissolved oxygen levels relative to the hypolimnion. The oxygenated water is pumped through the outlet piping to the distribution header, positioned at the same depth as the submersible pump, where the oxygenated water is injected back into the hypolimnion. Picture credit: (Gerling et al. 2014). Reproduced by permission

Oxygen nanobubbles can be combined with modified local soil materials, which can remove harmful algal blooms and remediate polluted sediment by modifying local soils with a minimal amount of natural products (usually less than 1% of the soil), such as chitosan, cationic starch, as well as the leaves, pods and kernels of the drumstick tree *Moringa oleifera*, and xanthan (Fig. 14.28).

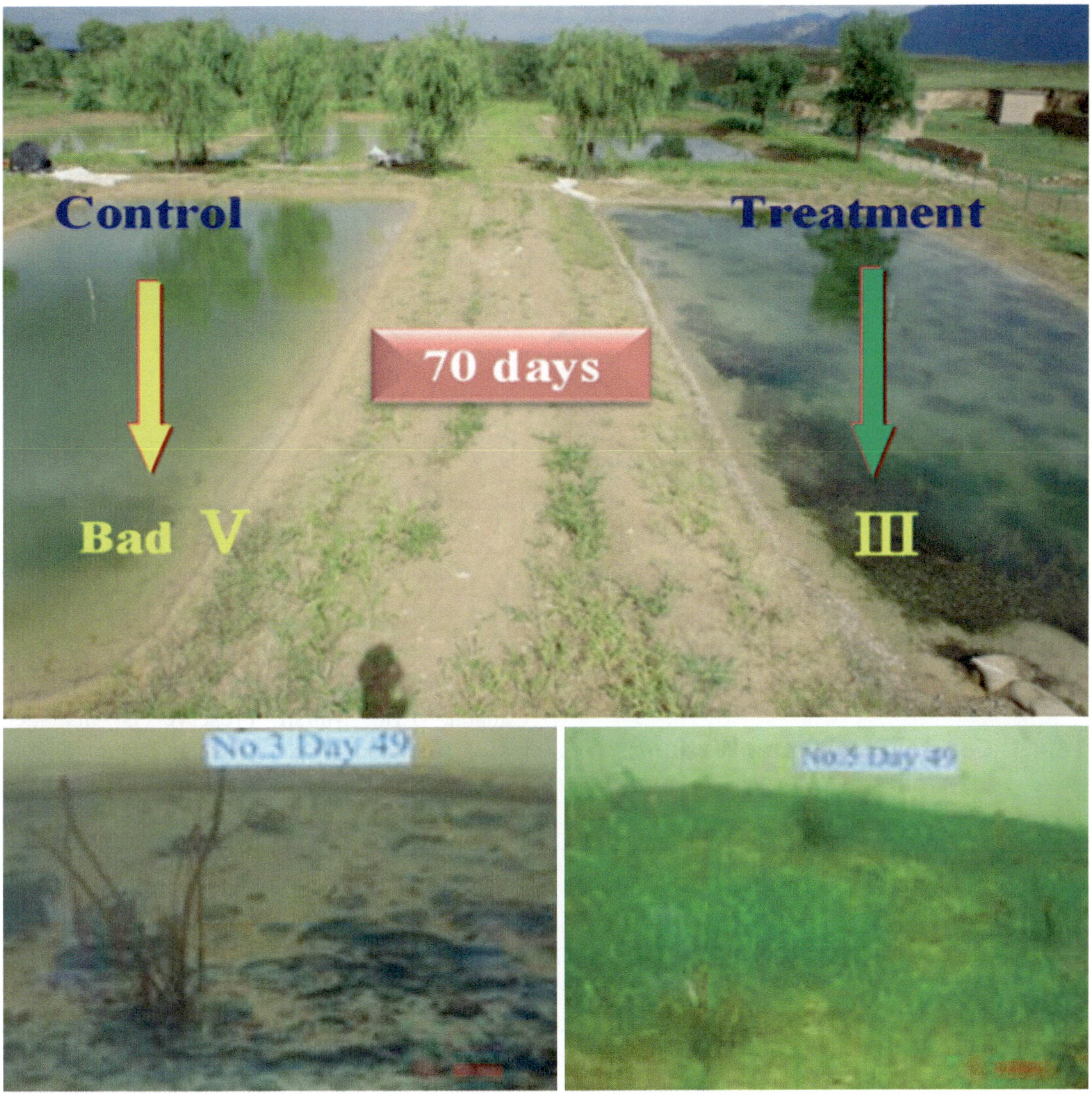

Fig. 14.27 Control (left) and treatment (right) ponds using oxygen nanobubbles at Cetian Reservoir, Datong City, North China. Credit: (Pan et al. 2019), licensed under CC BY 4.0

Excellent further readings are "Practices for Eutrophic Shallow Lake Water Remediation and Restoration: A Critical Literature Review" (Pereira and Mulligan 2023) and "Anthropogenic Impacts on Lake and Stream Ecosystems and Approaches to Restoration" (Søndergaard and Jeppesen 2007).

Exercises:

1. Watch the following three videos:
   - Inland Lake Shoreline Restoration https://youtu.be/3W8UBjGZp7o(approx. 5 min)
   - Natural Shoreline: Constructing Encapsulated Soil Lifts https://youtu.be/6PLSN9zhL9g (approx. 4:20)
   - Video 1—Natural Shorelines https://youtu.be/l7iZS_1AcOU (approx. 4:30).
2. Explain the benefits of lakeshore restoration. Describe the main techniques of lakeshore restoration in your own words.

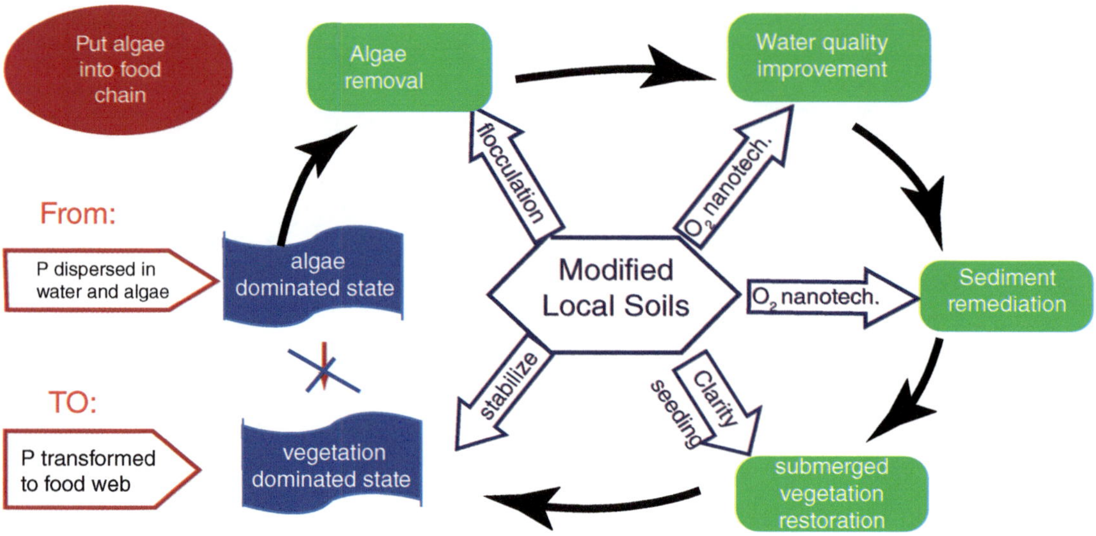

Fig. 14.28 Multidisciplinary principle of Modified Local Soil (MLS) technology. Credit: (Pan et al. 2019), CC BY 4.0

14.4 Wetland Restoration

14.4.1 Coastal Freshwater Wetlands: Erie Marsh

Coastal wetlands of inland lakes are one of the fastest declining ecosystems globally. The Great Lakes region lost over 50% of its coastal wetlands, and the remaining habitats are of poor quality (Gehring et al. 2020). In the Western basin of Lake Erie, less than 5% of coastal wetlands remain (Krieger et al. 1992) (Fig. 14.29). The Erie Marsh is one of Lake Erie's most extensive coastal wetlands. It constitutes 11% of all remaining coastal wetlands in SE Michigan. The Erie Marsh supports many animals and plants that would otherwise not exist in this area anymore. The habitat is also significant as a migratory and nesting area for birds, and it harbors some of Michigan's few remaining colonies of American lotus and swamp rose mallow, both listed as state-threatened. In 2006, Erie Marsh Preserve became a privately owned part of the Detroit River International Wildlife Refuge (https://www.fws.gov/refuge/detroit-river). The refuge protects habitat for 65 species of fish, 29 species of waterfowl, and 300 species of migratory birds in Michigan and Ontario, Canada. Erie Marsh Preserve is managed by a collaboration between The Nature Conservancy, US Fish & Wildlife Service, and other partners.

Only around 5% of the coastal wetlands in the Western Lake Erie basin persist from the 1950s, when pollution and dike construction reduced the water quality and flow into marshlands, rendering most of the remaining wetland areas barely restorable (Fig. 14.29). A $ 5,000,000 restoration project is on its way to restore the water flow in and out of one of the most extensive restorable coastal wetlands, the Erie Marsh (red circle in Fig. 14.29), and restore natural conditions to 946 of the 2217 acres. The execution of a four-phase restoration plan will gradually reestablish a mosaic of wetland types, such as emergent marsh, lake plain wet prairie, seasonally flooded moist soil wetlands, and deep water marsh, and includes effective control of invasive species. Details of this ambitious restoration project of coastal freshwater wetland ecosystems can be found in the project draft developed by Chris May for The Nature Conservancy, provided here: https://wassmer.org/freiland/MI_Erie_Marsh_EA.pdf.

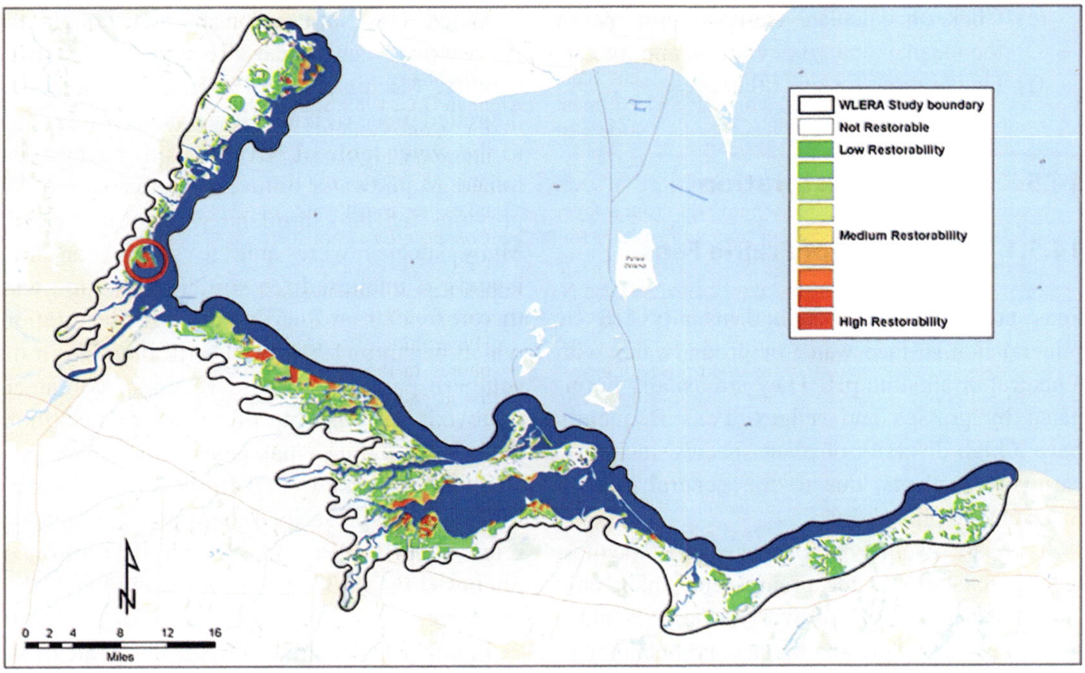

Fig. 14.29 Restorability of coastal wetlands in Western Lake Erie. The red circle marks the Erie Marsh. Picture credit: (NOAA Office for Coastal Management 2024). Reproduced by permission

Exercises:

1. Open or import the Google Earth KMZ file for Erie Marsh: https://tinyurl.com/eriemarsh1 in GoogleEarth Pro to see where the Erie Marsh is located.
2. Watch the following two videos:
   (a) SPOTLIGHT ON NATURE: Erie Marsh Preserve: https://youtu.be/7kjIqDJeFPE (about 3 min).
   (b) Restoring Erie Marsh https://youtu.be/vnzcffBfV3A (about 5 min).
3. Watch the Webinar Erie Marsh Preserve Coastal Wetland Restoration and Enhancement: https://digitalmedia.fws.gov/digital/collection/videos/id/293/rec/13 (about 42 min). You can follow the lecture with Chris May's original PowerPoint slides (Copyright: Chris May/TNC): https://tinyurl.com/eriemarsh2, which provide a better resolution.
4. Answer the following questions about the content of the webinar:
   (a) How does the new water pumping station keep out carp, and why are carp terrible news for a nature preserve?
   (b) What are the goals for a booming increase in species richness and abundance between pre- and post-restoration?
   (c) Name a snake species listed as a threatened species in the Erie Marsh.
5. Take a look at the Western Lake Erie Restoration Assessment at https://glcwra.wim.usgs.gov/wlera/
   (a) Select the checkbox WLERA Composite Area, select the submenu parameters, and unselect WaterMask.
   (b) Go to the area summary tool. Under the custom area, select draw.
   (c) Click on the corners of any area within the outlined study area boundary. You may want to zoom in and pan to the area of interest first.
   (d) When you have closed the area, click stop drawing. This will display your area. Take a screenshot.

(e) Click on calculate statistics and record the mean, standard deviation, and max.
(f) Do the same for one other area.

14.5 Peatland Restoration

14.5.1 Restoration of Prairie Fens

Fens are a type of peatland usually fed by mineral-rich surface water or groundwater with a neutral or alkaline pH. They are usually dominated by grasses and sedges. Fens frequently have a high diversity of plant species, including carnivorous plants, due to the general lack of nitrogen and phosphorus.

There are several types of fens. One of the most diverse fen types is the prairie fen, which only occurs in the glaciated Midwest of the USA and is distinguished by flora and fauna components typical for tallgrass prairies. Prairie fens arise when glacial deposits at outwash plains, moraines, and ice ridges conduct groundwater and bring it to the surface (Amon et al. 2002) (Fig. 14.30). For details, please refer to "Exploring the Prairie Fen Wetlands of Michigan" (Kost and Hyde 2009).

All over the world, swamps and fens are being drained for agricultural purposes. An example from northern Germany is shown in Fig. 14.31.

Prairie fens are critically endangered due to widespread delineation (draining) in the area of their occurrence (Fig. 14.32).

When The Nature Conservancy bought the Ives Road Fen (near Tecumseh, Lenawee County, Michigan) in 1987, it was a highly degraded piece of land. However, as it was close to the water table of River Raisin and had sustained groundwater influx, the conversion of the previous fen into farmland was not successful. Many species were able to survive on little islands of unaltered fen soil. The location was, therefore, extraordinarily suited for restoration, which began in 1994 and is still ongoing in the southern part of the preserve. First, volunteers removed vast thickets of European buckthorn that covered large patches of the abandoned farmland using saws and clippers (Fig. 14.33a), followed by prescribed burning. The manual removal of tiles (Fig. 14.33b) and the refilling of ditches (Fig. 14.33c) was extremely very labor intensive.

Figure 14.34 below shows a flowchart of the restoration process. The mostly community-carried restoration is one of the biggest success stories in wetland restoration in Michigan. For more information, refer to (May 2011)—provided at: https://tinyurl.com/IvesRdFen1

Exercises:

1. Open or import the Google Earth KMZ file of Ives Road Fen at https://tinyurl.com/IvesRdFen2 in GoogleEarth Pro to see where the Ives Road Fen is located.
2. Watch the following two videos:

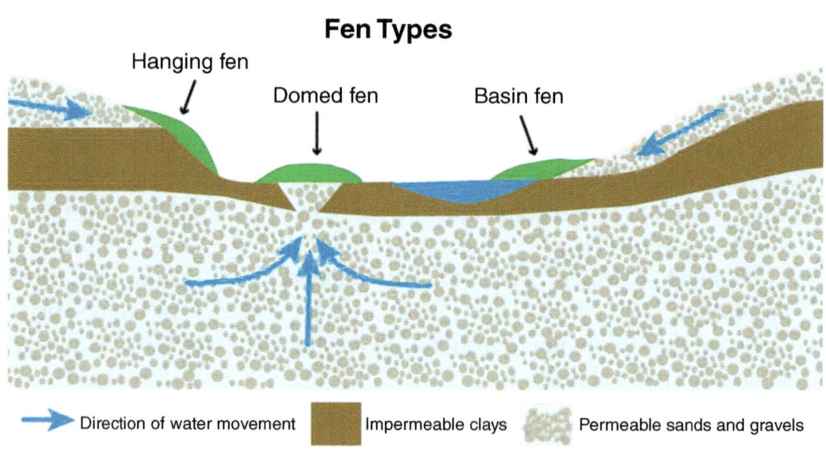

Fig. 14.30 Geologic settings for Midwestern USA fens. Permeable deposits transmit groundwater to the surface on hillsides to form slope fens (hanging fens), upward through openings in confining strata to form mound fens (domed fens), and to the perimeter of lakes to form lake edge fens (basin fens). Picture credit: Kost and Hyde (2009), reproduced by permission

   (a) Spotlight on Nature: Ives Road Fen https://youtu.be/ScFffG6AOD4 approx. 3:30.
   (b) Restoring and Protecting Tecumseh | Great Lakes Now Connect—Wild Places https://youtu.be/gb18kJE5sCw approx. 6:45.
3. Watch the slides: "Perseverance: Invasive Plant Control and Restoration at the Ives Road Fen Preserve." The presentation by Christopher May, The Nature Conservancy, at the RRIP-IT-UP Conference, Marquette, Michigan, on November 10, 2011,is provided at: https://tinyurl.com/IvesRdFen1.
4. Answer the following questions about the Ives Rd. Fen restoration.
   (a) What is the acreage of the Ives Road Fen preserve, and how big is the prairie fen?
   (b) Explain the geology and hydrology of prairie fens.
   (c) When did the TNC establish the preserve?
   (d) Why were drainage tiles and ditches put into the land before it became a preserve, and what did this cause in terms of hydrology and vegetation?
   (e) What are the four steps from a degraded area dominated by adult buckthorn to a fen community, and how can it be maintained?
   (f) How did they get rid of the ditches?
   (g) How do "Flammable Cover Ratings" and "Herbaceous Species Cover Ratings" measure restoration effectiveness, and what relationship exists between these two parameters?
   (h) What made this laborious restoration possible and successful?
   (i) Name three successes achieved by the restoration.
   (j) What is EDRR?
   (k) What is adaptive management?

Fig. 14.31 Wetland in Northern Germany that has been extensively drained for agricultural use

One problem for peatland restoration after agricultural use is the soil's excess nutrients. Peatlands must be low in nutrients. Excess nutrients can be reduced by grazing the area with cows or mowing the vegetation from the area. It would

Fig. 14.32 Farm tiles and ditches drained the water table of the naturally saturated fen soil, trying to dry up the soil for farming. Picture credit: Randall Schaetzl, Professor of Geography—Michigan State University. Reproduced by permission

Fig. 14.33 Volunteer-carried fen restoration at Ives Road Fen. (**a**) Heaps of removed buckthorn thickets. (**b**) Removal of drainage titles. (**c**) Refilling of ditches and blockage of drainage using temporary weirs made of wood. Picture credits: Chris May/The Nature Conservancy (May 2011). Reproduced by permission

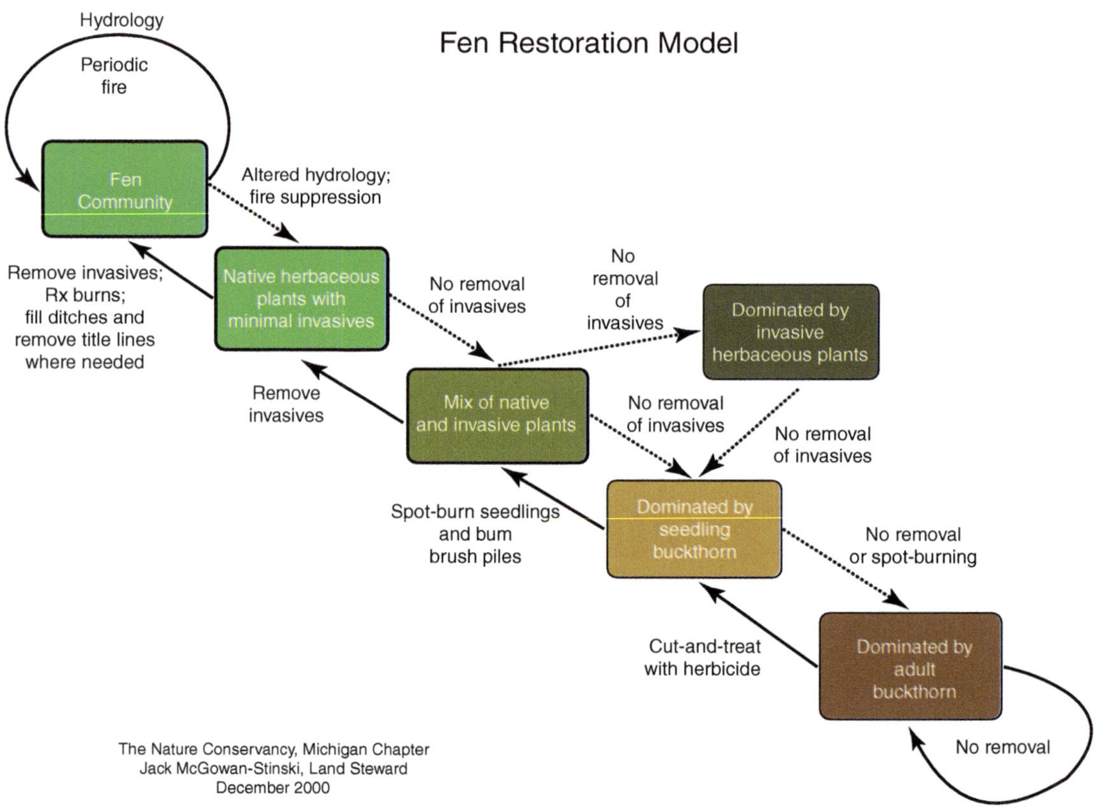

Fig. 14.34 Flowchart of the restoration process at Ives Road Fen. Reproduced by permission

be ideal if natural wetlands still existed in the neighborhood of the restoration areas. Seeds from the natural vegetation can then be transferred (see above). Soils contaminated with pollutants can be improved microbially by bioremediation if the soil conditions are suitable (Alpha-Omega Training and Compliance 2023).

14.5.2 Bog Restoration

Upland moors are peatlands that are only fed by rainwater (Fig. 14.35a). They are also classified as ombrotrophic bogs (Lindsay 1995, 2018). As mentioned above for fens, bogs are also endangered by drainage and peat extraction (Fig. 14.35c). Europe originally had large areas (up to 10%) of wetlands. Around 80% of these areas have been lost in the past millennium (Verhoeven 2014). Fens and bogs are common in the boreal, Atlantic, and temperate zones. In the United States, almost 50% of wetlands were lost (Mitsch and Gosselink 1986). The Ramsar Convention of 1971 (https://www.ramsar.org/sites/default/files/documents/library/current_convention_text_e.pdf) places wetlands of international importance under protection, particularly concerning waterfowl habitat. Peatlands are essential for climate mitigation as they are CO_2 sinks/reservoirs. A broad overview and potential

Fig. 14.35 Upland moors and bogs. (**a**) Bog in the National Park Harz, Germany. Bogs are peat-based ecosystems that are maintained by precipitation (**b**) Bog restoration in Harz National Park. Old drainage channels are backed up for rewetting. (**c**) Peat extraction in Rhön, Central Germany

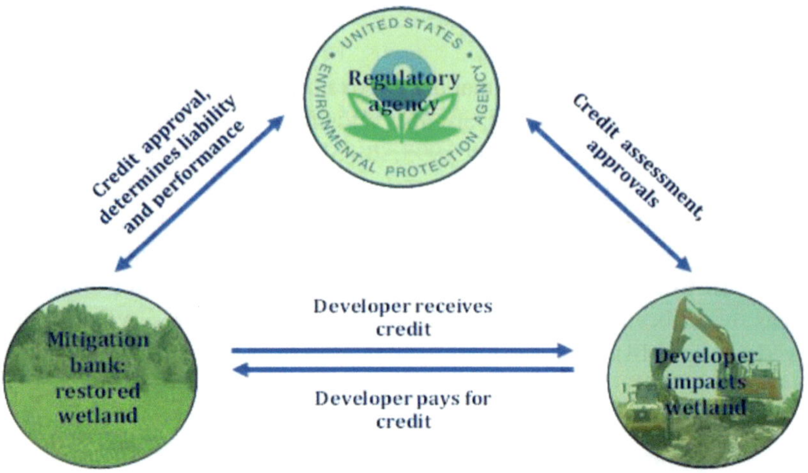

Fig. 14.36 A mitigation bank involves three entities: The mitigation bank/environmental engineering company, the regulators, and the developer/credit buyer. The basic transactions are with the wetland mitigation bank, which receives and sells mitigation credits

benefits of peatland restoration can be found in Bonn et al. (2016).

Farrell et al. (2021) describe how the condition of peatland and its restoration management should be accounted for across the European Union. Succow and Jeschke (1990) and Dewitz et al. (2023) describe the peatland distribution and conditions in Europe. Richert et al. (2003) emphasized the importance of the vegetation composition for the development of a bog.

We want to show an example of the restoration of an upland moor. Similar to the restoration of fens (Fig. 14.33c), drainage ditches or similar drainage points in bogs are dammed up using simple barriers, thus enriching the rainwater in the bog and rewetting the moor (Fig. 14.35b).

14.5.3 No Net Loss Policies

Wetland mitigation banking in Michigan requires the replacement of wetland functions as a condition of many construction and land use change permits issued under state law (Part 303, Wetlands Protection, of the Natural Resources and Environmental Protection Act, 1994 PA 451, as amended, (Michigan Department of Environment, Great Lakes, and Energy (EGLE) 2024) and federal law (Part 404 of the Clean Water Act) (US EPA 2015a). The rationale of wetland mitigation is the replacement of wetland functions that provide public benefits, such as flood storage, water quality protection, fish and wildlife habitat, and groundwater recharge. Related are the Wetland Conservation (also known as "Swampbuster") provisions of the Food Security Act (FSA) (US EPA 2015b), which hold back certain federal farm program benefits from farmers who convert or modify wetlands. The idea of wetland mitigation banking in all these provisions is to compensate the impacted or destroyed wetland function at the construction or conversion site, with an equal acreage and functionality at a different site (no net loss policy (Everard 2018, b)) (Fig. 14.36).

Despite the shortcomings of this approach in terms of the exact locality of the lost wetland and its connectivity to similar ecosystems, these programs have helped reduce the overall wetlands conversion rate by agriculture and other uses. Although restored wetlands need to be the same size and in the same watershed as the destroyed wetlands, their wetland functions and biodiversity are not necessarily replaced, especially in a local context (Fennessy and Dresser 2018; Harris and Roach 2021). One reason for caution is that restored wetlands are typically monitored for only 5 years, during which their biodiversity and functionality seem comparable to local reference wetlands but become less biodiverse and functional after the monitoring period has passed (Robertson et al. 2018; Stefanik and Mitsch 2012). A newer study by Tillman and Matthews (2023) found that wetland banks replaced only an average of 45% of the native plant species in impacted natural wetlands. The authors further state that policies intended to ensure only equivalence in acreage between impacted wetlands and banks do not safeguard biodiversity. It seems fair to conclude that the conservation instrument of wetland mitigation banks can help slow the overall decline of wetlands but will not automatically preserve biodiversity and all associated ecosystem services if they do not mimic the topographic and hydrologic habitat diversity, transplant local plant and animal species from the impacted to the compensation site, are as well or better connected to other wetlands in the watershed, and are monitored long-term.

References

Adler, J. (2024). Meet the Ecologist Who Wants You to Unleash the Wild on Your Backyard. *Smithsonian Magazine*. https://www.smithsonianmag.com/science-nature/meet-ecologist-who-wants-unleash-wild-backyard-180974372/ . Accessed 25 August 2024

Ahmed, A. K. A., & Marhaba, T. F. (2017). Review on river bank filtration as an in situ water treatment process. Clean Technologies and Environmental Policy, 19(2), 349–359. https://doi.org/10.1007/s10098-016-1266-0

Alexander, G. G., & Allan, J. D. (2006). Stream Restoration in the Upper Midwest, U.S.A. *Restoration Ecology*, *14*(4), 595–604. https://doi.org/10.1111/j.1526-100X.2006.00171.x

Allaby M. 2004 (ed.): *A Dictionary of Ecology*. 2nd ed. – Oxford University Press, 440 p.

Alpha-Omega Training & Compliance (AOTC) 2023: Types of Bioremediation – Applications, Benefits, and Examples. - Alpha-Omega Training & Compliance, Inc. 05.12.2023. https://a-otc.com › types-of-bioremediation

Amon, J. P., Thompson, C. A., Carpenter, Q. J., & Miner, J. (2002). Temperate zone fens of the glaciated Midwestern USA. *Wetlands*, *22*(2), 301–317. https://doi.org/10.1672/0277-5212(2002)022[0301:TZFOTG]2.0.CO;2

Ampoorter E, Barbaro L, Jactel H, Baeten L, Boberg J, Carnol M, Castagneyrol B, Charbonnier Y, Dawud SM, Deconchat M, De Smedt P, de Wandeler H, Guyot V, Hättenschwiler S, Joly F-X, Koricheva J, Milligan H, Muys B, Nguyen D, Ratcliffe S, Raulund-Rasmussen K, Scherer-Lorenzen M, van der Plas F, van Keer J, Verheyen K, Vesterdal L. & Allan E. 2020: Tree diversity is key for promoting the diversity and abundance of forest-associated taxa in Europe. – *Oikos* 129(2), 133–46. https://doi.org/10.1111/oik.06290

Atristain, M., Solagaistua, L., Larrañaga, A., von Schiller, D., & Elosegi, A. (2024). Slow drawdown, fast recovery: Stream macroinvertebrate communities improve quickly after large dam decommissioning. *Journal of Applied Ecology*, *n/a*(n/a). https://doi.org/10.1111/1365-2664.14656

Axelsson, E. P., Grady, K. C., Alloysius, D., Falck, J., Lussetti, D., Vairappan, C. S., et al. (2024). Lessons learned from 25 years of operational large-scale restoration: The Sow-A-Seed project, Sabah, Borneo. *Ecological Engineering*, *206*, 107282. https://doi.org/10.1016/j.ecoleng.2024.107282

Bayen, P., Lykke, A. M., Moussa, B. M., Bognounou, F., & Thiombiano, A. (2020). Effects of Three Different Planting Techniques on Soil Water Content, Survival, and Growth of Senegalia Seedlings on Semi-Arid Degraded Lands in Burkina Faso. *Tropical Conservation Science*, *13*, 1940082920972081. https://doi.org/10.1177/1940082920972081

Bellucci, L. G., Cassin, D., Giuliani, S., Botter, M., & Zonta, R. (2016). Sediment pollution and dynamic in the Mar Piccolo of Taranto (southern Italy): insights from bottom sediment traps and surficial sediments. *Environmental Science and Pollution Research*, *23*(13), 12554–12565. https://doi.org/10.1007/s11356-016-6738-6

Benstead, P., & JoseJosé, P. (2001). Wetlands Restoration. In S. A. Levin (Ed.), Encyclopedia of Biodiversity (pp. 805–822). New York: Elsevier. https://doi.org/10.1016/B0-12-226865-2/00283-2

Berger J.J. 1991: The federal mandate to restore: laws and policies on environmental restoration. – The Environmental Professional 13, 195-206

Bernhardt, E. S., Palmer, M. A., Allan, J. D., Alexander, G., Barnas, K., Brooks, S., et al. (2005). Synthesizing U.S. River Restoration Efforts. *Science*, *308*(5722), 636–637. https://doi.org/10.1126/science.1109769

Bernhardt, Emily S., Sudduth, E. B., Palmer, M. A., Allan, J. D., Meyer, J. L., Alexander, G., et al. (2007). Restoring Rivers One Reach at a Time: Results from a Survey of U.S. River Restoration Practitioners. *Restoration Ecology*, *15*(3), 482–493. https://doi.org/10.1111/j.1526-100X.2007.00244.x

Bertalan, L., Novák, T. J., Németh, Z., Rodrigo-Comino, J., Kertész, Á., & Szabó, S. (2018). Issues of Meander Development: Land Degradation or Ecological Value? The Example of the Sajó River, Hungary. *Water*, *10*(11), 1613. https://doi.org/10.3390/w10111613

Bonn Convention 1979: *Convention on the conservation of migratory species of wild animals*. Entered into force on November 1 1983.

Bonn A., Allott T., Evans M., Joosten H. & Stoneman R. (eds.) 2016: *Peatland Restoration and Ecosystem Services: Science, Policy and Practice*. – Ecological Reviews. Cambridge University Press, 493 p. https://doi.org/10.1017/CBO9781139177788

Boromisza, Z., Török, E. P., & Ács, T. (2014). Lakeshore-restoration-landscape ecology-land use: Assessment of shore-sections, being suitable for restoration, by the example of lake velence (Hungary). Carpathian

Journal of Earth and Environmental Sciences, 9, 179–188.
Braat L.C. & de Groot R. 2012: The ecosystem services agenda: bridging the worlds of natural science and economics, conservation and development, and public and private policy. – *Ecosystem Services* 1 (1), 4-15
Brancalion P. 2023: Tropical Ecological Restoration. Understanding the contributions of restored forests for nature and people: The NewFor Project. - https://mbgecologicalrestoration.wordpress.com/category/tropical-ecological-restoration/
Brancalion P., Niamir A., Broadbent E., Crouzeilles R., Barros F.S.M., Zambrano A.M.A., Baccini A., Aronson J., Goetz S., Reid J.L., Strassburg B.B.N., Wilson S. & Chazdon R.L. 2019: Global restoration opportunities in tropical rainforest landscapes. – *Science Advances* 5 (7), https://doi.org/10.1126/sciadv.aav3223
Braunss, M., Gücker, B., Wagner, C., Garcia, X.-F., Walz, N., & Pusch, M. T. (2011). Human lakeshore development alters the structure and trophic basis of littoral food webs. *Journal of Applied Ecology*, *48*(4), 916–925. https://doi.org/10.1111/j.1365-2664.2011.02007.x
Brettschneider, D. J., Spring, T., Blumer, M., Welge, L., Dombrowski, A., Schulte-Oehlmann, U., et al. (2023). Much effort, little success: causes for the low ecological efficacy of restoration measures in German surface waters. *Environmental Sciences Europe*, *35*(1), 31. https://doi.org/10.1186/s12302-023-00736-1
Burghardt, K. T., Tallamy, D. W., Philips, C., & Shropshire, K. J. (2010). Non-native plants reduce abundance, richness, and host specialization in lepidopteran communities. *Ecosphere*, *1*(5), art11. https://doi.org/10.1890/ES10-00032.1
Burrow, A. K., & Lance, S. (2022). Restoration of Geographically Isolated Wetlands: An Amphibian-Centric Review of Methods and Effectiveness. *Diversity*, *14*(10), 879. https://doi.org/10.3390/d14100879
Cardinale B.J., Primack R.B. & Murdoch J.D. 2020: *Conservation Biology*. – Sinauer Associates, Oxford University Press. 584 p.
Chazdon R.L., Brancalion P.H.S., Laestadius L., Bennett-Curry A., Buckingham K., Kumar C., Moll-Rocek J., Vieira I.C.G. & Wilson S.J. 2016: When is a forest a forest? Forest concepts and definitions in the era of forest and landscape restoration. – *Ambio* 45, 538–550. https://doi.org/10.1007/s13280-016-0772-y
Cheik, S., & Jouquet, P. (2020). Integrating local knowledge into soil science to improve soil fertility. *Soil Use and Management*, *36*(4), 561–564. https://doi.org/10.1111/sum.12656
Clewell A.F. & Aronson J. 2007: *Ecological Restoration. Principles, Values, and Structure of an Emerging Profession*. – Island Press, 216 p.
Crisp N., Dick J. & Mullins M. 2004: *Mongolia Forestry Sector*. Review, p. 61-68. – The World Bank, Victoria B.C.
Czech, B., Krausman, P. R., & Devers, P. K. (2000). Economic Associations among Causes of Species Endangerment in the United StatesAssociations among causes of species endangerment in the United States reflect the integration of economic sectors, supporting the theory and evidence that economic growth proceeds at the competitive exclusion of nonhuman species in the aggregate. *BioScience*, *50*(7), 593–601. https://doi.org/10.1641/0006-3568(2000)050[0593:EAACOS]2.0.CO;2
Dewitz I., Wenz K., Hüpperling S. & Peters J. (eds.) 2023: *Mooratlas 2023*. Daten und Fakten zu nassen Klimaschützern. – Heinrich Böll Stiftung, Bonifatius Druck, Paderborn
E. Dinerstein, Olson D.P., + 46 authors & Saleem M., *BioScience* 67, 534-545 (2017). https://doi.org/10.1093/biosci/bix014
Elias, J. E., & Meyer, M. W. (2003). Comparisons of undeveloped and developed shorelands, northern Wisconsin, and recommendations for restoration. *Wetlands*, *23*(4), 800–816. https://doi.org/10.1672/0277-5212(2003)023[0800:COUADS]2.0.CO;2
Ellenberg H. & Leuschner C. 2010: *Vegetation Mitteleuropas mit den Alpen: In ökologischer, dynamischer und historischer Sicht*. 6th ed. – UTB, Stuttgart. 1357 p.
Everard, M. (2018). No Net Loss: Overview. In C. M. Finlayson, M. Everard, K. Irvine, R. J. McInnes, B. A. Middleton, A. A. van Dam, & N. C. Davidson (Eds.), *The Wetland Book: I: Structure and Function, Management, and Methods* (pp. 821–825). Dordrecht: Springer Netherlands. https://doi.org/10.1007/978-90-481-9659-3_179
Falk D.A., Palmer M.A. & Zedler JB (eds.) 2006: *Foundations of Restoration Ecology*. – Island Press, 364 p.
Farrell C.A., Coleman L., Norton D., Kelly-Quinn M., Obst C., Eigenraam M., O'Donoghue C., Kinsella S., Smith F., Sheehy I. & Stout J.C. 2021: Developing peatland ecosystem accounts to guide targets for restoration. – *One Ecosystem* 6: e76838. https://doi.org/10.3897/oneeco.6.e76838
Fennessy, M. S., & Dresser, A. R. (2018). No Net Loss Case Study: Structural and Functional Equivalence of Mitigation Wetlands. In C. M. Finlayson, M. Everard, K. Irvine, R. J. McInnes, B. A. Middleton, A. A. van Dam, & N. C. Davidson (Eds.), *The Wetland Book: I: Structure and Function, Management, and Methods* (pp. 827–836). Dordrecht: Springer Netherlands. https://doi.org/10.1007/978-90-481-9659-3_177
Federal Interagency Stream Restoration Working Group (FISRWG). (1998). Stream Corridor Restoration: Principles, Processes, and Practices. Washington D.C.: Federal Interagency Stream Restoration Working Group.
Fletcher, J., Willby, N., Oliver, D. M., & Quilliam, R. S. (2020). Phytoremediation Using Aquatic Plants. In B. R. Shmaefsky (Ed.), *Phytoremediation: In-situ Applications* (pp. 205–260). Cham: Springer International Publishing. https://doi.org/10.1007/978-3-030-00099-8_7

Florentine, S., Gibson-Roy, P., Dixon, K. W., & Broadhurst, L. (Eds.). (2024). *Ecological Restoration: Moving Forward Using Lessons Learned* (2023rd edition.). Springer.

García-Haba, E., Hernández-Crespo, C., Martín, M., & Andrés-Doménech, I. (2023). The role of different sustainable urban drainage systems in removing microplastics from urban runoff: A review. *Journal of Cleaner Production*, *411*, 137197. https://doi.org/10.1016/j.jclepro.2023.137197

Gehring, T. M., Blass, C. R., Murry, B. A., & Uzarski, D. G. (2020). Great Lakes coastal wetlands as suitable habitat for invasive mute swans in Michigan. *Journal of Great Lakes Research*, *46*(2), 323–329. https://doi.org/10.1016/j.jglr.2019.12.013

Gerling, A. B., Browne, R. G., Gantzer, P. A., Mobley, M. H., Little, J. C., & Carey, C. C. (2014). First report of the successful operation of a side stream supersaturation hypolimnetic oxygenation system in a eutrophic, shallow reservoir. *Water Research*, *67*, 129–143. https://doi.org/10.1016/j.watres.2014.09.002

Girona M.M., Morin H., Gauthier S. and Bergeron Y. (eds.) 2023: *Boreal Forests in the Face of Climate Change*. Sustainable Management. - Advances in Global Change Research 74, Springer, 837 p. https://doi.org/10.1007/978-3-031-15988-6

Gonzalez-Abraham, C. E., Radeloff, V. C., Hammer, R. B., Hawbaker, T. J., Stewart, S. I., & Clayton, M. K. (2007). Building patterns and landscape fragmentation in northern Wisconsin, USA. *Landscape Ecology*, *22*(2), 217–230. https://doi.org/10.1007/s10980-006-9016-z

Grantham, H. S., Duncan, A., Evans, T. D., Jones, K. R., Beyer, H. L., Schuster, R., et al. (2020). Anthropogenic modification of forests means only 40% of remaining forests have high ecosystem integrity. Nature Communications, 11(1), 5978. https://doi.org/10.1038/s41467-020-19493-3

Greipsson S. 2011: *Restoration Ecology*. – Jones & Bartlett Learning, L.L.C., 408 p.

Groom M.J., Meffe G.K. & Carroll C.R. 2006: *Principles of Conservation Biology*. 3rd ed. – Sinauer Associates, Inc., 793 p.

Grove S.J. & Meggs J. 2003: Coarse woody debris, biodiversity and management: a review with particular reference to Tasmanian wet eucalypt forests. – *Australian Forestry* 66(4). https://doi.org/10.1080/00049158.2003.10674920

Hambler C. 2004: *Conservation*. – Cambridge University Press, 368 p.

Harker D., Libby G., Harker K., Evans S. & Evans M. 1999: *Landscape Restoration Handbook*. (2nd ed.). – Lewis Publishers, 884 p.

Harris, J., & Roach, B. (2021). *Environmental and Natural Resource Economics: A Contemporary Approach - International Student Edition* (5th edition.). New York, NY: Routledge.

Harrison R.D. & Swinfield T. 2015: Restoration of logged humid tropical forests: An experimental programme at Harapan Rainforest, Indonesia. – *Tropical Conservation Science* 8(1), 4-16. https://doi.org/10.1177/194008291500800103. Mongabay.com Open Access Journal

Heymann EW, Culot L, Knogge C, Smith AC, Tirade Herrera ER, Müller B, Stojan-Dolar M, Ferrer YL, Kubisch P, Kupsch D, Slana D, Koopmann ML, Ziegenhagen B, Bialozyt R, Mengel C, Hambuckers J, Heer K (2019): Small Neotropical primates promote the natural regeneration of anthropogenically disturbed areas. *Nature Scientific Reports*. http://nature.com/articles/s41598-019-46683-x

Hilderbrand, R. H., Bambakidis, T., & Crump, B. C. (2023). The Roles of Microbes in Stream Restorations. *Microbial Ecology*, *85*(3), 853–861. https://doi.org/10.1007/s00248-023-02179-w

Holl, K. (2020). *Primer of Ecological Restoration*. Washington: Island Press.

Homegrown National Park, Inc. (2024). Homegrown National Park. *Homegrown National Park*. https://homegrownnationalpark.org/. Accessed 25 August 2024

Huser, B. J., Bajer, P. G., Kittelson, S., Christenson, S., & Menken, K. (2022). Changes to water quality and sediment phosphorus forms in a shallow, eutrophic lake after removal of common carp (Cyprinus carpio). *Inland Waters*, *12*(1), 33–46. https://doi.org/10.1080/20442041.2020.1850096

Ielpi, A., Lapôtre, M. G. A., Gibling, M. R., & Boyce, C. K. (2022). The impact of vegetation on meandering rivers. *Nature Reviews Earth & Environment*, *3*(3), 165–178. https://doi.org/10.1038/s43017-021-00249-6

IUCN 2020: The Bonn Challenge. Restore our Future. – www.bonnchallenge.org

Jähnig, S. C., Lorenz, A. W., Hering, D., Antons, C., Sundermann, A., Jedicke, E., & Haase, P. (2011). River restoration success: a question of perception. *Ecological Applications*, *21*(6), 2007–2015. https://doi.org/10.1890/10-0618.1

Jenkinson, R. G., Barnas, K. A., Braatne, J. H., Bernhardt, E. S., Palmer, M. A., Allan, J. D., & Synthesis, T. N. R. R. S. (2006). Stream Restoration Databases and Case Studies: A Guide to Information Resources and Their Utility in Advancing the Science and Practice of Restoration. *Restoration Ecology*, *14*(2), 177–186. https://doi.org/10.1111/j.1526-100X.2006.00119.x

Johnson, M. F., Thorne, C. R., Castro, J. M., Kondolf, G. M., Mazzacano, C. S., Rood, S. B., & Westbrook, C. (2020). Biomic river restoration: A new focus for river management. *River Research and Applications*, *36*(1), 3–12. https://doi.org/10.1002/rra.3529

Kaboré, D., & Reij, C. (2004). The emergence and spreading of an improved traditional soil and water conservation practice in Burkina Faso: *EPTD discussion papers*. https://ideas.repec.org//p/fpr/eptddp/114.html. Accessed 25 July 2024

Kaiser, D. T. (2014). *Termites and Ants in Burkina Faso (West Africa): Taxonomic and Functional Diversity Along Land-use Gradients : Ecosystem Services of Termites in the Traditional Zaï-System* (Ph.D. Dissertation). Julius-Maximilians-Universität Universität, Würzburg.

Kaiser, D. T., Tra-Bi, C. S., Yeo, K., Konate, S., & Linsenmair, K. E. (2015). Species richness of termites (Blattoidea: Termitoidae) and ants (Hymenoptera: Formicidae) along disturbance gradients in semi-arid Burkina Faso (West Africa). *Bonn zoological Bulletin*, *64*, 16–31.

Kaiser, D. T., Lepage, M., Konate, S., & Linsenmair, K. E. (2017). Ecosystem services of termites (Blattoidea: Termitoidae) in the traditional soil restoration and cropping system Zai in northern Burkina Faso (West Africa). *Agriculture, Ecosystems & Environment*, *236*, 198–211. https://doi.org/10.1016/j.agee.2016.11.023

Kaufmann, P. R., Peck, D. V., Paulsen, S. G., Seeliger, C. W., Hughes, R. M., Whittier, T. R., & Kamman, N. C. (2014). Lakeshore and littoral physical habitat structure in a national lakes assessment. *Lake and Reservoir Management*, *30*(2), 192–215. https://doi.org/10.1080/10402381.2014.906524

Kibuye, F. A., Zamyadi, A., & Wert, E. C. (2021). A critical review on operation and performance of source water control strategies for cyanobacterial blooms: Part II-mechanical and biological control methods. *Harmful Algae*, *109*, 102119. https://doi.org/10.1016/j.hal.2021.102119

Koh L.P., Miettinen J., Liew S.C. & Ghazoul J. 2011: Remotely sensed evidence of tropical peatland conversion to oil palm. – *PNAS* 108 (12), 5127-5132. https://doi.org/10.1073/pnas.1018776108

Kost, M. A., & Hyde, D. A. (2009). *Exploring the Prairie Fen Wetlands of Michigan*. East Lansing, MI: Michigan State University Extension.

Kondor, A. C., Vancsik, A. V., Bauer, L., Szabó, L., Szalai, Z., Jakab, G., et al. (2024). Efficiency of the bank filtration for removing organic priority substances and contaminants of emerging concern: A critical review. Environmental Pollution, 340, 122795. https://doi.org/10.1016/j.envpol.2023.122795

Koszelnik, L. B. & P. (2015). Lakes and reservoirs restoration – Short description of the chosen methods. In *Progress in Environmental Engineering*. CRC Press.

Kramer, A. T., Crane, B., Downing, J., Hamrick, J. l., Havens, K., Highland, A., et al. (2019). Sourcing native plants to support ecosystem function in different planting contexts. *Restoration Ecology*, *27*(3), 470–476. https://doi.org/10.1111/rec.12931

Krieger, K. A., Klarer, D. M., Heath, R. T., & Herdendorf, C. E. (1992). A Call for Research on Great Lakes Coastal Wetlands. *Journal of Great Lakes Research*, *18*(4), 525–528. https://doi.org/10.1016/S0380-1330(92)71319-7

Leibnitz IZW 2019: New rainforest conservation initiative converts oil palm plantations into rainforest for the first time. – *News, Press Release* 28-11-2019.

Lindsay RA. 1995: *Bogs: the ecology, classification and conservation of ombrotrophic mires*. Battleby, Perth: Scottish Natural Heritage; 1995. Available from: http://roar.uel.ac.uk/3594/ .

Lindsay, R. 2018: Peatland (Mire Types): Based on Origin and Behavior of Water, Peat Genesis, Landscape Position, and Climate. In: Finlayson, C., Milton, G., Prentice, R., Davidson, N. (eds) The Wetland Book. Springer, Dordrecht. https://doi.org/10.1007/978-94-007-4001-3_279

Lindsay, A. R., Gillum, S. S., & Meyer, M. W. (2002). Influence of lakeshore development on breeding bird communities in a mixed northern forest. *Biological Conservation*, *107*(1), 1–11. https://doi.org/10.1016/S0006-3207(01)00260-9

Liu, H., Zhou, W., Li, X., Chu, Q., Tang, N., Shu, B., et al. (2020). How many submerged macrophyte species are needed to improve water clarity and quality in Yangtze floodplain lakes? *Science of The Total Environment*, *724*, 138267. https://doi.org/10.1016/j.scitotenv.2020.138267

Lorenz, S., Pusch, M. T., & Blaschke, U. (2015). Minimum shoreline restoration requirements to improve the ecological status of a north-eastern German glacial lowland lake in an urban landscape. *Fundamental and Applied Limnology*, 323–332. https://doi.org/10.1127/fal/2015/0647

Ly C. 2024: Reforestation initiatives in Africa may damage grassland and savannah. – *NewScientist* February 15 2024.

Lyu, T., Wu, S., Mortimer, R. J. G., & Pan, G. (2019). Nanobubble Technology in Environmental Engineering: Revolutionization Potential and Challenges. *Environmental Science & Technology*, *53*(13), 7175–7176. https://doi.org/10.1021/acs.est.9b02821

Manning A.D., Cunningham R.B. & Lindenmayer DB 2013: Bringing forward the benefits of coarse woody debris in ecosystem recovery under different levels of grazing and vegetation density. – *Biological Conservation* 157, 204-214. https://doi.org/10.1016/j.biocon.2012.06.028

May, C. A. (2011). Perseverance: Invasive plant control and restoration at the Ives Road Fen Preserve. (p. 41). Presented at the RRIP-IT-UP Conference, Marquette, Michigan: The Nature Conservancy.

Michigan Department of Environment, Great Lakes, and Energy. (2024, April 25). Bioengineered Shoreline Protection on Inland Lakes. *ArcGIS StoryMaps*. https://storymaps.arcgis.com/stories/63d101f3984d4c4e8720364d0fdd4431. Accessed 5 June 2024

Michigan Department of Environment, Great Lakes, and Energy (EGLE). (2024). Wetland Mitigation Banking. *Wetland Mitigation Banking*. https://www.michigan.gov/egle/about/organization/water-resources/wetlands/mitigation-banking. Accessed 8 June 2024

Michigan Natural Shoreline Partnership. (2024). Michigan Natural Shoreline Partnership. *Michigan*

Natural Shoreline Partnership. https://www.shorelinepartnership.org/. Accessed 5 June 2024

Mitsch W.J. & Gosselink J.G. 1986: *Wetlands*. – Van Nostrand Reinhold, New York. 539 p.

Mühlenberg M., Slowik J., Wöll H. & Waitkuwait W.E. 1995: Strategies for the restoration of tropical forest, that incorporate Wildlife Protection: an example from the Ivory Coast, West Africa. p. 413-417 – In.: Bissonette J.A. & Krausman P.R. (eds.): *Integrating people and Wildlife for a sustainable future*. Proc. 1st Internat. Wildl. Management Congr., The Wildlife Society, Bethesda

Mühlenberg M., Slowik J., Waltert M., Fermon H., Singho B. & Waitkuwait E. 1997: *Présentation des données du biomonitoring dans les forêts classées de l'est de la côte d'Ivoire*. – Rapport avec 7 annexes aux GTZ, Abidjan.

Mühlenberg M., Slowik J., Waltert M., Fermon H., Bamba S. & Waitkuwait W.E. 1999: Biomonitoring in Tropenwäldern. Ein Beitrag zur Bewertung von Sustainable Use. – *Göttinger Beiträge zur Land- und Forstwirtschaft in den Tropen und Subtropen* 133, 220–225

Mühlenberg M., Barkhishig T., Dashzeveg Ts., Drößler L., Neusel B. & Tsogtbaatar J. 2006: *Lessons from Tree Planting Initiatives in Mongolia*. – Discussion Papers, East Asia and Pacific. Environment and Social Development Department. Washington, DC: World Bank, 38 p. https://documents1.worldbank.org/curated/en/143891468059947579/pdf/377950MOG0less1ing0P09260901PUBLIC1.pdf

Müller V. 2015: Termiten machen es grün. – *Spektrum.de*, Hintergrund 15.09.2015

Murray-Stoker D. 2021: On the efficacy of restoration in stream networks: comments, critiques, and prospective recommendations. – *Peer Community Journal* (Section Ecology) 1, article no. e52. https://doi.org/10.24072/pcjournal.52

Naiman, R. J. (1996). Water, society and landscape ecology. *Landscape Ecology*, *11*(4), 193–196. https://doi.org/10.1007/BF02071809

Narango, Desiree L., Tallamy, D. W., & Marra, P. P. (2017). Native plants improve breeding and foraging habitat for an insectivorous bird. *BIOLOGICAL CONSERVATION*, *213*, 42–50. https://doi.org/10.1016/j.biocon.2017.06.029

Narango, Desirée L., Tallamy, D. W., & Marra, P. P. (2018). Nonnative plants reduce population growth of an insectivorous bird. *Proceedings of the National Academy of Sciences*, *115*(45), 11549–11554. https://doi.org/10.1073/pnas.1809259115

Narango, Desiree L., Tallamy, D. W., & Shropshire, K. J. (2020). Few keystone plant genera support the majority of Lepidoptera species. *Nature Communications*, *11*(1), 5751. https://doi.org/10.1038/s41467-020-19565-4

Natter A. 2018: Tropenforscher Karl Eduard Linsenmair und der Termiten-Trick. – *Main Post* 15.12.2018. aktualisiert: 07.04.2020

Natural Resources Conservation Service, U.S. Department of Agriculture. (1996). *National Engineering Handbook (Title 210), Part 650, Chapter 16, Streambank and Shoreline Protection* (No. PB98114358) (p. 143). Washington, DC.: USDA, NRCS. https://ntrl.ntis.gov/NTRL/dashboard/searchResults/titleDetail/PB98114358.xhtml. Accessed 4 June 2024

Newete, S. W., & Byrne, M. J. (2016). The capacity of aquatic macrophytes for phytoremediation and their disposal with specific reference to water hyacinth. *Environmental Science and Pollution Research*, *23*(11), 10630–10643. https://doi.org/10.1007/s11356-016-6329-6

NOAA Office for Coastal Management. (2024). Locating and Assessing Western Lake Erie's Restorable Wetlands. *Locating and Assessing Western Lake Erie's Restorable Wetlands*. https://coast.noaa.gov/digitalcoast/stories/lake-erie.html. Accessed June 7 2024

O'Connor N.A. 1991: The effects of habitat complexity on the macroinvertebrates colonising wood substrates in a lowland stream. – *Oecologia* 85, 504–512. https://doi.org/10.1007/BF00323762

Padonou, E. A., Djagoun, C. A. M. S., Akakpo, A. B., Ahlinvi, S., Lykke, A. M., Schmidt, M., et al. (2020). Role of termites in the restoration of soils and plant richness on bowé in West Africa. *African Journal of Ecology*, *58*(4), 828–835. https://doi.org/10.1111/aje.12778

Palmer, M.a., Bernhardt, E. s., Allan, J. D., Lake, P. s., Alexander, G., Brooks, S., et al. (2005). Standards for ecologically successful river restoration. *Journal of Applied Ecology*, *42*(2), 208–217. https://doi.org/10.1111/j.1365-2664.2005.01004.x

Palmer, M. A., Zedler, J. B., Falk, D. A., & PhD, D. K. H. (2016). *Foundations of Restoration Ecology* (Second Edition, New edition, Second Edition, New.). Washington: Island Press.

Pan, G., Miao, X., Bi, L., Zhang, H., Wang, L., Wang, L., et al. (2019). Modified Local Soil (MLS) Technology for Harmful Algal Bloom Control, Sediment Remediation, and Ecological Restoration. *Water*, *11*(6), 1123. https://doi.org/10.3390/w11061123

Pander, J., & Geist, J. (2013). Ecological indicators for stream restoration success. *Ecological Indicators*, *30*, 106–118. https://doi.org/10.1016/j.ecolind.2013.01.039

Pangau-Adam M., Slowik J., Trei J-N. & Waltert M. 2021: Negative Effects of Logging on Bird Dispersed Plants in Northern Papuan Lowland Forest, Indonesia. – Tropical Conservation Science 14, 1-9. https://doi.org/10.1177/19400829211031171

Catherine L. Parr, Mariska te Beest, Nicola Stevens. Conflation of reforestation with restoration is widespread. *Science*, 2024; 383 (6684): 698-701. https://doi.org/10.1126/science.adj0899

Pejchar, L., Reed, S. E., Bixler, P., Ex, L., & Mockrin, M. H. (2015). Consequences of residential development for biodiversity and human well-being. *Frontiers in Ecology and the Environment.*

13(3): 146-153., *13*(3), 146–153. https://doi.org/10.1890/140227

Pereira, A. C., & Mulligan, C. N. (2023). Practices for Eutrophic Shallow Lake Water Remediation and Restoration: A Critical Literature Review. *Water*, *15*(12), 2270. https://doi.org/10.3390/w15122270

Perleberg, D., Radomski, P., & Simon, S. (2012). *Score your shore - Citizen Shoreline Description Survey* (2nd ed.). St. Paul, MN: Minnesota Department of Natural Resources.

Polvi, L. E., Lind, L., Persson, H., Miranda-Melo, A., Pilotto, F., Su, X., & Nilsson, C. (2020). Facets and scales in river restoration: Nestedness and interdependence of hydrological, geomorphic, ecological, and biogeochemical processes. *Journal of Environmental Management*, *265*, 110288. https://doi.org/10.1016/j.jenvman.2020.110288

Poorter L., Craven D., Jakova CC, van der Sande M., et al. & Hérault B. 2021: Multi-dimensional tropical forest recovery. – *Science*, https://doi.org/10.1126/science.abh3629

Preece, E. P., Moore, B. C., Skinner, M. M., Child, A., & Dent, S. (2019). A review of the biological and chemical effects of hypolimnetic oxygenation. *Lake and Reservoir Management*, *35*(3), 229–246. https://doi.org/10.1080/10402381.2019.1580325

Radeloff, V. C., Hammer, R. B., & Stewart, S. I. (2005). Rural and suburban sprawl in the U.S. midwest from 1940 to 2000 in its relation to forest fragmentation. *Conservation Biology. 19(3): 1-13.*, *19*(3), 1–13.

Rahel, F. J. (2002). Homogenization of Freshwater Faunas. *Annual Review of Ecology, Evolution, and Systematics*, *33*(Volume 33, 2002), 291–315. https://doi.org/10.1146/annurev.ecolsys.33.010802.150429

Rahmani, T. A., Nurrochmat, D. R., Park, M. S., Boer, R., Ekayani, M., & Satria, A. (2022). Reconciling Conflict of Interest in the Management of Forest Restoration Ecosystem: A Strategy to Incorporate Different Interests of Stakeholders in the Utilization of the Harapan Rainforest, Jambi, Indonesia. *Sustainability*, *14*(21), 13924. https://doi.org/10.3390/su142113924

Reij, C., Smale, M., & Tappan, G. G. (2009a). Re-greening the Sahel: Farmer-led innovation in Burkina Faso and Niger, 53–58.

Reij C., Gary T. & Melinda S. 2009b. "Agroenvironmental transformation in the Sahel: Another kind of "Green Revolution"," IFPRI discussion papers 914, International Food Policy Research Institute (IFPRI).

Richard, M., Tallamy, D. W., & Mitchell, A. B. (2019). Introduced plants reduce species interactions. *Biological Invasions*, *21*(3), 983–992. https://doi.org/10.1007/s10530-018-1876-z

Richert M., Dietrich O., Koppisch D. & Roth S. 2003: The influence of vegetation development and decomposition in a degraded fen. – *Restoration Ecology* 8, 186-195

Riera, J., Voss, P. R., Carpenter, S. R., Kratz, T. K., Lillesand, T. M., Schnaiberg, J. A., et al. (2001). Nature, society and history in two contrasting landscapes in Wisconsin, USA: Interactions between lakes and humans during the twentieth century. *Land Use Policy*, *18*(1), 41–51. https://doi.org/10.1016/S0264-8377(00)00044-2

Rinaudo, T., McKenzie, S., Huynh, T.-B., & Sterrett, C. L. (2021). Farmer Managed Natural Regeneration: Community Driven, Low Cost and Scalable Reforestation Approach for Climate Change Mitigation and Adaptation. In Handbook of Climate Change Management (pp. 1–34). Springer, Cham. https://doi.org/10.1007/978-3-030-22759-3_281-1

Robertson, M., Galatowitsch, S. M., & Matthews, J. W. (2018). Longitudinal evaluation of vegetation richness and cover at wetland compensation sites: implications for regulatory monitoring under the Clean Water Act. *Wetlands Ecology and Management*, *26*(6), 1089–1105. https://doi.org/10.1007/s11273-018-9633-8

Rofkar, J. R., Dwyer, D. F., & Bobak, D. M. (2014). Uptake and Toxicity of Arsenic, Copper, and Silicon in Azolla caroliniana and Lemna minor. *International Journal of Phytoremediation*, *16*(2), 155–166. https://doi.org/10.1080/15226514.2012.759534

Roose, E., Kabore, V., & Guenat, C. (1999). Zai Practice: A West African Traditional Rehabilitation System for Semiarid Degraded Lands, a Case Study in Burkina Faso. *Arid Soil Research and Rehabilitation*, *13*(4), 343–355. https://doi.org/10.1080/089030699263230

Schmieder, K. (2004). European lake shores in danger — concepts for a sustainable development. *Limnologica*, *34*(1), 3–14. https://doi.org/10.1016/S0075-9511(04)80016-1

Schnaiberg, J. A., Riera, J., Turner, M. G., & Voss, P. R. (2002). Explaining Human Settlement Patterns in a Recreational Lake District: Vilas County, Wisconsin, USA. *Environmental Management*, *30*(1), 24–34. https://doi.org/10.1007/s00267-002-2450-z

Sinclair, J. S., Mademann, J. A., Haubrock, P. J., & Haase, P. (2023). Primarily neutral effects of river restoration on macroinvertebrates, macrophytes, and fishes after a decade of monitoring. *Restoration Ecology*, *31*(3), e13840. https://doi.org/10.1111/rec.13840

Søndergaard, M., & Jeppesen, E. (2007). Anthropogenic impacts on lake and stream ecosystems, and approaches to restoration. *Journal of Applied Ecology*, *44*(6), 1089–1094. https://doi.org/10.1111/j.1365-2664.2007.01426.x

Stefanik, K. C., & Mitsch, W. J. (2012). Structural and functional vegetation development in created and restored wetland mitigation banks of different ages. *Ecological Engineering*, *39*, 104–112. https://doi.org/10.1016/j.ecoleng.2011.11.016

Stokland J.N. 2021: Volume increment and carbon dynamics in boreal forest when extending the rotation length towards biologically old stands. – *Forest Ecology and Management* 488, 119017. https://doi.org/10.1016/j.foreco.2021.119017

Strayer, D. L., & Findlay, S. E. G. (2010). Ecology of freshwater shore zones. *Aquatic Sciences*, *72*(2), 127–163. https://doi.org/10.1007/s00027-010-0128-9

Succow M. and Jeschke L. 1990: Moore in der Landschaft – Enststehung, Haushalt, Lebewelt, Verbreitung, Nutzung und Erhaltung der Moore [Peatlands in the landscape – formation, ecology, biodiversity, distribution, use and conservation of peatlands]. Publisher Harri Deutsch, 268 p.

Sullivan T.P., Sullivan D.S., Lindgren P.M.F. & Ransome D.B. 2012: If we build habitat, will they come? Woody debris structures and conservation of forest mammals. – *Journal of Mammalogy* 93, (6), 1456–1468. https://doi.org/10.1644/11-MAMM-A-250.1

Tallamy, D. W. (2017). Creating Living Landscapes: Why We Need to Increase Plant/Insect Linkages in Designed Landscapes. https://doi.org/10.21273/HORTTECH03699-17

Tallamy, D. W., & Shriver, W. G. (2021). Are declines in insects and insectivorous birds related? *Ornithological Applications*, *123*(1), duaa059. https://doi.org/10.1093/ornithapp/duaa059

The River Restoration Centre. (2020). *Manual of River Restoration Techniques | The RRC*. Cranfield, UK: Cranfield University. https://www.therrc.co.uk/manual-river-restoration-techniques. Accessed 26 July 2024

Theobald, D. M., Miller, J. R., & Hobbs, N. T. (1997). Estimating the cumulative effects of development on wildlife habitat. *Landscape and Urban Planning*, *39*(1), 25–36. https://doi.org/10.1016/S0169-2046(97)00041-8

Tilch S. 2015: Es lebe der Boden - wie Mensch und Tier gemeinsam Wüstenböden urbar machen. – *Netzwerk-Forum zur Biodiversitätsforschung*, Pressemitteilung 16.04.2015

Tillman, S. C., & Matthews, J. W. (2023). Evaluating the ability of wetland mitigation banks to replace plant species lost from destroyed wetlands. *Journal of Applied Ecology*, *60*(6), 990–998. https://doi.org/10.1111/1365-2664.14391

Tip of the Mitt Watershed Council. (2007). *Understanding, Living With, & Controlling Shoreline Erosion: A Guidebook for Shoreline Property Owners* (3rd ed.). Petoskey, MI: Tip of the Mitt Watershed Council. https://watershedcouncil.org/wp-content/uploads/2023/04/Understanding-Shoreline-Erosion.pdf

Tomter S.M., Kuliešis A. & Gschwantner T. 2016: Annual volume increment of the European forests-description and evaluation of the national methods used. – *Annals of Forest Science* 73 (4), 849-856. DOI https://doi.org/10.1007/s13595-016-0557-2

U.S. Environmental Protection Agency. (2022). *National Lakes Assessment: The Third Collaborative Survey of Lakes in the United States* (No. EPA 841-R-22-002). U.S. Environmental Protection Agency, Office of Water and Office of Research and Development. https://nationallakesassessment.epa.gov/webreport/. Accessed 11 May 2024

University of Liverpool. "Reforestation programs could threaten vast area of tropical grasslands." *ScienceDaily*. ScienceDaily, February 15 2024. <www.sciencedaily.com/releases/2024/02/240215142140.htm>.

US Environmental Protection Agency. (2022). National Lakes Assessment: The Third Collaborative Survey of Lakes in the United States (No. EPA 841-R-22-002). US Environmental Protection Agency, Office of Water and Office of Research and Development. https://nationallakesassessment.epa.gov/webreport/. Accessed May 11 2024

US EPA, O. (2015a, August 20). Compensatory Mitigation Factsheet under CWA Section 404. Overviews and Factsheets. https://www.epa.gov/cwa-404/compensatory-mitigation-factsheet-under-cwa-section-404. Accessed 8 June 2024

US EPA, O. (2015b, December 3). CWA Section 404 and Swampbuster: Wetlands on Agricultural Lands. Overviews and Factsheets. https://www.epa.gov/cwa-404/cwa-section-404-and-swampbuster-wetlands-agricultural-lands. Accessed 8 June 2024

USDA NRCS. (1996). National Engineering Handbook (Title 210), Part 650, Chapter 16, Streambank and Shoreline Protection (No. PB98114358) (p. 143). Washington, DC.: Natural Resources Conservation Service, US Department of Agriculture. https://ntrl.ntis.gov/NTRL/dashboard/searchResults/titleDetail/PB98114358.xhtml. Accessed June 4 2024

Van Dyke F. 2008: Conservation Biology. Foundations, Concepts, Applications. 2nd ed. – Springer, 477 p.

Verhoeven J.T.A. 2014: Wetlands in Europe: Perspectives for restoration of a lost paradise. - Ecological Engineering 66, 6-9. https://doi.org/10.1016/j.ecoleng.2013.03.006

Wang, H., Zhang, H., & Cai, G. (2011). An Application of Phytoremediation to River Pollution Remediation. *Procedia Environmental Sciences*, *10*, 1904–1907. https://doi.org/10.1016/j.proenv.2011.09.298

Waters, S., Hamilton, D., Pan, G., Michener, S., & Ogilvie, S. (2022). Oxygen Nanobubbles for Lake Restoration—Where Are We at? A Review of a New-Generation Approach to Managing Lake Eutrophication. *Water*, *14*(13), 1989. https://doi.org/10.3390/w14131989

Wilcove, D. S., Rothstein, D., Dubow, J., Phillips, A., & Losos, E. (1998). Quantifying Threats to Imperiled Species in the United StatesAssessing the relative importance of habitat destruction, alien species, pollution, overexploitation, and disease. *BioScience*, *48*(8), 607–615. https://doi.org/10.2307/1313420

Wohl, E., Angermeier, P. L., Bledsoe, B., Kondolf, G. M., MacDonnell, L., Merritt, D. M., et al. (2005). River restoration. Water Resour. Res., 41(10), W10301. https://doi.org/10.1029/2005wr003985

Wohl, E., Lane, S. N., & Wilcox, A. C. (2015). The science and practice of river restoration. Water Resources Research, 51(8), 5974–5997. https://doi.org/10.1002/2014WR016874

Woodford, J. E., & Meyer, M. W. (2003). Impact of lakeshore development on green frog abundance. *Biological Conservation*, *110*(2), 277–284. https://doi.org/10.1016/S0006-3207(02)00230-6

Wright, K. K. (2021). Assessing stream restoration and the influence of scale, variable choice, and comparison sites. *Ecosphere*, *12*(3), e03440. https://doi.org/10.1002/ecs2.3440

Wu, S., Bashir, M. A., Raza, Q.-U.-A., Rehim, A., Geng, Y., & Cao, L. (2023). Application of riparian buffer zone in agricultural non-point source pollution control—A review. *Frontiers in Sustainable Food Systems*, *7*. https://doi.org/10.3389/fsufs.2023.985870

Zemp, D. C., Ehbrecht, M., Seidel, D., Ammer, C., Craven, D., Erkelenz, J., et al. (2019). Mixed-species tree plantings enhance structural complexity in oil palm plantations. *Agriculture, Ecosystems & Environment*, *283*, 106564. https://doi.org/10.1016/j.agee.2019.06.003

Zemp, D.C., Guerrero-Ramirez N., Brambach F., Darras K., Grass I., Potapov A., Röll A., Arimond I., Ballauff J., Behling H., Berkelmann D., Biagioni S., Buchori D., Craven D., Daniel R., Gailing O., Ellsäßer F., Fardiansah R., Hennings N., Irawan B., Khokthong W., Krashevska V., Krause A., Kückes J. & Kreft H. 2023: Tree islands enhance biodiversity and functioning in oil palm landscapes.—*Nature* 618, 316–321 (2023). https://doi.org/10.1038/s41586-023-06086-5

Zilli, FL & Fernández F. 2023: Factors explaining the diversity of invertebrates inhabiting woods in the Paraná River wetlands. – *Wetlands Ecol. Manage.* 31, 595–609. https://doi.org/10.1007/s11273-023-09933-4

15 Designing Data Collection

Abstract

Ecological observations are considered scientific evidence if you can show convincingly that they were not received by chance. Before starting a study, you should consider how to distribute and replicate your observations. An introduction to statistical methods is provided for this purpose. We do not provide or discuss the calculations, but we refer you to rapidly developing applications. Fundamental decisions that should be made before collecting data are explained, such as stratification, dependent and independent samples, and sample size.

Individual statistical tests are presented by explaining what test can be used for what data type. They can be used either for comparisons of sample series or for testing influencing factors on a selected variable. The latter analysis ranges from simple correlations to complex multivariate procedures. We also provide advice for the presentation of results. For some questions, descriptive indices are sufficient. A frequently used example is the description of diversity. The formulae for calculating some indices are also provided here.

15.1 Main Features of Statistics

In this chapter, we do not intend to provide an exhausting description of statistical methods but rather convey a necessary, basic understanding of statistics, which is the prerequisite of the quantitative approach in ecology. Detailed instructions can be found in many books written specifically about statistics, such as Sachs (1984), Hedderich and Sachs (2020), Poole (1974), Finney (1980), Sokal and Rohlf (2011), Flury and Riedwyl (1983), Herrmann (1984), Ehrenberg (1986), Weber (1986), Lorenz (1988), Fowler and Cohen (1996), Ludwig and Reynolds (1988), Krebs (1989), Dalgaard (2008), Timischl (2013), R Core Team (2018), Rudolf and Kuhlisch (2020), Harris (2020), Whitlock and Schluter (2020), Hector (2021), Schmuller (2022), Triola and Roy (2023), and Rasch and Verdooren (2023). Meanwhile, instructions for statistical tests and sometimes even data entry and calculations are offered on the internet for free. We recommend executing a search in any search engine using the appropriate phrases you will encounter below.

However, since ecological field data are subject to many factors, usually show high variability, and can often only be measured or observed inaccurately, a statistical data treatment is necessary for each evaluation. Since field observations are always laborious and usually not repeatable under identical conditions, a balanced experimental design is required before you start collecting data (see also Lehner 1979; Hurlbert 1984; Whitlock and Schluter 2020). This includes a rough preliminary investigation (pilot study) with a concise duration, so that representative variation coefficients, distribution patterns and numbers of species, and the number of individuals

M. Mühlenberg, T. Waßmer, *Field Ecology*, https://doi.org/10.1007/978-3-031-87547-2_15

one will be confronted with in the actual research can be estimated. The rough preliminary investigation determines the planned sample size and distribution (Alreck and Settle 2003; Conroy 2021).

When planning the experiment, one should also be clear about the variables to be observed (or measured). At this stage, i.e., before the actual observations (data collection, experiment), we recommend creating a variable list; see the example below. If observations are missing or are incomplete for a variable, then the later handling of this variable in the evaluation process becomes questionable.

The use of computers for evaluating data is standard and is based on a variety of statistical software offered (e.g., Afflerbach 1987). The five most commonly used packages in the health sciences are SPSS, SAS, Stata, Excel, and R (Masuadi et al. 2021). NYU DataServices (2023) and Wikipedia offer a list and compare statistical packages (https://en.wikipedia.org/wiki/Comparison_of_statistical_packages). Most of the current packages are highly comparable and can all compute any standard tests, so it is more a question of personal preference or what your institution may have already available or what software you already know how to use. The most flexible software by far is R. Due to its open-source architecture, all standard procedures and almost every specialized computation can be run in R. Lai et al. (2019) explored the extent of R usage in ecology and the frequencies of R packages reported for 10 years ending in 2017.

15.2 Samples: Formulation of Questions, Choice of Methods

The formulation of research questions and the knowledge of existing preliminary evidence (see samples for the pilot study) are decisive for the sample selection. If we want to investigate the spatial and temporal distribution of species, e.g., the distribution of ground beetle species in a heterogeneous meadow, systematic sampling is recommended. A systematic sample selection, for example, involves observing at a constant distance and time (e.g., every 30 m, or every 2 h).

If we are interested in accurate information about the population size of a species, for example, in the wet parts of a meadow, a random sample selection is recommended.

Random coordinates are used to determine sample locations in the field randomly. The study area is drawn on an accurate map within a coordinate system or divided into a checkerboard pattern by a grid. Pairs of random coordinates between 1 and 100 can also be taken directly from random tables (e.g., in Lewis and Taylor 1967). Random coordinates can also be generated with simple software programs or Microsoft Excel.

Randomly drawn samples have an advantage over systematic samples in that the results obtained from their basic statistical parameters (mean, standard deviation) only show inevitable random deviations compared to the corresponding values of the total population (Sachs 1984; Hedderich and Sachs 2020; Siegel and Castellan 1988; Zar 2009). The probability of a systematic (nonrandom) error is therefore significantly lower.

A systematic sample selection can contain methodological or systematic errors (*bias*), the size of which often cannot be precisely specified. If the sample contains systematic errors, it does not represent the total population (Sachs 1984; Hedderich and Sachs 2020; Siegel and Castellan 1988; Zar 2009). Compared to random samples, the disadvantages of comparatively less precise statements are compensated by better data distribution.

If the variable to be examined is homogeneous across the entire study area, then there is only a small difference between systematic and random sample selection.

Suppose a specific trend within a study area, e.g., due to a moisture gradient (see Sect. 10.15 Forest Edge). In that case, the systematic sample provides a more accurate result, e.g., regarding the distribution in the gradient. If there are periodic fluctuations, e.g., an increase in the terrain every 10 m with corresponding microhabitat changes, then a systematic sample can always lie in the depressions and give a false picture across the entire area. We would reduce such a systematic error if we drew several systematic samples with randomly changed starting points: e.g., 0, 10, 20, 30… and 3, 13, 23, 33... We also achieve the same goal through changes in the step size.

15.3 Stratification

Stratification of samples means dividing all samples into subunits, such as a population into subpopulations or a habitat into different areas (see Fig. 15.1a), in which fewer variates fluctuate.

Temporally variable factors (rain, seasonal management) or habitat constants (habitat type, altitude layer, altitude, rock base, dry and wet areas) can be used to assign samples to different areas. The main goal of stratification is to increase the precision, e.g., of a population density estimate, and keep the variance between individual samples low. The strata are chosen to make the population distribution within a stratum more homogeneous than the entire study area.

If we are mainly interested in one parameter in a study, e.g., a species, we will easily find a suitable stratification. With several parameters, we will have difficulties achieving an effective stratification. Prior knowledge about the distribution pattern of a species facilitates stratification, whether from the literature or your own systematic preliminary studies.

We expect a more homogeneous distribution of the animals within a stratum than when con-

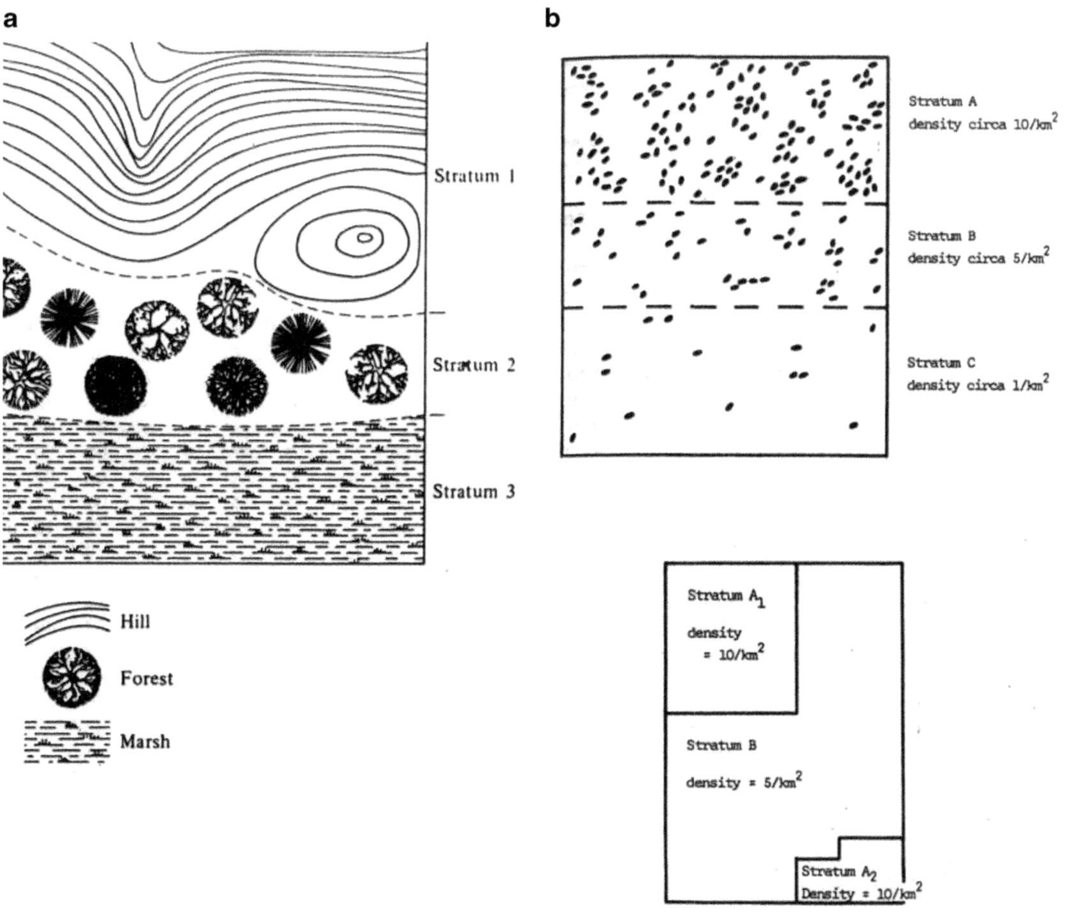

Fig. 15.1 (**a**) Scheme for stratification by habitats. (**b**) Scheme for stratification by abundances

sidering all areas together. Therefore, we should arrange the samples randomly within a stratum. This form of sample selection is called "stratified random sampling," a method most commonly used in ecological fieldwork (Southwood 1978; Sutherland 2006). If the areas are of different sizes, then we choose the number of samples taken in each stratum proportional to its size.

15.4 Independent and Dependent Samples

Just as we reduce the variance between samples for statistical evaluation by dividing the samples into strata, we can reduce the sampling error when comparing two samples, x and y, from individual measurements (x_1, x_2,... and y_1, y_2,...) by forming pairs and reducing the sampling error through so-called parallel samples. Instead of comparing weight measurements from randomly selected songbirds of population A with measurements from population B, the result would be more precise if we only compared adult ♂♂ (males) from population A with adult ♂♂ from population B. If we want to make a statement about the ♂♂ of a bird species (e.g., yellowhammer), then we would have independent samples when comparing populations A and B (the total population is all ♂♂ of a population, so we have two total populations, namely male yellowhammers of population A and male yellowhammers of population B). However, if we want to test an influencing factor on a bird species (yellowhammer) as a whole, i.e., make a general statement for yellowhammers, and this influence could have different effects on the sexes, then the sample is paired by the preselection (♂♂ A against ♂♂ B or ♀♀ A against ♀♀ B), i.e., we then have dependent (linked) samples in front of us.

A dependent sample also exists when comparing an observation of an animal before and after an event (possibly also after an experiment) (see also Hurlbert 1984).

For statistical tests to evaluate your data correctly, you must decide whether your samples are independent or dependent (linked). This choice often does not seem clear when comparing field data. In principle, samples from two different total populations are considered independent, whereas samples from one total population are dependent. If an unexamined factor is assumed to influence both samples simultaneously or in the same direction, then the samples will be treated as linked. Linked samples should also be taken as paired measurements, i.e., the simultaneous emptying of trap sets. Also linked are samples in a meadow, from which observation data are collected before and after a treatment, e.g., mowing. It must be assumed that other factors had also acted in the mowed investigation area "meadow" before and after. They influenced both sample collections, or the species community before and after mowing still contained many typical individuals. Another example of linked data are behavioral observations during feeding, carried out one after the other on the same bird (cf. Beal and Khamis 1990).

Trap catches or observations would be considered independent samples if they come from two areas presumably not influenced in the same direction by the same external factor: samples from a dry and wet meadow, for example, or from a meadow and a forest. A comparison of two bird species, for example, regarding the feeding behavior (choice of location, duration of seed or prey treatment) would also be independent samples.

15.5 Descriptive Statistics

For the initial evaluation and summary of the recorded ecological data, one calculates the average of a series of measurements or groups of numbers from a series of observations. Since it is uncertain whether all observation values are close to the average or vary greatly, information about the dispersion and distribution of the available data are also provided.

The arithmetic mean $\overline{x}$ is the most frequently calculated and helpful average for further comparisons. In skewed and one-sided distributions of the observation values (see below), the median (med(x), $\tilde{x}$, $\mu_{1/2}$, or M) can represent the observation values more sensibly than the arithmetic mean. The median is the central value in the middle of n sample values ordered by size, with even n being the arithmetic mean of the two central values. For example, the median value is practical when specifying animal group sizes: If only once a large herd was observed, this numerical value would strongly and one-sidedly influence the average group size calculated as the arithmetic mean. Similarly, when stating the average size of protected areas in Bavaria, the area value of a national park would mask the fact that most nature reserves are far too small for the area requirements of animal species (in the sense of securing their future population, see Sect. 10.10 population viability). One advantage of the median value is that it is hardly influenced by outlier values (see below). Not in all cases do extreme values belong to outlier values. For example, extreme values can be ecologically significant in the case of pollutant contamination within the lethal range.

The dispersion of measured values in a data set can be described using their deviations from the mean. The standard deviation (SD) formula can be found in any statistics book (e.g., Finney 1980). A mean value (arithmetic mean) should always be presented together with its standard deviation in the form $\overline{x} \pm SD$. The coefficient of variation (CV) expresses the standard deviation as a percentage of the mean value. The relative coefficient of variation V_F can only take values between 0% and 100%.

When you have calculated a value like the mean of a sample with *n* number of observed values, the observed dispersion is not the only interest but also the "reliability" of the parameter calculated from the sample. For this purpose, the confidence interval (CI, confidence range) is provided. The confidence interval around the arithmetic mean is only calculated when the observed values are approximately normally distributed. We also assume that we are dealing with random samples. The confidence limits of 95% indicate, for example, that in 95 out of 100 cases, you are correct about the assertion: within the limits of two standard deviations on either side of the sample mean lies the true but unknown mean of the so-called total population. If, as in most cases, the observations or sample results are not randomly distributed and there may be a sampling bias, computer programs that run simulations based on the results and carry out a so-called "bootstrapping" can help. This allows confidence limits to be determined more precisely (Sutherland 2006; Manly 2007).

For example, if one has tried 60 times on observation walks to determine species A and B and has encountered species A 60 times and species B only 15 times, we would conclude that species B is only 1/4 as common as species A, p (the frequency of certain observations) = 15/60 = 0.25 (if we want to make a statement about the ratio, we only count samples in which at least species A was seen). However, with a probability of 95%, we can only claim that species B is approximately 1/3 to 1/7 as frequent as species A: The confidence limits in this example are 0.14 and 0.36 (0.25 ± 0.11).

Under certain circumstances, excessively high or low values within a series of observations may be neglected. Prior knowledge can be used to determine if extreme values should be considered possible or rejected. Such extreme values could, for example, be due to measurement or observation errors or even calculation errors when transcribing values into tables. They would not belong to the same population, and their omission allows for much broader usability of the statistical tests.

As a rule of thumb, with at least ten individual values, a value can be discarded as an "outlier" if it lies outside the range of $\overline{x} \pm 4\ SD$, where the mean and standard deviation are calculated without the values of the suspected outliers. However, this rule of thumb can only be applied if the data are approximately normally distributed. In any case, any outlier elimination should be clearly stated to avoid any indication of data falsification.

A frequency distribution indicates how often the different values occur in measurements or observation values. The values can be more or less symmetrically distributed around a mean (normal distribution, parametric data) or occur in clusters at the upper or lower end of a distribution (skewed distribution, nonparametric data). There are mathematical descriptions for many distribution forms. A normal distribution describes data that follow a particular type of symmetric single-peaked distribution in the form of a "bell curve" in which 68% of the observation values in the hypothetical example fall within a simple standard deviation upwards and downwards $(\overline{x} \pm SD)$, and 95% fall within double the standard deviation $(\overline{x} \pm 2SD)$. Since the computational analysis and application of statistical tests compare observation values with theoretical distributions, it is essential to know the distribution of your data and accordingly select the correct statistical methods. If the observed distribution resembles a theoretical distribution, it can be assumed that the same laws apply.

15.6 Size and Number of Samples

15.6.1 How Large Must a Sample Be?

The desired accuracy generally determines the size of the sample. The greater the variability of the individual results of a sample, the more individual samples we must collect to achieve an equally precise result. The fluctuation of the measured values and, thus, the sample size is influenced, for example, by the population density of the species to be investigated, its distribution (clumped or uniform), and the standardization of the recording method. The number of samples will always be a compromise between the available resources (time, cost) and the desired sensitivity (how small a difference should still be detected?) or the size of the confidence interval to be determined (e.g., in 95 out of 100 cases, the true mean of the total population should be included, i.e., confidence interval 95%).

In the case of sample selection stratified by population density, the largest number of samples is taken in the most densely populated area to obtain the most information with the least possible effort. This means that more samples are taken where the highest individual density is expected (see Fig. 15.1b), which can, for example, provide additional results about, e.g., phenology and sex ratios with less statistical effort.

The question of sensitivity or the confidence interval also depends on the objective of the investigation. The determination of the population size can be interesting for (a) population management (how much can be harvested?), (b) population analysis (which characteristics determine population dynamics?), (c) testing of hypotheses (e.g., does competition limit population size?) and (d) population monitoring (how does the population change over time under a particular influence?) (see Eberhardt 1978).

Different calculations of the necessary sample size depend on whether the accuracy is determined for "estimating" a value (estimation methods serve for the precision of measurements and counts, test methods for testing hypotheses, Lorenz 1988) or comparing two groups. First, one chooses an acceptable confidence interval (statistical certainty), e.g., 95%, which tolerates that in 5% of the experimental or observation repetitions, the true mean of the total population lies outside the recorded range.

For approximately normally distributed samples, a deviation of 2 $SD/\sqrt{n}$ (SD = standard deviation, n = sample size) up and down from the mean is tolerated. To determine the sample size, one determines which absolute deviation up and down from the mean at the 95% confidence interval should be accepted, e.g., in the measurement of wing length, e.g., an accuracy of ±1 cm. In the second case, one determines the sensitivity, i.e., the size of the difference that should still be statistically reliably detected. The difference is given as a ratio, e.g., of the two mean values. For example, the first mean value is one and a half times as large as the second (Eberhardt 1978).

As a rule of the thumb, the simplest formula for estimating the minimum required number of samples (n) is:

$$n = 4 \cdot \frac{S^2}{d}$$

where S = Dispersion, indicated as the coefficient of variation C

C = standard deviation / mean value * 100 = (SD/$\overline{x}$) * 100.

d = Tolerance limit.

For example, with a coefficient of variation of 0.6 and tolerance limits of 20% (d = 0.2), n is calculated as follows:

$$n = 4 \cdot \frac{0.6^2}{0.2} = 4 \cdot 9 = 36$$

As a rule of the thumb, the simplest formula for estimating the minimum required sample size (n) when comparing two groups is:

$$n = \frac{21 \cdot C^2}{(k-1)^2}$$

where C = Coefficient of variation

k = Ratio between the two values of the comparison.

For example, with a coefficient of variation of C = 0.6 and a chosen difference accuracy of k = 1.5, n is calculated as follows:

$$n = \frac{21 \cdot (0.6)^2}{(1.5-1)^2} = 21 \cdot \frac{0.6^2}{0.5^2} \approx 30$$

This means that at least 30 sample values are needed for each group to demonstrate a difference in the ratio of 1.5 as significant. The number 21 contains the following two conventions: the critical probability of a false decision, (a) to reject a true null hypothesis is assumed to be 5%, and (b) to maintain a false null hypothesis is assumed to be 10% (= 90% test power) (Hedderich and Sachs 2020; Sachs 1984).

Coefficients of variation can be estimated from one's own previous studies or literature. According to Eberhardt (1978), the values of the coefficient of variation C for wildlife counts after evaluation of the literature for roadside counts are around 0.8, for call counts around 0.7, for transects (on foot) between 0.5 and 2.0, and for pellet counts around 1.0.

The necessary sample size for D-Vac insect collections (see Sect. 2.8) can be calculated as follows (for example, see Potts 1986):

$$\ln N = (\ln a - 2\ln p) - (2-b)\ln m$$

where

N = Sample size, dependent on the population density of the species under study and the degree of aggregation

a = Constant (species-specific, can be set to 1).

m = Average density (animals per m^2).

p = Degree of accuracy, as standard error/mean.

b = Index of clumping degree (b = 1, if randomly distributed; $b > 1$, if clumped; realistic values are around b = 2).

b is calculated according to the following equation:

$$\ln s^2 = \ln a + b \ln m$$

In addition to computational estimation of the minimum number of required sample values, one can often also graphically determine which sample size to settle for a desired result. The number of samples is plotted on the x-axis, and the desired parameter (e.g., number of species, each calculated standard error) is plotted on the y-axis. The curves usually have a hyperbolic-like course, from which the approximation to a value (desired parameter) is evident.

15.7 Setting Up a Variable List

In research based on trap catches, the animal material is first sorted and then determined to the lowest taxon possible. The first step in the evaluation is a tally list. Besides for the catch results of pitfalls, it is the basis for studies involving observations of a bird community, or the food items identified during an investigation.

When compiling the variable list, we consider which information is essential for answering our research questions. Only these variables are included in the list, which saves us time and storage space for data collection and subsequent data management. Do I want or need, for example, to separate subadults from juvenile stages in spider catches, and can I distinguish them in all species?

After we have asked ourselves what we want to compare between groups, we determine the sampling unit. It can be a single animal observation, but it can also stand for all observations in a specific space or time, the catch result of a pitfall trap, or the counting result of all animal individuals of a unit quadrat.

Depending on the statistical software, the sampling unit becomes the data record or a case when setting up the data file. The properties of these cases are labeled as fields or variables. They contain individual numbers of different species, ecological information, climate measurements, and general information like the date of the sampling unit, study area, and trap number. Such a file structure is common to most ecological studies, whether large mammals in African savannas or spiders in a forest in Europe are studied.

It is recommended that a detailed variable list is created in advance using a word processing or tabular data program. It can then be used more flexibly than if the corresponding information is only queried in a statistical program, as it can be printed separately and changed as desired.

For the creation of the variable list, you should consider the following recommendations:

- Only code data with numerical values that cannot be meaningfully captured or entered otherwise. Text strings reduce the probability of errors in data collection, entry, and correction. For long phrases, abbreviations make sense. For the few tests that cannot be performed on text strings, these can easily be transformed into numbers.
- For numerical coding, use decimal points and not commas.
- Depending on the question, include grouping variables to identify sample groups with a single variable. Without a grouping variable, multiple variables could only be identified by combining several pieces of information.
- Code similar things with adjacent numbers as it facilitates a later summary.

15.8 About the Selection of Statistical Tests

With a statistical test, you want to test if an observed result can reject a null hypothesis or if the sample has a statement value for the unknown totality. The research question and the distribution of the data determine the selection of statistical tests. Each test is tied to certain prerequisites, and the data must adhere to those.

Although it is common to recognize significance only from a particular threshold value (e.g., $p < 0.05$), one should not equate statistical significance with biological significance (Yoccoz 1991). A statistically significant difference depends primarily on the sample size. We should often ask ourselves before investigating how big a difference we can interpret biologically.

Table 15.1 provides a simple overview of the selection process for practical and frequently used tests in ecology, together with the most important prerequisites.

The data material must first be distinguished into discrete or continuous characteristics. Discrete data can only assume certain values, such as observed frequencies (numbers of individuals, number of particular food objects), and principally jump from value to value. Continuous data can (within an interval) assume any values, such as measurements of wing lengths, body weights, or durations of flower visits. Such data are called measurements with continuous sequences; their apparent jumps are measurement inaccuracies. Rank numbers are not measured values but numbers that replace observation values in ascending order of size with numbers from 1 to n. An example of rank numbers is given below (Tables 15.2 and 15.3).

The distribution of the data determines the selection of the appropriate statistical tests. Distributions were discussed in Sect. 15.5. If it is determined with a certain probability that your observation data come close to a discrete distribution, it can be assumed that the same laws apply to your sample values as to the specific theoretical distribution, such as a truly random

Table 15.1 Statistical tests for differences

<table>
<tr><th colspan="9">Statistical tests for differences</th></tr>
<tr><th rowspan="2">Data type</th><th colspan="2" rowspan="2">Distribution</th><th rowspan="2">1 sample</th><th colspan="3">2 Samples</th><th colspan="2">> 2 samples</th></tr>
<tr><th>Independent</th><th colspan="2">Linked (dependent)</th><th>Independent</th><th>Linked (dependent)</th></tr>
<tr><td>Frequencies</td><td colspan="2" rowspan="2">Discrete distributions</td><td>Chi-square test of goodness of fit</td><td>Chi-square test; median test</td><td rowspan="2">Wilcoxon-test</td><td rowspan="4">Sign test</td><td>Chi-square test</td><td>(Q-test)</td></tr>
<tr><td>Rank numbers</td><td>------------</td><td>Mann-Whitney-U-test</td><td rowspan="2">Kruskal-Wallis H-test</td><td rowspan="2">Friedman-test</td></tr>
<tr><td rowspan="2">Measured values</td><td rowspan="2">Continuous series of numbers</td><td>Any distribution</td><td>Kolmogorov-Smirnov-test</td><td>Kolmogorov-Smirnov-test</td><td rowspan="2">(t-test)</td></tr>
<tr><td>(Almost) normally distributed</td><td>t-test</td><td>F-test;
t-test</td><td colspan="2">One way analysis of variance (ANOVA)</td></tr>
</table>

Table 15.2 Examples of a frequency distribution of an animal in four habitats (A–D)

A	B	C	D
20	31	9	32
28	29	15	22
44	31	11	22
36	27	8	16
29	34	12	27
33		8	19
31		14	
34			
$\bar{x}$ = 31.9	30.4	11.0	23.0

Table 15.3 Transformation of number series from biotopes B and D into rank numbers

Habitat in the first column												
B						27	29	31	31		34	$n_1 = 5$
D	16	19	22	22	27					32		$n_2 = 6$
Rank number						5.5	7	8.5	8.5		11	$R_1 = 40.5$
	1	2	3.5	3.5	5.5					10		$R_2 = 25.5$

distribution. Chi-square tests compare your observation values with the theoretically expected values according to a particular distribution. Tests that assume normally or approximately normally distributed data are always the most sensitive. Approximately normally distributed data are a prerequisite for variance analyses and multivariate calculation methods (see below).

However, ecological field data rarely obey to a normal distribution and are often not transformable to an approximate normal distribution by a data transformation. Even a series of measured values are often not normally distributed, e.g., the duration of food intake. Therefore, distribution-free test methods (nonparametric tests) are primarily considered for the field ecologist. These tests do not make any demands on the shape of the distribution, but they are also not as effective in revealing differences as significant. One could usually compensate for such test weaknesses by increasing the sample size. However, this effort is limited by time, repeatability, or, in some cases, by procuring animals (possibly involving killing). Therefore, it is desirable to transform the data for normal distribution.

The nature of your research question dictates a division of your data into one or more samples to be statistically tested. One-sample tests are, in principle, nothing more than two-sample tests. There is no separate data series for the first series of observations (sample), only a predetermined parameter value. The distinction between independent and connected (dependent) samples is essential for comparing multiple samples. This separation must be strictly observed when applying statistical tests. In Sect. 15.4, it was explained in more detail when independent and connected samples are present.

As an example of a data transformation into rank numbers, the following observation series should serve:

In four biotopes, A, B, C, and D, the activity densities of carabid beetles were determined in several test areas. A table was created (Table 15.2). The observation data are not normally distributed. A similar-looking table would also be obtained if the number of visits by flower visitors to four flower plant species had been counted several times per time unit.

A rank test (U-test, see Sect. 15.9) can determine whether the average activity density or the average number of approaches per unit of time for B and D differ significantly.

The observation values are plotted in the order of the numbers and ranked (lowest to highest numbers). If identical values are present, numbering continues as normal, but the arithmetic mean of the rank places of the identical values is assigned to all identical values, as an example see Table 15.3.

Another form of data transformation involves forming a new variable from the difference between two variables for tests of paired samples (see, for example, the sign test in Sect. 15.9).

15.9 Examples of Statistical Tests

Let us consider some research questions to show examples of selecting the appropriate test to handle a particular problem. Let us assume we have observed blackbirds (*Turdus merula)* at different locations and also caught them alive to measure certain characteristics. It should be emphasized that the data given here are purely fictitious, but we have tried to keep them largely within realistic limits based on literature values. However, please do not interpret our examples in this chapter as results of scientific work!

Let us imagine we caught several blackbirds in a forest and measured their wing lengths. We know that, on the one hand, morphological features are adapted to life in particular habitats, and on the other hand, they can differ between populations, especially from different climatic regions. We, therefore, ask ourselves whether the blackbirds we caught come from a uniform subpopulation of the forest we studied.

If all our blackbirds belong to one subpopulation, we expect a unimodal distribution with only one mean value of our measured characteristic. The statistician would say that our blackbirds come from a basic population. We would see confirmation of this if our measurements were normally distributed.

To test for a parameter's normal distribution, we perform a descriptive statistical test, the **Kolmogorov-Smirnov test**. A test for normal distribution is also the basis for the correct choice of further test procedures. A unimodal distribution of the measured values indicates a normal distribution (comparing only males), and a bimodal distribution indicates a deviation from a normal distribution (in the example, if males and females are compared).

We want to know precisely: Do males and females from the same biotope differ significantly? (Statistically speaking, how large is the error probability if we reject the null hypothesis (males and females have the same wing lengths)? We could also be interested in the question of whether the males in the forest differ from the males of a city population in terms of wing length.

Since we can assume a normal distribution in our case, we first compare the variance of the measurements of our male animals with the female animals (or the forest animals with the city animals) with the **F-Test**. The result of the F-test is essential for the correct choice of the Student's *t*-test. The ***t*-test** compares two samples (basic populations) based on their mean values.

We now want to ask ourselves whether our blackbirds measured in Southern Germany show the same values as those in Denmark. We have no observations from that location, only a mean value from the literature. The suitable test in this case is the one-sample *t*-test. Its calculation is similar to the *t*-test for two connected samples. The other sample is the literature value, assigned to the values as a new variable.

Next, we wonder if blackbirds occur in the same densities in different habitats. In spring, we assume that we would have determined the number of breeding pairs in the forest and the city based on the number of singing males within the sample size (area in m^2). If we want to compare the blackbird density between three habitats in winter, then we cannot count the singing males as they do not announce territories in winter. However, we can carry out control walks of the same duration and the same distance and note how many blackbirds we encounter.

With the Chi-square goodness of fit test for one sample, we can compare the observed frequencies with expected values, which would be calculated based on whether the total counted breeding pairs or sightings were evenly distributed across the two habitats or over several habitats. A complete even distribution would exist if we had found the same number of breeding pairs per area in each habitat or, in our example, in each habitat, the same number of positive sightings per sample (= per fixed number of control walks) {(n sightings total)/(number of habitats)}.

Another problem arises when we want to compare different places and there are two manifestations of a variable in one place. A question could be whether males in individual moth species usually fly higher than females. The encounter frequencies in a "trap low" and a "trap high" are compiled for both "males" and "females" in a fourfold contingency (two-by-two) table. In our blackbird example, we determine the encounter frequency (observed *a* times at one location and not observed *b* times) and compare this between a cemetery, forest, and city center. Our general question is: Do we encounter the species in habitat A more often than in habitat B?

To solve this problem, we use the 2 by 2 **Chi-square test.** It is tied to the condition that *n* (all controls) > 20 and the so-called expected frequencies in the fields >3. The expected frequency of one of the four fields is the value expected according to the corresponding marginal sums, e.g., for field *a* $(a + c) \cdot (a + b)/n$.

If n < 20 and the expected frequencies are <3, Fisher's exact test is used "for weakly populated contingency tables."

Our test could show, for example, that we encounter more blackbirds in the forest in winter than in the city center.

If we want to describe how significant the difference in the encounter frequency is, we calculate the Pearson **correlation coefficient** (PCC). The calculation shows, for example, that the differences between the cemetery and the city center are greater than between the forest and the city center. If the differences are maximal, i.e., two fields (b) and (c) have zero values, the PCC in the hypothetical example reaches 0.7071.

The corrected correlation coefficient CC corr is more illustrative and suitable for comparing differently sized contingency tables. Its values lie between 0 and 1 and thus give an impression about the strength of the connection comparable to other correlation coefficients.

$$CC_{corr} = \frac{CC}{CC_{max}}$$

Above, we explained when the median value summarizes data meaningfully. Just as we can test arithmetic means for differences, there is also a statistical test for comparing median values of different observation series, the **Median test**. We are interested, for example, in the size of finch bird sleeping societies in winter in different habitats (groves, reed areas).

If we want to compare two observation series or measurement series with each other and the data are not normally distributed, the Mann-Whitney *U*-test (short **U-test**) is most commonly used. We imagine that we would have measured the song duration of blackbirds in the forest interior and at the forest edge. We hypothesize that longer stanzas express greater territorial defense readiness and that the defense strength changes towards the forest edge. So, we have non-normally distributed measurements (singing duration). The *U*-test requires transforming the data into rank numbers (see above).

In all previous tests presented above, we could assume independent samples. Connected samples exist when the same blackbird individuals are to be compared in different situations (see above). We can, for example, first ask whether the duration of the morning singing is different from the duration of the evening singing. Ideally, we would have two values per blackbird and thus paired samples.

Paired samples are always obtained in two ways: (1) when test repetitions are carried out with the same individuals or (2) when a preselection of the test objects is made and differences in characteristics independent of the overall population are investigated. With blackbirds, we would also make a preselection if we examined the males at the forest's edge concerning the question of singing duration in the morning/evening, males from the inner forest and other different, but in themselves ± homogeneous areas. A preselection would also lie in the distinction of age classes or gender with a behavioral characteristic (duration in the treatment of different fruits, are acorns cracked open faster than walnuts or better hidden?). External factors or different behavioral programs influence the pair to be tested similarly. Suppose the paired samples belong to a common total population (total population of birds at the edge of the forest, the males, ...). In that case, one uses the **Wilcoxon signed-rank test** as a statistical test (*Wilcoxon matched-pairs signed-rank test*).

The **sign test** extends the Wilcoxon signed-rank test because the individual pairs do not need to come from a common total population. It is a quick test that is mainly free of prerequisites. If the preselection is not definable and paired samples are present, this is the test of choice. The sign test is also suitable when we want to compare, for example, two types of traps in terms of their effectiveness or test whether a certain attractant attracts more animals to the trap. The animals probably belong to a total population in terms of their property of being caught or attracted by the trap. The two trap types, I and II (qualitative differences or with or without attractant), are exposed as paired samples under similar conditions. In the sign test, the differences are formed from the catch results in I and II. They turn out either positive or negative for n paired samples. If zero differences occur, they are disregarded, and the sample size is reduced accordingly. The test size h is the number of those differences that have occurred less frequently. If 11 positive and four negative differences have resulted from 15 samples, $h = 4$. h is then compared with a critical values table for the corresponding significance level. If h is equal to or less than the table value, the null hypothesis (H_0, there is no difference) is rejected.

The test corresponding to the Wilcoxon signed-rank test for paired samples with a normal distribution of the difference values is the paired samples t-test. Since this t-test is only about 5% more sensitive than the Wilcoxon signed-rank test, the latter is almost always recommended for practice, as it does not set conditions on the distribution of the differences. Deviations from the normal distribution are expected for most field ecological data.

If more than two samples are to be compared, then the corresponding tests first only check whether all samples belong to the same total population ("overall test"). If differences exist, it is unclear which samples differ significantly from others or where the differences lie. To find this out, each sample must be compared using the tests for two samples discussed above. The "overall tests" serve to prepare these tests for two samples. The **Friedmann test** corresponds to the Wilcoxon signed-rank test but handles more than two related samples simultaneously.

With the **Kruskal–Wallis test** by ranks (Kruskal–Wallis *H-test*), one investigates the same questions as with the **Mann-Whitney U-test** but for multiple observation series simultaneously. It is the appropriate test if we, for example, are interested in whether there are any differences in the song duration of blackbirds between different habitats (forest, cemetery, city, hedge landscape).

Finally, the Chi-square test for more than two samples is, in the same sense as the H-test, an extension of the Chi-square test for two samples (see above).

15.10 Examination of Influencing Factors

The statistical tests covered in the previous sections mainly dealt with parameter differences between different samples. Another problem arises when, in a sample, two variables are related to each other. The relationship between two quantities is measured or quantified by correlation analysis and described by regression analysis. An observed or measured variable (y-values) takes large and small values, and we are looking for how these fluctuations are mathematically influenced by another variable (x-values). We define y as the dependent variable (target value) and x as the independent variable (influencing factor). Table 15.4 provides an overview of correlation and regression tests, and Table 15.5 shows the corresponding mathematical analyses.

Table 15.4 Statistical tests for correlations between variables in a sample

Data type	Distribution		2 Variables	>2 Variables
Frequencies	Discrete distributions		Chi-square test + Pearson-correlation-coefficient	Chi-square test + Pearson-correlation-coefficient
Rank numbers			Spearman rank correlation	
Measured values	Continuous series of numbers	Any distribution		
		Almost normally distributed	Simple regression analysis	Multiple regression

Table 15.5 Overview of regression and variance analyses

		Independent variable (influencing variable)	
		One	Many
Dependent variable (influenced value)	One	Single-factor ANOVA	ANOVA
		Simple regression analysis	Multiple regression
	Many	---------------	MANOVA

Regression serves as a method (1) for predicting the size (y) of a given independent x-value and (2) for explaining the variation of y by changes of x. A possible question for this scenario is: how much water does an animal lose when the outside temperature increases in certain steps (problem 1)? If we suspect that larger animals lose less water than smaller ones, we are dealing with problem 2.

From the measured y-values, we would first form the average as usual and then consider all y-values as deviations from this average. We imagine the y-values in the coordinate system plotted to the corresponding x-values. Now, we are looking for the best fitting line, which makes the deviations smaller than the initially horizontal line through the average of the y-values would have done. This procedure is called the "**least squares**" method. However, even with a very good fit, there will still be deviations of the y-values from the fitted line. This is referred to as the residual scatter. Our regression result looks like this because the estimated line explains a certain proportion of the variation of y but not the whole variation. There is still the residual scatter.

The resulting line has a slope described by the **regression coefficient**. The regression coefficient b is the factor that indicates how much y changes when x is increased or decreased by a certain value. If $b = 0$, this means that x does not influence y. The significance of the regression coefficient indicates by how much b differs from zero. The 95% confidence limits state that with 95% certainty, the found straight line represents the true nature of the actual correlation between the variables.

We now ask how good the relationship between the two variables, y and x, is. How well does the straight line hit the points in the cloud of points? How well can we predict the y-values from the x-values? This question is answered by calculating the **correlation coefficient**. The correlation coefficient r measures the strength of the relationship between the two variables. It measures how large the scatter is in this link by giving us the relative size of this statistical dispersion.

The values of the correlation coefficient range between -1 and $+1$, with the value of 0 indicating no correlation. The significance of r indicates whether there is a relationship at all, a strong or weak one, and with what certainty r differs from zero (no relationship). A general disadvantage of r is that it does not say what the relationship looks like, i.e., how y varies with x.

If a significant relationship is found, this does not show a causal link! The two variables do not have to be directly causally linked but could be linked to a third variable, which causes the correlation. A causal connection is only indicated if the descriptive relationship between the two variables can be generalized and applied similarly to different data sets.

We have seen above that despite an indicated regression, a residual scatter remains unexplained. The scientific goal is to explain this residual scatter with another factor: "*The aim of science is to reduce the unexplained variances to zero*" (quote from Sokal and Rohlf 2011). In the above example, the water loss of an animal would not be determined solely by the outside temperature but also by the animal's nutritional state and probably other factors. If you try to explain the residual variance step by step by adding new factors, i.e., you adjust the straight line better so that the residual scatter becomes smaller, you perform a **multiple linear regression**.

Although multiple regressions can easily be calculated using computers, problems can arise because it is no longer easy to distinguish if a variable is dependent (y) or independent (x_1 and x_2 ...) when multiple variables are considered. In suitable computer programs, the variables are selected step by step in a multiple regression: the correlations of y with each x-variable are first calculated, and a simple regression is started with the highest of these correlations (e.g., y with x_5). Next, the residuals are calculated, and then the x-variables most highly correlated with these residuals are included in the regression. You stop adding new x-variables to the regression equation when this only causes a small or no reduction of the residual standard deviation. This is equivalent to the multiple correlation coefficient R, which measures the relationship between y and all x variables. The square of the multiple correlation coefficient (= R^2) represents the proportion of the variance of y that is "explained" by the regression equation.

Sometimes, various x-variables are strongly correlated with each other. In multiple regression, only one of these x-variables may be selected even though both x-variables can be biologically significant, as the other variable contributes nothing, is insignificant in reducing variation further, or explains the existing variation just as well. Several techniques have been developed that reduce many variables to a few influencing factors, e.g., the principal components. The main concern of the **principal component analysis** is to reduce a large number of partly correlated variables to a small number of significant, non-correlated variable groups, called the principal components (Shipley 2021). One result could be that the first principal component determines 60% of the variance, and the second principal component another 20%. The third principal component is responsible for another 10% of the variance. The principal component I is aligned so that its direction coincides with the greatest extent of the point cloud (= greatest variance). The principal component II is then orthogonal to it.

Factor analyses generally describe the measured variables as a function of hypothetical variables, not directly measurable on the object, the so-called factors. The aim is to have fewer variables with the hypothetical variables than with the multitude of observed or measured variables (Flury and Riedwyl 1983).

Before using statistical packages, one should critically engage with the interpretation possibilities from multivariate calculations (e.g., Manly 2004; Rudolf and Buse 2020, and Backhaus et al. 2023). Non-mathematicians often hardly understand indices from models; they are not easily interpretable ecologically. Different calculation methods lead to different results, as arbitrary decisions are included. In many cases, a detailed consideration (direct observation) and discussion of the original correlation matrix can be helpful. At least it will not lead you astray! A correlation matrix is a square matrix with the correlation coefficients between all pairs of variables.

The following will present some examples whose substantive questions require the calculation of correlations and regressions and which will be further analyzed with multivariate methods. One of the most common ecological problems is the relationship of a species to its habitat. The standard questions for conservation management are:

- What effect will a habitat change have on a particular species?
- What structural features are found in habitat A but not in habitat B that allow the species in question to live in A and not in B?
- How can I make habitat type B suitable for the species in question?
- How can I get more individuals of the species in question into habitat A?
- Can I also introduce this species into a habitat of type C?

Usually, many habitat variables are measured (see examples in Sect. 4.2), and you want to explain the occurrence of a species using at least some of them. The frequency of the species would be the y-variable, and the habitat characteristics are various independent variables (e.g., tree density x_1, shrub density x_2, number of herbaceous plants per unit area x_3,...). Many of the x-variables correlate, e.g., tree density with tree height or canopy closure with ground cover.

A similar problem arises when we want to investigate the egg size of a bird species, as it is known that stronger chicks with lower average mortality hatch from larger eggs. We define the egg size as the dependent variable y, and the influencing factors (x-variable) can be, e.g., the microclimate, the laying sequence of the eggs, the age of the females, and the location of the egg laying (at the edge or in the middle of a colony).

In a third example, we ask why a particular individual (i) dominates a group. Dominance can be measured as the proportion of interactions with another individual won in achieving resources such as food or females willing to mate. We define an animal as dominant, e.g., if it has won 75% or more of the interactions. A dominance index describes the proportion of dominated individuals. We now consider which variables we need to measure or observe to explain the variance of dominance, e.g., the size or weight of the individuals, their plumage, age, and location.

In our example, we start with a simple regression and correlation. We want to know if there is a relationship between the total weight measured in blackbirds and their respective wing lengths. Since the measured weight values were not normally distributed, we take the logarithmized values. Section 15.9 above referred to a test for normal distribution using the Kolmogorov-Smirnov test.

Variance analyses can be performed at several levels. They generally assume independent random samples with an approximate normal distribution. Simple variance analyses ("*one-way analysis of variance*," simplest form of an "***AN**alysis **Of** **VA**riance*," ANOVA) test whether the variances of multiple observation series are equal or unequal. They represent an extension of the F-test for two samples for this question. Also, several means can be tested for differences in their variances in an extension of the Student's *t*-test.

An example is the test for differences in the song duration of blackbirds at four different locations, which we introduced in Sect. 15.9 with the presentation of the *H*-test by Kruskal-Wallis.

Table 15.6 Table of data material for a variable (y) that is dependent on various influencing variables

Observation values of a dependent variable y (influenced size)	Influencing factor x_1	Influencing factor x_2
y_1	Expression 1	Variable 1
y_2	.	...
y_3	.	Variable 2
.	.	...
.	.	
y_1	Variable 2	Variable 1
y_2	.	...
y_3	.	Variable 2
.	.	...
.	.	

To understand the principle of ANOVA, we imagine the following tabular scheme for our data material (Table 15.6):

The influencing factor 1 (x_1) could be the pH of a particular breeding medium, which influences the size of a developing insect; the influencing factor 2 (x_2) could be the medium's sugar content. Similarly, the influencing factor x_1 could be a temperature gradient, and we observe the frequency of an insect species (y) depending on the temperature. The influencing factor x_2 could then be, for example, the humidity.

We must distinguish between two possible variances (a) and (b):

(a) the variance within the groups, i.e., the variance of the y-values for each of the variants of x_1 (= column "influencing factor x_1" in Table 15.6). We can calculate as many means and variances as groups ($x_1 1$, $x_2 1$).

(b) There is also the variance between the groups: Each group has a mean, and from these means 1, 2, ..., we calculate the variance of the means.

The schema of the data material thus allows two different variance estimations. If we assert with a null hypothesis that the different groups (x_1 1, x_2 1, ...) come from the same population, then the estimation of variance (a) is the same as the estimation of variance (b): (b) $\approx$ (a). However, if the influencing factor x_2 1, for example, makes a difference compared to $x_1 1$ on the observation series y, i.e., if there is a difference between the groups, then the variance (b), the variance between groups, becomes larger than the variances (a) within the groups: (b) > (a).

With the simple variance analysis, we test the influence of the different manifestations of the influencing factor and the differences between the manifestations of column x_1. Thus, we can replace the *t*-test.

The ANOVA is mathematically equivalent to the Student's *t*-test. Although an ANOVA assumes independent random samples from approximately normally distributed observation series, the procedure is robust against deviations from the normal distribution. This insensitivity is mainly because the means usually follow a normal distribution more closely than the distribution of variable values themselves. In principle, one should try to check one's data for normal distribution beforehand (see above) and, if possible, perform a data transformation to normal distribution if they are not normally distributed. The consequences of a deviation from the normal distribution are not too serious. However, they affect the significance level.

So far, we have discussed one variable (x_1) as an influencing factor with different manifestations (x_1 1, x_1 2, ...) and asked whether the variation in the observed target variable, the influenced (dependent) variable y, can be explained by the different manifestations of the influencing factor.

The ANOVA now checks whether each newly added factor (influencing factor) is responsible for a new variation. Thus, an ANOVA serves as an extended variance analysis for the quantitative investigation of several influencing factors on experimental results or observation values. The question for the analysis is about the causes of the variance in the results. Since many influencing factors must be considered, we are dealing with this similar to multiple regressions using multivariate methods.

Let us consider an example for the variance analysis: We have observed birds and counted them standardized at different locations, under different weather conditions (temperature measured), and at different times of day. What influences our different observation results?

Single-factor ANOVA tests x_1 (different locations) against y (recorded number of individuals). You can also test x_2 (different temperatures) against y and x_3 (different times of day).

The result of single-factor ANOVA would be that the location (x_1) influences the observed number of individuals but not the temperature (x_2) or the time of day (x_3).

Since we have found that different locations affect the frequency of blackbirds, we are also interested in which of the four locations are different.

"ANOVA" initially tests the influence of these three values x_1, x_2, and x_3 on y, like a one-factorial

ANOVA. The result up to this point is the same: only the variable x_1 (location) has an effect in our imagined example (indicated as the significance of *F*). Then, the extended variance analysis forms all possible combinations of two factors with the "*two-way interactions*": time combined with location (x_3, x_1), time combined with temperature (x_3, x_2), and location combined with temperature (x_1, x_2). It turns out that the combination of time and temperature has an influence, but not any of the other combinations. In our imagined example, birds are observed more often when it is cold in the morning or warm in the evening but less often when it is warm in the morning and cold in the evening. Thus, we have gained new information and possibly grasped a new biological content. If we were to test with a single-factor ANOVA (one-way ANOVA) instead of a multiway-ANOVA, we would miss out on this new information. ANOVA can now calculate the "*three-way interactions*." It turns out that the combination of location, time, and temperature does not influence the observed bird individual count.

If we deal with a problem field of many independent variables (influence sizes x_1, x_2,...) and several dependent variables (target sizes y_1, y_2,... as influenced sizes), then we need test procedures of a multiple ANOVA, the MANOVA. A MANOVA has many options for the calculation process, for which the corresponding manuals are needed. To handle a MANOVA, the questions must be well thought out. The same critical caution applies as was already hinted at in the principal component analysis above. James and Mc Culloch (1990), Alminagorta et al. (2021), Dray et al. (2012), and Podani (2005) provide an overview of the goals and limitations of various multivariate calculation methods.

Another problem arises when we compare ecological relationships in the form of similarities (see below, e.g., faunal similarity or niche calculations in Sect. 15.11). Our data again result in a matrix, in which, e.g., different species are all mutually compared concerning a variable (similar resource demand, niche overlap). A grouping of the species can also be done with multivariate methods, e.g., cluster analyses (Deichsel and Trampisch 1985; Legendre and Legendre 1986; Borcard et al. 2018; Everitt et al. 2011; Legendre and Legendre 2012). There are also free online courses on cluster analysis in R statistics (https://techvidvan.com/tutorials/cluster-analysis-in-r/, https://data-flair.training/blogs/clustering-in-r-tutorial/).

Several calculation methods are available for this type of multivariate analysis, which one must get familiar with before using them. **Cluster analyses** are merely descriptive methods that do not serve to explain relationships or the effect of factors. Different methods lead to different descriptions. A fundamental decision must be made before the cluster analysis: whether to group the variables or the cases. More often, the cases are lumped together. For example, the catch result of a pitfall trap or the observations in a quadrat can be considered a case. A variable cluster is used if we want to describe how certain observational data (variables) similarly affect the occurrence of animals: Do certain animal species correlate in their frequency in all observation quadrats?

Another decision that needs to be made is the form of representation. The result of a cluster analysis is usually presented in a so-called **dendrogram**. Should the tip of the dendrogram point upward or downward? The various methods split all data from a pool once, but more often, it is the agglomerative method, in which initially, the two types or cases that are most similar are singled out as the first pair, and then the next most similar ones are sought. One method is constructing the "minimal tree" (see Deichsel and Trampisch 1985; McGarigal et al. 2000a, b; Mukherjee et al. 2018; Williams 1983, Fig. 15.2).

Complex calculation methods can now calculate the distances between the species in a multidimensional space and then group them. The selection of different methods can lead to different dendrograms.

For the graphical representation of results, there are various graphic packages, the inexpensive MVSP by Kovach Computing Services (KCS, https://www.kovcomp.co.uk/mvsp/) (Kovach 2010) and the Java-based platform-inde-

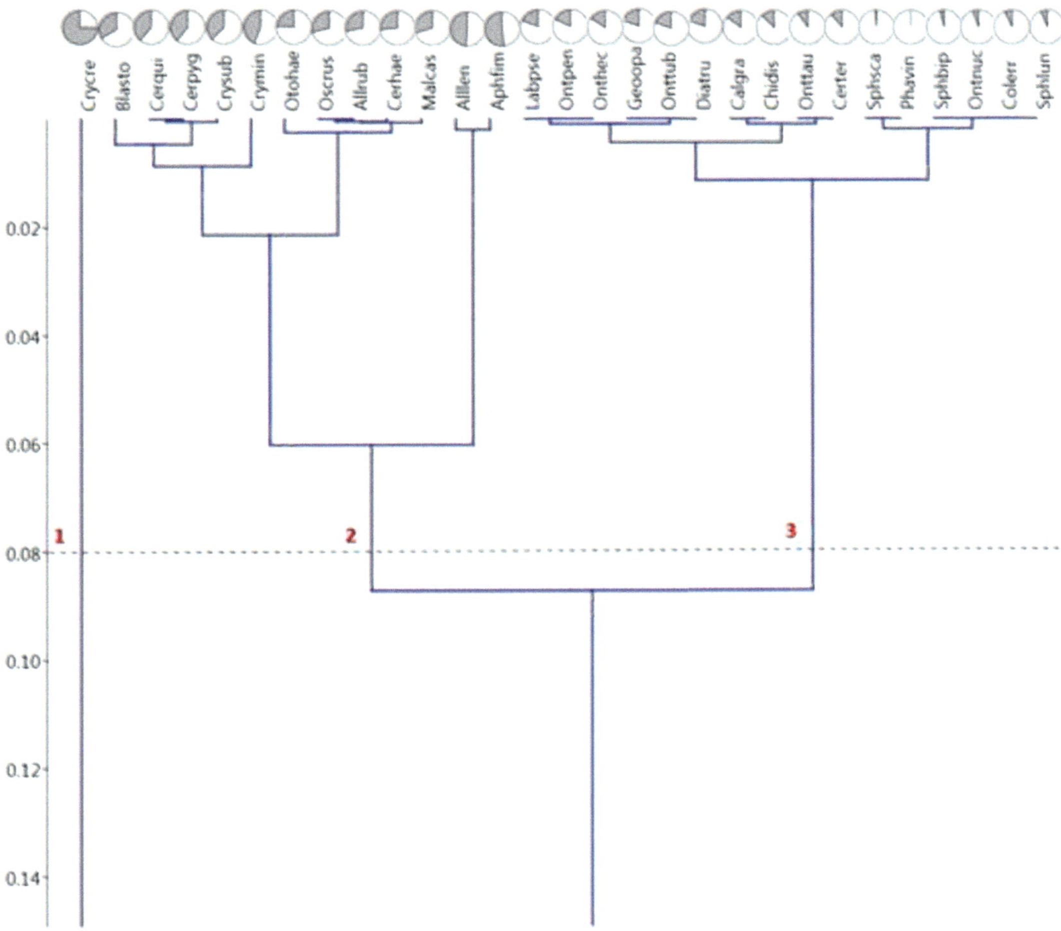

Fig. 15.2 Example of a cluster analysis of the similarities of coprophilous beetle species in terms of their successional appearance based on Pianka's niche overlap index (see below Sect. 15.11). For explanations, see Wassmer (2020)

pendent freeware Past by Øyvind Hammer (https://www.nhm.uio.no/english/research/resources/past/) (Hammer et al. 2001).

15.11 Use of Indices

This section presents indices that describe species communities (composition, distribution, similarity) and the distribution of resources among species. For a comprehensive coverage of indices and free software, see (Krebs 1998). Many indices can also be calculated using the above mentioned applications MVSP and Past (Hammer et al. 2001; Kovach 2010). In addition, you can use an internet search engine to locate Excel sheets to calculate many indices.

15.11.1 Faunal Similarity

There are many descriptive indices for the distribution of individuals among species within a species community (see also Table 10.6.1 in Sect. 10.6).

The **Dominance** describes the relative frequency of a species compared to the other species based on a particular habitat size.

$$D_i = \frac{\text{Number of individuals of the species } i \times 100}{\text{Total number of individuals in the species community.}}$$

The dominance index d expresses what share the most common species has in the community:

$$d = \frac{N_{\max}}{N_T}$$

N_{max} = Number of individuals of the most common species.

N_T = Total number of individuals of the species community.

Dominance is classified differently by various authors. In addition to a linear division (dominant species with more than 5% relative frequency), Engelmann (1978) suggested a logarithmic relation for the class formation:

Eudominant 32.0–100%
Dominant 10.0–31.9% "main species"
Subdominant 3.2–9.9%
Recedent 1.0–3.1%
Succedent 0.32–0.99% "accompanying species"
Sporadic: under 0.32%

Most species appear at the level of the subdominants, which are primarily suitable for differential diagnostic purposes. The classification is designed to include 85% of the recorded individuals within the "main species."

With the **dominance structure**, the species are ordered according to their relative frequency within a taxocoenosis or species community. Usually, the percentages for the individual species are represented as a histogram or pie charts.

The **Renkonen index**, also known as **percent similarity**, is defined as:

$$P = \sum_{p_1 i,\, p_2 i} \text{minimum}\left(p_1 i, p_2 i\right)$$

where:

P = Percentage similarity between samples 1 and 2.

$p_1 i$ Percentage of species i in community sample 1.

$P_2 i$ Percentage of species i in community sample 2.

It measures the congruence in the dominance relations of two species communities. This measure is not heavily influenced by sample size and species number. For calculation and comparison, see Wolda (1981)and (Krebs 1998, 2014). The index can also be calculated within R-packages: https://rdrr.io/github/wilsonfrantine/R4eco/man/renkonen.html.

The **Dice-Sørensen coefficient (DSC)**, also known as the **similarity coefficient**, only considers the number of common species and is used for a simple comparison of species communities:

$$DSC(\%) = \frac{2G}{S_A + S_B} \times 100$$

G = Number of species common to both areas.

S_A, S_B = Number of species in area A or B

The DSC ranges between 0% and 100%; the higher the value, the greater the similarity in the species composition.

According to Wainstein (1967), **Wainstein's Similarity Index**, K_w also considers the common species and their relative frequencies. The Wainstein index takes values between 0 and 100, with higher values indicating greater similarity.

Another form of description, how strongly two species, s_1 and s_2, are connected, is achieved through the **Pearson correlation coefficient** (PCC). It expresses the relationship within a 2 × 2 contingency table and was discussed in Sect. 15.8. For this analysis, the 2 × 2 contingency table comprises the fields: Species A present or missing and Species B present or missing. The disadvantage of the PCC as an association coefficient is that it does not reach the value of +1.

Huhta (1979) compared several indices for fauna comparison based on the description of natural succession and recommended **Kendall's tau** (Kendall rank correlation coefficient) as the best index. This coefficient takes values from −1 to +1, with 0 indicating an utterly indifferent association between the two species. Kendall's tau does not depend on the distribution of the initial values and is a non-parametric measure (see above Sect. 15.8). Its calculation can be found in statistics textbooks and Wolda (1981).

Applying the similarity index K_w applies, for example, when two species communities with their distributions of individuals are to be compared. The question arises, for example, when we examine the samples of individual pitfall traps for similarity or compare taxocoenoses between two differently moist meadows. We can also perform this comparison with only two species: To what extent is species B correlated with the occurrence of species A? We recommend calculating Kendall's tau using a statistics program. Calculating niche overlap is another way to quantitatively express the relationship between two species (see below).

Suppose we want to correlate many species with each other in their occurrence or compare many habitats for their fauna (taxocoenosis). In that case, collecting the correlation values or calculated similarity indices in a matrix is worth compiling. A more modern way to display similarity indices in a sorted manner is achieved with cluster analyses (see above Sect. 15.10).

15.11.2 Diversity

Many indices have been developed to calculate diversity. Only those measures that consider both the number of species and the distribution of individuals among the species are suitable. Peet (1974), Pielou (1975), and Magurran (1988) provide overviews with evaluations of the individual indices. Roswell et al. (2021) critically compare different measurement methods for recording species diversity.

It is practically impossible to record all existing species in a particular area. However, the higher the effort to record, the more species will be found. The estimated number of species depends on the "sample effort." This relationship is most easily described in a **cumulative species curve** (Fig. 15.3). It is important for practice, as it can be used to decide how much effort is necessary to estimate the number of species in an area.

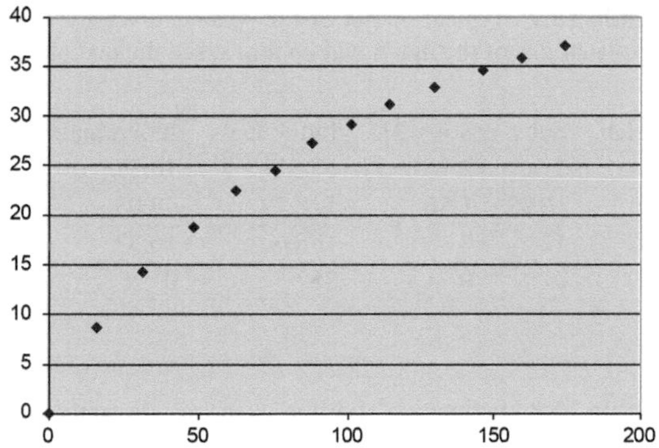

Fig. 15.3 Example of a cumulative species curve: The y-axis describes the number of species, and the x-axis the number of individuals. The number of individuals is directly correlated with the sample effort

Since the cumulative species curve asymptotically approaches a maximum value, it is possible to estimate the number of species with fewer samples through modeling. A software called "**EstimateS**" is offered for free online for this purpose: A user's guide and application of the estimation of species richness and related indices are provided by Robert K. Colwell (latest version 2023). EstimateS 9.1 is a free Windows or Mac OS software application. It can be downloaded from the website: https://www.robertkcolwell.org. Calculation methods for species diversity have also been developed for R (Hsieh et al. 2016).

A diversity calculation according to EstimateS is provided in a table (Table 15.7) as the values of the observed number of species (Sobs) and their standard deviation (SD), the average number of the recorded individuals in each species ("Individuals"), the number of species that were observed in only one individual ("Singletons"), the number of species which were recorded only once ("Uniques"), and the results of various models (ACE, ICE, Chao2) describing the cumulative species curve.

In another table (Table 15.8), the calculation results for various diversity indices are compiled.

Table 15.7 Typical output of the species diversity calculations for four samples (e.g., four habitats Hab A-D). For explanations of the individual columns, see the text above

Hab	Sobs	Sobs_SD	Individuals	Individuals_SD	Singletons	Uniques	ACE	ACE SD	ICE	Chao2
A	7.56	2.16	67.8	16.52	1.76	7.56	9.71	0.79	38.42	38.42
B	10.98	1.56	128.72	19.03	1.14	5.94	11.72	1.13	26.04	15.17
C	12	0	188.6	16.37	0.6	2.38	12.28	0.44	13.69	12.42
D	2	0	257	0	0	0	12	0	12	12

Table 15.8 Estimation of different species diversity indices, an example of the output of Estimates

Hab	Alpha	Alpha_SD	Shannon	Shannon_SD	Simpson	Simpson_SD
A	2.39	0.54	1.59	0.47	5.18	2.63
B	2.92	0.49	1.96	0.35	6.35	2.64
C	2.86	0.43	2.13	0.22	7.05	2.46
D	2.61	0.37	2.19		6.77	0

The preferable index depends on the actual sampling conditions. An explanation of all the indices is given in the user's guide (Colwell 2005, 2011–2023, https://spaces-cdn.owlstown.com/blobs/f7yfc31w8r1p8ruymhvfp4mcdb9w). While it is not difficult to learn how to apply the software program, it is crucial to collect the data in a way that is reasonable for statistical analyses. As the software is not developed anymore, the author recommends R-based applications from Anne Chao's website: http://chao.stat.nthu.edu.tw/wordpress/software_download.

To depict faunal similarity between habitat types, a measure of faunal description is used in a multidimensional scaling (StatSoft 2014), and the study sites are ordinated two-dimensionally. The multidimensional scaling is conducted with the abundance-based similarity index of Morisita Horn (Colwell 2005). The same result is shown by a two-dimensional ordination of the study sites, analyzing abundance data in a correspondence analysis.

Significant responses at the species level to variation in habitat are detected by one-way ANOVA. Although a significant difference between the habitats can be determined using ANOVA, it is unclear between which habitats the significant difference exists. Therefore, a test is usually added *post hoc*, e.g., the Tukey-HSD test for more than two observation groups (Gurvich and Naumova (2021). The results should be presented by graphs (an example is given in Waltert et al. 2004).

Ecologically, we must distinguish between **α- and β-diversity**. The α-diversity refers to the species richness in a habitat. It is a kind of "point diversity" and describes, e.g., the diversity of a sample within a ± homogeneous area (*within-habitat diversity)*. However, we can also be interested in how much the fauna changes from one habitat to the next and how many species disappear or are added. Such diversity is indicated by the β-diversity (*between-habitat diversity*). It can also be called "gradient diversity" as it describes the degree of change in species composition along an environmental gradient. A low α-diversity can be linked to a high β-diversity and vice versa.

The following will discuss individual indices describing species richness in more detail. These indices all refer to the point diversity. Again, (Krebs 1998, 2014) provides a comprehensive overview of diversity indices. The freeware PAST can calculate many diversity indices (Hammer et al. 2001) and the Excel-based Community Ecology Parameter Calculator is also very helpful (Drozd 2010).

The most commonly used index is the **Shannon index** H (also known in the literature as the Shannon-Weaver or Shannon-Wiener index, hereafter referred to as the Shannon index), whose formula comes from information theory and describes the mean degree of uncertainty of encountering any particular species of the S species in a random sampling.

The diversity value H increases with increasing numbers of species (**species richness**, S) and an increasingly even distribution of the existing individuals among the species. H reaches the following limit values: At $S = 1$, p_i equals 1, and the H_s value equals 0.

Maximum diversity is reached when all species in a community containing S species are in the same quantitative ratio. Since species in nature are never entirely evenly distributed, H_s values of real biocoenoses reach, at most values up to 4.5. However, most values fall in the range between 1.5 and 3.5.

The Shannon index is tied to a prerequisite that cannot be overlooked for its calculation: The sample must be randomly taken. Only if we can assume that our observation or catch of the different species with their counts of individuals is a random sample from the total population can we apply the Shannon(-Wiener) formula:

$$H' = -\sum_{i=1}^{s} p_i \ln(\mathrm{p_i})$$

where:

H'= Index of species diversity.

s = Number of species.

p_i = proportion of total sample belonging to ith species.

In cases where the species community to be described was not randomly sampled and was selectively recorded in some form, the **Brillouin index** is suggested for calculation, which is also based on information theory. However, no parametric statistics (variance and t-test, see above) can be applied. The Brillouin index is also suitable when the entire composition of the biocoenosis is known.

$$\mathrm{HB} = (\ln \mathrm{N}! - \Sigma \ln \mathrm{n}_i!)/\mathrm{N}$$

where:

$\hat{H}$ = Brillouin's index of species diversity.

N = Total number of individuals in the entire collection.

n_1 = Number of individuals belonging to species 1.

n_2 = Number of individuals belonging to species 2.

Ground eclector catches (see above Sect. 2.20.2) and catches with a biocoenometer represent random samples from a species community. They justify the use of the Shannon index. Light trap catches and, in most cases (at least for statements for an entire taxocoenosis), soil trap catches (like pitfall traps) are nonrandom samples, as the traps act selectively, i.e., they attract or capture the species of a taxocoenosis differently. Such samples are treated like total catches and are better described with the Brillouin index. Since the Brillouin index assumes that a total collection of the individuals and species of a collection is present, no statistical difference calculations can be carried out.

Since in comparing different ecosystems, the diversity index alone does not reveal whether its value is due to a high number of species with different numbers of individuals or an even distribution of individuals across a few species, the calculated **evenness** is used as a comparison measure. The diversity value is related to the maximum possible diversity value, resulting in the same number of species but with the greatest possible even distribution of individuals across the existing species (see above and Fig. 15.4). The evenness (E) is sometimes considered the "degree of diversity development."

$$E = \frac{\text{Calculated diversity value}}{\text{Maximum diversity value}}$$

The value of E_s ranges between 0 and + 1.

The Simpson index is a diversity index that indicates the probability that two randomly encountered individuals from an unlimitedly large species community belong to different species.

The logarithmic series well describes the form of the frequency distribution of species in a community. There are always a few species with many individuals and many with few individuals. This relationship is captured with the diversity

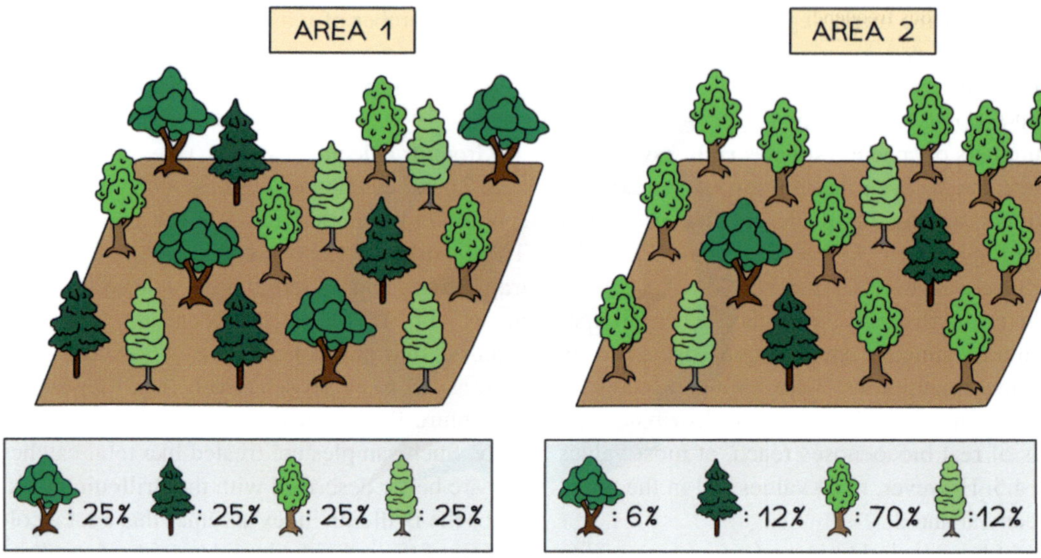

Fig. 15.4 Comparison of two habitats regarding tree species diversity. Habitat 1 has the highest diversity due to the equal distribution of species. Picture credit: Save My Exams, savemyexams.com. Reproduced by permission

measure "α (log series)." α has the best discrimination ability between different areas, according to Magurran (1988).

The choice of the diversity value depends on which influencing factor of diversity should be given more weight: The dominance of the most common species, the abundance or even distribution of neither dominant nor rare (recedent) species, or mainly the number of species. It is crucial to consider the sensitivity to sample size. Because recording the number of species depends heavily on the sample size, the diversity indices cannot be unaffected by the sample size.

If a diversity measure is also used for functional aspects of an ecosystem, then the different body sizes must be considered. Since predators and parasites usually also differ in their specialization on prey or hosts, their respective diversity should be calculated separately. With these considerations, the search for new diversity measures has begun (Cousins 1991).

15.11.3 Faunal Change

Different communities can be compared by separately calculating their respective diversities or with similarity indices (see above). The simplest way to describe the change in species composition is to calculate the turnover rate. It can be spatially related to changes along an environmental gradient or temporally related to different periods. Only the presence or absence of species is included in the consideration.

A formula for species turnover is:

$$T = \frac{J + E}{S_I + S_{II}}$$

J = Number of species added between Season I and II (or gradient sections I and II), E = Number of species that have disappeared between I and II, S_I = Number of species in Season I (or gradient Section I), S_{II} = Number of species in Season II (or gradient Section II)

The change in fauna can also be used to assess the stability of communities. Stability has two aspects: (a) uniformity over time ("constancy") and (b) regenerative capacity. In the second aspect (b), we must distinguish between the ability of an ecosystem to resist changes ("persistence") and the property of returning to the initial state after a disturbance ("elasticity," measured in

Table 15.9 Various frequency data (fictitious) demonstrating fluctuations between two species

		Population density		
		Species 1	Species 2	Total density
Time period	1	5	45	50
	2	10	40	50
	3	40	15	55
	4	20	25	45
	5	5	40	45

the time dimension). Here we only want to deal with the characterizing aspect (a), constancy.

As parameters of a species community, we can use the total density (total abundance), the number of species, the diversity, the evenness, and the frequencies of the existing species. The variability of these parameters describes the stability aspect constancy. For the calculation, the respective coefficient of variation would be suitable (Järvinen 1979). The turnover (formula above) expresses the change in species composition. Similar to the diversity indices (see above), each coefficient of variation proposed here is most strongly influenced by other factors: the total density and the frequency of individual species by the change of common species, the number of species, and the turnover rate by the variation in the population density of the rare species or their appearance or disappearance. With species diversity and evenness, the magnitude of the influence depends on the chosen index.

Since it is believed that a system's stability often results from the populations within the system showing a compensatory fluctuation, the path of documentation proposed by Järvinen (1979) is interesting. Suppose the ratio of the variance of species *i* with different population densities between different time periods to the variance of the total density for the different time periods is much greater than 1. In that case, there is a compensatory fluctuation between the species.

The following example illustrates compensatory fluctuations (Table 15.9):

15.11.4 Niche Calculations

Niche measurements refer to the use of various resources by individuals or species (see Sect. 12.1). Resource categories include food objects, microhabitats, and microclimate ranges. Using the same calculations, one could also classify behavioral patterns into feature classes and compare their occurrence in animals.

To calculate niche width and overlap, a resource matrix is constructed according to the following pattern (Fig.15.5):

We call the niche width of a species the special niche width (NBi).

It can be calculated using the abovementioned formulas based on the Simpson or Shannon formula.

Hanski and Koskela (1977) suggest a scaling of the niche width:

$$\frac{NB_i - NB_{\min}}{NB_{\max} - NB_{\min}}$$

where:

$NB_{\min}$ = smallest calculated niche width

$NB_{\max}$ = largest calculated niche width

With this expression, the species with the largest observed niche width receives the value 1.0 (largest observed generalist), and those with the smallest niche width receive the value 0 (largest observed specialist).

The specific niche overlap (degree of common resource use by two specific species or individuals) can be calculated according to Pianka (1973) or Colwell and Futuyma (1971). According to Colwell and Futuyma, the values are between 0 and 1. Pielou (1972) also calculated the average niche width and overlap using the corresponding formulae.

Species	Resource Classes RC1 RC2 RC3 ... RCr	
Species 1	N_{11} ··················· N_{1j} ··· N_{1r}	y_1
Species 2	· · ·	
Species 3	· · ·	
·	N_{i1} ···························· N_{ir}	y_i
·	· · ·	
·	· · ·	
Species S	N_{s1} ··················· N_{sj} ··· N_{sr}	y_s
	X_1 X_j X_r	Z

Fig. 15.5 Scheme of a resource matrix. N_{ij} = Number of observed occurrences of species i in resource class *j*, X_j = Total number of observed occurrences of all species in resource class *j*, Y_i = Total number of observed individuals of species *i*. *Z* = Sum of all observations

Critical caution is advised in the ecological interpretation of the above mentioned niche calculations. We usually measure only one or a few dimensions without knowing what proportion of the total niche we have captured (hypervolume). Another problem arises with the formation of the resource classes: are they used differently because they are available differently? Therefore, attempts have been made to evaluate the resource classes and to extend the calculations by weighing factors (e.g., Colwell and Futuyma 1971; Hanski 1978). The weighing ultimately considers the occupation of the respective resource class by the total number of individuals in this class as an expression of the availability or productivity of this resource class. Even complicated weighing factors correlate highly significantly with the respective sum of individuals (Hanski 1978). Schoener (1974) considers resource availability as the frequency of, e.g., prey. In practice, one may distinguish between (1) resource gradients, which provide a ± continuous spectrum, and (2) separable resource classes. Examples of (1) are climate gradients, spatial dimensions such as vertical ranges of foraging in birds (see Sect. 12.2 Prey objects) when essentially the size is responsible for niche separation, and ± homogeneous microhabitats like leaves and fungi. (2) includes, for example, macrohabitats. Weighting would only make sense for type (1) variables, whereas they are doubtful for type (2).

Vitt and Caldwell (2013) provide larger-scale examples of niche overlaps in species assemblages of reptiles and amphibians.

When trying to quantify multiple niche dimensions, it must be clarified whether the individual dimensions are independent or dependent on each other. If they are independent, a multiplication of the simple dimension values is mathematically appropriate. If they are entirely dependent on each other, a summation is appropriate. In the latter case, the average value from the individual dimension values would also represent a suitable estimate of the multidimensional niches. The actual situation in the field often lies between these two extreme possibilities. Hanski (1978) concludes that summation measures a better estimate of niche width, whereas the product measures a better estimate of niche overlap. Despite the limitations of niche measurements, the great benefit of structuring fieldwork accordingly and critically discussing the data remains undisputed.

The questions of resource use and availability are linked to an animal's (or a species') preference for a particular resource. This is particularly interesting in the choice of food. King (1980) has proposed a simple calculation for prey selectivity:

$$D = \frac{r - p}{r + p - 2rp}$$

D = "selection index".

r = fraction of a prey type in the diet.

p = fraction of this prey type within all prey populations.

D becomes 0 if no prey (or food objects) is selected.

15.11.5 Dispersion Indices

The spatial distribution pattern of most species corresponds to a clumped distribution. To describe the degree of clumping (measure for the aggregation), Ludwig and Reynolds (1988) recommend Green's index of dispersion. This index is independent of the total number of individuals in the samples. It varies between 0 (for random distribution) and + 1 (for maximum clumping).

15.11.6 Area-Species Relationships and Isolation Indices

The relationship between the sample area (*A*) and the number of species (*S*) was described by Mac Arthur and Wilson (1967) with the equation:

$$S = c \cdot A^z$$
$$\text{or} \log S = c + z \log A.$$

c = constant, which depends on the respective animal group, the biogeographical area, and the population density. When comparing a taxocoenosis between different islands, c does not need to be calculated as a constant. z = slope of the line in double logarithmic representation. The parameter z does not change significantly between animal groups and geographical regions. Most empirical values for biocoenoses on islands fluctuate around 0.3. On habitat islands of mainland areas, the z-values are lower (between 0.15 and 0.25, compare, for example, Feath and Kane 1978).

The slope of the area-species curve depends on the colonization ability of the studied animal group and the ratio of the equilibrium species number (S) of the island and the species number (P) in the species pool as a source of colonization (see Williamson 1981; Schoener 1976).

The size of the habitat and the degree of isolation influence the number of species on an island (or habitat island).

The indication of the distance to the nearest island or the mainland is not very informative when describing the degree of isolation. Power (1972) suggests the following simple calculation:

$$I = 1 - \sum_{i=1}^{k} \frac{1}{m}$$

m = distance in m or km, k = the number of islands or nearest mainland points within a certain radius from the considered island, e.g., in a 10 km, 50 km, or 200 km radius.

Two habitat islands can have the same area but very different shapes (e.g., from circular to narrow elongated). Since the shape of the habitat islands also influences their colonization, an easy index for describing the island shape can be given through the expression (F) (Feath and Kane 1978):

$$F = \frac{U}{2\sqrt{A\pi}}$$

F = measure for the shape of the island

A = area of the island (or habitat)

U = circumference of the island (or habitat)

For a circular area, F = 1, with increasing deviation from the circular area, F becomes larger.

References

Afflerbach L. 1987: *Statistik-Praktikum mit dem PC.* – Teubner, Stuttgart, 193 p.

Alminagorta, O., Loewen, C. J. G., de Kerckhove, D. T., Jackson, D. A., & Chu, C. (2021). Exploratory analysis of multivariate data: Applications of parallel coordinates in ecology. *Ecological Informatics, 64*, 101361. https://doi.org/10.1016/j.ecoinf.2021.101361

Alreck, P. L., & Settle, R. B. (2003). *The Survey Research Handbook, Third Edition* (3rd edition.). Boston: McGraw-Hill Education.

Backhaus K., Erichson B., Gensler S., Weiber R. & Weiber T. (2023). Multivariate Analysis: An Application-Oriented Introduction. Wiesbaden: Springer Gabler.

Beal K.G. & Khamis H.J. 1990: Statistical analysis of a problem data set: correlated observations. – *The Condor* 92, 248–251

Borcard, D., Gillet, F., & Legendre, P. (2018). *Numerical Ecology with R* (2nd ed. 2018th edition.). New York, NY: Springer.

Colwell R.K. 2005: Estimates: Statistical estimation of species richness and shared species from samples. User's guide and application. https://spaces-cdn.owlstown.com/blobs/f7yfc31w8r1p8ruymhvfp4mcdb9w

Colwell, R.K. 2011–2023. *EstimateS*: Statistical Estimation of Species Richness and Shared Species from Samples. Version 9.1. University of Colorado. Center for Open Science. Boulder, CO - USA

Colwell R.K. & Futuyma D.J. 1971: On the measurement of niche breadth and overlap. –*Ecology* 52, 567–576

Conroy, R. M. (2021). *The RCSI Sample size handbook*.https://doi.org/10.13140/RG.2.2.30497.51043

Cousins S.H. 1991: Species diversity measurement: choosing the right index. – *Trends in Ecology & Evolution* 6 (6), 190-192

Dalgaard P. 2008: *Introductory Statistics with R*. – Springer Nature, 364 p.

Deichsel G. & Trampisch H.J. 1985: Clusteranalyse und Diskriminanzanalyse. – G. Fischer Verl., Stuttgart, New York, 135 p.

Dray, S., Pélissier, R., Couteron, P., Fortin, M.-J., Legendre, P., Peres-Neto, P. R., et al. (2012). Community ecology in the age of multivariate multiscale spatial analysis. *Ecological Monographs*, *82*(3), 257–275. https://doi.org/10.1890/11-1183.1

Drozd, P. (2010). ComEcoPaC – Community Ecology Parameter Calculator. Ostrava, Czech Republic: University of Ostrava, Department of Biology. https://prf.osu.cz/kbe/8072/ecological-software/

Eberhardt L L. 1978: Appraising Variability in Population Studies. – *J. Wildl. Mangament* 42 (2), 207–238

Ehrenberg A.S.C. 1986: *Statistik oder der Umgang mit Daten. Eine praktische Einführung mit Übungen*. – VCH Verlagsgesellschaft, Weinheim, 344 p.

Engelmann H.-D. 1978: Zur Dominanzklassifizierung von Bodenarthropoden. – *Pedobiologia* 18, 378–380

Everitt, B. S., Landau, S., Leese, M., & Stahl, D. (2011). *Cluster Analysis* (5th edition.). Wiley.

Feath S.H. & Kane T.C. 1978: Urban biogeography. City parks as islands for Diptera and Coleoptera. – *Oecologia* 32, 127–133

Finney D.J. 1980: *Statistics for Biologists*. – Springer Dordrecht, 166 p.

Flury B. & Riedwyl H. 1983: *Angewandte multivariate Statistik*. Computergestützte Analyse mehrdimensionaler Daten. – G. Fischer-Verlag, Stuttgart, New York, 187 p.

Fowler J. & Cohen, L. 1996: *Statistics for Ornithologists*. – BTO Guide No. 22, 176 p.

Gurvich V. and Naumova M. 2021: Logical Contradictions in the One-Way ANOVA and Tukey–Kramer Multiple Comparisons Tests with More Than Two Groups of Observations. - *Symmetry* 13(8),1387; https://doi.org/10.3390/sym13081387

Hammer, Ø., Harper, D. A. T., & Ryan, P. D. (2001). Paleontological statistics software package for education and data analysis. *Palaeontologia Electronica*, *4*(1), 9pp.

Hanski I. 1978: Some Comments on the Measurement of Niche Metrics. – *Ecology* 59 (1), 168–174

Hanski I. & Koskela H. 1977: Niche relations among dung-inhabiting beetles. – *Oecologia* 28, 203–231

Harris J.K. 2020: *Statistics With R: Solving Problems Using Real-World Data*. – SAGE Publications, Inc., 784 p.

Harvard Graphics, from Software Publishing Corporation and initially called Harvard Presentation Graphics, is a graphing/plotting/presentation creation application for DOS. Harvard Graphics Advanced Presentations 3.0 is available for free download

Hector A. 2021: *The New Statistics with R. An Introduction for Biologists* (2nd ed.). – Oxford University Press, 288 p.

Hedderich, J., & Sachs, L. (2020). *Angewandte Statistik: Methodensammlung mit R*. Berlin, Heidelberg: Springer. https://doi.org/10.1007/978-3-662-62294-0

Herrmann D. 1984: *Wahrscheinlichkeitsrechnung und Statistik* – 30 BASIC-Programme. – Anwendung von Mikrocomputern; Bd. 2. – Vieweg-Verl., Braunschweig, 70 p.

Hsieh T.C., Ma K.H. & Chao A. iNEXT 2016: an R package for rarefaction and extrapolation of species diversity (Hill numbers). – *Methods Ecol. Evol.* 7, 1451–1456

Huhta V. 1979: Evaluation of Different Similarity Indices as Measures of Succession in Arthropod Communities of the Forest Floor After Clear-Cutting. – *Oecologia* (Berl.) 41, 11–23

Hurlbert S.H. 1984: Pseudoreplication and the design of ecological field experiments. – *Ecological Monographs* 54 (2), 187–211

James F.C. & Mc Culloch C.E. 1990: Multivariate analysis in Ecology and Systematics: Panacea or Pandora's Box? – *Ann. Rev. Ecol. Syst.* 21, 129–166

Järvinen O. 1979: Geographical Gradients of Stability in European Land Bird Communities. – *Oecologia* 38, 51–69

King C.M. 1980: The weasel *(Mustela nivalis)* and its prey in an English woodland. – *J. Anim. Ecol.* 49, 127–159

Kovach, W. L. (2010). MVSP - A MultiVariate Statistical Package for Windows. Pentraeth, Wales, U.K.: Kovach Computing Services.

Krebs C.J. 1989: *Ecological Methodology*. – Harper & Row Publishers, New York, 654 p.

Krebs, C. J. (1998). *Ecological Methodology* (2nd edition.). Menlo Park, Calif: Benjamin Cummings.

Krebs, C. J. (2014). *Ecological methodology* (3rd ed.). Vanvouver: Charles Krebs. https://www.zoology.ubc.ca/~krebs/books.html

Lai, J., Lortie, C. J., Muenchen, R. A., Yang, J., & Ma, K. (2019). Evaluating the popularity of R in ecology. *Ecosphere*, *10*(1), e02567. https://doi.org/10.1002/ecs2.2567

Legendre P. & Legendre L. 1986: *Developments in Numerical Ecology*. – NATO ASI Series G: Ecological Sciences, Vol. 14., Springer-Verlag, Berlin, Heidelberg, New York, 585 p.

Legendre, P., & Legendre, L. (2012). *Numerical Ecology* (3rd edition.). Amsterdam: Elsevier.

Lehner Ph.N. 1979: *Handbook of Ecological Methods*. – Garland STPM Press, New York & London, 403 p.

Lewis T. & Taylor L.R. 1967: *Introduction to Experimental Ecology*. – Academic Press, London, New York, 401.

Lorenz R.J. 1988: *Grundbegriffe der Biometrie*. 2. Aufl. – Gustav Fischer Verlag, Stuttgart, New York, 241 p.

Ludwig J.A. & Reynolds J.F. 1988: *Statistical Ecology. A primer on methods and computing*. – John Wiley & Sons, New York, Chichester, Brisbane, Toronto, Singapore, 337 p.

Mac Arthur R.H. & Wilson E.O. 1967: *The Theory of Island Biogeography*. – Princeton University Press, Princeton, N. J., 203 p.
Magurran A.E. 1988: *Ecological Diversity and Its Measurement*. – Croom Helm, Sydney, London, 179 p.
Manly B.F.J. 2004: *Multivariate Statistical Methods: A Primer*. – Chapman and Hall/CRC, 208 p.
Manly B.F.J. 2007: *Randomization, Bootstrap and Monte Carlo Methods in Biology* (3rd ed.). – London, Chapman and Hall, 480 p.
Masuadi, E., Mohamud, M., Almutairi, M., Alsunaidi, A., Alswayed, A. K., & Aldhafeeri, O. F. (2021). Trends in the Usage of Statistical Software and Their Associated Study Designs in Health Sciences Research: A Bibliometric Analysis. *Cureus*, *13*(1), e12639. https://doi.org/10.7759/cureus.12639
McGarigal, K., Stafford, S., & Cushman, S. (2000a). Cluster Analysis. In Multivariate Statistics for Wildlife and Ecology Research (pp. 81–128). Springer, New York, NY. https://doi.org/10.1007/978-1-4612-1288-1_3
McGarigal, K., Stafford, S., & Cushman, S. (2000b). Discriminant Analysis. In Multivariate Statistics for Wildlife and Ecology Research (pp. 129–187). Springer, New York, NY. https://doi.org/10.1007/978-1-4612-1288-1_4
Mukherjee, S.P., Sinha, B.K., & Chattopadhyay, A.K. (2018). Cluster and Discriminant Analysis. In Statistical Methods in Social Science Research (pp. 75–94). Springer, Singapore. https://doi.org/10.1007/978-981-13-2146-7_8
NYU Data Services 2023: Quantitative Analysis Guide: Which Statistical Software to Use? URL: https://guides.nyu.edu/quant
Peet R.K. 1974: The measurement of species diversity. – *Ann. Rev. Ecol. Syst.* 5, 285–307
Pianka E.R. 1973: The structure of lizard communities. – *Ann. Rev. Ecol. Syst.* 4, 53–74
Pielou E.C. 1972: Niche width and niche overlap: a method for measuring them. – *Ecology* 53 (4), 687–692
Pielou E.C. 1975: *Ecological Diversity*. – John Wiley & Sons, New York, London, Sydney, Toronto, 165 p.
Podani, J. (2005). Multivariate exploratory analysis of ordinal data in ecology: Pitfalls, problems and solutions. *Journal of Vegetation Science*, *16*(5), 497–510. https://doi.org/10.1111/j.1654-1103.2005.tb02390.x
Poole R.W. 1974: *An Introduction to Quantitative Ecology*. – McGraw-Hill Book, Company, New York, Toronto, 531 p.
Potts G.R. 1986: *The Partridge. Pesticides, Predation and Conservation*.—Collins, London, 274 p
Power D.M. 1972: Numbers of bird species on the California islands. – *Evolution* 26, 451–463
R Core Team 2018. R: A Language and Environment for Statistical Computing (R Foundation for Statistical Computing, 2018)
Rasch D. & Verdooren R. 2023: *Angewandte Statistik mit R für Agrarwissenschaften*. Versuchsplanung und -auswertung mit konkreten Beispielen und Übungsaufgaben. – Springer Spectrum, 280 p.
Roswell M., Dushoff J. & Winfree R. 2021: A conceptual guide to measuring species diversity. – *Oikos* 130, 321–338
Rudolf M. & Buse J. 2020: *Multivariate Methods*. A practice-oriented introduction with application examples. 3rd ed. – Hogrefe Publishing, 457 p.
Rudolf M. & Kuhlisch W. 2020: *Biostatistik*. Eine Einführung für Bio- und Umweltwissenschaftler. 2nd ed. – Pearson Studium – Biologie, Deutschland, 464 p.
Sachs, L. (1984). Applied Statistics. New York, NY: Springer. https://doi.org/10.1007/978-1-4612-5246-7
Schmuller J. 2022: *Statistik mit R für Dummies*. 2nd ed. – Wiley-VCH, 432 p.
Schoener T.W. 1974: Some methods for calculating competition coefficients from resource-utilization spectra. – *Am. Nat.* 108, 332–340
Schoener T.W. 1976: The species-area relation within archipelagos: models and evidence from Island land birds. – *Proc. 16th Int. Ornithol. Congr.*, 629–642
Shipley, B. (2021). *Ordination methods for biologists: a non-mathematical introduction using R*. BS Publishing.
Siegel, S., & Castellan, N. J. (1988). *Nonparametric Statistics for The Behavioral Sciences* (2nd edition.). Boston, Mass.: McGraw-Hill Humanities/Social Sciences/Languages.
Sokal R.R. & Rohlf F.J. 2011: *Biometry. The Principles and Practice of Statistics in Biological Research*. 4th ed. – W. H. Freeman and Company, New York, 937 p.
Southwood T.R.E. 1978: *Ecological Methods – with particular reference to the study of Insect Populations*. – Chapman and Hall, London, John Wiley & Sons, New York, 524 p.
StatSoft is the original developer of Statistica, acquired by Dell 2014. Statistica is an analytics software portfolio. (after Wikipedia.org)
Sutherland W.J. 2006: *Ecological Census Techniques*, a handbook. 2nd ed. – Cambridge University Press, 432 p.
Timischl W. 2013: *Angewandte Statistik*. Eine Einführung für Biologen und Mediziner. 3rd ed. – Springer Vienna, 550 p.
Triola M. & Roy J. 2023: *Biostatistics for the Biological and Health Sciences*, Global Edition. 3rd ed. – Pearson Education Limited.
Vitt, L.J. & Caldwell J.P. 2013: *Herpetology: An Introductory Biology of Amphibians and Reptiles*. 4th ed. – Burlington, Massachusetts, U.S.A.: Academic Press, 776 p.
Wainstein B.A. 1967: Some methods of evaluation of similarity of biocoenoses. – *Zool. Z.* 46, 981–986. (russ.)
Waltert M., Mardiastuti A. & Mühlenberg M. 2004: Effects of Land Use on Bird Species Richness in Sulawesi, Indonesia. – *Conservation Biology* 18 (5): 1339-1346

Wassmer, T. (2020). Attractiveness of Cattle Dung to Coprophilous Beetles (Coleoptera: Scarabaeoidea and Sphaeridiinae) and Their Segregation During the Initial Stages of the Heterotrophic Succession on a Pasture in Southeast Michigan. Journal of Insect Science, 20(3). https://doi.org/10.1093/jisesa/ieaa040

Weber E. 1986: *Grundriss der biologischen Statistik*. 9. Aufl. – Gustav Fischer Verlag, Stuttgart, New York, 652 p.

Whitlock, M., & Schluter, D. (2020). *The Analysis of Biological Data* (Third edition.). Austin: W. H. Freeman.

Williams, B.K. (1983). Some Observations of the Use of Discriminant Analysis in Ecology. Ecology, 64(5), 1283–1291. https://doi.org/10.2307/1937836

Williamson M. 1981: *Island Populations*. – Oxford University Press, Oxford, 286 p.

Wolda H. 1981: Similarity indices, sample size and diversity. – *Oecologia* 50, 296-302

Yoccoz N.G. 1991: Use, overuse and misuse of significance tests in evolutionary biology and ecology. – *Bulletin of the Ecological Society of America* 72 (2), 106–111

Zar, J. H. (2009). *Biostatistical Analysis* (5th edition.). Upper Saddle River, N.J: Pearson.

16 Writing Research and Grant Proposals

Abstract

Scientific results must be presented and published in slide presentations, manuscripts, or submitted publications. We provide some general guidelines for successful presentations and research papers, and present aspects that increase your chances that a grant application will be eligible for funding.

16.1 Recommendations for Writing Scientific Papers

Once a data collection or experiment was analyzed satisfactory, it should be communicated appropriately. This can be as a presentation for a meeting or a conference, a manuscript template, or a submitted scientific publication. For any of these types of disseminations, recommendations and general rules can be found, for example, in the following books: Davis (2005), Gustavii (2008), Matthews and Matthews (2008), Gastel and Day (2016), and McMillan (2016), Heard (2022), Hofmann (2023, 2025), ter Kuile (2024), Zaumanis (2021, 2023).

First, you should ask yourself what message or story you want to tell or convey. This also includes the search for literature on the subject to explain the background of the problem to the reader or listener. What facts and ideas must be addressed to classify your research? Are these facts presented in the correct order? Selecting tables, graphs, and photos supporting the results are essential. Conclusions must be formulated very clearly and must refer to the information presented. Finally, you should revise the paper to ensure it is conveyed concisely and precisely. Here, a colleague's review can help you by constructively criticizing your paper. Finally, it is also essential to address any remaining questions and make recommendations for further research. The bibliography must be checked for exact punctuation and format.

A written paper should follow the structure that is generally considered to be the standard for publications: *Title, Abstract, Introduction, Material and Methods, Results, Discussion and Conclusions, Acknowledgments, References, and possibly a Summary*. The title should not be too long (ideally, not more than 100 characters) and should catch the reader's interest without exaggeration or bragging.

16.2 On Structuring a Grant Proposal

Research work and conservation activities cost money. It is, therefore, always important to apply for support from potential funders (e.g., *grant agencies)*. Several books provide detailed instructions for grant applications, e.g., Reif-

M. Mühlenberg, T. Waßmer, *Field Ecology*, https://doi.org/10.1007/978-3-031-87547-2_16

Lehrer (2004), Friedland et al. (2018a, b), Glasman-deal (2020), Li and Marrongelle (2012), and Zaumanis (2023). For your applications to succeed, you must at least adhere to certain generally accepted rules for the structure and the form of the application. A general guideline will be outlined below.

Title: The title should sound interesting. It is also recommended that general objectives, e.g., from the Millennium Goals (United Nations 2024a) or the Sustainable Development Goals of the UN (United Nations 2024b), such as sustainability, climate adaptation, and poverty reduction, be included in the title. This can underline the relevance of the project.

Background: What do we already know about the topic? What do we not know? Why should we be interested in expanding our knowledge about this topic? Justification of the project. An overview of relevant literature should be given here.

State of the art: The latest science and corresponding technology should be discussed here.

Research questions and hypotheses: The research project's goal should be divided into individual steps and the corresponding questions. Now, you can speculate and then set up more precise and testable hypotheses.

Objectives: Here, the overall goal of the project is first presented. It also points out what methodological or technical achievements are to be expected.

Activities: Describe each activity planned to process the individual steps (see above research questions). Also, indicate when and for how long a measure will be carried out. If it is planned to involve additional employees or local specialists, this must also be stated (see below). The methodology to be applied must also be described in detail. For the reader, it must become apparent that individual goals can be achieved with the methods mentioned. One can also argue that the methods applied in your proposed research can lead to better results than previously used techniques. It is important to emphasize that the described goals are achievable and their application is feasible. Possible challenges, bottlenecks or failures (pitfalls) of individual work steps should also be addressed.

Expected output: What are the expected results? How will the results fit into existing knowledge? What will be new? In what way do the results complement known relationships? This section also refers to the justification of the application. One should highlight which preliminary data are available and how the project ties in with them. If you have already contributed to the preliminary results through your previous work, this underlines the credibility that the applicant is well suited to achieve the stated goals.

Implications for conservation: Special conclusions that are useful for conservation should be highlighted. Conservation may also be one of the research project's goals. You must describe in detail how the results can be implemented for conservation. It can be pointed out that practical conservation always needs a scientific basis.

Beneficiaries: Who can benefit from the possible results? The focus is on general science, possible partners, local authorities, or specialists. Could the new results make resource management more sustainable? Will a local community benefit from the application of the results?

Potential project partners: As a rule, networking of activities with other partners or institutions is always preferred. If the work is planned in a host country, it is usually necessary to include partners from the host country in the application in advance. This also clarifies questions about research permits in foreign countries. In some countries, a research permit must be applied for in advance. If partners are included in the project, you may motivate them by promising them that they will be co-authors in the publications of the results.

Target area/region: With the description of the project region, you also indicate that you are already familiar with the local conditions and have adapted your methods to them.

Expected duration—schedule: Once again, a concrete schedule is presented. It is better to point out at the beginning that if some of the goals may possibly take more time.

Expected costs—detailed plan of expenses: The more detailed the expenditure plan is compiled, the more convincing it appears. Even local transport costs and possibly local remuneration for local assistance should be listed. Planned trips due to project interruptions or laboratory costs incurred must be calculated. Ensuring that your institution or partner institutions will cover certain incurred costs is advantageous. Corresponding agreements should be presented in writing. With a *memorandum of understanding* (MoU), you can document what services the partners provide to the cooperation.

Overall, one should consider several times which idea or question in the *proposal* is new, and then one must *justify* the application convincingly. It usually depends on the specifications of the *granting agency* as to where in the application an abstract should be placed (at the beginning, often after the title page, or at the end). Since the *reviewers* are usually overwhelmed with applications, often only the abstract is scanned in the first reading. It must, therefore, be concise, precise, and convincing. It must contain the research objectives and the justification for the planned investigations. This requires the basic methodology and the anticipated results with significant conclusions to be presented. These specifications are usually written after the application is completed and should be reviewed several times. It is best practice if people not working in the same field can understand the *self-explanatory* application instantly.

References

Davis M. 2005: *Scientific papers and presentations* (2nd ed.). – Elsevier, Academic Press, 356 p.

Friedland, A. J., Folt, C. L., & Mercer, J. L. (2018). *Writing Successful Science Proposals* (Third edition.). New Haven: Yale University Press, 261 p.

Gastel B. & Day R.A. 2016: *How to Write and Publish a Scientific Paper* (8th ed.). – Greenwood, 326 p.

Glasman-deal, H. (2020). *Science Research Writing: For Native And Non-native Speakers Of English.* New Jersey London Singapore Beijing Shanghai Tapei Chennai Tokio: WSPC.

Gustavii B. 2008 (2nd ed.): *How to Write and Illustrate Scientific Papers.* – Cambridge University Press, 168 p.

Heard, S. B. (2022). *The Scientist's Guide to Writing, 2nd Edition: How to Write More Easily and Effectively throughout Your Scientific Career* (2nd edition.). Princeton: Princeton University Press.

Hofmann, Angie. (2022). *Scientific Writing and Communication* (5th edition.). New York: Oxford University Press.

Hofmann, Angelika. (2024). *Writing in the Biological Sciences: A Comprehensive Guide to Scientific Communication* (5th edition.). New York: Oxford University Press.

Kuile, B. ter. (2024). *Life Sciences Research and Scientific Writing* (2024th edition.). Cham: Springer.

Li, P., & Marrongelle, K. (2012). *Having Success with NSF: A Practical Guide* (1st edition.). Hoboken, N.J: Wiley-Blackwell.

Matthews J.R. & Matthews R.W. 2008 (3rd ed.): *Successful Scientific Writing. A step-by-step guide for the biological and medical sciences.* – Cambridge University Press, 240 p.

McMillan V.E. 2016: *Writing Papers in the Biological Sciences* (6th ed.). – Bedford/St. Martin's, 264 p.

Reif-Lehrer L. 2004: *Grant Application Writer's Handbook.* (4th ed.) – Jones & Bartlett Publ., 362 p.

United Nations. (2024a). United Nations Millennium Development Goals. *Millennium Development Goals.* United Nations. https://www.un.org/millennium-goals/. Accessed 14 September 2024

United Nations. (2024b). THE 17 GOALS | Sustainable Development. *THE 17 GOALS.* https://sdgs.un.org/goals. Accessed 24 July 2024

Zaumanis, M. (2021). *Write an impactful research paper: A scientific writing technique that will shape your academic career.* Las Vegas: Independently published.

Zaumanis, M. (2023). *Write a Winning Research Proposal: How to Generate Grant Ideas and Secure Funding Using Research Project Canvas.* Independently Published By Peer.

Subject Index

M. Mühlenberg, T. Waßmer, *Field Ecology*, https://doi.org/10.1007/978-3-031-87547-2

Taxonomic Index

M. Mühlenberg, T. Waßmer, *Field Ecology*, https://doi.org/10.1007/978-3-031-87547-2

D

E

F

GPSR Compliance

The European Union's (EU) General Product Safety Regulation (GPSR) is a set of rules that requires consumer products to be safe and our obligations to ensure this.

If you have any concerns about our products, you can contact us on ProductSafety@springernature.com

In case Publisher is established outside the EU, the EU authorized representative is:

Springer Nature Customer Service Center GmbH
Europaplatz 3
69115 Heidelberg, Germany

Batch number: 10731383

Printed by Printforce, the Netherlands